TETRAHEDRON ORGANIC CHEMIS

Series Editors: **J-E Bäckvall, J E Baldwin, F**

D1760384

VOLUME 26

U

D
Palladium in
erocyclic Chemistry

This b
self

le for the Synthetic Chemist

Second Edition

Related Titles of Interest

BOOKS

Tetrahedron Organic Chemistry Series:
CARRUTHERS: Cycloaddition Reactions in Organic Synthesis
CLARIDGE: High-Resolution NMR Techniques in Organic Chemistry
CLAYDEN: Organolithiums: Selectivity for Synthesis
FINET: Ligand Coupling Reactions with Heteroatomic Compounds
GAWLEY & AUBÉ: Principles of Asymmetric Synthesis
HASSNER & STUMER: Organic Syntheses Based on Name Reactions
and Unnamed Reactions (2nd Edition)
LI & GRIBBLE: Palladium in Heterocyclic Chemistry
McKILLOP: Advanced Problems in Organic Reaction Mechanisms
OBRECHT & VILLALGORDO: Solid-Supported Combinatorial and
Parallel Synthesis of Small-Molecular-Weight Compound Libraries
PERLMUTTER: Conjugate Addition Reactions in Organic Synthesis
PIETRA: Biodiversity and Natural Product Diversity
PIRRUNG: Molecular Diversity and Combinatorial Chemistry
SESSLER & WEGHORN: Expanded, Contracted & Isomeric Porphyrins
TANG & LEVY: Chemistry of *C*-Glycosides
WONG & WHITESIDES: Enzymes in Synthetic Organic Chemistry

Studies in Natural Products Chemistry *(series)*
Strategies and Tactics in Organic Synthesis *(series)*
Rodd's Chemistry of Carbon Compounds *(series)*
Best Synthetic Methods *(series)*

JOURNALS

Bioorganic & Medicinal Chemistry
Bioorganic & Medicinal Chemistry Letters
Tetrahedron
Tetrahedron Letters
Tetrahedron: Asymmetry

Full details of all Elsevier publications are available on www.elsevier.com or from your nearest Elsevier office

Palladium in Heterocyclic Chemistry

A Guide for the Synthetic Chemist

Second Edition

Edited by

JIE JACK LI

Michigan Laboratories
Pfizer Global Research & Development
Ann Arbor, USA

GORDON W. GRIBBLE

Department of Chemistry
Dartmouth College
Hanover, USA

ELSEVIER

Amsterdam - Boston - Heidelberg - London - New York - Oxford - Paris
San Diego - San Francisco - Singapore - Sydney - Tokyo

ELSEVIER

The Boulevard, Langford Lane, Kidlington, Oxford OX5 1GB, UK
Radarweg 29, PO Box 211, 1000 AE Amsterdam, The Netherlands

First edition 2007

British Library Cataloguing in Publication Data
A catalogue record for this book is available from the British Library

Library of Congress Cataloging-in-Publication Data
A catalog record for this book is available from the Library of Congress

ISBN-13:   978-0-08-045116-9 and ISBN-10: 0-08-045116-0 (hardbound)
ISBN-13:   978-0-08-045117-6 and ISBN-10: 0-08-045117-9 (paperback)
ISSN: 1460-1567 (series)

For information on all Elsevier publications
visit our website at books.elsevier.com

Printed and bound in The United Kingdom

07 08 09 10 11   10 9 8 7 6 5 4 3 2 1

Contributing authors

Nadia M. Ahmad
School of Chemistry
University of Nottingham
University Park
Nottingham
NG7 2RD, UK

Dr. Marudai Balasubramanian
Research Informatics
Pfizer Global Research & Development
2800 Plymouth Road
Ann Arbor, MI 48105

Dr. Paul Galatsis
Department of Chemistry
Pfizer Global Research & Development
2800 Plymouth Road
Ann Arbor, MI 48105

Prof. Gordon W. Gribble
Department of Chemistry
6128 Burke Laboratory
Dartmouth College
Hanover, NH 03755

Dr. Jie Jack Li
Department of Chemistry
Pfizer Global Research & Development
2800 Plymouth Road
Ann Arbor, MI 48105

Dr. Chris Limberakis
Chemistry Department
Pfizer Global Research & Development
2800 Plymouth Road
Ann Arbor, MI 48105

Prof. Bert U.W. Maes
Department of Chemistry
University of Antwerp
Groenenborgerlaan 171
B-2020 Antwerp, Belgium

**Prof. Richard J. Mullins
and Adam M. Azman**
Department of Chemistry
Xavier University
3800 Victory Parkway
Cincinnati, OH 45207-4221

Dr. Michael Palucki
Department of Process Research
Merck & Co., Inc.
Rahway, NJ 07065-0900

Prof. Kevin M. Shea
Department of Chemistry
Clark Science Center
Smith College
Northampton, MA 01063

Prof. John P. Wolfe
Department of Chemistry
University of Michigan
930 N. University Avenue
Ann Arbor, MI 48109-1055

Preface to the second edition

The first edition of "Palladium in Heterocyclic Chemistry – A Guide for the Synthetic Chemist," published five years ago, was warmly received by many readers. Five years is a fleeting interval in scientific history, but palladium chemistry in general, and palladium in heterocyclic chemistry in particular, has matured considerably. In order to make the second edition more thorough and up-to-date, we are fortunate to have recruited a group of stellar authors to take on the challenge. As a consequence, not only have we attempted to improve and update each chapter, but we have also added three new chapters: Chapter 12: Quinolines; Chapter 13: Pyridazines; Chapter 14: Industrial scale palladium chemistry. We are indebted to all authors for their contributions.

We welcome your critique.

Jack Li and Gordon Gribble, March 2006

Preface to the first edition

Palladium chemistry, despite its immaturity, has rapidly become an indispensable tool for synthetic organic chemists. Today, palladium-catalyzed coupling is the method of choice for the synthesis of a wide range of biaryls and heterobiaryls. The number of applications of palladium chemistry to the syntheses of heterocycles has grown exponentially.

Then, is there a need for a monograph dedicated solely to the palladium chemistry in heterocycles? The answer is a resounding "yes!":

1. Palladium chemistry of heterocycles has its "idiosyncrasies" stemming from their different structural properties from the corresponding carbocyclic aryl compounds. Even activated chloroheterocycles are sufficiently reactive to undergo Pd-catalyzed reactions. As a consequence of α and γ activation of heteroaryl halides, Pd-catalyzed chemistry may take place regioselectively at the activated positions, a phenomenon rarely seen in carbocyclic aryl halides. In addition, another salient peculiarity in palladium chemistry of heterocycles is the so-called "heteroaryl Heck reaction." For instance, while intermolecular Pd-catalyzed arylations of carbocyclic arenes are rare, Pd-catalyzed arylations of azoles and many other heterocycles readily take place. Therefore, the principal aim of this book is to highlight important Pd-mediated reactions of heterocycles with emphasis on the unique characteristics of individual heterocycles.

2. A myriad of heterocycles are biologically active and therefore of paramount importance to medicinal and agricultural chemists. Many heterocycle-containing natural products (they are highlighted in boxes throughout the text) have elicited great interest from both academic and industrial research groups. Recognizing the similarities between the palladium chemistry of arenes and heteroarenes, a critical survey of the accomplishments in heterocyclic chemistry will keep readers abreast of such a fast-growing field. We also hope this book will spur more interest and inspire ideas in such an extremely useful area.

We have compiled important preparations of heteroaryl halides, boranes, and stannanes for each heterocycle. The large body of data regarding Pd-mediated polymerization of heterocycles in material chemistry is not focused here; neither is coordination chemistry involving palladium and heterocycles.

We are much indebted to Susan E. Hagan, Douglas S. Johnson, Michael Palucki, Howard W. Roark, Roderick J. Sorenson, Peter L. Toogood, Sharon Ward, and Kim Werner for proofreading the manuscript. We are grateful to Professor Louis S. Hegedus of Colorado State University who read the entire manuscript and offered many invaluable comments and suggestions. Professor Rick L. Danheiser of Massachusetts Institute of Technology also read part of the manuscript and provided very insightful suggestions. Any remaining errors and omissions are, of course, entirely our own.

Last, but not least, we wish to thank Wendy O. Berryman of Dartmouth College for typing part of the manuscript and Sharon Ward of Elsevier Science for editorial assistance throughout the project.

Jack Li and Gordon Gribble, June 2000

Contents

Preface to the second edition   vii

Preface to the first edition   ix

Abbreviations   xix

1   An introduction to palladium catalysis   1
John P. Wolfe and Jie Jack Li

1.1   Oxidative coupling/cyclization 3
1.2   Cross-coupling reactions with organometallic reagents 5
     1.2.1   The Negishi coupling 6
     1.2.2   The Suzuki coupling 7
     1.2.3   The Stille coupling 10
     1.2.4   The Stille–Kelly coupling 11
     1.2.5   The Kumada coupling 12
     1.2.6   The Hiyama coupling 13
1.3   The Sonogashira reaction 14
1.4   The Heck, intramolecular Heck, and heteroaryl Heck reactions 15
1.5   Carbonylation reactions 19
1.6   The Pd-catalyzed C−P bond formation 20
1.7   Palladium-catalyzed C−N bond and C−O bond-forming reactions 21
     1.7.1   The Buchwald–Hartwig Pd-catalyzed aryl
                 C−N bond formation 21
     1.7.2   Palladium-catalyzed aryl C−O bond formation 22
     1.7.3   Palladium-catalyzed heteroatom-Heck reactions 23
     1.7.4   Palladium-catalyzed carboamination and carboetherification
                 of alkenes 24
     1.7.5   Palladium-catalyzed C−H bond oxidation/halogenation 25
1.8   The Tsuji–Trost reaction 25
1.9   The Wacker-type reactions 26
1.10   Mori–Ban, Hegedus, and Larock indole syntheses 27
1.11   References 29

2 Pyrroles   **37**
Gordon W. Gribble

    2.1   Synthesis of pyrrolyl halides 38
    2.2   Oxidative coupling/cyclization 41
    2.3   Coupling reactions with organometallic reagents 46
           2.3.1   Grignard reagents (the Kumada coupling) 46
           2.3.2   Organozinc reagents (the Negishi coupling) 46
           2.3.3   Organoboron reagents (the Suzuki coupling) 48
           2.3.4   Stille coupling 54
    2.4   Sonogashira reaction. 59
    2.5   Heck and intramolecular Heck reactions 61
    2.6   Carbonylation 66
    2.7   C−N bond formation reactions 67
    2.8   Miscellaneous 72
    2.9   References 74

3 Indoles   **81**
Gordon W. Gribble

    3.1   Synthesis of indolyl halides 83
    3.2   Oxidative coupling/cyclization 90
    3.3   Coupling reactions with organometallic reagents 96
           3.3.1   Kumada coupling 96
           3.3.2   Negishi coupling. 97
           3.3.3   Suzuki coupling 102
           3.3.4   Stille coupling 113
    3.4   The Sonogashira coupling 124
    3.5   Heck couplings. 128
           3.5.1   Heck coupling reaction. 128
           3.5.2   The Mori–Ban indole synthesis 140
           3.5.3   The Larock indole synthesis 147
    3.6   Carbonylation. 150
    3.7   C−N bond formation reactions. 155
    3.8   Miscellaneous 165
    3.9   References 168

4 Pyridines   **189**
Paul Galatsis

    4.1   Synthesis of halopyridines 192
           4.1.1   Direct metalation followed by quenching with halogens 192
           4.1.2   Metal–halogen exchange 195
           4.1.3   Halogen–halogen exchange 196
           4.1.4   Dehydroxy–halogenation. 199
    4.2   Coupling reactions with organometallic reagents 200
           4.2.1   Kumada coupling. 200
           4.2.2   Negishi coupling 201
           4.2.3   Suzuki coupling 205

	4.2.4	Stille coupling	216
	4.2.5	Hiyama coupling	223
4.3		Sonogashira reaction	225
4.4		Heck and intramolecular Heck reactions	231
4.5		Buchwald–Hartwig aminations (C–N bond formation)	236
4.6		Direct C–C bond formation	240
4.7		Summary	242
4.8		References	243

5 Thiophenes and benzo[b]thiophenes   **251**
Chris Limberakis

5.1		Preparation of halothiophenes and halobenzothiophenes	252
	5.1.1	Direct halogenation	252
	5.1.2	Quenching lithiothiophenes with halogens	255
5.2		Oxidative and reductive coupling reactions	256
5.3		Cross-coupling with organometallic reagents	256
	5.3.1	Kumada coupling	256
	5.3.2	Negishi coupling	258
	5.3.3	Suzuki coupling	261
		5.3.3.1 Preparation of boronic acids, boronates, and trifluoroborates	261
		5.3.3.2 Couplings with aryl and heteroaryl substrates	263
		5.3.3.3 Regioselectivity	270
	5.3.4	Stille coupling	272
		5.3.4.1 Preparation of stannylthiophenes	272
		5.3.4.2 Alkyl-, vinyl-, and alkynylthiophenes	274
		5.3.4.3 Arylthiophenes	277
		5.3.4.4 Heteroarylthiophenes	279
		5.3.4.5 Thiophene-containing condensed heteroaromatics	282
		5.3.4.6 Cu(I) thiophene-2-carboxylate (CuTC)	283
	5.3.5	Hiyama coupling	283
5.4		Sonogashira reaction	284
5.5		Heck and intramolecular Heck reactions	287
5.6		Carbonylation reactions	290
5.7		Buchwald–Hartwig aminations	291
5.8		Miscellaneous	294
	5.8.1	Palladium-catalyzed C–P bond formation	294
	5.8.2	Palladium-catalyzed cycloisomerization	294
5.9		References	295

6 Furans and benzo[b]furans   **303**
Kevin M. Shea

6.1		Synthesis of halofurans and halobenzo[b]furans	303
	6.1.1	Halofurans	303
	6.1.2	Halobenzofurans	305
6.2		Oxidative coupling/cyclization	306

6.3   Coupling reactions with organometallic reagents 308
        6.3.1   Kumada coupling. 308
        6.3.2   Negishi coupling 308
        6.3.3   Suzuki coupling 312
                6.3.3.1   Furylboronic acids 312
                6.3.3.2   Furans as electrophiles. 314
        6.3.4   Stille coupling 316
                6.3.4.1   Furan motif as a nucleophile (stannane). 316
                6.3.4.2   Furan motif as an electrophile (halide or triflate). . . . 320
        6.3.5   Stille–Kelly reaction 322
        6.3.6   Hiyama coupling 322
6.4   Sonogashira reaction 322
6.5   Heck, intramolecular Heck, and heteroaryl Heck reactions 324
        6.5.1   Intermolecular Heck reaction 324
        6.5.2   Intramolecular Heck reaction 325
        6.5.3   Heteroaryl Heck reaction. 327
6.6   Heteroannulation 328
        6.6.1   Propargyl carbonates 328
        6.6.2   Alkynols 330
        6.6.3   Dibromoalkenols 333
        6.6.4   Alkynones 333
        6.6.5   Enones 335
        6.6.6   Allenones 336
6.7   Carbonylation and C−N and C−O bond formation 336
6.8   References 337

7   Thiazoles and benzothiazoles   345
      Richard J. Mullins and Adam M. Azman

7.1   Synthesis of halothiazoles 345
        7.1.1   Direct halogenation. 346
        7.1.2   Sandmeyer reaction 347
7.2   Coupling reactions with organometallic reagents 347
        7.2.1   Negishi coupling 347
                7.2.1.1   Thiazole as Negishi nucleophile 347
                7.2.1.2   Thiazole as Negishi electrophile 349
        7.2.2   Suzuki coupling 352
        7.2.3   Stille coupling 354
                7.2.3.1   Synthesis of stannylthiazoles 354
                7.2.3.2   Thiazole as Stille nucleophile 356
                7.2.3.3   Thiazole as Stille electrophile 361
7.3   Sonogashira reaction 363
7.4   Heck and heteroaryl Heck reactions 367
7.5   Carbonylation. 370
7.6   C−N bond formation 371
7.7   Site selective coupling reactions   371
7.8   References 374

8   Oxazoles and benzoxazoles   **379**
Marudai Balasubramanian

  8.1   Introduction. 379
  8.2   Synthesis of halooxazoles and halobenzoxazoles 383
         8.2.1   Direct halogenation. 383
         8.2.2   Metalation and halogen quench 383
         8.2.3   Sandmeyer reaction 384
  8.3   Coupling reactions with organometallic reagents 385
         8.3.1   Negishi coupling 386
         8.3.2   Suzuki coupling 388
         8.3.3   Suzuki–Miyaura coupling 389
         8.3.4   Stille coupling 390
         8.3.5   Sonogashira coupling. 396
  8.4   Heck and heteroaryl Heck reactions 398
  8.5   Carbonylation. 400
  8.6   Palladium-catalyzed amination. 402
  8.7   Carbopalladation of nitriles 402
  8.8   Quaterfuran and quinquifuran 403
  8.9   Summary 404
  8.10   References 404

9   Imidazoles   **407**
Richard J. Mullins and Adam M. Azman

  9.1   Synthesis of haloimidazoles 407
  9.2   Homocoupling reaction 410
  9.3   Coupling reactions with organometallic reagents 411
         9.3.1   Negishi coupling 411
         9.3.2   Suzuki coupling 414
         9.3.3   Stille coupling 417
  9.4   Sonogashira reaction 422
  9.5   Heck and heteroaryl Heck reactions 424
  9.6   Tsuji–Trost reaction 428
  9.7   Phosphonylation 429
  9.8   References 430

10 Pyrazines and quinoxalines   **435**
Marudai Balasubramanian

  10.1   Pyrazines 435
  10.2   Coupling reactions with organometallic reagents 437
         10.2.1   Kumada reaction 437
         10.2.2   Negishi coupling 439
         10.2.3   Stille coupling. 440
         10.2.4   Suzuki coupling. 444
         10.2.5   Sonogashira coupling 451
  10.3   Palladium-catalyzed amination. 455
  10.4   Heck reaction. 456

10.5   Carbonylation reactions 458
10.6   Cyanation of pyrazines. 460
10.7   Deoxygenation of heteroamine-*N*-oxide 460
10.8   Quinoxalines 461
    10.8.1   Coupling reactions with organometallic reagents 461
        10.8.1.1   Stille coupling 461
        10.8.1.2   Sonogashira reaction 463
    10.8.2   Intramolecular Heck reaction. 466
    10.8.3   Carbonylation reactions. 467
    10.8.4   Hydrogenation 467
    10.8.5   Amination 467
    10.8.6   Cyanation of pyridopyrazine 468
    10.8.7   Pyrrolopyrazine 468
    10.8.8   Lumazines. 469
10.9   References 469

11 Pyrimidines   **475**
Michael Palucki

11.1   Synthesis of pyrimidinyl halides and triflates 476
11.2   Coupling reactions with organometallic reagents 478
    11.2.1   Negishi coupling 478
    11.2.2   Suzuki coupling. 482
    11.2.3   Stille coupling. 488
        11.2.3.1   The pyrimidine motif as a nucleophile 488
        11.2.3.2   The pyrimidine motif as an electrophile. 492
    11.2.4   Organozirconium, organoaluminum, and
        organoindium reagents 495
11.3   Sonogashira reaction 496
11.4   Heck reaction. 502
11.5   The carbonylation reaction. 504
11.6   Heteroannulation 505
11.7   References 506

12 Quinolines   **511**
Nadia M. Ahmad

12.1   Synthesis of quinoline electrophiles 512
    12.1.1   Halogenation of quinolones. 512
    12.1.2   Direct halogenation of quinolines 513
    12.1.3 S_NAr reaction 514
    12.1.4   Halogen-dance reaction. 515
    12.1.5   Vilsmeier–Haack reaction 515
    12.1.6   Miscellaneous syntheses of haloquinolines. 516
12.2   Synthesis of quinoline nucleophiles 517
12.3   Cross-coupling reactions with organometallic reagents 519
    12.3.1   Negishi coupling 519
    12.3.2   Suzuki coupling. 520

		12.3.3	Stille coupling	523
		12.3.4	Hiyama coupling	527
	12.4	Sonogashira reaction		528
	12.5	Heck reaction		530
		12.5.1	Intermolecular Heck reaction	530
		12.5.2	Intramolecular Heck reaction	532
	12.6	Miscellaneous reactions mediated by palladium		533
		12.6.1	Oxidative cyclization	533
		12.6.2	Reductive cyclization	534
		12.6.3	Heteroannulation	535
		12.6.4	Cyanation	535
		12.6.5	Homocoupling	536
		12.6.6	Phosphination	536
		12.6.7	Oxidative C—C_{Ar} bond formation	536
	12.7	References		537

13 Pyridazines   **541**
Bert U.W. Maes

	13.1	Synthesis of (pseudo)halopyridazines and (pseudo)halopyridazin-3(2*H*)-ones		543
		13.1.1	Ring synthesis of halopyridazin-3(2*H*)-ones	543
		13.1.2	Direct halogenation of pyridazines and pyridazin-3(2*H*)-ones	544
		13.1.3	From pyridazine *N*-oxides via reaction with $POCl_3$	545
		13.1.4	Deoxy-halogenation of pyridazinones	545
		13.1.5	Pseudohalopyridazines and -pyridazin-3(2*H*)-ones by esterification	546
		13.1.6	Direct metalation of pyridazines followed by quenching with halogens	547
		13.1.7	From metalopyridazines by reaction with an electrophilic halogen source	548
		13.1.8	Diazotization—S_N1 reaction of aminopyridazines	549
		13.1.9	Transhalogenation on chloropyridazines and chloropyridazin-3(2*H*)-ones	549
	13.2	Coupling reactions with organometallic reagents		550
		13.2.1	Negishi coupling	550
		13.2.2	Suzuki coupling	551
			13.2.2.1 *(Pseudo)halopyridazines*	551
			13.2.2.2 *(Pseudo)halopyridazin-3(2H)-ones*	556
		13.2.3	Stille coupling	562
	13.3	Sonogashira reaction		566
	13.4	Heck and intramolecular Heck reactions		574
	13.5	Carbonylation reactions		577
		13.5.1	Alkoxycarbonylation	577
		13.5.2	Aminocarbonylation	579
	13.6	C—N bond formation		579
	13.7	References		582

14 Industrial scale palladium chemistry   **587**
Marudai Balasubramanian

   14.1   Introduction. 587
   14.2   Pharmaceutical products 588
       14.2.1   Heck coupling. 588
       14.2.2   Carbonylations 591
       14.2.3   Suzuki coupling. 592
       14.2.4   Sonogashira coupling 594
       14.2.5   Negishi coupling 595
       14.2.6   Amination. 595
   14.3   Cosmaceuticals. 599
       14.3.1   Heck reaction 599
   14.4   Agrochemical products. 599
       14.4.1   Matsuda–Heck reaction. 599
       14.4.2   Carbonylations 600
       14.4.3   Amination. 601
   14.5   Material sciences. 601
   14.6   Polymer chemistry 603
   14.7   New catalyst developments in fine chemical synthesis 607
       14.7.1   Heck reactions 607
       14.7.2   Suzuki coupling. 608
       14.7.3   Suzuki–Miyaura coupling. 610
       14.7.4   Sonogashira coupling 611
       14.7.5   Negishi coupling 611
       14.7.6   Kumada coupling 612
   14.8   Amination 612
   14.9   Carbonylation. 612
   14.10 Amidocarbonylation 613
   14.11 References 613

Index   **621**

Abbreviations

AIBN	2,2′-azobisisobutylonitrile
AMPA	amino-3-hydroxy-5-methyl-4-isoxazolepropionic acid
9-BBN	9-borabicyclo[3.3.1]nonane
BHT	2,6-di-*tert*-butyl-4-methylphenol
BINAP	2,2′-bis(diphenylphosphino)-1,1′-binaphthyl
Boc	*tert*-butyloxycarbonyl
t-Bu	*tert*-butyl
cAMP	adenosine cyclic 3′,5′-monophosphate
Cbz	benzyloxycarbonyl
CCK-A	cholecystokinin-A
CNS	central nerve system
m-CPBA	*m*-chloroperoxybenzoic acid
CuTC	copper thiophene-2-carboxylate
DABCO	1,4-diazabicyclo[2.2.2]octane
dba	dibenzylideneacetone
DBH	1,3-dibromo-5,5-dimethylhydantoin
DCE	dichloroethane
Δ	solvent heated under reflux
DDQ	2,3-dichloro-5,6-dicyano-1,4-benzoquinone
DIBAL	diisobutylaluminum hydride
DMA	*N,N*-dimethylacetamide
DME	1,2-dimethoxyethane
DMF	dimethylformamide
DMI	1,3-dimethyl-2-imidazolidinone
DMPU	*N,N*-dimethylpropyleneurea
	(1,3-dimethyl-3,4,5,6-tetrahydro-2-(1*H*)-pyrimidinone)
DMSO	dimethyl sulfoxide
DMT	dimethoxytrityl
DNA	deoxyribonucleic acid
dppb	1,4-bis(diphenylphosphino)butane
dppe	1,2-bis(diphenylphosphino)ethane
dppf	1,1′-bis(diphenylphosphino)ferrocene
dppp	1,3-bis(diphenylphosphino)propane

ECE	endothelin conversion enzyme
EDC	1-(3-dimethylaminopropyl)3-ethylcarbodiimide hydrochloride
GABA	γ-aminobutyric acid
HMG-CoA	hydroxymethylglutaryl coenzyme A
HMPA	hexamethylphosphoric triamide
HT	hydroxytryptamine (serotonin)
LDA	lithium diisopropylamide
LHMDS	lithium hexamethyldisilazane
LTMP	lithium 2,2,6,6-tetramethylpiperidine
NAD H	reduced nicotinamide adenine dinucleotide
NBS	*N*-bromosuccinimide
NCS	*N*-chlorosuccinimide
NIS	*N*-iodosuccinimide
NK	neurokinin
NMP	1-methyl-2-pyrrolidinone
PMB	*para*-methoxybenzyl
PPA	polyphosphoric acid
RaNi	Raney Nickel
Red-Al®	sodium bis(2-methoxyethoxy)aluminum hydride
RNA	ribonucleic acid
SEM	2-(trimethylsilyl)ethoxymethyl
$S_N Ar$	nucleophilic substitution on an aromatic ring
TBAF	tetrabutylammonium fluoride
TBDMS	*tert*-butyldimethylsilyl
TBS	*tert*-butyldimethylsilyl
Tf	trifluoromethanesulfonyl (triflyl)
TFA	trifluoroacetic acid
TFP	tri-*o*-furylphosphine
THF	tetrahydrofuran
TIPS	triisopropylsilyl
TMEDA	*N,N,N′,N′*-tetramethylethylenediamine
TMG	tetramethyl guanidine
TMP	tetramethylpiperidine
TMS	trimethylsilyl
Tol-BINAP	2,2′-bis(di-*p*-tolylphosphino)-1,1′-binaphthyl
TTF	tetrathiafulvalene

Chapter 1

An introduction to palladium catalysis

John P. Wolfe and Jie Jack Li

Over the past 30–40 years, organopalladium chemistry has found widespread use in organic synthesis, and has been described in detail in a number of useful and informative books [1]. Palladium catalysts facilitate unique transformations that cannot be readily achieved using classical techniques, and in many cases palladium-catalyzed reactions proceed under mild reaction conditions and tolerate a broad array of functional groups. As such, the use of palladium catalysts for the synthesis of important, biologically active heterocyclic compounds has been the focus of a considerable amount of research [2].

This chapter describes the fundamentals of palladium catalysis in the context of heterocyclic chemistry, including the basic mechanisms of many useful transformations along with a number of new synthetic and mechanistic developments. The majority of the Pd-catalyzed reactions described in this book proceed via catalytic cycles that are comprised of eight fundamental organopalladium transformations shown below [1]. Most of these transformations can occur via more than one mechanistic pathway, and in some instances the precise mechanisms have not been fully elucidated.

The basic reactions are:

(1) oxidative addition, in which a Pd(0) complex undergoes insertion into a (usually) polarized σ-bond to afford a Pd(II) complex;

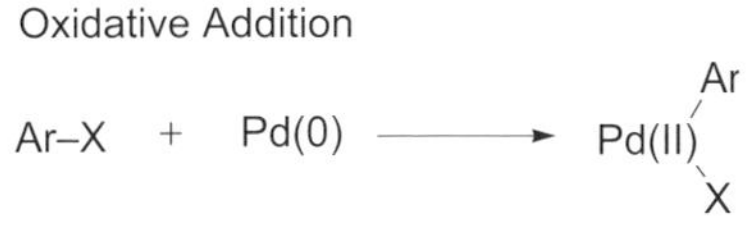

Oxidative Addition

Ar–X + Pd(0) ⟶ Pd(II) $\begin{smallmatrix} Ar \\ X \end{smallmatrix}$

(2) reductive elimination, which is the microscopic reverse of oxidative addition, and leads to formation of a σ-bond with concomitant formal reduction of Pd(II) to Pd(0);

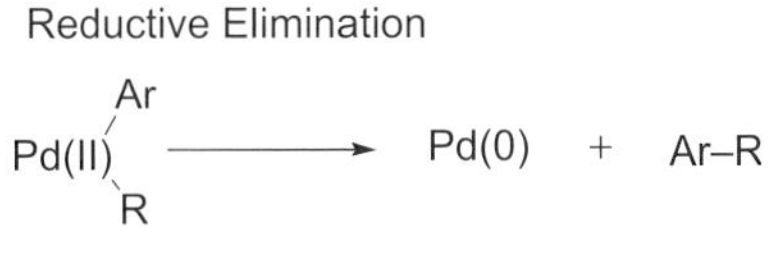

Reductive Elimination

Pd(II) $\begin{smallmatrix} Ar \\ R \end{smallmatrix}$ ⟶ Pd(0) + Ar–R

1

(3) migratory insertion, which involves the *syn*-addition of a palladium–carbon or palladium-heteroatom bond across an alkene with no change in metal oxidation state;

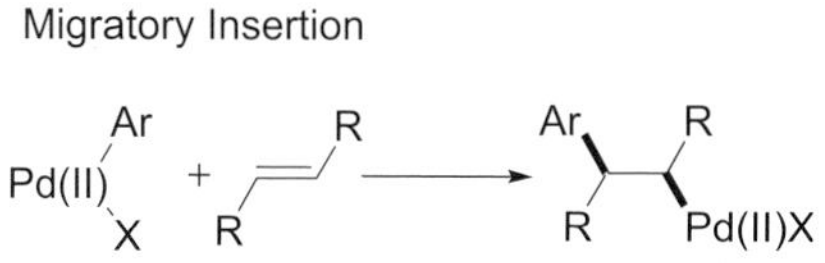

Migratory Insertion

(4) β-hydride elimination, which is the *syn*-elimination of a hydrogen atom and Pd(II) from a palladium alkyl complex with no change in oxidation state;

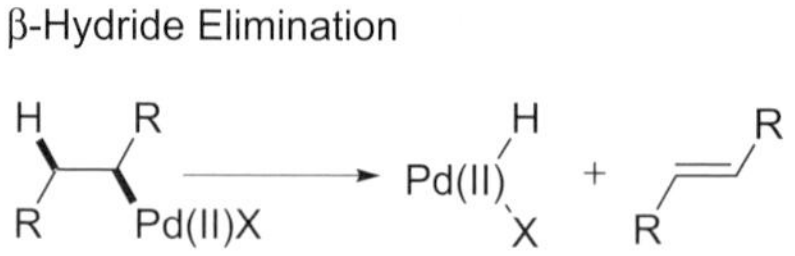

β-Hydride Elimination

(5) Wacker-type addition, which is the *anti*-addition of (most commonly) a heteroatom and a Pd(II) species across a C–C double bond;

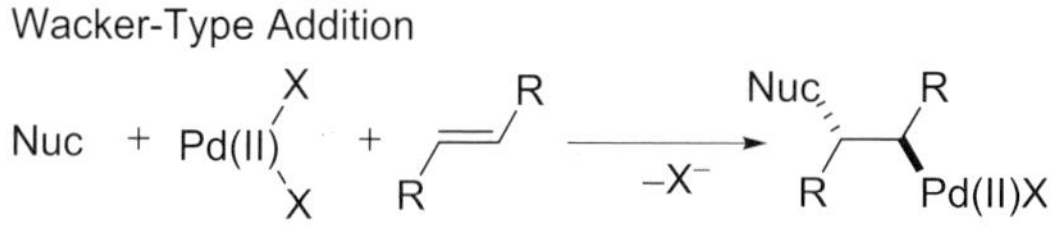

Wacker-Type Addition

(6) electrophilic palladation, in which a C–H σ-bond is exchanged for a C–Pd bond with loss of one equivalent of acid;

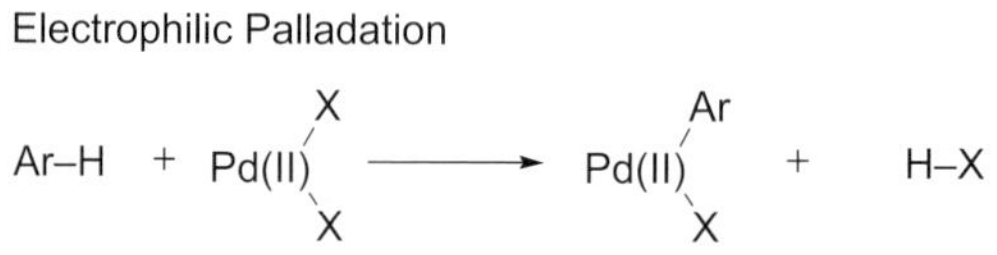

Electrophilic Palladation

(7) transmetalation, which involves the exchange of an R–M bond with a Pd–X bond to form Pd–R and M–X;

Transmetalation

(8) formation and trapping of π-allylpalladium species (formally a type of oxidative addition/reductive elimination sequence). The linking of these individual steps together

in synthetically useful catalytic cycles is described throughout the course of this chapter.

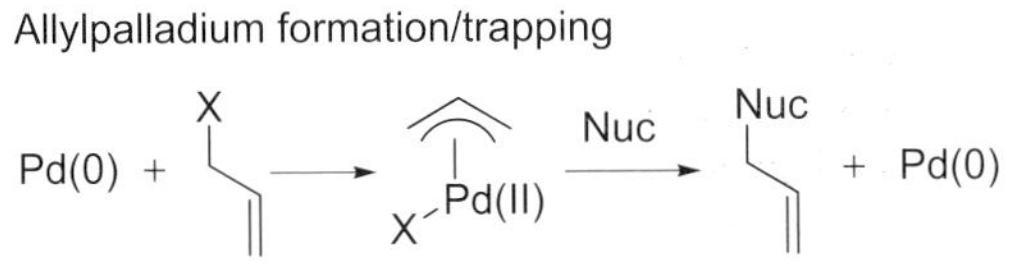

Allylpalladium formation/trapping

$$Pd(0) \; + \; \overset{X}{\underset{\underset{X^{\diagup}Pd(II)}{\nwarrow}}{\big|}} \quad \longrightarrow \quad \overset{\frown}{\underset{X^{\diagup}Pd(II)}{\big|}} \quad \overset{Nuc}{\longrightarrow} \quad \overset{Nuc}{\underset{\big\|}{\big|}} \; + \; Pd(0)$$

In most of the mechanistic schemes described below, the ligands on palladium have been omitted for the sake of clarity and simplicity. However, the nature of the ligands is often crucial for the reactivity and selectivity of palladium catalysts. For example, in many instances Pd-catalyzed amination reactions of aryl halides provide low yields with PPh_3-ligated palladium complexes but proceed in excellent yields when catalysts bearing bulky electron-rich ligands are employed (see Section 1.7.1 below). Thus, the choice of the appropriate catalyst/ligand is often crucial for success in these reactions.

Palladium chemistry involving heterocycles has many unique characteristics stemming from the inherently different structural and electronic properties of heterocyclic molecules in comparison to the corresponding aromatic carbocycles. One salient feature of heterocycles is the marked activation at positions α and γ to the heteroatom. For N-containing heterocycles, the presence of the N-atom polarizes the aromatic ring, thereby activating the α and γ positions, making them more prone to nucleophilic attack. For example, the order of S_NAr displacement of heteroaryl halides with EtO⁻ is [3]:

chloropyrimidine > chloroquinoline > chloropyridine >> chlorobenzene
7×10^5 3×10^2 1   no reaction

The order of reactivities observed in S_NAr displacement reactions often parallels the order of reactivity of aryl halides in oxidative additions to Pd(0). Likewise, the ease with which the oxidative addition occurs for heteroaryl halides can often be predicted on the basis of S_NAr reactivity of a given substrate. In addition, α- and γ-chloroheteroarenes are sufficiently activated for use in Pd-catalyzed reactions with a variety of different catalysts, whereas Pd-catalyzed reactions of unactivated aryl chlorides (e.g. chlorobenzene) typically require large, electron-rich phosphine or N-heterocyclic carbene ligands [4].

The α- and γ-position activation has a remarkable impact on the regiochemical outcome for the Pd-catalyzed reaction of heterocycles. For example, Pd-catalyzed reactions of 2,5-dibromopyridine take place regioselectively at the C(2) position [5], whereas lithium–halogen exchange takes place at C(5) [6]. Palladium-catalyzed reactions of 2,4- or 2,6-dichloropyrimidines take place at C(4) and C(6) more readily than at C(2) [7].

1.1. Oxidative coupling/cyclization

The oxidative coupling/cyclization reaction is the intramolecular union of two arenes with formal loss of H_2 promoted by a Pd(II) species (typically $Pd(OAc)_2$). In an early example of this transformation, treatment of diphenylamines **1** with $Pd(OAc)_2$ in acetic acid yielded carbazoles **2** [8]. The role of acetic acid in such oxidative cyclization processes is to protonate one of the acetate ligands, which affords a more electrophilic cationic Pd(II) species, thereby promoting the initial electrophilic palladation of the aromatic ring.

X = H, CH$_3$; R = CH$_3$, CH$_3$O, Cl, Br, NO$_2$, CO$_2$H

Presumably, the oxidative cyclization of **1** commences with direct palladation at the *ortho*-position, forming σ-arylpalladium(II) complex **3** *in a fashion analogous to a typical electrophilic aromatic substitution* (this notion is useful in predicting the regiochemistry of oxidative cyclizations). The mechanism of the second formal C–H bond functionalization step is not fully elucidated, but may occur either via (a) an intramolecular carbopalladation reaction (migratory insertion) followed by *anti*-β-hydride elimination from **4** (Path A); (b) by σ-bond metathesis (through a four-centered transition state) followed by reductive elimination (Path B); (c) by electrophilic aromatic substitution followed by C–C bond-forming reductive elimination (Path C) [9].

Overall, this transformation leads to the conversion of Pd(II) to Pd(0), which consumes one equivalent of expensive Pd(OAc)$_2$ in most cases. However, progress has been made towards the development of catalytic versions of this transformation, in which catalytic turnover is effected by employing a second oxidant that serves to convert Pd(0) back to Pd(II). For example, Knölker described the oxidative cyclization of **5** using *catalytic* Pd(OAc)$_2$ to afford indole derivative **6** [10]. The reoxidation of Pd(0) to Pd(II) was accomplished using excess cupric acetate in a manner analogous to the Pd-catalyzed Wacker reaction (see Section 1.9) [11].

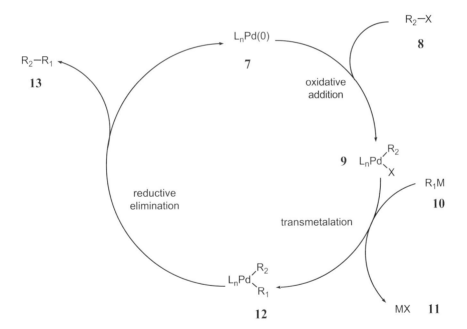

1.2. Cross-coupling reactions with organometallic reagents

One of the most frequently employed methods for the construction of Csp^2–Csp^2 bonds is the Pd-catalyzed cross-coupling of aryl or vinyl halides with main-group organometallic reagents. These reactions generally proceed with retention of regiochemistry and/or olefin geometry, and can be effected with a broad range of substrates. A simplified mechanism for these transformations (shown below) typically commences with oxidative addition of the aryl/vinyl halide to a Pd(0) complex to afford intermediate **9**. The organopalladium halide complex **9** can then undergo transmetalation with the main-group coupling partner (R_1M, **10**) to afford the diorganopalladium species **12**. Carbon–carbon bond-forming reductive elimination from **12** affords the desired cross-coupling product **13** and regenerates the Pd(0) catalyst [12].

The turnover-limiting step in this catalytic cycle depends on the steric and electronic properties of both the organohalide and the organometallic reagent as well as the nature of the main-group metal, and can also be affected by the structure of the metal catalyst. The order of halide reactivity in oxidative addition processes is: I > Br ≈ OTf > Cl, and as noted above, the relative rate of oxidative addition of various aromatic halides is roughly proportional to the relative rates of S_NAr transformations of these substrates. The trans-metalation step [13] is often faster with more nucleophilic/electron-rich organic frag-ments (R_1), and is inhibited by the steric bulk of both coupling partners. When there is more than one group attached to metal M, the order of transmetalation for different substituents is:

$$\text{alkynyl} > \text{vinyl} > \text{aryl} > \text{allyl} \sim \text{benzyl} >> \text{alkyl}$$

Reductive elimination can often be facilitated by the use of catalysts bearing bulky, monodentate phosphine ligands, and is believed to be most rapid when the two coupling partners have opposite electronic properties (i.e. one electron-rich and one electron-poor) [14]. The rates of all three steps in the catalytic cycle are believed to be maximized by employing conditions that favor the formation of intermediates bearing a single phos-phine ligand [15].

In some cases, the active Pd(0) complex is generated *in situ* via reduction of a Pd(II) precatalyst **14** by an organometallic reagent R_1M (**10**). The transmetalation product **15** undergoes a reductive elimination, giving rise to Pd(0) species **7**, along with the homo-coupling product R_1-R_1. This is one of the reasons why the organometallic coupling partners are often used in a slight excess relative to the electrophilic partners.

$$L_nPd\begin{smallmatrix}X\\X\end{smallmatrix} + R_1M \xrightarrow{\text{transmetalation}} L_nPd\begin{smallmatrix}R_1\\R_1\end{smallmatrix} \xrightarrow[\text{elimination}]{\text{reductive}} R_1-R_1 + L_nPd(0)$$

$$\quad\quad\ \mathbf{14}\quad\quad\ \mathbf{10} \quad\quad\quad\quad\quad\quad\quad\quad\quad \mathbf{15}\quad\quad\quad\quad\quad\quad\quad\quad\quad\quad\ \mathbf{7}$$

Cross-coupling reactions are generally divided into subcategories based on the choice of main-group metal. The most commonly employed main-group organometallics are organostannanes (Stille coupling), organoboranes (Suzuki coupling), organosilanes (Hiyama coupling), and organozinc reagents (Negishi coupling). Organostannanes and organozinc reagents will undergo direct transmetalation with the intermediate Pd(Ar)(X) complexes, whereas organoboranes and organosilanes require the presence of base or fluoride to facilitate the transmetalation process via formation of anionic four- or five-coordinate "ate" complexes (see below) [12].

1.2.1. The Negishi coupling

The Negishi reaction is the Pd-catalyzed cross-coupling of organozinc reagents with organohalides or triflates [16]. It is compatible with many functional groups including ketones, esters, amines, and nitriles. The organozinc reagents are usually generated and used *in situ* by transmetalation of Grignard or organolithium reagents with $ZnCl_2$ [17]. In addition, some halides may oxidatively add to Zn(0) to give the corresponding organozinc reagents. In one case, Knochel's group has developed a facile method to prepare organozinc reagents via direct oxidative addition of organohalides to Zn(0) dust [18]. This approach is more advantageous than the usual transmetalation using organolithium or Grignard reagents because of better tolerance of functional groups. The conversion of alkyl halides to alkylzinc reagents often requires the addition of catalytic amounts of NaI and the use of zinc that has been activated via sequential treatment with 1,2-dibromoethane and chlorotrimethylsilane [19]. However, Huo has recently developed a method for gen-eration of alkylzinc reagents from alkyl halides using zinc activated with I_2 in DMA [20].

The high functional group tolerance of these transformations is nicely illustrated by the coupling shown below. Evans and Bach prepared organozinc reagent **17** that contains both ester and amide functional groups via quantitative metalation of iodide **16** with activated Zn/Cu couple. Treatment of 2-iodoimidazole **18** with three equivalents of **17** in the presence of catalytic PdCl$_2$(Ph$_3$P)$_2$ provided adduct **19** in 79% yield [21].

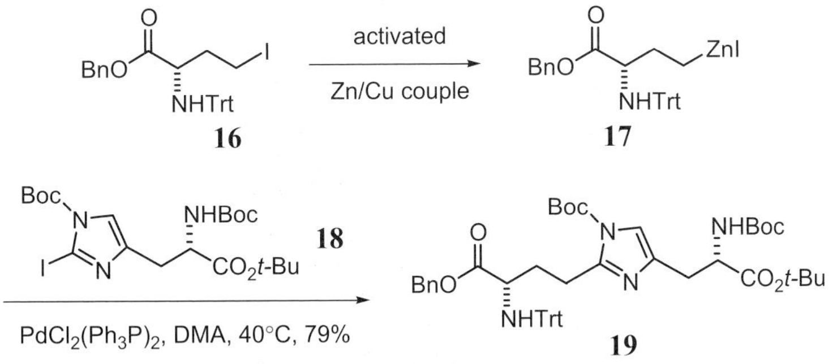

The successful Negishi coupling between *alkylzinc reagent* **17** and 2-iodoimidazole **18** also illustrates that primary aliphatic alkyl groups can be effectively cross-coupled using this method. Although competing β-hydride elimination of the intermediate (aryl)(alkyl)Pd(II) complex can lead to side products (typically the aryl halide is reduced and the alkyl nucleophile is oxidized to an olefin), there are many examples of Negishi couplings of alkylzinc reagents that indicate C–C bond-forming reductive elimination supercedes β-hydride elimination side reactions [16]. Cross-coupling of alkyl organometallics can also be achieved using Suzuki or Stille reactions (described below).

1.2.2. The Suzuki coupling

The Suzuki reaction is one of the most frequently employed Pd-catalyzed C–C bond-forming reactions, and involves the cross-coupling of organoboron reagents with organohalides (or triflates) [22]. This transformation has two significant advantages over related Stille couplings of organostannanes (see below). First of all, organoboron reagents have very low toxicity, especially compared to aryl(trialkyl)tin derivatives. In addition, the boron-containing side products of these transformations are usually easily removed through a simple alkali workup, whereas the trialkyltin halide byproducts of Stille coupling reactions are much more difficult to remove. For these reasons, Suzuki coupling is a more attractive choice than Stille coupling for syntheses conducted on a large scale.

The main limitations of using organoboronic acids in cross-coupling chemistry are three-fold. These compounds are quite reactive towards many nucleophiles and oxidants, which prohibits installation of the boronic acid functional group early in a synthetic sequence. Moreover, purification of boronic acids is often problematic. Finally, since the C–B bond is relatively nonpolar, transmetalation of an organoboron reagent to Pd(II) will not occur without coordination of an anionic base or a fluoride ion to the boron atom to afford a four-coordinate "ate"-complex (e.g. **20** shown below). This requirement poses some limitations with respect to transformations of base- or fluoride-sensitive substrates, although these limitations are usually not insurmountable obstacles and problems can frequently be solved with a judicious choice of reaction conditions.

20

Recently, there has been considerable progresses in the development of new organoboron reagents for cross-coupling chemistry that overcome some of the limitations described above. Most notably, Molander has described the use of air- and moisture-stable alkyl, aryl, and vinyl potassium trifluoroborate salts in Suzuki coupling reactions [23]. For example, the cross-coupling of **21** with **22** afforded biaryl **23** in 64% yield. These salts are easier to purify and handle than the analogous boronic acid derivatives, and they have proven to be stable to epoxidation reactions that allow for the preparation and cross-coupling of functionalized organoboron derivatives [24]. Although there are considerably fewer known heteroarylboron reagents than the related heteroarylstannanes, a growing number of these compounds are appearing in the literature.

As each of the ensuing chapters will entail, pyrrolyl-, indolyl-, pyridinyl-, pyrimidinyl-, thienyl-, pyrazinyl-, and furylboron reagents have been documented, although those derived from thiazoles, oxazoles, and imidazoles have yet to be described in the primary literature. The most popular methods for synthesizing heteroarylboron reagents include: (a) halogen–metal exchange of a heteroaryl halide followed by treatment with a borate; and (b) direct metalation of a heteroarene followed by quenching with a trialkylborate [25]. As a representative example, the conversion of bromopyrimidine **24** to the corresponding boronic acid (**25**) followed by subsequent Suzuki coupling is described below.

The preparation of 5-pyrimidineboronic acid is not trivial because the requisite lithiopyrimidine is prone to react with another substrate molecule by undergoing addition to the azomethine carbon [26]. However, the analogous transformation of 2,4-di-*tert*-butoxy-5-bromopyrimidine (**24**) can be effected in good yield. The halogen-metal exchange can be conducted at –75°C, and the nucleophilic attack of the corresponding organolithium species on the azomethine carbon of a second molecule is retarded due to steric hindrance [26]. The organolithium reagent generated upon lithium–halogen exchange is then treated with tri-*n*-butylborate, followed by basic hydrolysis and acidification to afford 2,4-di-*t*-butoxy-5-pyrimidineboronic acid (**25**). The Suzuki coupling of boronic acid **25** with a variety of heteroaryl halides was conducted to prepare 5-substituted heteroarylpyrimidines such as **26**, which were then hydrolyzed to 5-substituted uracils for investigation of potential antiviral properties [27].

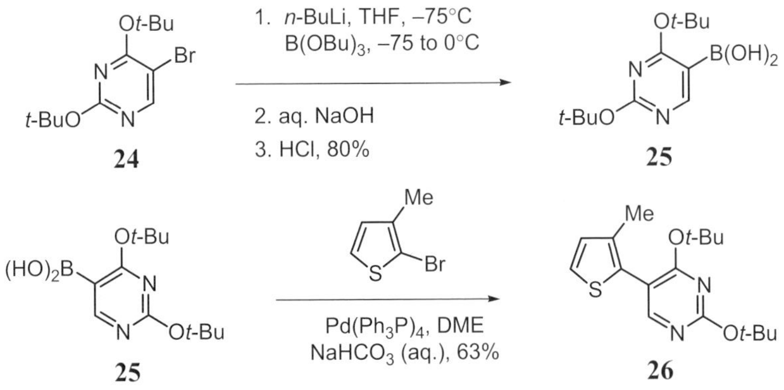

24 → **25**

1. *n*-BuLi, THF, −75°C
   B(OBu)$_3$, −75 to 0°C
2. aq. NaOH
3. HCl, 80%

25 + (3-methyl-2-bromothiophene) → **26**

Pd(Ph$_3$P)$_4$, DME
NaHCO$_3$ (aq.), 63%

Recognizing the distinct difference in reactivity for each site of *N*-protected 2,4,5-triiodoimidazole **27**, Ohta's group executed the regioselective arylation of this substrate [28]. As shown below, coupling of **27** with one equivalent of 3-indolylboronic acid **28** provided imidazolylindole **29**; the Suzuki reaction occurred regioselectively at C(2) of the imidazole ring. Product **29** was subsequently employed in the total synthesis of nor-topsentin C.

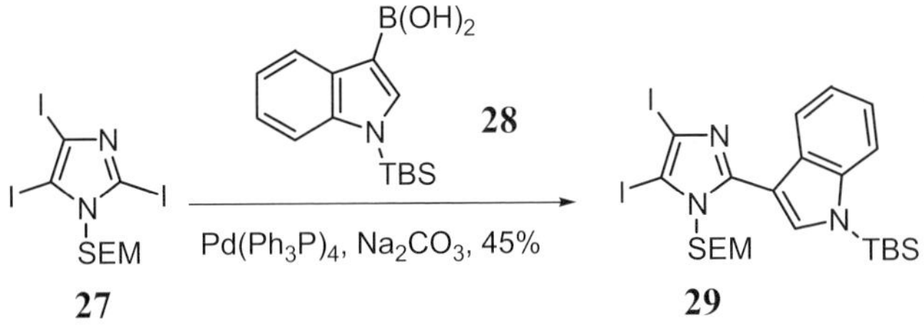

27 + **28** → **29**

Pd(Ph$_3$P)$_4$, Na$_2$CO$_3$, 45%

In recent years, a number of groups have been involved in the development of new, highly active catalysts for Suzuki coupling reactions [29]. For example, the groups of Buchwald and Fu have developed catalysts for the coupling of aryl chlorides with arylboronic acids [30]. With many catalysts (e.g. Pd(PPh$_3$)$_4$), chlorobenzene is very unreactive because the oxidative addition of aryl chlorides to Pd(0) is slow (relative to analogous oxidative additions of aryl iodides, bromides, and triflates). However, in the presence of sterically hindered, electron-rich phosphine ligands (e.g. P(*t*-Bu)$_3$ or 2-dicyclohexylphosphinobiphenyl derivatives), enhanced reactivity is acquired presumably because the oxidative addition of an aryl chloride is more facile with a more electron-rich palladium complex. Extensions of these methods to Suzuki couplings using inexpensive heteroaryl chlorides have also been described [23, 31]. For example, the Pd(OAc)$_2$/**31** catalyzed Suzuki coupling of phenylboronic acid with 4-chloroquinoline (**30**) provided **32** in 97% yield [31]. This methodology should be amenable to large-scale synthesis as the required ligands are now commercially available, the transformations can be effected with low catalyst loadings, and the organic chlorides are less expensive and more readily available than other organic halides.

As in the Negishi reaction, various *alkylboron reagents* have also been successfully coupled with electrophilic partners [32]. For example, Suzuki *et al.* coupled 1-bromo-1-phenylthioethene with 9-[2-(3-cyclohexenyl)ethyl]-9-BBN (34), prepared by a simple addition of 9-borabicyclo[3.3.1]nonane (9-BBN) to 4-vinyl-1-cyclohexene (33), to furnish 4-(3-cyclohexenyl)-2-phenylthio-1-butene (35) in good yield [33].

1.2.3. The Stille coupling

The Stille reaction is the Pd-catalyzed cross-coupling between an organostannane and an electrophile to form a new C–C single bond [34]. It is regarded as the most versatile method in Pd-catalyzed cross-coupling reactions with organometallic reagents for two reasons. First, the organostannanes are readily prepared, purified, and stored. Second, the Stille conditions tolerate a wide variety of functional groups. In contrast to the Suzuki, Kumada, and Heck reactions that are run under basic conditions, the Stille reactions are usually run under neutral conditions. The pitfall of the Stille reaction is the toxicity of stannanes, which makes this process unsuitable for large-scale synthesis.

Stille coupling reactions of organic triflates are often facilitated by added LiCl when conducted in THF solvent. Presumably the chloride anion replaces the triflate ligand on palladium following oxidative addition, which leads to increased stability of the Pd(Ar)(X) species [35]; this additive may also assist in the transmetalation step. However, LiCl is not needed when the Stille reactions are carried out in polar solvents like NMP [36].

Many Stille coupling reactions are improved by the use of added Cu(I) co-catalysts [37]. Farina and Liebeskind postulated that Cu(I) performs dual roles in these transformations [34d, 38]. In ethereal solvents, Cu(I) acts as a phosphine ligand scavenger to facilitate formation of the coordinatively unsaturated Pd(II) intermediate needed to effect transmetalation. In addition, Cu(I) may also react with the organostannane to form a more reactive organocopper reagent in highly polar solvents such as NMP in the presence of soft ligands (e.g. $AsPh_3$).

Heteroarylstannanes are more prevalent than their heteroarylboron counterparts. As the following chapters will entail, all of the heteroarylstannanes are known with the one exception of quinoxalines bearing tin at the C(2) or C(3) positions. The exception could be simply a result of lack of necessity—as this aromatic group could be installed through cross-coupling of the available quinoxalinyl halides.

In comparison to the Stille couplings of simple aryl halides, those involving heteroaryl substrates have their own characteristics [34d]. First of all, many heterocycles have activated positions such as the α and γ positions of pyridines. As a consequence, many heteroaryl chlorides are suitable precursors for Stille couplings. Second, also due to activation of different positions on a heteroarene, regioselective coupling is often possible. Such a phenomenon is not commonly seen in the Stille couplings of simple carbocyclic aryl substrates. A good example is illustrated in the Stille coupling of 2,3-dibromofuran **36**. It is known that nucleophilic substitutions on 2,3-dibromofurans bearing an acceptor substituent at C(5) occur preferentially at C(2) [39]. As the trend of oxidative addition is similar to that of nucleophilic substitution, the first Stille coupling of **36** with an allylstannane occurred predominantly at C(2), giving rise to **37** [40]. Under more forcing conditions, the second Stille coupling then took place to afford 2,3,5-trisubstituted furan **38**.

1.2.4. The Stille–Kelly coupling

The Stille–Kelly reaction is the Pd-catalyzed intramolecular cross-coupling of bis(aryl halide) substrates using distannane reagents [41]. In 1990, *en route* to the total synthesis of pradimicinone, Kelly *et al.* synthesized tricyclic compound **41** by treating dibromide **39** with hexamethylditin in the presence of Pd(PPh$_3$)$_4$. The intermediacy of monostannane **40** was confirmed by three experiments: (1) Pd(PPh$_3$)$_4$ alone does not promote the cyclization of **39** to **41**; (2) in the presence of catalytic Pd(PPh$_3$)$_4$, independently prepared **40** was converted to **41** in the absence of Me$_3$Sn–SnMe$_3$; (3) workup of the Me$_3$Sn–SnMe$_3$-mediated cyclization after partial conversion reveals the presence of the monostannane **40**. The cyclization reaction was also effective with the analogous bis(iodide) and bis(triflate) substrates. The Stille–Kelly reaction offers a means for predetermining the regiochemical outcome of the cyclization of unsymmetrical stilbenes; such control is not always available through photochemical methods or oxidative cyclization.

In Grigg's approach to hippadine (**44**), the biaryl carbon–carbon bond was generated via a Stille–Kelly reaction [42]. Treatment of **42** with hexamethyl ditin in the presence of a Pd(0) catalyst resulted in cyclization to afford lactam **43**. Oxidation of the indoline moiety in **43** using 2,3-dichloro-5,6-dicyano-1,4-benzoquinone (DDQ) then delivered hippadine (**44**). Analogously, the Stille–Kelly coupling of dibromide **45** led directly to hippadine (**44**) [43].

Applications of the aforementioned methodology are also found in the total synthesis of plagiochin D to link a 16-membered biaryl system [44], as well as to the intramolecular cyclization of dibenzyl halides [42, 45]. Additional examples include the synthesis of dithienothiophene (**47**) from dithienyl dibromide **46** [46] and the preparation of carbazole **49** from diaryldibromide **48** [47].

1.2.5. The Kumada coupling

The Kumada coupling is the Pd-catalyzed cross-coupling of a Grignard reagent with an electrophile such as an alkenyl-, aryl-, or heteroaryl halide or triflate [48]. One significant advantage of this reaction is that numerous Grignard reagents are commercially available. Those that are not commercially available may be readily prepared from the corresponding halides. Another advantage is that the reaction can often be run at room temperature or lower. However, a major limitation of this method is the low functional group tolerance of these processes due to the high reactivity and basicity of Grignard reagents. Nonetheless, the presence of functional groups that are not irreversibly transformed by Grignard reagents is usually not problematic. For example, in the synthesis of thienylbenzoic acid **50**, the carboxylic acid moiety survived under the coupling conditions [49]. Although deprotonation of the acid likely occurs, the resulting carboxylate salt does not undergo undesired side reactions, and also does not interfere with the cross-coupling.

In analogy to other Pd-catalyzed cross-couplings of heteroaryl dihalides described above, regioselective Kumada coupling can often be achieved. For example, 2,5-dibromopyridine was mono-heteroarylated regioselectively at C(2) with one equivalent of a 2-indolyl Grignard reagent affording indolylpyridine **51**. Subsequent Kumada coupling of **51** with a second Grignard reagent, 2-thienylmagnesium bromide, under more rigorous conditions gave heterotriaryl **52** [50].

1.2.6. The Hiyama coupling

Hiyama-type cross-coupling reactions employ organosilanes as the main-group organometallic component [51]. As is the case in Suzuki reactions of organoboron derivatives, transmetalation of an organosilicon reagent does not occur under normal Pd-catalyzed cross-coupling conditions due to the relatively nonpolar nature of the C—Si bond. However, the C—Si bond can be activated by a nucleophile such as F^- or HO^- through formation of a pentacoordinated silicate, which weakens the C—Si bond by enhancing the polarization. As a result, the transmetalation becomes more facile and the cross-coupling proceeds readily. One of the advantages of the Hiyama coupling is the nontoxic nature of organosilicon reagents. Other advantages include better tolerance of functional groups in comparison to other strong nucleophilic organometallic reagents, and the low reactivity of the organosilicon reagents towards many standard organic reaction conditions. The combination of these characteristics makes the Hiyama coupling an attractive alternative to other Pd-catalyzed cross-couplings. Therefore, this reaction has much promise for future exploration and exploitation.

The Hiyama couplings of heterocycles are still being developed to their full potential. Nevertheless, several heteroaryl halides have been cross-coupled with aryl or heteroaryl silicon reagents [52]. For example, in the presence of catalytic π-allylpalladium chloride dimer and two equivalents of KF, the cross-coupling of ethyl(2-thienyl)difluorosilane (**53**) and methyl 3-iodo-2-thiophenecarboxylate led to bis-thiophene **54** under relatively forcing conditions [53].

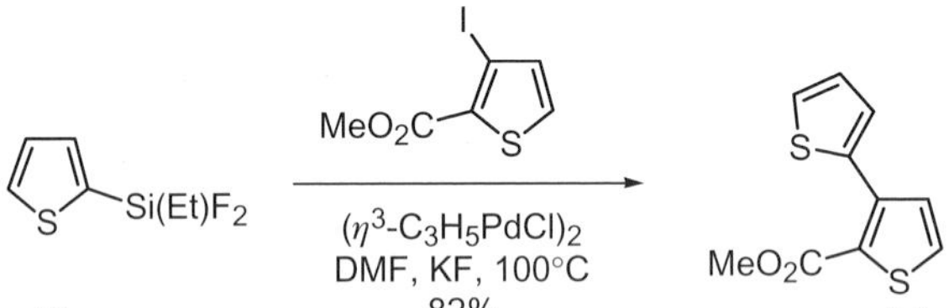

The Hiyama coupling of aryltrimethoxysilane **55** and 3-bromopyridine provided arylpyridine **56** with the aid of TBAF [54]. The aryltrimethoxysilane reagents are generally less susceptible to hydrolysis than the analogous chlorosilanes.

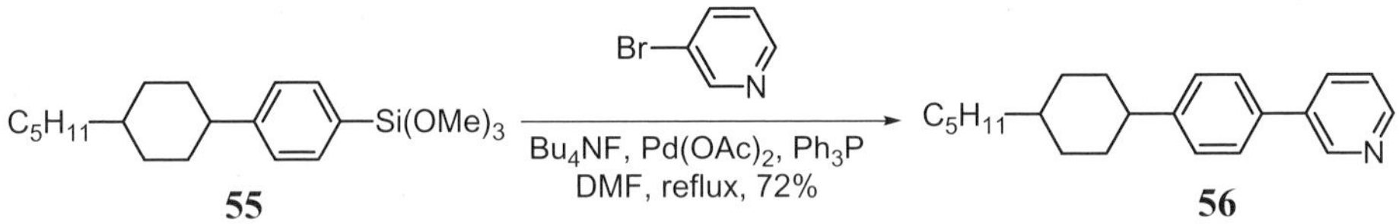

1.3. The Sonogashira reaction

The Stephens–Castro reaction is the cross-coupling reaction of copper(I) arylacetylides with iodoalkenes [55]. The scope of this process is sometimes limited by the vigorous reaction conditions and by the difficulty in preparing cuprous acetylides. However, the mechanistically related Sonogashira reaction [56] allows for the direct cross-coupling of terminal alkynes with aryl halides under mild conditions through use of catalytic bis(triphenylphosphine)-palladium dichloride and CuI as the co-catalyst in the presence of an aliphatic amine. The formation of diphenylacetylene exemplifies the original Sonogashira reaction conditions:

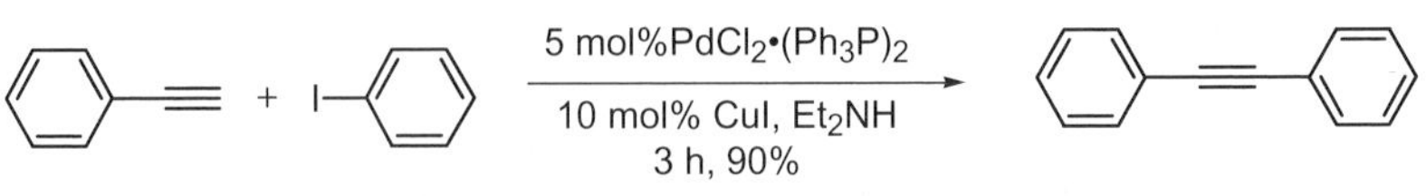

It is speculated that an alkynylcopper species, which undergoes the transmetalation process more readily, is generated during the reaction with the aid of an amine. The aliphatic amine also serves as a reducing agent to generate Pd(0) from the Pd(II) precatalyst.

The Sonogashira reaction has enjoyed tremendous success in the synthesis of almost all types of heteroarylacetylenes because of the extremely mild reaction conditions and great tolerance of nearly all types of functional groups. A representative reaction with conditions similar to those described in Sonogashira's original report [57]:

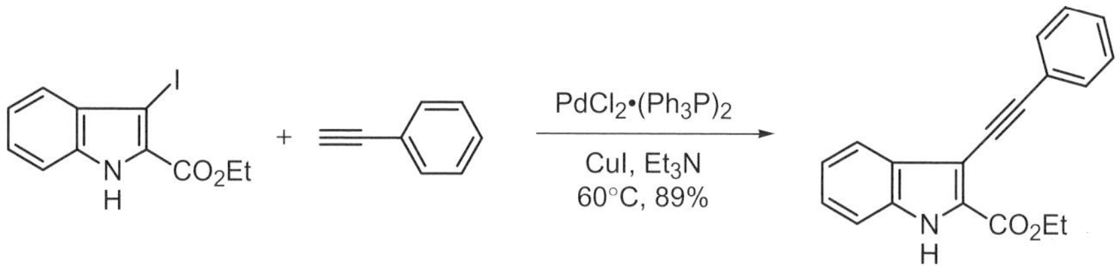

Quite different from the original Sonogashira conditions, the following cross-coupling was achieved using the Jeffery's "ligand-free" protocol (see Section 1.4 below) in the absence of cuprous iodide [58]. Much recent work has focused on the development of other conditions for "copper-free" Sonogashira couplings [59].

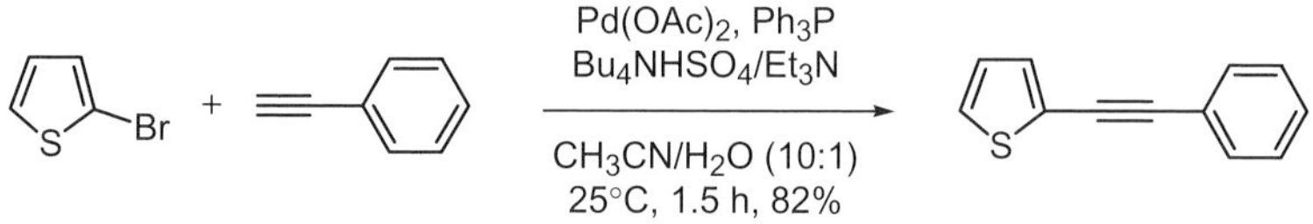

1.4. The Heck, intramolecular Heck, and heteroaryl Heck reactions

The Heck reaction, first disclosed independently by Mizoroki and Heck in the early 1970s [60], is the Pd-catalyzed coupling reaction of organohalides (or triflates) with olefins. In recent years, this transformation has become an indispensable tool for organic chemists [61]. Inevitably, many applications to heterocyclic chemistry have been pursued and successfully executed. In a representative example, Ohta *et al.* treated 2-chloro-3,6-dimethylpyrazine (**57**) with styrene in the presence of KOAc and catalytic Pd(PPh$_3$)$_4$ to furnish (*E*)-2,5-dimethyl-3-styrylpyrazine (**58**) [62]. The high *E*-selectivity is a consequence of the reaction mechanism (described below).

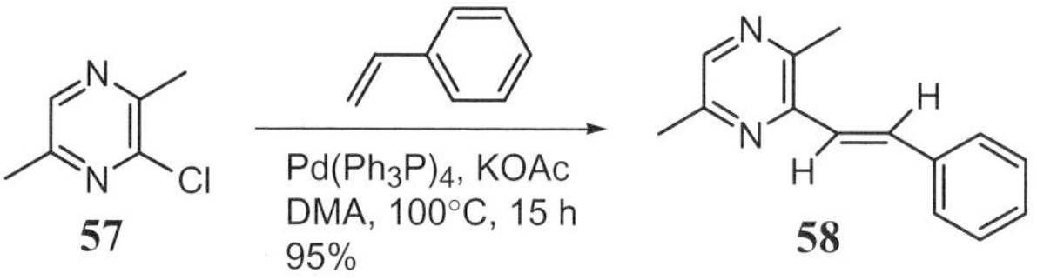

The mechanism of this transformation likely commences via oxidative addition of the 2-chloropyrazine **57** to coordinatively unsaturated Pd(0), giving rise to σ-pyrazinylpalladium(II) complex **59** [61]. Subsequently, complex **59** undergoes a *syn*-insertion of styrene into the Pd–C bond to afford alkylpalladium(II) complex **60**. Steric effects govern the regiochemistry in this particular case, with the arylation occurring at the less hindered, unsubstituted olefinic carbon. However, electronic effects can control regioselectivities with other substrate combinations. The intermediate **60** is transformed via a σ-bond rotation to conformational isomer **61**, which then undergoes *syn*-β-hydride elimination to generate the (alkene)Pd(H)(Cl) complex **62**. Because the σ-bond rotation and -β-hydride elimination are thermodynamically controlled and reversible, the thermody-namically more stable *E*-olefin **58** is ultimately generated, and hydridopalladium(II) chloride (**63**) is released. With the aid of a base, reductive elimination of HCl then takes place, regenerating Pd(0) to close the catalytic cycle [63].

The rate of the alkene insertion step is quite sensitive to steric effects, and intermolecular reactions are generally limited to mono- and disubstituted alkenes. However, intramolecular reactions of tri- and tetra-substituted olefins are well precedented [61b].

H–Pd(II)–Cl $\xrightarrow{\text{base}}$ Pd(0) + HCl

In 1984, Jeffery discovered that under "ligand-free" conditions, Pd-catalyzed vinylation of organic halides proceeds at or near room temperature, whereas normal Heck reactions require higher temperatures [64]. Jeffery's "ligand-free" conditions have been broadly applied to a variety of Heck arylations that were not feasible using more standard reaction conditions. For example, efforts to use classical Heck-reaction conditions for the conversion of *N*-allyl-*N*-benzyl(3-bromoquinoxalin-2-yl)amine (**64**) to 1-benzyl-3-methylpyrrolo[2,3-*b*]quinoxaline (**65**), resulted in slow reactions and low yields [65], which may be attributed to the poisoning of the palladium catalyst via complexation to the aminoquinoxalines. In contrast, the Jeffery conditions afforded the desired product in 83% yield. The enhanced reactivity and yield under Jeffery's "ligand-free" conditions may be due to the coordination/solvation of the palladium intermediates by bromide ions present in the reaction mixture, which presumably prevents the precipitation of Pd(0).

The intramolecular version of the Heck reaction has been particularly useful, enabling elegant syntheses of many complex molecules [61b]. For example, Rawal and Iwasa carried out an intramolecular Heck cyclization of pentacyclic lactam **66** with a pendant vinyl iodide moiety using Jeffery's "ligand-free" conditions to afford hexacyclic *strychnan* alkaloid precursor **67** with complete stereochemical control [66]. The Mori–Ban indole synthesis (Section 1.10, see below) is another good example of the utility of this process.

Overman's group [67] enlisted an intramolecular Heck reaction to form a quaternary center in their synthesis of gelsemine. When the cyclization precursor **68** was subjected to "ligandless" conditions [Pd₂(dba)₃, Et₃N] in the weakly coordinating solvent toluene, the quaternary center was formed with ~9:1 diastereoselectivity (**70:69** = 89:11). In contrast, use of the coordinating solvent THF and a stoichiometric amount of an added silver salt completely reversed the sense of asymmetric induction in the cyclization reaction (**70:69** = 3:97).

There has also been considerable interest and progress in the development of *asymmetric Heck reactions* (AHR) [68]. This methodology allows for the preparation of enantiomerically enriched products from achiral substrates using a *catalytic* amount of a chiral palladium complex, making the process practical and economical. For example, treatment of triflate **71** with catalytic Pd(OAc)₂/(R)-BINAP provided oxindole **72** in 96% yield and 88% ee [69].

One class of transformations that illustrate the striking difference in reactivity between heteroarenes and carbocyclic arenes is the heteroaryl Heck reaction, in which an aryl or heteroaryl halide is coupled directly with a heteroaromatic compound to afford a biaryl product (formally a C−H bond functionalization process). Intermolecular Heck reactions involving the functionalization of aromatic carbocycles with aryl/heteroaryl halides are rare [70], whereas heterocycles including thiophenes, furans, thiazoles, oxazoles, imidazoles, pyrroles, and indoles, etc. are excellent substrates for these types of transformations. For example, the heteroaryl Heck reaction of 2-chloro-3,6-diethylpyrazine (**73**) and benzoxazole leads to arylation of the benzoxazole C(2) position afford pyrazinylbenzoxazole **74** [71].

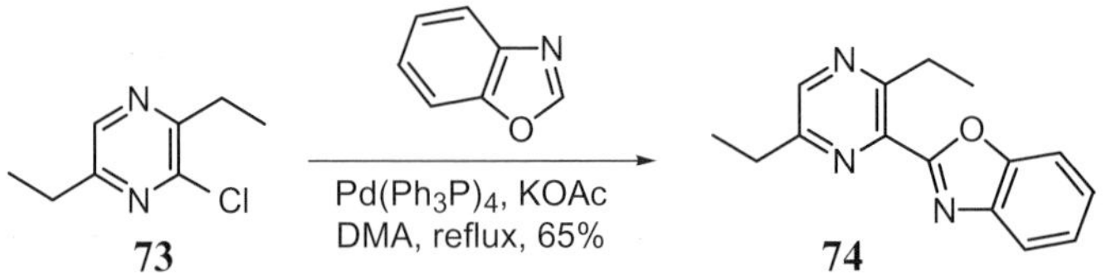

Pd(Ph$_3$P)$_4$, KOAc
DMA, reflux, 65%

73   **74**

A plausible mechanism for the above reaction involves oxidative addition of the heteroaryl chloride to Pd(0) to provide Pd(II) intermediate **75**, which subsequently inserts into benzoxazole to form the arylpalladium(II) complex **76**. β-Hydride elimination of **76** would afford **77** with concomitant regeneration of Pd(0) for the next catalytic cycle.

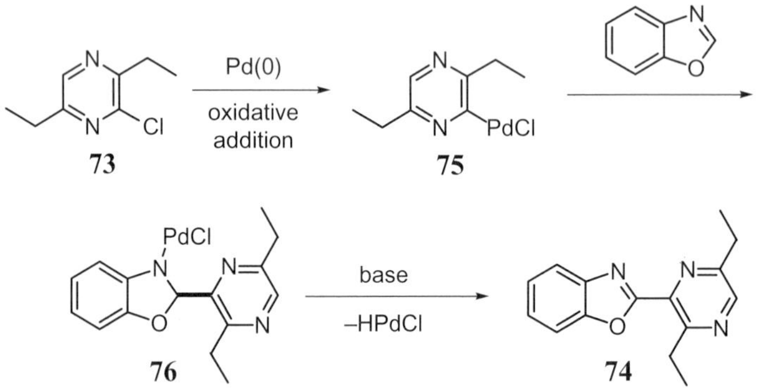

Pd(0)
oxidative
addition

73   **75**

base
–HPdCl

76   **74**

An intramolecular heteroaryl Heck reaction was the pivotal step in the synthesis of 5-butyl-1-methyl-1*H*-imidazo[4,5-*c*]quinolin-4(5*H*)-one (**77**), a potent antiasthmatic agent [72]. The optimum yield was obtained under Jeffery's "ligand-free" conditions.

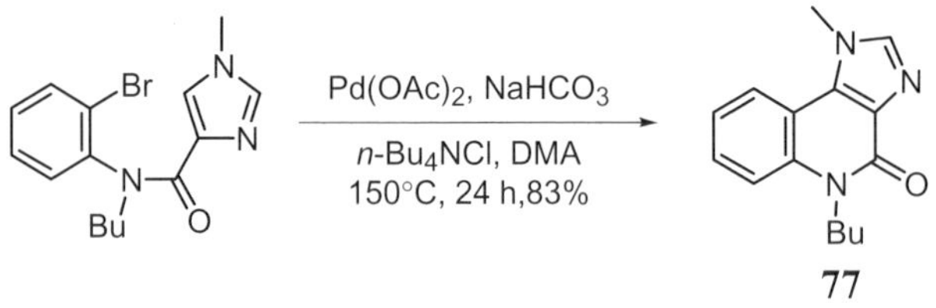

Pd(OAc)$_2$, NaHCO$_3$

n-Bu$_4$NCl, DMA
150°C, 24 h, 83%

77

Although intermolecular Heck-type reactions of carbocyclic arenes are uncommon, related intramolecular reactions are well precedented as illustrated by the following two examples [73, 74].

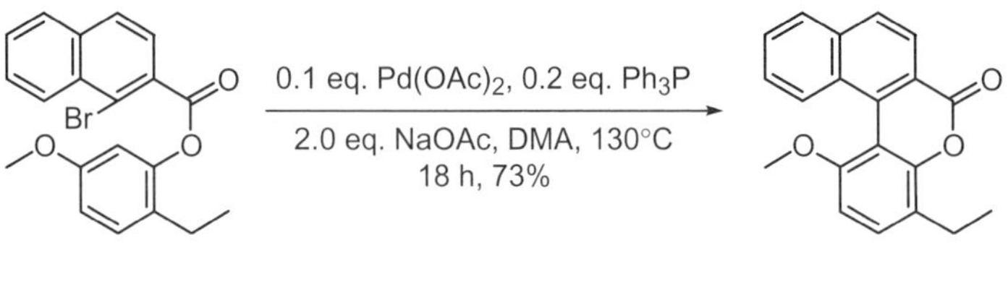

0.1 eq. Pd(OAc)$_2$, 0.2 eq. Ph$_3$P

2.0 eq. NaOAc, DMA, 130°C
18 h, 73%

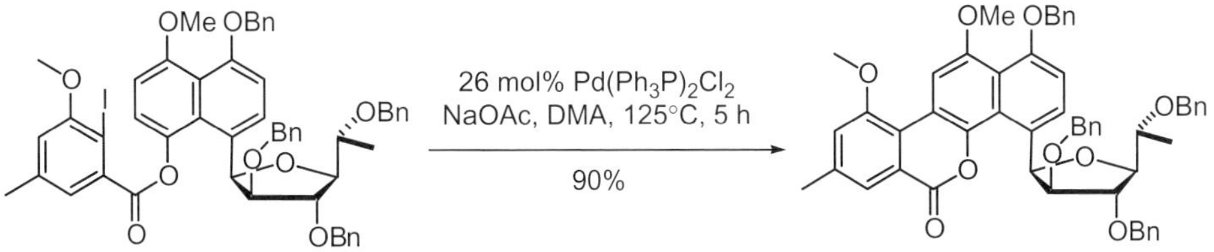

The development of new catalysts for Heck arylations of unactivated aryl chloride substrates has also been a recent focus of research in this area. In late 1990s, Reetz [75] and Fu [76] successfully conducted intermolecular Heck reactions using aryl chlorides as substrates, as exemplified by the conversion of *p*-chloroanisole to adduct **78** [76]. A few examples of Heck reactions involving unactivated heteroaryl chlorides have also been described [77].

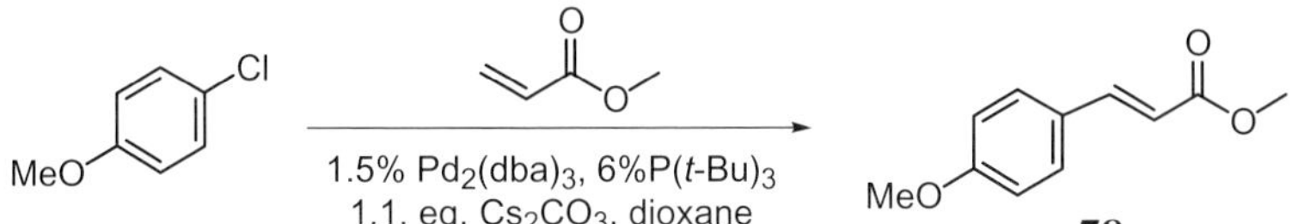

1.5. Carbonylation reactions

Palladium-catalyzed carbonylation of heteroaryl halides provides a quick entry to heteroaryl carbonyl compounds such as heteroaryl aldehydes, carboxylic acids, ketones, esters, amides, α-keto esters, and α-keto amides. In addition, Pd-catalyzed alkoxycarbonylation and aminocarbonylation are compatible with many functional groups, and therefore have advantages over conventional methods for preparing esters and amides [78].

Representative examples of Pd-catalyzed heterocycle carbonylation reactions are shown below. Alkoxycarbonylation of 2,3-dichloro-5-(methoxymethyl)pyridine (**79**) took place regioselectively at C(2) to give ester **80** [79]. Aminocarbonylation of 2,5-dibromo-3-methylpyridine also proceeded preferentially at C(2) to give amide **81** despite the steric hindrance of the 3-methyl group [80].

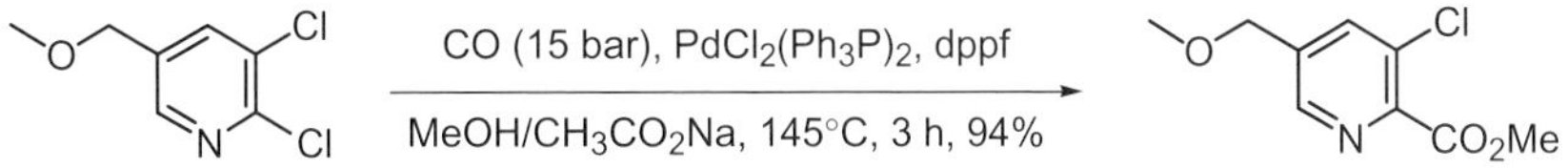

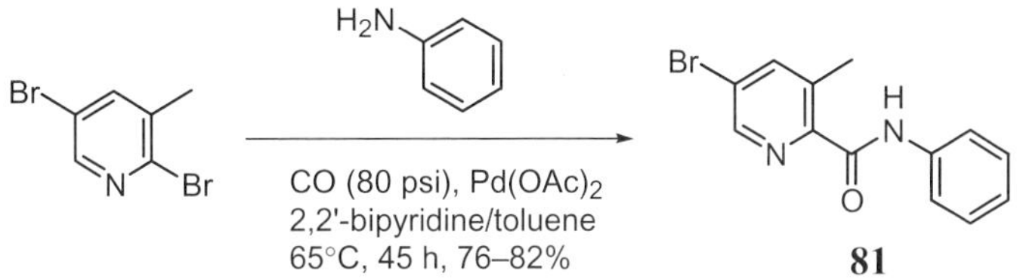

1.6. The Pd-catalyzed C–P bond formation

The formation of Csp^2–P bonds is readily achievable using the Michaelis–Arbuzov reaction. Although this method is not applicable to the formation of heteroaryl Csp^2–P bonds, Pd-catalyzed reactions provide suitable approaches to such compounds [81].

The first example of Pd-catalyzed C–P bond formation was described by Hirao *et al.* [82]. As shown below, diethyl-3-pyridylphosphonate was prepared via Pd-catalyzed phosphonation of 3-bromopyridine with diethyl phosphite [82b,c].

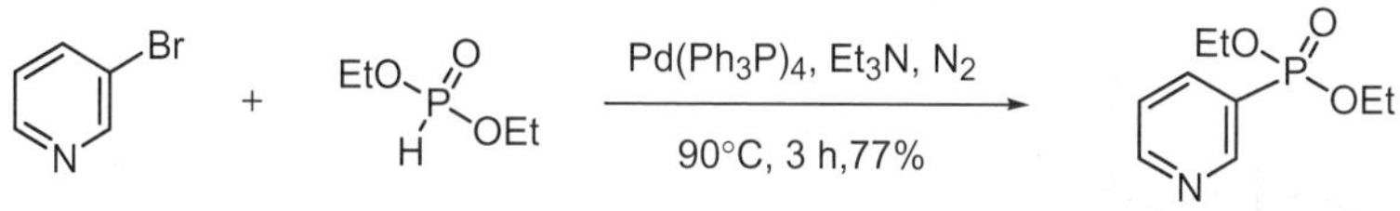

Xu *et al.* have further expanded the scope of Pd-catalyzed Csp^2–P bond formation reactions to include a variety of different organophosphorous coupling partners [83]. For example, the coupling of 2-bromothiophene with *n*-butyl phenylphosphite was effected to form *n*-butyl diarylphosphinate **82**. This chemistry also allows for the synthesis of alkyl arylphenylphosphine oxides, functionalized alkyl arylphenylphosphinates, alkenyl arylphenylphosphinates, alkenylbenzyl-phosphine oxides, as well as chiral, nonracemic isopropyl arylmethylphosphinates [84]. Intramolecular Pd-catalyzed Csp^2–P bond formation has also been reported [84].

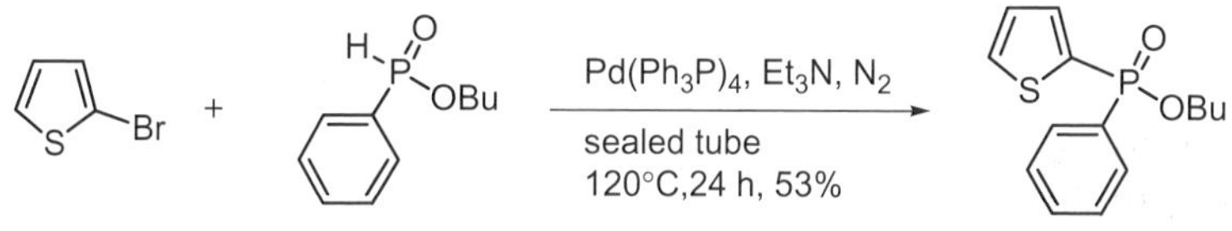

A plausible mechanism for the Pd-catalyzed Csp^2–P bond forming reactions proposed by Xu *et al.* is shown below. Oxidative addition of 2-bromothiophene to Pd(0) results in Pd(II) intermediate **83**, which then undergoes a reaction with *n*-butyl phenylphosphite and triethylamine to afford **84**. Finally, reductive elimination of **84** furnishes unsymmetrical alkyl arylphosphinate **82**, regenerating Pd(0). Although this mechanism seems quite reasonable based on more recent studies of Pd-catalyzed carbon-heteroatom bond-forming processes (see below), alternative mechanisms have also been proposed [81, 82c].

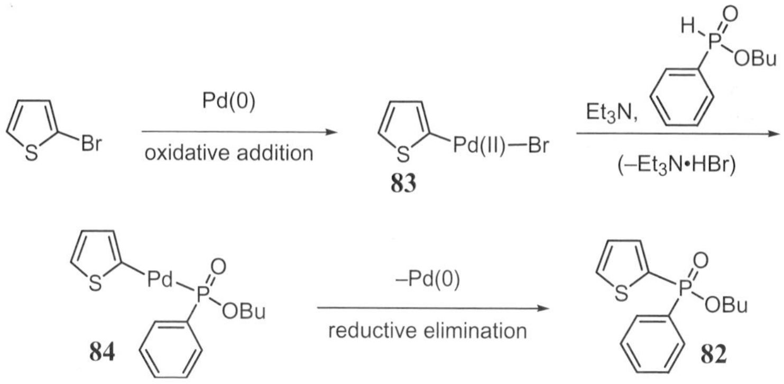

Aryl triflates have also proven to be viable substrates for the Pd-catalyzed Csp^2–P bond formation reactions [85]. Intriguingly, phosphorylation can be achieved from the Pd-catalyzed coupling of alkenyl triflates with not only dialkylphosphites, but also with *hypophosphorous acid* [85b]. Thus, phosphinic acid **86** was obtained when triflate **85** was treated with hypophosphorous acid in the presence of Pd(Ph$_3$P)$_4$. Due to the broad availability of alkenyl triflates and the mild reaction conditions employed in their coupling reactions, alkenyl triflates have certain advantages over the corresponding alkenyl halides as substrates for Pd-catalyzed phosphorylations to make alkenyl phosphonates or phosphinates.

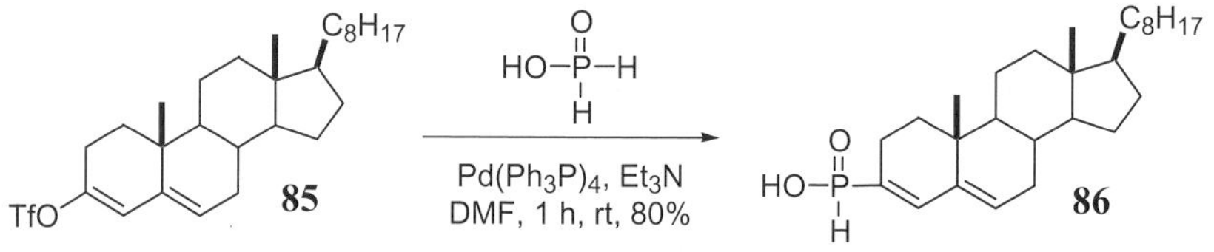

1.7. Palladium-catalyzed C—N bond and C—O bond-forming reactions

1.7.1. The Buchwald–Hartwig Pd-catalyzed aryl C—N bond formation

The first example of a Pd-catalyzed aryl C—N bond forming-reaction was described by the Migita group in 1983; the cross-coupling of bromobenzene with N,N-diethylaminotributyltin in the presence of catalytic Cl$_2$Pd[P(o-tol)$_3$]$_2$ led to the formation of N,N-diethylaniline [86]. Although the Migita reaction employs only a catalytic amount of palladium, this method suffers from the use of toxic and moisture-sensitive aminostannanes. In 1984, a Pd-mediated C—N bond formation was carried out via the intramolecular amination of an aryl halide in Boger's synthesis of lavendamycin [87], although 1.5 equivalents of Pd(Ph$_3$P)$_4$ were employed.

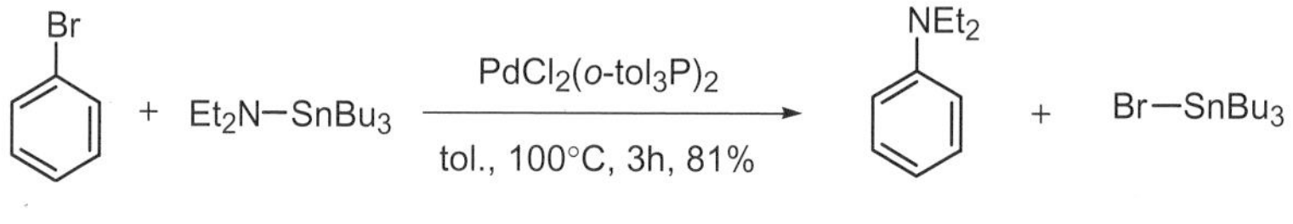

In 1995, Buchwald and Hartwig independently discovered the direct Pd-catalyzed cross-coupling of aryl halides with amines in the presence of a stoichiometric amount of base [88]. This field is rapidly expanding, and many reviews covering the scope and limitations of this amination have been published since the seminal 1995 papers [89]. In the context of heterocycle synthesis, one example is given to showcase the utility and mechanism of this reaction. As shown below, the Pd(OAc)$_2$/**87** catalyzed reaction of 3-bromopyridine with N-methylaniline afforded **88** in 94% yield [90]. Applications of this chemistry to the synthesis of specific classes of heterocycles may be found in their respective chapters.

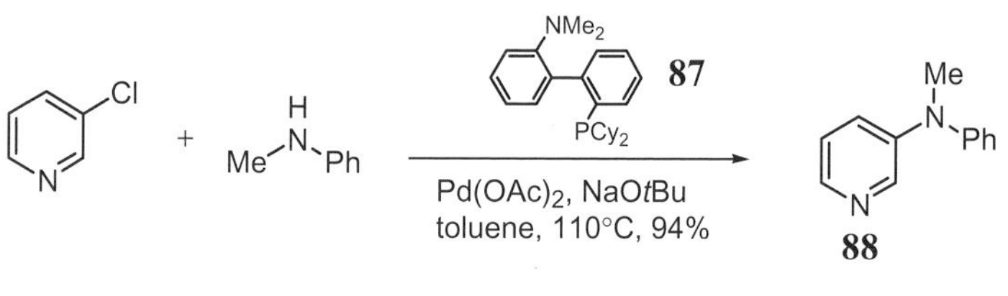

A simplified version of the mechanism of these reactions is shown below, and commences with the oxidative addition of the aryl bromide to Pd(0), giving rise to the palladium complex **89** [91]. This complex is subsequently converted to amido complex **90** via reaction with the amine and the base. Reductive elimination of **90** then gives the amination product **88** with concomitant regeneration of the Pd(0) catalyst. If the amine has β-hydrogen atoms (in amido complex **90**), β-hydride elimination is a potential competing pathway that leads ultimately to the reduction of the aryl bromide. However, reductive elimination is faster than β-hydride elimination in most cases. Although the ligands have been omitted from the mechanistic scheme shown below, they are absolutely critical to the success of these transformations. Use of nonoptimal ligands frequently leads to low yields and considerable amounts of side products from aryl bromide reduction.

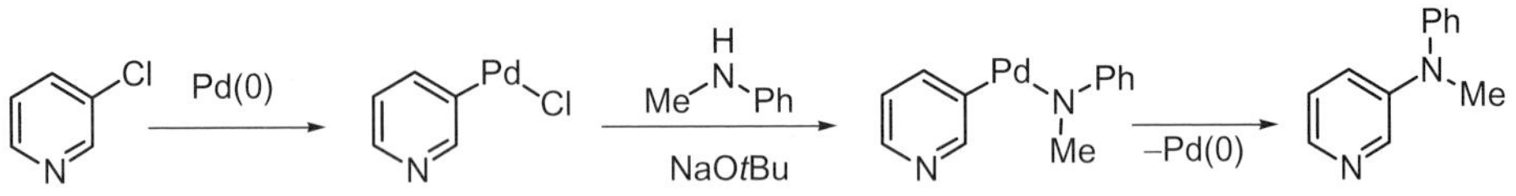

The electrophiles in catalytic amination reactions can be either aryl halides or triflates possessing electron-rich, neutral, or electron-poor ring systems, whereas amines can range from aliphatic to aromatic and primary to secondary amines. The Pd-catalyzed C—N bond formation works both inter- and intramolecularly.

1.7.2. Palladium-catalyzed aryl C—O bond formation

In comparison to amines and thiols, alcohols are much less nucleophilic. Consequently, the reductive elimination process to form the C—O bond [91b, 92] is much slower than the analogous processes that form C—N and C—S bonds [89]. In 1996, Palucki *et al.* described the first intramolecular Pd-catalyzed synthesis of cyclic aryl ethers from *o*-haloaryl-substituted alcohols [93]. For example, 3-(2-bromophenyl)-2-methyl-2-butanol (**91**) was converted to 2,2-dimethylchroman (**92**) under the agency of catalytic Pd(OAc)$_2$ in the presence (*S*)-(–)-2,2′-bis(di-*p*-tolylphosphino)-1,1′-binaphthyl (Tol-BINAP) as the ligand and K$_2$CO$_3$ as the base. The catalysts originally employed in this method worked well for the tertiary alcohols, and moderately well for cyclic secondary alcohols, but not for acyclic secondary alcohols. However, the use of new, di-*tert*-butylphosphinobiaryl ligands has significantly expanded the scope of this process [94].

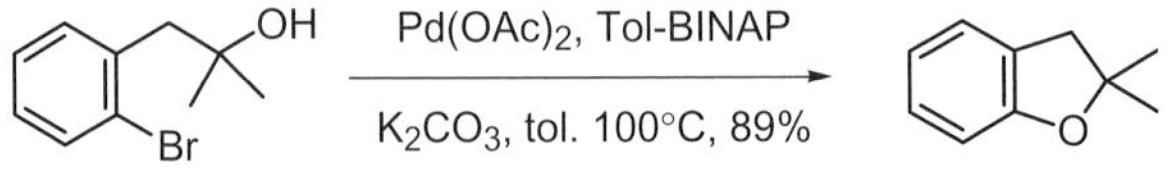

The Pd-catalyzed intermolecular C—O bond formation has also been achieved, and is also most effective with bulky, electron-rich phosphines [91b, 95]. In addition, Buchwald has shown that these types of ligands facilitate the Pd-catalyzed diaryl ether formation [96]. When 2-(di-*tert*-butylphosphino)biphenyl (**95**) was used as the ligand, the coupling of triflate **93** and phenol **94** afforded diaryl ether **96** in the presence of Pd(OAc)$_2$ and K$_3$PO$_4$. The methodology also worked for electron-poor, neutral, and electron-rich aryl halides.

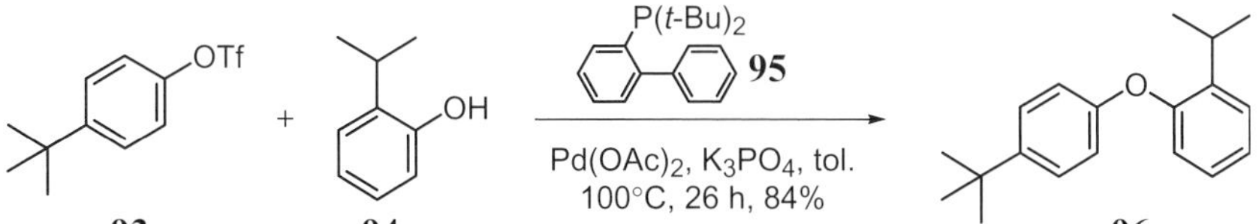

The mechanism for Pd-catalyzed C–O bond formation is believed to be similar to that of C–N bond formation [92]. Applications of this method to heterocyclic chemistry are uncommon [95e], partially because the S_NAr displacements of many heteroaryl halides with alkoxides are facile without the aid of palladium.

1.7.3. Palladium-catalyzed heteroatom-Heck reactions

As described above in Section 1.4, intramolecular Heck-reactions proceed via *syn*-alkene insertion into Pd–C bonds of intermediate (Ar)Pd(X) complexes. Although alkene insertions into late-transition metal–carbon bonds are common and well precedented, the analogous *syn*-insertion of alkenes into palladium-heteroatom bonds of $(R_2N)Pd(X)$ complexes are quite rare. However, the known examples of these transformations (heteroatom-Heck reactions) provide useful routes into an array of nitrogen heterocycles.

The pioneering studies in this area were reported in 1999 by Narasaka, who demonstrated intramolecular heteroatom Heck-type reactions of *O*-pentafluorobenzoyl oximes [97]. As shown below, treatment of unsaturated substrate **97** with a catalytic amount of $Pd(PPh_3)_4$ in the presence of triethyl amine provided pyrrole **98** upon workup with chlorotrimethylsilane. The mechanism of this reaction proceeds via oxidative addition of the N–O bond to afford **99**, which undergoes alkene insertion into the Pd–N bond to provide alkyl-palladium complex **100**. The exo-methylene product **101** is generated by β-hydride elimination from **100**, and isomerization to the desired pyrrole **98** occurs when chlorotrimethylsilane is added.

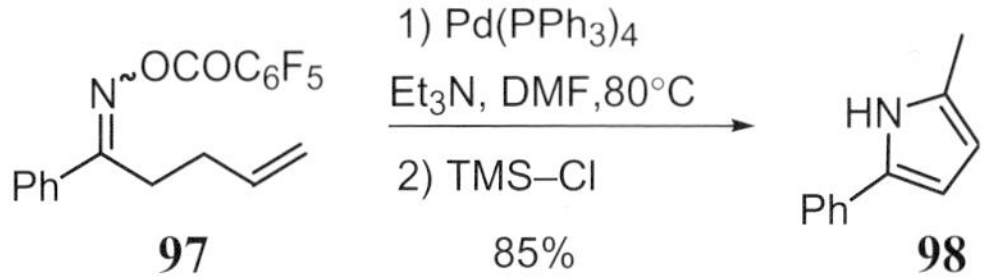

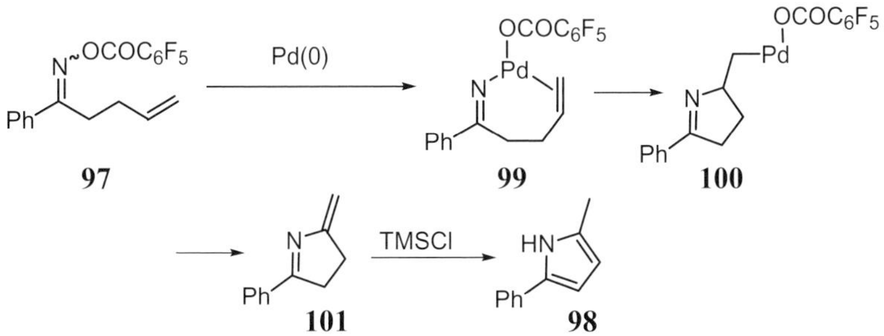

This type of transformation has been employed for the synthesis of several different nitrogen heterocycles [98], and an analogous reaction of unsaturated *N*-chloroamines that generates pyrrolidine products has also been described [99].

1.7.4. Palladium-catalyzed carboamination and carboetherification of alkenes

The first Pd-catalyzed carboamination and carboetherification reactions of γ-amino or γ-hydroxy alkenes with aryl halides were described by Wolfe in 2004 [100]. As shown below, treatment of these substrates with aryl or heteroaryl bromides in the presence of NaO*t*Bu and a palladium catalyst provides substituted tetrahydrofuran or pyrrolidine products.

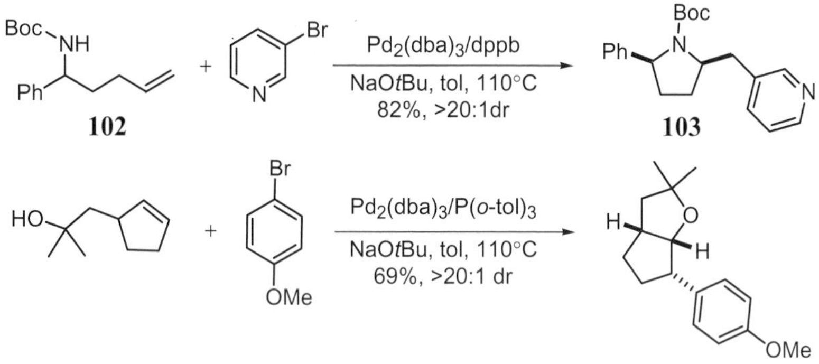

The carboamination has also been utilized in a tandem process that effects the conversion of 2-allylanilines to *N*-aryl-2-benzylindolines in a one-pot transformation [101].

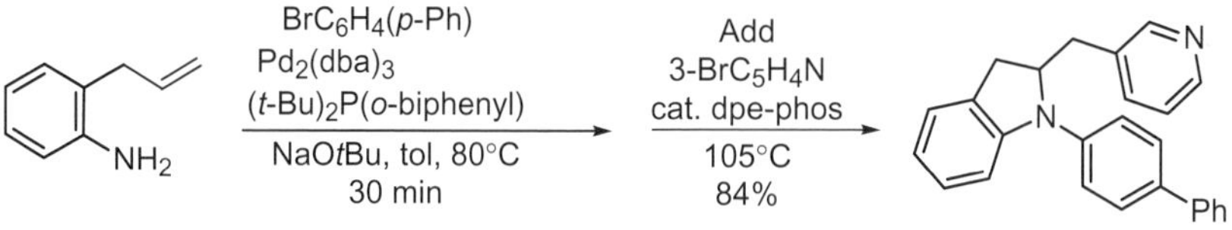

The proposed mechanism for the pyrrolidine-forming reactions is shown below [102], and commences with oxidative addition of the aryl halide to Pd(0) to afford **104**. This intermediate is converted to palladium(aryl)(amido) complex **105**, which undergoes intramolecular *syn*-insertion of the alkene into the Pd–N bond to generate **106**. Complex **106** then liberates the product **103** via C–C bond-forming reductive elimination with concomitant regeneration of the Pd(0) catalyst. Although complexes similar to **105** are well documented in Pd-catalyzed arylamination processes (Section 1.7.1), the carboamination reactions are the first examples of transformations involving alkene insertion into these intermediates. The mechanism for tetrahydrofuran formation is believed to proceed in an analogous manner.

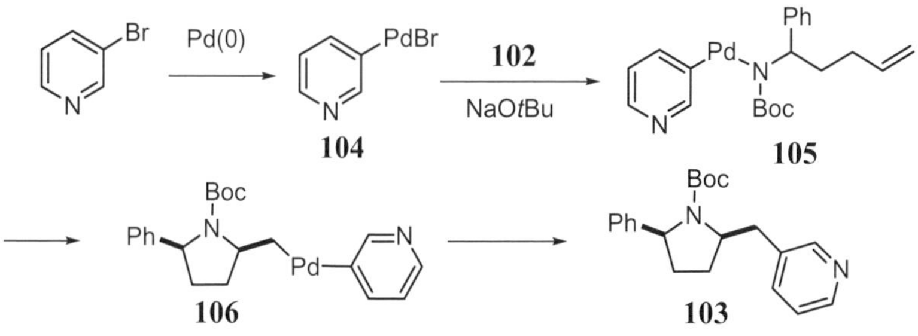

1.7.5. Palladium-catalyzed C—H bond oxidation/halogenation

The Pd-catalyzed oxidation of unfunctionalized C—H bonds has recently been described by Sanford. These reactions lead to the direct, regioselective installation of hydroxyl groups or halogen atoms onto aromatic and heteroaromatic ring systems. For example, benzo[*h*]quinoline is selectively converted to 10-chlorobenzo[*h*]quinoline upon treatment with catalytic Pd(OAc)$_2$ and NCS [103]. As shown below, these transformations are also effective for the installation of oxygenated functional groups including acetates and alkyl ethers. The oxidative functionalization of *sp*^{3}C—H bonds has also been achieved [104].

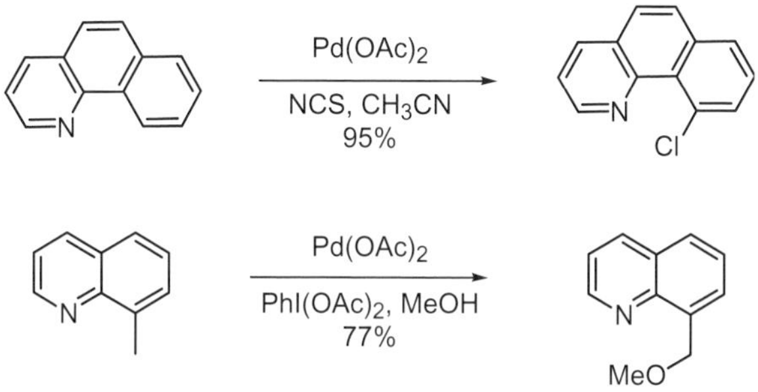

These reactions are believed to proceed via an unusual Pd(II)–Pd(IV)–Pd(II) catalytic cycle that is initiated by directed palladation of the substrate to afford metallacycle **107**. This Pd(II) complex then is oxidized to Pd(IV) intermediate **108**, which then undergoes carbon-heteroatom bond-forming reductive elimination to generate the product and regenerate the Pd(II) catalyst.

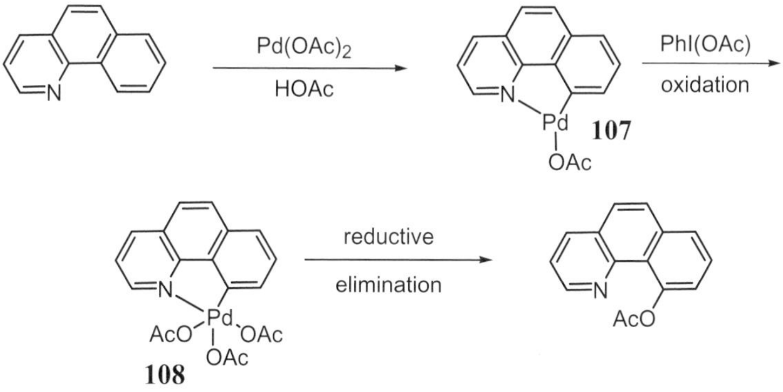

1.8. The Tsuji–Trost reaction

The Tsuji–Trost reaction is the Pd-catalyzed allylation of nucleophiles [105] with allylic halides, acetates, carbonates, etc. This transformation proceeds via intermediate allylpalladium complexes (e.g. **110**), and typically proceeds with overall retention of stereochemistry. In addition, the trapping of the intermediate allylpalladium complex usually occurs at the least hindered carbon. A representative example of this transformation is shown below in an application to the formation of an *N*-glycosidic bond. Treatment of 2,3-unsaturated hexopyranoside **109** with imidazole in the presence of a Pd(0) catalyst afforded *N*-glycopyranoside **111** [106]. The oxidative addition of allylic substrate **109** to Pd(0) occurs with inversion of configuration to provide the π-allyl complex **110**. Nucleophilic attack of imidazole on the least hindered allylic carbon also proceeds with inversion of configuration to give **111**. More recent studies have expanded the scope of these glycosidation reactions [107] and allow for catalyst-based control of anomeric stereochemistry [107c].

1.9. The Wacker-type reactions

Wacker-type reactions are Pd(II)-catalyzed transformations involving heteroatom nucleophiles and alkenes or alkynes as electrophiles [108]. In most of these reactions, the Pd(II) catalyst is converted to an inactive Pd(0) species in the final step of the process, and use of stoichiometric oxidants is required to effect catalytic turnover. For example, the synthesis of furan **113** from α-allyl-β-diketone **112** is achieved via treatment of the substrates with a catalytic amount of $Pd(OAc)_2$ in the presence of a stoichiometric amount of $CuCl_2$ [109]. This transformation proceeds via Pd(II) activation of the alkene to afford **114**,

which undergoes nucleophilic attack of the enol oxygen onto the alkene double bond to provide alkylpalladium complex **115**. β-Hydride elimination of **115** gives **116**, which undergoes isomerization and loss of Pd(H)(Cl) to yield the furan product. The Pd(H)(Cl) complex undergoes reductive elimination of HCl to generate Pd(0), which is subsequently reoxidized to Pd(II) by the $CuCl_2$.

The synthesis of a variety of other heterocycles has been achieved using similar methodology [108]. For example, benzopyran derivatives are readily prepared from 2-allyl phenols [110]; note that the Pd(0) precatalyst is oxidized to Pd(II) by air before the reaction commences. Nitrogen heterocycles are also accessible from Wacker-type transformations, as demonstrated by the Hegedus indole synthesis described below in Section 1.10.

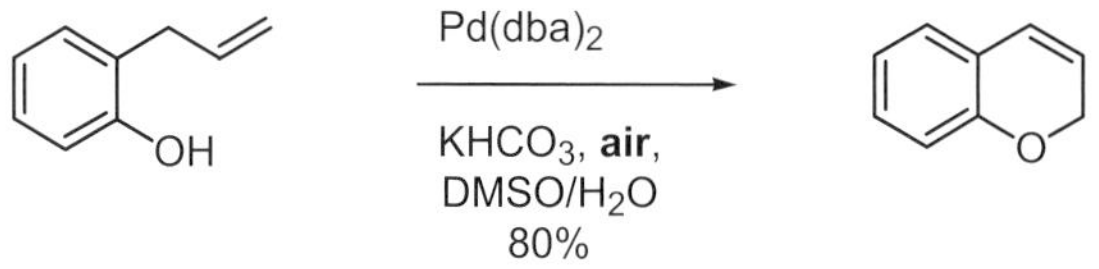

1.10. Mori–Ban, Hegedus, and Larock indole syntheses

The final section in this chapter will discuss the application of some of the transformations described above to the synthesis of indoles. This section serves to illustrate some of the differences between Pd(0) and Pd(II)-catalyzed reactions, and also shows that many mechanistically different, but synthetically complementary pathways can be employed for the Pd-catalyzed synthesis of heterocycles.

Among early reported Pd-catalyzed reactions, the Mori–Ban indole synthesis has proven to be very useful for pyrrole annulation. In 1977, based on their success with the nickel-catalyzed synthesis of indole from 2-chloro-*N*-allylaniline, the group led by Mori and Ban disclosed Pd-catalyzed intramolecular reactions of aryl halides with pendant olefins [111]. Compound **117**, easily prepared from 2-bromo-*N*-acetylaniline and methyl bromocrotonate, was adopted as a cyclization precursor. Treatment of **117** with Pd(OAc)$_2$ (2 mol%), Ph$_3$P (4 mol%), and NaHCO$_3$ in DMF provided indole **118** via an intramolecular Heck reaction followed by olefin isomerization to afford the fully aromatic product. Although yields from the initial report were moderate, they have been greatly improved over the last two decades [112].

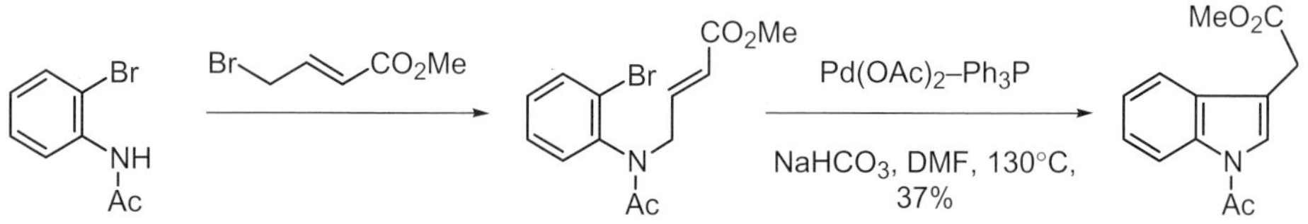

The *in situ* reduction of Pd(OAc)$_2$ to Pd(0) is effected by PPh$_3$ and proceeds with concomitant formation of triphenylphosphine oxide and acetic acid as shown below [113].

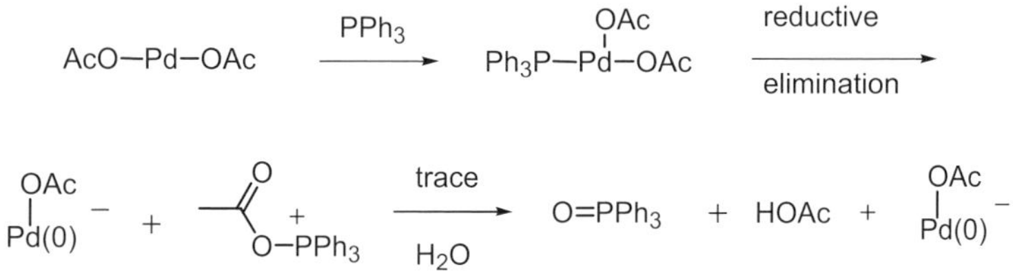

In 1976, Hegedus *et al.* described the synthesis of indoles using a Pd-assisted intramolecular amination of olefins, which tolerated a range of functionalities [8a,114]. For example, the requisite *o*-allylaniline **119** was prepared in high yield by the reaction of 5-methoxycarbonyl-2-bromoaniline with π-allylnickel bromide. Addition of **119** to a suspension of stoichiometric PdCl$_2$(CH$_3$CN)$_2$ in THF produced a yellow precipitate (the putative intermediate **121**), which upon treatment with Et$_3$N gave rise to indole **120** and deposited metallic palladium.

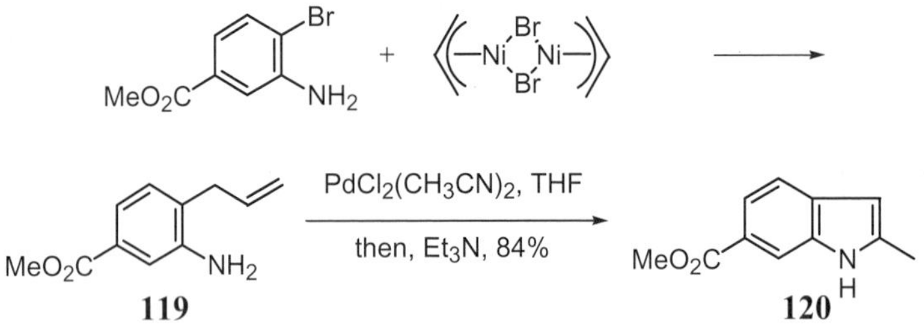

Hegedus proposed that the mechanism of this transformation proceeds through a Wacker-type reaction mechanism that is promoted by Pd(II). As shown below, coordination of the olefin to Pd(II) results in precipitate **121**, which upon treatment with Et$_3$N undergoes intramolecular *trans*-aminopalladation to afford intermediate **122**. As expected, the nitrogen atom attack occurs in a *5-exo-trig* fashion to afford **123**. β-Hydride elimination of **123** gives rise to exocyclic olefin **124**, which rearranges to indole **120**. The final step of this mechanism leads to the formation of catalytically inactive Pd(0). However, addition of oxidants such as benzoquinone allows for catalytic turnover.

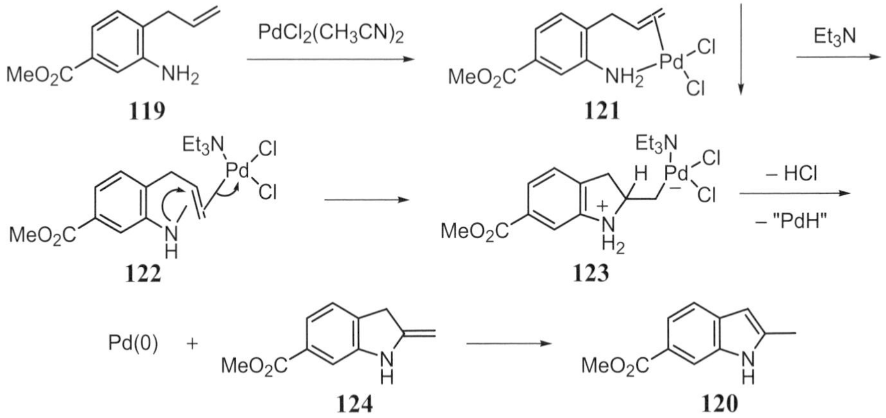

In 1991 [115], Larock *et al.* described the synthesis of indoles via a Pd-catalyzed heteroannulation of internal alkynes using *o*-iodoaniline and its derivatives, as exemplified by the Pd-catalyzed reaction between *o*-iodoaniline and propargyl alcohol **125** to give indole **126**. The pronounced directive effect of neighboring alcohol groups on the regiochemistry of alkyne insertion appears to be the result of coordination of the alcohol to the palladium during the insertion step (see intermediate **128** in the mechanistic scheme below). Generally, the annulation of unsymmetrical alkynes has proven to be highly regioselective, providing one regioisomer in most cases. The more sterically bulky group ends up adjacent to the nitrogen atom in the indole product.

The mechanism of the Larock indole synthesis is postulated as follows: reduction of Pd(OAc)$_2$ with small amounts of Me$_2$NH impurities present in DMF generates Pd(0), which is subsequently solvated by chloride to afford the anionic Pd(0) complex L$_2$PdCl$^-$. Oxidative addition of *o*-iodoaniline to Pd(0) gives rise to the palladium intermediate **127**, which then coordinates the internal alkyne, affording complex **128**. Regioselective *syn*-insertion of the alkyne into the arylpalladium bond furnishes vinylpalladium inter-mediate **129**, which forms a six-membered palladacyclic intermediate **130** from the nitro-gen displacement of the halide ligand. Finally, reductive elimination of **130** provides indole **126**, with concomitant regeneration of Pd(0).

1.11. References

1. (a) *Metal-catalyzed Cross-coupling Reactions*; Diederich, F.; Stang, P. J. Eds.; Wiley-VCH: Weinhein, Germany, **1998**. (b) Farina, V.; Krishnamurthy, V.; Scott, W. J. *The Stille Reaction*: Wiley, New York, NY, **1998**. (c) Tsuji, J. *Perspectives in Organopalladium Chemistry for the 21st Century*: Elsevier, Lausanne, Switzerland, **1999**. (d) Tsuji, J. *Palladium Reagents and Catalysts: Innovations in Organic Synthesis*: Wiley, Chichester, UK, **1995**. (e) Hegedus, L. S. *Transition Metals in the Synthesis of Complex Organic Molecules* 2nd Ed.,: University Science Books, Mill Valley, USA, **1999**. (f) Malleron, J.-L.; Fiaud, J.-C.; Legros, J.-Y. *Handbook of Palladium-catalyzed Organic Reactions*: Academic Press, San Diego, USA, **1997**.
2. (a) Wolfe, J. P.; Thomas, J. S. *Curr. Org. Chem.* **2005**, *9*, 625–55. (b) Nakamura, I.; Yamamoto, Y. *Chem Rev.* **2004**, *104*, 2127–98. (c) Zeni, G.; Larock, R. C. *Chem. Rev.* **2004**, *104*, 2285–309. (d) Rubin, M.; Sromek, A. W.; Gevorgyan, V. *Synlett* **2003**, 2265–91.
3. Chapman, N. B.; Russell-Hill, D. Q. *J. Chem. Soc.* **1956**, 1563–72.
4. For a review, see: Littke, A. F.; Fu, G. C. *Angew. Chem. Int. Ed.* **2002**, *41*, 4176–211.
5. Tilley, J. W.; Zawoiski, S. *J. Org. Chem.* **1988**, *53*, 386–90.

6. Parham, W. E.; Piccirili, R. M. *J. Org. Chem.* **1977**, *42*, 257–60.

7. Simkovsky, N. M.; Ermann, M.; Roberts, S. M.; Parry, D. M.; Baxter, A. D. *J. Chem. Soc., Perkin Trans. 1* **2002**, 1847–9.

8. (a) For an early review, see: Hegedus, L. S. *Angew. Chem., Int. Ed. Engl.* **1988**, *27*, 1113–126. (b) For a recent review, see: Fujiwara, Y.; Jia, C. in *Handbook of Organopalladium Chemistry for Organic Synthesis*, Negishi, E.-i. Ed.; John Wiley and Sons: Hoboken, **2002**, vol. 2, pp. 2859–62. (c) Åkermark, B.; Eberson, L.; Jonsson, E.; Pettersson, E. *J. Org. Chem.* **1975**, *40*, 1365–7. (d) Knölker, H.-J.; Fröhner, W. *J. Chem. Soc., Perkin Trans. 1* **1998**, 173–6. (e) Knölker, H.-J.; O'Sullivan, N. *Tetrahedron* **1994**, *50*, 10893–908.

9. For mechanistic discussions of related transformations, see: (a) Hennessy, E. J.; Buchwald, S. L. *J. Am. Chem. Soc.* **2003**, *125*, 12084–5. (b) Reference 8b.

10. Knölker, H.-J.; Reddy, K. R.; Wagner, A. *Tetrahedron Lett.* **1998**, *39*, 8267–70.

11. For related transformations that employ other oxidants for the conversion of Pd(0) to Pd(II), see: (a) Åkermark, B.; Oslob, J. D.; Heuschert, U. *Tetrahedron Lett.* **1995**, *36*, 1325–6. (b) Hagelin, H.; Oslob, J. D.; Åkermark, B. *Chem. Eur. J.* **1999**, *5*, 2413–16.

12. For recent reviews, see: (a) Echavarren, A. M.; Cardenas, D. J. in *Metal-Catalyzed Cross-Coupling Reactions*, de Meijere, A.; Diederich, F., Eds.; 2nd Ed., Wiley-VCH: Weinheim, **2004**, pp. 1–40. (b) Hassan, J.; Sevignon, M.; Gozzi, C.; Schulz, E.; Lemaire, M. *Chem. Rev.* **2002**, *102*, 1359–469.

13. (a) Espinet, P.; Echavarren, A. M. *Angew. Chem. Int. Ed.* **2004**, *43*, 4704–34. (b) Miyaura, N. *J. Organomet. Chem.* **2002**, *653*, 54–7.

14. Shekhar, S.; Hartwig, J. F. *J. Am. Chem. Soc.* **2004**, *126*, 13016–27.

15. Christmann, U.; Vilar, R. *Angew. Chem. Int. Ed.* **2005**, *44*, 366–74.

16. (a) Negishi, E.-i.; Baba, S. *J. Chem. Soc., Chem. Commun.* **1976**, 596–7. (b) Negishi, E.-i. *Acc. Chem. Res.* **1982**, *15*, 340–8. (c) Negishi, E.-i. in *Handbook of Organopalladium Chemistry for Organic Synthesis*, Negishi, E.-i. Ed.; John Wiley and Sons: Hoboken, **2002**, vol. 1, pp. 229–47.

17. Knochel, P.; Singer, R. D. *Chem. Rev.* **1993**, *93*, 2117–88.

18. Prasad, A. S.; B.; Stevenson, T. M.; Citineni, J. R.; Nyzam, V.; Knochel, P. *Tetrahedron* **1997**, *53*, 7237–54.

19. Jubert, C.; Knochel, P. *J. Org. Chem.* **1992**, *57*, 5425–31.

20. Huo, S. *Org. Lett.* **2003**, *5*, 423–5.

21. Evans, D. A.; Bach, T. *Angew. Chem., Int. Ed. Engl,* **1993**, *32*, 1326–7.

22. (a) Suzuki, A. in *Modern Arene Chemistry*, Astruc, D., Ed.; Wiley-VCH: Weinheim, **2002**, pp. 53–106. (b) Miyaura, N. in *Metal-Catalyzed Cross-Coupling Reactions*, de Meijere, A.; Diederich, F., Eds.; 2nd Ed., Wiley-VCH: Weinheim, **2004**, pp. 41–123. (c) Kotha, S.; Lahiri, K.; Kashinath, D. *Tetrahedron* **2002**, *58*, 9633–95.

23. (a) Molander, G. A.; Yun, C.-S.; Ribagorda, M.; Biolatto, B. *J. Org. Chem.* **2003**, *68*, 5534–9. (b) Molander, G. A.; Biolatto, B. *J. Org. Chem.* **2003**, *68*, 4302–14.

24. Molander, G. A.; Ribagorda, M. *J. Am. Chem. Soc.* **2003**, *125*, 11148–9.

25. Tyrrell, E.; Brookes, P. *Synthesis* **2003**, 469–83.

26. Gronowitz, S.; Hörnfeldt, A.-B.; Kristjansson, V.; Musil, T. *Chem. Scr.* **1986**, *26*, 305–9.

27. (a) Peters, D.; Hörnfeldt, A.-B.; Gronowitz, S. *J. Heterocycl. Chem.* **1990**, *27*, 2165–73. (b) Wellmar, U.; Hörnfeldt, A.-B.; Gronowitz, S. *J. Heterocycl. Chem.* **1995**, *32*, 1159–63.

28. (a) Kawasaki, I.; Yamashita, M.; Ohta, S. *Chem. Pharm. Bull.* **1996**, *44*, 1831–9, and references cited therein. (b) Kawasaki, I.; Katsuma, H.; Nakayama, Y.; Yamashita, M.; Ohta, S. *Heterocycles* **1998**, *48*, 1887–901.

29. For a review, see: Bellina, F.; Carpita, A.; Rossi, R. *Synthesis* **2004**, 2419–40.

30. (a) Barder, T. E.; Walker, S. D.; Martinelli, J. R.; Buchwald, S. L. *J. Am. Chem. Soc.* **2005**, *127*, 4685–96. (b) Wolfe, J. P.; Singer, R. A.; Yang, B. H.; Buchwald, S. L. *J. Am. Chem. Soc.* **1999**, *121*, 9550–61. (c) Littke, A. F.; Dai, C.; Fu, G. C. *J. Am. Chem. Soc.* **2000**, *122*, 4020–8.

31. Walker, S. D.; Barder, T. E.; Martinelli, J. R.; Buchwald, S. L. *Angew. Chem. Int. Ed.* **2004**, *43*, 1871–6.

32. For a review, see: Chemler, S. R.; Trauner, D.; Danishefsky, S. J. *Angew. Chem. Int. Ed.* **2001**, *40*, 4544–68.

33. Ishiyama, T.; Nishijima, K.; Miyaura, N.; Suzuki, A. *J. Am. Chem. Soc.* **1993**, *115*, 7219–25.

34. (a) Milstein, D.; Stille, J. K. *J. Am. Chem. Soc.* **1978**, *100*, 3636–8. (b) Kosugi, M.; Shimizu, Y.; Migita, T. *J. Organomet. Chem.* **1977**, *129*, C36–C38. (c) Stille, J. K. *Angew. Chem., Int. Ed. Engl.* **1986**, *25*, 508–24. (d) Farina, V.; Krishnamurthy, V.; Scott, W. J. *Org. React.* **1997**, *50*, 1–652. (e) For an excellent review on the intramolecular Stille reaction, see, Duncton, M. A. J.; Pattenden, G. *J. Chem. Soc., Perkin Trans. 1* **1999**, 1235–46.

35. Echavarren, A. M.; Stille, J. K. *J. Am. Chem. Soc.* **1987**, *109*, 5478–86.

36. Farina, V.; Krishnan, B.; Marshall, D. R.; Roth, G. P. *J. Org. Chem.* **1993**, *58*, 5434–44.

37. Han, X.; Stoltz, B. M.; Corey, E. J. *J. Am. Chem. Soc.* **1999**, *121*, 7600–5 and references cited therein.

38. Liebeskind, L. S.; Fengl, R. W. *J. Org. Chem.* **1990**, *55*, 5359–64.

39. Kada, R.; Knoppova, V.; Kovac, J.; Cepec, P. *Coll. Czech. Chem. Commun.* **1984**, *49*, 984–91.

40. Bach, T.; Krüger, L. *Synlett* **1998**, 1185–6.

41. Kelly, T. R.; Li, Q.; Bhushan, V. *Tetrahedron Lett.* **1990**, *31*, 161–4.

42. Grigg, R.; Teasdale, A.; Sridharan, V. *Tetrahedron Lett.* **1991**, *32*, 3859–62.

43. Sakamoto, T.; Yasuhara, A.; Kondo, Y.; Yamanaka, H. *Heterocycles* **1993**, *36*, 2597–600.

44. Fukuyama, Y.; Yaso, H.; Nakamura, K.; Kodama, M. *Tetrahedron Lett.* **1999**, *40*, 105–8.

45. Seiders, T. J.; Baldridge, K. K.; Elliott, E. L.; Grube, G. H.; Siegel, J. S. *J. Am. Chem. Soc.* **1999**, *121*, 7439–40.

46. Iyoda, M.; Miura, M.-i; Sasaki, S.; Kabir, S. M. H.; Kuwatani, Y.; Yoshida, M. *Tetrahedron Lett.* **1997**, *38*, 4581–2.

47. Iwaki, T.; Yasuhara, A.; Sakamoto, T. *J. Chem. Soc., Perkin Trans. 1* **1999**, 1505–10.

48. (a) Tamao, K.; Sumitani, K.; Kiso, Y.; Zembayashi, M.; Fujioka, A.; Kodama, S.-i.; Nakajima, I.; Minato, A.; Kumada, M. *Bull. Chem. Soc. Jpn.* **1976**, *49*, 1958–69. (b) Murahashi, S.-i. *J. Organomet. Chem.* **2002**, *653*, 27–33.

49. Amatore, C.; Jutand, A.; Negri, S.; Fauvarque, J. F. *J. Organomet. Chem.* **1990**, *390*, 389–98.

50. Minato, A.; Suzuki, K.; Tamao, K.; Kumada, M. *J. Chem. Soc., Chem. Commun.* **1984**, 511–13.

51. (a) Hiyama, T. *J. Organomet. Chem.* **2002**, *653*, 58–61. (b) Denmark, S. E.; Sweis, R. F. in *Metal-Catalyzed Cross-Coupling Reactions*, de Meijere, A.; Diederich, F., Eds.; 2nd Ed., Wiley-VCH: Weinheim, **2004**, pp. 163–216.

52. Pierrat, P.; Gros, P.; Fort, Y. *Org. Lett.* **2005**, *7*, 697–700 and references cited therein.

53. Hatanaka, Y.; Fukushima, S.; Hiyama, T. *Heterocycles* **1990**, *30*, 303–6.

54. Shibata, K.; Miyazawa, K.; Goto, Y. *Chem. Commun.* **1997**, 1309–10.

55. Stephens, R. D.; Castro, C. E. *J. Org. Chem.* **1963**, *28*, 3313–15.

56. (a) Sonogashira, K.; Tohda, Y.; Hagihara, N. *Tetrahedron Lett.* **1975**, 4467–70. (b) Rossi, R.; Carpita, A.; Bellina, F. *Org. Prep. Proc. Int.* **1995**, *27*, 129–60. (c) Campbell, I. B. in *Organocopper Reagents*, Taylor, R. J. K. Ed.; IRL Press: Oxford, UK, **1994**, pp. 217–35. (d) Sonogashira, K. *J. Organomet. Chem.* **2002**, *653*, 46–9.

57. Sakamoto, T.; Nagano, T.; Kondo, Y.; Yamanaka, H. *Chem. Pharm. Bull.* **1988**, *36*, 2248–52.

58. (a) Jeffery, T. *Synthesis* **1987**, 70–1. (b) Nguefack, J.-F.; Bolitt, V.; Sinou, D. *Tetrahedron Lett.* **1996**, *37*, 5527–30.

59. (a) Soheili, A.; Albaneze-Walker, J.; Murry, J. A.; Dormer, P. G.; Hughes, D. L. *Org. Lett.* **2003**, *5*, 4191–4. (b) Fukuyama, T.; Shinmen, M.; Nishitani, S.; Sato, M.; Ryu, I. *Org. Lett.* **2002**, *4*, 1691–4. (c) Ma, Y.; Song, C.; Jiang, W.; Wu, Q.; Wang, Y.; Liu, X.; Andrus, M. B. *Org. Lett.* **2003**, *5*, 3317–19. (d) Park, S. B.; Alper, H. *Chem. Commun.* **2004**, 1306–7. (e) Gelman, D.; Buchwald, S. L. *Angew. Chem. Int. Ed.* **2003**, *42*, 5993–6. (f) Arques, A.; Aunon, D.; Molina, P. *Tetrahedron Lett.* **2004**, *45*, 4337–40. (g) Cheng, J.; Sun, Y.; Wang, F.; Guo, M.; Xu, J.-H.; Pan, Y.; Zhang, Z. *J. Org. Chem.* **2004**, *69*, 5428–32. (h) Alonso, D. A.; Najera, C.; Pacheco, M. C. *Adv. Synth. Catal.* **2003**, *345*, 1146–58.

60. (a) Mizoroki, T.; Mori, K.; Ozaki, A. *Bull. Chem. Soc. Jpn.* **1971**, *44*, 581. (b) Heck, R. F.; Nolley, J. P., Jr. *J. Org. Chem.* **1972**, *37*, 2320–2.

61. (a) Braese, S.; de Meijere, A. in *Metal-Catalyzed Cross-Coupling Reactions*, de Meijere, A.; Diederich, F., Eds.; 2nd Ed., Wiley-VCH: Weinheim, **2004**, pp. 217–315. (b) Link, J. T. *Org. React.* **2002**, *60*, 157–534.

62. Akita, Y.; Inoue, A.; Mori, Y.; Ohta, A. *Heterocycles* **1986**, *24*, 2093–7.

63. For recent mechanistic studies on this step, see: Hills, I. D.; Fu, G. C. *J. Am. Chem. Soc.* **2004**, *126*, 13178–9.

64. (a) Jeffery, T. *Tetrahedron* **1996**, *52*, 10113–30. (b) Jeffery, T. *J. Chem. Soc., Chem. Commun.* **1984**, 1287–9. (c) Jeffery, T. *Tetrahedron Lett.* **1999**, *40*, 1673–6. (d) Jeffery, T.; Galland, J.-C. *Tetrahedron Lett.* **1994**, *35*, 4103–6. (e) Jeffery, T. *Adv. Met.-Org. Chem.* **1996**, *5*, 153–260. (f) Jeffery, T.; David, M. *Tetrahedron Lett.* **1998**, *39*, 5751–4.

65. Li, J. J. *J. Org. Chem.* **1999**, *64*, 8425–7.

66. Rawal, V. H.; Iwasa, S. *J. Org. Chem.* **1994**, *59*, 2685–6.

67. (a) Earley, W. G.; Oh, T.; Overman, L. E. *Tetrahedron Lett.* **1988**, *29*, 3785–8. (b) Madin, A.; Overman, L. E. *Tetrahedron Lett.* **1992**, *33*, 4859–62. (c) Madin, A.; O'Donnell, C. J.; Oh, T.; Old, D. W.; Overman, L. E.; Sharp, M. J. *Angew. Chem. Int. Ed.* **1999**, *38*, 2934–6.

68. (a) Dounay, A. B.; Overman, L. E. *Chem. Rev.* **2003**, *103*, 2945–63. (b) Shibasaki, M.; Vogl, E. M.; Ohshima, T. *Adv. Synth. Catal.* **2004**, *346*, 1533–52.

69. Dounay, A. B.; Hatanaka, K.; Kodanko, J. J.; Oestreich, M.; Overman, L. E.; Pfeifer, L. A.; Weiss, M. M. *J. Am. Chem. Soc.* **2003**, *125*, 6261–71.

70. Intermolecular arylation of benzanilides: (a) Kametani, Y.; Satoh, T.; Miura, M.; Nomura, M. *Tetrahedron Lett.* **2000**, *41*, 2655–8. (b) Daugulis, O.; Zaitsev, V. G.

Angew. Chem. Int. Ed. **2005**, *44*, 4046–8. Intermolecular arylation of phenolates: (c) Satoh, T.; Kawamura, Y.; Miura, M.; Nomura, M. *Angew. Chem., Int. Ed. Engl.* **1997**, *36*, 1740–2. (d) Kawamura, Y.; Satoh, T.; Miura, M.; Nomura, M. *Chem. Lett.* **1999**, 961–2. Intermolecular arylation of biarylmethanols: (e) Terao, Y.; Wakui, H.; Nomoto, M.; Satoh, T.; Miura, M.; Nomura, M. *J. Org. Chem.* **2003**, *68*, 5236–43. Intermolecular arylation of benzaldehydes: (f) Gürbüz, N.; Özdemir, I.; Cetinkaya, B. *Tetrahedron Lett.* **2005**, *46*, 2273–7.

71. Aoyagi, Y.; Inoue, A.; Koizumi, I.; Hashimoto, R.; Tokunaga, K.; Gohma, K.; Komatsu, J.; Sekine, K.; Miyafuji, A.; Kunoh, J. Honma, R. Akita, Y.; Ohta, A. *Heterocycles* **1992**, *33*, 257–72.

72. Kuroda, T.; Suzuki, F. *Tetrahedron Lett.* **1991**, *32*, 6915–18.

73. (a) Bringmann, G.; Heubes, M.; Breuning, M.; Göbel, L.; Ochse, M.; Schöner, B.; Schupp, O. *J. Org. Chem.* **2000**, *65*, 722–728. (b) Hosoya, T.; Takashiro, E.; Matsumoto, T.; Suzuki, K. *J. Am. Chem. Soc.* **1994**, *116*, 1004–15.

74. For additional recent examples, see: (a) Campeau, L.-C.; Thansandote, P.; Fagnou, K. *Org. Lett.* **2005**, *7*, 1857–60. (b) Campeau, L.-C.; Parisien, M.; Leblanc, M.; Fagnou, K. *J. Am. Chem. Soc.* **2004**, *126*, 9186–7. (c) Huang, Q.; Fazio, A.; Dai, G.; Campo, M. A.; Larock, R. C. *J. Am. Chem. Soc.* **2004**, *126*, 7460–1.

75. Reetz, M. T.; Lohmer, G.; Schwickardi, R. *Angew. Chem., Int. Ed.* **1998**, *37*, 481–3.

76. Littke, A. F.; Fu, G. C. *J. Org. Chem.* **1999**, *64*, 10–11.

77. Littke, A. F.; Fu, G. C. *J. Am. Chem. Soc.* **2001**, *123*, 6989–7000.

78. (a) Tsuji, J. *Palladium Reagents and Catalysts: Innovations in Organic Synthesis*: Wiley: Chichester, UK, **1995**, pp. 340–5. (b) Kumar, K.; Zapf, A.; Michalik, D.; Tillack, A.; Heinrich, T.; Böttcher, H.; Arlt, M.; Beller, M. *Org. Lett.* **2004**, *6*, 7–10 and references cited therein. (c) Albaneze-Walker, J.; Bazaral, C.; Leavey, T.; Dormer, P. G.; Murry, J. A. *Org. Lett.* **2004**, *6*, 2097–100 and references cited therin.

79. Bessard, Y.; Roduit, J. P. *Tetrahedron* **1999**, *55*, 393–404.

80. Wu, G. G.; Wong, Y.; Poirier, M. *Org. Lett.* **1999**, *1*, 745–7.

81. (a) Schwan, A. L. *Chem. Soc. Rev.* **2004**, *33*, 218–24. (b) Montchamp, J.-L. *J. Organomet. Chem.* **2005**, *690*, 2388–406.

82. (a) Hirao, T.; Masunaga, T.; Ohshiro, Y.; Agawa, T. *Tetrahedron Lett.* **1980**, *21*, 3595–8. (b) Hirao, T.; Masunaga, T.; Ohshiro, Y.; Agawa, T. *Synthesis* **1981**, 56–7. (c) Hirao, T.; Masunaga, T.; Yamada, N.; Ohshiro, Y.; Agawa, T. *Bull. Chem. Soc. Jpn.* **1982**, *55*, 909–13.

83. (a) Xu, Y.; Li, Z.; Xia, J.; Guo, H.; Huang, Y. *Synthesis* **1983**, 377–8. (b) Xu, Y.; Zhang, J. *Synthesis* **1984**, 778–80.

84. Xu, Y.; Wei, H.; Zhang, J.; Huang, G. *Tetrahedron Lett.* **1989**, *30*, 949–52 and references cited therein.

85. (a) Petrakis, K. S.; Nagabhushan, T. L. *J. Am. Chem. Soc.* **1987**, *109*, 2831–3. (b) Holt, D. A.; Erb, J. M. *Tetrahedron Lett.* **1989**, *30*, 5393–6. (c) Kurz, L.; Lee, G.; Morgan, D., Jr.; Waldyke, M. J.; Ward, T. *Tetrahedron Lett.* **1990**, *31*, 6321–4. (d) Uozumi, Y.; Tanahashi, A.; Lee, S. Y.; Hayashi, T. *J. Org. Chem.* **1993**, *58*, 1945–8.

86. Kosugi, M.; Kameyama, M.; Migita, T. *Chem. Lett.* **1983**, 927–8.

87. Boger, D. L.; Panek, J. S. *Tetrahedron Lett.* **1984**, *25*, 3175–8.

88. (a) Guram, A. S.; Rennels, R. A.; Buchwald, S. L. *Angew. Chem. Int. Ed.* **1995**, *34*, 1348–50. (b) Louie, J.; Hartwig, J. F. *Tetrahedron Lett.* **1995**, *36*, 3609–12.

89. (a) Hartwig, J. F. *Angew. Chem., Int. Ed.* **1998**, *37*, 2046–67. (b) Wolfe, J. P.; Wagaw, S.; Marcoux, J.-F.; Buchwald, S. L. *Acc. Chem. Res.* **1998**, *31*, 805–18. (c) Muci, A. R. *Top. Curr. Chem.* **2002**, *219*, 131–209. (d) Hartwig, J. F. in *Modern Arene Chemistry* Astruc, D., Ed.; Wiley-VCH: Weinheim, **2002**, pp. 107–68. (e) Schlummer, B.; Scholz, U. *Adv. Synth. Catal.* **2004**, *346*, 1599–626.

90. Wolfe, J. P.; Tomori, H.; Sadighi, J. P.; Yin, J.; Buchwald, S. L. *J. Org. Chem.* **2000**, *65*, 1158–74.

91. For detailed mechanistic studies, see: (a) Singh, U. K.; Strieter, E. R.; Blackmond, D. G.; Buchwald, S. L. *J. Am. Chem. Soc.* **2002**, *124*, 14104–14. (b) Mann, G.; Hartwig, J. F. *J. Am. Chem. Soc.* **1996**, *118*, 13109–10. (c) Guari, Y.; van Strijdonck, G. P. F.; Boele, M. D. K.; Reek, J. N. H.; Kamer, P. C. J.; van Leeuwen, P. W. N. M. *Chem. Eur. J.* **2001**, *7*, 475–82.

92. Widenhoefer, R. A.; Buchwald, S. L. *J. Am. Chem. Soc.* **1998**, *120*, 6504–11.

93. Palucki, M.; Wolfe, J. P.; Buchwald, S. L. *J. Am. Chem. Soc.* **1996**, *118*, 10333–4.

94. Kuwabe, S.-i.; Torraca, K. E.; Buchwald, S. L. *J. Am. Chem. Soc.* **2001**, *123*, 12202–6.

95. (a) Palucki, M.; Wolfe, J. P.; Buchwald, S. L. *J. Am. Chem. Soc.* **1997**, *119*, 3395–6. (b) Mann, G.; Incarvito, C.; Rheingold, A. L.; Hartwig, J. F. *J. Am. Chem. Soc.* **1999**, *121*, 3224–5. (c) Watanabe, M.; Nishiyama, M.; Koie, Y. *Tetrahedron Lett.* **1999**, *40*, 8837–40. (d) Parrish, C. A.; Buchwald, S. L. *J. Org. Chem.* **2001**, *66*, 2498–500. (e) Vorogushin, A. V.; Huang, X.; Buchwald, S. L. *J. Am. Chem. Soc.* **2005**, *127*, 8146–9 and references cited therein.

96. Aranyos, A.; Old, D. W.; Kiyomori, A.; Wolfe, J. P.; Sadighi, J. P.; Buchwald, S. L. *J. Am. Chem. Soc.* **1999**, *121*, 4369–78.

97. Tsutsui, H.; Narasaka, K. *Chem. Lett.* **1999**, 45–6.

98. (a) Chiba, S.; Kitamura, M.; Saku, O.; Narasaka, K. *B. Chem. Soc. Jpn.* **2004**, *77*, 785–96. (b) Zaman, S.; Kitamura, M.; Narasaka, K. *B. Chem. Soc. Jpn.* **2003**, *76*, 1055–62.

99. Helaja, J.; Göttlich, R. *Chem. Commun.* **2002**, 720–1.

100. (a) Wolfe, J. P.; Rossi, M. A. *J. Am. Chem. Soc.* **2004**, *126*, 1620–1. (b) Ney, J. E.; Wolfe, J. P. *Angew. Chem. Int. Ed.* **2004**, *43*, 3605–8. (c) Bertrand, M. B.; Wolfe, J. P. *Tetreahedron* **2005**, *61*, 6447–59.

101. Lira, R.; Wolfe, J. P. *J. Am. Chem. Soc.* **2004**, *126*, 13906–7.

102. Ney, J. E.; Wolfe, J. P. *J. Am. Chem. Soc.* **2005**, *127*, 8644–51.

103. Dick, A. R.; Hull, K. L.; Sanford, M. S. *J. Am. Chem. Soc.* **2004**, *126*, 2300–1.

104. Desai, L. V.; Hull, K. L.; Sanford, M. S. *J. Am. Chem. Soc.* **2004**, *126*, 9542–3.

105. For recent reviews, see: (a) Tsjui, J. in *Handbook of Organopalladium Chemistry for Organic Synthesis*, Negishi, E.-i. Ed.; John Wiley and Sons: Hoboken, **2002**, vol. 2, pp. 1669–87. (b) Acemoglu, L.; Williams, J. M. J. in *Handbook of Organopalladium Chemistry for Organic Synthesis*, Negishi, E.-i. Ed.; John Wiley and Sons: Hoboken, **2002**, vol. 1, pp. 1689–705.

106. Bolitt, V.; Chaguir, B.; Sinou, D. *Tetrahedron Lett.* **1992**, *33*, 2481–4.

107. (a) Comely, A. C.; Eelkema, R.; Minnaard, A. J.; Feringa, B. L. *J. Am. Chem. Soc.* **2003**, *125*, 8714–5. (b) Babu, R. S.; O'Doherty, G. A. *J. Am. Chem. Soc.* **2003**, *125*, 12406–7. (c) Kim, H.; Men, H.; Lee, C. *J. Am. Chem. Soc.* **2004**, *126*, 1336–7.

108. For recent reviews: (a) Balme, G.; Bouyssi, D.; Lomberget, T.; Monteiro, N. *Synthesis* **2003**, 2115–34. (b) Reference 2c.

109. Han, X.; Widenhoefer, R. A. *J. Org. Chem.* **2004**, *69*, 1738–40.

110. Larock, R. C.; Wei, L.; Hightower, T. R. *Synlett* **1998**, 522–4.

111. (a) Mori, M.; Chiba, K.; Ban, Y. *Tetrahedron Lett.* **1977**, *18*, 1037–40; (b) Ban, Y.; Wakamatsu, T.; Mori, M. *Heterocycles* **1977**, *6*, 1711–15.

112. For a recent review on Pd-catalyzed indole-forming reactions, see: Cacchi, S.; Fabrizi, G. *Chem. Rev.* **2005**, *105*, 2873–920.

113. (a) Amatore, C.; Carre, E.; Jutand, A.; M'Barki, M. A.; Meyer, G. *Organometallics* **1995**, *14*, 5605–14. (b) Amatore, C.; Carre, E.; Jutand, A.; M'Barki, M. A. *Organometallics* **1995**, *14*, 1818–26. (c) Amatore, C.; Jutand, A.; M'Barki, M. A. *Organometallics* **1992**, *11*, 3009–13. (d) Amatore, C.; Azzabi, M; Jutand, A. *J. Am. Chem. Soc.* **1991**, *113*, 8375–84.

114. (a) Hegedus, L. S.; Allen, G. F.; Waterman, E. L. *J. Am. Chem. Soc.* **1976**, *98*, 2674–6. (b) Hegedus, L. S.; Allen, G. F.; Bozell, J. J.; Waterman, E. L. *J. Am. Chem. Soc.* **1978**, *100*, 5800–7.

115. (a) Larock, R. C.; Yum, E. K. *J. Am. Chem. Soc.* **1991**, *113*, 6689–90. (b) Larock, R. C.; Yum, E. K.; Refvik, M. D. *J. Org. Chem.* **1998**, *63*, 7652–62. (c) Larock, R. C. *J. Organomet. Chem.* **1999**, *576*, 111–24.

Chapter 2

Pyrroles

Gordon W. Gribble

The pyrrole ring is widely distributed in nature. It occurs in both terrestrial and marine plants and animals [1–3]. Examples of simple pyrroles include the *Pseudomonas* metabolite pyrrolnitrin, a recently discovered seabird hexahalogenated bipyrrole [4], and an ant trail pheromone. An illustration of the abundant complex natural pyrroles is konbu'acidin A, a sponge metabolite that inhibits cyclin dependent kinase 4. The enormous reactivity of pyrrole in electrophilic substitution reactions explains the occurrence of more than 100 naturally occurring halogenated pyrroles [2, 3].

| pyrrolnitrin | seabird compound | ant trail pheromone |

| laughine | catuabine I | konbu'acidin A |

Following the discovery of the unique electronic properties of polypyrrole, numerous polymers of pyrrole have been crafted. A copolymer of pyrrole and pyrrole-3-carboxylic acid is used in a glucose biosensor, and a copolymer of pyrrole and *N*-methylpyrrole operates as a redox switching device. Self-doping, low-band gap, and photorefractive pyrrole polymers have been synthesized, and some examples are illustrated [1, 5].

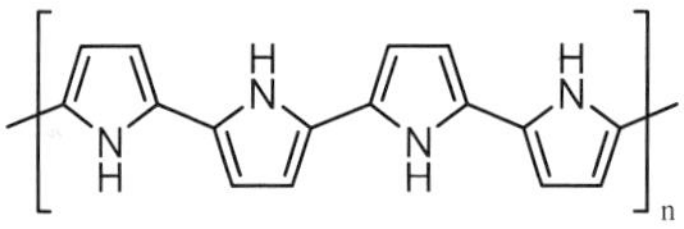

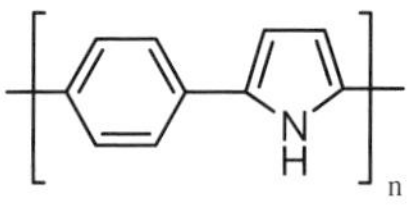

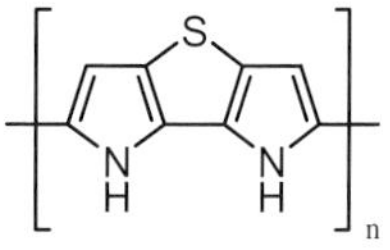

polypyrrole

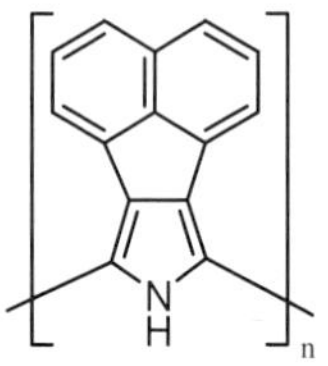

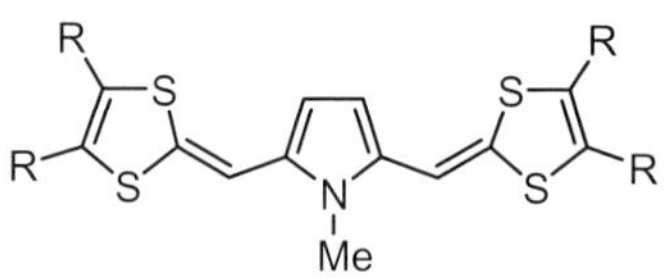

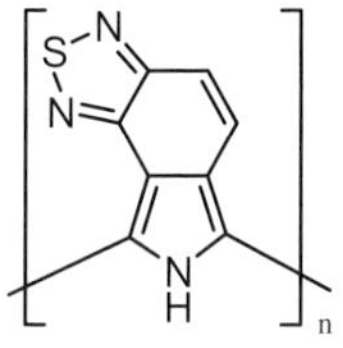

The pyrrole ring has found great use in the design and development of pharmaceuticals. Lipitor is the leading cholesterol lowing drug and other bioactive pyrroles shown below include the opioid antagonist norbinaltorphimine, a broad-spectrum insecticide, a sodium-independent dopamine receptor antagonist, a DNA cross-linking agent, and an antipsychotic agent.

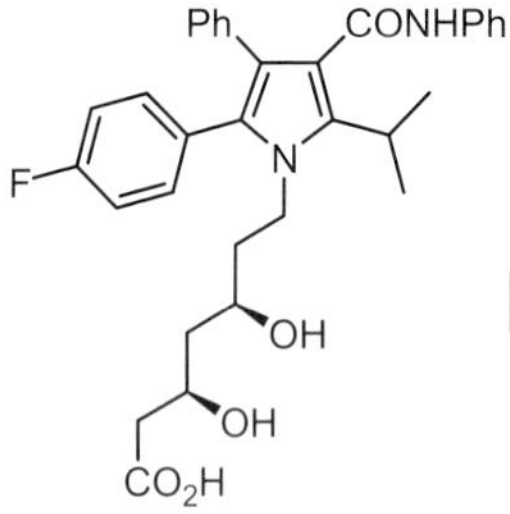

Lipitor 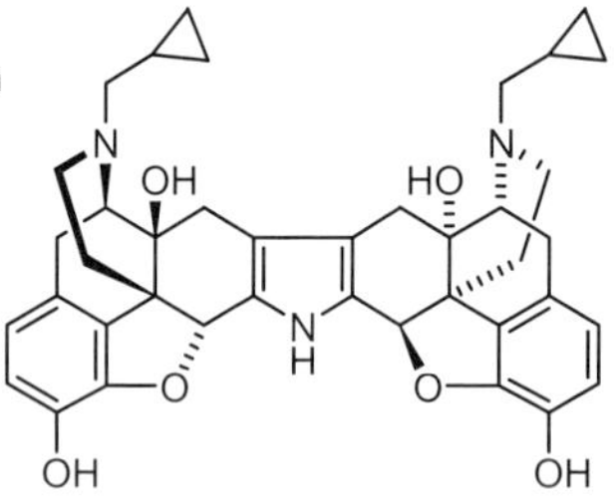   norbinaltorphimine 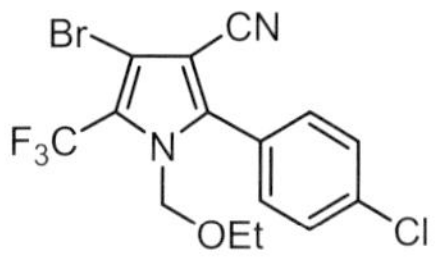   insecticide

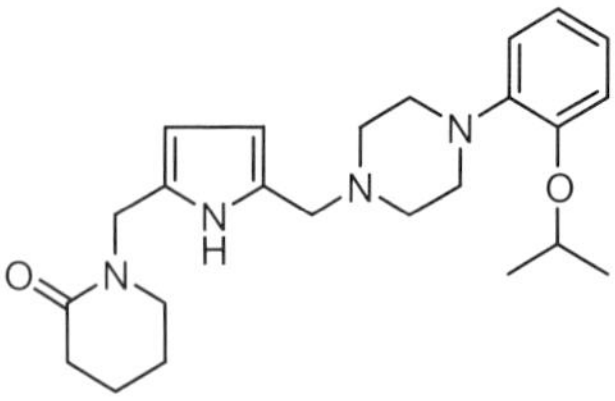

dopamine receptor antagonist 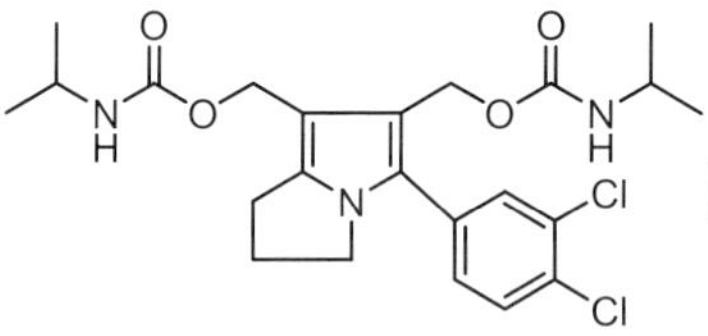   DNA cross-linking agent 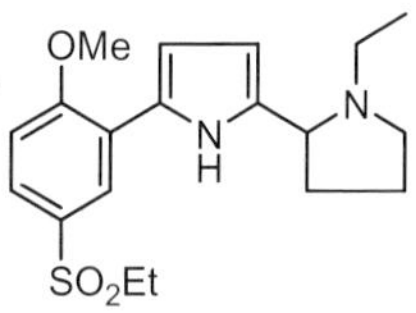   antipsychotic

2.1. Synthesis of pyrrolyl halides

Although pyrrolyl halides are well-known compounds, their instability to acid, alkali, and heat precludes their commercial availability. Since pyrrole is a very reactive π-excessive heterocycle, it undergoes halogenation extremely readily [6, 7]. For example, the labile 2-bromopyrrole, which decomposes above room temperature, is a well-known compound, as are *N*-alkyl-2-halopyrroles, readily prepared by direct halogenation, usually with NBS for the synthesis of bromopyrroles [8, 9]. The 2-halopyrrole is usually the kinetic product but the 3-halopyrrole is often the thermodynamic product, and this property of halopyrroles can be exploited in synthesis. For example, N-benzylpyrrole (1) can be dibrominated to give 2 as the kinetic product, which rearranges to 3 upon treatment with acid [10, 11]. Other N-alkyl-2,5-dibromopyrroles are available in this fashion.

Bromination of 1 with Br_2 (CCl_4, 0°C) affords 1-benzyl-3-bromopyrrole directly in 55–66% yield [12]. Since these bromopyrroles are still very labile, other N-protecting groups, which are electron-withdrawing, have been employed to provide access to more robust halogenated pyrroles. The bromination of N-BOC-pyrrole, which is readily available [13] with 1,3-dibromo-5,5-dimethylhydantoin (NBH) or NBS, can be adjusted to afford either the 2-bromo derivative [14] or the 2,5-dibromo derivative in good yields [14, 15]. Not surprisingly, a bulky N-protecting group directs halogenation to the C-3 position of pyrrole. Thus, the bromination of N-tritylpyrrole (4) affords either mono- (5) or 3,4-dibromination products depending on conditions [16]. The trityl protecting group is readily removed under Birch conditions.

A more versatile N-protection strategy is seen in Muchowski's N-TIPS-pyrrole (6) [17–19]. Whereas bromination of N-trimethylsilylpyrrole gives almost exclusively C-2 attack, the NBS bromination of 6 affords the C-3 product 7 in a 96:4 ratio (C-3/C-2) in 93% yield. Deprotection is readily achieved with fluoride to give the unstable 8. The C-3, C-4 dibromo derivative can be prepared by tribromination of 6 and then selective C-2 bromine–lithium exchange. Similar chemistry affords 3-iodo-N-TIPS pyrrole.

The power of Muchowski's method is seen by the fact that these bromopyrroles can be subjected to bromine–lithium exchange to afford the versatile 3-lithio species that can be quenched with a variety of electrophiles in good to excellent yields [18–21]. This is illustrated by a synthesis of verrucarin E (**11**) [19].

Triple bromination of *N*-TIPS-pyrrole (**6**) with NBS affords tribromopyrrole **12** (97%) [22], which can either be deprotected to afford the marine acorn worm metabolite 2,3,4-tribromoindole **13** [22, 23] or lithiated selectively at C-2 to give, after quenching with carbon dioxide, pyrrole carboxylic acid **14** [24].

Although labile, 2-bromopyrrole (**15**) can be prepared from pyrrole and *N*-protected *in situ* to give 2-bromo-1-tosylpyrrole (**16**) in 80% yield [25].

Depending on the reaction conditions, 1-methylpyrrole (**17**) can be brominated at C-2 with NBS to give 2-bromo-1-methylpyrrole (**18**) or at C-3 with NBS and catalytic PBr_3 to give 3-bromo-1-methylpyrrole (**19**). Both reactions are essentially quantitative, but both **18** and **19** decompose on silica gel [26].

If the pyrrole is substituted with an electron-withdrawing group, then more vigorous halogenation conditions are required, but the products are usually more stable than simple halogenated pyrroles. For example, the bromination of pyrrole-2-carboxylic acid (**20**) yields the 4,5-dibromo isomer (**21**) in excellent yield [27]. Similarly, the bromination of 4-chloropyrrole-2-carboxylic acid furnishes 5-bromo-4-chloropyrrole-2-carboxylic acid in 90% yield [28].

Various iodinated pyrroles have been prepared by direct iodination [19, 29] or via thallation [30]. For example, 3-iodo-*N*-TIPS-pyrrole is prepared in 61% yield from **6** [19], and 3,4-diiodo-2-formyl-1-methylpyrrole is available in 54% yield via a bis-thallation reaction [30]. Although *N*-protected 2-lithiopyrroles are readily generated and many types are known [6, 31], these intermediates have not generally been employed to synthesize halogenated pyrroles. One exception is the synthesis of the two natural seabird hexahalogenated bipyrroles **23** and **25**, which were prepared as shown from bipyrrole **22** [32]. The presumed electron-withdrawing inductive effect of the nitrogen atoms is responsible for the regiochemistry of the halogen–metal exchange reaction in going from **23** to **25**.

Other halopyrroles and related triflates will be cited as appropriate in the subsequent sections.

2.2. Oxidative coupling/cyclization

One of the earliest examples of the use of palladium in pyrrole chemistry was the Pd(OAc)2-induced oxidative coupling of *N*-methylpyrrole with styrene to afford a mixture of olefins **26** and **27** in low yield based on palladium acetate [33].

26 (11%)   **27** (3%)

A similar reaction of pyrroles **28** with acrylates provides the C-2 substituted α-alkenyl derivatives **29** in 24–91% yield [34]. The 2,6-dichlorobenzoyl protecting group is noteworthy as it prevents cyclization of the phenyl group onto the pyrrole ring (*vide infra*).

R$_1$ = 2,6-dichlorobenzoyl, SO$_2$Ph
R$_2$ = H, Me
R$_3$ = H, Ac
R$_4$ = H, Me, CHO
R$_5$ = Me, Et

28   **29**

Itahara has also found that the phenylation of *N*-aroylpyrroles can be achieved using Pd(OAc)$_2$ [35, 36]. Although *N*-benzoylpyrrole (**30**) yields a mixture of diphenylpyrrole **31**, cyclized pyrrole **32**, and bipyrrolyl **33** as shown, 1-(2,6-dichlorobenzoyl)pyrrole **34** gives the diphenylated pyrrole **35** in excellent yield. The *N*-aroyl groups are readily cleaved with aqueous alkali, and the arylation reaction also proceeds with *p*-xylene and *p*-dichlorobenzene. Unfortunately, *N*-methyl-, *N*-acetyl-, and *N*-(phenoxycarbonyl) pyrroles give complex mixtures of products.

30   **31** (25%)   **32** (20%)   **33** (8%)

34   **35**

As expected, reaction of *N*-aroylpyrroles **36** in the absence of added arene affords the bipyrroles **37** or cyclized product **38** [37, 38]. Bipyrrole **39** was prepared via this oxidative coupling reaction [37].

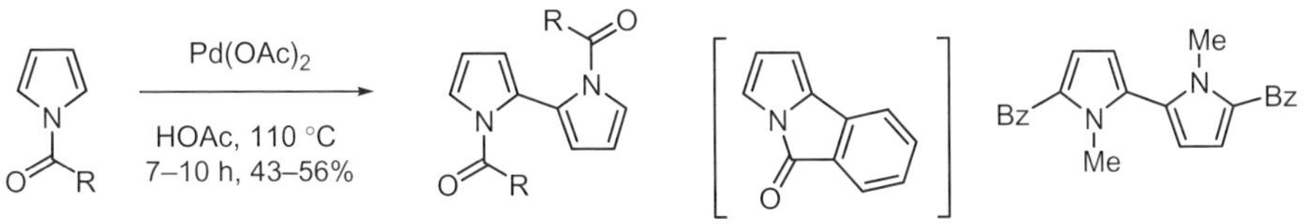

R = Ph, 4-MePh, 1-naphthyl

36   **37**   **38**   **39**

Itahara has extended these stoichiometric Pd(OAc)$_2$-induced reactions to the coupling of N-(phenylsulfonyl)pyrrole (**40**) and 1,4-naphthoquinone (**41**) to afford **42** [39].

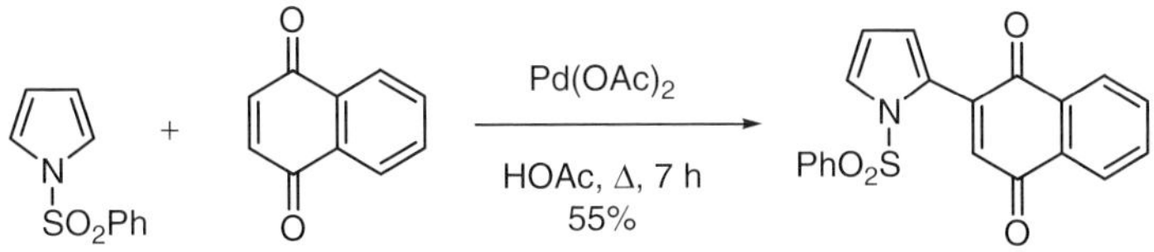

40   **41**   **42**

A key step in Boger's synthesis of prodigiosin and related compounds is the oxidative cyclization of dipyrrolyl ketones such as **43** to give **44** in excellent yield [40, 41]. Noteworthy is the use of polymer-supported Pd(OAc)$_2$ in this chemistry.

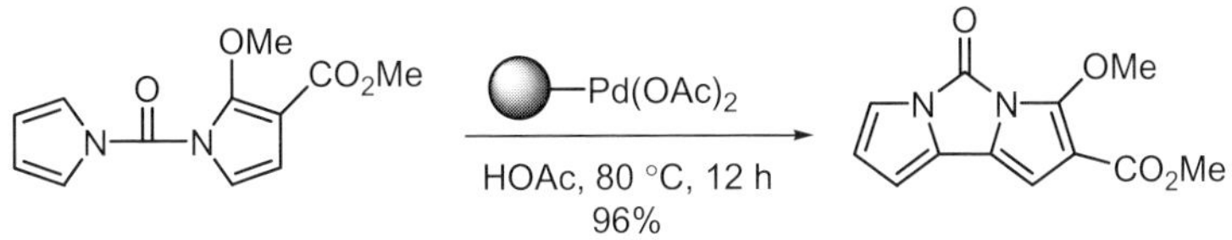

43   **44**

Depending on the reaction conditions, pyrrole amide **45** cyclizes either to pyrrolo[1,2-*a*]pyrazin-1-ones (**46**) or to a mixture of pyrrolo[2,3-*c*]pyridin-7-ones (**47**) and pyrrolo[3,2-*c*]pyridin-4-ones (**48**) [42]. Isomers **47** and **48** are distinguishable by the lower chemical shift of the N−H proton (10.21 ppm) for **47** (*syn* to the carbonyl) relative to that observed for **48** (9.08 ppm) and other spectral data. The cyclization leading to **47-48** is proposed to involve spiro intermediate **49**.

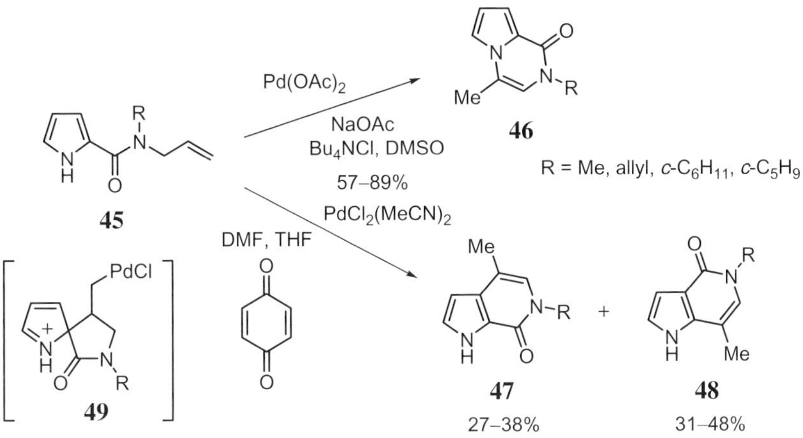

45

46

R = Me, allyl, *c*-C$_6$H$_{11}$, *c*-C$_5$H$_9$

49

47   **48**

27–38%   31–48%

A synthesis of the marine alkaloid (+)-dragmacidin F featured the oxidative cyclization of pyrrole **50** to give **51** in good yield [43]. Further elaboration to the natural product involved a Suzuki coupling (Section 2.3.3).

During a synthesis of lamellarin and related pyrrole alkaloids, Iwao *et al.* observed the Pd(OAc)$_2$-mediated decarboxylative cyclization of pyrrole acid **52** to **53** [44]. Lesser amounts (12%) of the product from decarboxylation of **52** were isolated.

Yamamoto and co-workers have reported the novel pyrrole ring synthesis shown for **56**, involving the Pd-induced union of methyleneaziridines **54** and 2-acetylpyridine (**55**) [45]. Similar reactions of 3- and 4-acetylpyridine give pyrroles **57** and **58**, respectively.

R = Bn, *n*-C$_6$H$_{13}$, PhCHMe, (MeO)$_2$CH$_2$CH$_2$, MeO(CH$_2$)$_3$, MeCH(*c*-C$_6$H$_{11}$)

Pyrrole rings frequently serve as precursors to indole rings [46] and $PdCl_2$ induces the oxidative cyclization of pyrrole **59** to a mixture of **60** and **61** [47]. Since the oxidation of tetrahydroindoles to indoles, such as **60** to **61**, is usually straightforward, this transformation can be viewed as a novel and efficient indole ring synthesis.

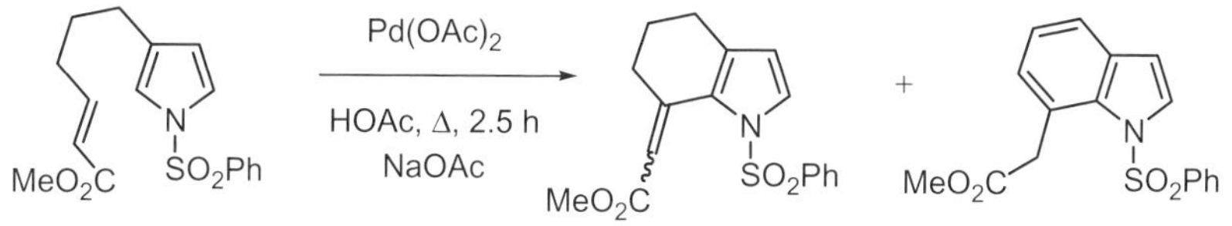

59 **60** (40%) **61** (27%)

A new pyrrole ring synthesis developed by Arcadi involves the addition of ammonia or benzylamine to 4-pentynones, the latter of which are conveniently prepared via a palladium oxidative coupling sequence as shown below for the synthesis of **62** [48, 49].

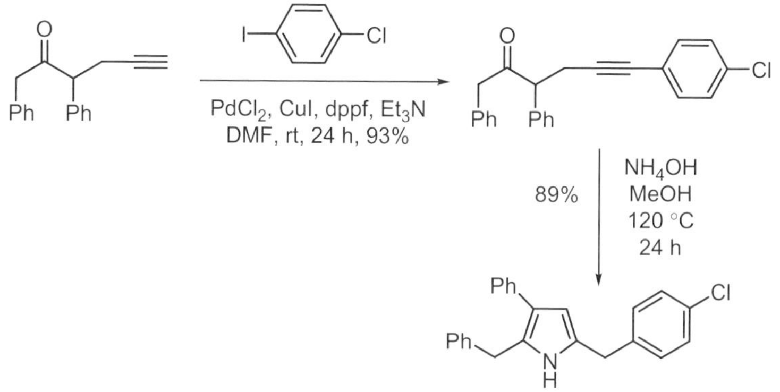

62

Ohta has developed a facile and efficient synthesis of pyrroles **64** that involves the Pd-catalyzed oxidative cyclization of hydroxy enamines such as **63** [50]. Fused pyrroles **65** and **66** were also synthesized in similar fashion.

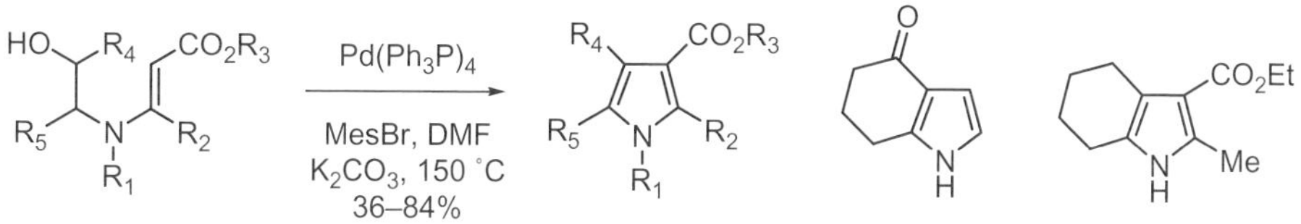

63 **64** **65** (63%) **66** (85%)

R_1 = H, Me, R_2 = Me, Ph, R_3 = Me, Et, R_4 = H, Me, Ph, R_5 = H, Me, Bn, Ph, *i*-Pr

2.3. Coupling reactions with organometallic reagents

2.3.1. Grignard reagents (the Kumada coupling)

Pyrrole Grignard reagents can be generated either by bromine–magnesium exchange on bromopyrroles [51–53] or by transmetalation of lithiopyrroles with magnesium bromide [54–56]. The resulting pyrrole Grignard reagents undergo a variety of coupling reactions (the Kumada coupling [57, 58]) with aryl, alkyl, and heteroaryl halides as catalyzed by palladium [51, 52, 54–56]. Several examples are shown below leading to pyrroles 67–69 [52, 54, 55].

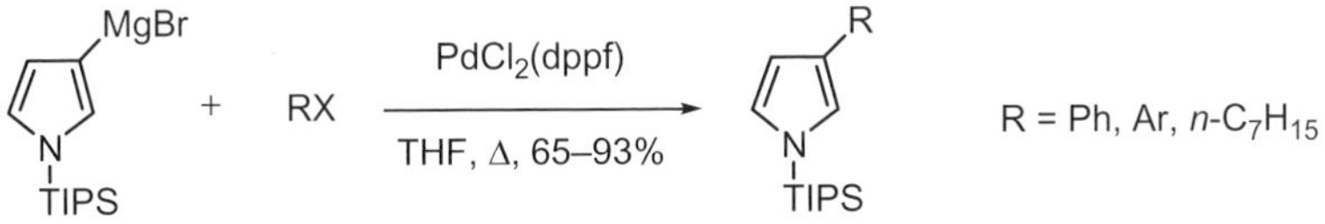

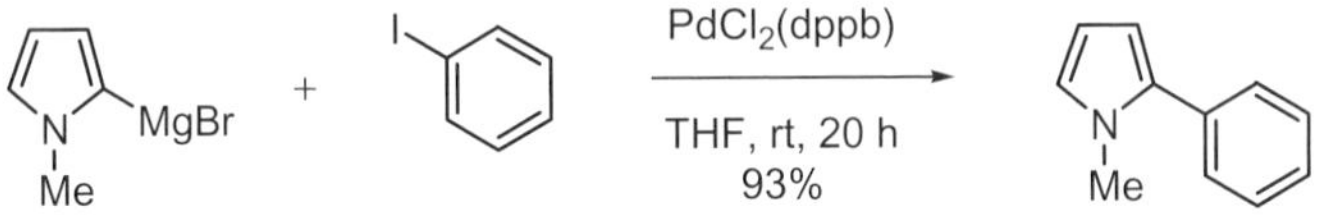

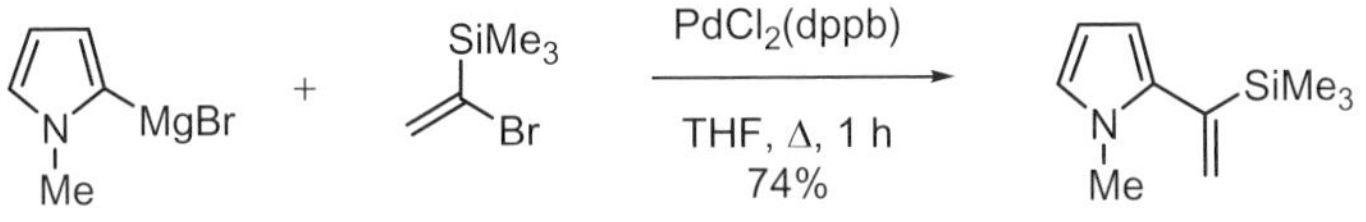

Noteworthy is the fact that one can utilize the appropriate bromopyrrole (e.g. 70) in conjunction with the desired Grignard reagent in a one-step operation to afford the corresponding substituted pyrroles (e.g. 71) [52]. The mixed pyrrole–pyridine heterocycle 72 was made in this fashion [54].

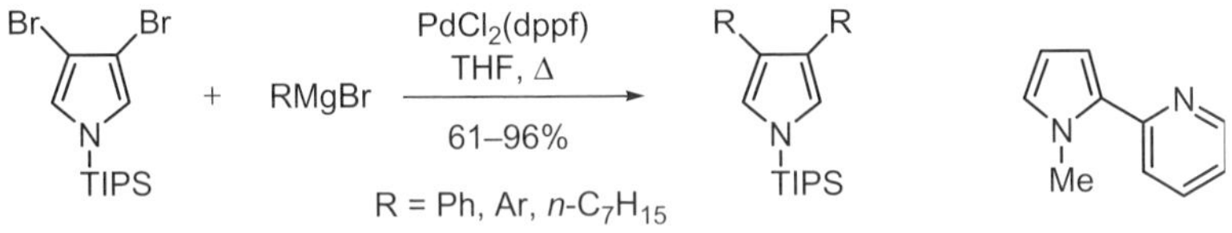

2.3.2. Organozinc reagents (the Negishi coupling)

Pyrrolylzinc reagents are normally generated by transmetalation of lithiopyrroles with zinc chloride. As shown in the examples below, the resulting pyrrolylzinc species undergo the Negishi coupling with a range of aryl and acyl halides leading to pyrroles 68, 73, and 74 [28, 54, 59, 60]. Pyrrole 73 was synthesized in a search for novel near-infrared dyes [54].

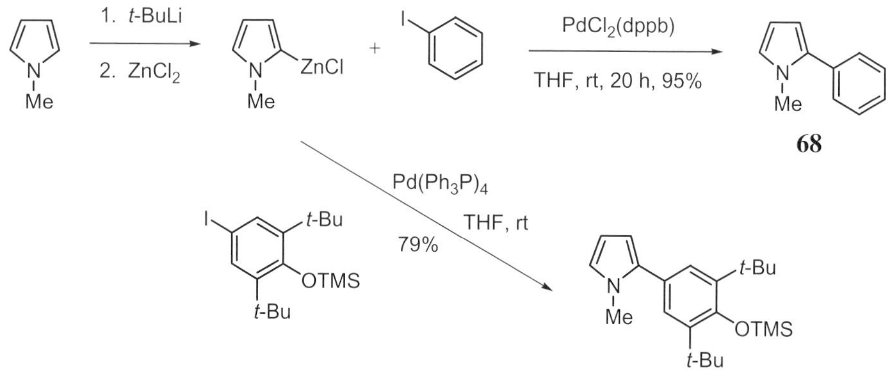

68

73

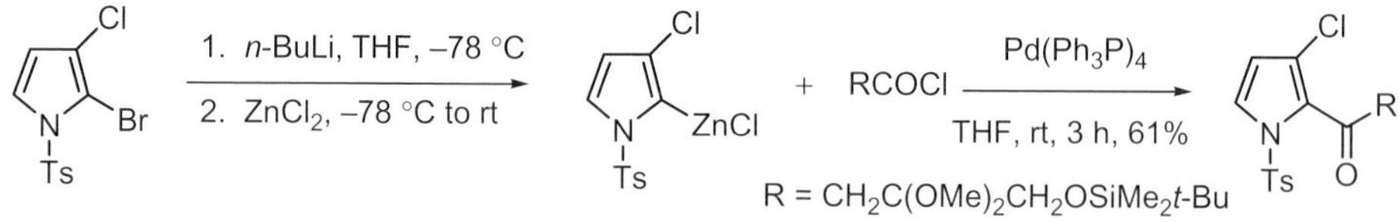

74

The *N*-pyrrolylzinc chloride **75** undergoes Pd-catalyzed coupling with perfluoroalkyl iodides to afford the 2-substituted pyrroles **76** in good yield [61]. Smaller amounts (15–20%) of 2-perfluoroalkanoyl pyrroles, which presumably arise by hydrolysis of the benzylic difluoromethylene group, are also formed. This reaction, which is performed in one pot, also affords 2-phenylpyrrole (75%) and 3-phenylpyrrole (5%) with iodobenzene. Some biphenyl (15%) is also formed.

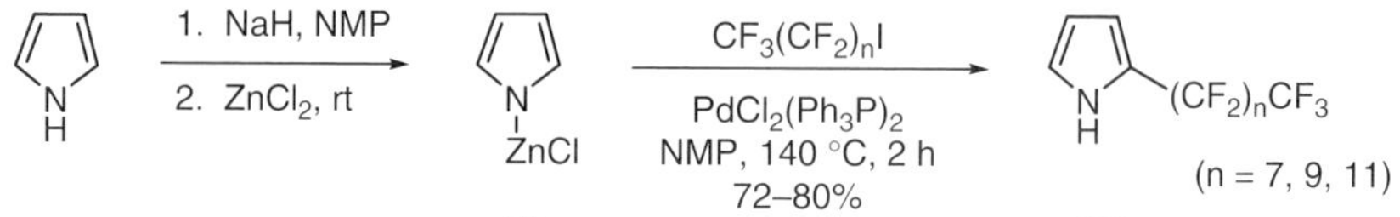

75   **76**

The *N*-pyrrolylzinc chloride (**75**) can be generated and coupled *in situ* with aryl bromides to give the expected 2-arylated pyrroles **77** in good to excellent yields [62]. Some aryl chlorides and iodides also couple with **75**.

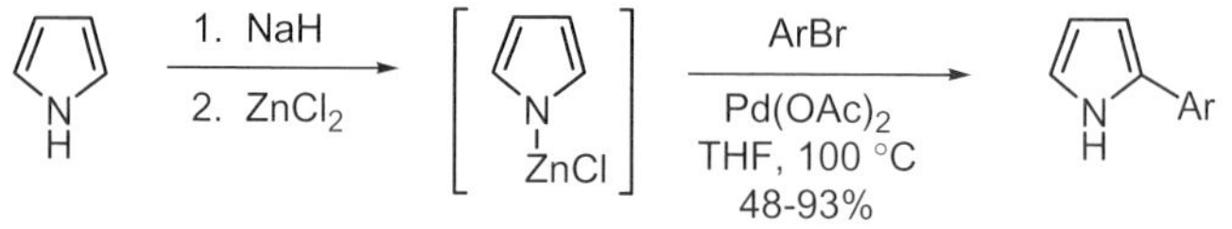

75   **77**

Ar = Ph, Mes, *i*-Pr₃Ph, 3-quinolinyl, 2-MeOPh, 4-Me₂NPh, 3,5-diMeOPh

A solid-phase synthesis of the lamellarin alkaloids Q and O featured the preparation of scaffold **79** and subsequent Suzuki coupling [24]. Dibromopyrrole **78** was synthesized as shown earlier for **14**.

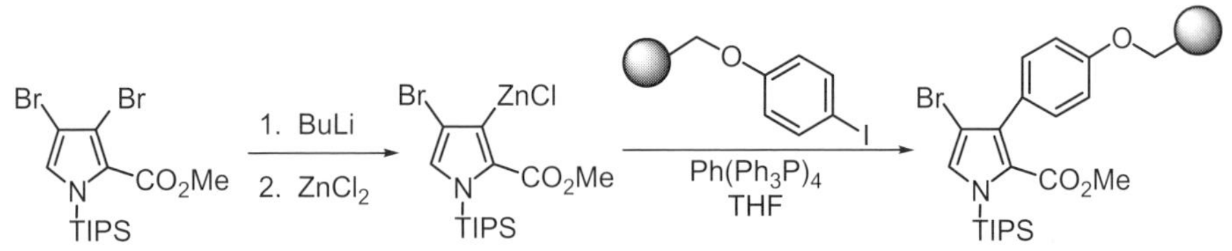

78   **79**

Bromopyrrole **80** undergoes bromine–lithium exchange, transmetalation, and Negishi coupling with some pyridyl halides to give **81** [63].

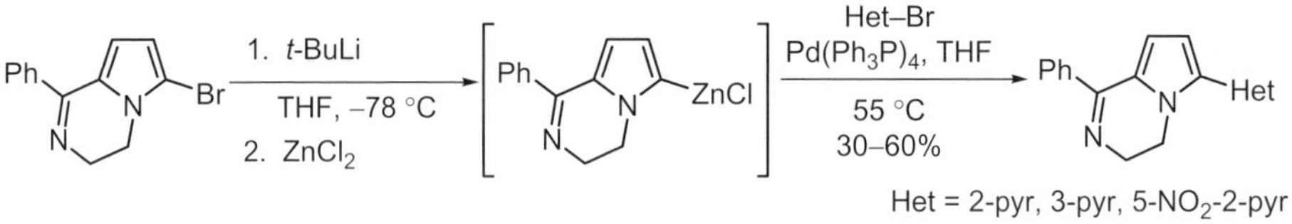

Het = 2-pyr, 3-pyr, 5-NO$_2$-2-pyr

80   **81**

Dihydropyrrole triflate **82** is smoothly converted to the 2-phenyl derivative **83** under Negishi conditions [64]. This sequence represents a useful alternative to conventional pyrrole Negishi couplings.

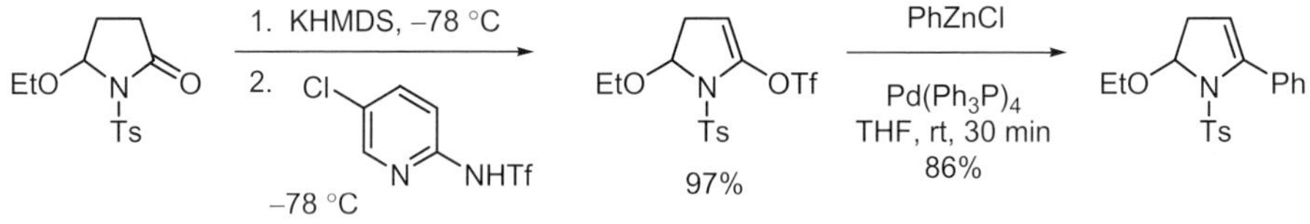

82   **83**

The Negishi coupling of halogenated porphyrins has also been described [65, 66].

2.3.3. Organoboron reagents (the Suzuki coupling)

As we have seen in Chapter 1, the Suzuki coupling reaction is a powerful method for preparing biaryls and several applications in pyrrole chemistry have been described. Schlüter reported the first pyrrole boronic acid **84**, but, surprisingly, the related pyrrole borate **85** could not be hydrolyzed [15].

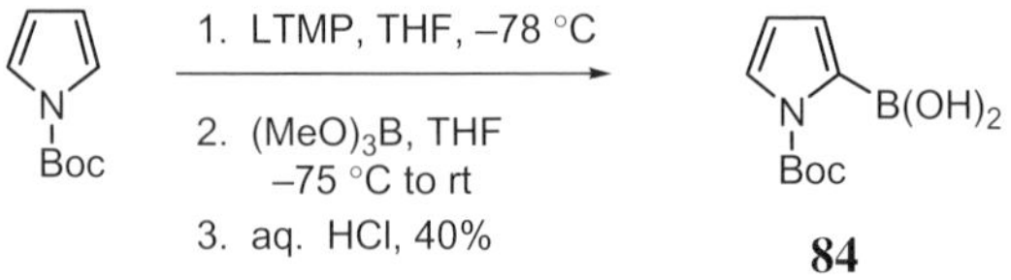

84

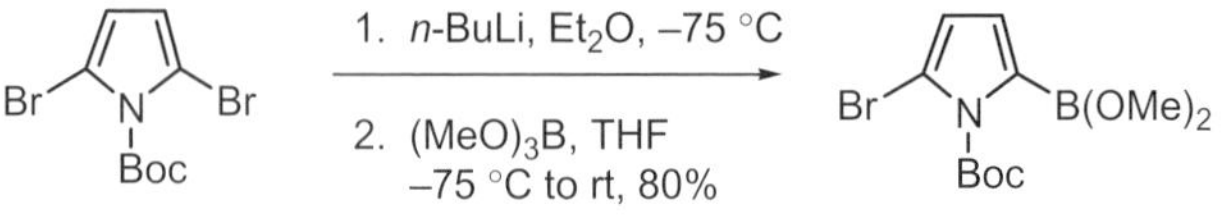

85

Gallagher has effected the Suzuki coupling reaction of **84** with a wide variety of aryl and heteroaryl halides to give pyrroles **86** as shown [67]. In all cases, the major side product is the bipyrrole. Iodobenzene gives a higher yield than bromobenzene, but all of the other examples are aryl and heteroaryl bromides. This study included the synthesis of new, selective dopamine D_3 receptor antagonist intermediates **87** and **88**, among others. The BOC group is readily cleaved in these compounds with methoxide in methanol at room temperature.

A similar sequence with 5-substituted pyrrole boronic acids **89** affords a range of 2-aryl- and 2-hetarylpyrroles **90** [68]. Boronic acids **89** were prepared as shown for **84**. This study included the synthesis of bis-pyrrole **91** from **89** (R = Et) and 3-iodobromobenzene (71%). Thermal deblocking affords **92**.

R = Bn, *i*-Pr, Et; Ar = Ph, 4-MePh, 4-CNPh, 3-pyr, 4-AcPh, 4-NO$_2$Ph

The unsubstituted pyrrole-2-boronic acid **93** was coupled with pyridyl bromide **94** on a Wang resin to afford nicotinic acid derivative **95** after cleavage [69].

93   **94**   **95**

A synthesis of the immunosuppressive agent undecylprodigiosine involved a Suzuki coupling between **84** and triflate **96** to give tripyrrole **97** [70]. The corresponding Stille reaction was unsuccessful.

96   **97**

Ketcha had prepared 1-(phenylsulfonyl)pyrrole-2-boronic acid (**99**) in low yield and effected Suzuki coupling reactions to afford the corresponding 2-aryl derivatives **100** [71].

Ar = Ph, 1-naphthyl, 4-NO$_2$Ph, 4-AcPh, 4-MeOPh, X = Br, I

98   **99**   **100**

The pyrrole-3-boronic acid **102** has been prepared from the 3-iodopyrrole **101** by Muchowski and subjected to a range of Suzuki couplings to afford 2-arylpyrroles [72]. Subsequent fluoride deblocking to give **103** occurs in excellent yield.

6   **101**   **102**

103

Ar = Ph, 4-MePh, 4-MeOPh, 4-MeCONHPh, 4-ClPh, 4-NO$_2$Ph, 4-CO$_2$MePh, 4-CHOPh, 4-CNPh, 2-MePh, 3-MeOPh, 3-pyridyl; X = Br, I

Swager has utilized pyrrole-3-boronic acid **102**, which was prepared from **6** by NBS bromination (88%), lithiation (BuLi), and boronation (B(OMe)$_3$; aq MeOH) (50% overall), in Suzuki reactions to synthesize bis-pyrroles **104–106**, *en route* to novel conductive electrochromic polymers [73].

Banwell and co-workers have converted diiodopyrrole **107** [19] to bis-boronate **108**, which without purification was successfully coupled to afford bis-quinoline pyrrole **109** [74].

Similarly, pyrrole boronate **110**, which was prepared from **51**, coupled regioselectively with bromopyrazine **111** to give **112** (77%), *en route* to the synthesis of (+)-dragmacidin F [43].

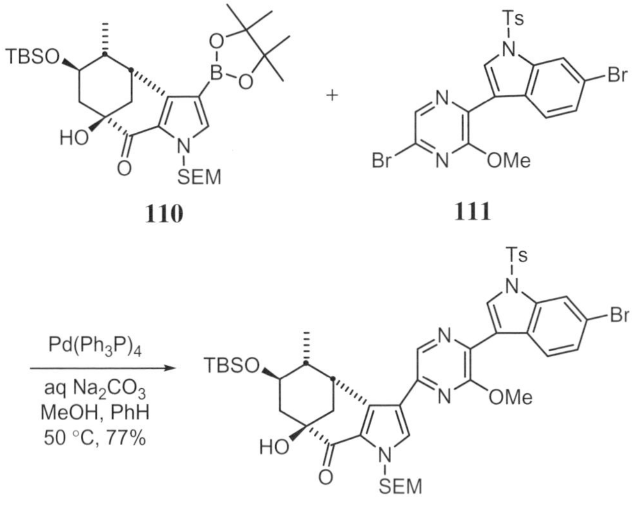

110   +   **111**

Pd(Ph₃P)₄

aq Na₂CO₃
MeOH, PhH
50 °C, 77%

112

The pyrrole component can also be employed as the aryl halide in Suzuki coupling with aryl boronic acids. Thus, Chang has effected several such reactions using phenylboronic acid and halopyrroles such as **113** and **114** [75].

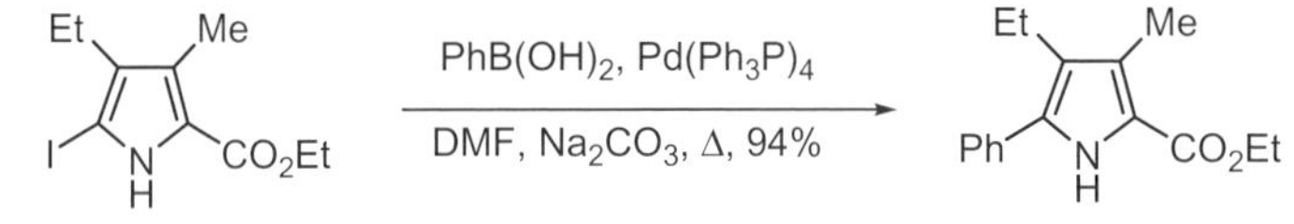

PhB(OH)₂, Pd(Ph₃P)₄

DMF, Na₂CO₃, Δ, 94%

113

PhB(OH)₂, Pd(Ph₃P)₄

DMF, Na₂CO₃, Δ, 95%

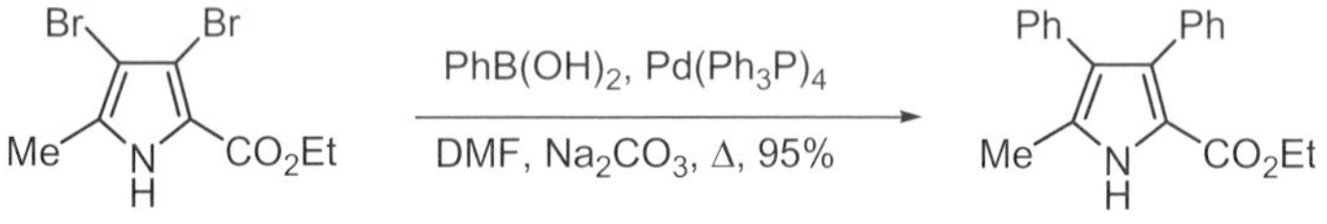

114

Using 2-bromopyrrole **115**, Burgess has synthesized 2-arylpyrroles **116** in excellent yield [76]. The resulting hydrolyzed pyrroles **117** were used to prepare 3,5-diaryl BODIPY@dyes.

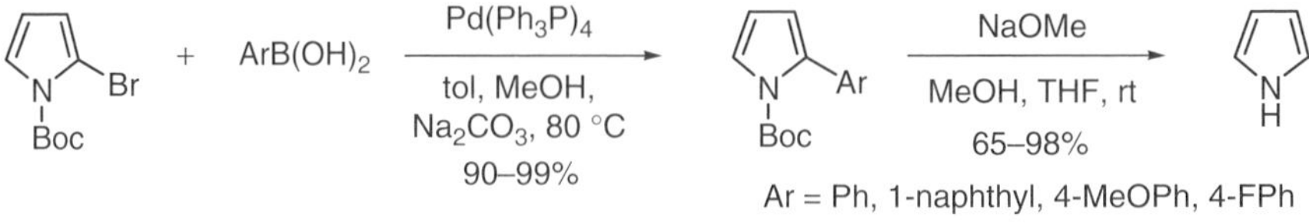

+   ArB(OH)₂

Pd(Ph₃P)₄

tol, MeOH,
Na₂CO₃, 80 °C
90–99%

NaOMe

MeOH, THF, rt
65–98%

Ar = Ph, 1-naphthyl, 4-MeOPh, 4-FPh

115   **116**   **117**

Likewise, 2-bromo-1-methyl-5-phenylpyrrole (**118**) undergoes Suzuki couplings with arylboronic acids to afford **119** [77].

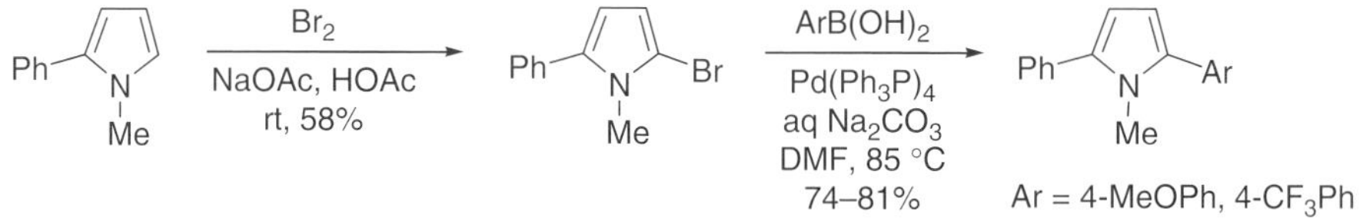

118 **119**

Other bromopyrroles that engage in Suzuki coupling reactions are **16** [25], **79** [24], **80** [63], **120** [78], **121** [78], **122** [79], and **123** [80].

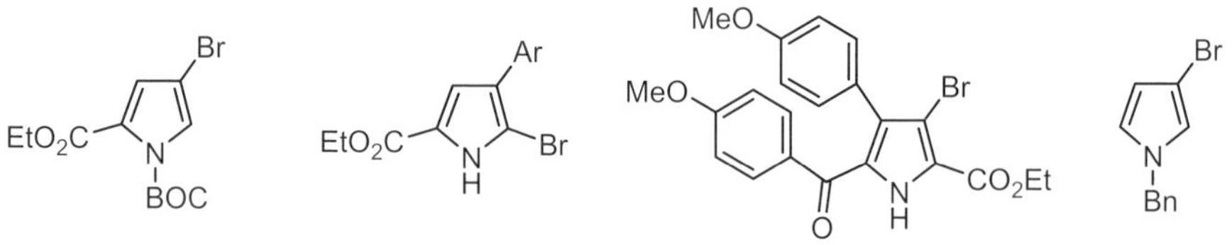

120 **121** **122** **123**

Dibromide **124** undergoes regioselective and sequential Suzuki reactions to afford **125** and then **126** [63]. A double coupling of **124** with 3-thienylboronic acid yields **127** (79%).

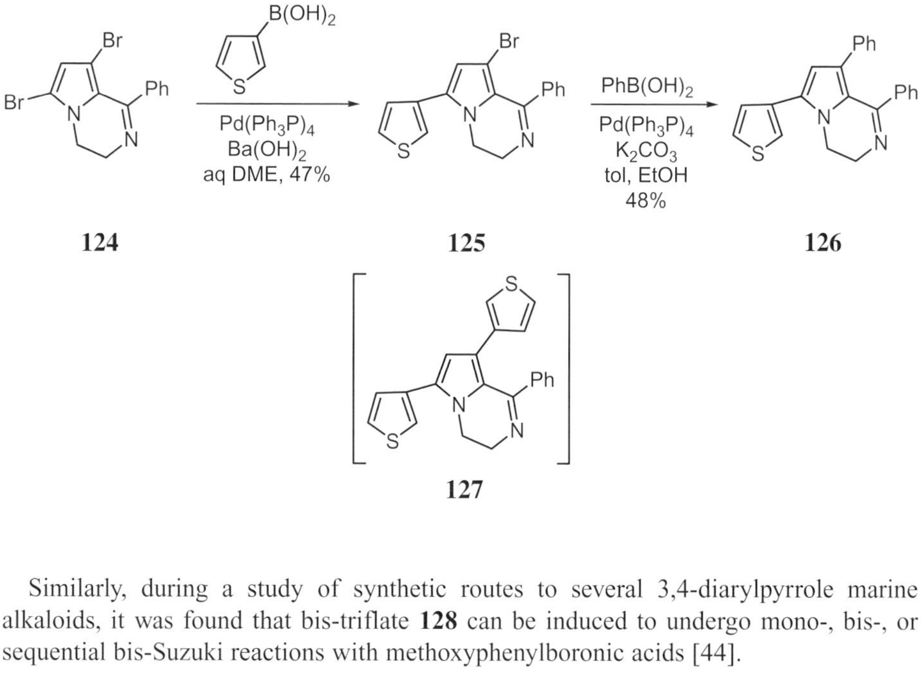

124 **125** **126**

127

Similarly, during a study of synthetic routes to several 3,4-diarylpyrrole marine alkaloids, it was found that bis-triflate **128** can be induced to undergo mono-, bis-, or sequential bis-Suzuki reactions with methoxyphenylboronic acids [44].

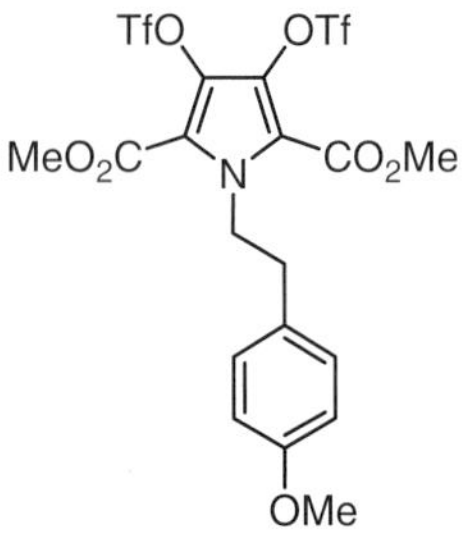

128

The Suzuki coupling has been utilized to craft β-octasubstituted tetramesitylpor-phyrins using various arylboronic acids [81], and Schlüter has adopted this reaction to prepare phenyl–pyrrole mixed polymers **129** [82]. The BOC group is easily removed by heating [83] and polymers with molecular weights of up to 23,000 were synthesized. These polymers are potentially interesting for their electrical and nonlinear optical properties [84].

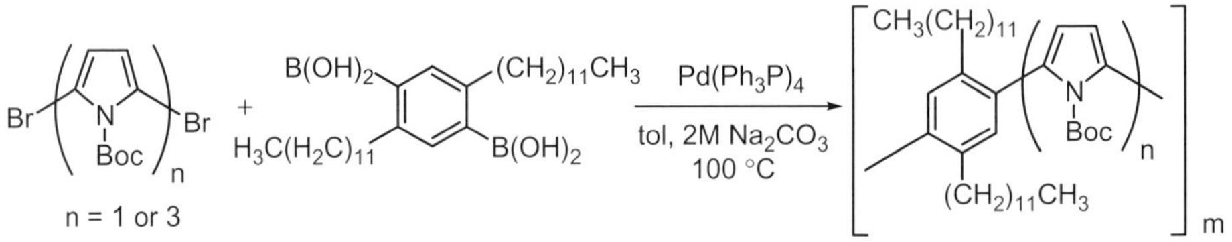

129

2.3.4. Stille coupling

Bailey described the first application of the Stille coupling to pyrroles, and one of the earliest examples of any such reaction involving heterocycles [85]. Lithiation of *N*-methylpyrrole and quenching with trimethylstannyl chloride gives 2-(trimethylstannyl) pyrrole (**130**), and Pd-catalyzed coupling with iodobenzene affords 1-methyl-2-phenylpyrrole (**68**) in good yield. The coupling of 1-methyl-2-(tri-*n*-butylstannyl)pyrrole with iodobenzene under similar conditions (Pd(Ph$_3$P)$_2$Cl$_2$, THF, 85°C) gives **68** in 82% yield [77].

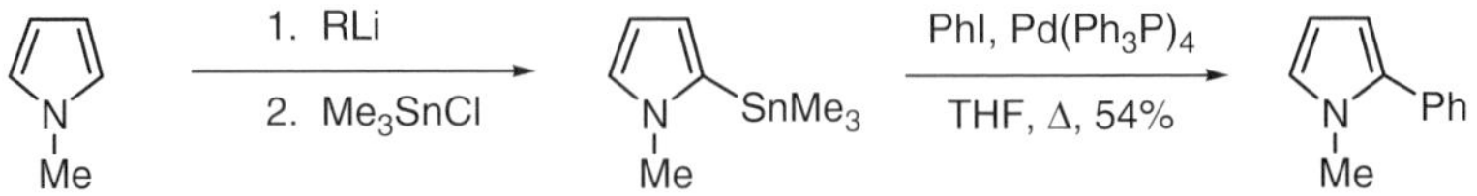

130   **68**

Mono- or bis-bromine–lithium exchange on dibromopyrrole **131** affords stan-nylpyrroles **132** or **133**, respectively, and *N*-BOC-2-trimethylstannylpyrrole is obtained in 75% yield from *N*-BOC pyrrole by lithiation with LTMP and quenching with Me$_3$SnCl [15]. This latter stannane couples smoothly with *o*-bromostilbenes [86].

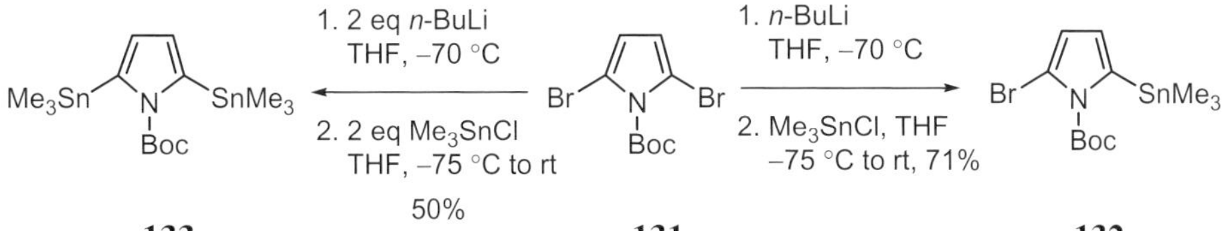

133   **131**   **132**

The bis-stannylpyrrole **134** is obtained from 1-methylpyrrole via 2,5-dilithiation. Subsequent Stille coupling affords **135**, which was used to craft novel diguanidine antifungal agents [87].

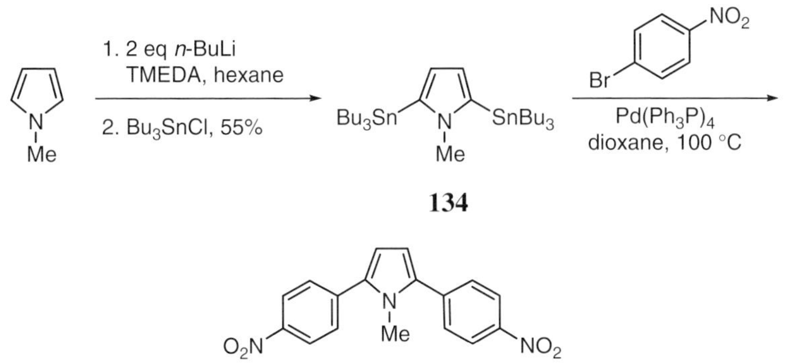

134

135

A Stille coupling between stannylpyrrole **136** and 3-bromoindole **137** gave the expected **138** albeit in low yield [63]. Pyrrole **136** was prepared from **80** in the usual manner (1. *n*-BuLi; 2. Bu$_3$SnCl; 65%).

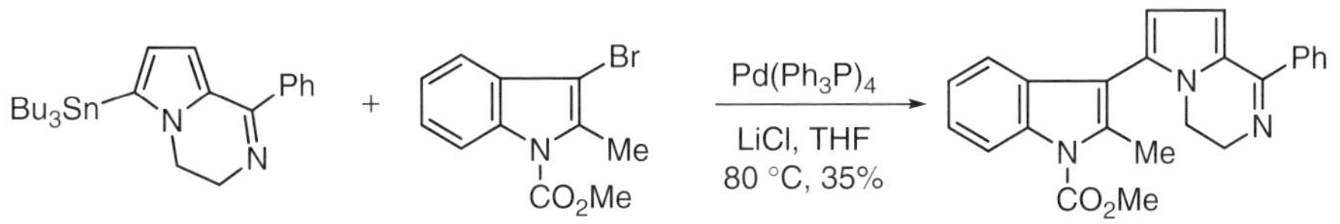

136   **137**   **138**

Direct lithiation of *N*-(dimethylamino)pyrrole (**139**) and subsequent quenching with trimethylstannyl chloride affords **140** in excellent yield [88]. The *N*-dimethylamino group can be removed with Cr$_2$(OAc)$_4$·2H$_2$O.

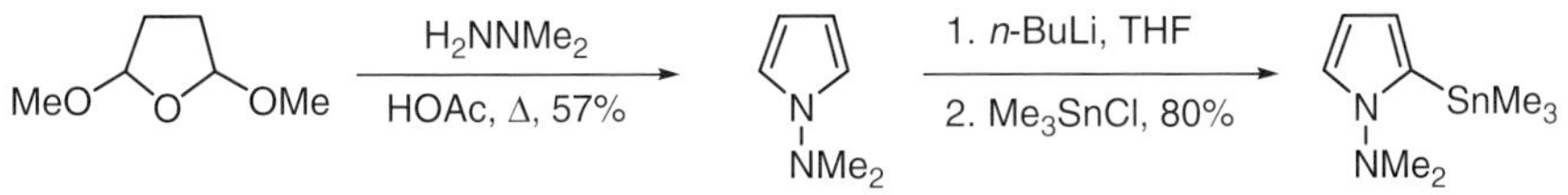

139   **140**

Although the stannylated pyrroles are normally obtained via lithiation, two other methods to prepare these Stille precursors have been devised. Caddick has found that the addition of tri-*n*-butylstannyl radical to pyrrole **141** affords stannylpyrrole **142** in good yield [89].

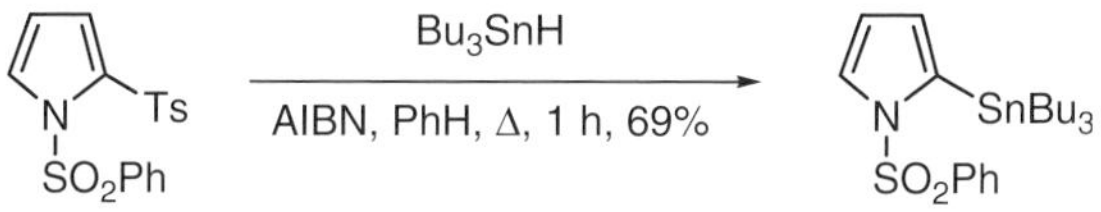

141   **142**

The second alternative synthesis of stannylpyrroles involves the versatile van Leusen pyrrole ring synthesis, illustrated below for the synthesis of **144** [90, 91]. Several examples were prepared in these studies, although at present the scope of this reaction is limited to pyrroles having an electron-withdrawing group at the 3-position. The structure of a BOC derivative of **144** (R_1 = Ph, R_2 = Bz) was established by X-ray crystallography [90].

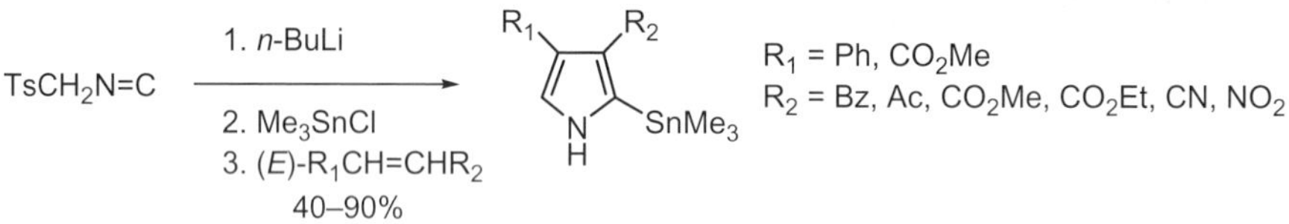

143   **144**

Stille couplings on N-protected derivatives of **144** proceed well, as shown below for the synthesis of pyrroles **145** [91].

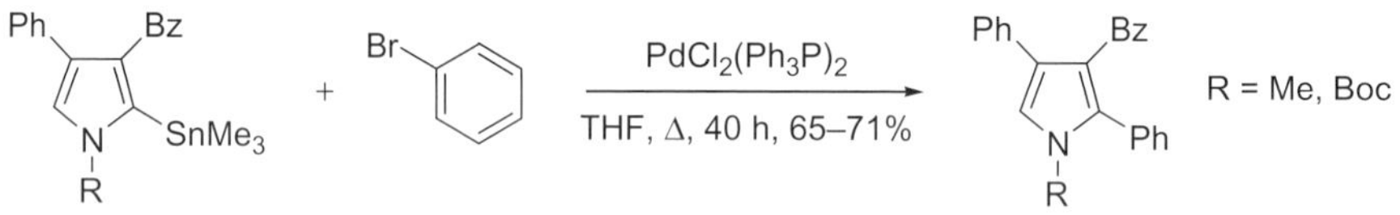

145

Dubac has employed a Stille coupling reaction to synthesize the pyrroles **147** and **148** from stannylpyrrole aldehyde **146** [92]. The latter tin compound was prepared as shown, and related stannylpyrroles were synthesized similarly [93] or using Muchowski's 6-dimethylamino-1-azafulvene dimer lithiation methodology [94].

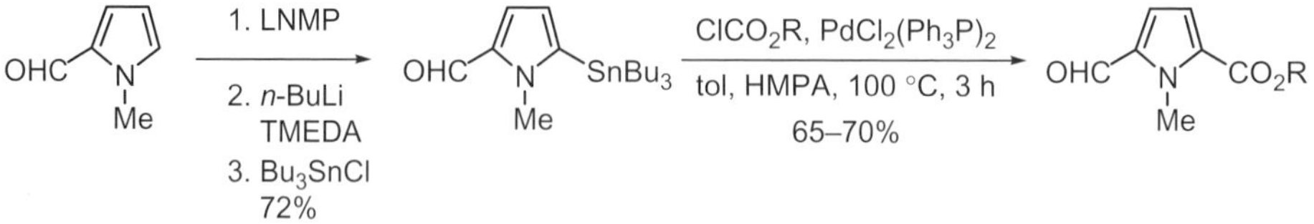

146

147, R = n-octyl
148, R = NMePh

The related stannylpyrrole **149**, which was reported by Dubac [93], has been used to synthesize the sponge metabolite mycalazol 11 (**150**) and related compounds, which have activity against the P388 murine leukemia cell line [95].

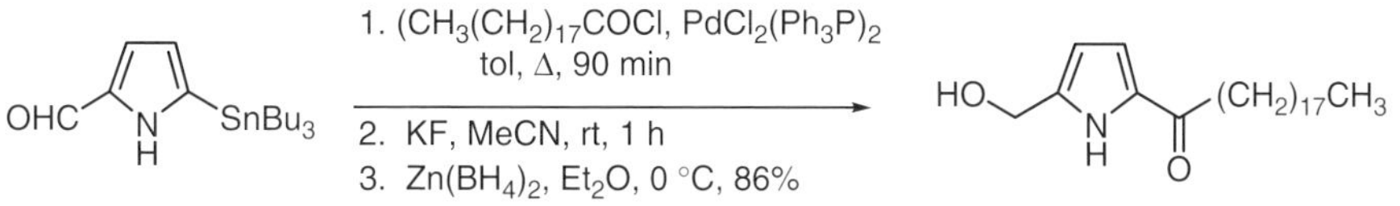

149 → **150**

Stille couplings between stannylpyrrole **151** and siloles **152** and **153** have afforded the silole–pyrrole **154** [96]. These workers also prepared dimers and trimers of **154**.

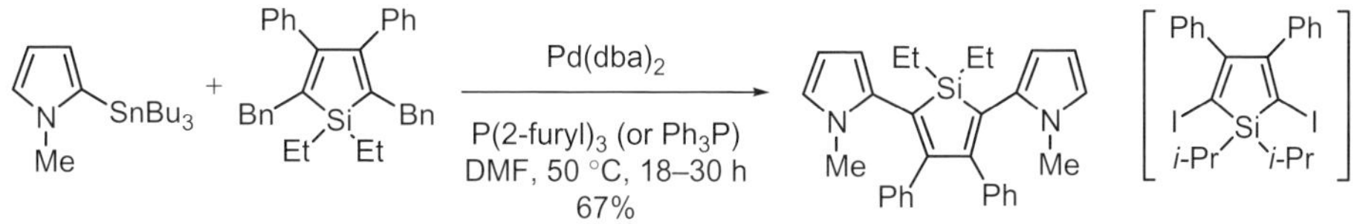

151   **152**   **154**   **153**

The Pd-catalyzed cross-coupling reaction of 3-stannylated pyrroles is also known. Muchowski has thus prepared and utilized **155** to effect Stille couplings leading to **156** [72].

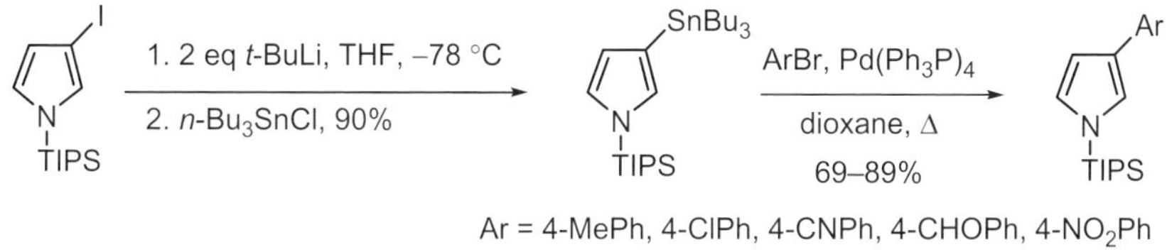

Ar = 4-MePh, 4-ClPh, 4-CNPh, 4-CHOPh, 4-NO$_2$Ph

101   **155**   **156**

Pyrrole **157** has been employed in Stille couplings with dibromobenzoquinone **158** [97]. The product **159** can be subjected to a second Stille coupling to afford unsymmetrical diheteroarylquinones. Similar couplings between **158** and 2,3-dibromo-1,4-naphthoquinone were also described.

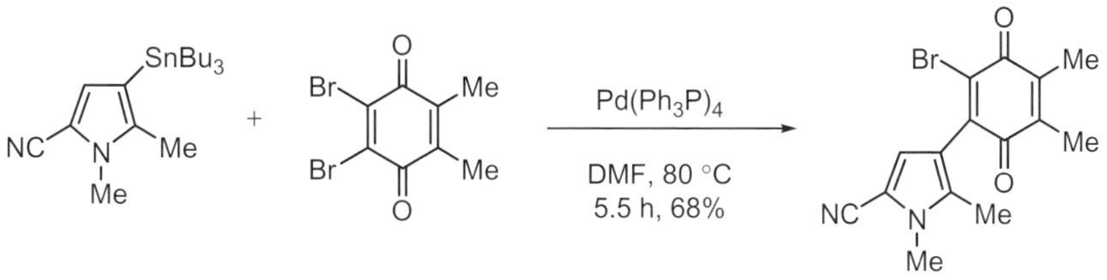

157   **158**   **159**

Halogenated pyrroles can serve as the aryl halide in Stille couplings with organotin reagents. Scott has used this idea to prepare a series of 3-vinylpyrroles, which are important building blocks for the synthesis of vinyl-porphyrins, bile pigments, and indoles [98]. Although 3-chloro- and 3-bromopyrroles fail completely or fared poorly in this chemistry, 3-iodopyrroles **160** work extremely well to yield 3-vinylpyrroles **161**.

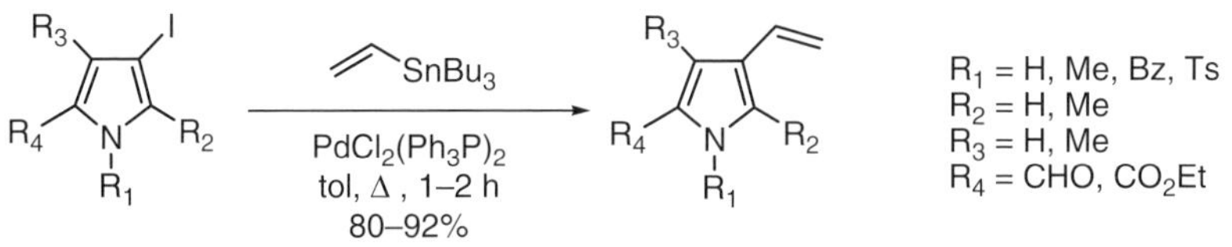

SnBu3
PdCl2(Ph3P)2
tol, Δ, 1–2 h
80–92%
R1 = H, Me, Bz, Ts
R2 = H, Me
R3 = H, Me
R4 = CHO, CO2Et

160 **161**

These workers have also used 3-iodopyrrole **162** to prepare the corresponding tin derivative **163** for Stille couplings to furnish 3-arylpyrroles **164** [99]. These pyrrole derivatives are important for the synthesis of β-aryl-substituted porphyrins for studies of heme catabolism. Also synthesized in this study were compounds **165** and **166** [99].

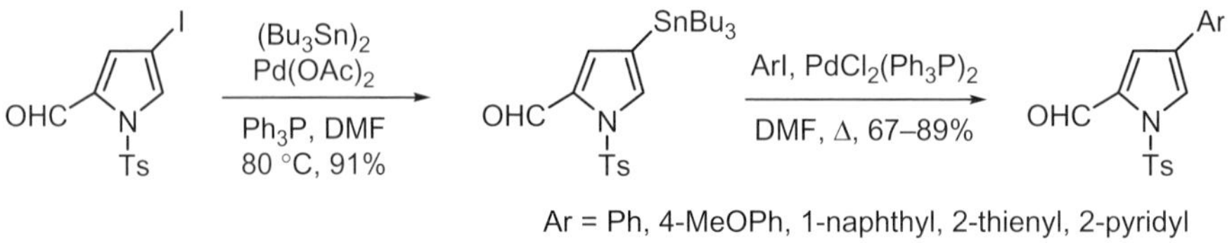

(Bu3Sn)2
Pd(OAc)2
Ph3P, DMF
80 °C, 91%
ArI, PdCl2(Ph3P)2
DMF, Δ, 67–89%
Ar = Ph, 4-MeOPh, 1-naphthyl, 2-thienyl, 2-pyridyl

162 **163** **164**

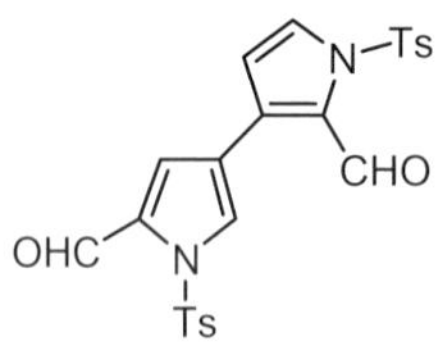

165 (70%)

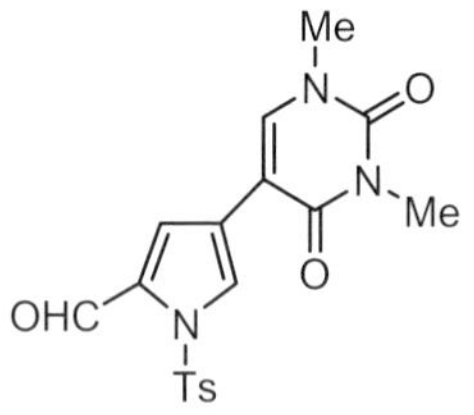

166 (62%)

Similarly, Hibino *et al.* achieved excellent yields of 2-vinylpyrroles **168** via Stille coupling of pyrroles **167** and tributylvinylstannane [100].

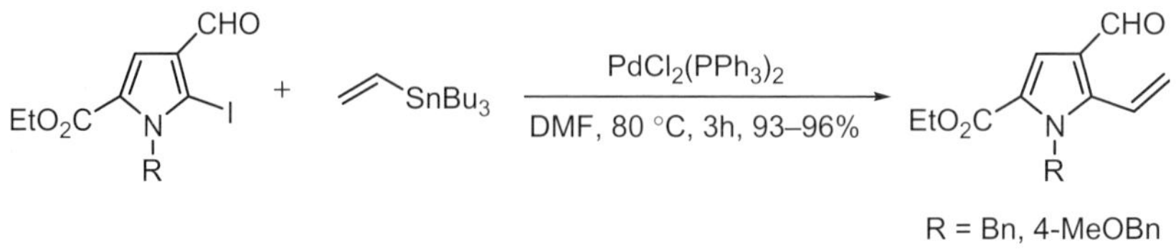

PdCl2(PPh3)2
DMF, 80 °C, 3h, 93–96%
R = Bn, 4-MeOBn

167 **168**

Tour has utilized dibromopyrrole **169** to prepare zwitterionic diiodopyrrole **171**, which in turn was employed in a synthesis of diphenyl derivative **172** and pyrrole polymers [10, 11].

Stille has employed the Pd-catalyzed coupling that bears his name in syntheses of anthramycin and analogues [101, 102]. Thus, enol triflate **173** is smoothly coupled with acrylates to provide **174-176**.

Dihydropyrrole triflate **82**, which was mentioned in Section 2.3.2, undergoes facile Stille couplings to yield, for example, 2-acylpyrrole derivative **177** after hydrolysis. Also prepared in these studies were the corresponding 2-vinyl (77%) and 2-phenyl (56%) derivatives [64, 103].

The Stille Pd-catalyzed cross-coupling has been employed in the synthesis of modified porphyrins [65, 66, 104]. For example, the union of dihaloporphyrins with tri-*n*-butylvinylstannane affords protoporphyrin IX in excellent yield [104].

Pyridine **178** undergoes a Stille coupling with 1-methyl-2-trimethylstannylpyrrole (**130**) to give **179** in 36% yield [105], and the same stannane has been joined with bromopyrimidines [106].

2.4. Sonogashira reaction

The Sonogashira reaction frequently serves as a platform for the construction of indoles, and we will explore this application in Chapter 3, but it also is a valuable method for the preparation of alkynyl pyrroles.

Muchowski has utilized N-TIPS-3-iodopyrrole (**101**) to prepare a series of 3-alkynyl pyrroles **180** using standard Sonogashira conditions [72]. Pyrroles **181** were obtained after fluoride cleavage of the TIPS group. Similarly, 3,4-bis(alkynyl)pyrroles were prepared from N-TIPS-3,4-diiodopyrrole [72].

The coupling of trimethylsilylacetylene with 2,5-diiodo-1,3,4-trimethylpyrrole (**182**) affords the corresponding bis-acetylene after cleavage of the TMS groups [107].

Tandem Sonogashira reactions starting with iodopyrrole **183** have furnished the dipyrrolylacetylene **185**, and dimerization of **184** gives **186** [108]. The Sonogashira coupling has also been reported for enol triflate **82** reacting with an N-protected propargyl amine [103].

In their studies towards the synthesis of lukianol A, Wong *et al.* observed the regioselective *ipso*-iodination of pyrrole **187** followed by Sonogashira coupling to give **189** in quantitative yield overall [109]. Iodopyrrole **188** also undergoes Suzuki couplings with arylboronic acids.

Dolphin has employed the Sonogashira coupling of terminal alkynes with zinc(II)-10-iodo-5,15-diphenylporphyrin (**190**) to prepare a series of alkynyl derivatives **191** [110], and similar syntheses of β-alkynyl porphyrins have been described [111].

2.5. Heck and intramolecular Heck reactions

As will be seen in this chapter and in the rest of the book, the Heck reaction and its numerous variations represent a fantastically powerful set of tools available to the heterocycle chemist.

Although most Heck chemistry that involves pyrroles is intramolecular or entails synthesis of the pyrrole ring, a few intermolecular Heck reactions of pyrroles are known. Simple pyrroles (pyrrole, *N*-methylpyrrole, *N*-(phenylsulfonyl)pyrrole) react with 2-chloro-3,6-dialkylpyrazines under Heck conditions to give mixtures of C-2 and C-3 pyrrole-substituted pyrazines in low yields [112]. For example, chloropyrazine **192** reacts with *N*-(phenylsulfonyl)pyrrole to give a 5:2 mixture of coupled products (**193**, **194**). After alkali cleavage of the phenylsulfonyl group in **193** and **194**, the isomeric heterocycles could ultimately be separated.

Pyrrole fails to undergo a Heck reaction with 1-bromoadamantane [113], but does couple with iodobenzene (Pd(OH)$_2$, Ph$_3$P, MgO, dioxane, 150°C) to give 2-phenylpyrrole in 86% yield [114]. Dihydropyrroles such as **195** and **196** undergo Heck reactions with ease, although the yields are variable [115–117]. Some examples are illustrated below. Since dihydropyrroles can be oxidized to pyrroles with a variety of reagents [118, 119], these Heck reactions of dihydropyrroles should constitute viable routes to pyrroles.

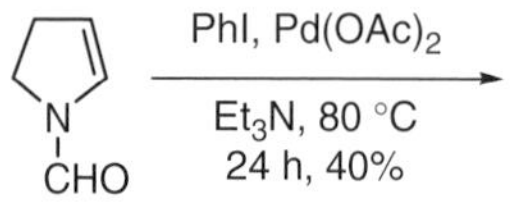

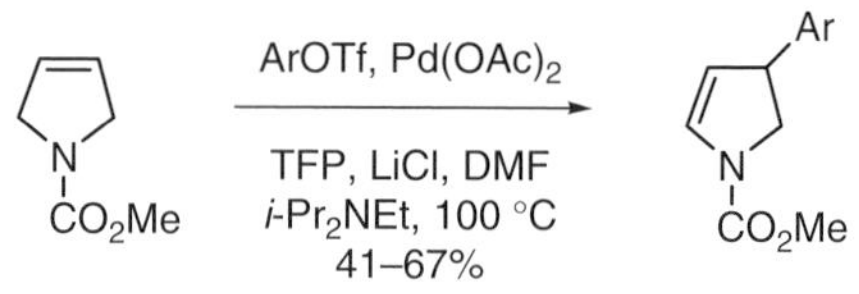

Ar = Ph, 3-CF$_3$Ph, 2-MeOPh, 4-MeOPh, 3-pyridyl, 1-cyclohexyl, 1-naphthyl

195   **196**

In these Heck reactions, some degree of enantioselectivity (up to 83% ee) is achieved in the presence of (*R*)-BINAP, although the yields of Heck products are often very low in the highest degree of enantioselectivity (e.g. 19% isolated yield at 83% ee) [117]. An example of a tandem Heck reaction is shown below involving the arylation of dihydropyrrole **196** with 1-naphthyl triflate (**197**) [116]. Complete chirality transfer is observed for the arylation of **198** to **199**.

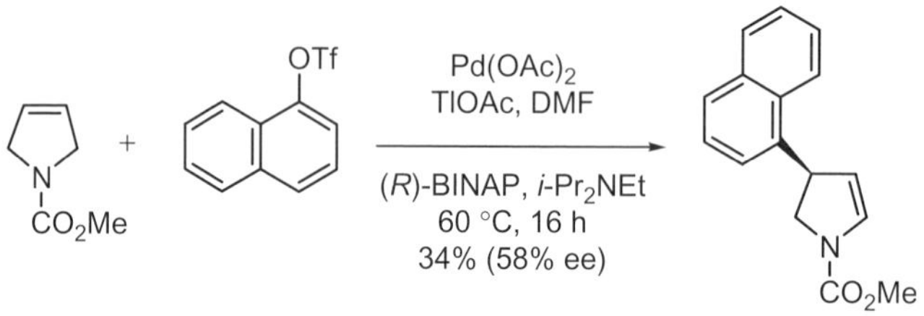

196   **197**   **198**

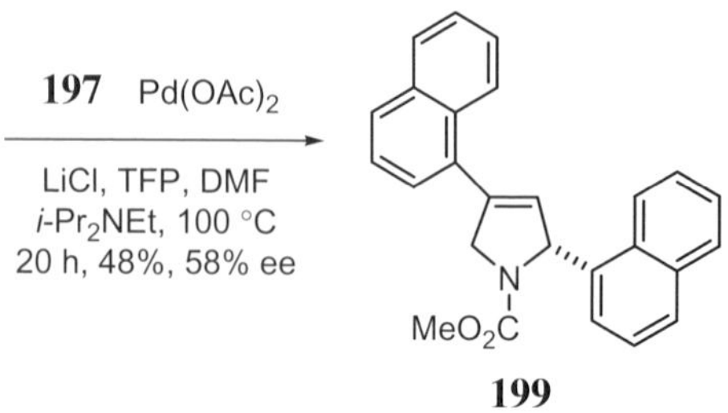

199

Stille employed a Heck reaction on triflate **200** to install the acrylamide side chain in his synthesis of anthramycin and analogues, as illustrated below [101, 102]. Ironically, a Stille reaction was less efficient in this transformation.

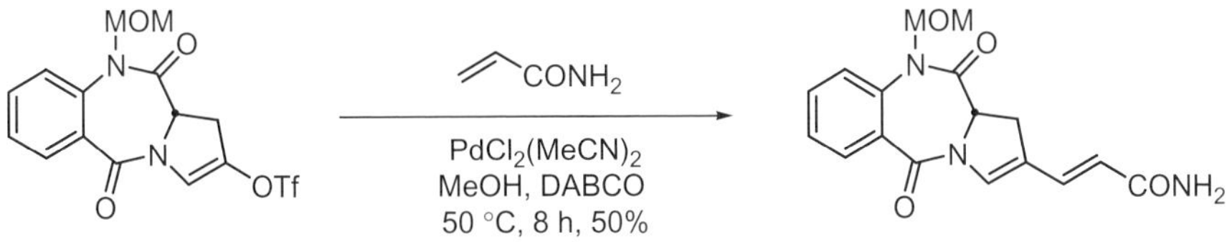

200

In a series of papers, rich in chemistry, Natsume has parlayed the intramolecular Heck reaction with **201** and **202** into an elegant construction of (+)-duocarmycin SA and related compounds [120–123].

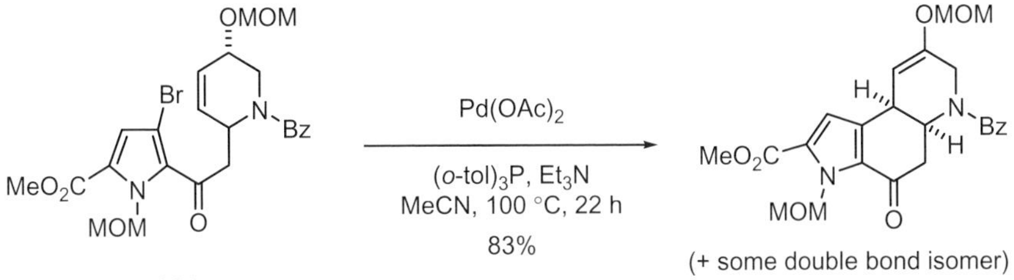

201

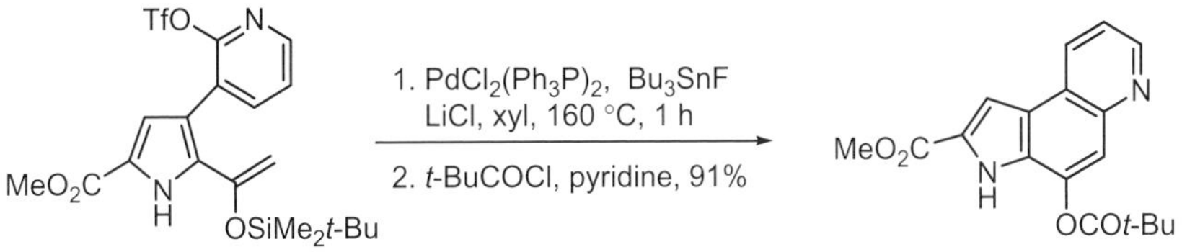

(+ some double bond isomer)

202

Grigg has utilized the Heck reaction in several ways, from the simple cyclization of *N*-(2-iodobenzoyl)pyrrole (**203**) to afford tricyclic lactam (**204**) [124] to the complex cascade transformation of pyrrole **205** with diphenylacetylene to tetracycle **208** [125]. The initial intermediate **206** undergoes a second Heck reaction with the pendant alkene to furnish **207**. A final insertion into the appropriate phenyl C–H bond completes the sequence.

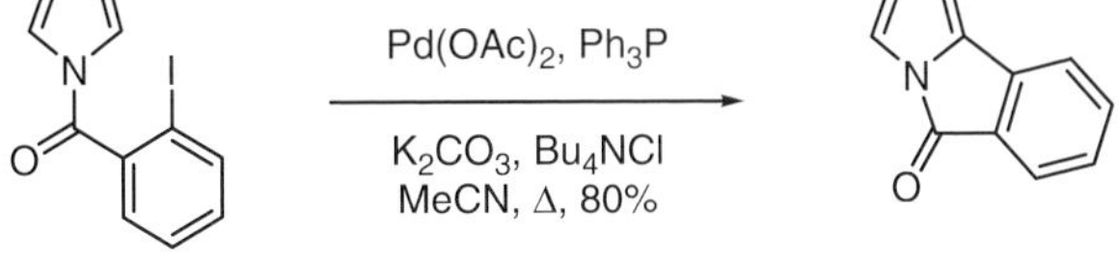

203   **204**

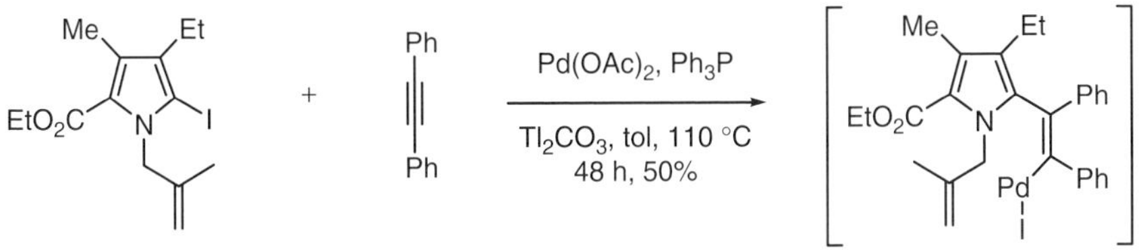

205   **206**

A related tandem Heck reaction is seen in the conversion of **209** to **210**, wherein the pyrrole ring is the site of termination [126].

We will encounter more of these fantastic tandem cascade Pd-induced cyclizations in Chapter 3. Using microwave heating, Beccalli and co-workers performed an intramolecular Heck cyclization with pyrrole **211** leading to tricycle **212** in excellent yield [127].

Several pyrrole ring syntheses have been developed that utilize an intramolecular Heck reaction, but, since these transformations involve C–N bond formation, they are covered in Section 2.7.

Grigg has employed the intramolecular Heck reaction to craft a series of spiro-pyrrolidines (**213**) [128], and Gronowitz has prepared several thienopyrroles, such as **214**, via a Heck strategy [129]. The *N*-BOC group in **214** is readily removed on mild heating after adsorption on silica gel (92% yield).

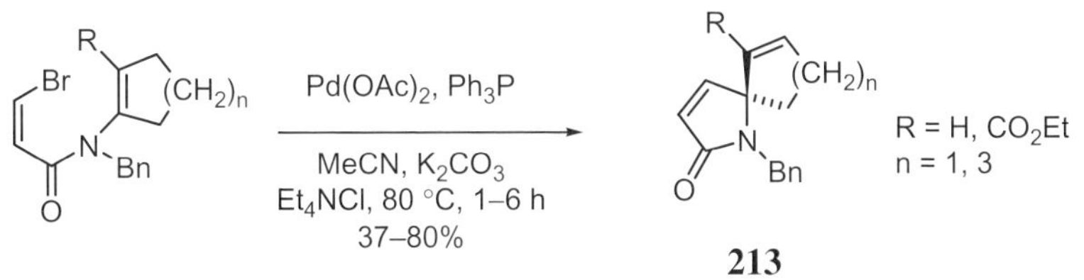

213

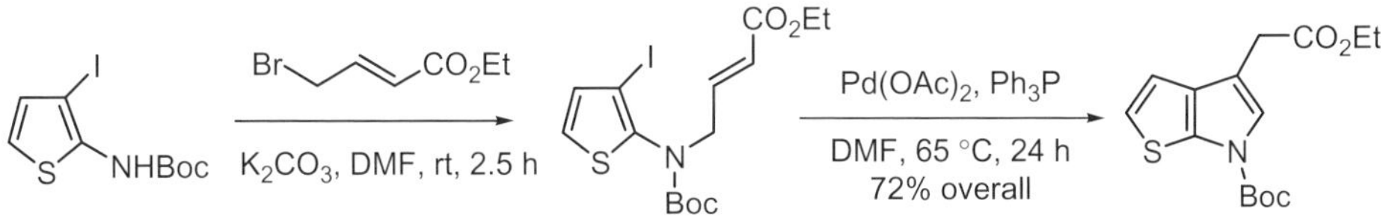

214

Since Heck reactions on metalated substrates are known (e.g. with organomercurials [130]), applications of these transformations to pyrrole chemistry have been reported. For example, mercuration of pyrrole **215** followed by exposure to methyl acrylate under Heck reaction conditions leads to **216** [131]. This Heck variation has been extended by Smith to mercurated porphyrins [132]. Pyrrolylacrylates like **216** have also been made using conventional Heck reactions on 3-iodopyrroles [133, 134].

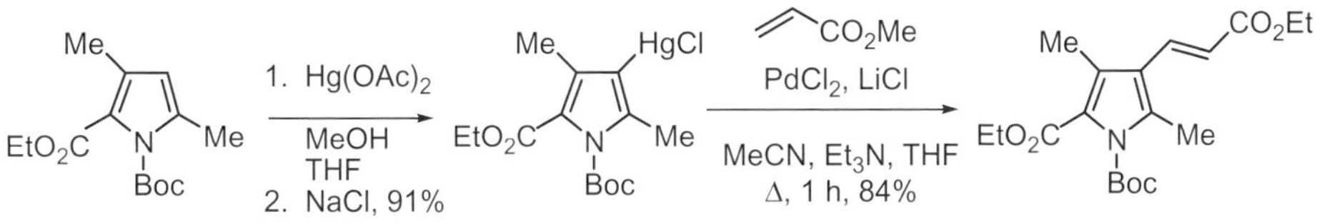

215 **216**

Pyrrole **217** can be thallated [135] and subjected to Heck (and Sonogashira) conditions to afford the anticipated products **218** [136].

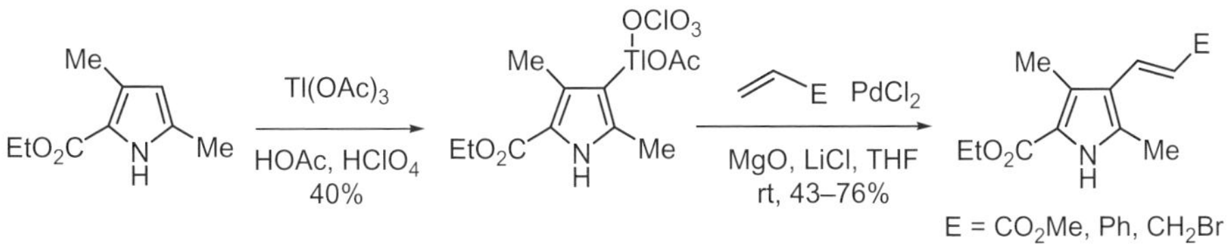

E = CO$_2$Me, Ph, CH$_2$Br

217 **218**

A Heck coupling of 2-lithio-1-methylpyrrole and bromohydrin **219** affords **220** [137].

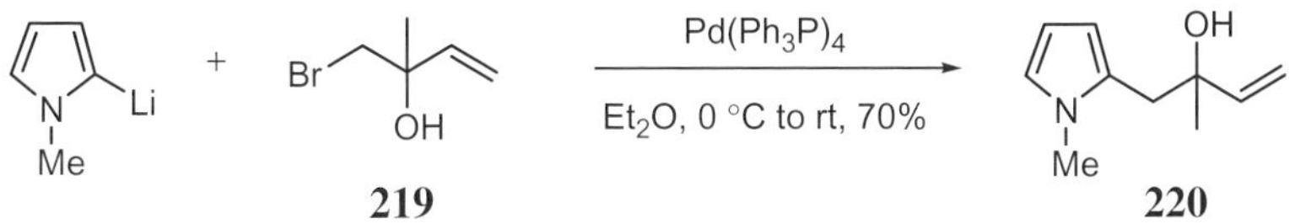

219 **220**

2.6. Carbonylation

As will be seen throughout the book, carbon monoxide has become an important player in the extension of Heck chemistry to carbonylation reactions. And, since most of these involve CO at atmospheric pressure, this variation is quite accessible to the synthetic chemist.

Edstrom has utilized the carbonylation Heck variation to engineer new routes to 3-substituted 4-hydroxyindoles, indolequinones, and mitosene analogs [138, 139]. For example, triflate **221** is converted to methyl ester **222** in high yield [139]. Subsequent oxidation affords indole **223**.

221   **222**   **223**

Dihydropyrrole triflate **82**, which we have encountered earlier in this chapter, undergoes a Pd-catalyzed carbonylation reaction to give ester **224** [64, 103]. A similar carbonylation sequence in the presence of the hydride donor n-Bu₃SnH gives the corresponding aldehyde in 56% yield [103].

82   **224**

Under mercuration conditions, pyrrole itself reacts with a mixture of $Hg(OAc)_2$, $PdCl_2$, LiBr, CO, EtOH, and $Cu(OAc)_2$ to give 2-(ethoxycarbonyl)pyrrole, but in only 4% yield [140]. In contrast, using the thallation-palladium modification of the Heck reaction, Monti and Sleiter have prepared pyrrole ester **225** in high yield [136].

225

Cyclocarbonylation of pyrrole **226** leads in modest yield to 4-acetoxyindole **227** [141].

226   **227**

An interesting Pd-induced diyne cyclization–carbonylation sequence has been discovered by Chiusoli [142]. Thus, diyne **228** is transformed as shown into a mixture of (*E*)- and (*Z*)-bis(alkylidene)pyrrolidines **229** and **230**.

Directed-carbonylation of 2-(dimethylaminomethyl)-1-(phenylsulfonyl)pyrrole (**231**) leads to the isolable palladium complex **232**. Exposure of **232** to CO yields pyrrole ester **235** in good yield [143].

2.7. C–N bond formation reactions

Palladium-catalyzed reactions of alkenes containing nitrogen nucleophiles have proven to be a powerful methodology for C–N bond formation leading to pyrroles.

An early example of this strategy is the palladium black catalyzed conversion of (*Z*)-2-buten-1,4-diol with primary amines (cyclohexyl amine, 2-aminoethanol, *n*-hexyl amine, aniline) at 120°C to give *N*-substituted pyrroles in 46–93% yield [144]. Trost extended this amination to the synthesis of a series of *N*-benzyl amines **235** from the readily available α-acetoxy-α-vinylketones **234** [145]. This methodology allowed for the facile preparation of pyrrolo-fused steroids.

Fürstner has employed the Trost pyrrole synthesis in the first total synthesis of roseophilin, wherein this *N*-benzylpyrrole ring forming step occurred in 70% yield [28]. Bäckvall has found that primary amines react with dienes under the guidance of Pd(II) to form pyrroles **236** in variable yields [146]. The intermediate π-allyl-palladium complexes are quite stable.

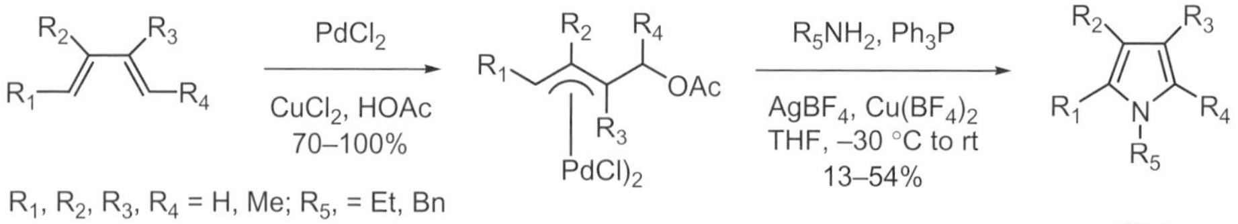

236

Most of the Pd-induced pyrrole ring forming reactions involve precursors having the requisite nitrogen nucleophile and alkene in the same molecule. Thus, as part of a general study of the amination of allylic systems, Genet and Bäckvall have observed that amine **237** cyclizes to pyrrole **238** [147]. Similar conditions and appropriate substrate design leads to bicyclic systems such as **239** and **240** where an amide nitrogen is a nucleophile [148].

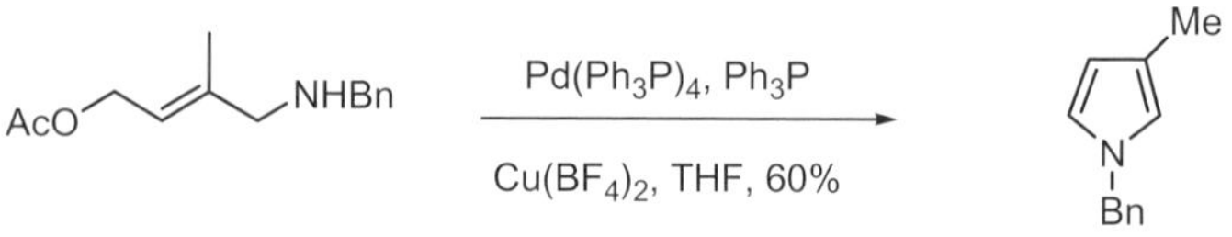

237   **238**

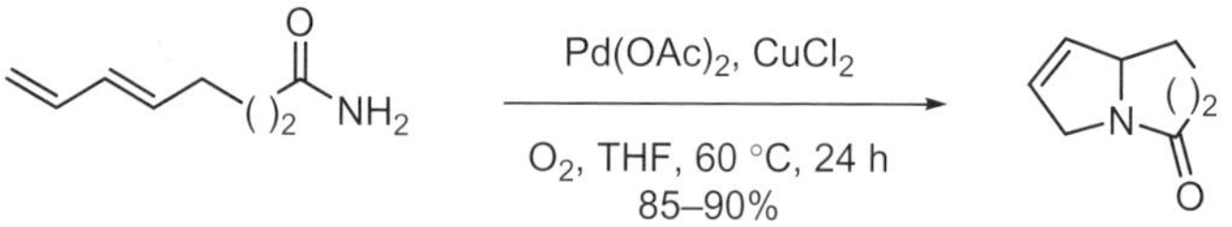

239 (n = 1)
240 (n = 2)

Likewise, the nitrogen in sulfonamides, such as **241** and **242**, serves very well in pyrrole forming reactions, as evidenced by the two examples illustrated below [149, 150]. In the second reaction, the hydroxyl group is essential for success, as is chloride (i.e. $Pd(OAc)_2$ does not work) [150]. It would seem that additional work is necessary to tame this latter reaction.

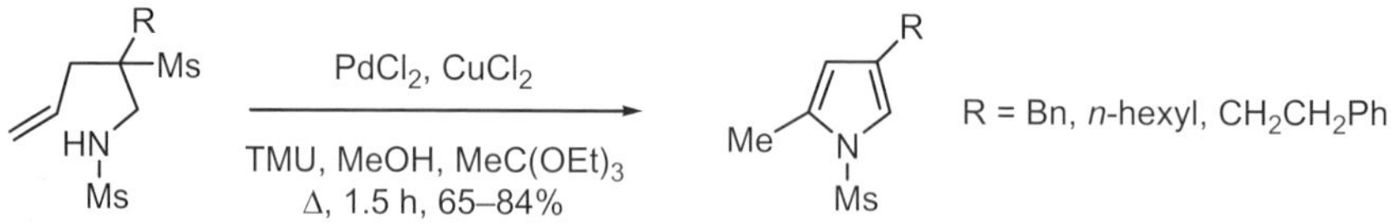

241

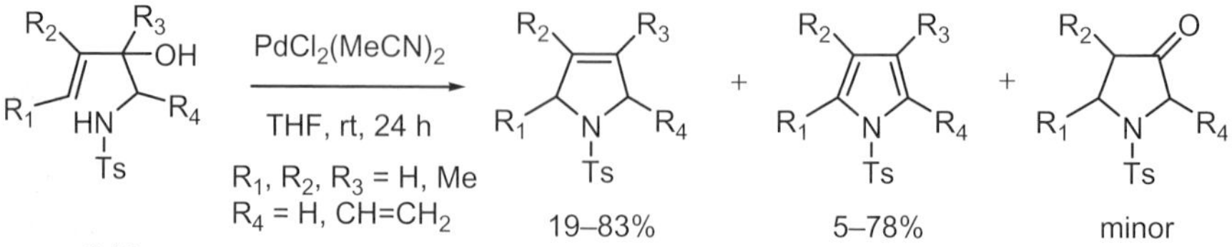

242

Cyclization of amino alcohol **243** leads to efficient chirality transfer into product dihydropyrrole **244** [151].

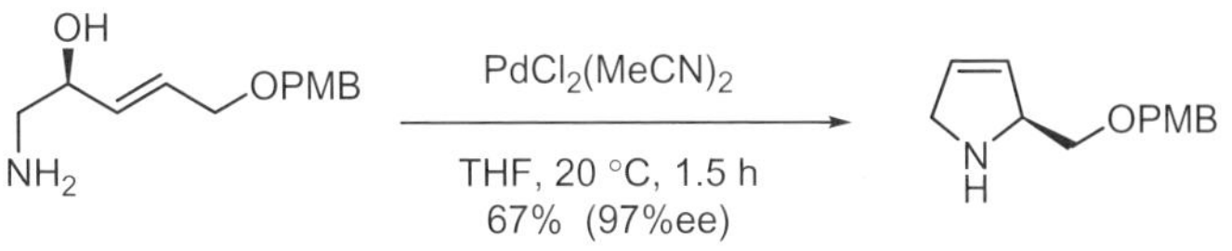

243   **244**

Alkynes can also serve as the recipient of cycloamination protocols. For example, 2,4-disubstituted pyrroles **246** are formed in high yields from Pd-catalyzed cyclization of aminoalkynes **179** [152]. Less effective is Pd(PPh$_3$)$_4$ but Pd(OAc)$_2$ works as well as PdCl$_2$.

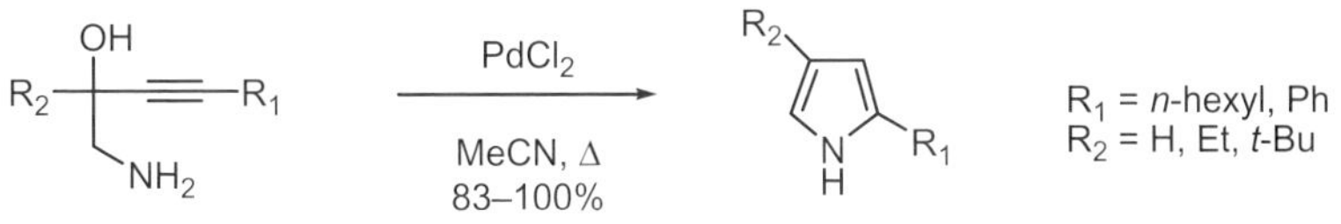

245   **246**

Gabriele *et al.* have described the cycloamination of aminoalkynes **247** leading to pyrroles **248** [153, 154], and in the presence of carbon monoxide to pyrrole esters **249** [155].

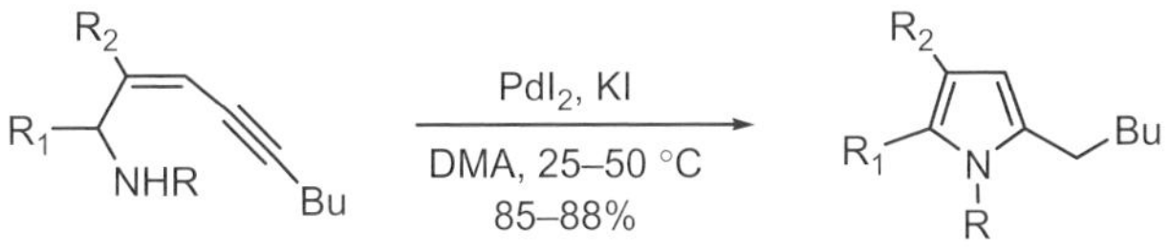

R = Bu, Bn; R$_1$ = H, Bu; R$_2$ = Et, Ph, H

247   **248**

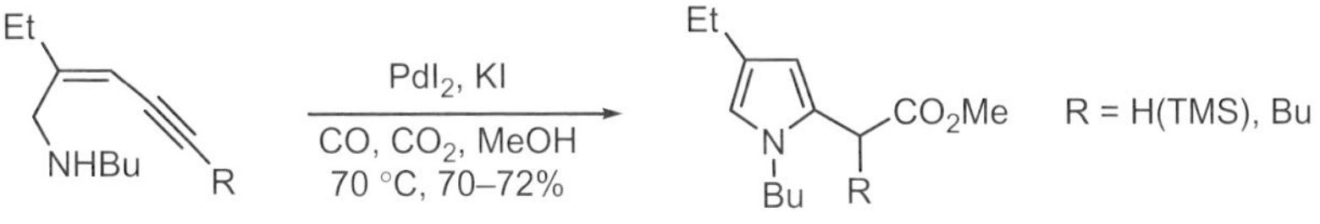

249

Arcadi has described a series of pyrrole ring-forming reactions leading to *N*-aminopyrrole derivatives **251–253** from common precursor **250**, depending on conditions [156]. Since *N*-aminopyrroles are widely used as pharmaceutical precursors, this chemistry could find wide appeal for the synthesis of medicinal agents.

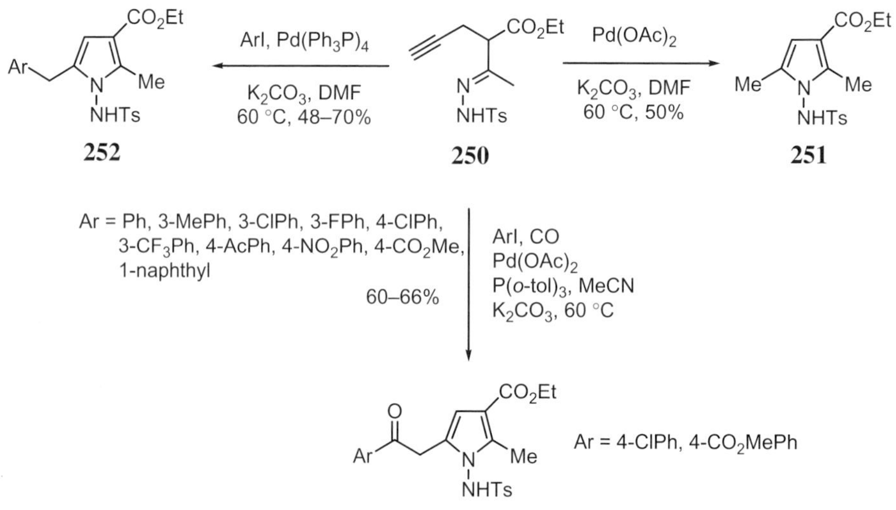
CO2Et
Ar
Me
N
NHTs
ArI, Pd(Ph3P)4
K2CO3, DMF
60 °C, 48–70%
Pd(OAc)2
K2CO3, DMF
60 °C, 50%
Ar = Ph, 3-MePh, 3-ClPh, 3-FPh, 4-ClPh,
3-CF3Ph, 4-AcPh, 4-NO2Ph, 4-CO2Me,
1-naphthyl
60–66%
ArI, CO
Pd(OAc)2
P(o-tol)3, MeCN
K2CO3, 60 °C
Ar = 4-ClPh, 4-CO2MePh

252 **250** **251** **253**

The oxime nitrogen can also participate in cycloamination reactions to give pyrroles. Thus, treatment of oxime esters such as **254** with $Pd(PPh_3)_4$ readily affords **255** [157]. The pentafluorophenyl group is necessary for good results; otherwise a Beckmann rearrangement can unfavorably enter the picture. The oxime stereochemistry makes no difference on the outcome of the reaction. In addition to **255**, pyrroles **256** and **257** were also prepared in this study (among others) [157].

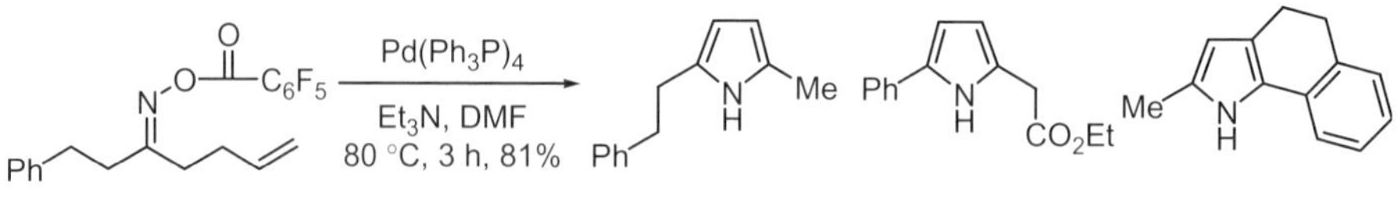
C6F5
Ph
Pd(Ph3P)4
Et3N, DMF
80 °C, 3 h, 81%
Me
CO2Et

254 **255** **256** (88%) **257** (78%)

A related amino-Heck cyclization of hydrazonium salts has been discovered by Narasaka. For example, **258** is smoothly converted to pyrrole **259** in 81% yield [158].

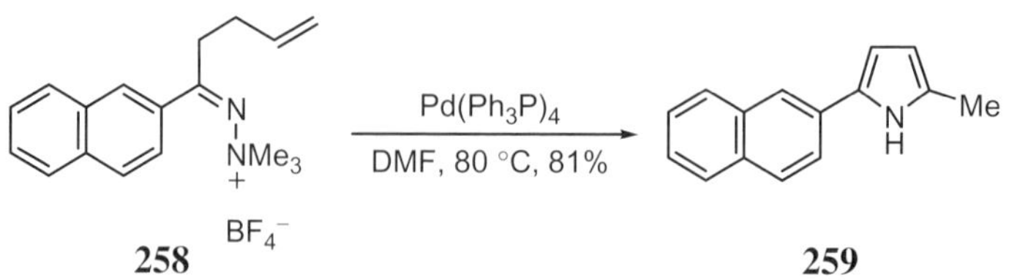
NMe3
BF4−
Pd(Ph3P)4
DMF, 80 °C, 81%
Me

258 **259**

Larock has described the reaction of diphenylacetylene with iodosulfonamides **260** to give alkylidene dihydropyrroles **261** [159]. This ring-forming reaction is similar to the large number of related indole syntheses we will see in the next chapter. For example, **262** affords **263** under these conditions.

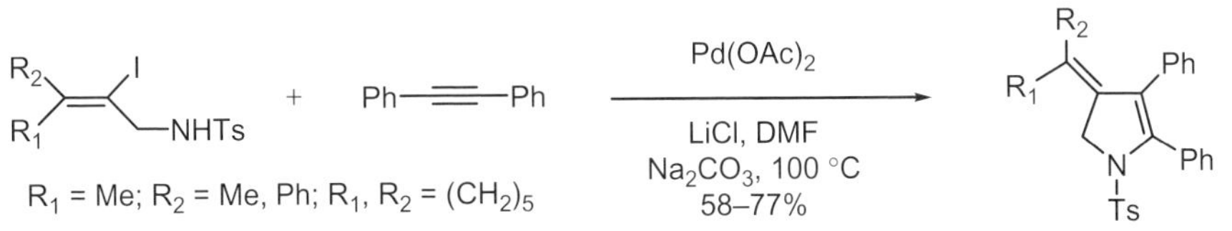

260   **261**

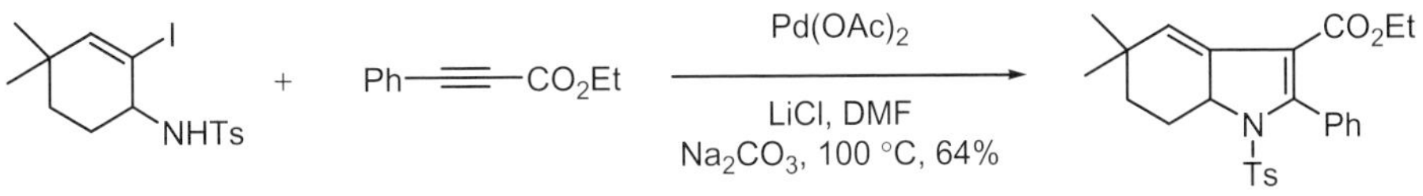

262   **263**

Using a similar Larock strategy, Gronowitz has synthesized a series of thienopyrroles, such as **264**, and other heterocyclic-fused pyrroles [160, 161]. Related reactions leading to indoles will be presented in the next chapter.

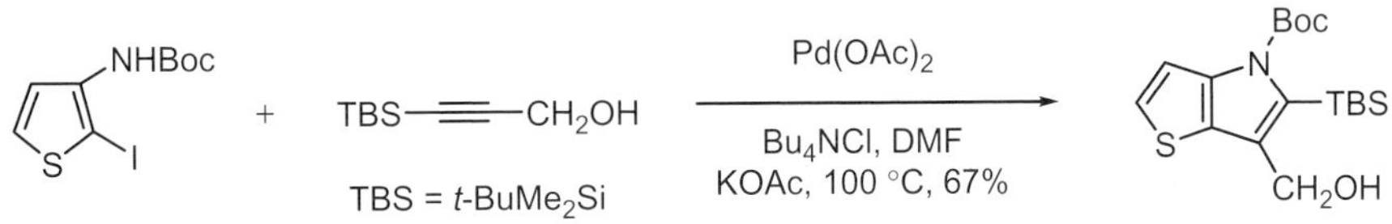

264

Two groups simultaneously found that trimethylsilyl cyanide reacts with disubstituted acetylenes in the presence of a Pd catalyst to form silylated 2-amino-5-cyanopyrroles **265** or **266** [162, 163]. These reactions are run neat and a variety of Pd species are successful in this transformation [162]. In the case of unsymmetrical diaryl acetylenes, the reaction is not regioselective [163].

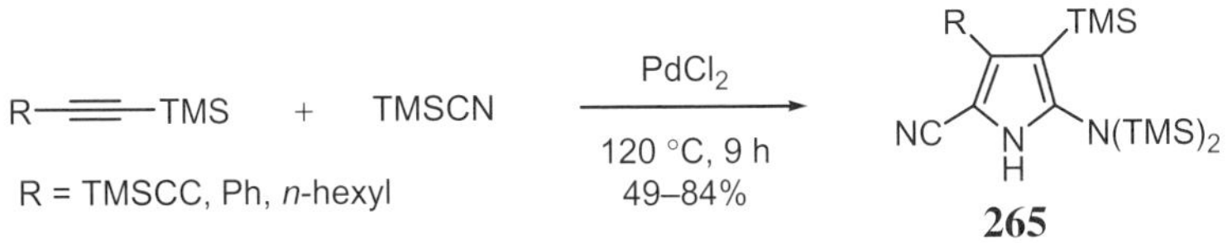

265

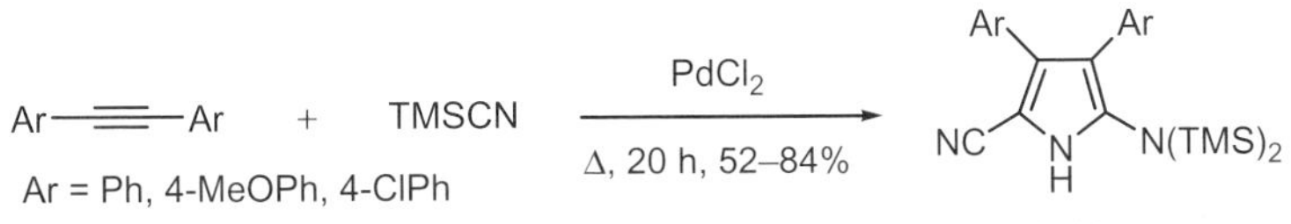

266

The powerful Buchwald–Hartwig aryl amination methodology [164–170] has been applied by Hartwig to the synthesis of *N*-arylpyrroles (**267**) [171, 172].

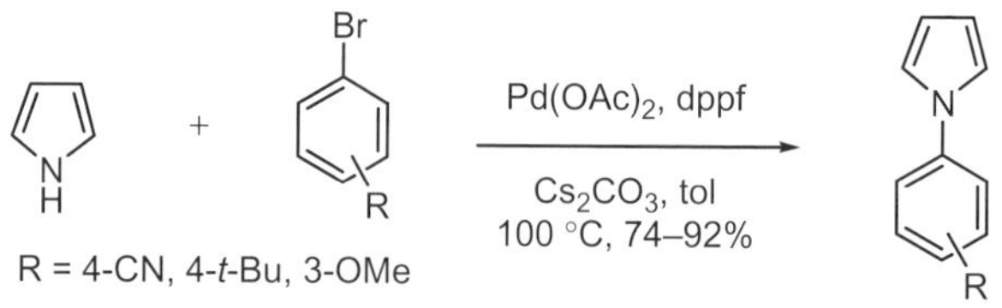

267

Likewise, the amination of bromopyrroles **268** and **80** with a wide range of primary and secondary amines has been disclosed [63, 173].

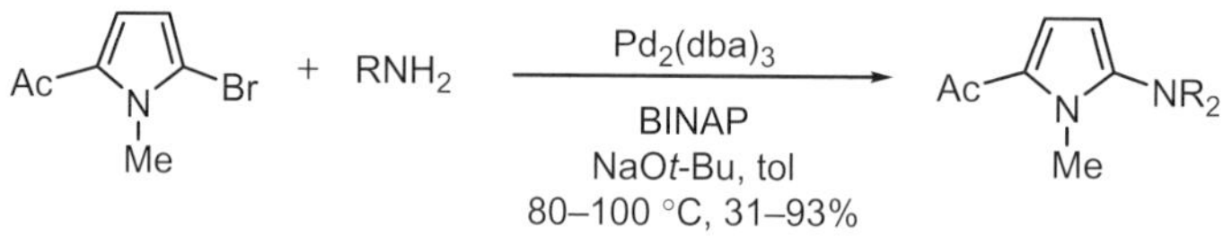

268

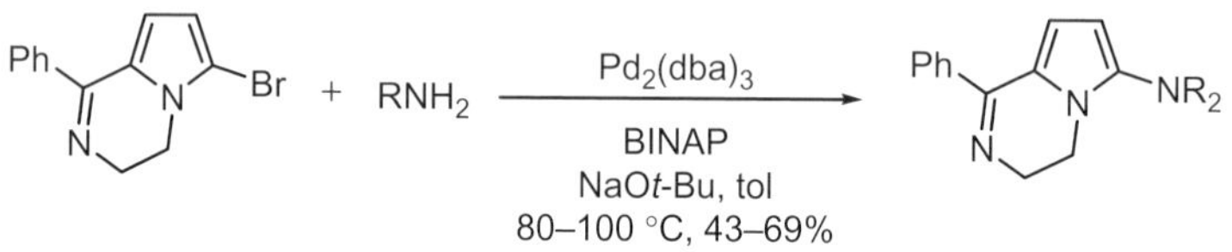

80

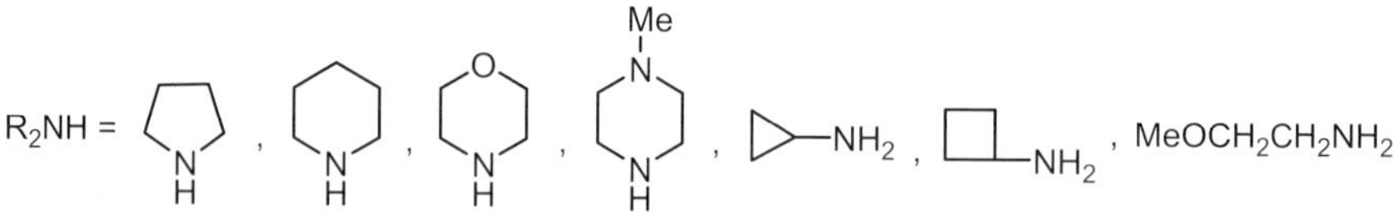

2.8. Miscellaneous

Several reactions and syntheses of pyrroles, which involve Pd catalysis, do not fall into the previous categories in this chapter, and therefore are presented here.

With regard to the Pd-catalyzed arylation of dihydropyrroles presented in Section 2.5, it is noteworthy that the readily available 2,5-dihydropyrroles **269** can be smoothly isomerized to 2,3-dihydropyrroles **270** under the influence of Pd [174].

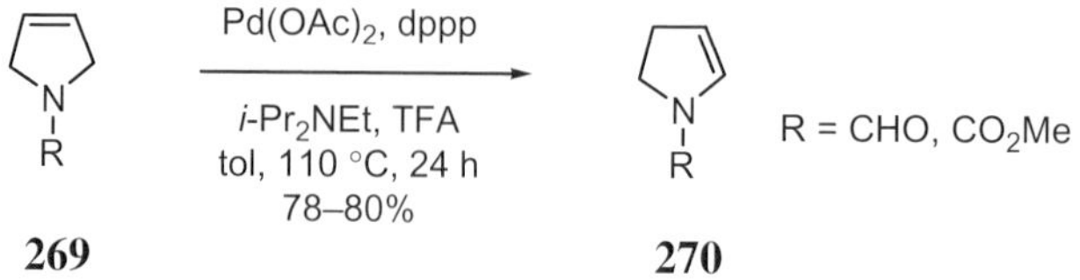

269   **270**

Iodopyrroles **271** can be conveniently deiodinated with formate as the hydride donor in the presence of Pd(0) [175]. This transformation is particularly important in the synthesis of dipyrromethanes for porphyrins and for linear pyrroles. Interestingly, no reaction occurs in refluxing THF.

271

R₁ = Bn, Et, *t*-Bu
R₂ = Me, Et, CH₂CH₂CO₂Me

It is frequently necessary in synthesis to operate on the nonheterocycle portion of a molecule, and palladium technology often succeeds admirably in this regard. For example, *N*-(4-iodophenyl)pyrrole (**272**) is cyanated to **273** in excellent yield [176].

272   **273**

An efficient and novel route to 3,3′-bipyrroles, such as **275**, involves the oxidative-cyclization of cyclodiyne **274** [177].

274   **275**

The use of Pd/C to effect carbonylation reactions, which has long been known, is equally successful with pyrrole aldehydes, e.g. **276** → **277** [133].

276   **277**

A three-component coupling of imines **278**, alkynes **279**, and acid chlorides **280** affords pyrroles **281** [178]. The active catalyst is proposed to be **282**, which leads to the generation of mesoionic münchnones **283**. An example is the synthesis of **284**.

278   **279**   **280**   **281**

R$_1$ = Et, Ar, Bn; R$_2$ = Ar, 4-MePh, 3-indolyl;
R$_3$ = H, Ar, CO$_2$Me, Bz; R$_4$ = H, Ar, CO$_2$Me,
Bz; R$_5$ = Ph, 2-furyl, 4-MePh, c-C$_6$H$_{11}$

282   **283**

284

2.9. References

1. Gribble, G. W. *Pyrroles and their Benzo Derivatives* in *Comprehensive Heterocyclic Chemistry* — 2nd Edition, vol. 2; Eds. Katritzky, A. R.; Rees, C. W.; Scriven, E. F. V.; Pergamon: New York; **1996,** Chapter 2.04.

2. Gribble, G. W. *Prog. Chem. Org. Nat. Prod.* **1996,** *68*, 1–498.

3. (a) Gribble, G. W. *Chem. Soc. Rev.* **1999,** *28*, 335–46. (b) Gribble, G. W. *Environ. Sci. Pollut. Res.* **2000,** *7*, 37–49.

4. Tittlemier, S. A.; Simon, M.; Jarman, W. M.; Elliott, J. E.; Norstrom, R. J. *Environ. Sci. Technol.* **1999,** *33*, 26–33.

5. For a leading reference, see Martina, S.; Enkelmann, V.; Wegner, G.; Schlüter, A.-D. *Synth. Metals* **1992,** *51*, 299–305.

6. For a review of pyrrole chemistry, see Jones, R. A.; Bean, G. P. *The Chemistry of Pyrroles,* Academic Press: New York, **1977.**

7. Anderson, H. J.; Loader, C. E. *Synthesis* **1985,** 353–64.

8. Cordell, G. A. *J. Org. Chem.* **1975,** *40*, 3161–9.

9. Gilow, H. M.; Burton, D. E. *J. Org. Chem.* **1981,** *46*, 2221–5.

10. Brockmann, T. W.; Tour, J. M. *J. Am. Chem. Soc.* **1995,** *117*, 4437–47.

11. Choi, D.-S.; Huang, S.; Huang, M.; Barnard, T. S.; Adams, R. D.; Seminario, J. M.; Tour, J. M. *J. Org. Chem.* **1998,** *63*, 2646–55.

12. Anderson, H. J.; Griffins, S. J. *Can. J. Chem.* **1967,** *45*, 2227–34.

13. Grehn, L.; Ragnarsson, U. *Angew. Chem., Int. Ed. Eng.* **1984,** *23*, 296–7.

14. Chen, W.; Cava, M. P. *Tetrahedron Lett.* **1987,** *28*, 6025–6.

15. Martina, S.; Enkelmann, V.; Wegner, G.; Schlüter, A.-D. *Synthesis* **1991**, 613–15.
16. Chadwick, D. J.; Hodgson, S. T. *J. Chem. Soc., Perkin Trans. 1* **1983**, 93–102.
17. Muchowski, J. M.; Solas, D. R. *Tetrahedron Lett.* **1983**, *24*, 3455–6.
18. Muchowski, J. M.; Naef, R. *Helv. Chim. Acta* **1984**, *67*, 1168–72.
19. Bray, B. L.; Mathies, P. H.; Naef, R.; Solas, D. R.; Tidwell, T. T.; Artis, D. R.; Muchowski, J. M. *J. Org. Chem.* **1990**, *55*, 6317–28.
20. Kozikowski, A. P.; Cheng, X.-M. *J. Org. Chem.* **1984**, *49*, 3239–40.
21. Stefan, K.-P.; Schuhmann, W.; Parlar, H.; Korte, F. *Chem. Ber.* **1989**, *122*, 169–74.
22. John, E. A.; Pollet, P.; Gelbaum, L.; Kubanek, J. *J. Nat. Prod.* **2004**, *67*, 1929–31.
23. Emrich, R.; Weyland, H.; Weber, K. *J. Nat. Prod.* **1990**, *53*, 703–5.
24. Marfil, M.; Albericio, F.; Álvarez, M. *Tetrahedron* **2004**, *60*, 8659–68.
25. Knight, L. W.; Huffman, J. W.; Isherwood, M. L. *Synlett* **2003**, 1993–6.
26. Dvornikova, E.; Kamieńska-Trela *Synlett* **2002**, 1152–4.
27. Ponasik, J. A.; Conova, S.; Kinghorn, D.; Kinney, W. A.; Rittschof, D.; Ganem, B. *Tetrahedron* **1998**, *54*, 6977–86.
28. Fürstner, A.; Weintritt, H. *J. Am. Chem. Soc.* **1998**, *120*, 2817–25.
29. Farnier, M.; Fournari, P. *Bull. Soc. Chim. Fr.* **1973**, 351–9.
30. Hollins, R. A.; Colnago, L. A.; Salim, V. M.; Seidl, M. C. *J. Heterocycl. Chem.* **1979**, *16*, 993–6.
31. For an excellent summary, see Gharpure, M.; Stoller, A.; Bellamy, F.; Firnau, G.; Snieckus, V. *Synthesis* **1991**, 1079–82.
32. Gribble, G. W.; Blank, D. H.; Jasinski, J. P. *Chem. Commun.* **1999**, 2195–6.
33. Asano, R.; Moritani, I.; Fujiwara, Y.; Teranishi, S. *Bull. Chem. Soc. Jpn.* **1973**, *46*, 663–4.
34. Itahara, T.; Kawasaki, K.; Ouseto, F. *Bull. Chem. Soc. Jpn.* **1984**, *57*, 3488–93.
35. Itahara, T. *J. Chem. Soc., Chem. Commun.* **1981**, 254–5.
36. Itahara, T. *J. Org. Chem.* **1985**, *50*, 5272–5.
37. Itahara, T. *J. Chem. Soc., Chem. Commun.* **1980**, 49–50.
38. Itahara, T. *Heterocycles* **1986**, *24*, 2557–62.
39. Itahara, T. *J. Org. Chem.* **1985**, *50*, 5546–50.
40. Boger, D. L.; Patel, M. *Tetrahedron Lett.* **1987**, *28*, 2499–502.
41. Boger, D. L.; Patel, M. *J. Org. Chem.* **1988**, *53*, 1405–15.
42. Beccalli, E. M.; Broggini, G.; Martinelli, M.; Paladino, G. *Tetrahedron* **2005**, *61*, 1077–82.
43. (a) Garg, N. K.; Caspi, D. D.; Stoltz, B. M. *J. Am. Chem. Soc.* **2004**, *126*, 9552–3. (b) Garg, N. K.; Caspi, D. D.; Stoltz, B. M. *J. Am. Chem. Soc.* **2005**, *127*, 5970–8.
44. Iwao, M.; Takeuchi, T.; Fujikawa, N.; Fukuda, T.; Ishibashi, F. *Tetrahedron Lett.* **2003**, *44*, 4443–6.
45. Siriwardana, A. I.; Kathriarachchi, K. K. A. D. S.; Nakamura, I.; Gridnev, I. D.; Yamamoto, Y. *J. Am. Chem. Soc.* **2004**, *126*, 13898–9.
46. For reviews of indole ring syntheses, see (a) Gribble, G. W. *Contemp. Org. Synth.* **1994**, *1*, 145–72. (b) Gribble, G. W. *Perkin Trans. 1* **2000**, 1045–75.
47. Yokoyama, Y.; Suzuki, H.; Matsumoto, S.; Sunaga, Y.; Tani, M.; Murakami, Y. *Chem. Pharm. Bull.* **1991**, *39*, 2830–6.
48. Arcadi, A.; Rossi, E. *Synlett* **1997**, 667–8.
49. Arcadi, A.; Rossi, E. *Tetrahedron* **1998**, *54*, 15253–72

50. Aoyagi, Y.; Mizusaki, T.; Ohta, A. *Tetrahedron Lett.* **1996**, *37*, 9203–6.
51. Bumagin, N. A.; Sokolova, A. F.; Beletskaya, I. P.; Wolz, G. *Russ. J. Org. Chem.* **1993**, *29*, 136–7.
52. Bumagin, N. A.; Nikitina, A. F.; Beletskaya, I. P. *Russ. J. Org. Chem.* **1994**, *30*, 1619–29.
53. Abarbri, M.; Dehmel, F.; Knochel, P. *Tetrahedron Lett.* **1999**, *40*, 7449–53.
54. Minato, A.; Tamao, K.; Hayashi, T.; Suzuki, K.; Kumada, M. *Tetrahedron Lett.* **1981**, *22*, 5319–22.
55. Minato, A.; Suzuki, K.; Tamao, K.; Kumada, M. *Tetrahedron Lett.* **1984**, *25*, 83–6.
56. Minato, A.; Suzuki, K.; Tamao, K.; Kumada, M. *J. Chem. Soc., Chem. Commun.* **1984**, 511–13.
57. Hayashi, T.; Konishi, M.; Kumada, M. *Tetrahedron Lett.* **1979**, 1871–4.
58. Widdowson, D. A.; Zhang, Y.-Z. *Tetrahedron* **1986**, *42*, 2111–16.
59. Sakamoto, T.; Kondo, Y.; Takazawa, N.; Yamanaka, H. *Heterocycles* **1993**, *36*, 941–2.
60. Takahashi, K.; Gunji, A. *Heterocycles* **1996**, *43*, 941–4.
61. Filippini, L.; Gusmeroli, M.; Riva, R. *Tetrahedron Lett.* **1992**, *33*, 1755–8.
62. Rieth, R. D.; Mankad, N. P.; Calimano, E.; Sadighi, J. P. *Org. Lett.* **2004**, *6*, 3981–3.
63. Castellote, I.; Vaquero, J. J.; Fernández-Gadea, J.; Alvarez-Builla, J. *J. Org. Chem.* **2004**, *69*, 8668–75.
64. Luker, T.; Hiemstra, H.; Speckamp, W. N. *Tetrahedron Lett.* **1996**, *37*, 8257–60.
65. DiMagno, S. G.; Lin, V. S.-Y.; Therien, M. J. *J. Org. Chem.* **1993**, *58*, 5983–93.
66. DiMagno, S. G.; Lin, V. S.-Y.; Therien, M. J. *J. Am. Chem. Soc.* **1993**, *115*, 2513–15.
67. Johnson, C. N.; Stemp, G.; Anand, N.; Stephen, S. C.; Gallagher, T. *Synlett* **1998**, 1025–7.
68. Paulus, O.; Alcaraz, G.; Vaultier, M. *Eur. J. Org. Chem.* **2002**, 2565–72.
69. Fernàndez, J.-C.; Solé-Feu, L.; Fernández-Forner, D.; de la Figuera, N.; Forns, P.; Albericio, F. *Tetrahedron Lett.* **2005**, *46*, 581–5.
70. D'Alessio, R.; Rossi, A. *Synlett* **1996**, 513–14.
71. Grieb, J. G.; Ketcha, D. M. *Synth. Commun.* **1995**, *25*, 2145–53.
72. Alvarez, A.; Guzmán, A.; Ruiz, A.; Velarde, E.; Muchowski, J. M. *J. Org. Chem.* **1992**, *57*, 1653–6.
73. Nadeau, J. M.; Swager, T. M. *Tetrahedron* **2004**, *60*, 7141–6.
74. Banwell, M. G.; Bray, A. M.; Edwards, A. J.; Wong, D. J. *J. Chem. Soc., Perkin Trans. 1* **2002**, 1340–3.
75. Chang, C. K.; Bag, N. *J. Org. Chem.* **1995**, *60*, 7030–2.
76. Thoresen, L. H.; Kim, H.; Welch, M. B.; Burghart, A.; Burgess, K. *Synlett* **1998**, 1276–8.
77. Vachal, P.; Toth, L. M. *Tetrahedron Lett.* **2004**, *45*, 7157–61.
78. Handy, S. T.; Zhang, Y.; Bregman, H. *J. Org. Chem.* **2004**, *69*, 2362–6.
79. Gupton, J. T.; Miller, R. B.; Krumpe, K. E.; Clough, S. C.; Banner, E. J.; Kanters, R. P. F.; Du, K. X.; Keertikar, K. M.; Lauerman, N. E.; Solano, J. M.; Adams, B. R.; Callahan, D. W.; Little, B. A.; Scharf, A. B.; Sikorski, J. A. *Tetrahedron* **2005**, *61*, 1845–54.
80. de Koning, C. B.; Michael, J. P.; Pathak, R.; van Otterlo, W. A. L. *Tetrahedron Lett.* **2004**, *45*, 1117–19.
81. Zhou, X.; Zhou, Z.; Mak, T. C. W.; Chan, K. S. *J. Chem. Soc., Perkin Trans. 1* **1994**, 2519–20.

82. Martina, S.; Schlüter, A.-D. *Macromolecules* **1992**, *25*, 3607–8.
83. Rawal, V. H.; Cava, M. P. *Tetrahedron Lett.* **1985**, *26*, 6141–4.
84. Brédas, J. L.; Chance, R. R., Eds. Conjugated Polymeric Materials: Opportunities in Electronics, Optoelectronics, and Molecular Electronics; Kluwer Academic Publishers: Dordrecht, The Netherlands, **1990**.
85. Bailey, T. R. *Tetrahedron Lett.* **1986**, *27*, 4407–10.
86. Basaric, N.; Mariniç, Z.; Sindler-Kulyk, M. *Tetrahedron Lett.* **2003**, *44*, 7337–40.
87. Jana, G. H.; Jain, S.; Arora, S. K.; Sinha, N. *Bioorg. Med. Chem. Lett.* **2005**, *15*, 3592–5.
88. Martinez, G. R.; Grieco, P. A.; Srinivasan, C. V. *J. Org. Chem.* **1981**, *46*, 3760–1.
89. Caddick, S.; Joshi, S. *Synlett* **1992**, 805–6.
90. Meetsma, A.; Dijkstra, H. P.; Ten Have, R.; van Leusen, A. M. *Acta Cryst.* **1996**, *C52*, 2747–50.
91. Dijkstra, H. P.; Ten Have, R.; van Leusen, A. M. *J. Org. Chem.* **1998**, *63*, 5332–8.
92. Jousseaume, B.; Kwon, H.; Verlhac, J.-B.; Denat, F.; Dubac, J. *Synlett* **1993**, 117–18.
93. Denat, F.; Gaspard-Iloughmane, H.; Dubac, J. *J. Organomet. Chem.* **1992**, *423*, 173–82.
94. Muchowski, J. M.; Hess, P. *Tetrahedron Lett.* **1988**, *29*, 770–80.
95. Nabbs, B. K.; Abell, A. D. *Bioorg. Med. Chem. Lett.* **1999**, *9*, 505–8.
96. Tamao, K.; Ohno, S.; Yamaguchi, S. *Chem. Commun.* **1996**, 1873–4.
97. Yoshida, S.; Kubo, H.; Saika, T.; Katsumura, S. *Chem. Lett.* **1996**, 139–40.
98. Wang, J.; Scott, A. I. *Tetrahedron Lett.* **1995**, *36*, 7043–6.
99. Wang, J.; Scott, A. I. *Tetrahedron Lett.* **1996**, *37*, 3247–50.
100. Hirayama, M.; Choshi, T.; Kumemura, T.; Tohyama, S.; Nobuhiro, J.; Hibino, S. *Heterocycles* **2004**, *63*, 1765–70.
101. Peña, M. R.; Stille, J. K. *Tetrahedron Lett.* **1987**, *28*, 6573–6.
102. Peña, M. R.; Stille, J. K. *J. Am. Chem. Soc.* **1989**, *111*, 5417–24.
103. Bernabé, P.; Rutjes, F. P. J. T.; Hiemstra, H.; Speckamp, W. H. *Tetrahedron Lett.* **1996**, *37*, 3561–4.
104. Minnetian, O. M.; Morris, I. K.; Snow, K. M.; Smith, K. M. *J. Org. Chem.* **1989**, *54*, 5567–74.
105. Massa, M. A.; Patt, W. C.; Ahn, K.; Sisneros, A. M.; Herman, S. B.; Doherty, A. *Bioorg. Med. Chem. Lett.* **1998**, *8*, 2117–22.
106. Peters, D.; Hörnfeldt, A.-B.; Gronowitz, S. *J. Heterocycl. Chem.* **1990**, *27*, 2165–73.
107. Ortaggi, G.; Scarsella, M.; Scialis, R.; Sleiter, G. *Gazz. Chim. Ital.* **1988**, *118*, 743–4.
108. (a) Cho, D. H.; Lee, J. H.; Kim, B. H. *J. Org. Chem.* **1999**, *64*, 8048–50. (b) Mártire, D. O.; Jux, N.; Aramendía, P. F.; Negri, R. M.; Lex, J.; Braslavsky, S. E.; Schaffner, K.; Vogel, E. *J. Am. Chem. Soc.* **1992**, *114*, 9969–78.
109. Liu, J.-H.; Yang, Q.-C.; Mak, T. C. W.; Wong, H. N. C. *J. Org. Chem.* **2000**, *65*, 3587–95.
110. Boyle, R. W.; Johnson, C. K.; Dolphin, D. *J. Chem. Soc., Chem. Commun.* **1995**, 527–8.
111. Ali, H.; van Lier, J. E. *Tetrahedron* **1994**, *50*, 11933–44.
112. Aoyagi, Y.; Inoue, A.; Koizumi, I.; Hashimoto, R.; Tokunaga, K.; Gohma, K.; Komatsu, J.; Sekine, K.; Miyafuji, A.; Kunoh, J.; Honma, R.; Akita, Y.; Ohta, A. *Heterocycles* **1992**, *33*, 257–72.
113. Bräse, S.; Waegell, B.; de Meijere, A. *Synthesis* **1998**, 148–52.
114. Sezen, B.; Sames, D. *J. Am. Chem. Soc.* **2003**, *125*, 5274–5.

115. Nilsson, K.; Hallberg, A. *J. Org. Chem.* **1990**, *55*, 2464–70.

116. Sonesson, C.; Larhed, M.; Nyqvist, C.; Hallberg, A. *J. Org. Chem.* **1996**, *61*, 4756–63.

117. Ozawa, F.; Hayashi, T. *J. Organometal. Chem.* **1992**, *428*, 267–77.

118. Shim, Y. K.; Youn, J. I.; Chun, J. S.; Park, T. H.; Kim, M. H.; Kim, W. J. *Synthesis* **1990**, 753–4.

119. Yagi, T.; Aoyama, T.; Shioiri, T. *Synlett* **1997**, 1063–4.

120. Muratake, H.; Abe, I.; Natsume, M. *Tetrahedron Lett.* **1994** *35*, 2573–6.

121. Muratake, H.; Abe, I.; Natsume, M. *Chem. Pharm. Bull.* **1996**, *44*, 67–79.

122. Muratake, H.; Tonegawa, M.; Natsume, M. *Chem. Pharm. Bull.* **1996**, *44*, 1631–3.

123. Muratake, H.; Tonegawa, M.; Natsume, M. *Chem. Pharm. Bull.* **1998**, *46*, 400–12.

124. Grigg, R.; Sridharan, V.; Stevenson, P.; Sukirthalingam, S.; Worakun, T. *Tetrahedron* **1990**, *46*, 4003–18.

125. Grigg, R.; Loganathan, V.; Sridharan, V. *Tetrahedron Lett.* **1996**, *37*, 3399–402.

126. Grigg, R.; Fretwell, P.; Meerholtz, C.; Sridharan, V. *Tetrahedron* **1994**, *50*, 359–70.

127. Beccalli, E. M.; Broggini, G.; Martinelli, M.; Paladino, G.; Zoni, C. *Eur. J. Org. Chem.* **2005**, 2091–6.

128. Grigg, R.; Sridharan, V.; Stevenson, P.; Sukirthalingam, S. *Tetrahedron* **1989**, *45*, 3557–68.

129. Wensbo, D.; Annby, U.; Gronowitz, S. *Tetrahedron* **1995**, *51*, 10323–42.

130. Heck, R. F. *J. Am. Chem. Soc.* **1971**, *93*, 6896–901.

131. Ganske, J. A.; Pandey, R. K.; Postich, M. J.; Snow, K. M.; Smith, K. M. *J. Org. Chem.* **1989**, *54*, 4801–7.

132. Morris, I. K.; Snow, K. M.; Smith, N. W.; Smith, K. M. *J. Org. Chem.* **1990**, *55*, 1231–6.

133. Demopoulos, B. J.; Anderson, H. J.; Loader, C. E.; Faber, K. *Can. J. Chem.* **1983**, *61*, 2415–22.

134. Faber, K.; Anderson, H. J.; Loader, C. E.; Daley, A. S. *Can. J. Chem.* **1984**, *62*, 1046–50.

135. Monti, D.; Sleiter, G. *Gazz. Chim. Ital.* **1990**, *120*, 587–90.

136. Monti, D.; Sleiter, G. *Gazz. Chim. Ital.* **1994**, *124*, 133–6.

137. Araki, S.; Ohmura, M.; Butsugan, Y. *Bull. Chem. Soc. Jpn.* **1986**, *59*, 2019–20.

138. Edstrom, E. D. *Synlett* **1995**, 49–50.

139. Edstrom, E. D.; Yu, T.; Jones, Z. *Tetrahedron Lett.* **1995**, *36*, 7035–8.

140. Jaouhari, R. Dixneuf, P. H.; Lécolier, S. *Tetrahedron Lett.* **1986**, *27*, 6315–18.

141. Iwasaki, M.; Kobayashi, Y.; Li, J.-P.; Matsuzaka, H.; Ishii, Y.; Hidai, M. *J. Org. Chem.* **1991**, *56*, 1922–7.

142. Chiusoli, G. P.; Costa, M.; Masarati, E.; Salerno, G. *J. Organomet. Chem.* **1983**, *255*, C35–8.

143. Cartoon, M. E. K.; Cheeseman, G. W. H. *J. Organomet. Chem.* **1982**, *234*, 123–36.

144. Murahashi, S.; Shimamura, T.; Moritani, I. *J. Chem. Soc., Chem. Commun.* **1974**, 931–2.

145. Trost, B. M.; Kernan, E. *J. Org. Chem.* **1980**, *45*, 2741–6.

146. Bäckvall, J.-E.; Nyström, J.-E. *J. Chem. Soc., Chem. Commun.* **1981**, 59–60.

147. Genet, J. P.; Balabane, M.; Bäckvall, J. E.; Nyström, J. E. *Tetrahedron Lett.* **1983**, *24*, 2745–8.
148. Anderson, P. G.; Bäckvall, J.-E. *J. Am. Chem. Soc.* **1992**, *114*, 8696–8.
149. Igarashi, S.; Haruta, Y.; Ozawa, M.; Nishide, Y.; Kinoshita, H.; Inomata, K. *Chem. Lett.* **1989**, 737–40.
150. Kimura, M.; Harayama, H.; Tanaka, S.; Tamaru, Y. *J. Chem. Soc., Chem. Commun.* **1994**, 2531–3.
151. Saito, S.; Hara, T.; Takahashi, N.; Hirai, M.; Moriwake, T. *Synlett* **1992**, 237–8.
152. Utimoto, K.; Miwa, H.; Nozaki, H. *Tetrahedron Lett.* **1981**, *22*, 4277–8.
153. Gabriele, B.; Salerno, G.; Fazio, A. *J. Org. Chem.* **2003**, *68*, 7853–61.
154. Gabriele, B.; Salerno, G.; Costa, M. *Synlett* **2004**, 2468–83.
155. Gabriele, B.; Salerno, G.; Fazio, A.; Campana, F. B. *Chem. Commun.* **2002**, 1408–9.
156. Arcadi, A.; Anacardio, R.; D'Anniballe, G.; Gentile, M. *Synlett* **1997**, 1315–7.
157. Tsutsui, H.; Narasaka, K. *Chem. Lett.* **1999**, 45–16.
158. Kitamura, M.; Yanagisawa, H.; Yamane, M.; Narasaka, K. *Heterocycles* **2005**, *65*, 273–7.
159. Larock, R. C.; Doty, M. J.; Han, X. *Tetrahedron Lett.* **1998**, *39*, 5143–6.
160. Wensbo, D.; Eriksson, A.; Jeschke, T.; Annby, U.; Gronowitz, S.; Cohen, L. A. *Tetrahedron Lett.* **1993**, *34*, 2823–6.
161. Wensbo, D.; Gronowitz, S. *Tetrahedron* **1996**, *52*, 14975–88.
162. Kusumoto, T.; Hujama, T.; Ogata, K. *Tetrahedron Lett.* **1986**, *27*, 4197–200.
163. Chatani, N.; Hanafusa, T. *Tetrahedron Lett.* **1986**, *27*, 4201–4.
164. Hartwig, J. F. *Synlett* **1997**, 329–40.
165. (a) Marcoux, J.-F.; Wagaw, S.; Buchwald, S. L. *J. Org. Chem.* **1997**, *62*, 1568–9. (b) Sadighi, J. P.; Harris, M. C.; Buchwald, S. L. *Tetrahedron Lett.* **1998**, *39*, 5327–30.
166. Hartwig, J. F. *Angew. Chem. Int. Ed.* **1998**, *37*, 2046–67.
167. (a) Wolfe, J. P.; Wagaw, S.; Marcoux, J.-F.; Buchwald, S. L. *Acc. Chem. Res.* **1998**, *31*, 805–18. (b) Yang, B. H.; Buchwald, S. L. *J. Organometal. Chem.* **1999**, *576*, 125–46.
168. Hartwig, J. F. *Angew. Chem. Int. Ed.* **1998**, *37*, 2090–2.
169. Belfield, A. J.; Brown, G. R.; Foubister, A. J. *Tetrahedron* **1999**, *55*, 11399–428.
170. Huang, J.; Grasa, G.; Nolan, S. P. *Org. Lett.* **1999**, *1*, 1307–9.
171. Mann, G.; Hartwig, J. F.; Driver, M. S.; Fernandez-Rivas, C. *J. Am. Chem. Soc.* **1998**, *120*, 827–8.
172. Hartwig, J. F.; Kawatsura, M.; Hauck, S. I.; Shaughnessy, K. H.; Alcazar-Roman, L. M. *J. Org. Chem.* **1999**, *64*, 5575–80.
173. Castellote, I.; Vaquero, J. J.; Alvarez-Builla, J. *Tetrahedron Lett.* **2004**, *45*, 769–72.
174. Sonesson, C.; Hallberg, A. *Tetrahedron Lett.* **1995**, *36*, 4505–6.
175. Leung, S. H.; Edington, D. G.; Griffith, T. E.; James, J. J. *Tetrahedron Lett.* **1999**, *40*, 7189–91.
176. Anderson, B. A.; Bell, E. C.; Ginah, F. O.; Harn, N. K.; Pagh, L. M.; Wepsiec, J. P. *J. Org. Chem.* **1998**, *63*, 8224–8.
177. Gleiter, R.; Ritter, J. *Tetrahedron* **1996**, *52*, 10383–8.
178. Dhawan, R.; Arndtsen, B. A. *J. Am. Chem. Soc.* **2004**, *126*, 468–9.

Chapter 3

Indoles

Gordon W. Gribble

Indole is perhaps the single most visible heterocycle in all of chemistry. It is embodied in a myriad of natural products, pharmaceutical agents, and a growing list of polymers. In addition to the hundreds of well-known indole plant alkaloids (e.g. yohimbine, reserpine, strychnine, ellipticine, lysergic acid, physostigmine), the indole ring is present in an array of other organisms. The indigo analog Tyrian Purple is the ancient Egyptian dye produced by Mediterranean molluscs, and a bromine-containing triindole is found in an *Orina* sponge. Sciodole is a *Tricholoma* fungal product and the plant growth hormone 4-chloroindole-3-acetic acid (4-Cl-IAA) is utilized by peas, beans, lentils, and other plants. Another chlorine-containing fungal metabolite is pennigritrem, produced by *Penicillium nigricans*. The funnel-web spider employs the toxic argiotoxin 659 in chemical defense.

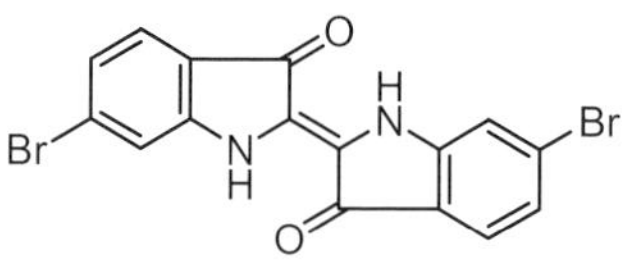

Tyrian Purple

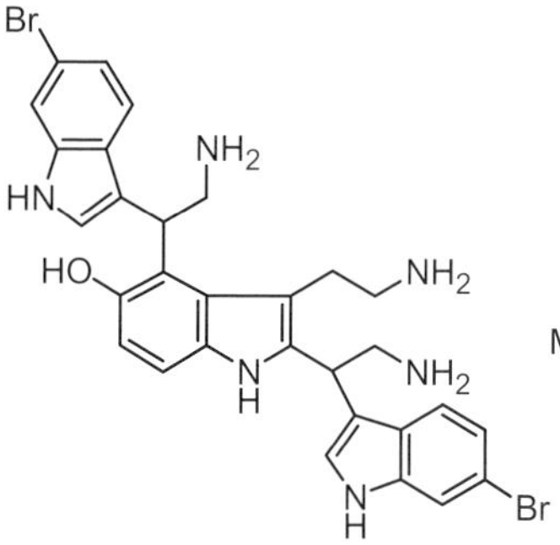

Orina metabolite

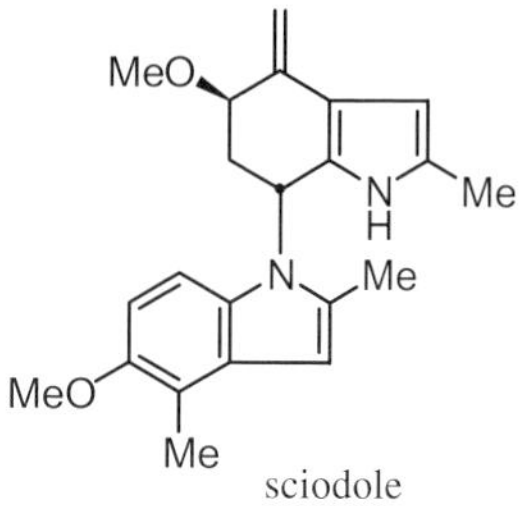

sciodole

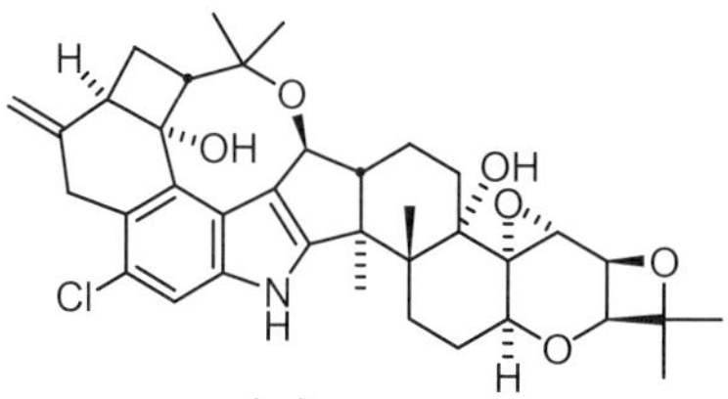

pennigritrem

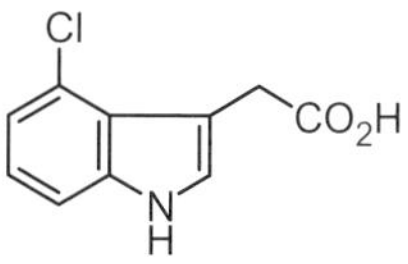

4-Cl-IAA

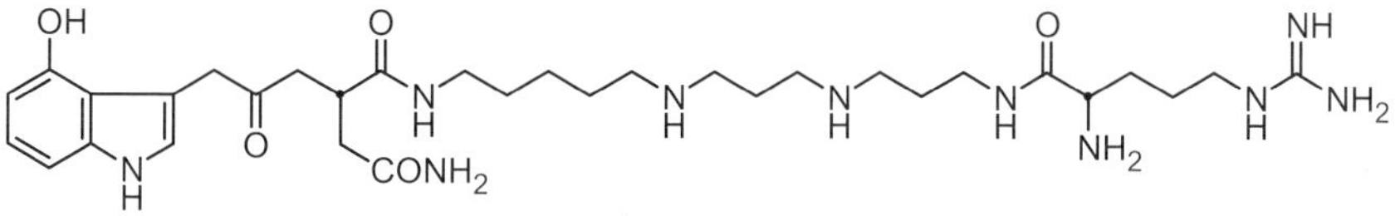

argiotoxin 659

The central importance of the indole derivatives serotonin and the amino acid trypto-phan in living organisms has inspired chemists to design and synthesize thousands of indole-containing pharmaceuticals. Some of the most recent indole-containing drugs are shown here. Three new antimigraine drugs—competitors of the highly effective Sumatriptan—are Naramig, Zomig, and Maxalt. The antiemetics Nasea and Anzemet are potent and highly selective 5-HT$_3$ receptor antagonists for the treatment of chemother-apy-induced nausea and vomiting. Rescriptor is a new HIV-1 reverse transcriptase inhibitor for HIV-positive individuals, and Serdolect is a neuroleptic for acute and chronic schizophrenia. Lescol is an HMG-CoA reductase inhibitor, and Somatuline (not shown) is an indole cyclic peptide for the treatment of acromegaly by acting as a growth hormone inhibitor. Accolate is a new antiasthma drug, and the simple oxindole ReQuip is effective in the treatment of early-stage Parkinson's disease.

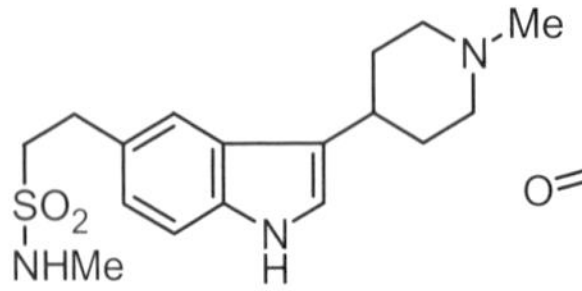

Naramig 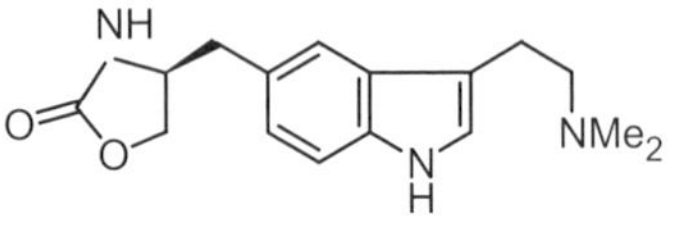   Zomig 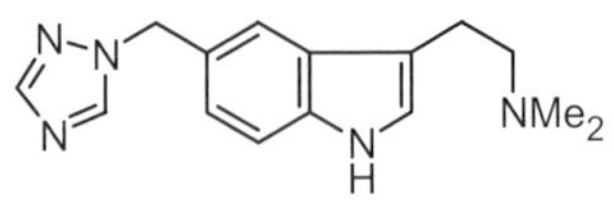   Maxalt

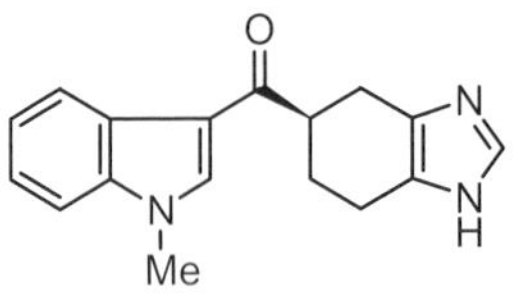

Nasea 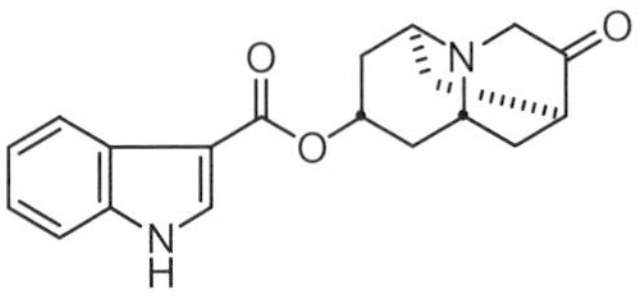   Anzemet

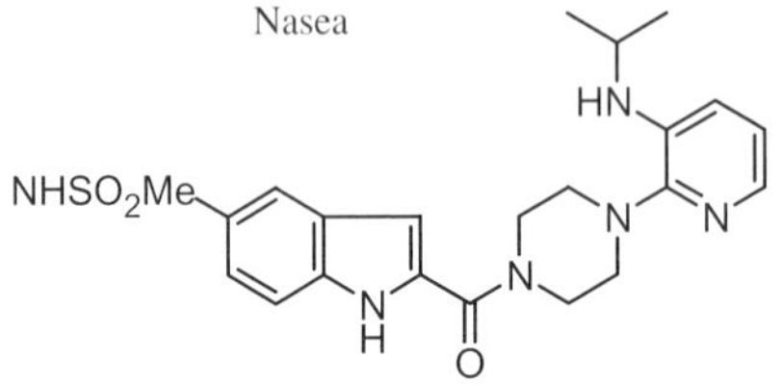

Rescriptor 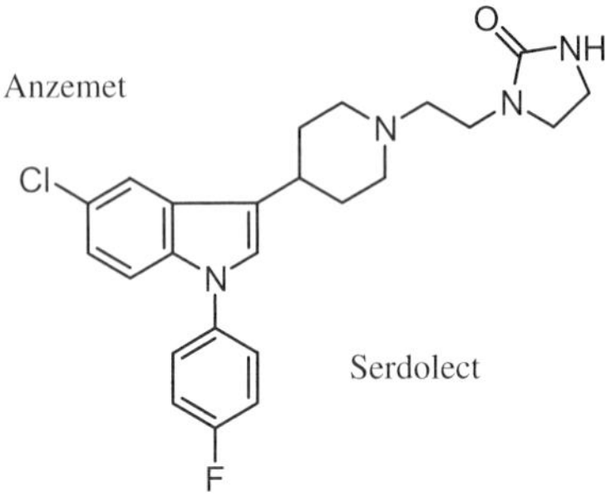   Serdolect

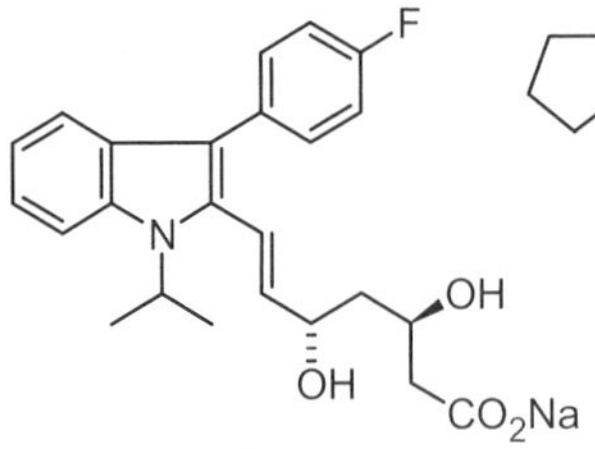

Lescol 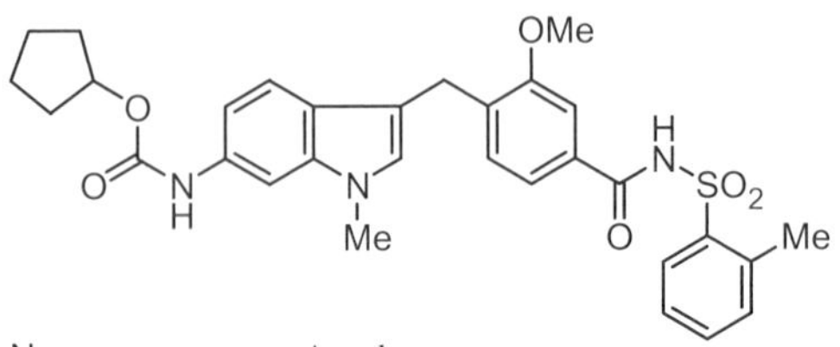   Accolate 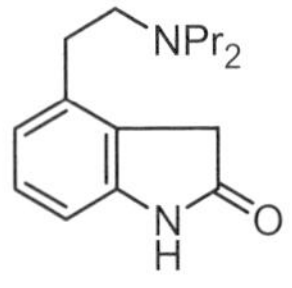   ReQuip

Although much less than pyrrole polymers, indole polymers are beginning to be syn-thesized and studied as new materials. Electropolymerized films of indole-5-carboxylic acid are well-suited for the fabrication of micro-pH sensors and they have been used to measure ascorbate and NADH levels. The three novel pyrroloindoles shown have been electrochemically polymerized, and the polymeric pyrrolocarbazole has similar physical properties to polyaniline.

3.1. Synthesis of indolyl halides

Like pyrrole, indole is a very reactive π-excessive heterocycle and reacts with halogens and other electrophiles extremely rapidly [1]. Nature has exploited this property to produce more than 300 halogenated indoles, mainly bromoindoles in marine organisms. A few examples are illustrated below. Chapter 2 contains references to reviews of these natural organohalogens.

X = Cl, Y = H	X = H, Y = Br	R = Me, X = H, Y = Br
X = Br, Y = H	X = Cl, Y = H	R = Me, X = Br, Y = H
X = Cl, Y = Br	X = Y = Br	R = H, X = Y = Br
X = Y = Br	X = Y = Cl	R = Me, X = Y = Br
marine acorn worm	New Zealand red alga	Caribbean red alga

Early syntheses of haloindoles involved direct reaction of indoles with chlorine, bromine, or iodine. In some cases, this approach was reasonably successful, but the instability of the resulting 3-haloindoles made product isolation and further chemistry difficult. For example, although attempted preparations of 3-chloro-, 3-bromo-, and 3-iodoindole were described in the early 1900s [2], only recently have practical syntheses of these compounds and their *N*-protected derivatives become available. For example, 3-bromoindole (**2**) can be prepared in 64% yield by the treatment of indole (**1**) with pyridinium bromide perbromide. The product decomposes at 65°C and should be stored at – 20°C [2]. Likewise, the unstable 3-iodoindole can be formed by treating **1** with I$_2$, KI, NaOH, MeOH [3]. The product was converted to indoxyl acetate (AgOAc, HOAc) in 28% overall yield.

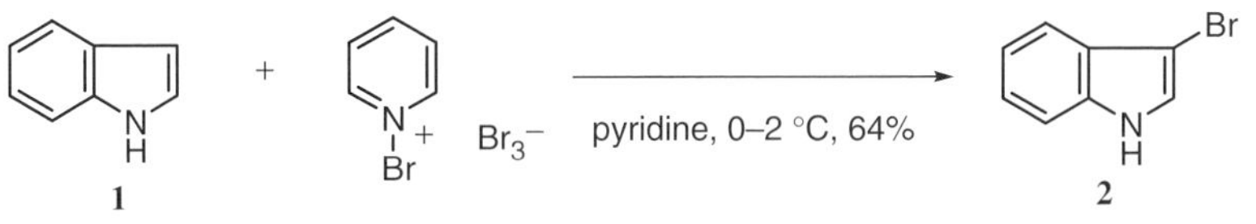

A vast improvement for the synthesis of both 3-bromo- and 3-iodoindole by using DMF as solvent was described by Bocchi and Palla, as summarized below [4]. This appears to be the method of choice for the preparation of simple 3-bromo- and 3-iodoindoles.

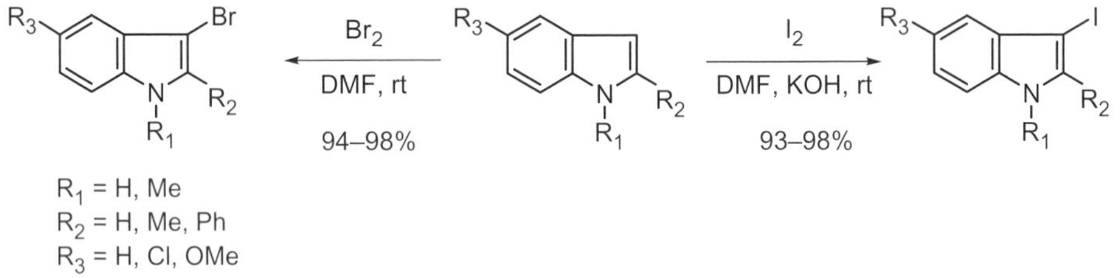

R_1 = H, Me
R_2 = H, Me, Ph
R_3 = H, Cl, OMe

Erickson extended these reactions to useful preparations of both 3-chloroindole and several 2,3-dihaloindoles, many of which occur naturally [5]. When the C-3 position is already substituted, halogenation usually occurs at C-2. A summary of these halogenations is shown below. Erickson was able to improve Piers' synthesis of **2** to a yield of 82%. Interestingly, the action of sulfuryl chloride on 3-iodoindole gives the *ipso* product 3-chloroindole in 84% yield.

The notoriously unstable 2-chloroindole was first synthesized in pure form by Powers [6], and this procedure was later extended to the preparation of 2-bromoindole by Erickson [5]. The method involves reaction of oxindole with either $POCl_3$ or $POBr_3$ but yields are very low in both cases (26% for 2-chloroindole and 15% for 2-bromoindole). Powers also synthesized the more stable 1-benzyl-2-chloroindole from *N*-benzyloxyindole and $POCl_3$, but discontinued all work in this area after developing a severe skin rash from these haloindoles. This toxicity is consistent with the role of natural halogenated indoles in chemical defense by marine organisms.

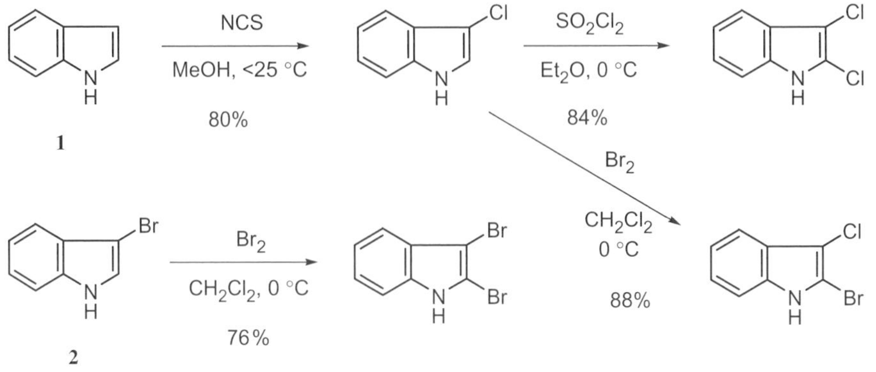

The most efficient route to *N*-unsubstituted 2-haloindoles is that developed by Bergman involving a Katritzky indole C-2 lithiation protocol [7] and quenching with selected sources of halogen as illustrated [8]. Bromine and iodine gave lower yields. Furthermore, subjecting 2-iodoindole to the Bocchi and Palla procedure with iodine affords 2,3-diiodoindole in 82% yield. Bergman observed that the order of stability of 2-haloindoles at –20°C is 2-I >> 2-Cl > 2-Br. Nevertheless, simple *N*-unsubstituted 2- and 3-haloindoles would seem to be too labile to undergo consistently successful Pd-catalyzed chemistry.

Several methods for synthesizing *N*-protected (usually with electron-withdrawing groups) 2- and 3-haloindoles have been developed and the resulting haloindoles are much less prone to decomposition than the unsubstituted compounds. Bromination of *N*-(phenylsulfonyl)indole (**3**), which is readily available via lithiation [9, 10] or phase-transfer chemistry [11, 12], affords 3-bromo-1-(phenylsulfonyl)indole (**4**) in nearly quantitative yield [12].

Bromination of *N-tert*-butyldimethylsilylindoles **5** proceeds under mild conditions with NBS [13].

The synthesis of both 3-iodo-1-(phenylsulfonyl)indole (**6**) and 2,3-diiodo-1-(phenyl-sulfonyl)indole (**7**) can be achieved in excellent overall yields as illustrated [10]. The preparation of **6** is done in one pot.

The lithiation of 3-haloindoles represents an excellent method for the preparation of other 2,3-dihaloindoles. For example, treatment of **4** with LDA followed by quenching with CNBr affords 2,3-dibromoindole **8** in good yield [14], and quenching with iodine furnishes 3-bromo-2-iodoindole **9** [15].

Likewise, quenching the 2-lithio species derived from **6** with CNBr gives 2-bromo-3-iodo-1-(phenylsulfonyl)indole in 80% yield. Lithiation of *N*-(phenylsulfonyl)indole (**3**) with LDA followed by quenching with CNBr or benzenesulfonyl chloride gives the corresponding 2-bromo and 2-chloro derivatives in 82 and 93% yields, respectively [15]. Similar lithiation methods have been used to prepare 2-iodo-1-methylindole [16] and 1-Boc-2-iodoindole, the latter of which can be converted to 2-iodotryptamine, and its 5-methoxy derivative can be similarly crafted [17]. Widdowson prepared 2- and 3-fluoro-1-(4-toluenesulfonyl)indoles from the corresponding trimethylstannylindoles by reaction with either cesium fluoroxysulfate or Selectfluor™ [18]. Unfortunately, the C-F bond is too strong to undergo oxidative addition of Pd so these fluoroindoles will have to await the development of new technology to find applications in synthesis. Treatment of 2-(trimethylsilyl)azaindoles with ICl furnishes the corresponding 2-iodoazaindoles [19].

An entirely different approach to 3-haloindoles involves a mercuration/iodination sequence, which has been adopted by Hegedus to prepare 4-bromo-3-iodo-1-(4-toluene-sulfonyl)indole for use in the synthesis of ergot alkaloids [20, 21]. We will discuss this chemistry later.

The synthesis of the benzene-ring substituted haloindoles entails more conventional aryl ring chemistry or *de novo* synthesis of the indole ring [1, 22–24]. As expected, these haloindoles are far more stable than the C-2 and C-3 haloindoles.

Hegedus also described an efficient synthesis of 4-bromoindole (**10**) and 4-bromo-1-(4-toluenesulfonyl)indole (**11**) starting from 2-amino-6-nitrotoluene [20]. The synthesis is lengthy but yields are good, and the method involves the Hegedus indole ring synthesis discussed later in this chapter.

Conventional aryldiazonium salt chemistry on 4-aminoindole provides 4-iodo-1-(4-toluenesulfonyl)indole (13), 4-iodoindole (14), and 1-(*tert*-butyldimethylsilyl)-4-iodoindole (15) in excellent yields as shown [25, 26].

An excellent synthesis of 5-bromo- (18) and 5-iodoindole (19) involves protecting the indole double bond as sulfonate 16, acetylation to 17, and halogenation [27]. Indoline itself undergoes bromination at C-4 and C-7 [28].

Carrera and Sheppard improved upon a Leimgruber–Batcho indole synthesis [24] to prepare 6-bromoindole (20) in excellent overall yield from 4-bromo-2-nitrotoluene [29a], and Rapoport utilized this method to synthesize 4-, 5-, 6-, and 7-bromoindole [29b].

1. $(MeO)_2CHNMe_2$, DMF
   pyrrolidine, 110 °C

2. H_2, RaNi, tol

74%

20

The Bartoli indole synthesis [24] is an excellent method for preparing 7-bromoindole (21) [29a, 30].

THF, −70 °C

53%

21

In addition, due to the proximity and inductive effect of the indoline nitrogen, a metalation strategy provides for a facile synthesis of 7-bromoindoline, as well as other 7-substituted indolines. Thus, Iwao and Kuraishi find that N-Boc-indoline (22) is smoothly lithiated at C-7 to give 7-lithio 23. Quenching 23 with suitable electrophiles gives 7-haloindolines 24–26, and quenching with tri-n-butylstannyl chloride affords the corresponding tin derivative in 65% yield [31].

s-BuLi, TMEDA

THF or Et_2O, −78 °C

22   23   24

I_2

59%

C_2Cl_6

61%

$BrCH_2CH_2Br$

57%

26   25

Meyers adopted and refined this method to prepare 1-benzyl-7-bromoindoline (60% overall yield from 22) for use in Grignard coupling chemistry, and obtained a higher yield of 7-bromoindoline 25 by quenching with 1,2-dibromo-1,1,2,2-tetrafluoroethane (68%) [32, 33]. It is important to note that several methods are available for the oxidation of indolines to indoles [22–24], including one involving $PdCl_2$ as developed by Kuehne [34]. Interestingly, in the presence of Ph_3P, indoline reacts with $PdCl_2$ to give the stable complex 27, confirmed by X-ray crystallography [35]. In any event, conditions can usually be found to oxidize halo-substituted and other indolines to the indoles (Pd/C, MnO_2, DDQ, NBS, etc.).

A third efficient synthesis of 7-bromoindole (**21**) involves the Stille stannylation of 2,6-dibromoaniline to give enol ether **28**, which, after hydrolysis and cyclization, affords **21** in 96% overall yield [36].

Widdowson expanded his hexacarbonylchromium chemistry to the synthesis and lithiation of Cr(CO)$_3$-*N*-TIPS indole (**29**), leading to 4-iodoindole **30** after oxidative decomplexation [37]. Stannylation at C-4 could also be achieved using this method (62% yield), and comparable chemistry with 3-methoxymethylindole leading to C-4 substitution was described.

Another general approach to benzene-ring haloindoles involves thallation chemistry. Hollins and co-workers demonstrated that C-4 thallation occurs readily in a series of 3-acylindoles **31** affording 4-iodoindoles **32** following treatment of the thallated intermediates with KI [38].

R_1 = H, Me; R_2 = H, Me; R_3 = H, Me

Somei improved this methodology by quenching the appropriate thallated intermediate with I$_2$/CuI/DMF to give 4-iodoindole-3-carboxaldehyde in 94% yield [39], and he extended this method to achieve efficient syntheses of C-7 haloindoles [40, 41]. For example, 7-iodoindole (**33**) was prepared in good overall yield from *N*-acetylindoline as illustrated. Thallation at C-5 is a minor (5%) pathway.

The invention of the triflate (trifluoromethylsulfonyl) group—a superior leaving group—has led to its use in palladium chemistry [42]. Conway and Gribble described the synthesis of 3-indolyl triflate **34** [12] and 2-indolyl triflate **35** from oxindole [43]. Mérour synthesized the *N*-phenylsulfonyl derivative **36** by employing a Baeyer–Villiger oxidation of the appropriate indolecarboxaldehyde [44].

3.2. Oxidative coupling/cyclization

Most of the early applications of palladium to indole chemistry involved oxidative coupling or cyclization using stoichiometric Pd(II). Åkermark first reported the efficient oxidative coupling of diphenyl amines to carbazoles **37** with Pd(OAc)$_2$ in refluxing acetic acid [45]. The reaction is applicable to several ring-substituted carbazoles (Br, Cl, OMe, Me, NO$_2$), and 20 years later Åkermark and colleagues made this reaction catalytic in the conversion of arylaminoquinones **38** to carbazole-1,4-quinones **39** with *tert*-butylhydroperoxide or oxygen as the oxidant [46]. This oxidative cyclization is particularly useful for the synthesis of benzocarbazole-6,11-quinones (e.g. **40**).

38 → **39**

R$_1$ = H, OMe
R$_2$ = H, Me, OMe
R$_3$ = H, Me
R$_4$ = H, Me
R$_5$ = H, Me
R$_4$, R$_5$ = benzo

40 (74%)

Such oxidative cyclizations with Pd(OAc)$_2$ have been used to synthesize carbazole and carbazolequinone alkaloids [47], ellipticines [48], 6H-pyrazolo[4,3-c]carbazole [49], and a series of 3-acetyl-1-methoxy-2-methylcarbazoles [50]. Knölker and co-workers exploited this reaction in the synthesis of numerous naturally occurring carbazolequinones including prekinamycin analogs (benzo[b]carbazolequinones) [51, 52], carbazomycins G and H [53], carbazoquinocin C [54, 55], (±)-carquinostatin A [55], and carbazomadurin A [56]. Noteworthy is the use of Cu(OAc)$_2$ as a reoxidant to make these oxidative cyclizations catalytic [53, 54]. Illustrative is the synthesis of **41**, the acetate derivative of (±)-carquinostatin A [55]. In other cases, Knölker achieved yields of carbazolequinones higher than 90%.

41

Trost was the first to apply a Pd-catalyzed cyclization reaction to the synthesis of indole alkaloids [57–59]. Thus, exposure of indole isoquinuclidine **42** to PdCl$_2$(MeCN)$_2$ gave ibogamine (**44**) (racemic and optically active) after reductive cleavage of the Pd–C bond in **43** [58]. Other iboga alkaloids were similarly crafted by Trost, and Williams adopted this strategy in a key step in his synthesis of (+)-paraherquamide B [60]. By reducing the palladium intermediate with NaBD$_4$, Trost deduced that this cyclization mechanism involves initial indole C-2 palladation followed by ring closure to **43** [58].

42 → **43**

44

Stoltz has reported the first oxidative indole annulations that are catalytic in palladium, and two examples are illustrated below [61]. The ligand is ethyl nicotinate.

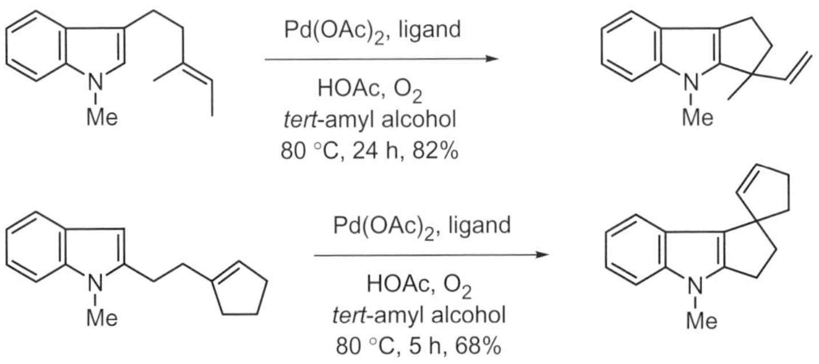

A similar Pd-catalyzed cyclization–carboalkoxylation of several alkenyl indoles has been described by Widenhoefer, one of which is shown [62].

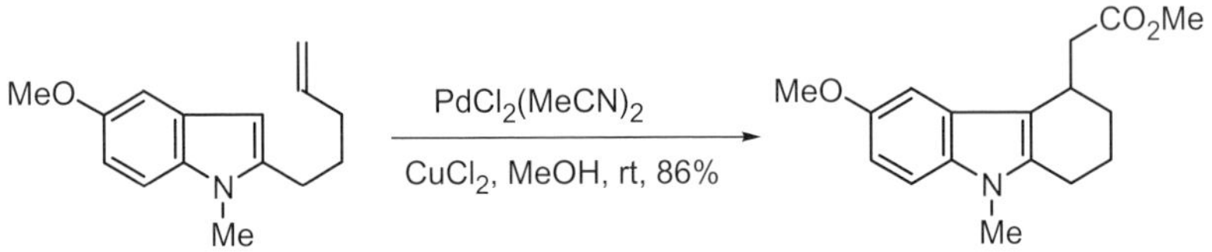

In a series of papers, Itahara established the utility of Pd(OAc)$_2$ in the oxidative cyclization of C- and N-benzoylindoles, and two examples are shown [63]. Itahara also found that the cyclization of 3-benzoyl-1,2-dimethylindole proceeds to the C-4 position (31% yield) [63a]. Under similar conditions, both 1-acetylindole and 1-acetyl-3-methylindole are surprisingly intermolecularly arylated at the C-2 position by benzene and xylene (22–48% yield) [64, 65].

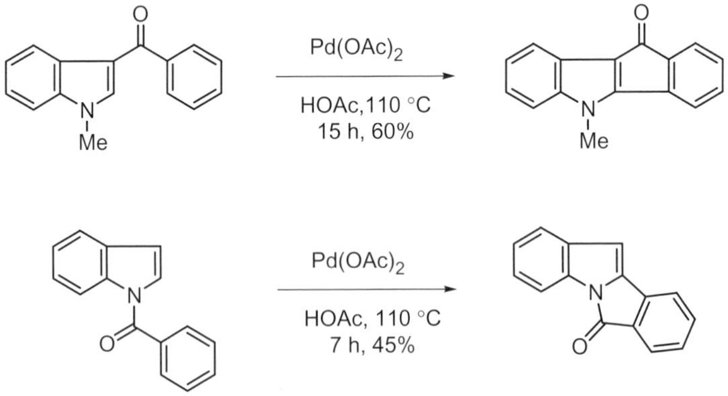

Itahara's work has paved the way for several Pd(OAc)$_2$-oxidative cyclizations leading to indole alkaloids. Thus, Black synthesized a series of pyrrolophenanthridone alkaloids (e.g. **45**→**46**) [66] and, following the pioneering work of Bergman on the parent system [16], he effected the cyclization of diindole urea **47** to **48** [67]. The presence of a 3-methyl group in each indole ring blocks cyclization. N-acylindolines corresponding to

45 also undergo this cyclization, and DDQ introduces the indole double bond quantitatively [66a]. However, the cyclization of *N*-benzylindoles corresponding to **45** fails.

Depending on the reaction conditions, indole 2-carboxamide **49** undergoes cyclization to afford either carbolinones or pyrazino[1,2-*a*]indoles [68].

Hill described the Pd(OAc)$_2$-oxidative cyclization of bisindolylmaleimides (e.g. **50**) to indolo[2,3-*a*]pyrrolo[3,4-*c*]carbazoles (e.g. **51**) [69], which is the core ring system in numerous natural products, many of which have potent protein kinase activity [70]. Other workers employed this Pd-induced reaction to prepare additional examples of this ring system [71, 72]. Ohkubo found that PdCl$_2$/DMF was necessary to prevent acid-induced decomposition of benzene-ring substituted benzyloxy analogs of **50**, and the yields of cyclized products under these conditions are 85–100% [71].

Srinivasan found that the typical stoichiometric Pd(OAc)$_2$ conditions effect cyclization of 2-(*N*-arylaminomethyl)indoles to aryl-fused β-carbolines in low yield [e.g. **52**] [73]. Similar to the chemistry observed with *N*-(phenylsulfonyl)pyrrole, 1,4-naphthoquinone also undergoes Pd(OAc)$_2$ oxidative coupling with N-(phenylsulfonyl)indole to give **53** in 68% yield [74].

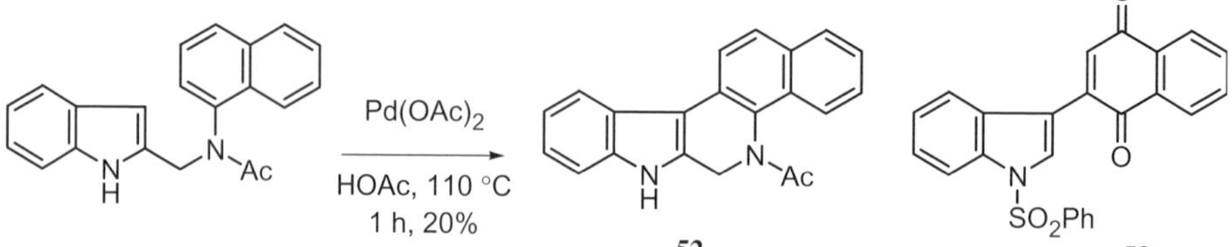

Intermolecular Pd oxidative-couplings with indoles are well established, although initial results were unpromising. For example, Billups found that indole reacts with allyl acetate (Pd(acac)/Ph$_3$P/HOAc) to give a mixture of 3-allyl- (54%), 1-allyl- (7%), and 1,3-diallylindole (11%) [75]. Allyl alcohol is also successful in this reaction but most other allylic alcohols fail. Likewise, methyl acrylate reacts with N-acetylindole (Pd(OAc)$_2$/HOAc) to give only a 20% yield of methyl (E)-3-(1-acetyl-3-indolyl)acrylate and a 9% yield of N-acetyl-2,3-bis(carbomethoxy)carbazole [76]. Itahara improved these oxidative couplings by employing both N-(2,6-dibenzoyl)indoles (e.g. **54**, **55**) and N-(phenylsulfonyl)indole as substrates [77]. Reaction occurs at C-3 unless this position is blocked. The coupling can be made catalytic using AgOAc or other reoxidants [77b]. Some examples are shown below and E-stereochemistry is the major or exclusive isomer. Acrylonitrile also reacts with **54** under these conditions (52%; E/Z = 3/1) [77a], and methyl vinyl ketone, ethyl (E)-crotonate, and ethyl α-methylacrylate react with N-(phenylsulfonyl)indole under these oxidative conditions [77b]. Interestingly, an N-indole 2-pyridylmethyl substituent leads to C-2 alkenylation with methyl acrylate, acrylonitrile, and phenyl vinyl sulfone under typical conditions (Pd(OAc)$_2$, Cu(OAc)$_2$, HOAc, dioxane, 70°C) [78].

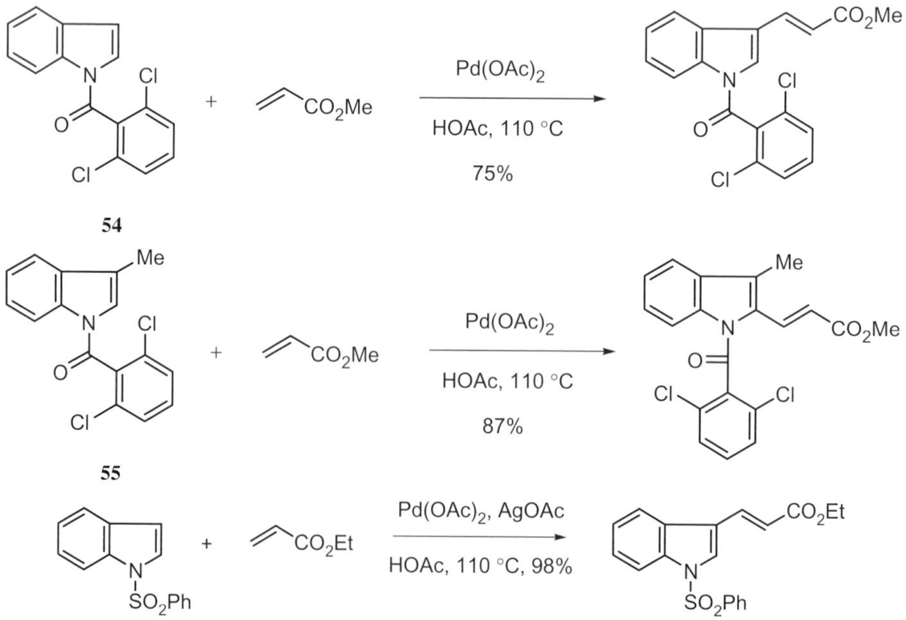

Hegedus found that 4-bromo-1-(4-toluenesulfonyl)indole (**11**) reacts with methyl acrylate to form the C-3 product in low yield under stoichiometric conditions [20]. Heck reactions of this substrate and related haloindoles are discussed in Section 3.5. Yokoyama, Murakami, and co-workers also utilized **11** in total syntheses of clavicipitic acid and costaclavine, one key step of which is the oxidative coupling of **11** with **56** to give dehydrotryptophan derivative **57** [79]. The use of chloranil as a reoxidant to recycle Pd(O) to Pd(II) greatly improves the coupling over earlier conditions [80, 81]. For example, chloranil was more effective than DDQ, MnO_2, Ag_2CO_3, $Co(salen)_2/O_2$, and $Cu(OAc)_2$. In the absence of chloranil, the yield of **57** is 31%.

The reaction of *N*-protected dehydroalanine methyl esters (e.g. **56**, **59**) with other indoles **58** can also be effected to give the corresponding dehydrotryptophans **60**, invariably as the *Z*-isomers [81]. Murakami, Yokoyama, and co-workers also studied oxidative couplings of acrylates, acrylonitrile, and enones with 2-carboethoxyindole, 1-benzylindole, and 1-benzyl-2-carboethoxyindole and $PdCl_2$ and $CuCl_2$ or $Cu(OAc)_2$ to give C-3 substitution in 50–84% yields [82, 83].

R_1 = H, Ts
R_2 = H, CO_2Et
R_3 = H, 4-Br, 4-Me, 4-CO_2Me, 5-NO_2, 5-Br, 5-OMe, 7-Br

These workers have also found that 3-alkenylpyrroles, such as **61**, are cyclized to indoles [84].

In addition to examining the vinylation of 1-methyl-2-indolecarboxaldehyde with methyl acrylate ($Pd(OAc)_2$/HOAc/AgOAc) to give methyl (*E*)-3-(2-formyl-1-methyl-3-indolyl)acrylate in 60% yield, Pindur found that similar reactions of methyl 3-(1-methyl-2-indolyl)acrylate afford bis(carbomethoxy)carbazoles albeit in low yield [85]. Fujiwara discovered that the combination of catalytic $Pd(OAc)_2$ with benzoquinone and *t*-butylhydroperoxide serves to couple indole with methyl acrylate to give methyl (*E*)-3-(3-indolyl)acrylate in 52% yield [86].

Abdrakhmanov and co-workers observed the cycloaddition of *N*-(1-methyl-2-butenyl)aniline or 2-(1-methyl-2-butenyl)aniline with PdCl$_2$/DMSO to give a 69% yield of a mixture of 2-ethyl-3-methylindole and 2,4-dimethylquinoline [87]. The authors propose that a Claisen rearrangement is initially involved. A similar oxidative cyclization of a 5-amino-indoleacrylate was the starting point for syntheses of CC-1065 and related compounds [88]. In an elegant mechanistic analysis, Sames has developed a general Pd-catalyzed arylation of indoles, at C-2 or C-3 depending on the reaction conditions [89].

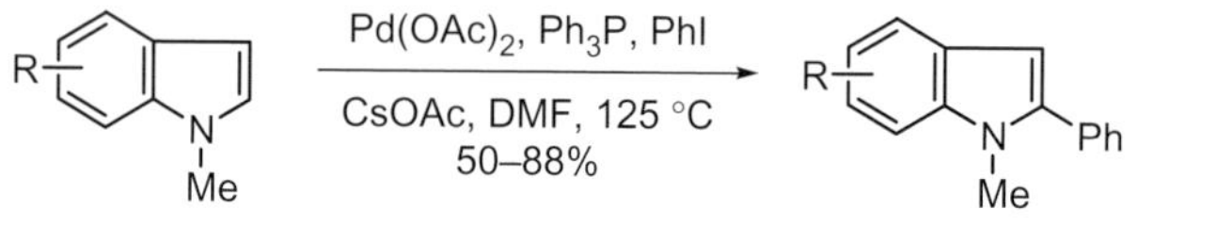

R = H, 5-CN, 5-NO$_2$, 4-MeO, 6-NO$_2$, 6-MeO, 6-Me, 6-CO$_2$Me, 5-NHSO$_2$Ph

The Pd-catalyzed C-3 alkylation of indoles via nucleophilic allylic substitution on allylic carbonates and acetates has been described [90]. Two clever indole ring syntheses involving oxidative cyclization are illustrated below [91].

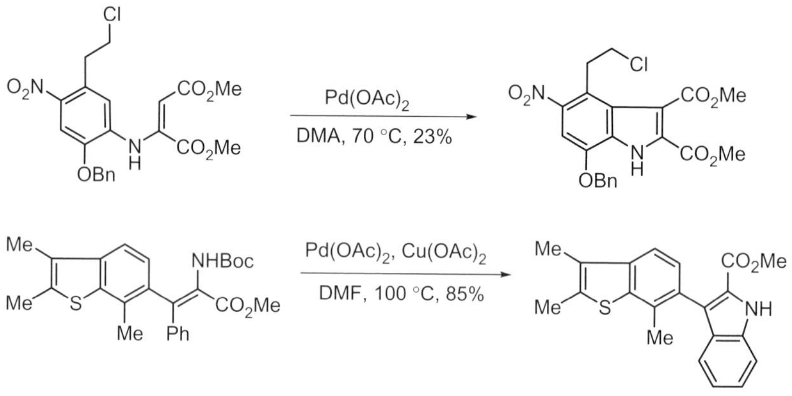

3.3. Coupling reactions with organometallic reagents

3.3.1. Kumada coupling

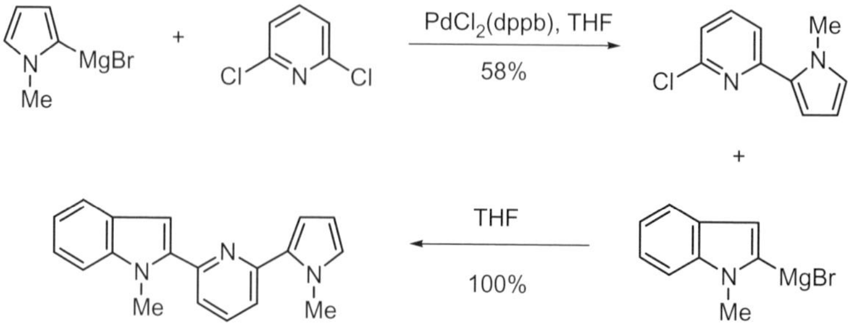

Of all the Pd-catalyzed coupling reactions, the Kumada coupling has been applied least often in indole chemistry. However, this Grignard-Pd cross-coupling methodology has been used to couple 1-methyl-2-indolylmagnesium bromide with iodobenzene and α-bromovinyltrimethylsilane to form 1-methyl-2-phenylindole and 1-methyl-2-(1-trimethyl-silyl)vinylindole in 79 and 87% yields, respectively [92a, b]. Kumada constructed the tri-heterocycle **62** using a tandem version of his methodology [92c].

Kondo employed the Kumada coupling using the Grignard reagents derived from 2- and 3-iodo-1-(phenylsulfonyl)indole to prepare the corresponding phenyl derivatives in 50% yield [93]. Widdowson expanded the scope of the Kumada coupling and provided some insight into the mechanism [94].

3.3.2. Negishi coupling

Although the Negishi coupling has been less frequently used in indole synthetic manipulations than either Suzuki or Stille couplings, we will see in this chapter that Negishi chemistry is often far superior to other Pd-catalyzed cross-coupling reactions involving indoles. One of the first such examples is Pichart's coupling of 1-methyl-2-indolylzinc chloride (**63**) with iodopyrimidine **64** to give **65** [95].

Sakamoto and co-workers studied extensively the generation and couplings of *N*-protected 2- and 3-indolylzinc halides [96]. The best *N*-protecting groups are Boc, SO_2Ph, and CO_2^-, and the zinc species can either be generated by transmetalation of a lithiated indole with $ZnCl_2$ or by oxidative addition of active Zn to iodoindoles. The latter technique is required for successful generation of 3-indolylzinc halides since 1-(phenylsulfonyl)-3-lithioindoles are known to rearrange to the more stable 2-lithioindoles at room temperature [10]. A selection of these reactions is shown that illustrates the range of compatible functional groups. These workers also generated 3-carbomethoxy-1-(phenylsulfonyl)-4-indolylzinc iodide and coupled it with iodobenzene under the same conditions to give the phenylated derivative in 43% yield.

25–74%   Ar = Ph, 4-MeOPh, 4-NO_2Ph, 4-CO_2EtPh,
3-BzPh, 2-pyridyl, 2-thienyl, 3-$CONMe_2$Ph

Independently and simultaneously, Bosch developed similar Negishi indole chemistry [97] to that described above by Sakamoto. An important discovery is that 1-(*tert*-butyl-dimethylsilyl)-3-lithioindole is stable at room temperature and does not rearrange to the 2-lithio isomer. Therefore, a halogen–lithium exchange sequence followed by transmetalation with $ZnCl_2$ conveniently generates the 3-indolylzinc species. This species and 1-(phenylsulfonyl)-2-indolylzinc chloride smoothly couple with halopyridines under Negishi conditions to give 3- and 2-(2-pyridyl)indoles, respectively, as illustrated below for the former (**66** to **67** to **68**) [97b,c]. The *tert*-butyldimethylsilyl group is readily removed with TsOH. Bosch and co-workers also coupled **67** with a range of other heteroaryl halides (2- and 3-bromothiophene, 3-bromofuran, 2-chloropyrazine, and 3-bromo-1-(phenylsulfonyl)indole) [97b,c].

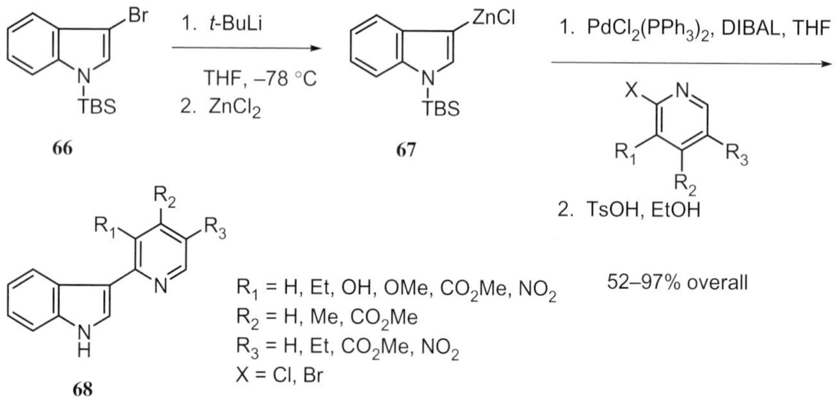

Danieli extended the Pd-catalyzed coupling of 2-indolylzinc chlorides to a series of halopyridin-2-ones and halopyran-2-ones [98]. This Negishi coupling is more efficient than a Suzuki approach but is not as good as a Stille coupling. An example of the latter will be shown in Section 3.3.4. These workers also generated zinc reagents from 5-iodopyridin-2-one and 5-bromopyran-2-one but Negishi couplings were sluggish. Since direct alkylation of a 2-lithioindole failed, Fisher and co-workers utilized a Negishi protocol to synthesize 2-benzylindole **69** as well as the novel CNS agent **70** [99].

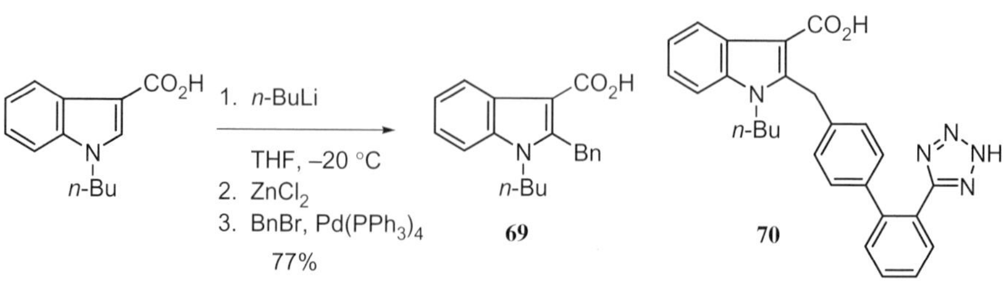

Cheng and Cheung also employed a 2-indolylzinc chloride **72** to couple with indole **71** in a synthesis of "inverto-yuehchukene" **73** [100]. Other Pd catalysts were no better in this low-yielding process.

Maas and co-workers coupled 1-methyl-2-indolylzinc chloride (63) with several (Z)-bromostyrenes (e.g. 74) to give the corresponding 2-(2-arylethenyl)-1-methylindoles (e.g. 75) [101]. Although the (Z)-olefin is the kinetic product, this compound can be isomerized to the (E)-olefin. The β-bromostyrenes are conveniently synthesized by treating the requisite dibromostyrene with Pd(PPh₃)₄/Bu₃SnH (61–79%). In addition to 74, 1-[(Z)-2-bromo-ethenyl)]-4-nitrobenzene and 2-[(Z)-2-bromoethenyl)]thiophene were prepared using this method and successfully coupled with 63.

In a synthesis of polyketides, Kocienski crafted indole 78 from 2-iodo-1-methylindole and the appropriate organozinc reagent 77 derived from the corresponding stannane (76), which itself was reluctant to undergo a Stille coupling [102].

In his general studies aimed at the synthesis of ergot alkaloids, Hegedus effected the coupling of dihaloindole 79 with the zinc reagent 80, generated *in situ* from 1-methoxy-1, 2-propadiene, to afford the expected indole ketene [21]. A Negishi coupling between

5-bromoindole and commercially available 3-cyanopropylzinc bromide to give **81** was the first step *en route* to a potential non-peptidic $\alpha_v\beta_3$ antagonist [103].

Negishi methodology can also be used to achieve the 3-acylation of indoles. Thus, Faul used this tactic to prepare a series of 3-acylindoles **83** from indole **82** [104]. Indole **82** could also be cleanly iodinated at C-3 with *N*-iodosuccinimide (78%).

R = Me, Bn, Pr, Ph, CH₂Cl

Bergman described indole C-3 acylation with acid chlorides via 1-indolylzinc chloride in the *absence* of palladium [105]. Davidsen and co-workers synthesized **86**, which is a potent antagonist of platelet activating factor-mediated effects, using this Bergman acylation sequence as shown [106]. As will be discussed in the next section, a Suzuki coupling was used to prepare **84**.

In continuation of his extraordinarily versatile and efficient directed-metalation technology, Snieckus employed indole **87** to selectively lithiate C-4 and to effect a Negishi coupling with 3-bromopyridine to give **88** in 90% yield [107]. In contrast, a Suzuki protocol gave **88** in only 19% yield (with loss of the TBS group).

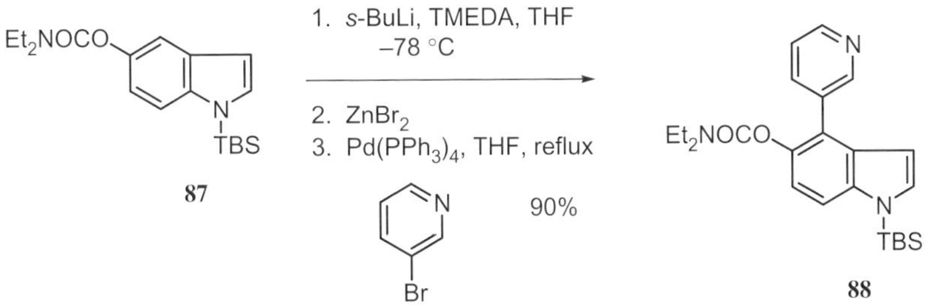
Et$_2$NOCO
TBS
1. s-BuLi, TMEDA, THF
−78 °C
2. ZnBr$_2$
3. Pd(PPh$_3$)$_4$, THF, reflux
N
Br
90%
Et$_2$NOCO
N
TBS

87 **88**

1,9-Dilithio-β-carboline (**89**), which was generated from 1-bromo-β-carboline, undergoes the Negishi coupling with 2-chloroquinoline to form the alkaloid nitramarine (**90**) [108].

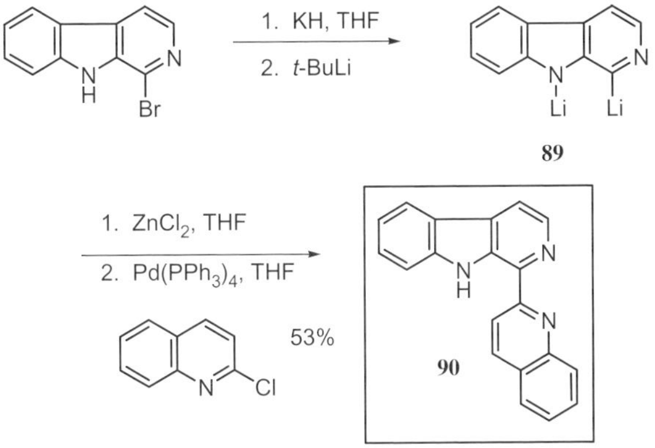
N
H
Br
1. KH, THF
2. t-BuLi
Li
Li
1. ZnCl$_2$, THF
2. Pd(PPh$_3$)$_4$, THF
N
Cl
53%

89

90

Grigg employed organozinc chemistry to construct 3-alkylidenedihydroindoles such as **91** via a tandem Pd-catalyzed cyclization–cross-coupling sequence [109]. A similar route to such compounds was reported by Luo and Wang, e.g. **92** [110].

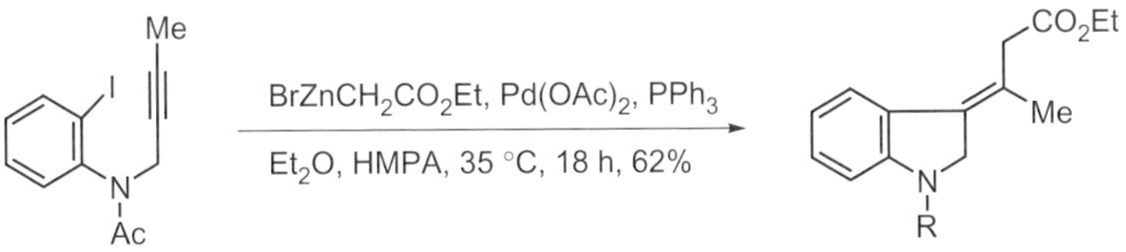
Me
I
N
Ac
BrZnCH$_2$CO$_2$Et, Pd(OAc)$_2$, PPh$_3$
Et$_2$O, HMPA, 35 °C, 18 h, 62%
CO$_2$Et
Me
R

91 (R = Ac, H; 1.4:1)

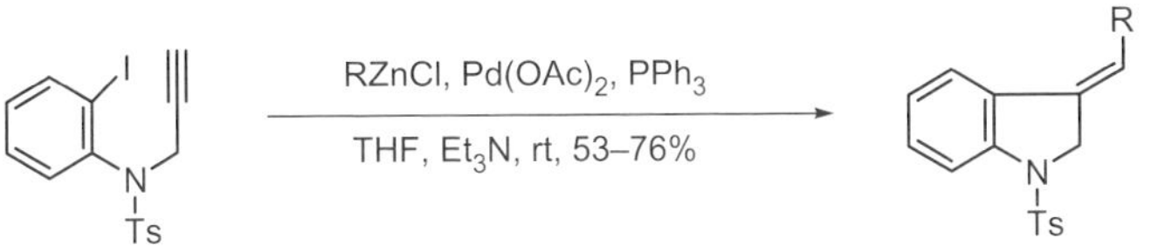
I
N
Ts
RZnCl, Pd(OAc)$_2$, PPh$_3$
THF, Et$_3$N, rt, 53–76%
R

92 (*Z* only)

R = Ph, 2-furyl, 2-thienyl, 2-pyridyl, *n*-Bu, *n*-BuCC, TMSCC, PhC(CH_2)=CH_2

Knochel has found that transmetalation of aryl iodide **93** to the corresponding arylz-inc species followed by Negishi coupling with ethyl 4-iodobenzoate (ArI) gives a highly functionalized indole [111].

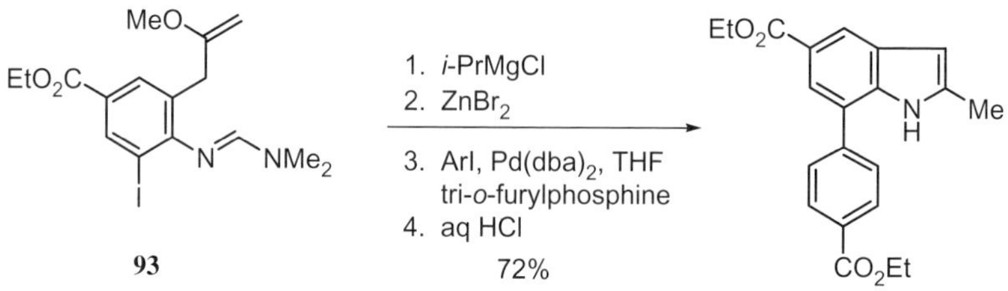

3.3.3. Suzuki coupling

Although the first report of an indoleboronic acid was by Conway and Gribble in 1990, this compound (**94**) was not employed in Suzuki coupling, but rather it was utilized *en route* to 3-indolyl triflate **34** as described in Section 3.1 [12].

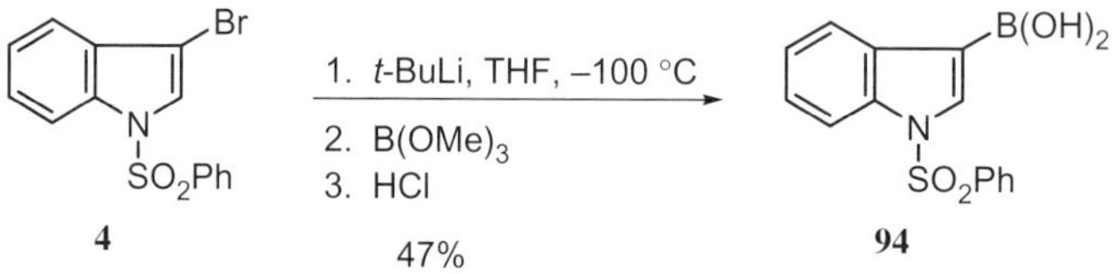

In the intervening years, indoleboronic acids substituted at all indole carbon positions have found use in synthesis. For example, Claridge and co-workers employed **94** in a synthesis of isoquinoline **95** under standard Suzuki conditions in high yield [112]. Compound **95** was subsequently converted to the new Pd-ligand 1-methyl-2-diphenylphosphino-3-(1′-isoquinolyl)indole.

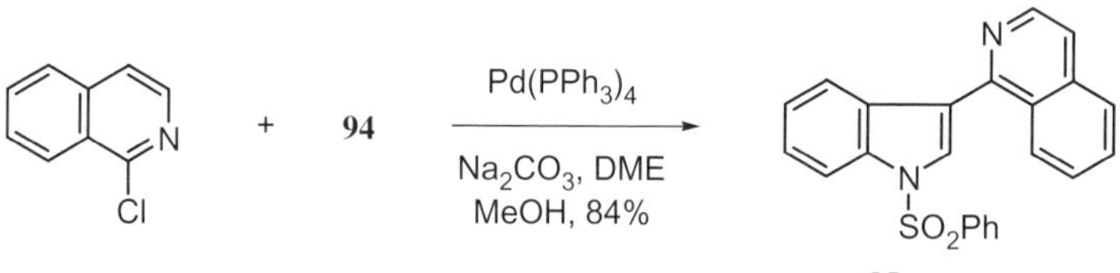

Martin utilized indoleboronic acids in Pd-catalyzed coupling to great effect, and has improved upon the halogen–metal exchange route to indole-3-boronic acids by adopting a mercuration–boronation protocol as illustrated below for the preparation of **96** and **97** [113].

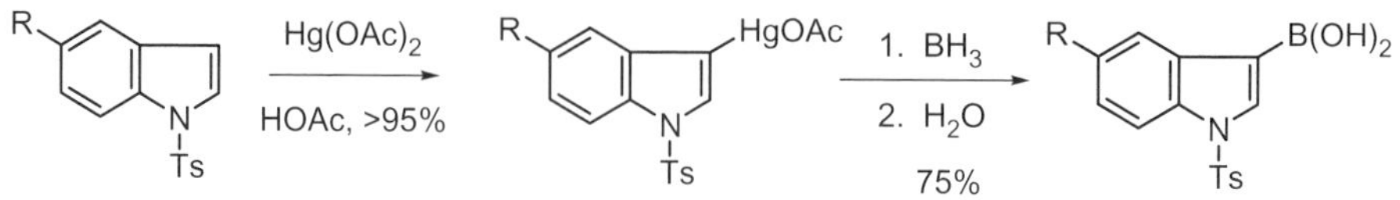

96, R = H
97, R = OMe

Boronic acids **96** and **97** couple very well with vinyl triflates **98** and **99** under typical Suzuki conditions (Pd(PPh$_3$)$_4$/Na$_2$CO$_3$/LiCl/DME) to give indoles **100** and **101**, respectively, in 76–92% yield [113]. Enol triflates **98** and **99** were prepared in good yield (73–86%) from *N*-substituted 3-piperidones, wherein the direction of enolization (LDA/THF/–78°C; PhNTf$_2$) is dictated by the *N*-substituent.

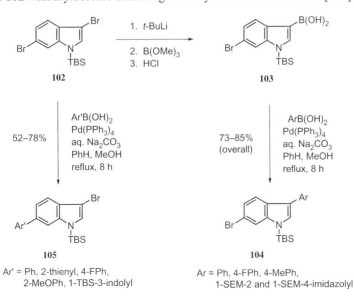

98, R = Me, Bn

100, R = Me, Bn; R' = H, OMe

99, R = Cbz, Boc

101, R = Cbz, Boc; R' = H, OMe

Boronic acid **96** has been employed by Jiang to synthesize both a series of novel indolylpyrimidines and indolylpyrazines [114a] and the dragmacidin D skeleton [114b]. This research group has also employed *N*-TBS-indole-3-boronic acids in the latter dragmacidin D studies [114b], as well as in the preparation of 2,1,3-benzothiadiazole bisindoles as novel fluorescent materials [114c]. Stoltz has made great use of indole-3-boronic acids (and others) in his several syntheses of the dragmacidin alkaloids [115]. For example, the 6-bromo derivative of **96** couples smoothly with halogenated pyrazines [115].

In the course of their successful syntheses of the marine alkaloids nortopsentins A–D, Kawasaki and co-workers were able to prepare selectively boronic acid **103** from 1-(*tert*-butyldimethylsilyl)-3,6-dibromoindole (**102**) and effect Suzuki couplings to give 3-arylindoles **104** in good yields [116]. Complementary to this chemistry is the direct Pd-catalyzed reaction of **102** with arylboronic acids to give 6-aryl-3-bromoindoles **105** [116].

102

1. *t*-BuLi
2. B(OMe)$_3$
3. HCl

103

52–78%

Ar'B(OH)$_2$
Pd(PPh$_3$)$_4$
aq. Na$_2$CO$_3$
PhH, MeOH
reflux, 8 h

73–85%
(overall)

ArB(OH)$_2$
Pd(PPh$_3$)$_4$
aq. Na$_2$CO$_3$
PhH, MeOH
reflux, 8 h

105

104

Ar' = Ph, 2-thienyl, 4-FPh,
2-MeOPh, 1-TBS-3-indolyl

Ar = Ph, 4-FPh, 4-MePh,
1-SEM-2 and 1-SEM-4-imidazolyl

Hoerrner utilized *N*-TIPS boronic acid **106** to prepare dehydro-β-methyltryptophan **107**, *en route* to a synthesis of β-(2*R*,3*S*)-methyltryptophan by asymmetric hydrogenation of **107** [117].

Nishida employed boronic acid **106** and the corresponding *N*-TBS boronic acid in Suzuki reactions with a series of heteroaromatic halides [118]. The *N*-TIPS group proved superior to *N*-TBS. 3-Indoleboronic acid **96** was employed by Neel to prepare bis(indolyl)maleimides such as **109** [119]. However, since the standard Suzuki conditions failed (triflate **108** apparently decomposing under the reaction conditions), the use of a phosphine-free Pd catalyst [120] and cesium fluoride [121] was necessary and gave **109** in an acceptable yield of 55%.

Merlic synthesized a series of *N*-substituted 2-iodoindoles (*N*-methyl, *N*-allyl, *N*-benzyl, *N*-phenylsulfonyl) and converted them to the corresponding boronic acids. Two of the latter (*N*-methyl and *N*-benzyl) undergo a Suzuki coupling with 2-iodoindole to give 2,2′-biindolyls in 51–67% yield *en route* to indolocarbazoles [122]. By comparison, a Negishi protocol failed and a Stille coupling gave 2,2′-biindolyls in lower yields. Gallagher prepared *N*-Boc-2-indoleboronic acid **110** for use in Suzuki coupling with aryl bromides and iodobenzene in a search for new selective dopamine D$_3$ receptor antagonists [123]. Indoleboronic acid **110** and benzene ring substituted derivatives can also be prepared by a non-cryogenic *in situ* lithiation and quench protocol [124]. Snieckus has constructed 2-indoleboronopinacolates and 2-indoledihaloboranes by *ipso*-borodesilylation reactions, and the subsequent Suzuki couplings proceed in good yield [125].

110

Ar = Ph, 4-ClPh, 3-pyridyl
(2-thienyl failed)

Martin prepared indole-5-boronic acid from 5-bromoindole by halogen–lithium exchange (44% yield) and performed Suzuki couplings with a wide range of aryl bromides (mainly) and heteroaryl bromides to give the expected compounds in 52–94% yield under the standard conditions [126]. In addition to the typical substituted aryl iodides (fluoro, methoxy, nitro), the range of iodo partners included pyridines, pyrimidines, pyrazines, furans, thiophenes, thiazoles, and isoxazoles. Roussi and co-workers executed Suzuki couplings with a variety of 5-, 6-, and 7-indoleboronic acids and aryl bromides, and also with arylboronic acids and 5-, 6-, and 7-bromoindoles to give the biaryl products in yields up to 60% [127]. The first combination gave higher yields of coupling products. Unfortunately, these extensive model studies did not translate into a successful intramolecular version that the authors had devised as a synthesis of chloropeptin and kistamycin model compounds [127b]. Thus, neither **111** nor **112** underwent the desired intramolecular Suzuki coupling (although an Ni(O)-mediated approach was successful).

111   **112**

Several other groups have reported the synthesis and Suzuki reactions of an *N*-methylindolyl-3-carboxamido-2-boronic acid for the synthesis of benzo[*a*]carbazoles [128], an *N*-Boc-5-sulfonamidoindolyl-2-boronic acid for the synthesis of novel KDR kinase inhibitors [129], indolyl-4-boronic acid in a new synthesis of lysergic acid [130], and 5-, 6-, and 7-indolylboronic acids for the synthesis of aryl-substituted indoles [131]. Carbazole-2,7-bis (boronates) have been employed to construct diindolocarbazoles [132].

The "reverse" Suzuki coupling of indolyl halides (and triflates) with aryl- and heteroarylboronic acids has been pursued by a number of investigators. The choice of which coupling partner is the boronic acid and which is the halide (or triflate) is governed by relative substrate accessibility and availability. Zembower and Ames incorporated a Suzuki coupling into a general synthesis of 5-substituted tryptophans **115** via the iodinated cyclic tautomer **113** as illustrated [133]. The Suzuki product **114** (R = Ph) and **113** were unraveled by a sequence of H_2SO_4 (to regenerate the indole) (85–91%), TMSI carbamate cleavage (75–83%), and base hydrolysis of the *N*-acetyl and methyl ester (61–69%)

to yield **115**. These Suzuki couplings were performed under conventional conditions with
$RB(OH)_2$ or with the modified reagents derived from 9-BBN.

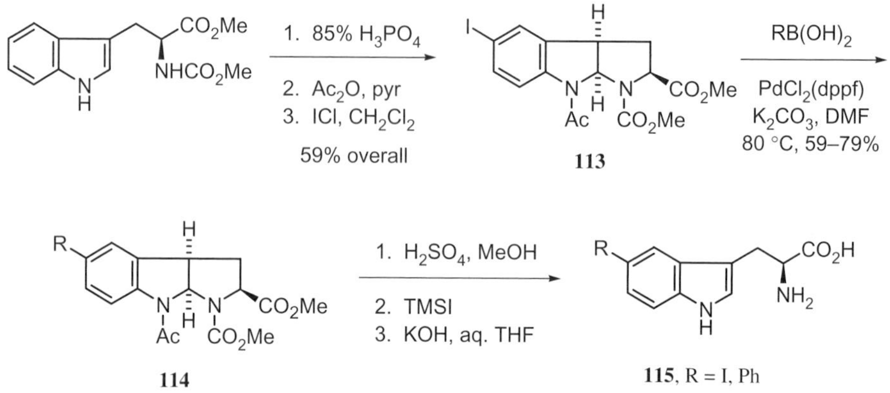

113

114

115, R = I, Ph

R= Ph, 3-NO_2Ph, CH_2CH_2Ph, CH_2CH_2-4-ClPh, n-C_8H_{17}

Snieckus described short syntheses of ungerimine (**121**) and hippadine by Suzuki
couplings of boronic acid **118** with 7-bromo-5-(methylsulfonyloxy)indoline (**116**) and
7-iodoindoline (**117**), respectively [134]. Cyclization and aerial oxidation also occur.
Treatment of **119** with Red-Al gave ungerimine (**121**) in 54% yield, and oxidation of **120**
with DDQ afforded hippadine in 90% yield. Indoline **116** was readily synthesized from
5-hydroxyindole in 65% overall yield by mesylation, reduction of the indole double bond,
and bromination. Indoline **117** was prepared in 67% yield from N-acetylindoline by thallation–iodination and basic hydrolysis.

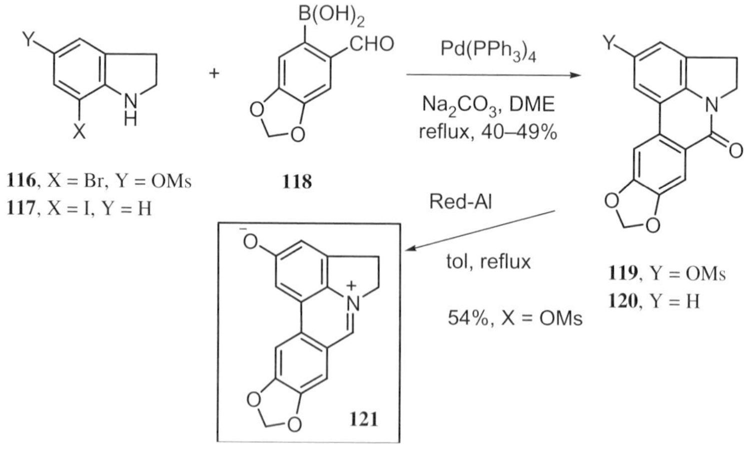

116, X = Br, Y = OMs
117, X = I, Y = H

118

119, Y = OMs
120, Y = H

121

Martin effected the synthesis of several 3,5-diarylated indoles by a tandem
Stille–Suzuki sequence [135]. The latter reaction involves exposure of 3-(3-pyridyl)-5-
bromo-1-(4-toluenesulfonyl)indole with arylboronic acids (aryl = 3-thienyl, 2-furyl,
phenyl) under typical conditions to give the expected products in 86–98% yield [135].

Carrera engaged 6- and 7-bromoindole in Pd-catalyzed couplings with 4-fluoro- and 4-methoxyphenylboronic acids to prepare 6- and 7-(4-fluorophenyl)indole (90 and 74% yield) and 6-(4-methoxyphenyl)indole (73% yield) [29]. Banwell and co-workers employed 7-bromoindole in a Suzuki coupling with 3,4-dioxygenated phenylboronic acids *en route* to the synthesis of Amaryllidaceae alkaloids [136]. Yields of 7-arylated indoles are 93–99%. Moody successfully coupled 4-bromoindole **122** with boronic acid **123**, which was derived from the corresponding bromo compound, to give **124** in high yield [137].

This reversal of coupling partners was used by Giralt to prepare aryl-substituted indoles [131a], Leclerc in the synthesis of diindolocarbazoles [132], Viaud–Massuard to craft dimers of indole and 7-azaindole as melatonin analogs [138], Schultz to prepare both a library of 5-aryltryptamines by coupling solid-supported 5-bromotryptamine with arylboronic acids [139] and 2,3,5-trisubstituted indoles [140]. Zhang also utilized the solid phase to generate a library of 2-aryl-3-alkylindoles and 2,3-diarylindoles from the corresponding haloindoles and arylboronic acids [141]. Gribble also reported the one-pot synthesis of symmetrical 2,3-diarylindoles in Suzuki reactions with 2,3-dihaloindoles [142]. The power of these methods is seen with the reported syntheses of compounds **125–128** [143–146] from the appropriate arylboronic acid and the 2- or 3-haloindole.

Terashima employed diethyl-(3-pyridyl)borane as the boron partner in a Suzuki coupling with both 5-bromo-1-(4-toluenesulfonyl)indole and 3-bromo-1-(mesitylenesulfonyl)indole to give the corresponding pyridylindoles in modest yields (47%, 39%) [147].

Somei effected the coupling of phenyl-, 2-furyl-, and 1-hexenylboronic acids with 4-thallated indole-3-carboxaldehyde (Pd(OAc)$_2$/DMF) to give 4-substituted 3-formylindoles [148]. Regioselective thallation of indole-3-carboxaldehyde is achieved using thallium tris-trifluoroacetate in 77% yield. Indole **129**, which is available by the Buchwald zirconium indoline synthesis, was used by Buchwald to synthesize **130** via a Suzuki protocol [149]. Boronate ester **130** is prepared by the hydroboration of 3-methyl-1-butyne with catechol borane. Indole **131** had been used in earlier studies to synthesize the clavicipitic acids.

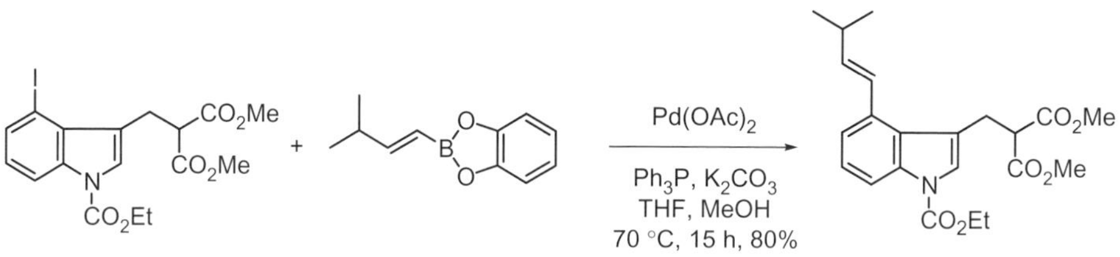

129   **130**   **131**

The medicinal importance of 2-aryltryptamines led Chu and co-workers to develop an efficient route to these compounds (**134**) via a Pd-catalyzed cross-coupling of protected 2-bromotryptamines **132** with arylboronic acids **133** [150]. Several Suzuki conditions were explored and only a partial listing of the arylboronic acids is shown here. In addition, boronic acids derived from naphthalene, isoquinoline, and indole were successfully coupled with **132**. The C-2 bromination of the protected tryptamines was conveniently performed using pyridinium hydrobromide perbromide (70–100%). Other groups have employed 2- and 5-halotryptamines (and homotryptamines) in Suzuki coupling to prepare novel inhibitors of 15-lipoxygenase [151] and selective 5-HT receptor agonists [152]. 2-Phenyl-5-(and 7-)azaindoles have been prepared via a Suzuki coupling of the corresponding 2-iodoazaindoles [19].

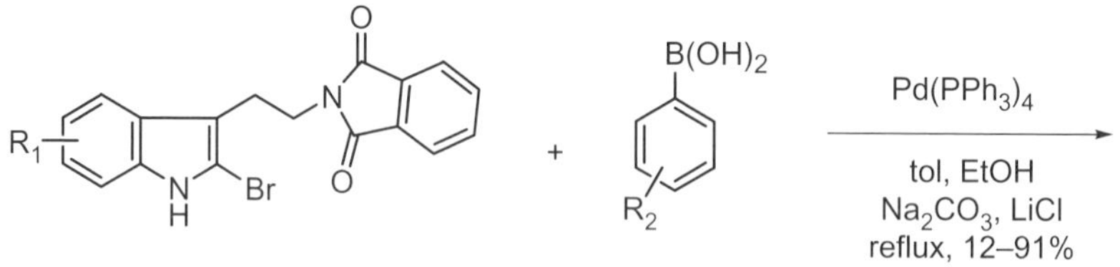

132   **133**

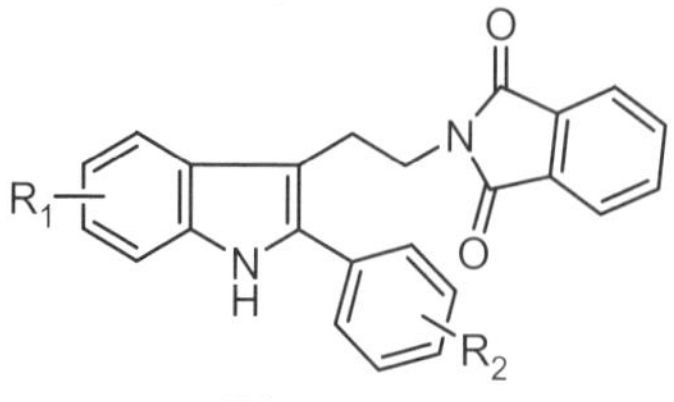

R$_1$ = H, 5-OBn, 5-OMe, 5-Cl, 5-Me, 6-F, 7-Me
R$_2$ = H, 2-Me, 3-Me, 4-Me, 3-NO$_2$, 4-F, 4-Cl, 3-OMe
        3,5-diMe, 2,4-diCl, 3,5-diOMe, 3,5-diCF$_3$, 3,5-diCl

134

Carini *et al.* converted 8-bromobenzo[*c*]carbazole to the corresponding aryl derivatives **135**, which are selective inhibitors of cyclin-dependent kinase 4 [153], and Nicolaou
employed a 4-bromoindole to craft **136** in a model study towards the synthesis of diazonamide A [154].

135

136

Indolyltriflates have been used in Suzuki couplings by Mérour [155] and others [152].
Thus, the readily available 1-(phenylsulfonyl)indol-2-yl triflate (**137**) smoothly couples
with arylboronic acids in 65–91% yield. Similarly, Pd-catalyzed cross-coupling of
phenylboronic acid with 1-benzyl-2-carbomethoxyindol-3-yl triflate affords the 3-phenyl
derivative (62% yield) [155b].

Ar = Ph, 2-CHO-Ph, 4-Br-Ph, 3-thienyl, 2-benzofuryl
5-indolyl, *N*-Boc-3-indolyl

Doi and Mori prepared the enol triflate of 1-(4-toluenesulfonyl)-4-oxo-4,5,6,7-tetrahy-
droindole which couples with 3,4-methylenedioxyphenylboronic acid ($PdCl_2(PPh_3)_2$) to
give the 4-substituted derivative in 52% yield. Dehydrogenation furnishes the indole in
59% yield [156]. Likewise, carbazole triflates have served as Suzuki partners with arylboronic acids and 9-alkyl-9-BBN derivatives in the synthesis of carazostatin, hyellazole,
and carbazoquinocins B-F [157], in the preparation of diindolocarbazoles [132], and in
the construction of several 4-aryl and 4-heteroaryl pyrrolo[3,4-*c*]carbazoles as models
for future combinatorial studies of new protein kinase C inhibitors [158]. An aryl triflate
was employed in the early stages of a Suzuki coupling route to novel naltrindoles [159].
Bracher and Hildebrand utilized 1-chloro-β-carboline in Suzuki methodology with
the appropriate arylboronic acids to prepare 1-phenyl-β-carboline and the alkaloids
komaroine (**139**) and perlolyrine (**140**) [160], and Mérour engineered the synthesis
of benzo[5,6]cyclohepta[*b*]indole derivatives, such as **141**, via a Suzuki coupling of
5-methyl-11-bromo-5,6-dihydrobenzo[5,6]cyclohepta[*b*]indol-6-one with 2-thienyl-
boronic acid [161].

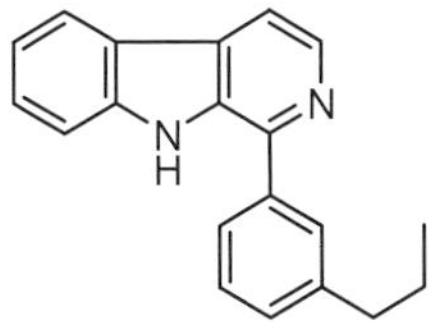

139 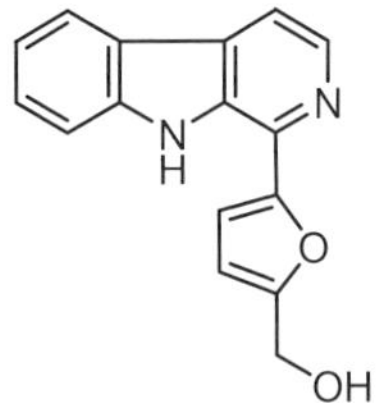   140 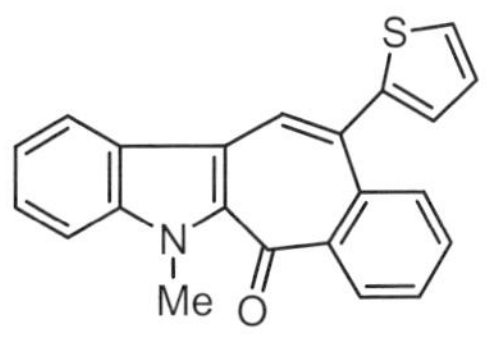   141

Ishikura and co-workers have done extensive work on the utility of indolylborates such as lithium triethyl(1-methylindol-2-yl)borate (142), prepared as shown from 1-methylindole, in Suzuki-like Pd-catalyzed reactions [162]. For example, 142 couples smoothly with aryl halides to afford 2-arylindoles 143 [162a]. The amount of 2-ethyl-1-methylindole by-product, formed by ethyl group migration, can be minimized by refluxing the mixture. At room temperature, 2-ethyl-1-methylindole is the major product. More recent work by Ishikura extended these couplings to the (removable) N-Boc analog of 142 with comparable yields to those obtained with 142 [162m].

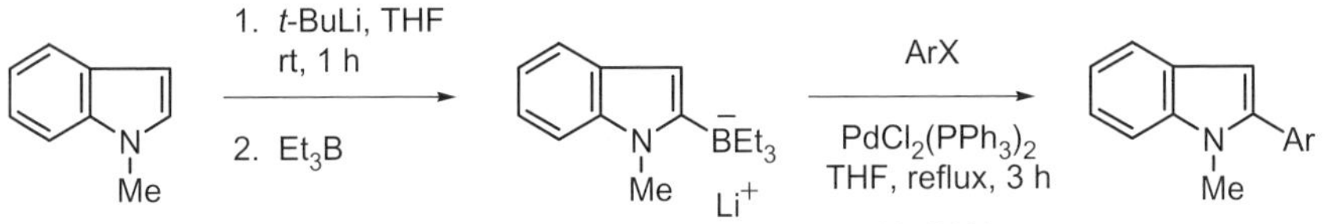

142

60–80%   143

Ar = Ph, 2- and 3-pyridyl, 3-thienyl, 2-styryl
X = Br, I

The extraordinary power of the Ishikura Pd-catalyzed couplings of indolylborates is revealed by the several examples shown below [162c,g,i–k]. The carbonylation version is discussed in Section 3.6. The formation of allenylindoles 145 vis-á-vis alkynylindoles 147 apparently depends on the equilibrium between an allenylpalladium complex and a propargylpalladium complex, and S_N2-like attack on the latter by 142 to give 147 is favored by the Ph_3P-ligated Pd complex.

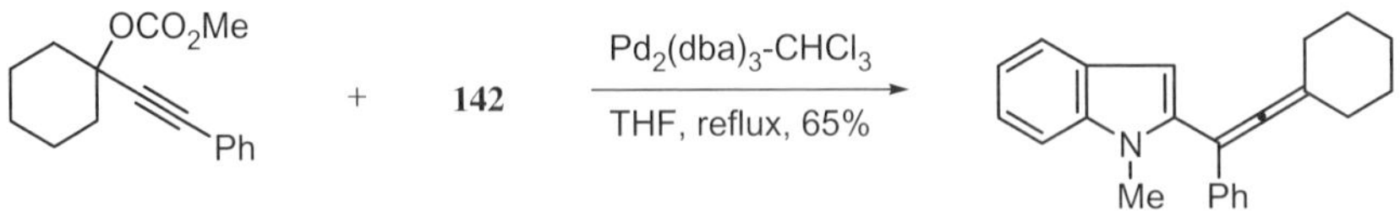

144   145

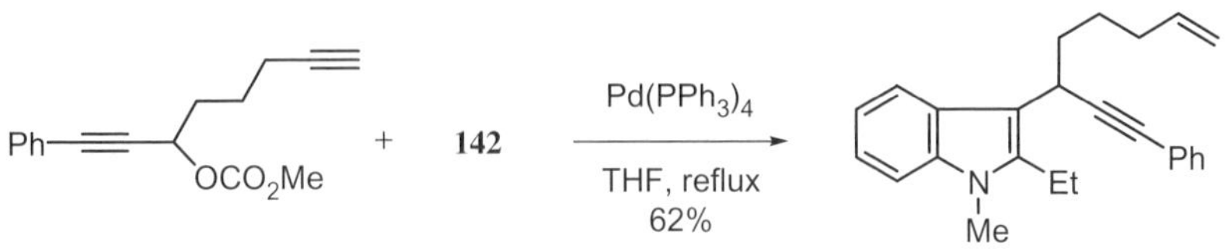

146   147

Murase and co-workers generated the *N*-methoxyindolylborate **153** and effected coupling with several indoles to give **154**, and, by reductive cleavage of the *N*-methoxyl group, arcyriacyanin A [25].

Molander has effected a Suzuki reaction between 7-bromoindole and potassium trifluoroborates (e.g. **155**) to give the corresponding coupled product **156** [163].

Palladium-catalyzed reactions of arylboronic acids have been utilized to craft precursors for constructing indole rings. Suzuki found that tris(2-ethoxyethenyl)borane (**157**) and catechol-derived boranes **158** readily couple with *o*-iodoanilines to yield **159**, which easily cyclize to indoles **160** with acid [164]. Kumar and co-workers used this method to prepare 5-(4-pyridinyl)-7-azaindoles from 6-amino-5-iodo-2-methyl-3,4′-bipyridyl [165].

A similar scheme with catechol-vinyl sulfide boranes also leads to indoles [166]. A Suzuki protocol has been employed by Sun and co-workers to synthesize a series of 6-aryloxindoles [167].

Abell utilized a Suzuki cross-coupling reaction on resin **161**. Subsequent acid treatment effected cyclization to indole **162**, which was readily cleaved with amines and alcohols to form potential libraries of amides and esters, respectively [168].

Quéguiner made great use of Suzuki methodology to prepare heterobiaryls for use in carboline synthesis [169]. This chemistry is discussed in Chapter 4. As explored in more detail in Section 3.5, Grigg developed several Pd-catalyzed tandem cyclization–anion capture processes, and these include organoboron anion transfer agents [109, 170, 171]. Two examples of this methodology are shown.

3.3.4. Stille coupling

Following the pioneering work of Stille, despite the well-documented toxicity of organotin compounds, the use of these reagents in Pd-catalyzed cross-coupling reactions continues unabated. Indolylstannanes are usually prepared either by treating the appropriate lithioindole with a trialkyltin halide or by halogen–tin exchange with, for example, hexamethylditin. Typical procedures for the generation of (1-(4-toluenesulfonyl)indol-2-yl)trimethylstannane (**163**) and (1-(4-toluenesulfonyl)indol-3-yl)trimethylstannane (**164**) are illustrated [18]. Bosch described an excellent route to the *N*-TBS-3-trimethyl-stannylindole [172].

The indolyltributylstannanes, which are more robust than their trimethylstannyl counterparts, are prepared similarly [173, 174]. Labadie and Teng synthesized the *N*-Me, *N*-Boc, and *N*-SEM (indol-2-yl)tributylstannanes [174], and Beak prepared the *N*-Boc trimethyl- and tributyltin derivatives in high yield [173]. Caddick and Joshi found that tributylstannyl radical reacts with 2-tosylindoles to give the corresponding indole tin compounds as illustrated [175].

Stannylindoles that are substituted in the benzene ring can either be prepared by halogen–metal exchange or, in the case of the C-4 and C-7 positions, by directed lithiation. For example, *N*-TIPS-indole chromium complex **29** can be treated with trimethylstannyl chloride to give the C-4 substituted stannane (69% yield) [37]. Oxidative removal of the Cr(CO)$_3$ yields the *N*-TIPS indolylstannane (90% yield). Although indolylstannanes are often prone to premature destannylation, in some cases they can be manipulated prior to Pd-catalyzed cross-coupling reactions. For example, reaction of (1-methylindol-2-yl) trimethylstannane with tosyl isocyanate gives rise to **165** in high yield and a similar reaction with ethoxycarbonyl isothiocyanate gives **166**. In contrast, phenyl isocyanate gives *ipso* substitution [176]. Other syntheses of indolyltrialkylstannanes, including the elegant methodology of Fukuyama, will be presented later in this section.

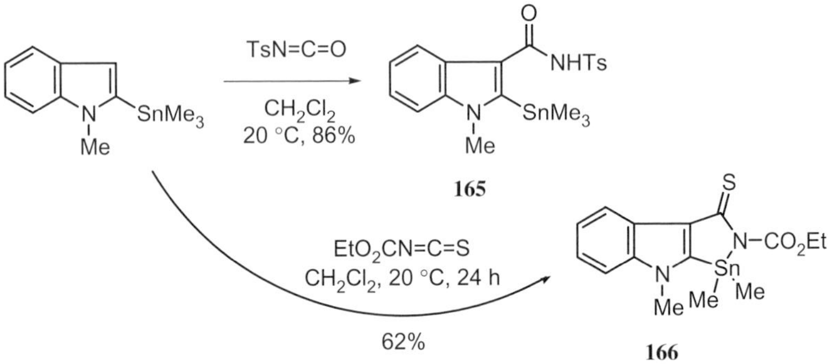

165

166

Palmisano and Santagostino first reported Stille reactions of indole ring stannylindoles with their detailed studies of *N*-SEM stannane **167** [177]. Thus **167**, which is readily prepared by C-2 lithiation of *N*-SEM indole and quenching with Bu₃SnCl (88%), couples under optimized Pd(O)-catalyzed conditions to give an array of cross-coupled products **168**. Some other examples and yields are shown. Stannane **167** could also be coupled with a (chloromethyl)cephem derivative (95%) and with a lysergic acid derivative (94%).

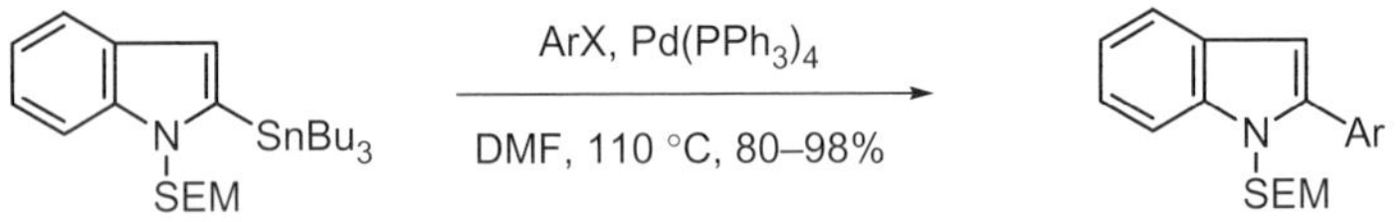

167   **168**

Ar= Ph, 2-NO₂-Ph, 2-Me-Ph, 4-OMe-Ph, 4-Ac-Ph, 2-pyridyl, 2-thienyl, Bn, allyl
X = Br, I

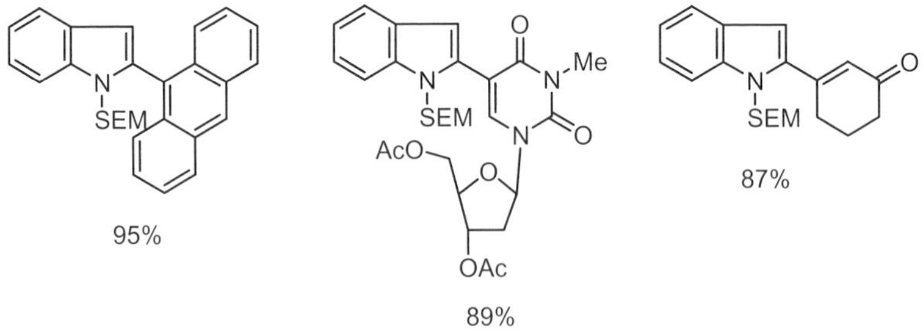

95%   89%   87%

These workers also synthesized tryptamine stannane **169** and effected Stille couplings with this compound, including the intramolecular reaction **170** to **171** [178]. Eight- and nine-membered rings could also be fashioned in this manner. Other Pd catalysts were much less successful. The *N*-tosyl derivative of **170** was similarly prepared and used in Stille chemistry.

In a study complementary to that of Palmisano and Santagostino, Labadie and Teng published a thorough and careful exploration of the Pd-catalyzed cross-coupling ability of *N*-protected 2-tributylstannylindoles (*N* = Me, Boc, SEM) [174]. Although the *N*-Me and *N*-SEM stannanes are unstable on silica gel, these can be purified by distillation. These two independent studies clearly establish the power of Stille Pd-catalyzed cross-coupling chemistry for synthesizing 2-substituted indoles. Using the Katritzky method of C-2 lithiation of indole-1-carboxylic acid, Hudkins synthesized 1-carboxy-2-tributylstannylindole and achieved cross-coupling reactions with an array of aromatic and heterocyclic halides using PdCl$_2$(Ph$_3$P)$_2$ [179]. This stannane is stable for one month at –20°C. A synthesis of arcyriacyanin A features the coupling of *N*-Boc-2-trimethylstannylindole with 4-bromo-1-(4-toluenesulfonyl)indole (**11**) [180]. Fraley and co-workers used *N*-Boc-5-TBS-2-trimethylstannylindole in a selective (for iodine) Stille coupling with 2-chloro-3-iodoquinoline to afford KDR kinase inhibitors [181], and Snieckus designed a Stille coupling to craft a synthesis of a blue–green alga indolo[2,3-*a*]carbazole [128]. Joule employed stannane **167** in a Pd-catalyzed coupling with 2-iodonitrobenzene to set the stage for a synthesis of isocryptolepine [182]. Although 1-tosyl-2-tributylstannylindole did couple with 2-bromonitrobenzene, subsequent chemistry was less successful than that involving **167**. Danieli made use of **167** to effect couplings with 5-bromo-2-pyridones and 5-bromo-2-pyranone to give the C-2 coupled products in 44–71% yields [98]. Use of 1-(4-methoxyphenylsulfonyl)-2-tributylstannylindole led only to indole dimer. Fukuyama devised a novel tin-mediated indole ring synthesis leading directly to 2-stannylindoles that can capture aryl and alkyl halides in a Pd-catalyzed cross-coupling termination reaction [183]. The presumed pathway is illustrated and involves initial tributylstannyl radical addition to the isonitrile **172**, cyclization, and final formation of stannylindole **173**.

Moreover, the *in situ* reaction of **173** under Stille conditions affords a variety of cou-pled products **174**, which have been employed in a synthesis of (–)-vindoline [183d].

R = CO$_2$Me, CH$_2$OTHP, *n*-Bu
R' = Ph, 4-Ac-Ph, Bn, 1-cyclohexenyl, 1-hexenyl
X = Br, I, OTf

The potential power of Fukuyama's method is illustrated by the synthesis of biindolyl **176** which was used in a synthesis of indolocarbazoles [183b,c]. The isonitriles (e.g. **175**) are generally prepared by dehydration of the corresponding formamides with POCl$_3$.

An application of Stille couplings to the solid phase using a traceless *N*-glycerol linker with 2-stannylindoles has been developed [184]. Only a few examples of the use of 3-stannylindoles in Stille reactions have been described. Ortar and co-workers pre-pared **177** and effected Pd-catalyzed cross-coupling reactions with several aryl, het-eroaryl, and vinyl substrates (bromides, iodides, triflates) to give the expected products **178** in high yields [185]. Enol triflates behave exceptionally well under the Ortar con-ditions, e.g. **179** to **180**.

177a, R = H
177b, R = OMe

R' = 4-MeO-Ph, 4-CO₂Me-Ph, 2-thienyl,
2-pyridyl, 2-naphthyl

178

179   180

Murakami generated 3-tributylstannylindoles *in situ* (but also isolable) using 3-bromoindole **181**, allylic acetates and carbonates, and hexamethyl tin [186]. A typical procedure is illustrated for the synthesis of **182**. The corresponding 5-bromo analog is allylated to the extent of 59%. 3-Stannylindoles couple smoothly in tandem fashion with 2,3-dibromo-5,6-dimethylbenzoquinone under Stille conditions [187].

181   182

A number of investigators employed 2- or 3-haloindoles (or triflates) in combination with organotin compounds to effect Pd-catalyzed cross-coupling. Somei used a Stille coupling of 2-bromo-3-indolecarboxaldehyde, which can be prepared from oxindole (77% yield), with several organotin reagents to synthesize the corresponding 2-substituted indoles [188]. One such reagent, (3-hydroxy-3-methyl-1-buten-1-yl)tributylstannane, was used in a synthesis of borrerine. Other substituents are methyl, phenyl, 2- and 3-pyridyl, and methyl acryloyl. Palmisano effected a similar coupling of the *N*-SEM derivative of 2-iodoindole with a stannylated pyrimidine (uridine) nucleoside [189]. As cited earlier, Fukuyama successfully coupled methyl 2-iodoindol-3-ylacetate with methyl 2-tributylstannylacrylate in total syntheses of (±)-vincadifformine and (−)-tabersonine [183b,c]. 2-Vinylazaindoles have been prepared in this fashion [19].

Indole triflates have proven to be very compatible with Pd-catalyzed methodologies, and Mérour applied triflate **36** in a Stille-type synthesis of 2-vinylindole **183** [155a].

Gribble and Conway observed that the corresponding indole-3-triflate **34** gives the 3-vinylindole in 62% yield under similar conditions using $Pd(Ph_3P)_2Cl_2$ [190]. Mérour also studied similar Stille couplings of 2-carboethoxy-1-methyl-3-indolyltriflate with vinylstannane and ethoxyvinylstannane [155b].

36   **183**

Sakamoto and co-workers employed a Stille coupling to prepare several 3-(2-ethoxyvinyl)indoles from (Z)-1-ethoxy-2-tributylstannylethene and 3-bromoindoles [191]. For example, the reaction of 3-bromo-1-(methylsulfonyl)indole with this tin reagents affords the (Z)-indole in 83% yield. Although Z-isomer is the kinetic product, the E-isomer is often obtained after longer reaction periods. These workers also coupled 2-ethoxy-trialkylstannylacetylenes (alkyl = Me, Bu) with 3-iodoindole **6** to yield 3-ethoxyethynyl-1-(phenylsulfonyl)indole, which upon acid hydrolysis provides ethyl 1-(phenylsulfonyl)-3-indolylacetate (70%) [192]. Hibino and co-workers pursued similar chemistry with 2-formyl-3-iodo-1-tosylindole and vinylstannanes during their syntheses of several carbazole alkaloids (hyellazole, carazostatin, carbazoquinocins B-F) [157]. The N-MOM indole protecting group was also employed in these studies, as was the unprotected indole. For example, 3-ethenyl-1-(methoxymethyl)indole-2-carboxaldehyde was prepared in 94% yield. In a tandem Stille–Suzuki operation, Martin was able to cross-couple 5-bromo-3-iodo-1-(4-toluenesulfonyl)indole with 3-pyridyl, 5-pyrimidyl-, and 2-pyrazinyltrimethylstannanes to give the corresponding 5-bromo-3-heteroarylindoles in 55–62% yields [135]. In a beautifully engineered total synthesis of the marine alkaloid (+)-hapalindole Q, Albizati achieved the coupling of 3-bromoindole **184** with bromide **185** to give the key intermediate indole **186** [193]. No other Pd conditions were successful.

184   **185**   **186** (+10% exo)

As a synthetic route to the grossularine natural products, Hibino and Potier independently studied the Stille coupling of ethyl 3-iodo-2-indolylcarboxylate, and N-protected analogs, with imidazolylstannanes [194, 195]. An example is illustrated for **187** to **189** [195]. During the course of their studies in this area, Hibino and co-workers discovered an interesting case of cine-substitution, which seems to be the first such example in heterostannane reactions [194c]. These workers also employed similar Stille reactions in syntheses of the antioxidant antiostatins, and carbazoquinocins [194d].

Palladium-catalyzed cross-coupling reactions involving the benzene ring positions in indoles have been the target of several investigations. Appropriately enough, Stille first studied the reactions of organotin compounds with C-4 substituted indoles. Thus, reaction of 1-tosyl-4-indolyltriflate (**190**) with stannanes **191** gives the coupled products **192** [196].

Buchwald effected a Stille reaction of 4-iodoindole **129** with vinyltributyltin to give the corresponding 4-vinylindole in 87% yield [149]. Widdowson and co-workers converted 4-iodo-1-triisopropylsilylindole (**30**) into the 4-methoxycarbonylmethylthio derivative in 98% yield by reaction with $Me_3SnSCH_2CO_2Me$ and $Pd(Ph_3P)_4$, at the beginning of a synthesis of (±)-chuangxinmycin methyl ester [197]. Somei employed thallated indole **193** in a cross-coupling reaction with pyridylstannane **194** to give **195** [198]. However, coupling of **193** with tetrabutylstannane was less successful (13% yield).

An approach to various 3,4-enynoindoles and 3,4-carbocycloindoles made use of Stille couplings with N-TIPS-4-iodogramine and vinylstannane [199], and a synthesis of

5-substituted tryptamines made use of 5-stannylindoles [152]. Doi and Mori made excellent use of dihydroindole triflate **196** in Pd-catalyzed cross-coupling reactions. This compound was discussed earlier in the Suzuki section, and it also undergoes Stille couplings as illustrated below [156]. A final dehydrogenation completes the sequence to indoles.

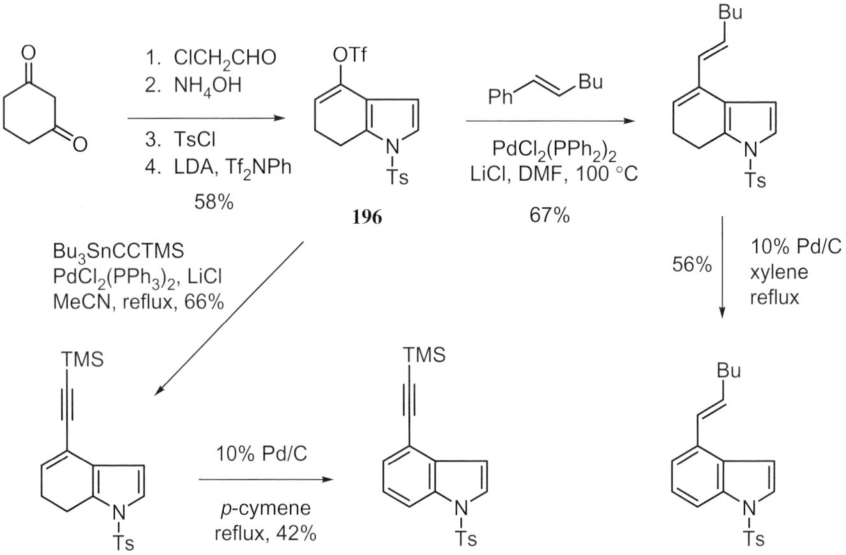

Martin prepared 5-trimethylstannylindole and effected coupling with bromobenzene to give 5-phenylindole [126a]. In a search for new cAMP phosphodiesterase inhibitors, Pearce prepared the furylindole **197** from 5-bromoindole and 5-*tert*-butoxy-2-trimethylstannylfuran [200a]. Benhida and co-workers explored Stille couplings of 6-bromo- and 6-iodoindole, and methyl 6-iodoindol-2-ylacetate with a variety of heteroarylstannanes and vinylstannanes [200b].

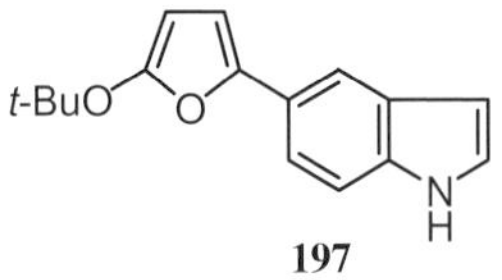

t-BuO ... **197** 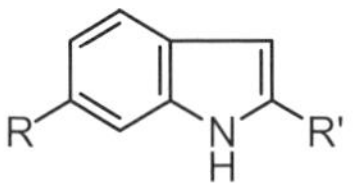 ... **198** ... R = allyl, 2-thienyl, 2-indolyl, 2-thiazoyl, 2-benzothiazoyl, (others)
R' = H, CH$_2$CO$_2$Me

Miki effected Pd-catalyzed cross-coupling between dimethyl 7-bromoindole-2,3-dicarboxylate and both tributylvinyltin and tributyl-1-ethoxyvinyltin to yield the expected 7-vinylindoles [201]. Hydrolysis of the crude reaction product from using tributyl-1-ethoxyvinyltin gave the 7-acetylindole. Sakamoto used dibromide **199**, which was prepared by acylation of 7-bromoindole, in a very concise and efficient synthesis of hippadine [36]. The overall yield from commercial materials is 39%. Somewhat earlier, Grigg employed the same strategy to craft hippadine from the diiodoindoline version of **199** using similar cyclization reaction conditions ((Me$_3$Sn)$_2$/Pd(OAc)$_2$), followed by DDQ oxidation (90%) [202].

199

The propensity for *N*-protected indoles to undergo metalation at C-7 has inspired two research groups to pursue Stille couplings using this tactic. Somei and co-workers allowed tetramethyltin to react with (1-acetylindolin-7-yl)thallium bis(trifluoroacetate) to give a low yield of 1-acetyl-7-methylindoline (32%) [203]. Similarly, 1-acetyl-7-phenylindoline was obtained in 35% yield by reaction with tetraphenyltin. Iwao and co-workers were able to lithiate and then stannylate the C-7 position of 1-*tert*-butoxycarbonylindoline. Coupling of this stannane with 6-bromopiperonal and 6-bromoveratraldehyde yielded the anticipated 7-arylindolines, which were converted into the pyrrolophenanthridone alkaloids hippadine, oxoassoanine, kalbretorine, pratosine, and anhydrolycorin-7-one [204]. Stille couplings have also been reported for carbazoles, carbolines, and other fused indoles. For example, Bracher and Hildebrand employed tributyl-1-ethoxyvinyltin in a reaction with 1-chloro-β-carboline (PdCl$_2$(Ph$_3$P)$_2$) to synthesize nitramarine and annomontine via the hydrolysis product 1-acetyl-β-carboline [205]. These workers also reported the reaction of tributylvinyltin with 1-chloro-β-carboline to give 1-vinyl-β-carboline in modest yield [160]. Bosch described tin couplings of a 1-bromonauclefine with tetraethyltin and tributylvinyltin to give the ethyl- and vinyl-substituted compounds in 87 and 95% yields, respectively [206]. In his researches towards the synthesis of grossularine analogs, Potier effected the Stille coupling of α-carboline triflate **200** with tributyl-4-methoxyphenyltin (**201**) to afford **202** [195]. Triflate **200** also couples readily with 1-(phenylsulfonyl)-3-tributylstannylindole under the same conditions (72% yield).

200　　　　**201**　　　　**202**

McCort employed Stille technology to join a SEM-protected 4-bromopyrrolo [3,4-*c*]carbazole to the 3-tributylstannyl derivatives of pyridine and quinoline [158]. Mérour effected coupling of tributylvinylstannane and tributylallylstannane with bromide **203** to afford the expected products **204** in excellent yield [161].

203　　　　**204**

Conde prepared indole vinylstannane **205** and used it to synthesize an osteoclast inhibitor derived from diene **206** [207].

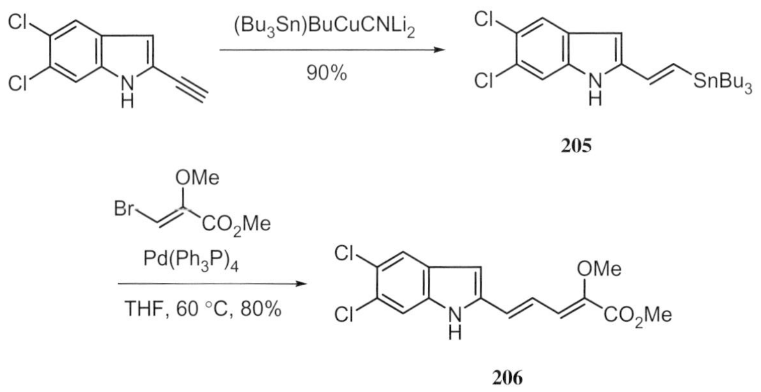

The Stille tin-coupling protocol was employed by Rice and co-workers to fashion novel naltrindole derivatives as potential new δ opioid receptor antagonists [159]. Thus, the triflate of naltrexone was allowed to react with tetramethyltin, tributylvinyltin, and tributyl-2-furyltin to give the requisite 3-desoxynaltrindole precursors. The reaction with trimethylphenyltin was less satisfactory. Several tin-mediated reactions leading directly to indole ring precursors have been described. Stille effected Pd-catalyzed couplings of *o*-bromoacetanilide (**207**) with tributylalkynyltins. The resulting alkynylanilines **208** smoothly cyclize to indoles **209** under the influence of $PdCl_2(MeCN)_2$ [208]. We will encounter more of these Pd-catalyzed indole ring formations in Section 3.5. Likewise, many ring-substituted indoles (5-Me, 5-Cl, 5-OTf, 6-OMe, 6-CO_2Me) were prepared in this study. Much less successful were the *N*-Ts and *N*-$COCF_3$ analogs of **207**.

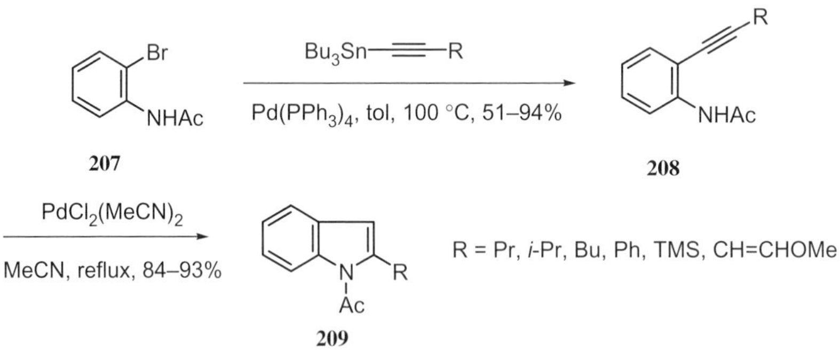

Sakamoto described similar reactions of *o*-bromoaniline derivatives with (*Z*)-tributyl-2-ethoxyvinyltin and subsequent cyclization of the coupled product with TsOH to yield, for example, *N*-acetylindole (29% yield overall) [191b]. This research group also used this methodology to synthesize a series of azaindoles, an example of which is illustrated below [209]. Halonitropyridines were particularly attractive as coupling partners with tributyl-2-ethoxyvinyltin and precursors to azaindoles. Although the (*Z*)-isomer of **210** is obtained initially, it isomerizes to the (*E*)-isomer which is the thermodynamic product.

This strategy represents a powerful method for the synthesis of all four azaindoles (1*H*-pyrrolopyridines). In fact, this method, starting with 2,6-dibromoaniline, is one of the best ways to synthesize 7-bromoindole (96% overall yield) [36].

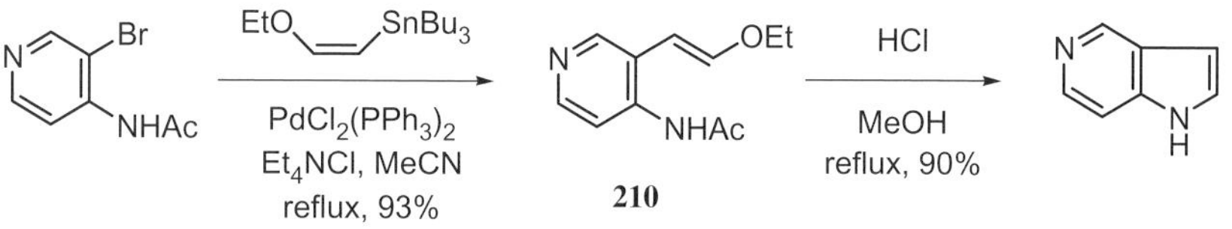

Iwao and co-workers utilized tin-chemistry to construct a series of 2-amino-biphenyls for use in carbazole synthesis via arynic cyclization [210]. For example, treating *N*-(*tert*-butoxycarbonyl)aniline with *t*-BuLi followed by stannylation with Bu_3SnCl gives the corresponding stannane. Coupling of this with 2-bromochlorobenzene yields *N*-(*tert*-butoxycarbonyl)-2-(2′-chlorophenyl)aniline (74% yield). Subsequent treatment with excess KNH_2 affords carbazole in 99% yield. This protocol was successful in synthesizing glycozolinine and glycozolidine. Similarly, Quéguiner coupled heteroaryl-stannanes with phenylpyridines to give precursors for α-substituted δ-carbolines [169c]. A representative procedure is shown below.

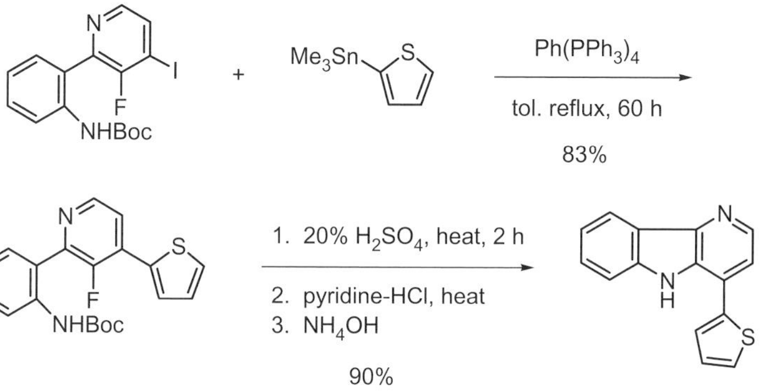

Grigg employed a tandem Pd-catalyzed cyclization to synthesize 3,3′-biindole **212** from symmetrical alkyne **211** [202]. The presumed intermediate exocyclic alkene is not detected.

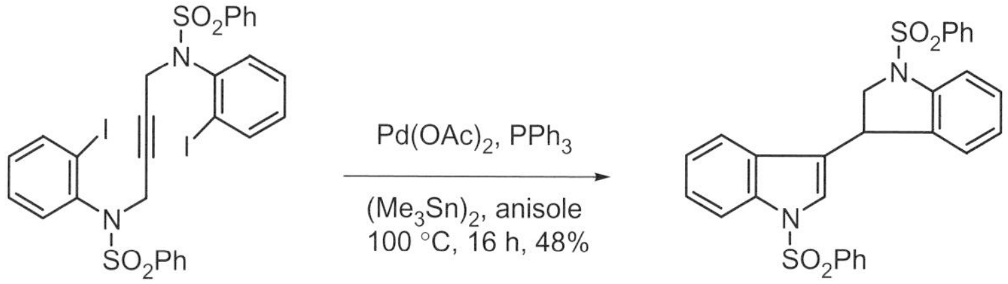

The combination of *o*-iodoaniline **213**, tributyl-2-thienylstannane (**214**), and bisalkyne **215** provides oxindole **216** in the presence of a Wilkinson's catalyst/Pd(O) system [171].

214   **215**

216

3.4. The Sonogashira coupling

The Sonogashira coupling is the Pd-catalyzed coupling of aryl halides and terminal alkynes [211], which, in the appropriate cases, can be followed by the spontaneous, or easily induced, cyclization to an indole ring. It is a sequel to the Castro acetylene coupling and subsequent cyclization to indoles in the presence of copper [212]. For example, Castro and co-workers found that copper acetylides react with *o*-iodoaniline to form 2-substituted indoles often in high yield. In the intervening years, the Pd-catalyzed cyclization of *o*-alkynylanilines to indoles has become a powerful indole ring construction. The related Larock indole ring synthesis is discussed in Section 3.5.

Yamanaka and co-workers were the first to apply the Sonogashira coupling reaction to an indole synthesis when they coupled trimethylsilylacetylene with *o*-bromonitrobenzene $(PdCl_2(Ph_3P)_2/Et_3N)$. Treatment with NaOEt/EtOH gives *o*-(2,2-diethoxyethyl)nitrobenzene (39% overall), and hydrogenation and acid treatment affords indole (87%, two steps) [213]. The method is applicable to a variety of ring-substituted indoles and, particularly, to the synthesis of 4- and 6-azaindoles (pyrrolopyridines) from halonitropyridines. Taylor coupled thallated anilides **217** with copper(I) phenylacetylide to afford the corresponding *o*-alkynylanilides **218**. In the same pot catalytic $PdCl_2$ is then used to effect cyclization to *N*-acylindoles **219** [214]. Hydrolysis to the indoles **220** was achieved by base.

217   **218**

219 R_1 = H, 6-Me, 5-Me, 6-Cl, 5,6-diMe   **220**
              R_2 = Me, Ph

Tischler and Lanza effected coupling of several substituted *o*-chloro- and *o*-bromo-nitrobenzenes with trimethylsilylacetylene to give the *o*-alkynylnitrobenzenes **221** [215]. Further manipulation affords the corresponding indoles **222** in good to excellent yield.

Yamanaka and co-workers studied Pd-catalyzed couplings of *o*-haloaniline derivatives with terminal alkynes, but cyclization to give indoles occurred upon treatment with mild base or CuI, often in one pot [213c, 216]. These workers also described the Pd-induced coupling of 1- and 2-substituted 2- and 3-iodoindoles with terminal alkynes to give the corresponding 2- and 3-alkynylindoles [217]. Similarly, Gribble and Conway coupled phenylacetylene with 3-indolyltriflate **34** to give the corresponding 3-(2-phenylethynyl)indole in 81% yield [190]. Monodendrons based on 9-phenylcarbazole were crafted by Zhu and Moore using the Sonogashira reaction. For example, **223** was prepared in high yield by these workers [218].

The combination of Pd-catalyzed coupling of terminal acetylenes with *o*-alkynylanilines or *o*-alkynylnitrobenzenes followed by base or CuI cyclization to an indole has been used in many situations with great success. Arcadi employed this methodology to prepare a series of 2-vinyl-, 2-aryl-, and 2-heteroarylindoles from 2-aminophenylacetylene and a subsequent elaboration of the acetylenic terminus. A final Pd-catalyzed cyclization completes the scheme [219].

Other indoles that have been prepared using the Sonogashira coupling and cyclization sequence include 5,7-difluoroindole and 5,6,7-trifluoroindole [220], 4-, 5-, and 7-methoxyindoles and 5-, 6-, and 7-(triisopropylsilyl)oxyindoles [221], the 5,6-dichloroindole SB 242784, a compound in development for the treatment of osteoporosis [207, 222], 5-azaindoles [223], 7-azaindoles [166], 2,2′-biindolyls [224, 183b,c], 2-octylindole for use in a synthesis of carazostatin [225], chiral indole precursors for syntheses of carbazoquinocins A and D [226], a series of 5,7-disubstituted indoles [227], a pyrrolo[2,3-*e*]indole [227], an indolo[7,6-*g*]indole [228], pyrrolo[3,2,1-*ij*]quinolines from 4-arylamino-8-iodoquinolines [229], optically active indol-2-ylarylcarbinols [230], 2-alkynylindoles [183b,c], 7-substituted indoles via the lithiation of the intermediate 2-alkynylaniline derivative [231], a variety of 2,5,6-trisubstituted indoles [145, 232] and 2,3,5-trisubstituted indoles [140], and indole-2-carboxylates for use in a synthesis of duocarmycin SA [233]. One study employs tetrabutylammonium fluoride, instead of CuI or alkoxide, to effect the final cyclization of **224** to indoles **225** [232].

A new, water-soluble palladium catalyst was used in the Sonogashira reaction (Pd(OAc)$_2$ triphenylphosphine-trisulfonate sodium salt) [234], and several groups adapted the Sonogashira coupling and subsequent cyclization to the solid-phase synthesis of indoles. Bedeschi and co-workers used this method to prepare a series of 2-substituted-5-indolecarboxylic acids [235]. Collini and Ellingboe extended the technique to 1,2,3-trisubstituted-6-indolecarboxylic acids [236]. Zhang and co-workers used the solid phase to prepare a series of 2-substituted-3-aminomethyl-5-indolecarboxamides, and, by manipulation of the resin-bound Mannich reaction intermediates, to synthesize 3-cyanomethyl-5-indole-carboxamide and other products of nucleophilic substitution [237]. This research team also employed a sulfonyl linker, as summarized below, to provide a series of substituted indoles [238, 141]. The advantages of this particular approach are that the sulfonyl linker is "traceless," since it disappears from the final indole product, and the polystyrene sulfonyl chloride resin is commercially available.

R$_1$ = H, 5-CO$_2$Me, 6-F, 6-OMe
R$_2$ = Ph, 4-Me-Ph, 4-F-Ph, 4-MeO-Ph, CH$_2$OMe, Bu, 2-pyridyl, others

The Sonogashira coupling has been used to prepare both *o*-alkynylformamides, for the synthesis of the corresponding isonitriles for the Fukuyama indole ring formation [184, 239], and several 4-alkynylindoles from the corresponding 4-iodoindoles [240]. Similarly, ethyl 3-triflyloxy-1-methylindole-2-carboxylate reacts with propargylic alcohol to form the corresponding indolyl-3-propynyl alcohol in 80% yield [155b]. The Sonogashira coupling was also used to prepare arylalkyne substrates for subsequent cyclization under Heck–Hegedus–Mori or Larock conditions, and these are discussed in the appropriate sections later in this chapter.

While the power and beauty of the Sonogashira cross-coupling reaction lies mainly in indole ring synthesis (vide supra), recent work has revealed that a wide range of acetylenic indoles is available using this methodology. For example, Rossi [241] and Larock [242] have shown that 2- and 3-haloindoles are readily coupled with terminal acetylenes to give acetylenic indoles that can be converted to carbolines by subsequent amination and cyclization. A Sonogashira reaction on a 2-iodoindole is a key step in Fukuyama's synthesis of (−)-aspidophytine [243], and 5-bromo-3-iodoindoles undergo regioselective Pd-catalyzed cross-coupling with terminal acetylenes at C-3 [143]. Higher temperatures effect a Sonogashira reaction at the C-5 bromine. A series of *N*-propargylindoles and indolines was employed in Sonogashira reactions to construct annulated carbazoles [244]. Illustrative of the versatility of this chemistry are the syntheses of the indole–pyrrole **226** [245] and oligoindoles **227** [246].

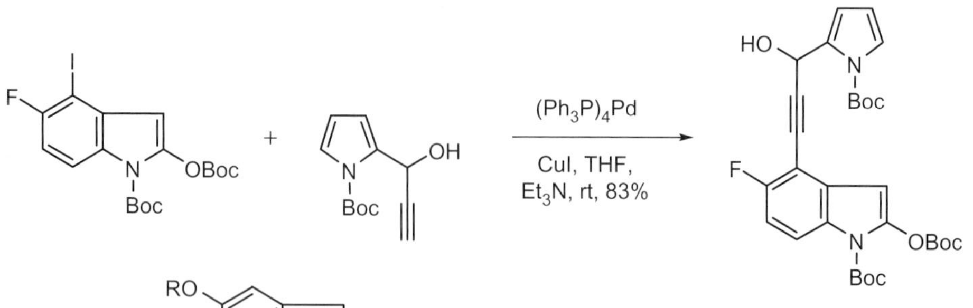

226

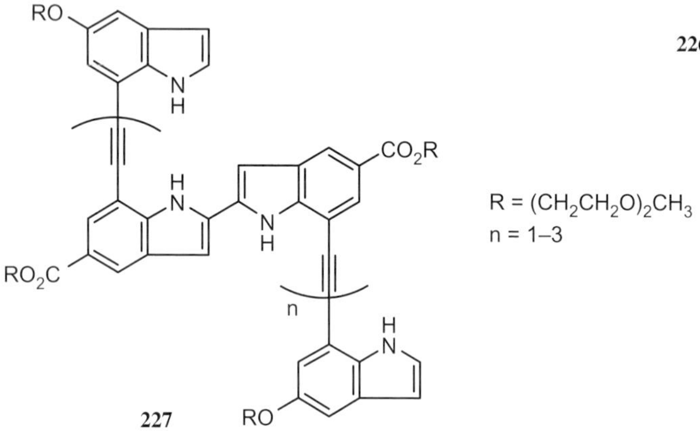

$R = (CH_2CH_2O)_2CH_3$
$n = 1–3$

227

3.5. Heck couplings

The incredibly powerful and versatile Heck coupling reaction has found enormous utility in the indole ring synthesis and in the elaboration of this important heterocycle. Due to the enormity of this topic, the section is divided into Heck reactions of indoles, the synthesis of the indole ring as developed by Hegedus, Mori–Ban, and Heck, and the Larock indole ring synthesis.

3.5.1. Heck coupling reaction

Both inter- and intramolecular Heck reactions of indoles have been pursued and these will be considered in turn. Appropriately, Heck and co-workers were the first to use Pd-catalyzed vinyl substitution reactions with haloindoles [247]. Thus, 1-acetyl-3-bromoindole (**228**) gave a 50% yield of 3-indolylacrylate **229**. A similar reaction with 5-bromoindole yielded (*E*)-methyl 3-(5-indolyl)acrylate (53% yield), but 3-bromoindole gave no identifiable product.

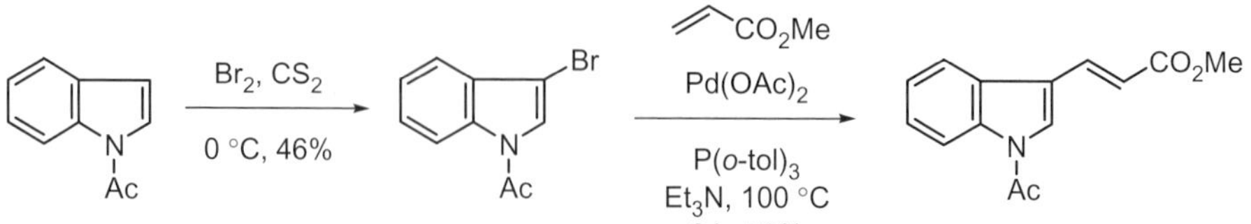

Somei and co-workers made extensive use of the Heck reaction with haloindoles in their synthetic approaches to ergot and other alkaloids [26, 40, 41, 248]. Thus, 4-bromo-1-carbomethoxyindole (69%) [26], 7-iodoindole (91%) (but not 7-iodoindoline or 1-acetyl-7-iodoindoline) [40, 41], and 1-acetyl-5-iodoindoline (96%) [41] undergo coupling with methyl acrylate under standard conditions (Pd(OAc)$_2$/Ph$_3$P/Et$_3$N/DMF/100°C) to give the corresponding (E)-indolylacrylates in the yields indicated. The Heck coupling of methyl acrylate with thallated indoles and indolines is productive in some cases [41, 248b,g]. For example, reaction of (3-formylindol-4-yl)thallium bis-trifluoroacetate (193) affords acrylate 230 in excellent yield [248b]. Similarly, this one-pot thallation–palladation operation from 3-formylindole and methyl vinyl ketone was used to synthesize 4-(3-formylindol-4-yl)-3-buten-2-one (86% yield).

However, the most important of Somei's contributions in this area are Heck reactions of haloindoles with allylic alcohols [26, 248a,c–f,h,i]. For example, reaction of 4-iodo-3-indolecarboxaldehyde with 2-methyl-3-buten-2-ol afforded alcohol 231 in high yield. This could be subsequently transformed to (±)-6,7-secoagroclavine (232) [248c]. Interestingly, the one-pot thallation–palladation protocol failed in this case.

Somei adapted this chemistry to syntheses of (±)-norchanoclavine-I, (±)-chanoclavine-I, (±)-isochanoclavine-I, (±)-agroclavine, and related indoles [248d–f,i]. Extension of this Heck reaction to 7-iodoindoline and 2-methyl-3-buten-2-ol led to a synthesis of the alkaloid annonidine A [248h]. In contrast to the uneventful Heck chemistry of allylic alcohols with 4-haloindoles, reaction of thallated indole 193 with 2-methyl-4-trimethylsilyl-3-butyn-2-ol affords an unusual 1-oxa-2-sila-3-cyclopentene indole product [248j]. Hegedus was also an early pioneer in exploring Heck reactions of haloindoles [249]. Thus, reaction of 4-bromo-1-(4-toluenesulfonyl)indole (11) under Heck conditions affords 4-substituted indoles 233 [249a]. Murakami described the same reaction with ethyl acrylate [83], and 2-iodo-5-(and 7-) azaindoles undergo a Heck reaction with methyl acrylate [19].

Hegedus also found that mercuration of **11** followed by a Heck reaction with methyl acrylate gives the corresponding 3-indolylacrylate in 70% yield [249a]. More relevant to Hegedus's goal of ergot alkaloid synthesis was the observation that the appropriate dihaloindole reacted regioselectively at C-3 to give exclusively the (Z)-isomer **234** [249a,b].

This reaction was used and extended by Murakami to the preparation of ring-substituted derivatives of **234** [80, 81], to similar Heck reactions leading to the C-4 (Z)-products [80], and to syntheses of clavicipitic acid and costaclavine [79]. Semmelhack improved upon this Heck reaction with *N*-protected dehydroalanine esters by adding chloride to the reaction mixture in his synthesis of the 4-fluoro analog of **234** (77 *vs.* 40% yield) [250]. Using his methodology and a Suzuki coupling, Hegedus and co-workers synthesized the ergot alkaloid (±)-aurantioclavine [249c] and the *N*-acetyl methyl ester of (±)-clavicipitic acid [249b]. The key intramolecular Heck cyclization utilized in the latter synthesis is presented later. Also worthy of mention is the Heck reaction of 5-bromo-1-tosylindole with an *N*-vinyloxazolone leading to 5-(2-aminoethyl)indoline [251], an approach to lysergic acid involving vinylation of 4-bromo-1-tosyl-3-indolecarboxaldehyde [252], a synthesis of (−)-6-*n*-octylindolactam-V featuring a Heck reaction on the corresponding 6-bromoindolactam [253], and reaction of a 3-iodo-1-methylindole (but not the NH analog) with methyl acrylate as part of Sundberg's approach to iboga alkaloids [254]. The intramolecular version of this strategy is presented later. Blechert and co-workers employed a Heck reaction between 1-benzyl-7-bromo-4-ethylindole and methyl crotonate in the early stages of a synthesis of (±)-*cis*-trikentrin [255].

Triflates also undergo Heck reactions, and Gribble and Conway reported several such couplings of 1-(phenylsulfonyl)indol-3-yl triflate (**34**) to afford 3-vinylindoles **235** [190]. Cyclohexene, allyl bromide, and methyl propiolate failed to react under these conditions, but triphenylphosphine afforded **236** in excellent yield (93%), and divinyl carbinol yielded the rearranged enal **237** (82% yield).

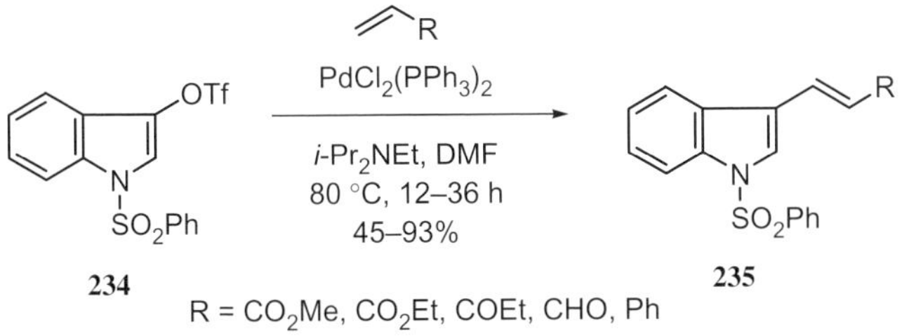
OTf
R
PdCl2(PPh3)2
i-Pr2NEt, DMF
80 °C, 12–36 h
45–93%
N
SO2Ph
R = CO2Me, CO2Et, COEt, CHO, Ph

234 235

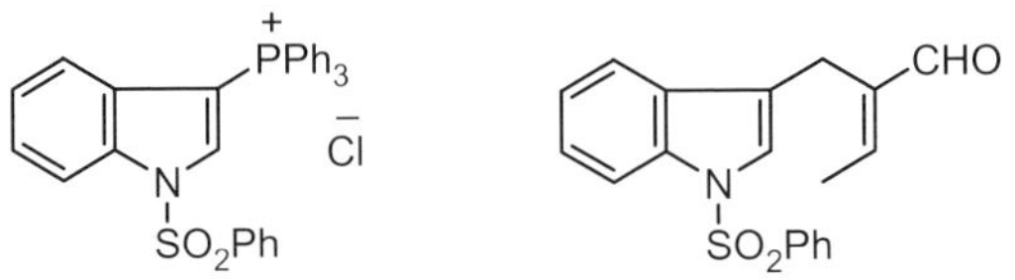
PPh3
Cl
CHO
N
SO2Ph

236 237

Mérour also explored the Heck reactions of indolyl triflates with allylic alcohols [155b, 256]. For example, reaction of triflate **238** with allyl alcohol gives the rearranged allylic alcohol **239** [155b].

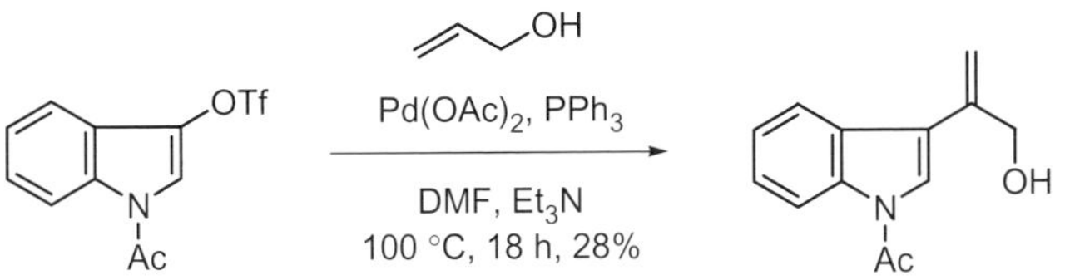
OH
OTf
Pd(OAc)2, PPh3
DMF, Et3N
100 °C, 18 h, 28%
N
Ac
OH

238 239

The vinyl triflate of Kornfeld's ketone has been subjected to Heck reactions with methyl acrylate, methyl methacrylate, and methyl 3-(*N*-*tert*-butoxycarbonyl-*N*-methyl)amino-2-methylenepropionate leading to a formal synthesis of lysergic acid [257]. A similar Heck reaction between 1-(phenylsulfonyl)indol-5-yl triflate and dehydroalanine methyl ester was described by this research group [258]. Chloropyrazines undergo Heck couplings with both indole and 1-tosylindole, and these reactions are discussed in Chapter 10 (pyrazines) [259]. Rajeswaran and Srinivasan described an interesting arylation of bromomethyl indole **240** with arenes [260]. Subsequent desulfurization and hydrolysis furnishes 2-arylmethylindoles **241**. Bis-indole **242** was also prepared in this study. Aryl iodides couple with indole under basic conditions to afford 2-arylindoles [261].

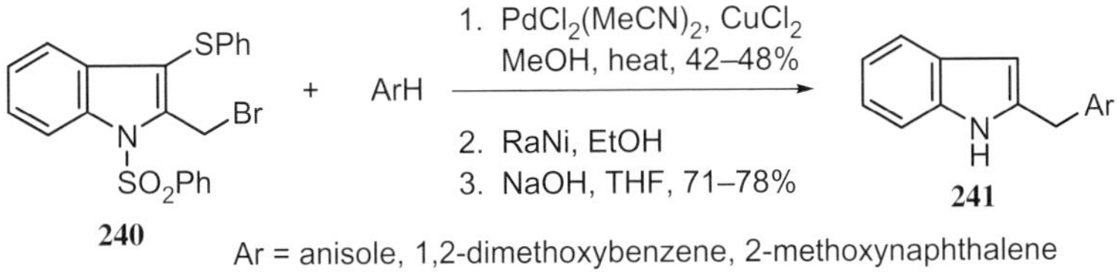
SPh
Br
N
SO2Ph
+ ArH
1. PdCl2(MeCN)2, CuCl2
MeOH, heat, 42–48%
2. RaNi, EtOH
3. NaOH, THF, 71–78%
Ar
N
H
Ar = anisole, 1,2-dimethoxybenzene, 2-methoxynaphthalene

240 241

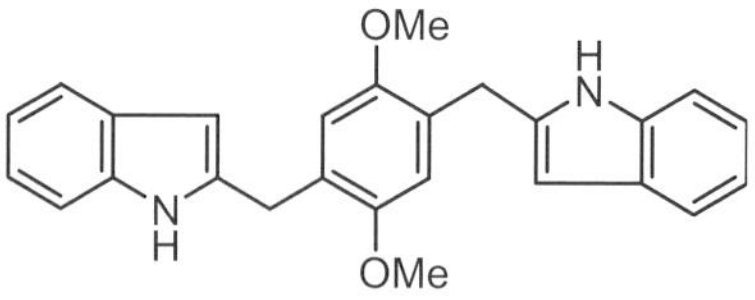

242

The Heck-mercuration modification, which was presented earlier, has been adapted to the solid phase by Zhang and co-workers, as illustrated below [238].

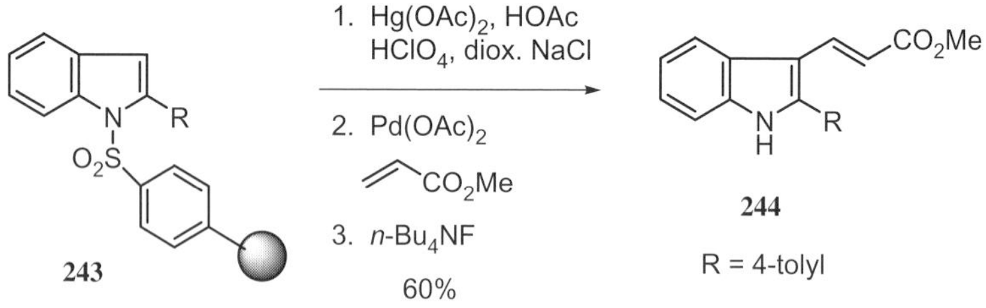
1. $Hg(OAc)_2$, HOAc
$HClO_4$, diox. NaCl
2. $Pd(OAc)_2$
CO_2Me
3. n-Bu_4NF
60%
CO_2Me
R = 4-tolyl

243 **244**

The intramolecular Heck reaction as applied to indoles has led to several spectacular synthetic achievements. As alluded to in the previous section, both Hegedus and Murakami exploited intramolecular Heck reactions to synthesize ergot alkaloids. In model studies, Hegedus noted that 3-allyl-4-bromo-1-tosylindole (**245**) cyclizes to **246** in good yield [249a], and Murakami's group observed that, for example, **247** cyclizes to **248** [262]. Roberts effected similar cyclizations leading to 7- and 8-membered ring tryptophan surrogates [263], and Snieckus used similar intramolecular Heck reactions to prepare *seco*-C/D ring analogs of ergot alkaloids [264].

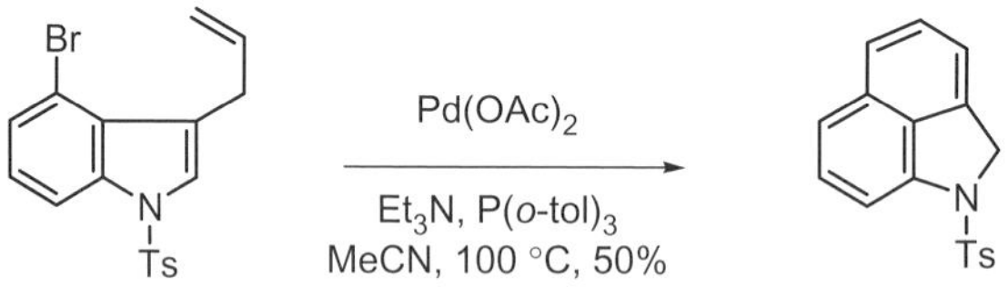
Br
Ts
$Pd(OAc)_2$
Et_3N, $P(o\text{-}tol)_3$
MeCN, 100 °C, 50%
Ts

245 **246**

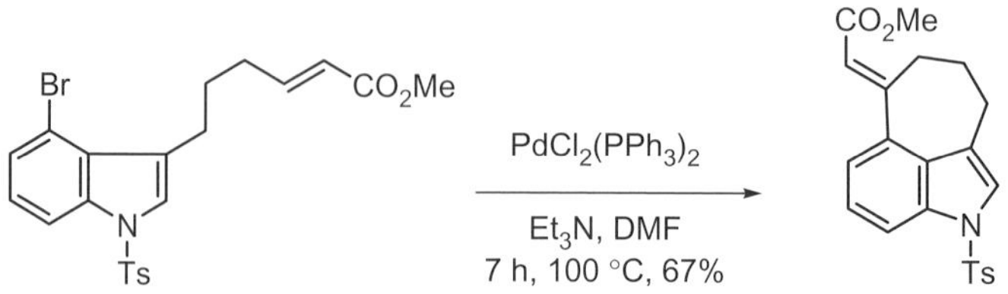
Br
CO_2Me
Ts
$PdCl_2(PPh_3)_2$
Et_3N, DMF
7 h, 100 °C, 67%
CO_2Me
Ts

247 **248**

Hegedus' synthesis of (±)-clavicipitic acid *N*-acetyl methyl ester culminated in the Pd-induced cyclization of **249** to **250**, the latter of which was reduced to the target mixture [249b]. Substrate **249** was prepared via a Heck reaction with the corresponding 4-bromo compound **234** and 2-methyl-3-buten-2-ol (83%). The cyclization also occurs with tosic acid (97%).

249   PdCl$_2$(MeCN)$_2$ MeCN, reflux, 2 h, 95%   **250**

In a related approach, Murakami synthesized clavicipitic acid and costaclavine [79], and later extended this chemistry to a synthesis of chanoclavine-I featuring the intramolecular Heck vinylation **251** to **252** [265]. The corresponding enone failed to cyclize under these conditions. Noteworthy is that radical cyclizations, which often compete successfully with Heck reactions, were poor in this system. Martin developed a novel route to the ergot skeleton that features a Heck cyclization of bromo alkyne **253** [266].

251   PdCl$_2$(bppp), Ag$_3$PO$_4$ CaCO$_3$, DMF, 100 °C   5 h, 77%   **252**   **253**

Several investigators have studied intramolecular Heck reactions on other alkene-tethered haloindoles. Black prepared several 1-allyl-7-bromoindoles and found that they undergo cyclization in the presence of palladium as shown for **254** to **255** [267]. Although this new synthesis of pyrroloquinolines is reasonably general, some of the products are unstable. Substrate **254** was prepared by bromination of 4,6-dimethoxy-2,3-diphenylindole (92%) and *N*-alkylation.

254   Pd(OAc)$_2$, P(*o*-tol)$_3$ Et$_3$N, MeCN, 100 °C, 15 h, 96%   **255**

Gilchrist examined the cyclization of *N*-alkenyl-2-iodoindoles with palladium [268, 269]. For example, reaction of *N*-pentenylindole **256** under Heck conditions affords a mixture of **257** and **258** in very good yield. In the absence of TlOAc, **258** is the major product. Further exposure of **257** to Pd(OAc)$_2$ gives **258**. Reaction of 1-(4-butenyl)-2-iodoindole under similar conditions affords the pyrrolo[1,2-*a*]indole ring system in modest yield (35%).

In his synthetic approaches to iboga alkaloids, Sundberg pursued several Heck cycliza- tion strategies but found the best one to be **259** to **260** [254].

Merlic discovered the novel benzannulation of biindole **261** to **262** during studies to synthesize indolocarbazoles [122b]. Several unsymmetrical biindoles were also prepared and their reactions with dimethyl acetylenedicarboxylate and related alkynes were stud- ied. Yields of indolocarbazoles were 51–88% and some regioselectivity was observed in unsymmetrical cases (up to 80:20).

A particularly elegant domino Heck reaction involving 4-bromoindole and bromo(indolyl)maleimide **263** to give *N*-methylarcyriacyanin A (**264**) in one operation was reported by Steglich [180]. This alkaloid could also be prepared from triflate **265** in higher yield in a single Heck reaction.

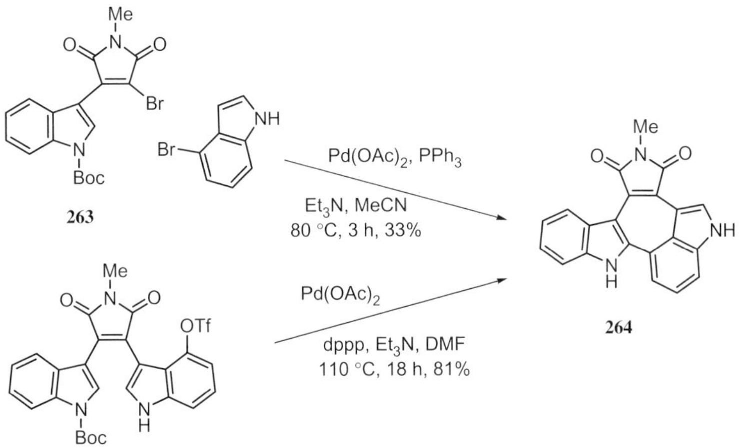

263

265

264

Several intramolecular Heck reactions involve aryl halides cyclizing onto indole rings. Grigg first described the simple Heck cyclizations of **266** and **267** [270], and this was followed by similar Heck reactions reported by Kozikowski and Ma on the bromide corresponding to **266** and the *N*-benzylindole **268** [271]. These investigators also observed cyclization to the C-3 position in a Heck reaction of indole **269**, and they prepared a series of peripheral-type benzodiazepine receptors **270** using this chemistry. For example, **270** (n = 3, R = *n*-Pr) is obtained in 81% yield.

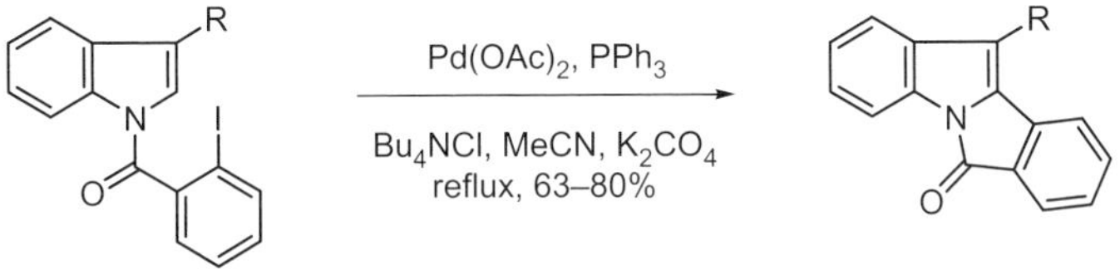

266, R = H
267, R = Me

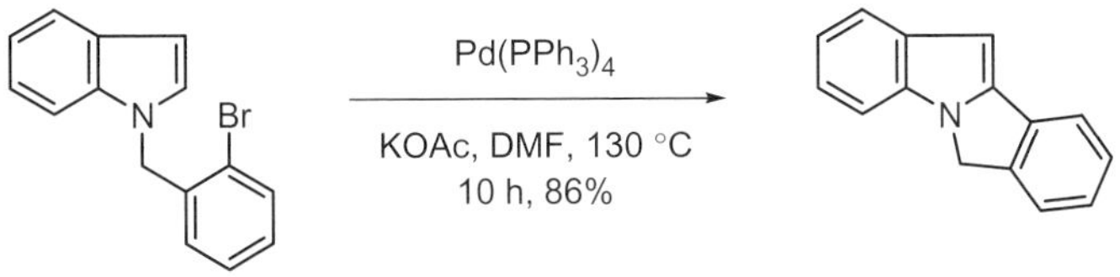

268

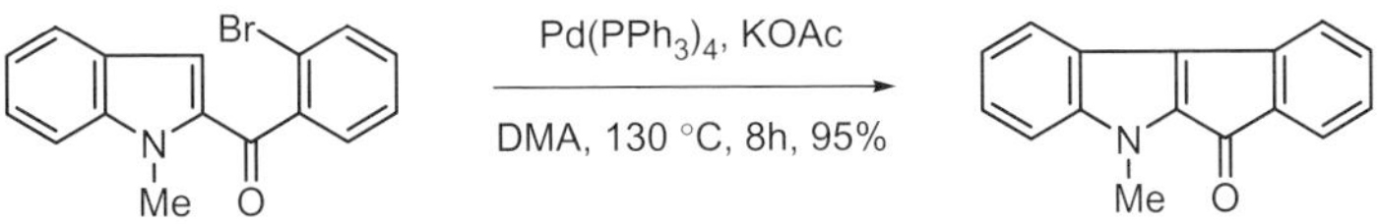

269

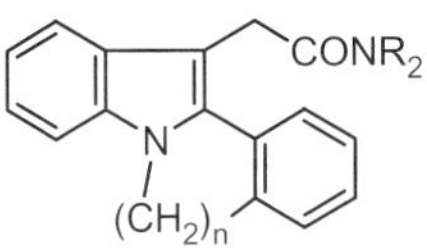

270

n = 1, 3
R = *n*-Pr, *n*-hexyl

Kraus found that a Pd-catalyzed cyclization is superior to those involving tin-initiated radical cyclizations in the construction of pyrrolo[1,2-*a*]indoles such as **272** [272]. The bromide corresponding to **271** cyclizes in 48% yield, and *N*-(2-bromo-1-cyclohexenecarbonyl)indole-3-carboxaldehyde cyclizes in 60% yield. In contrast, the corresponding radical reactions afford these products in 35–53% yields. Substrate **273** failed to cyclize under these Heck conditions, as did **274** as reported by Srinivasan [273]. However, radical cyclization of **274** did afford the desired 3,4-benzocarbolines.

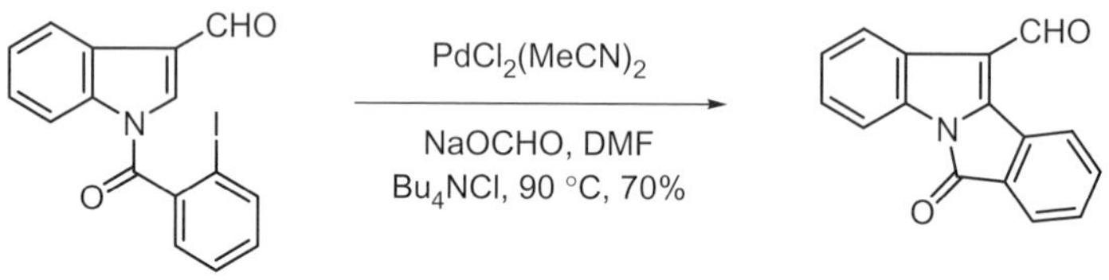

271   **272**

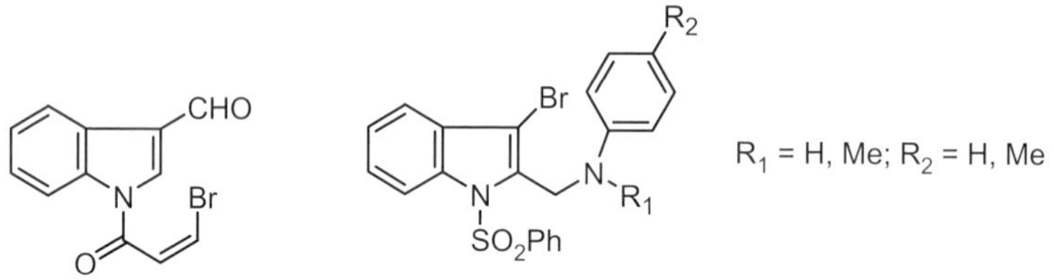

R_1 = H, Me; R_2 = H, Me

273   **274**

Mérour synthesized a series of indolo[2,1-*a*]isoquinolines and pyrrolophenanthridines via Heck cyclizations onto the C-2 and C-7 positions of indole, respectively [274]. Two examples are shown.

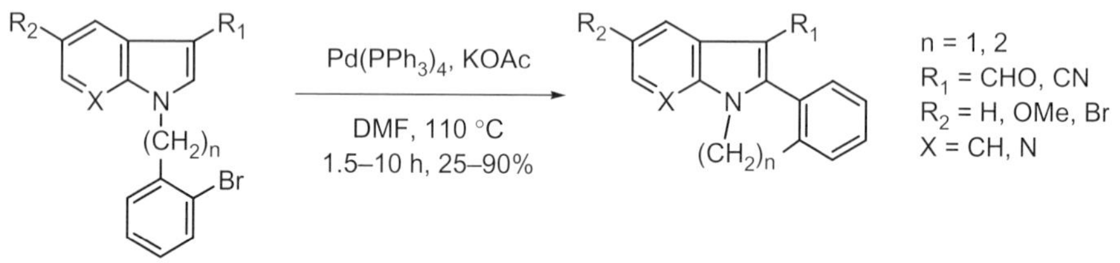

n = 1, 2
R_1 = CHO, CN
R_2 = H, OMe, Br
X = CH, N

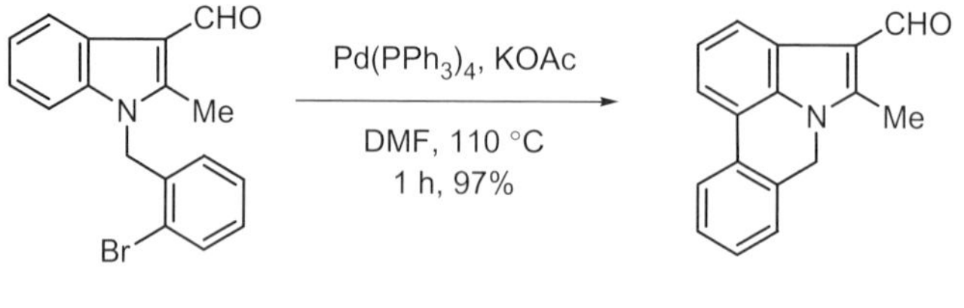

275

This indole C-7 Heck cyclization strategy was employed by Shao and Cai in a synthesis of anhydrolycorine-7-one from the requisite *N*-aroylindoline [275], by Miki in syntheses of pratosine and hippadine from substrates like **275** [276], by Rigby to synthesize anhydrodehydrolycorine from an *N*-benzylhydroindolone [277], and by Harayama to synthesize several of the pyrrolophenanthridine alkaloids [278]. Thal and co-workers constructed examples of the new ring systems, pyrido[2′,3′-*d*′]pyridazino[2,3-*a*]indole (**277**) and pyrido[2′,3′-*d*′]diazepino[1,6,7-*h*,*i*]indole (**278**), by effecting Heck cyclizations on the appropriate 2-bromopyridine precursors (e.g. **276**) at C-2 or C-7, respectively [279]. Compound **277** undergoes oxidative-addition with methyl acrylate at the C-3 position. This resulting product (not shown) can also be obtained from **276** in a tandem Heck sequence with methyl acrylate (62% yield).

Harayama and co-workers have synthesized rutaecarpine (**280**) via Heck cyclization of 2-bromoindole **279** [280].

Several other Heck cyclizations that do not involve the indole ring directly have been developed. Kelly employed the Heck cyclization of **281** to synthesize (and revise the structure of) maxonine (**282**) [281]. The product resulting from attack at C-8 was also obtained.

Gilchrist reported the conversion **283** to **284** [268, 269], and Grigg described syntheses of **285** and **286** using Heck conditions [270, 282].

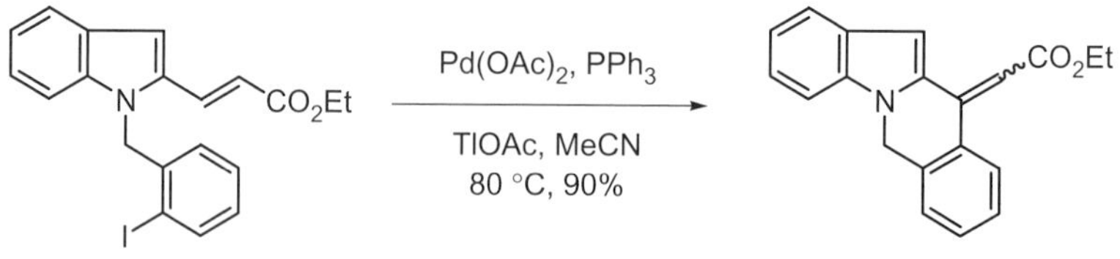
CO2Et
N
I
Pd(OAc)2, PPh3
TlOAc, MeCN
80 °C, 90%
CO2Et
N

283 284

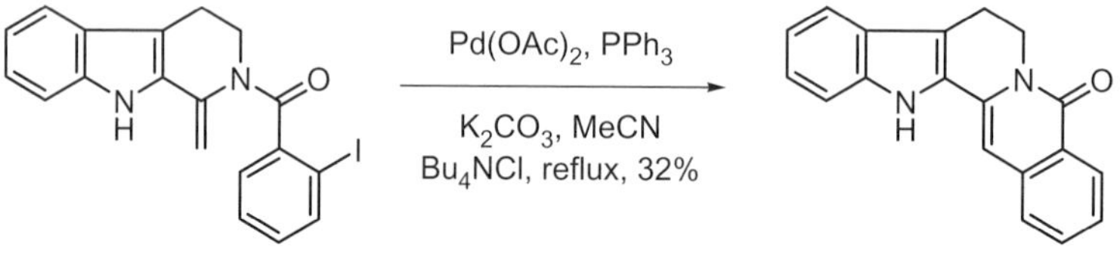
N
H
N
O
I
Pd(OAc)2, PPh3
K2CO3, MeCN
Bu4NCl, reflux, 32%
N
H
N
O

285

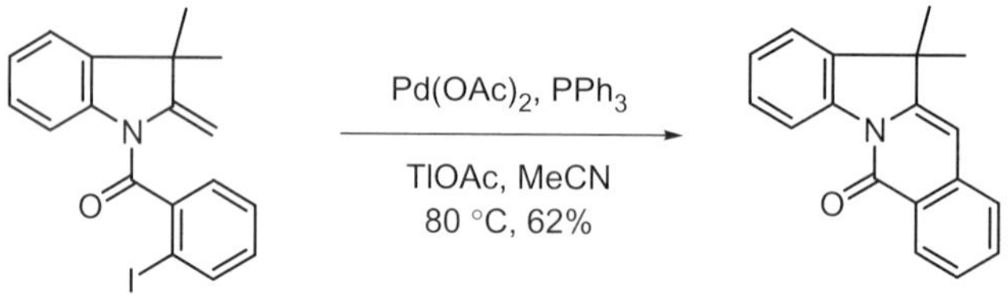
N
O
I
Pd(OAc)2, PPh3
TlOAc, MeCN
80 °C, 62%
N
O

286

Mérour and co-workers achieved in excellent yield the cyclization of **287** to benzo[4,5]cyclohepta[*b*]indole **288** [283].

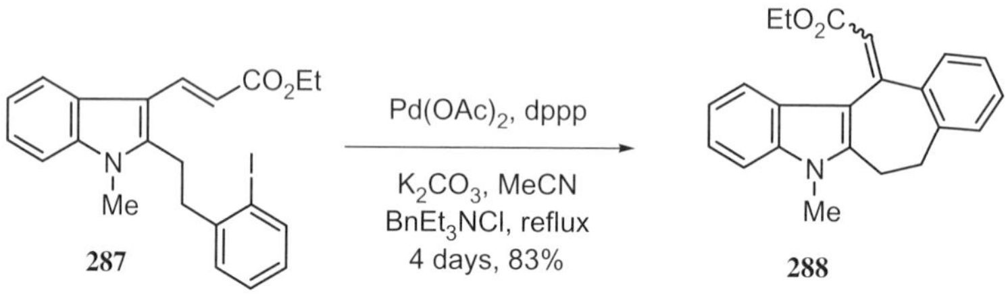
CO2Et
N
Me
I
Pd(OAc)2, dppp
K2CO3, MeCN
BnEt3NCl, reflux
4 days, 83%
EtO2C
N
Me

287 288

The naphth[3,2,1-*cd*]indole ring system, e.g. **290**, was constructed by Miki in very good yield via an intramolecular Heck reaction of triflate **289** [284]. The corresponding bromide furnished **290** in 74% yield. Interestingly, attempted radical cyclization of the bromide with Bu_3SnH failed.

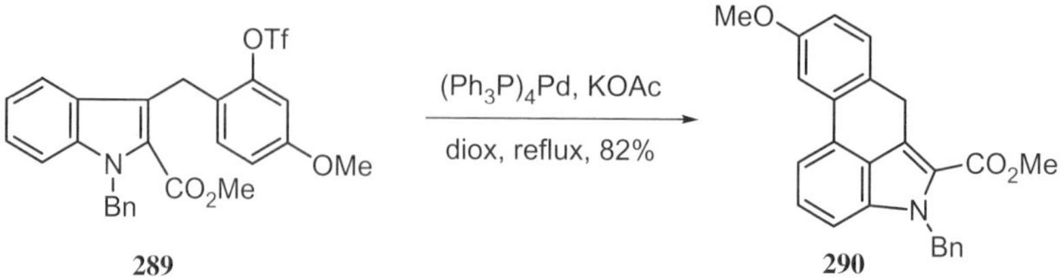
OTf
N
Bn
CO2Me
OMe
(Ph3P)4Pd, KOAc
diox, reflux, 82%
MeO
CO2Me
N
Bn

289 290

In a study of magallanesine analogs, Kurihara and co-workers effected the synthesis of **291** in low yield via Heck methodology [285].

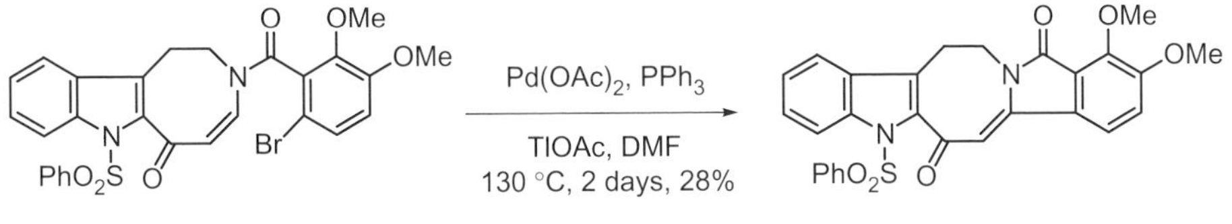

291

Larock has described a new carbazole synthesis that features palladation of 3-iodoindole **292**, insertion into an alkyne, and Heck cyclization to afford carbazole **293** after double bond isomerization [286]. Yields are higher with electron-deficient alkynes such as ethyl phenylpropiolate (91%; 60:40 regioisomers).

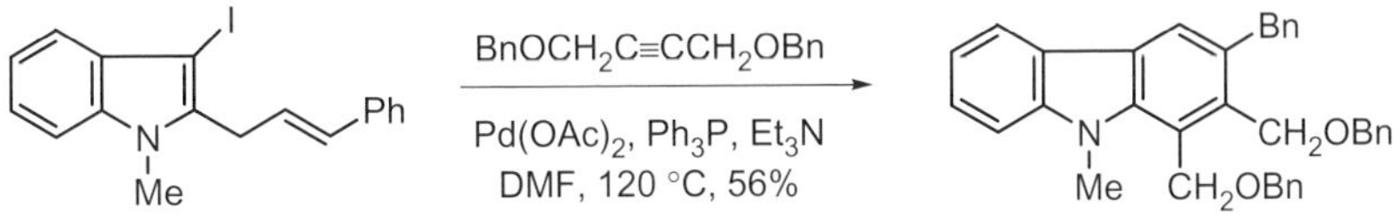

292   **293**

Rawal applied the Heck cyclization in elegant fashion to the construction of indole alkaloids. His route to geissoschizine alkaloids features a novel ring D formation, **294** to **295**–**296** [287]. Whereas classical Heck conditions favor the isogeissoschizal (**296**) product, the "ligand-free" modification of Jeffrey favors the geissoschizal (**295**) stereochemistry.

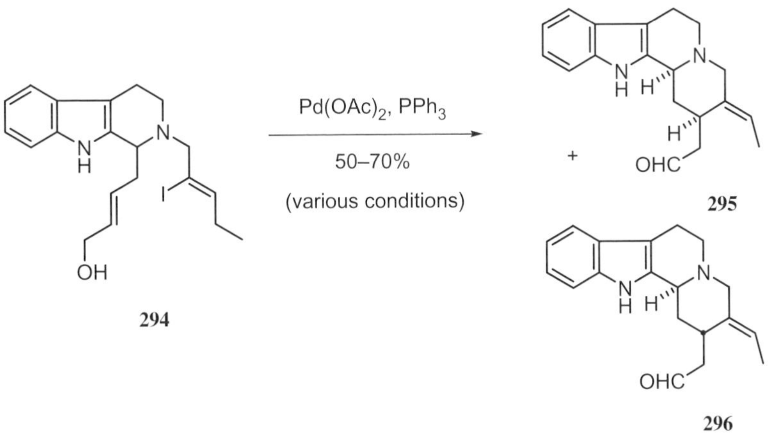

294

295

296

Following the application of a Heck cyclization to a concise synthesis of the *Strychnos* alkaloid dehydrotubifoline [288], and earlier model studies [289], Rawal employed a similar strategy to achieve a remarkably efficient synthesis of strychnine [290]. Thus, pentacycle

297 is smoothly cyclized and deprotected to isostrychnine (**298**) in 71% overall yield—an appropriate finale to this section!

3.5.2. The Mori–Ban indole synthesis

The application of Heck cyclizations to the synthesis of indoles, indolines, and oxindoles was discovered independently by the groups of Mori–Ban [291] and Heck [292]. These investigators found that Pd can effect the cyclization of *o*-halo-*N*-allylanilines to indoles under Heck conditions [293]. The cyclization of *o*-halo-*N*-allylanilines to indoles is a general and efficient methodology, especially with the Larock improvements in which he cyclized *o*-halo-*N*-allylanilines and *o*-halo-*N*-acryloylanilides into indoles and oxindoles [294]. For example, the conversion of **299** to **300** can be performed at lower temperature, shorter reaction time, and with less catalyst to give 3-methylindole (**300**) in 97% yield. Larock's improved conditions, which have been widely adopted, are catalytic (2%) Pd(OAc)$_2$, *n*-Bu$_4$NCl, DMF, base (usually Na$_2$CO$_3$), 25°C, 24 h. Larock extended his work in several ways [295, 296], particularly with regard to Pd-catalyzed cross-coupling of *o*-allylic and *o*-vinylic anilides with vinyl halides and triflates to produce 2-vinylindoles [296]. The related "Larock indole synthesis" is discussed separately in the next section.

Genet showed that a water-soluble palladium catalyst system (Pd(OAc)$_2$-trisodium-phosphinetriyltribenzene-sulfonate) converts *o*-iodo-*N*-allylaniline to 3-methylindole in 97% yield [297]. The novel catalysts prepared from Pd(OAc)$_2$ and a fluorinated phosphine (e.g. (C$_6$F$_{13}$CH$_2$CH$_2$)$_2$PPh) in supercritical CO$_2$ also accomplish this cyclization to give 3-methylindole [298]. Hegedus extended his original work on the cyclization of *N*-allyl-*o*-bromoanilines to the synthesis of indoloquinones [299]. In a program to synthesize CC-1065 analogs, Sundberg prepared indole **301** in excellent yield [300]. Silver carbonate and sodium carbonate were less effective than triethylamine.

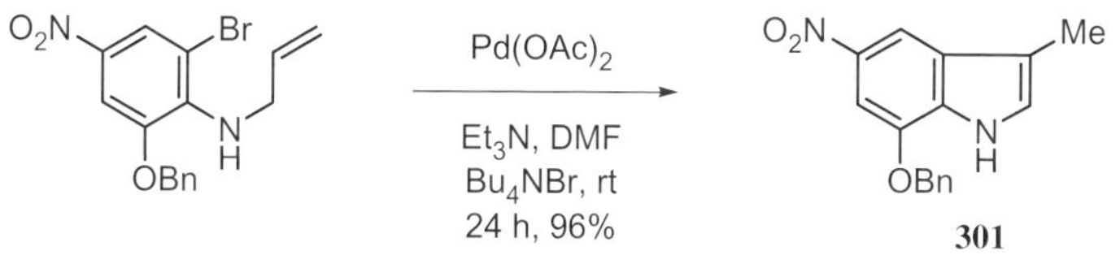

301

Likewise, Sakamoto has synthesized the CC-1065/duocarmycin pharmacophore **304** via the cyclization of **302** [301]. Silver carbonate prevented unwanted isomerization of the exocyclic double bond in **303**. Tietze and co-workers took full advantage of the power of these *N*-allyl-*o*-haloaniline Pd-catalyzed cyclizations in developing syntheses of the A-unit of CC-1065 and analogs [302].

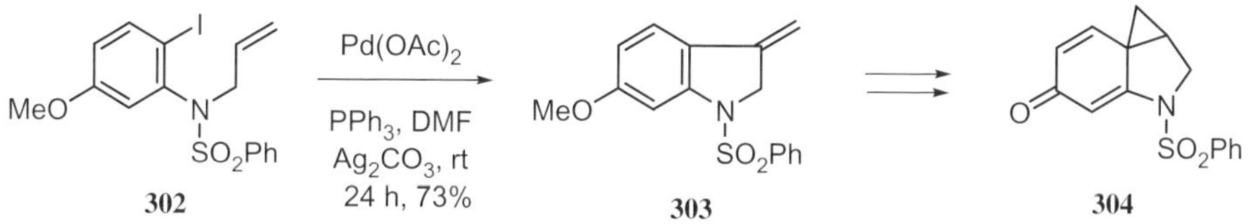

302 **303** **304**

Hoffmann and co-workers crafted the desoxyeserolin precursor **306** from the *N*-pyrrolidone aniline derivative **305** [303].

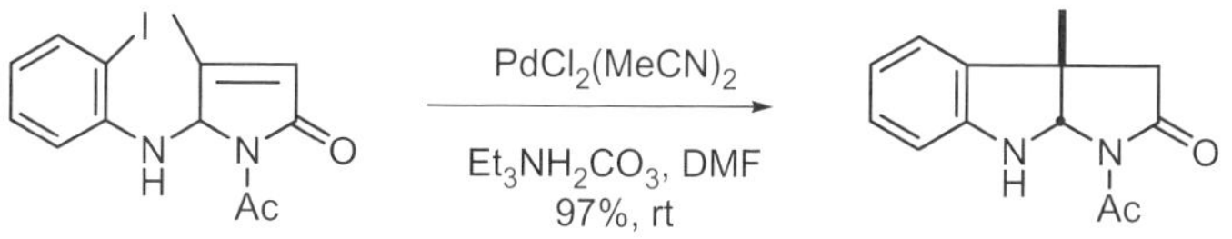

305 **306**

Macor exploited this methodology to synthesize several antimigraine analogs of Sumatriptan, e.g. **307** [304], and homotryptamines as potent and selective serotonin reuptake inhibitors [305].

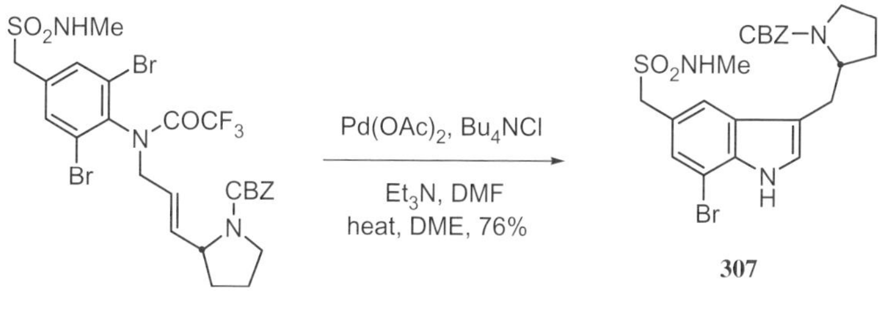

307

Gronowitz adapted this technology to one-pot syntheses of indole-3-acetic acids and indole-3-pyruvic acid oxime ethers from *N*-BOC protected *o*-iodoanilines [306]. Rawal employed the Pd-catalyzed cyclization of *N*-(*o*-bromoallyl)anilines to afford 4- and 6-hydroxyindoles, and a 4,6-dihydroxyindole [307], and Yang and co-workers have used a similar cyclization to prepare δ-carbolines **308** and **309** as illustrated by the two examples shown [308]. The apparent extraneous methyl group in **309** is derived from triethylamine.

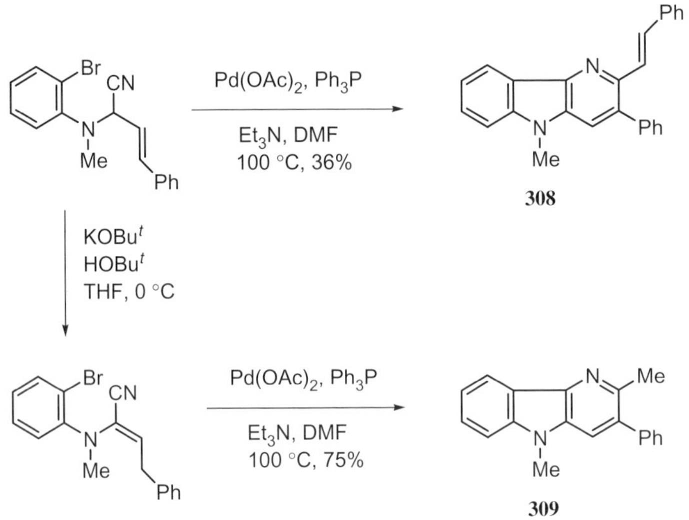

308

309

   The cyclization of *N*-allyl-*o*-haloanilines was adapted to the solid phase for both indoles [309, 310] and oxindoles [311]. For example, as illustrated below (**310**), a library of 1-acyl-3-alkyl-6-hydroxyindoles is readily assembled from acid chlorides, allylic bromides, and 4-bromo-3-nitroanisole [309]. Zhang and Maryanoff used the Rink amide resin to prepare *N*-benzylindole-3-acetamides and related indoles via Heck cyclization [310], and Balasubramanian employed this technology to the synthesis of oxindoles via the palladium cyclization of *o*-iodo-*N*-acryloylanilines [311]. This latter cyclization route to oxindoles is presented later in this section. Caddick has described Heck indole ring syntheses from 2-chloroanilines using Pd/imidazolium salts [312], and Anslyn uses the Heck cyclization of **299** to **300** as a fluorescence sensor [313].

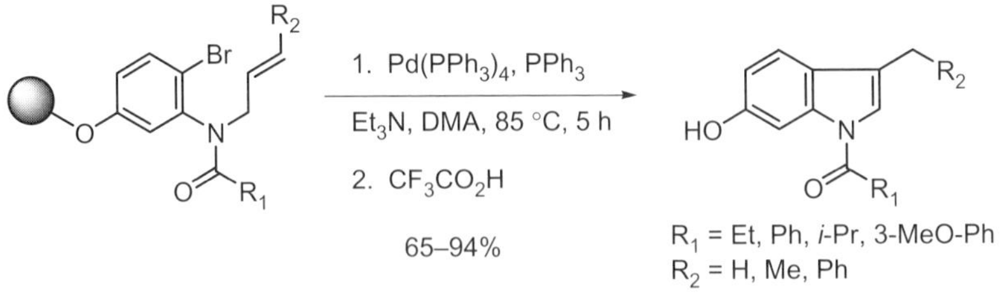

R_1 = Et, Ph, *i*-Pr, 3-MeO-Ph
R_2 = H, Me, Ph

310

   Whereas Hegedus [314] and Danishefsky [315] were the first to discover a tandem Heck reaction from *o*-allyl-*N*-acryloylanilines leading to tricyclic pyrrolo[1,2-*a*]indoles or pyridino[1,2-*a*]indoles [315], it has been the fantastic work of Grigg to unleash the enormous potential of this chemistry. Grigg and his co-workers parlayed their Pd-catalyzed tandem polycyclization-anion capture sequence into a treasure trove of syntheses starting with *N*-allyl-*o*-haloanilines [316]. Diels-Alder and olefin metathesis reactions can be interwoven into the sequence or can serve as the culmination step, as can a wide variety of nucleophiles. An example of the transformation of **311** to **312** is shown below in which indole is the terminating nucleophile [316d].

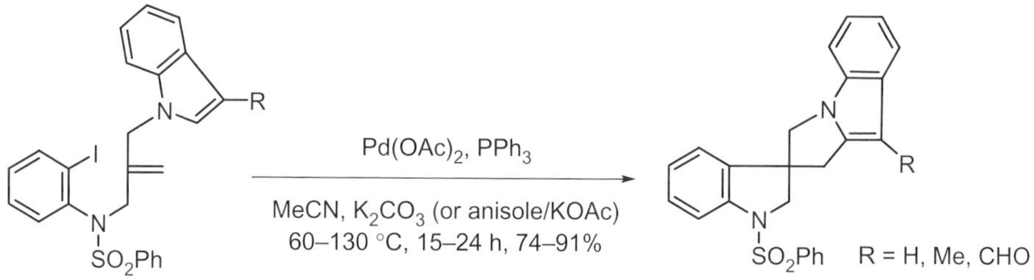

Grigg discovered that a 5-*exo-dig* Pd-catalyzed cyclization of *N*-acetylenic-*o*-haloanilines **313** to give 3-*exo*-alkylidene indolines **314** occurs in the presence of a hydride source, such as formic acid [317]. The reaction is stereoselective and regiospecific.

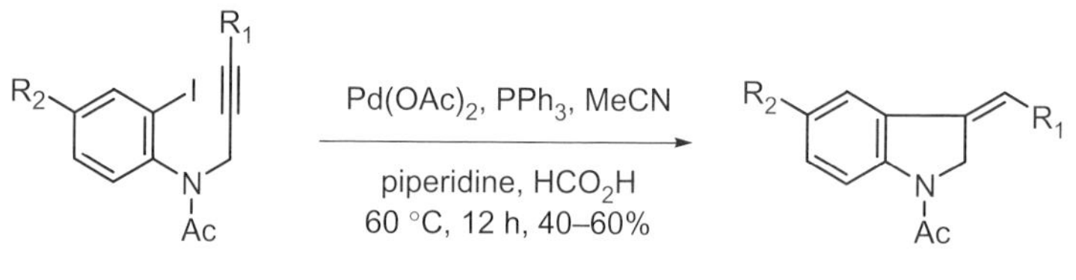

Grigg extended this alkyne cyclization to trapping with stannanes to give 3-*exo*-dienes [318], alkynes to afford tetracycles [316b, 319], and alkenes leading to cyclopropanes [320], an example of which is illustrated. In his studies, Grigg and co-workers have found that thallium and silver salts suppress direct capture of these palladium intermediates prior to capture [321].

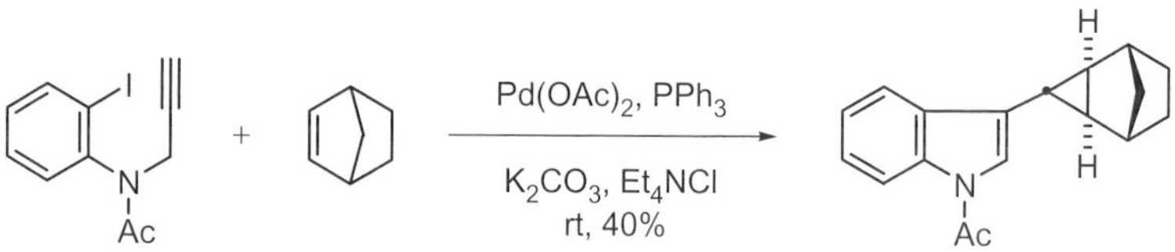

Overman's exhaustive study of the Pd-catalyzed cyclization of *o*-halo-*N*-acryloylanilines leading to spirooxindoles and related compounds has paid great dividends in advancing the art of organic synthesis [322]. Overman and his co-workers have developed this chemistry for the asymmetric synthesis of spirooxindoles leading to either enantiomer of physostigmine (**315**) [322e,i] and physovenine [322i], for gelsemine studies [322b,c], and, via a spectacular bis-Pd-catalyzed cyclization, to total syntheses of chimonan-thine and calycanthine [322e], as summarized in the transformation of **316** to **317**. Hiemstra, Speckamp and co-workers have pursued similar studies of Pd-catalyzed cyclizations to spirooxindoles, culminating in total syntheses of (±)-gelsemine and (±)-21-oxogelsemine [323].

It has been known for sometime that 2-(2-bromoanilino)enones undergo Heck-type cyclizations to form indoles and carbazoles. Thus, Kibayashi reported the synthesis of 4-keto-1,2,3,4-tetrahydrocarbazoles in this manner [324], and Rapoport employed this reaction (**318** to **319**) to achieve an improved synthesis of 7-methoxymitosene [325]. A series of related mitosene analogs has been crafted using Pd chemistry by Michael and co-workers [326]. They found that P(o-tol)$_3$ was far superior to PPh$_3$ in conjunction with Pd(OAc)$_2$.

Kasahara [327] and Sakamoto [328] have both shown that this Pd-cyclization is an excellent synthesis of a variety of 3-acylindoles, as shown for **320** to **321** [327]. Substrates **320** are also prepared using palladium. Sakamoto showed that 2-substituted 3-acylindoles are available in this manner [328].

R₁ = Ac, CO₂Et
R₂ = H, 6-OMe, 4-, 5-, 6-CO₂Me

Related Pd-cyclizations have been applied to the synthesis of 3-carboethoxy-2-trifluoromethylindoles [329], 2-carbobenzyloxy-4-hydroxymethyl-3-methylindoles, a unit that is present in the antibiotic nosiheptide, from a 2-(2-iodoanilino) unsaturated ester [330], 2- and 3-indolecarboxylates on solid phase [331], stephacidin A [332], an indolylquinoline KDR kinase inhibitor [333], tricyclic indoles from (2-iodophenyl)alkyl allenes [334], and 3-cyanoindoles [335]. A nice variation utilizes the *in situ* synthesis of 2-iodoanilino enamines and subsequent cyclization as shown for the preparation of indoles **322** [336].

This *in situ* generation of enamines and subsequent Heck cyclization to afford indoles has been adopted by several groups for the synthesis of indoles (and azaindoles) from 2-chloroanilines [337], tricyclic inhibitors of 5′-inosine monophosphate dehydrogenase [145], medium-ring fused indoles [338], and tetrahydrocarbazole in an *Organic Synthesis* preparation [339]. Some mechanistic insights have been offered [340]. The use of aryl-propynamides and *N*-alkynyl-2-haloanilides under Heck conditions affords oxindoles [341] and 2-aminoindoles [342], respectively. An example of the latter is illustrated for **323** to **324**.

This indole synthesis has been extended to β-tetrahydrocarbolines (**325**) [343], azaketotetrahydrocarbazoles [344], carbolines, carbazoles, and pyrido[1,2-*a*]benzimidazoles [345].

Examples of the former two reaction types are illustrated. An early Heck cyclization of 2-carboxy-2'-iododiphenylamine to 1-carbazolecarboxylic acid (73% yield) [346] has been generally overlooked by subsequent investigators. A microwave method is useful for the synthesis of azaindoles from aminopyridines and ketones [347], and a variety of cyclopent[b]indolones and carbazolones are available using this Heck chemistry [348, 349]. The requisite enamines can also be generated in situ by the addition of 2-haloanilines to alkynes with TiCl$_4$ [350]. The aryl Heck variation affords carbazoles [351–353] including novel indolo[3,2-b]benzo[b]thiophenes [353].

325   R$_1$ = Bn, CO$_2$Et
       R$_2$, R$_3$ = H, Me

A few examples of Pd-catalyzed cyclization of o-allylanilines to indoles have been reported [354, 355], but only in the case of the cyclization of N-alkyl-o-siloxyallylanilines leading to 3-alkoxyindoles is the method useful [355]. Palladium has been often used to prepare o-vinylaniline derivatives for subsequent (non-Pd) cyclization. Although Pd may not be involved in the indole-ring forming step, these reactions are still of interest to the palladium organic chemist. The Pd-catalyzed reaction of o-iodoacetanilide with ethyl α-methoxyacrylate affords o-vinylacetanilide 326 [356]. Acid treatment gives 2-carboethoxyindole.

326

A related Heck reaction of substituted o-bromoacetanilides with styrenes followed by selenium-induced cyclization of the resulting o-styrylacetanilides gives 2-arylindoles [357]. Substituted o-bromonitrobenzenes react with ethyl vinyl ether under the influence of Pd(OAc)$_2$ to give the corresponding o-ethoxyethenylnitrobenzenes. Zinc reduction then yields indoles [358]. The one-step Pd-catalyzed conversion of o-bromoanilines to indoles 327 with enamines (or with N-vinyl-2-pyrrolidone) has been reported [359].

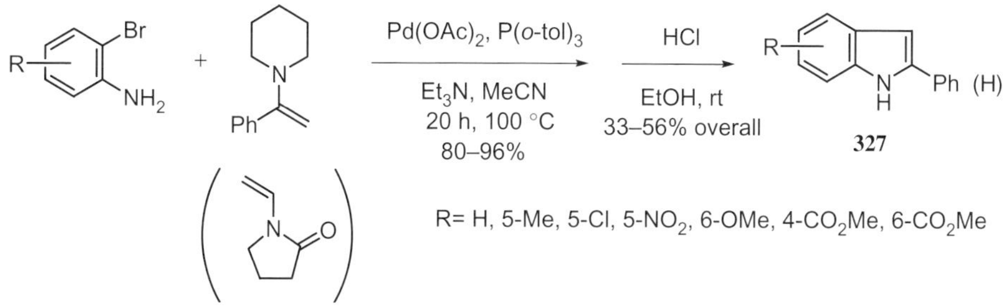

Ogasawara employed a Heck reaction of *o*-iodoaniline derivatives with dihydrodimethoxyfuran and vinylene carbonate to give intermediates that are readily cyclized to indoles with acid [360]. An example is shown below [360a].

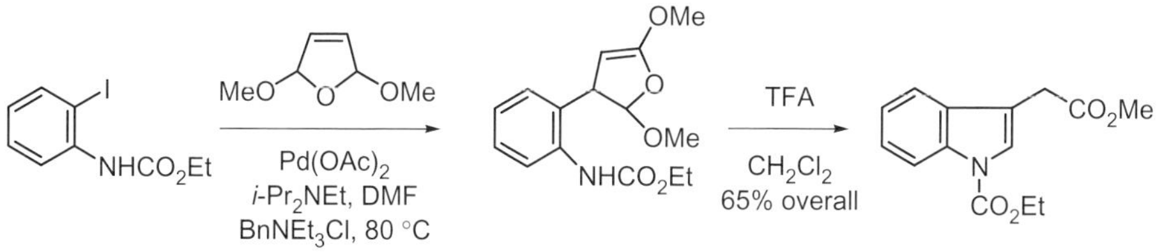

Carlström and Frejd described the one-step conversion of *o*-diiodobenzene (or iodobenzene) with *N*-Boc dehydroalanine methyl ester to give *N*-Boc-2-carbobenzyloxyindole in 30% yield [361].

3.5.3. The Larock indole synthesis

Larock and co-workers described the one-step Pd-catalyzed reaction of *o*-haloanilines with internal alkynes to give indoles [362]. This excellent reaction, which is shown for the synthesis of indoles **328**, involves oxidative addition of the aryl halide (usually iodide) to Pd(O), *syn*-insertion of the alkyne into the ArPd bond, nitrogen displacement of the Pd in the resulting vinyl–Pd intermediate, and final reductive elimination of Pd(O).

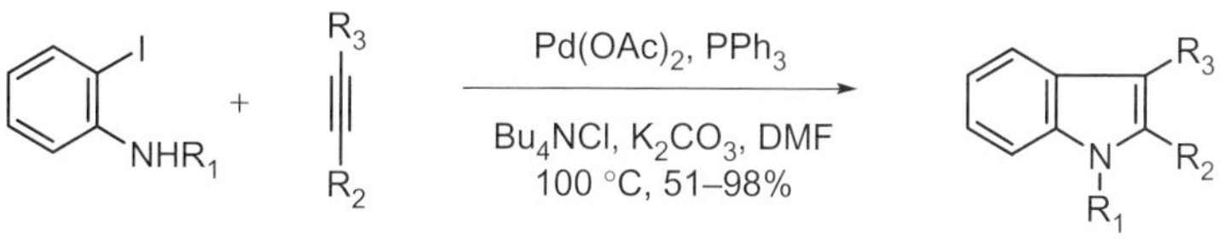

R$_1$ = H, Me, Ts
R$_2$, R$_3$ = *n*-Pr, *t*-Bu, cyclohexyl, TMS, Ph, CH$_2$OH, C(Me)=CH$_2$, (CH$_2$)$_2$OH, CMe$_2$OH

The reaction can be regioselective with unsymmetrical alkynes, and this is particularly true with silylated alkynes wherein the silyl group always resides at the C-2 indole position in the product. This is noteworthy because silyl-substituted indoles are valuable substrates for other chemistry (halogenation, Heck coupling). Gronowitz used the appropriate silylated

alkynes with *o*-iodoanilines to fashion substituted tryptophans following desilylation
with AlCl$_3$ [363]. Similarly, a series of 5-, 6-, and 7-azaindoles was prepared by
Ujjainwalla and Warner from *o*-aminoiodopyridines and silylated (and other internal)
alkynes using PdCl$_2$dppf [364]. Yum and co-workers also used a Larock indole synthesis
to prepare 7-azaindoles **329** [365a] and, from 4-amino-3-iodoquinolines, pyrrolo[3,2-
c]quinolines **330**, which have a wide spectrum of biological activity [365b].

R$_1$ = Me, Bn, H, *p*-MeOBn
R$_2$ = TMS, Ph, Pr
R$_3$ = CH$_2$OH, CH$_2$CH$_2$OH, Me, Ph, Pr

R$_1$ = aryl, Bn, *n*-Bu
R$_2$ = TMS, Ph, *n*-Pr
R$_3$ = Me, CH$_2$OH, CH$_2$CH$_2$OH, *n*-Pr = H, Me
R$_4$ = H, Me
R$_5$ = OMe, OCF$_3$

    The Larock synthesis was used by Chen and co-workers to synthesize the 5-(triazolyl-
methyl)tryptamine MK-0462, a potent 5-HT$_{1D}$ receptor agonist, as well as a metabolite
[366]. Larock employed his methodology to prepare tetrahydroindoles [367], and
Maassarani used this method for the synthesis of *N*-(2-pyridyl)indoles [368]. The latter
study features the isolation of cyclopalladated *N*-phenyl-2-pyridylamines. Rosso and co-
workers have employed this method for the industrial-scale synthesis of an antimigraine
drug candidate **331**. In this paper removal of spent palladium was best effected by trimer-
captotriazine (**332**) although many techniques were explored [369].

    The Larock indole synthesis was adapted to the solid phase both for the synthesis of
1,2,3-trisubstituted indole-5-carboxamides [370a] and, as illustrated, for the "traceless"
synthesis of 2,3-disubstituted indoles **333** [370b]. As seen earlier, the trimethylsilyl group
is fastened to C-2 with complete regioselectivity. The TMS group is cleaved under the resin
cleavage conditions. The original Larock conditions were not particularly successful.

R$_1$ = Ph, TMS, *n*-Pr, *t*-Bu
R$_2$ = Me,Et, Ph, CH$_2$CH$_2$OH, *n*-Pr

The current popularity and undisputed power of the Larock indole synthesis [371] is evidenced by applications to the preparation of tryptophans [372], tryptamines [373], tryptophols [146, 373c, 374], pyrrolo[2,3-*b*]pyrazines [375], 2- and 3-fluoroalkyl indoles [376], and a variety of other 2,3-disubstituted indoles [141, 145, 377].

Palladium-catalyzed chemistry between *o*-iodoanilines and 1,3-dienes leading to 2-vinylindolines is also known, having been first described by Dieck and co-workers [378]. This reaction, which is shown for the synthesis of **334**, was discovered before Larock's work in this area. The same reaction with 1,3-cyclohexadiene gives the corresponding tetrahydrocarbazole in 70% yield.

334

Larock extended this Pd-catalyzed diene heteroannulation to other dienes and anilines [379], including functionalized dienes leading to, for example, ketotetrahydrocarbazoles [380]. Back has employed 1-sulfonyl-1,3-dienes in this 2-vinylindoline synthesis [381], and the use of 1,3-dienes in constructing indolines has been adapted to the solid phase by Wang [382]. Interestingly, Larock has shown that the electronically-related vinylcyclopropanes undergo a similar cyclization with *o*-iodoanilines to form 2-vinylindolines, e.g. **335** [383]. Vinylcyclobutane also reacts in a comparable manner.

335

Larock found that allenes (1,2-dienes) undergo Pd-catalyzed reactions with *o*-iodoanilines to afford 3-alkylidene indolines, including examples using cyclic dienes, e.g. to give **336** [384], and ones leading to asymmetric induction, e.g. to give **337** [385]. The highest enantioselectivities ever reported for any Pd-catalyzed intramolecular allylic substitution reactions were observed in this study. Mérour modified this reaction for the synthesis of 7-azaindolinones, following ozonolysis of the initially formed *exo*-methylene indoline [386].

336

337

Prior to his work with internal alkynes, Larock found that *o*-thallated acetanilide undergoes Pd-catalyzed reactions with vinyl bromide and allyl chloride to give *N*-acetylindole and *N*-acetyl-2-methylindole each in 45% yield [387]. In an extension to reactions of internal alkynes with imines of *o*-iodoaniline, Larock reported a concise synthesis of isoindolo[2,1-*a*]indoles **338** and **339** [388]. The regioselectivity was excellent with unsymmetrical alkynes.

R$_1$ = Ph, Et, *n*-Bu, CO$_2$Et, (CH$_2$)$_4$OH
R$_2$ = H, Me, CF$_3$
R$_3$ = H, OMe, CO$_2$Et

338   **339** (93%)

3.6. Carbonylation

The insertion of carbon monoxide into σ-alkylpalladium(II) complexes followed by attack by either alcohols or amines is a powerful acylation method. This carbonylation reaction has been applied in several different ways to the reactions and syntheses of indoles. Hegedus and co-workers converted *o*-allylanilines to indoline esters **340** in yields up to 75% [389]. In most of the examples in this section, CO at atmospheric pressure was employed.

R$_1$ = Me, Ac
R$_2$ = H, Me

340

Hegedus applied this chemistry to the conversion of 1-lithio-3-methylindole to 1-(2-carbomethoxyethyl)-3-methylindole with CO and ethylene (53% yield) [390]. This interesting C–N bond formation reaction is revisited in Section 3.7. The amide precursor used by Overman in his gelsemine studies was prepared by a Pd-catalyzed carbonylation reaction between a vinyl triflate and *o*-bromoaniline [322b]. A similar amination scheme was used to synthesize 21-oxoyohimbine from 1-(2-bromobenzyl)-1,2,3,4-tetrahydro-β-carboline [391]. Mori and co-workers employed a similar strategy to prepare the quinazoline alkaloid rutecarpine and related compounds [392]. Edstrom expanded his studies on the carbonylation of pyrroles (Chapter 2) to the methoxycarbonylation of 5-azaindolones leading to **341** [393]. Doi and Mori performed a similar carbonylation on a dihydroindole 4-triflate leading, after dehydrogenation, to 4-carbomethoxy-1-tosylindole [156]. The direct carbonylation of *N*-acetyl- and *N*-benzoylindole leads to the corresponding C-3 carboxylic acids [394], and this carbonylation method was used to synthesize *N*-TBS-6-trifluoromethylindole-3-carboxylic acid [395].

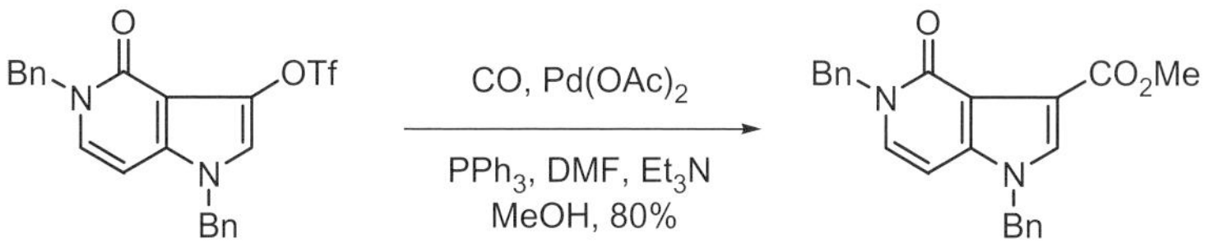

341

The tandem indolization–carbonylation of *o*-alkynylanilines has been explored by several groups. Kondo and co-workers have effected the synthesis of 3-carbomethoxyindoles **342** [396], and Arcadi has modified this reaction to prepare 3-acylindoles **343** [397]. This latter research group has extended this protocol to a synthesis of 12-acylindolo [1,2-*c*]quinazolines [398]. Oxindoles are also accessible via carbonylation of 2-ethynylanilines [399].

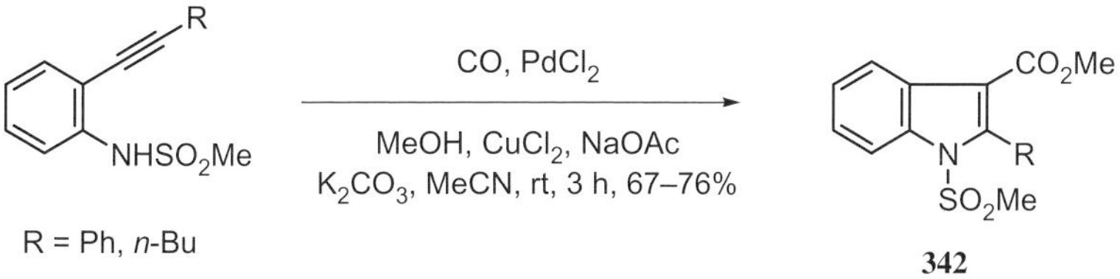

342

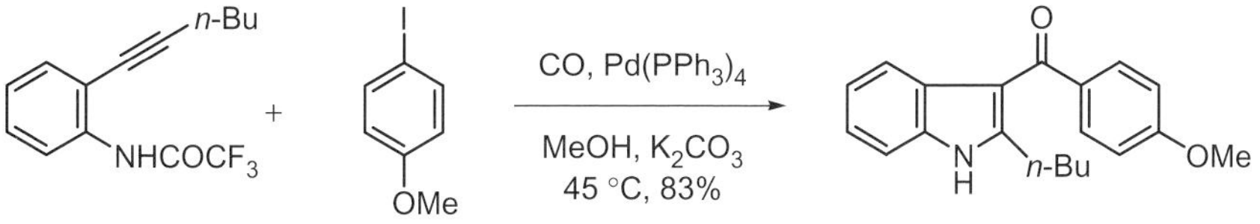

343

Herbert and McNeil have shown that the appropriate 2-iodoindole can be carbonylated in the presence of primary and secondary amines to afford the corresponding 2-indolecarboxamides in 33–97% yield. Further application of this protocol leads to amide **344**, which is a CCK-A antagonist (Lintitript) [400]. Carbonylation of unprotected 4-, 5-, 6-, and 7-bromoindoles with nucleophiles affords the corresponding amides, esters, and carboxylic acids [401].

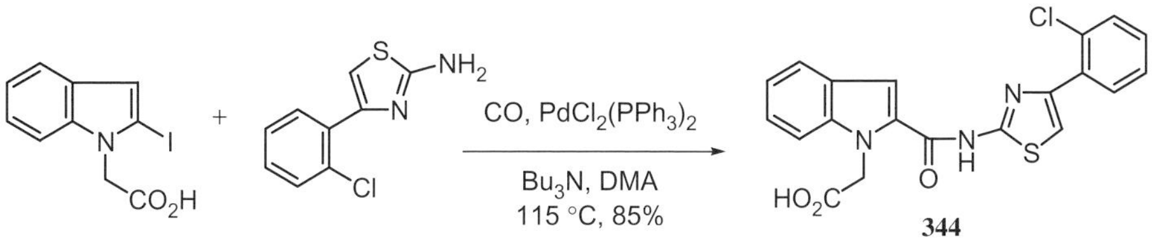

344

Fukuyama employed a vinyltin derivative in the carbonylation of 3-carbomethoxymethyl-2-iodoindole to afford **345** [183b]. Buchwald effected the carbonylation of 4-iodoindole **346** to give lactam **347** [149], and a similar carbonylation reaction on 4-iodoindole malonate **129** gives ketone **348** in 68% yield.

Somei carbonylated 7-thallated *N*-acetylindoline to give 7-carbomethoxy-*N*-acetylindoline as the major product $(CO/Pd(OAc)_2/MeOH/Cr(CO)_6/52\%)$ [203]. The carbonylation of C-2 in gramines and tryptamines has been achieved by Cenini and co-workers [402]. Thus, treatment of gramine (3-dimethylaminomethylindole) with Li_2PdCl_4 ($PdCl_2$ + LiCl) followed by treatment of the palladated intermediate (isolable, 93% yield) with methanol and CO afforded 2-carbomethoxygramine in 92% yield. In their grossularine synthetic studies, Hibino and co-workers effected the carbonylation of an α-carboline triflate leading to methyl ester **349** [194b]. The same triflate with an arylboronic acid and CO gave a low yield of an aryl ketone [194a].

In the presence of tetramethyltin, 1-bromonauclefine reacts with CO in a Pd-catalyzed carbonylation to give the alkaloid naucletine [206]. Dong and Busacca effected a new synthesis of tryptamines and tryptophols via Rh-catalyzed hydroformylation of functionalized anilines that are prepared by a standard Heck reaction, as shown for the preparation of tryptamine sulfonamide **350** [403]. This reaction is applicable to ring-substituted tryptamines (Cl, Br, F, OMe, CF_3). Likewise, the Rh-catalyzed carbonylation of *o*-alkynyl-lanilines, which were prepared by a Pd-catalyzed Sonogashira coupling, leads to oxindoles (60–86% yields) [404].

Hidai and co-workers found that 3-vinylindole **351** undergoes cyclocarbonylation to afford 1-acetoxycarbazole **352** [405]. The reaction of indole with allene and CO in the presence of catalytic Pd(O) leads to *N*-acylation (**353**) in good yield [406]. An analogous reaction with 5-hydroxyindole affords *N*- and *O*-acylation products (47% yield).

Grigg expanded his Pd-catalyzed cascade cyclization reactions to include carbonyla-tion as the termination step [407]. Thus, indoline **354** is obtained in excellent yield and the spiroindoline **355** is secured as a single diastereomer. Thallium acetate results in sig-nificant improvement in these reactions by allowing for low pressure carbonylation.

3-Spiro-2-oxindoles, such as **356**, are readily crafted from the Pd-catalyzed reactions of *o*-haloanilines with vinyl halides and triflates in the presence of CO [408]. The *o*-iodo enamide is presumed to form initially, followed by Heck cyclization.

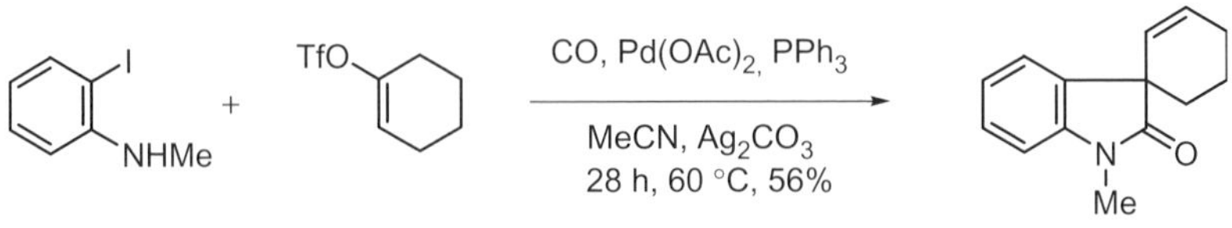

356

Grigg also extended these carbonylation reactions to a one- and two-pot protocols culminating in olefin metathesis [316h]. For example, substrate **357** is converted to **359**, via **358**, under these conditions. *N*-Tosylindolines were constructed in like fashion.

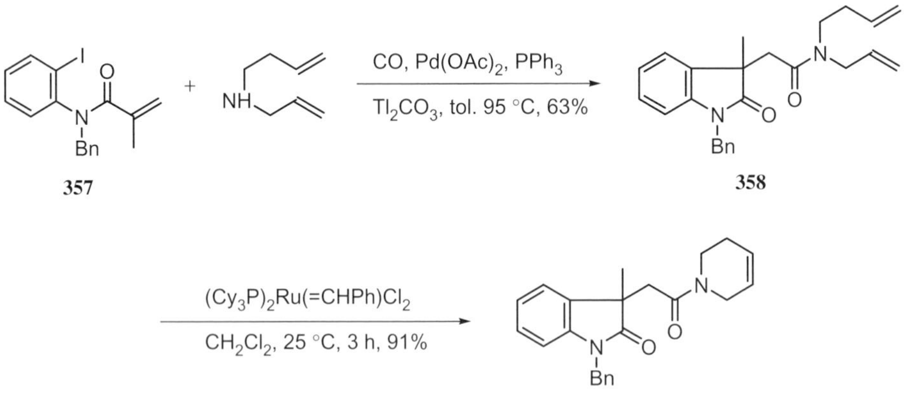

357   **358**

359

Ishikura has adapted his Pd-catalyzed cross-coupling methodology involving indolylborates to include carbonylation reactions [162f,h,m, 409]. For example, **142** reacts with enol triflates in the presence of CO and Pd to give 2-acylindoles such as **360** [409]. This particular ketone was cyclized to the indole C-3 position and the resulting cyclopentanone was converted to yuehchukene analogs. Borate **142** also reacts with oxindole precursor **361** to give 2-acylindole–oxindole **362** [162f]. Noteworthy is that Ishikura also used the removable *N*-Boc indolylborates in this chemistry [162m].

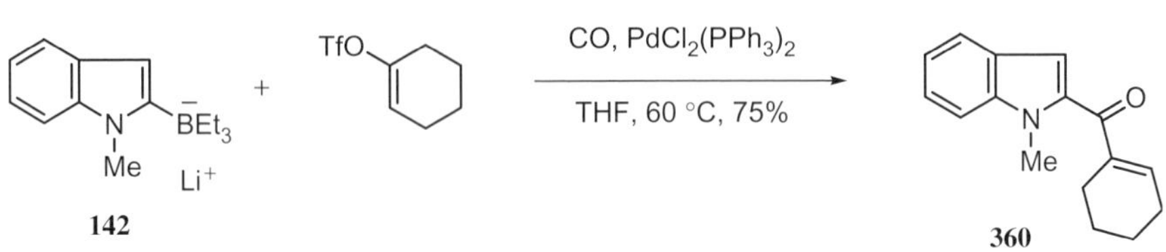

142   **360**

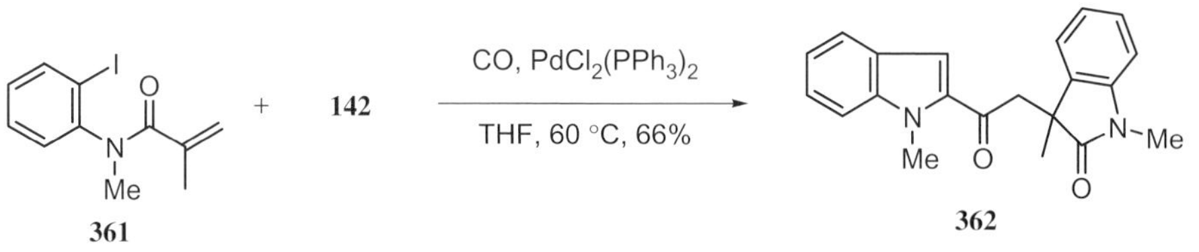

3.7. C–N bond formation reactions

While the Mori–Ban indole synthesis is catalyzed by a Pd(0) species, the Hegedus indole synthesis is catalyzed by a Pd(II) complex. In addition, the Mori–Ban indole synthesis is accomplished via a Pd-catalyzed vinylation (a Heck reaction), whereas the Hegedus indole synthesis established the pyrrole ring via a Pd(II)-catalyzed amination (a Wacker-type process). Hegedus conducted the Pd-induced amination of alkenes [410] to an intramolecular version leading to indoles from *o*-allylanilines and *o*-vinylanilines [249a,b, 389, 411, 412, 413]. Three of the original examples from the work of Hegedus are shown below.

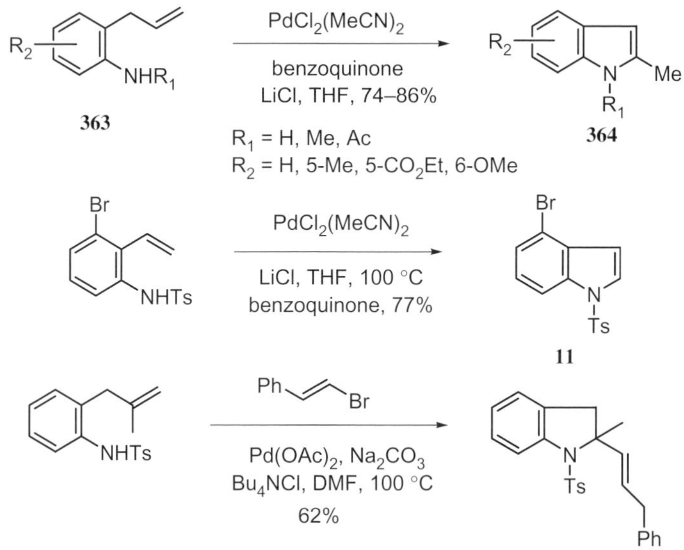

The Hegedus indole synthesis can be stoichiometric or catalytic and a range of indoles **365–367** was synthesized from the respective *o*-allylanilines in modest to very good yields (31–89%) [412].

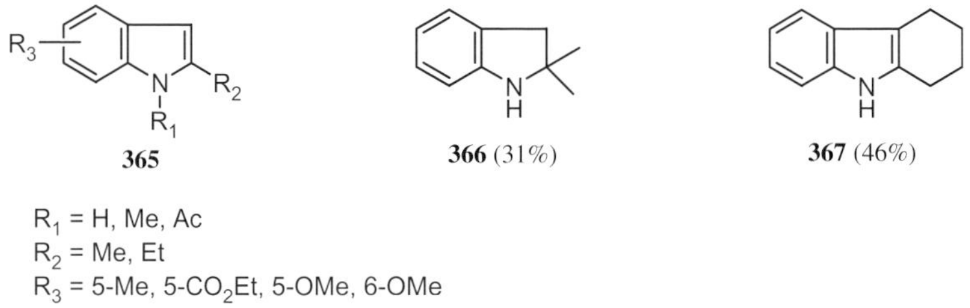

In addition, Hegedus and co-workers extended this chemistry to the N-alkylation of 1-lithioindole with alkenes (PdCl$_2$(MeCN)$_2$/THF/HMPA/Et$_3$N/then H$_2$) to afford N-alkylindoles in 28–68% yields [390]. Moreover, a similar reaction of N-allylindole **368** with nitriles leads either to **369** or **370** depending on how the intermediate Pd-alkyl (or acyl) complexes are treated [414]. The formation of these pyrazino[1,2-a]indoles is similar to a nitrile-Ritter reaction.

The Pd-catalyzed cyclization of o-vinylanilines to indoles has only been rarely utilized in synthesis, but Stille found this reaction to afford a variety of N-tosylindoles [196]. The requisite o-vinylaniline derivatives were conveniently prepared by a Stille coupling using tributylvinyltin and aryl bromides. A similar *ortho*-vinylation using SnCl$_4$–Bu$_3$N and acetylene and subsequent Pd-catalyzed cyclization to give 5- and 7-methylindole, and 1,4- and 1,6-dimethylindole was described by Yamaguchi [415]. Kasahara reported the vinylation of o-bromoacetanilides **371** and the Pd-catalyzed cyclization of the resulting o-vinylacetanilides **372** to N-acetylindoles **373** [416].

R= H, 4-Me, 4-CO$_2$Me, 5-OMe, 6-Cl, 7-CO$_2$Me

Several variations on the synthesis and cyclization of o-vinyl- and o-allylanilines using a palladium catalyst in one or both steps have been described, including azaindoles and pyrroloquinolines [417], 6-chloro-7-fluoroindole, the heterocyclic core of the 5-HT$_{2c}$

receptor agonist Ro60-0175 [418], and 5-(sulfamoylmethyl)indoles [419]. The cycliza-tion of *ortho-gem*-dihalovinylanilines **374** affords 2-substituted indoles **375** [420, 421], and a tandem sequence of boronic acid vinylation and organolithium addition-cyclization yields indoles **376** [422].

R = Ph, Ar, 2-thienyl, hexyl, alkenyl

R₁ = H, F, OMe
R₂ = *t*-Bu, *n*-Bu
R₃ = Ph, Ac, *t*-Bu, 2-thienyl

The cyclization of *o*-alkynylanilines to indoles, which usually does not require palla-dium, has been described in Section 3.4. In view of their extensive research with this transformation, this reaction is often referred to as the Sakamoto–Yamanaka indole syn-thesis [213c, 216, 217, 221, 231, 232]. Although the cyclization of *o*-alkynylanilines, which are often obtained by the Sonogashira coupling (Section 3.4), is usually accom-plished with base, Kundu used Pd(OAc)₂ to effect the conversion of **377** to **378** [423].

Trost used a Sonogashira coupling, followed by reduction of the nitro group, to pre-pare **379**, which in turn was converted to tricyclic lactam **380** by Pd-catalyzed cyclization (yield unreported) [424].

380

Illustrative of the mild conditions involved in these cyclizations, Fukuyama effected the transformation of **381** to biindole **382** [183b].

381   **382**

As discussed in Section 3.4, the Sonogashira coupling, and subsequent cyclization, plays a major role in the synthesis of indoles, a method not unrelated to the Larock indole synthesis. Indeed, recent years have seen an enormous number of reports utilizing this method, only some of which can be described. These applications include syntheses of 6-trifluoromethylindole [395], 5,6-difluoroindole [425], 5-amino-7-azaindole [426], pyrrolo[2,3-*b*]pyrazines [427], complex 2-arylindoles [428–430], novel tryptophans [372, 431], nitroindoles [432], biindoles [433], indole libraries [434], and naturally occurring indoles (±20*R*-dihydrocleavamine [435], asterriquinones [436], and furostifoline [437]). Although the cyclization of *o*-alkynylanilines normally proceeds well under simple alkoxide conditions, some improved conditions include a one-pot Sonogashira coupling–cyclization protocol using Pd/C in water (e.g. **383** to **384**) [438], and the use of tetra-*n*-butylammonium fluoride [439], potassium or cesium bases [440], or indium tribromide [441] as cyclization reagents. Cross-coupling and subsequent cyclization of 2-carboxamidoaryl triflates may also be employed in this indole ring synthesis [442], and solventless, microwave-enhanced coupling/cyclization conditions have been described [443]. A variation on this general theme features amination of *o*-alkynylhaloarenes and subsequent cyclization (e.g. **385** to **386**) [444], including an intramolecular version [444b].

383   **384**

385 + R$_3$NH$_2$ → **386**

Pd(OAc)$_2$

HIPrCl, tol
base, 105 °C
66–99%

R$_1$ = H, CF$_3$; R$_2$ = Ph, *t*-Bu, *n*-Hex; R$_3$ = *p*-Tol, Bn, Mes, *n*-Hex, *n*-Oct, PMB, 4-EtCO$_2$Ph
Base = KO*t*-Bu, K$_3$PO$_4$

Several groups have intercepted the indole–palladium complex that is initially obtained on cyclization by a subsequent Heck or other reaction. As will be seen, this can be a powerful elaboration of indoles. In the first example of this concept, Utimoto and co-workers ambushed intermediate **387** with a series of allylic chlorides to give **388**. Normal acid workup yields the corresponding C-3 unsubstituted indoles (52–83%) [445].

388

R$_1$ = Bu, *t*-Bu, Ph, CH=CHMe
R$_2$ = H, Ac, CO$_2$Me
R$_3$ = H, Me, Cl, CH$_2$Cl
R$_4$ = H, Me, Cl, CH$_2$Cl
R$_5$ = H, Me, Et, CH$_2$Cl, CH=CH$_2$

The research group of Cacchi made extensive use of these tandem cyclization–Heck reactions to prepare a wide variety of indoles [446]. For example, vinyl triflates react with *o*-aminophenylacetylene to afford an array of 2-substituted indoles in excellent yield, e.g. **389** to **390** [446b], and a similar reaction of **391** with aryl iodides leads to an excellent synthesis of 3-arylindoles **392** [446c]. The method is applicable to 2-substituted 3-alkylindoles [446e], indolo[1,2-*c*]quinazolines [446f], pyrrolo[2,3-*b*]quinoxalines [446g], and 2-substituted 3-alkynyl- and 3-acylindoles [446i]. For a review see reference 447.

1. Pd(PPh$_3$)$_4$, CuI
   Et$_2$NH, rt

2. PdCl$_2$, Bu$_4$NCl
   CH$_2$Cl$_2$, aq. HCl
   rt, 98%

389

390

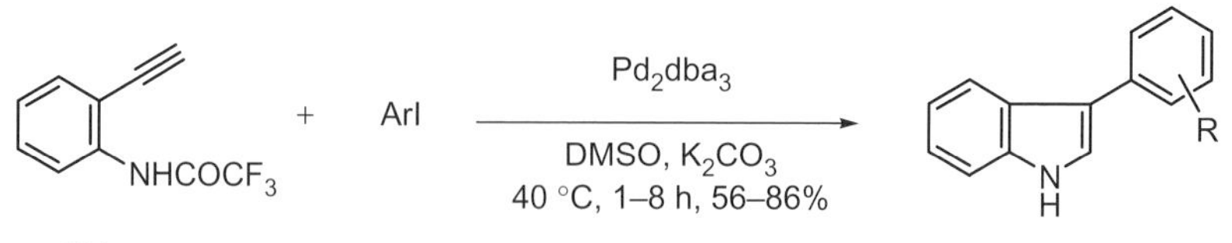

391   **392**

R = H, 4-NHAc, 4-F, 3-F, 4-Cl, 3-CF$_3$, 3-Ac, 4-CO$_2$Et, 3-CO$_2$Et, 3-NO$_2$, 4-Me

Saulnier expanded this reaction to an elegant synthesis of indolo[2,3-*a*]carbazole **393**, featuring a polyannulation from the diacetylene as shown [448].

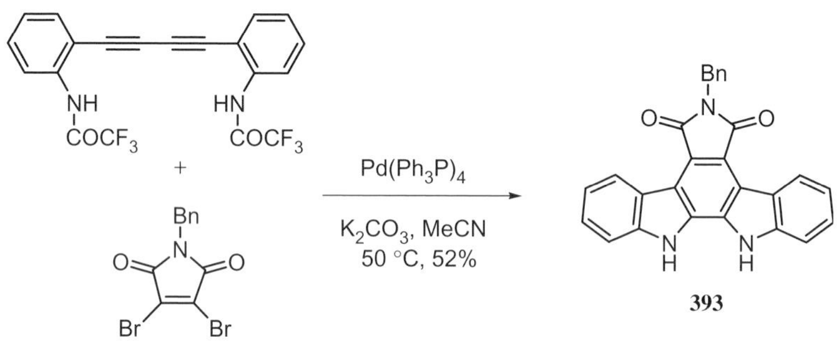

393

Yasuhara applied this methodology to the synthesis of 3-vinylindoles **394** [449].

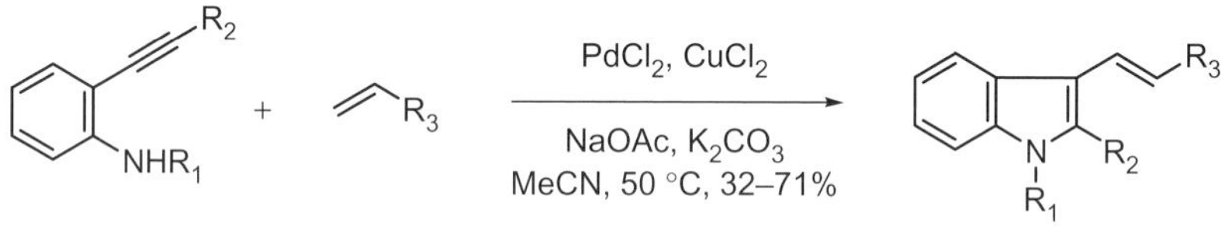

R$_1$ = Ms, SO$_2$Ph, CO$_2$Et
R$_2$ = Ph, *n*-hexyl, TMS
R$_3$ = CO$_2$Et, CHO, Ac

394

Larock [450], Knight [451], and Barluenga [452] described a Pd-catalyzed Sonogashira coupling of 2-haloanilines with terminal alkynes followed by iodocyclization to give 3-iodoindoles. Related indole syntheses that employ various *o*-alkynylanilines include the free-radical cyclization of *o*-alkynylarylisonitriles [453], the copper-catalyzed cyclization of 2-alkynyl-*N*-arylideneanilines [454], Pd–Cu-catalyzed cyclization of *o*-(alkyl)phenylisocyanates [455], and *o*-(alkyl)arylisocyanides [456]. For example, the latter chemistry provides *N*-cyanoindoles **395** [456a].

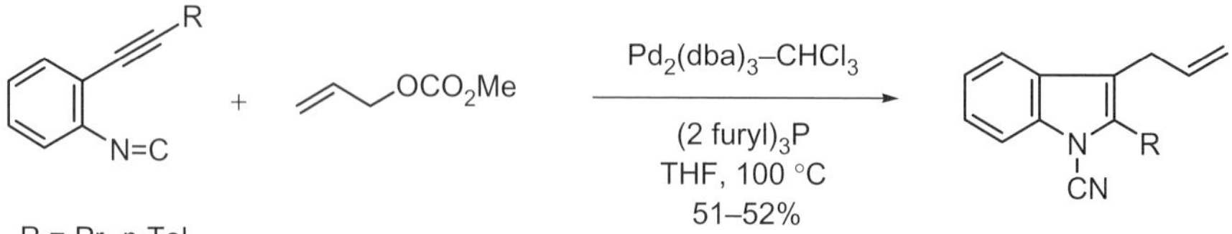

R = Pr, *p*-Tol

395

In a clever application of the hetero-Cope rearrangement, Martin used a Pd-catalyzed coupling of *N*-arylhydroxamates **396** with vinyl acetate to set up the [3,3]sigmatropic rearrangement **397** to **398** and final cyclization **398** to **399** [457]. Applications of this novel indole ring construction to the synthesis of the toxic fava bean metabolite 4-chloro-6-methoxyindole [457c], marine acorn worm 4,6-dibromo- and 3,4,6-tribromoindoles [457b,c], CC-1065 subunits, and PDE-I and -II have been achieved. *N*-Arylhydroxamates **396** are readily prepared from the appropriate nitrobenzene by partial reduction and acylation.

The aforementioned heteroannulation of *o*-haloanilines has been adapted by various means to the synthesis of carbazoles by several groups [458–462], and Larock has extended this to a carboline ring construction as shown for **400** to **401** [462].

Boger and co-workers were the first to report the intramolecular amination of aryl halides in their synthesis of lavendamycin [463]. Thus, biaryl **402** is smoothly cyclized under the action of palladium to β-carboline **403**, which comprises the CDE rings of lavendamycin.

Similarly, carbazoles can be synthesized via a double *N*-arylation of primary amines [464], and comparable tactics lead to indoles, as shown for **404** to **405** [465].

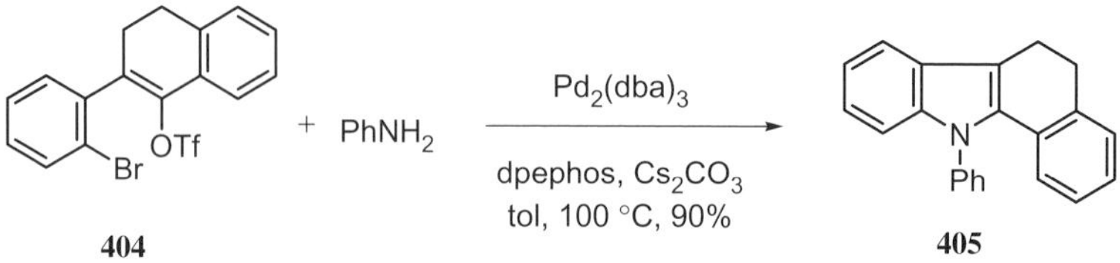

Buchwald parlayed the powerful Hartwig–Buchwald aryl amination technology [466] into a simple and versatile indoline synthesis [467]. For example, indole **406**, which has been employed in total syntheses of the marine alkaloids makaluvamine C and damirones A and B, was readily forged via the Pd-mediated cyclization shown below [467a]. This intramolecular amination is applicable to the synthesis of *N*-substituted optically active indolines [467c], and *o*-bromobenzylic bromides can be utilized in this methodology, as illustrated for the preparation of **407** [467d]. Furthermore, this Pd-catalyzed amination reaction has been applied to the synthesis of arylhydrazones, which are substrates for the Fischer indole synthesis [468]. Related chemistry has been employed to prepare 1-aminoindoles [469], *N*-arylindole-2-carboxylates [470], and indole-2-carboxylate libraries [471].

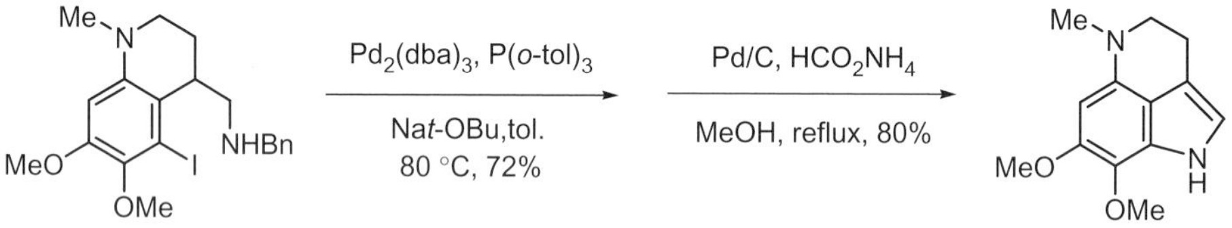

Snieckus and co-workers applied the Hartwig–Buchwald amination to the synthesis of *o*-carboxamido diarylamines, which can be elaborated to oxindoles [472]. Dodd synthesized α-carboline **408** via an intramolecular amination protocol [473]. These α-carbolines (pyrido[2,3-*b*]indoles) have been found to be modulators of the GABA$_A$ receptor, and this ring system is found in several natural products (grossularines, mescengricin). Snider achieved a similar cyclization of a 2-iodoindole leading to syntheses of (−)-asperlicin and (−)-asperlicin C as illustrated for the model reaction giving **409** [474]. The requisite 2-iodoindole was readily synthesized by a mercuration sequence (Hg(OCOCF$_3$)$_2$, KI/I$_2$/82%).

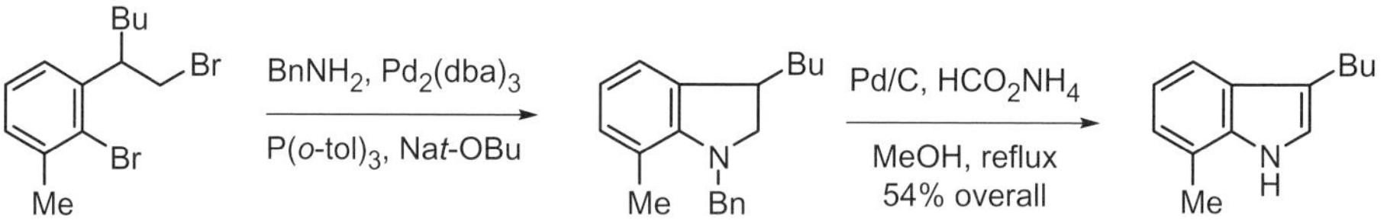

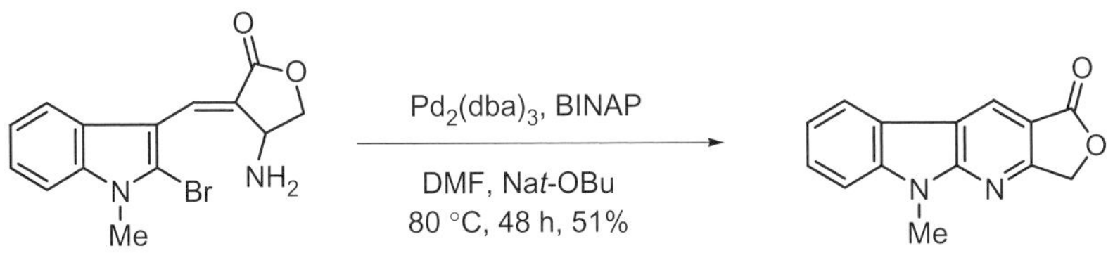

408

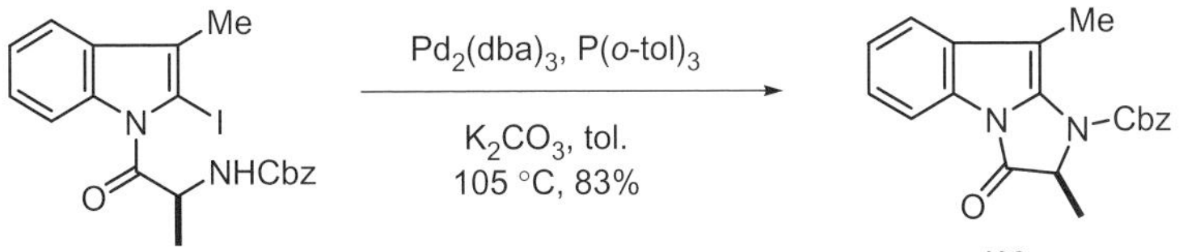

409

The intermolecular Pd-catalyzed amination of haloindoles [475] and indoleboronic acids [476] and the *N*-arylation [477] and *N*-vinylation of indoles [478] have been described. A method for a tandem carbon–hydrogen functionalization/amide arylation leads to carbazoles, e.g. **410** [479].

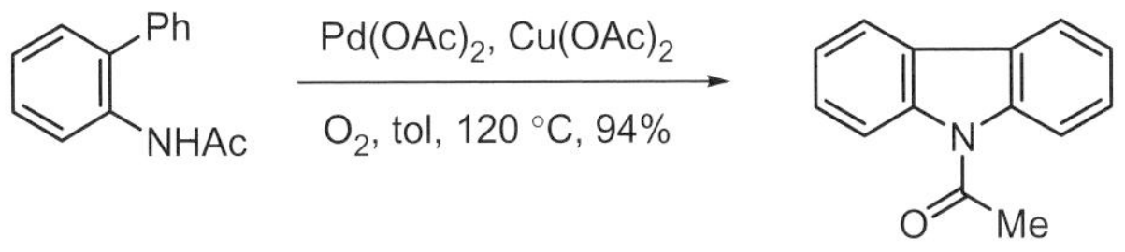

410

Several investigators have developed the reductive cyclization of *o*-nitrostyrenes into an efficient synthesis of indoles. Thus, research by the groups of Watanabe [480], Söderberg [481], and Cenini [482] have established this reductive Pd-catalyzed *N*-heteroannulation reaction as a viable and powerful route to simple indoles and fused indoles (**411**) as shown below. In addition, Söderberg has shown the application of this method to carbazolones [481d], natural β-carbolines [481e], and the carbazole alkaloid murrayaquinone A [481f]. Ohta described the related Pd-catalyzed cyclization of *o*-aminophenethyl alcohol to indole in 78% yield [483].

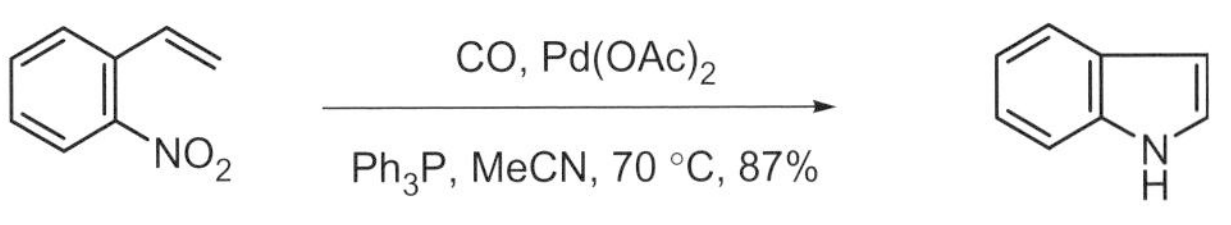

Also: 4-OH, 4-OMe, 5-OMe, 6-OMe, 4-Br, 4-NO$_2$, 4-CO$_2$Me,
5-CO$_2$Me, 6-CO$_2$Me, 7-CO$_2$Me, 2-Ph, 2-Me, and others

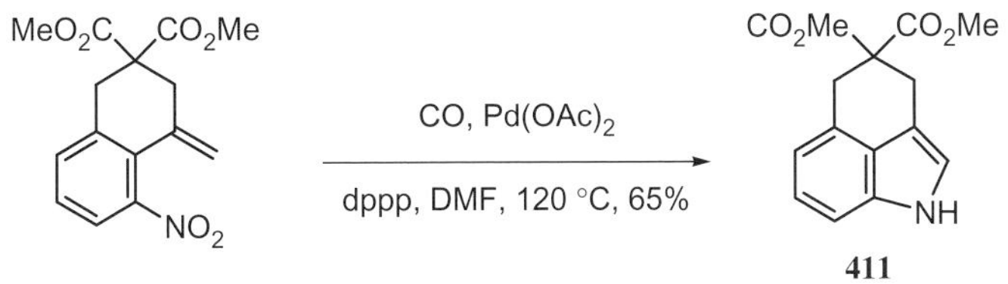

411

Kuethe and co-workers have used the reductive cyclization of *o*-nitrostyrenes [484] to construct indolo[2,3-*a*]carbazoles, including the natural tjipanazoles [484a], indolylquinolinones, a ring system found in potent and selective KDR kinase inhibitors [333, 484b], carbazoles and related tricyclic indoles, e.g. **412** [484c], and 2,2′-biindoles [484d].

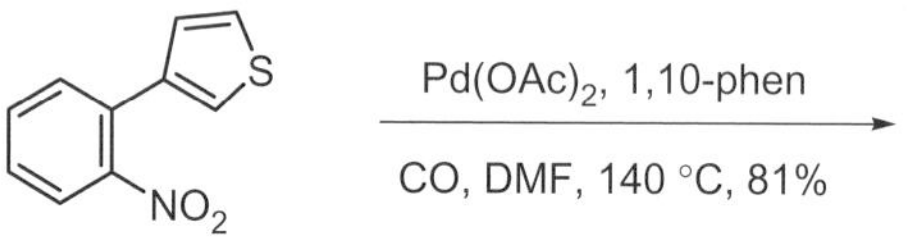

412

Izumi used Pd(OAc)$_2$ to effect the bis-alkoxylation of *o*-nitrostyrenes to form the corresponding *o*-nitrophenylacetaldehyde acetals, which, upon treatment with Fe/HOAc/HCl, give a variety of indoles (4-OMe, 5-OMe, 4-Me, 4-Cl, 4-CO$_2$Me, 6-Me) in 63–86% yields [485]. Mori and co-workers employed a titanium-isocyanate complex to construct tetrahydroindolones **413** from cyclohexane-1,3-diones [486].

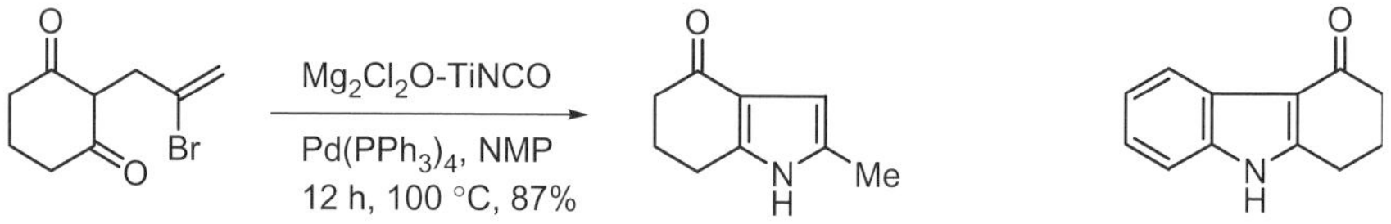

413a   **413b**

Mérour studied the reaction of indole triflates with diamines to afford pyrazino[2,3-*b*]indoles **414** and indolo[2,3-*b*]quinoxalines [487]. In the absence of palladium, the yield of **414** is only 31% after 15 h. In some cases spiroindoxyls are formed.

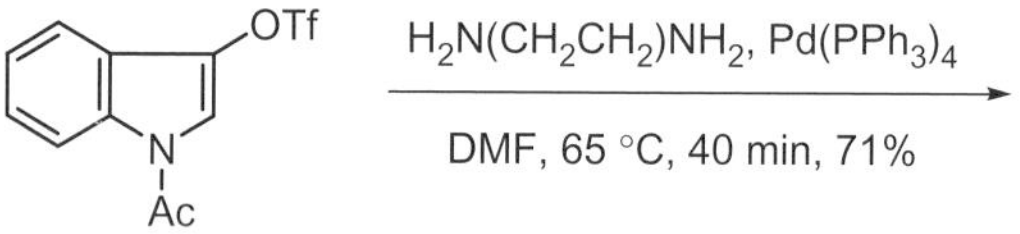

414

As we have seen earlier in this chapter, palladium is often employed to effect *N*-alkylation of indoles. Trost and Molander found that indole reacts with vinyl epoxide **415** to give indole **416** [488]. The utility of such *N*-alkylations remains to be established.

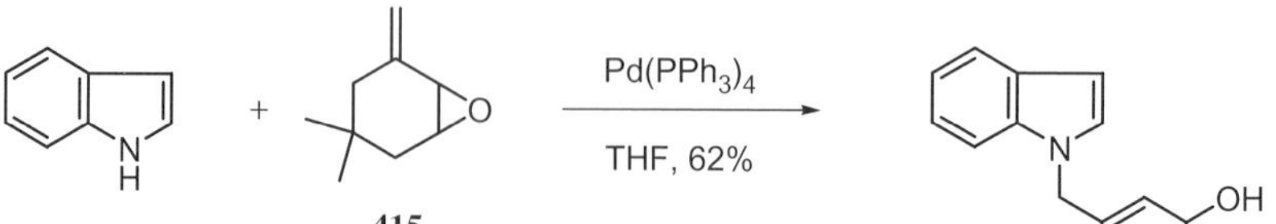

415

416

The Buchwald–Hartwig aryl amination methodology cited above in this section was engaged by Hartwig and others to synthesize *N*-arylindoles **417** [489]. Carbazole can be *N*-arylated under these same conditions with *p*-cyanobromobenzene (97% yield). Aryl chlorides also function in this reaction. The power of this amination method is seen by the facile synthesis of tris-carbazole **418** [489c].

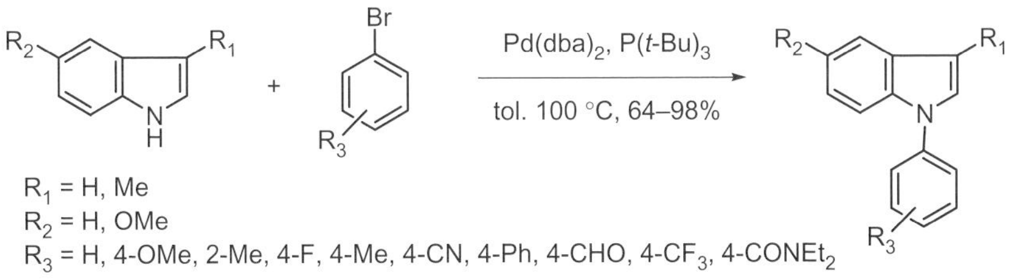

R$_1$ = H, Me
R$_2$ = H, OMe
R$_3$ = H, 4-OMe, 2-Me, 4-F, 4-Me, 4-CN, 4-Ph, 4-CHO, 4-CF$_3$, 4-CONEt$_2$

417

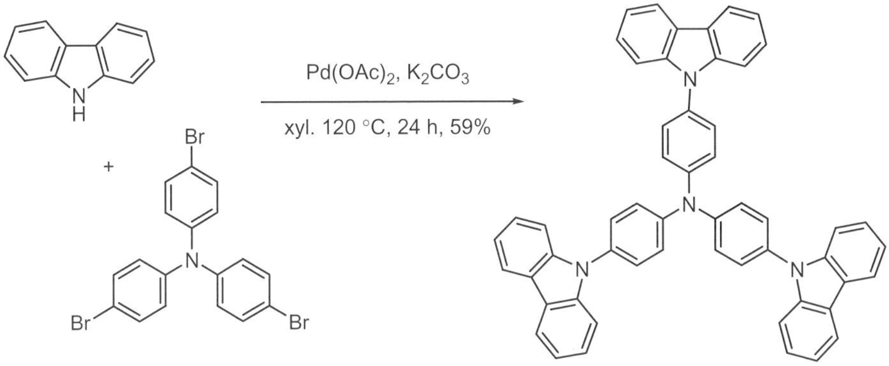

418

Although examples are sparse, a Pd-catalyzed carbon–sulfur bond formation leading to **419** was the penultimate reaction in a synthesis of (±)-chuangxinmycin [197, 490]. Earlier, Widdowson described the thiolation of 3-acetyl-4-iodoindole (MeO$_2$CCH$_2$SSnMe$_3$/Pd(Ph$_3$P)$_4$/83%) [491], and the coupling of 5-bromoindole with thiophenol (90%) has been reported by Itoh [492].

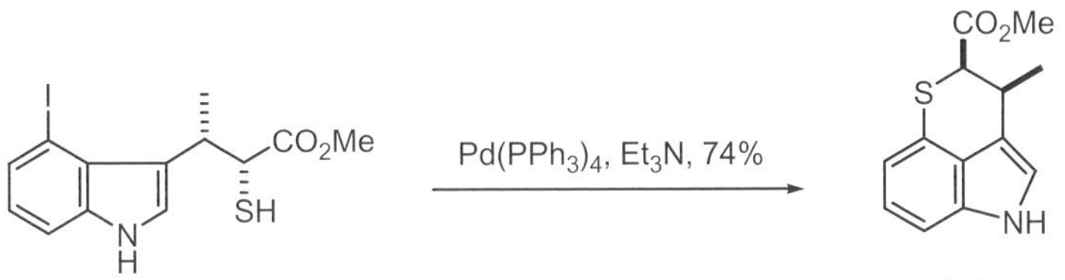

419

3.8. Miscellaneous

One of the more difficult transformations in organic synthesis is the deoxygenation of phenols to arenes. However, the use of palladium offers an attractive solution to this problem.

The phenolic group in naltrindole can be deoxygenated via the derived triflate by treatment with formic acid and PdCl$_2$(Ph$_3$P)$_2$ in good yield [159]. Triflate **420** is smoothly deoxygenated and further transformed to koumidine (**421**) under similar conditions by Sakai [493]. Noteworthy is the Pd-catalyzed isomerization of the ethylidene side chain. Sakai described a similar deoxygenation in the koumine series [494].

1. Pd(OAc)$_2$, dppf
   HCO$_2$H, Et$_3$N
   60 °C, 2 h, 98%

2. Mg, PdCl$_2$, PPh$_3$
   MeOH, rt, 50 h, 48%

420

421(+ 34% E)

Halogen-containing indoles can be subjected to other nucleophilic substitution reactions under the guidance of palladium. Thus, 1-bromonauclefine is smoothly debrominated to nauclefine with Pd(Ph$_3$P)$_4$ (96%) [206], and 5-haloindoles are converted to 5-cyanoindole with cyanide and palladium catalysis [495]. The selective Pd-catalyzed 6-debromination of a 4,6-dibromoindole has been reported [496]. A reaction using potassium cyanide and the triflate of 1-hydroxy-β-carboline gives 1-cyano-β-carboline, which was transformed into the marine alkaloid eudistomin T upon reaction with benzyl Grignard reagent [497]. Palladium-catalysis has also been employed to deprotect *N*-allylindolines prepared by the Bailey–Liebeskind indoline synthesis [498] using the procedure of Genet [499]. Genet has also described the "ALLOC" deprotection of 4-formyl-*N*-(allyloxycarbonyl)indole using Pd(OAc)$_2$ and a water-soluble sulfonated Ph$_3$P [500]. Takacs has employed indole as a trap for a Pd-mediated tetraene carbocyclization leading to **422** [501].

Pd(OAc)$_2$, P(*o*-tol)$_3$

CH$_2$Cl$_2$, 40 °C
19 h, 91%

422

Fukuyama employed a Pd(OAc)$_2$ carbamate cleavage to trigger the final cyclization to (±)-catharanthine [502]. Godleski utilized palladium to close the D-ring in the key step of a synthesis of alloyohimbone [503]. In a search for novel rigid tryptamines, Vangveravong and Nichols used palladium acetate to catalyze the cyclopropanation of a 3-vinylindole chiral sultam leading to asymmetric syntheses of both enantiomers of trans-2-(indol-3-yl)cyclopropylamines and the corresponding cyclopropane-carboxylic acids [504]. Matsumoto and co-workers cyclized diazoindoles **423** to **424** with Pd(OAc)$_2$ in a system where Rh$_2$(OAc)$_4$ gives only benzene-ring cyclized products [505].

R_1 = H, Me, i-Bu
R_2 = H, Me
R_3 = Me, Et

Pd(OAc)$_2$, MeOH

0 °C, 45 min, 55–77%

423   **424**

Although the yields are quite variable (6–100%), phenyl-1-azirines and allyl-1-azirines can be ring-enlarged to indoles under the action of Pd [506, 507]. In the example shown, **425** to **426**, the intermediate Pd-complex **427** was isolated [506]. The reaction of 2-phenylazirine with Pd(Ph$_3$P)$_4$ (CO/BnNEt$_3$Cl/NaOH) gives 2-styrylindole in 29% yield [507]. Interestingly, an atmosphere of N$_2$ is detrimental to the success of this reaction and CO is normally used.

PdCl$_2$(PhCN)$_2$

CO, PhH, 30 °C, 100%

425   **426**   **427**

R = H, Me, Ph

Yang described the Pd-induced cyclization of an aryl bromide onto a pendant cyano group leading to γ-carbolines and related compounds [508]. Genet studied the use of chiral palladium complexes in the construction of the C-ring of ergot alkaloids, a study that culminated in a synthesis of (–)-chanoclavine I [509]. For example, nitroindole **428** is cyclized to **429** in 57% yield and with enantioselectivities of upto 95% using Pd(OAc)$_2$ and (S)-(–)-BINAP.

Pd(OAc)$_2$, (S)$^-$(–)-BINAP

K$_2$CO$_3$, THF, rt, 6 h, 57%

428   **429**

The Pd-catalyzed Michael addition of indoles to enones occurs readily in ionic liquids, e.g. **430** to **431** [510]. Related to the Hiyama coupling is Denmark's development of the cross-coupling of 2-indolyldimethylsilanols **432** with aryl halides to give 2-arylindoles **433** [511].

430

R = H, 5-Br, 5-MeO, 5-NO$_2$, 7-NO$_2$

431

432

433

R = H, 4-NO$_2$, 3-NO$_2$, 4-CF$_3$, 4-CO$_2$*t*-Bu, 2-Me, 2-OMe, 4-MeO, 1-naphthyl

A fitting way to end this chapter is with Sakai's biomimetic syntheses of 11-methoxyk-oumine and koumine [494, 512]. Thus, the presumed biogenetic intermediate 18-hydroxytaberpsychine (**434**), which was synthesized from 18-hydroxygardnutine, was acetylated and transformed into koumine (**435**) [494].

434

1. Ac$_2$O, pyr, 96%

2. NaH, DMF, rt
   Pd(OAc)$_2$, PPh$_3$
   1 h, 90 °C, 80%

435

Sakai's elegant application of Pd-induced nucleophilic reactions of allylic acetates pro-vides the first experimental support for the biogenesis of the koumine alkaloid skeleton and is an excellent concluding illustration of the power of palladium in indole chemistry.

In conclusion, the fantastically diverse chemistry of indole has been significantly enriched by Pd-catalyzed reactions. The accessibility of all of the possible halogenated indoles and several indolyl triflates has resulted in a wealth of synthetic applications as witnessed by the length of this chapter. In addition to the standard Pd-catalyzed reactions such as Negishi, Suzuki, Heck, Stille, and Sonogashira, which have had great success in indole chemistry, oxidative coupling, amination, and cyclization are powerful routes to a variety of carbazoles, carbolines, indolocarbazoles, and other fused indoles.

3.9. References

1. For a review of haloindoles, see Powers, J. C. in *The Chemistry of Heterocyclic Compounds*; Ed. Houlihan, W. J.; J. Wiley and Sons: New York, **1972**; vol. 25, Part 2, p. 128.

2. For a leading reference to the early literature, see Piers, K.; Meimaroglou, C.; Jardine, R. V.; Brown, R. K. *Can. J. Chem.* **1963**, *41*, 2399–401.

3. Arnold, R. D.; Nutter, W. M.; Stepp, W. L. *J. Org. Chem.* **1959**, *24*, 117–8.

4. Bocchi, V.; Palla, G. *Synthesis* **1982**, 1096–7.

5. Brennan, M. R.; Erickson, K. L.; Szmalc, F. S.; Tansey, M. J.; Thornton, J. M. *Heterocycles* **1986**, *24*, 2879–85.

6. Powers, J. C. *J. Org. Chem.* **1966**, *31*, 2627–31.

7. Katritzky, A. R.; Akutagawa, K. *Tetrahedron Lett.* **1985**, *26*, 5935–8.

8. Bergman, J.; Venemalm, L. *J. Org. Chem.* **1992**, *57*, 2495–7.

9. Sundberg, R. J.; Russell, H. F. *J. Org. Chem.* **1973**, *38*, 3324–30.

10. Saulnier, M. G.; Gribble, G. W. *J. Org. Chem.* **1982**, *47*, 757–61.

11. Illi, V. O. *Synthesis* **1979**, 136.

12. Conway, S. C.; Gribble, G. W. *Heterocycles* **1990**, *30*, 627–33.

13. Yang, C.-G.; Liu, G.; Jiang, B. *J. Org. Chem.* **2002**, *67*, 9392–6.

14. Conway, S. C.; Gribble, G. W. *Heterocycles* **1992**, *34*, 2095–108.

15. Gribble, G. W.; Allison, B. D.; Conway, S. C.; Saulnier, M. G. *Org. Prep. Proc. Int.* **1992**, *24*, 649–54.

16. Bergman, J.; Eklund, N. *Tetrahedron* **1980**, *36*, 1439–43.

17. Kline, T. *J. Heterocycl. Chem.* **1985**, *22*, 505–9.

18. (a) Hodson, H. F.; Madge, D. J.; Widdowson, D. A. *Synlett* **1992**, 831–2. (b) Hodson, H. F.; Madge, D. J.; Slawin, A. N. Z.; Widdowson, D. A.; Williams, D. J. *Tetrahedron* **1994**, *50*, 1899–906.

19. Chi, S. M.; Choi, J.-K.; Yum, E. K.; Chi, D. Y. *Tetrahedron Lett.* **2000**, *41*, 919–22.

20. Harrington, P. J.; Hegedus, L. S. *J. Org. Chem.* **1984**, *49*, 2657–62.

21. Hegedus, L. S.; Sestrick, M. R.; Michaelson, E. T.; Harrington, P. J. *J. Org. Chem.* **1989**, *54*, 4141–6.

22. Sundberg, R. J. *The Chemistry of Indoles*, Academic Press: New York, **1970**.

23. Sundberg, R. J. *Indoles*, Academic Press: New York, **1996**.

24. For reviews of indole ring synthesis, see (a) Gribble, G. W. *Cont. Org. Syn.* **1994**, *1*, 145–172. (b) Gribble, G. W. *Perkin Trans. 1* **2000**, 1045–75.

25. Murase, M.; Watanabe, K.; Kurihara, T.; Tobinaga, S. *Chem. Pharm. Bull.* **1998**, *46*, 889–92.

26. Somei, M.; Tsuchiya, M. *Chem. Pharm. Bull.* **1981**, *29*, 3145–57.

27. (a) Thesing, J.; Semler, G.; Mohr, G. *Chem. Ber.* **1962**, *95*, 2205–11. (b) Russell, H. F.; Harris, B. J.; Hood, D. B.; Thompson, E. G.; Watkins, A. D.; Williams, R. D. *Org. Prep. Proc. Int.* **1985**, *17*, 391–9.

28. Miyake, Y.; Kikugawa, Y. *J. Heterocycl. Chem.* **1983**, *20*, 349–52.

29. (a) Carrera, Jr., G. M.; Sheppard, G. S. *Synlett* **1994**, 93–4. (b) Moyer, M. P.; Shiurba, J. F.; Rapoport, H. *J. Org. Chem.* **1986**, *51*, 5106–10.

30. Harrowven, D. C.; Lai, D.; Lucas, M. C. *Synthesis* **1999**, 1300–2.

31. Iwao, M.; Kuraishi, T. *Heterocycles* **1992**, *34*, 1031–8.

32. Meyers, A. I.; Hutchings, R. H. *Tetrahedron Lett.* **1993**, *34*, 6185–8.

33. Hutchings, R. H.; Meyers, A. I. *J. Org. Chem.* **1996**, *61*, 1004–13.

34. Kuehne, M. E.; Hall, T. C. *J. Org. Chem.* **1976**, *41*, 2742–6.

35. Chen, S.; Vasquez, L.; Noll, B. C.; Rakowski DuBois, M. *Organometallics* **1997**, *16*, 1757–64.

36. Sakamoto, T.; Yasuhara, A.; Kondo, Y.; Yamanaka, H. *Heterocycles* **1993**, *36*, 2597–600.

37. Beswick, P. J.; Greenwood, C. S.; Mowlem, T. J.; Nechvatal, G.; Widdowson, D. A. *Tetrahedron* **1988**, *44*, 7325–34.

38. Hollins, R. A.; Colnago, L. A.; Salim, V. M.; Seidl, M. C. *J. Heterocycl. Chem.* **1979**, *16*, 993–6.

39. Somei, M.; Yamada, F.; Kunimoto, M.; Kaneko, C. *Heterocycles* **1984**, *22*, 797–801.

40. Somei, M.; Saida, Y. *Heterocycles* **1985**, *23*, 3113–4.

41. Somei, M.; Saida, Y.; Funamoto, T.; Ohta, T. *Chem. Pharm. Bull.* **1987**, *35*, 3146–54.

42. Ritter, K. *Synthesis* **1993**, 735–62.

43. Conway, S. C.; Gribble, G. W. *Synth. Commun.* **1992**, *22*, 2987–95.

44. Bourlot, A. S.; Desarbre, E.; Mérour, J. Y. *Synthesis* **1994**, 411–6.

45. Åkermark, B.; Eberson, L.; Jonsson, E.; Pettersson, E. *J. Org. Chem.* **1975**, *40*, 1365–7.

46. (a) Åkermark, B.; Oslob, J. D.; Heuschert, U. *Tetrahedron Lett.* **1995**, *36*, 1325–6. (b) Hagelin, H.; Oslob, J. D.; Åkermark, B. *Chem. Eur. J.* **1999**, *5*, 2413–6.

47. (a) Furukawa, H.; Yogo, M.; Ito, C.; Wu, T.-S.; Kuoh, C.-S. *Chem. Pharm. Bull.* **1985**, *33*, 1320–2. (b) Furukawa, H.; Ito, C.; Yogo, M.; Wu, T. *Chem. Pharm. Bull.* **1986**, *34*, 2672–5. (c) Yogo, M.; Ito, C.; Furukawa, H. *Chem. Pharm. Bull.* **1991**, *39*, 328–34. (d) Bittner, S.; Krief, P.; Massil, T. *Synthesis* **1991**, 215–6. (e) Benavides, A.; Peralta, J.; Delgado, F.; Tamariz, J. *Synthesis* **2004**, 2499–2504. (f) Jacquelin, C.; Saettel, N.; Hounsou, C.; Teulade-Fichou, M.-P. *Tetrahedron Lett.* **2005**, *46*, 2589–92.

48. (a) Miller, R. B.; Moock, T. *Tetrahedron Lett.* **1980**, *21*, 3319–22. (b) Oliveira-Campos, A. M. F.; Queiroz, M.-J. R. P.; Rapso, M. M. M.; Shannon, P. V. R. *Tetrahedron Lett.* **1995**, *36*, 133–4.

49. Morel, S.; Boyer, G.; Coullet, F.; Galy, J.-P. *Synth. Commun.* **1996**, *26*, 2443–7.

50. Mandal, A. B.; Delgado, F.; Tamariz, J. *Synlett* **1998**, 87–9.

51. Knölker, H.-J.; O'Sullivan, N. *Tetrahedron Lett.* **1994**, *35*, 1695–8.

52. Knölker, H.-J.; O'Sullivan, N. *Tetrahedron* **1994**, *50*, 10893–908.

53. Knölker, H.-J.; Fröhner, W. *J. Chem. Soc., Perkin Trans. 1* **1998**, 173–5.

54. (a) Knölker, H.-J.; Reddy, K. R.; Wagner, A. *Tetrahedron Lett.* **1998**, *39*, 8267–70. (b) Knölker, H.-J.; Fröhner, W.; Reddy, K. R. *Synthesis* **2002**, 557–64.

55. Knölker, H.-J.; Reddy, K. R. *Synlett* **1999**, 596–8.

56. Knölker, H.-J.; Knöll, J. *Chem. Commun.* **2003**, 1170–1.

57. Trost, B. M.; Genet, J. P. *J. Am. Chem. Soc.* **1976**, *98*, 8516–7.

58. Trost, B. M.; Godleski, S. A.; Genet, J. P. *J. Am. Chem. Soc.* **1978**, *100*, 3930–1.

59. (a) Trost, B. M.; Godleski, S. A.; Belletire, J. L. *J. Org. Chem.* **1979**, *44*, 2052–4. (b) Trost, B. M.; Fortunak, J. M. D. *Organometallics* **1982**, *1*, 7–13.

60. Cushing, T. D.; Sanz-Cervera, J. F.; Williams, R. M. *J. Am. Chem. Soc.* **1996**, *118*, 557–79.

61. Ferreira, E. M.; Stoltz, B. M. *J. Am. Chem. Soc.* **2003**, *125*, 9578–9.

62. Liu, C.; Widenhoefer, R. A. *J. Am. Chem. Soc.* **2004**, *126*, 10250–1.

63. (a) Itahara, T.; Sakakibara, T. *Synthesis* **1978**, 607–8. (b) Itahara, T. *Synthesis* **1979**, 151–2. (c) Itahara, T. *Heterocycles* **1986**, *24*, 2557–62.

64. Itahara, T. *J. Chem. Soc., Chem. Commun.* **1981**, 254–5.

65. Itahara, T. *J. Org. Chem.* **1985**, *50*, 5272–5.

66. (a) Black, D. St C.; Keller, P. A.; Kumar, N. *Tetrahedron Lett.* **1989**, *30*, 5807–8. (b) Black, D. St C.; Keller, P. A.; Kumar, N. *Tetrahedron* **1993**, *49*, 151–64.

67. Black, D. St C.; Kumar, N.; Wong, L. C. H. *J. Chem. Soc., Chem. Commun.* **1985**, 1174–5.
68. Abbiati, G.; Beccalli, E. M.; Broggini, G.; Zoni, C. *J. Org. Chem.* **2003**, *68*, 7625–8.
69. Harris, W.; Hill, C. H.; Keech, E.; Malsher, P. *Tetrahedron Lett.* **1993**, *34*, 8361–4.
70. Gribble, G. W.; Berthel, S. J. *Stud. Nat. Prod. Chem.* Ed. Atta-ur-Rahman; Elsevier: New York, **1993**; vol. 12, p. 365.
71. Ohkubo, M.; Nishimura, T.; Jona, H.; Honma, T.; Morishima, H. *Tetrahedron* **1996**, *52*, 8099–112.
72. Faul, M. M.; Winneroski, L. L.; Krumrich, C. A. *J. Org. Chem.* **1998**, *63*, 6053–8.
73. Jeevanandam, A.; Srinivasan, P. C. *Synth. Commun.* **1995**, *25*, 3427–34.
74. Itahara, T. *J. Org. Chem.* **1985**, *50*, 5546–50.
75. Billups, W. E.; Erkes, R. S.; Reed, L. E. *Synth. Commun.* **1980**, *10*, 147–54.
76. Fujiwara, Y.; Maruyama, O.; Yoshidomi, M.; Taniguchi, H. *J. Org. Chem.* **1981**, *46*, 851–5.
77. (a) Itahara, T.; Ikeda, M.; Sakakibara, T. *J. Chem. Soc., Perkin Trans. 1* **1983**, 1361–3. (b) Itahara, T.; Kawasaki, K.; Ouseto, F. *Synthesis* **1984**, 236–7.
78. Capito, E.; Brown, J. M.; Ricci, A. *Chem. Commun.* **2005**, 1854–6.
79. (a) Yokoyama, Y.; Matsumoto, T.; Murakami, Y. *J. Org. Chem.* **1995**, *60*, 1486–7. (b) Osanai, K.; Yokoyama, Y.; Kondo, K.; Murakami, Y. *Chem. Pharm. Bull.* **1999**, *47*, 1587–90.
80. Yokoyama, Y.; Takahashi, M.; Kohno, Y.; Kataoka, K.; Fujikawa, Y.; Murakami, Y. *Heterocycles* **1990**, *31*, 803–4.
81. Yokoyama, Y.; Takahashi, M.; Takashima, M.; Kohno, Y.; Kobayashi, H.; Kataoka, K.; Shidori, K.; Murakami, Y. *Chem. Pharm. Bull.* **1994**, *42*, 832–8.
82. Murakami, Y.; Yokoyama, Y.; Aoki, T. *Heterocycles* **1984**, *22*, 1493–6.
83. Yokoyama, Y.; Takashima, M.; Higaki, C.; Shidori, K.; Moriguchi, S.; Ando, C.; Murakami, Y. *Heterocycles* **1993**, *36*, 1739–42.
84. Yokoyama, Y.; Suzuki, H.; Matsumoto, S.; Sunaga, Y.; Tani, M.; Murakami, Y. *Chem. Pharm. Bull.* **1991**, *39*, 2830–6.
85. Pindur, U.; Adam, R. *Helv. Chim. Acta* **1990**, *73*, 827–38.
86. Jia, C.; Lu, W.; Kitamura, T.; Fujiwara, Y. *Org. Lett.* **1999**, *1*, 2097–100.
87. Abdrakhmanov, I. B.; Mustafin, A. G.; Tolstikov, G. A.; Fakhretdinov, R. N.; Dzhemilev, U. M. *Chem. Heterocycl. Cpds.* **1986**, 262–4.
88. Fukuda, Y.; Furuta, H.; Shiga, F.; Asahina, Y.; Terashima, S. *Heterocycles* **1997**, *45*, 2303–8.
89. (a) Lane, B. S.; Sames, D. *Org. Lett.* **2004**, *6*, 2897–2900. (b) Lane, B. S.; Brown, M. A.; Sames, D. *J. Am. Chem. Soc.* **2005**, *127*, 8050–7.
90. (a) Bandini, M.; Melloni, A.; Umani-Ronchi, A. *Org. Lett.* **2004**, *6*, 3199–202. (b) Ma, S.; Yu, S. *Tetrahedron Lett.* **2004**, *45*, 8419–22.
91. (a) Daniell, K.; Stewart, M.; Madsen, E.; Le, M.; Handl, H.; Brooks, N.; Kiakos, K.; Hartley, J. A.; Lee, M. *Bioorg. Med. Chem. Lett.* **2005**, *15*, 177–80. (b) Abreu, A. S.; Ferreira, P. M. T.; Queiroz, M.-J. R. P.; Ferreira, I. C. F. R.; Calhelha, R. C.; Estevinho, L. M. *Eur. J. Org. Chem.* **2005**, 2951–7.
92. (a) Minato, A.; Tamao, K.; Hayashi, T.; Suzuki, K.; Kumada, M. *Tetrahedron Lett.* **1981**, *22*, 5319–22. (b) Minato, A.; Suzuki, K.; Tamao, K.; Kumada, M. *Tetrahedron Lett.* **1984**, *25*, 83–6. (c) Minato, A.; Suzuki, K.; Tamao, K.; Kumada, M. *J. Chem. Soc., Chem. Commun.* **1984**, 511–13.

93. Kondo, Y.; Yoshida, A.; Sato, S.; Sakamoto, T. *Heterocycles* **1996**, *42*, 105–8.

94. Widdowson, D. A.; Zhang, Y.-Z. *Tetrahedron* **1986**, *42*, 2111–6.

95. Vincent, P.; Beaucourt, J. P.; Pichart, L. *Tetrahedron Lett.* **1984**, *25*, 201–2.

96. (a) Sakamoto, T.; Kondo, Y.; Takazawa, N.; Yamanaka, H. *Heterocycles* **1993**, *36*, 941–2. (b) Sakamoto, T.; Kondo, Y.; Takazawa, N.; Yamanaka, H. *Tetrahedron Lett.* **1993**, *34*, 5955–6. (c) Sakamoto, T.; Kondo, Y.; Takazawa, N.; Yamada, H. *J. Chem. Soc., Perkin Trans. 1* **1996**, 1927–34. (d) Kondo, Y.; Takazawa, N.; Yoshida, A.; Sakamoto, T. *J. Chem. Soc., Perkin Trans. 1* **1995**, 1207–8.

97. (a) Amat, M.; Hadida, S.; Bosch, J. *Tetrahedron Lett.* **1993**, *34*, 5005–6. (b) Amat, M.; Hadida, S.; Bosch, J. *Tetrahedron Lett.* **1994**, *35*, 793–6. (c) Amat, M.; Hadida, S.; Pshenichnyi, G.; Bosch, J. *J. Org. Chem.* **1997**, *62*, 3158–75.

98. Danieli, B.; Lesma, G.; Martinelli, M.; Passarella, D.; Peretto, I.; Silvani, A. *Tetrahedron* **1998**, *54*, 14081–8.

99. Fisher, L. E.; Labadie, S. S.; Reuter, D. C.; Clark, R. D. *J. Org. Chem.* **1995**, *60*, 6224–5.

100. Cheng, K.-F.; Cheung, M.-K. *J. Chem. Soc., Perkin Trans. 1* **1996**, 1213–8.

101. Herz, H.-G.; Queiroz, M. J. R. P.; Maas, G. *Synthesis* **1999**, 1013–6.

102. Pimm, A.; Kocienski, P.; Street, S. D. A. *Synlett* **1992**, 886–8.

103. Leonard, K.; Pan, W.; Anaclerio, B.; Gushue, J. M.; Guo, Z.; DesJarlais, R. L.; Chaikin, M. A.; Lattanze, J.; Crysler, C.; Manthey, C. L.; Tomczuk, B. E.; Marugan, J. J. *Bioorg. Med. Chem. Lett.* **2005**, *15*, 2679–84.

104. Faul, M. M.; Winneroski, L. L. *Tetrahedron Lett.* **1997**, *38*, 4749–52.

105. Bergman, J.; Venemalm, L. *Tetrahedron* **1990**, *46*, 6061–6.

106. Davidsen, S. K.; Summers, J. B.; Albert, D. H.; Holms, J. H.; Heyman, H. R.; Magoc, T. J.; Conway, R. G.; Rhein, D. A.; Carter, G. W. *J. Med. Chem.* **1994**, *26*, 4423–9.

107. Griffen, E. J.; Roe, D. G.; Snieckus, V. *J. Org. Chem.* **1995**, *60*, 1484–5.

108. Bracher, F.; Hildebrand, D. *Tetrahedron* **1994**, *50*, 12329–36.

109. Burns, B.; Grigg, R.; Sridharan, V.; Stevenson, P.; Sukirthalingam, S.; Worakun, T. *Tetrahedron Lett.* **1989**, *30*, 1135–8.

110. Luo, F.-T.; Wang, R.-T. *Heterocycles* **1991**, *32*, 2365–72.

111. Lindsay, D. M.; Dohle, W.; Jensen, A. E.; Kopp, F.; Knochel, P. *Org. Lett.* **2002**, *4*, 1819–22.

112. Claridge, T. D. W.; Long, J. M.; Brown, J. M.; Hibbs, D.; Hursthouse, M. B. *Tetrahedron* **1997**, *53*, 4035–50.

113. (a) Zheng, Q.; Yang, Y.; Martin, A. R. *Tetrahedron Lett.* **1993**, *34*, 2235–8. (b) Zheng, Q.; Yang, Y.; Martin, A. R. *Heterocycles* **1994**, *37*, 1761–72.

114. (a) Jiang, B.; Yang, C.-G.; Xiong, W.-N.; Wang, J. *Bioorg. Med. Chem.* **2001**, *9*, 1149–54. (b) Yang, C.-G.; Liu, G.; Jiang, B. *J. Org. Chem.* **2002**, *67*, 9392–6. (c) Fang, Q.; Xu, B.; Jiang, B.; Fu, H.; Chen, X.; Cao, A. *Chem. Commun.* **2005**, 1468–70. (d) See also: de Koning, C. B.; Michael, J. P.; Pathak, R.; van Otterlo, W. A. L. *Tetrahedron Lett.* **2004**, *45*, 1117–19.

115. (a) Garg, N. K.; Sarpong, R.; Stoltz, B. M. *J. Am. Chem. Soc.* **2002**, *124*, 13179–84. (b) Garg, N. K.; Caspi, D. D.; Stoltz, B. M. *J. Am. Chem. Soc.* **2004**, *126*, 9552–3. (c) Garg, N. K.; Caspi, D. D.; Stoltz, B. M. *J. Am. Chem. Soc.* **2005**, *127*, 5970–8. (d) Garg, N. K.; Stoltz, B. M. *Tetrahedron Lett.* **2005**, *46*, 2423–6.

116. Kawasaki, I.; Yamashita, M.; Ohta, S. *Chem. Pharm. Bull.* **1996**, *44*, 1831–9.

117. Hoerrner, R. S.; Askin, D.; Volante, R. P.; Reider, P. J. *Tetrahedron Lett*. **1998**, *39*, 3455–8.
118. Nishida, A.; Miyashita, N.; Fuwa, M.; Nakagawa, M. *Heterocycles* **2003**, *59*, 473–6.
119. Neel, D. A.; Jirousek, M. R.; McDonald III, J. H. *Bioorg. Med. Chem. Lett*. **1998**, *8*, 47–50.
120. Wallow, T. I., Novak, B. M. *J. Org. Chem*. **1994**, *59*, 5034–7.
121. Wright, S. W.; Hageman, D. L.; McClure, L. D. *J. Org. Chem*. **1994**, *59*, 6095–7.
122. (a) Merlic, C. A.; McInnes, D. M.; You, Y. *Tetrahedron Lett*. **1997**, *38*, 6787–90. (b) Merlic, C. A.; McInnes, D. M. *Tetrahedron Lett*. **1997**, *38*, 7661–4.
123. Johnson, C. N.; Stemp, G.; Anand, N.; Stephen, S. C.; Gallagher, T. *Synlett* **1998**, 1025–7.
124. Vazquez, E.; Davies, I. W.; Payack, J. F. *J. Org. Chem*. **2002**, *67*, 7551–2.
125. Zhao, Z.; Snieckus, V. *Org. Lett*. **2005**, *7*, 2523–6.
126. (a) Yang, Y.; Martin, A. R.; Nelson, D. L.; Regan, J. *Heterocycles* **1992**, *34*, 1169–75. (b) Yang, Y.; Martin, A. R. *Heterocycles* **1992**, *34*, 1395–8.
127. (a) Carbonnelle, A-C.; Gonz·lez-Zamora, E.; Beugelmans, R.; Roussi, G. *Tetrahedron Lett*. **1998**, *39*, 4467–70. (b) Carbonnelle, A.-C.; Gonz·lez-Zamora, E.; Beugelmans, R.; Roussi, G. *Tetrahedron Lett*. **1998**, *39*, 4471–2.
128. Cai, X.; Snieckus, V. *Org. Lett*. **2004**, *6*, 2293–5.
129. (a) Fraley, M. E.; Arrington, K. L.; Buser, C. A.; Ciecko, P. A.; Coll, K. E.; Fernandes, C.; Hartman, G. D.; Hoffman, W. F.; Lynch, J. J.; McFall, R. C.; Rickert, K.; Singh, R.; Smith, S.; Thomas. K. A.; Wong, B. K. *Bioorg. Med. Chem. Lett*. **2004**, *14*, 351–5. (b) See also: Payack, J. F.; Vazquez, E.; Matty, L.; Kress, M. H.; McNamara, J. *J. Org. Chem*. **2005**, *70*, 175–8.
130. Hendrickson, J. B.; Wang, J. *Org. Lett*. **2004**, *6*, 3–5.
131. (a) Prieto, M.; Zurita, E.; Rosa, E.; Muñoz, L.; Lloyd-Williams, P.; Giralt, E. *J. Org. Chem*. **2004**, *69*, 6812–20. (b) However, see: Leadbeater, N. E.; Marco, M. *J. Org. Chem*. **2003**, *68*, 5660–7.
132. Bouchard, J.; Wakim, S.; Leclerc, M. *J. Org. Chem*. **2004**, *69*, 5705–11.
133. Zembower, D. E.; Ames, M. M. *Synthesis* **1994**, 1433–6.
134. Siddiqui, M. A.; Snieckus, V. *Tetrahedron Lett*. **1990**, *31*, 1523–6.
135. Yang, Y.; Martin, A. R. *Synth. Commun*. **1992**, *22*, 1757–62.
136. Banwell, M. G.; Bissett, B. D.; Busato, S.; Cowden, C. J.; Hockless, D. C. R.; Holman, J. W.; Read, R. W.; Wu, A. W. *J. Chem. Soc., Chem. Commun*. **1995**, 2551–2.
137. Moody, C. J.; Doyle, K. J.; Elliott, M. C.; Mowlem, T. J. *J. Chem. Soc. Perkin Trans.1* **1997**, 2413–9.
138. Guillard, J.; Larraya, C.; Viaud-Massuard, M.-C. *Heterocycles* **2003**, *60*, 865–77.
139. Wu, T. Y. H.; Schultz, P. G. *Org. Lett*. **2002**, *4*, 4033–6.
140. Wu, T. Y. H.; Ding, S.; Gray, N. S.; Schultz, P. G. *Org. Lett*. **2001**, *3*, 3827–30.
141. Zhang, H.-C.; Ye, H.; White, K. B.; Maryanoff, B. E. *Tetrahedron Lett*. **2001**, *42*, 4751–4.
142. Liu, Y.; Gribble. G. W. *Tetrahedron Lett*. **2000**, *41*, 8717–21.
143. Witulski, B.; Azcon, J. R.; Alayrac, C.; Arnautu, A.; Collot, V.; Rault, S. *Synthesis* **2005**, 771–80.
144. (a) de Koning, C. B.; Michael, J. P.; Rousseau, A. L. *Tetrahedron Lett*. **1998**, *39*, 8725–8. (b) de Koning, C. B.; Michael J. P.; Nhlapo, J. M.; Pathak, R.; van Otterlo, W. A. L. *Synlett* **2003**, 705–7.

145. Watterson, S. H.; Dhar, T. G. M.; Ballentine, S. K.; Shen, Z.; Barrish, J. C.; Cheney, D.; Fleener, C. A.; Rouleau, K. A.; Townsend, R.; Hollenbaugh, D. L.; Iwanowicz, E. J. *Bioorg. Med. Chem. Lett.* **2003**, *13*, 1273–6.
146. Walsh, T. F.; Toupence, R. B.; Ujjainwalla, F.; Young, J. R.; Goulet, M. T. *Tetrahedron* **2001**, *57*, 5233–41.
147. Ishikura, M.; Kamada, M.; Terashima, M. *Synthesis* **1984**, 936–8.
148. Somei, M.; Amari, H.; Makita, Y. *Chem. Pharm. Bull.* **1986**, *34*, 3971–3.
149. Tidwell, J. H.; Peat, A. J.; Buchwald, S. L. *J. Org. Chem.* **1994**, *59*, 7164–8.
150. Chu, L.; Fisher, M. H.; Goulet, M. T.; Wyvratt, M. J. *Tetrahedron Lett.* **1997**, *38*, 3871–4.
151. Weinstein, D. S.; Liu, W.; Gu, Z.; Langevine, C.; Ngu, K.; Fadnis, L.; Combs, D. W.; Sitkoff, D.; Ahmad, S.; Zhuang, S.; Chen, X.; Wang, F.-L.; Loughney, D. A.; Atwal, K. S.; Zahler, R.; Macor, J. E.; Madsen, C. S.; Murugesan, N. *Bioorg. Med. Chem. Lett.* **2005**, *15*, 1435–40.
152. Meng, C. Q.; Rakhit, S.; Lee, D. K. H.; Kamboj, R.; McCallum, K. L.; Mazzocco, L.; Dyne, K.; Slassi, A. *Bioorg. Med. Chem. Lett.* **2000**, *10*, 903–5.
153. Carini, D. J.; Kaltenback III, R. F.; Liu, J.; Benfield, P. A.; Boylan, J.; Boisclair, M.; Brizuela, L.; Burton, C. R.; Cox, S.; Grafstrom, R.; Harrison, B. A.; Harrison, K.; Akamike, E.; Markwalder, J. A.; Nakano, Y.; Seitz, S. P.; Sharp, D. M.; Trainor, G. L.; Sielecki, T. M. *Bioorg. Med. Chem. Lett.* **2001**, *11*, 2209–11.
154. Nicolaou, K. C.; Snyder, S. A.; Simonsen, K. B.; Koumbis, A. E. *Angew. Chem. Int. Ed.* **2000**, *39*, 3473–8.
155. (a) Joseph, B.; Malapel, B.; Mérour, J.-Y. *Synth. Commun.* **1996**, *26*, 3289–95. (b) Malapel-Andrieu, B.; Mérour, J.-Y. *Tetrahedron* **1998**, *54*, 11079–94.
156. Doi, K.; Mori, M. *Heterocycles* **1996**, *42*, 113–6.
157. (a) Choshi, T.; Sada, T.; Fujimoto, H.; Nagayama, C.; Sugino, E.; Hibino, S. *Tetrahedron Lett.* **1996**, *37*, 2593–6. (b) Choshi, T.; Sada, T.; Fujimoto, H.; Nagayama, C.; Sugino, E.; Hibino, S. *J. Org. Chem.* **1997**, *62*, 2535–43. (c) Choshi, T.; Kuwada, T.; Fukui, M.; Matsuya, Y.; Sugino, E.; Hibino, S. *Chem. Pharm. Bull.* **2000**, *48*, 108–13.
158. McCort, G.; Duclos, O.; Cadilhac, C.; Guilpain, E. *Tetrahedron Lett.* **1999**, *40*, 6211–15.
159. Kubota, H.; Rothman, R. B.; Dersch, C.; McCullough, K.; Pinto, J.; Rice, K. C. *Bioorg. Med. Chem. Lett.* **1998**, *8*, 799–804.
160. Bracher, F.; Hildebrand, D. *Liebigs Ann. Chem.* **1992**, 1315–19.
161. Joseph, B.; Alagille, D.; Rousseau, C.; Mérour, J.-Y. *Tetrahedron* **1999**, *55*, 4341–52.
162. (a) Ishikura, M.; Terashima, M. *J. Chem. Soc., Chem. Commun.* **1989**, 135–6. (b) For a review, see Ishikura, M.; Agata, I. *Recent Res. Devel. Org. Chem.* **1997**, *1*, 145–57. (c) Ishikura, M.; Terashima, M.; Okamura, K.; Date, T. *J. Chem. Soc., Chem. Commun.* **1991**, 1219–21. (d) Ishikura, M.; Terashima, M. *Tetrahedron Lett.* **1992**, *33*, 6849–52. (e) Ishikura, M.; Terashima, M. *J. Org. Chem.* **1994**, *59*, 2634–7. (f) Ishikura, M. *J. Chem. Soc., Chem. Commun.* **1995**, 409–10. (g) Ishikura, M.; Agata, I. *Heterocycles* **1996**, *43*, 1591–5. (h) Ishikura, M.; Matsuzaki, Y.; Agata, I. *Chem. Commun.* **1996**, 2409–10. (i) Ishikura, M.; Yaginuma, T.; Agata, I.; Miwa, Y. Yanada, R.; Taga, T. *Synlett* **1997**, 214–16. (j) Ishikura, M.; Hino, A.; Yaginuma, T.; Agata, I.; Katagiri, N. *Tetrahedron* **2000**, *56*, 193–207. (k) Ishikura, M.; Matsuzaki, Y.; Agata, I.; Katagiri, N. *Tetrahedron* **1998**,

54, 13929–42. (l) Ishikawa, M.; Agata, I.; Katagiri, N. *J. Heterocycl. Chem.* **1999**, *36*, 873–9. (m) Ishikura, M.; Matsuzaki, Y.; Agata, I. *Heterocycles* **1997**, *45*, 2309–12.

163. (a) Molander, G. A.; Bernardi, C. R. *J. Org. Chem.* **2002**, *67*, 8424–9. (b) Molander, G. A.; Biolatto, B. *J. Org. Chem.* **2003**, *68*, 4302–14.

164. Satoh, M.; Miyaura, N.; Suzuki, A. *Synthesis* **1987**, 373–7.

165. Kumar, V.; Dority, J. A.; Bacon, E. R.; Singh, B.; Lesher, G. Y. *J. Org. Chem.* **1992**, *57*, 6995–8.

166. Gridnev, I. D.; Miyaura, N.; Suzuki, A. *J. Org. Chem.* **1993**, *58*, 5351–4.

167. Sun, L.; Tran, N.; Liang, C.; Tang, F.; Rice, A.; Schreck, R.; Waltz, K.; Shawver, L. K.; McMahon, G.; Tang, C. *J. Med. Chem.* **1999**, *42*, 5120–30.

168. Todd, M. H.; Oliver, S. F.; Abell, C. *Org. Lett.* **1999**, *1*, 1149–51.

169. (a) Rocca, P.; Marsais, F.; Godard, A.; Quéguiner, G.; Adams, L.; Alo, B. *J. Heterocycl. Chem.* **1995**, *32*, 1171–5. (b) Trécourt, F.; Mongin, F.; Mallet, M.; Quéguiner, G. *Synth. Commun.* **1995**, *25*, 4011–24. (c) Arzel, E.; Rocca, P.; Marsais, F.; Godard, A.; Quéguiner, G. *J. Heterocycl. Chem.* **1997**, *34*, 1205–10.

170. Grigg, R.; Sansano, J. M.; Santhakumar, V.; Sridharan, V.; Thangavelanthum, R.; Thornton-Pett, M.; Wilson, D. *Tetrahedron* **1997**, *53*, 11803–26.

171. Grigg, R.; Sridharan, V.; Zhang, J. *Tetrahedron Lett.* **1999**, *40*, 8277–80.

172. Amat, M.; Hadida, S.; Sathyanarayana, S.; Bosch, J. *J. Org. Chem.* **1994**, *59*, 10–1.

173. Beak, P.; Lee, W. K. *J. Org. Chem.* **1993**, *58*, 1109–17.

174. Labadie, S. S.; Teng, E. *J. Org. Chem.* **1994**, *59*, 4250–4.

175. Caddick, S.; Joshi, S. *Synlett* **1992**, 805–6.

176. Arnswald, M.; Neumann, W. P. *J. Org. Chem.* **1993**, *58*, 7022–8.

177. Palmisano, G.; Santagostino, M. *Helv. Chim. Acta* **1993**, *76*, 2356–66.

178. Palmisano, G.; Santagostino, M. *Synlett* **1993**, 771–3.

179. Hudkins, R. L.; Diebold, J. L.; Marsh, F. D. *J. Org. Chem.* **1995**, *60*, 6218–20.

180. Brenner, M.; Mayer, G.; Terpin, A.; Steglich, W. *Chem. Eur. J.* **1997**, *3*, 70–4.

181. Fraley, M. E.; Hoffman, W. F.; Arrington, K. L.; Hungate, R. W.; Hartman, G. D.; McFall, R. C.; Coll, K. E.; Rickert, K.; Thomas, K. A.; McGaughey, G. B. *Curr. Med. Chem.* **2004**, *11*, 709–19.

182. Murray, P. E.; Mills, K.; Joule, J. A. *J. Chem. Res. (S)* **1998**, 377.

183. (a) Fukuyama, T.; Chen, X.; Peng, G. *J. Am. Chem. Soc.* **1994**, *116*, 3127–8. (b) Kobayashi, Y.; Fukuyama, T. *J. Heterocycl. Chem.* **1998**, *35*, 1043–55. (c) Kobayashi, Y.; Peng, G.; Fukuyama, T. *Tetrahedron Lett.* **1999**, *40*, 1519–22. (d) Kobayashi, S.; Ueda, T.; Fukuyama, T. *Synlett* **2000**, 883–6.

184. Kraxner, J.; Arlt, M.; Gmeiner, P. *Synlett* **2000**, 125–7.

185. Ciattini, P. G.; Morera, E.; Ortar, G. *Tetrahedron Lett.* **1994**, *35*, 2405–8.

186. (a) Yokoyama, Y.; Ito, S.; Takahashi, Y.; Murakama, Y. *Tetrahedron Lett.* **1985**, *26*, 6457–60. (b) Yokoyama, Y.; Ikeda, M.; Saito, M.; Yoda, T.; Suzuki, H.; Murakami, Y. *Heterocycles* **1990**, *31*, 1505–11.

187. Yoshida, S.; Kubo, H.; Saika, T.; Katsumura, S. *Chem. Lett.* **1996**, 139–40.

188. Somei, M.; Sayama, S.; Naka, K.; Yamada, F. *Heterocycles* **1988**, *27*, 1585–7.

189. Palmisano, G.; Santagostino, M. *Tetrahedron Lett.* **1993**, *49*, 2533–42.

190. Gribble, G. W.; Conway, S. C. *Synth. Commun.* **1992**, *22*, 2129–41.

191. (a) Sakamoto, T.; Kondo, Y.; Yasuhara, A.; Yamanaka, H. *Heterocycles* **1990**, *31*, 219–21. (b) Sakamoto, T.; Kondo, Y.; Yasuhara, A.; Yamanaka, H. *Tetrahedron* **1991**, *47*, 1877–86.

192. Sakamoto, T.; Yasuhara, A.; Kondo, Y.; Yamanaka, H. *Chem. Pharm. Bull.* **1994**, *42*, 2032–5.

193. Vaillancourt, V.; Albizati, K. F. *J. Am. Chem. Soc.* **1993**, *115*, 3499–502.

194. (a) Choshi, T.; Yamada, S.; Sugino, E.; Kuwada, T.; Hibino, S. *Synlett* **1995**, 147–8. (b) Choshi, T.; Yamada, S.; Sugino, E.; Kuwada, T.; Hibino, S. *J. Org. Chem.* **1995**, *60*, 5899–904. (c) Choshi, Y.; Yamada, S.; Nobuhiro, J.; Mihara, Y.; Sugino, E.; Hibino, S. *Heterocycles* **1998**, *48*, 11–14. (d) Choshi, T.; Fujimoto, H.; Sugino, E.; Hibino, S. *Heterocycles* **1996**, *43*, 1847–54.

195. Achab, S.; Guyot, M.; Potier, P. *Tetrahedron Lett.* **1995**, *36*, 2615–18.

196. Krolski, M. E.; Renaldo, A. F.; Rudisill, D. E.; Stille, J. K. *J. Org. Chem.* **1988**, *53*, 1170–6.

197. Dickens, M. J.; Mowlem, T. J.; Widdowson, D. A.; Slawin, A. M. Z.; Williams, D. J. *J. Chem. Soc., Perkin Trans. 1* **1992**, 323–5.

198. Somei, M.; Yamada, F.; Naka, K. *Chem. Pharm. Bull.* **1987**, *35*, 1322–5.

199. Pérez-Serrano, L.; Casarrubios, L.; Domínguez, G.; Freire, G.; Pérez-Castells, J. *Tetrahedron* **2002**, *58*, 5407–15.

200. (a) Pearce, B. C. *Synth. Commun.* **1992**, *22*, 1627–43. (b) Benhida, R.; Lecubin, F.; Fourrey, J.-L.; Rivas Castellanos, L.; Quintero, L. *Tetrahedron Lett.* **1999**, *40*, 5701–3.

201. Miki, Y.; Matsushita, K.; Hibino, H.; Shirokoshi, H. *Heterocycles* **1999**, *51*, 1585–91.

202. Grigg, R.; Teasdale, A.; Sridharan, V. *Tetrahedron Lett.* **1991**, *32*, 3859–62.

203. Somei, M.; Kawasaki, T.; Ohta, T. *Heterocycles* **1988**, *27*, 2363–5.

204. Iwao, M.; Takehara, H.; Obata, S.; Watanabe, M. *Heterocycles* **1994**, *38*, 1717–20.

205. Bracher, F.; Hildebrand, D. *Liebigs Ann. Chem.* **1993**, 837–9.

206. Lavilla, R.; Gullón, F.; Bosch, J. *J. Chem. Soc., Chem. Commun.* **1995**, 1675–6.

207. Conde, J. J.; McGuire, M.; Wallace, M. *Tetrahedron Lett.* **2003**, *44*, 3081–4.

208. Rudisill, D. E.; Stille, J. K. *J. Org. Chem.* **1989**, *54*, 5856–66.

209. Sakamoto, T.; Satoh, C.; Kondo, Y.; Yamanaka, H. *Heterocycles* **1992**, *34*, 2379–84.

210. Iwao, M.; Takehara, H.; Furukawa, S.; Watanabe, M. *Heterocycles* **1993**, *36*, 1483–8.

211. Sonogashira, K.; Tohda, Y.; Hagihara, N. *Tetrahedron Lett.* **1975**, 4467–70.

212. (a) Castro, C. E.; Stephens, R. D. *J. Org. Chem.* **1963**, *28*, 2163. (b) Stephens, R. D.; Castro, C. E. *J. Org. Chem.* **1963**, *28*, 3313–15. (c) Castro, C. E.; Gaughan, E. J.; Owsley, D. C. *J. Org. Chem.* **1966**, *31*, 4071–8. (d) Castro, C. E.; Havlin, R.; Honwad, V. K.; Malte, A.; Mojé, S. *J. Am. Chem. Soc.* **1969**, *91*, 6464–70.

213. (a) Sakamoto, T.; Kondo, Y.; Yamanaka, H. *Heterocycles* **1984**, *22*, 1347–50. (b) Sakamoto, T.; Kondo, Y.; Yamanaka, H. *Chem. Pharm. Bull.* **1986**, *34*, 2362–8. (c) For a review of the use of palladium catalysis in heterocycle synthesis, with a good summary of the authors' work, see Sakamoto, T.; Kondo, Y.; Yamanaka, H. *Heterocycles* **1988**, *27*, 2225–49.

214. Taylor, E. C.; Katz, A. H.; Salgado-Zamora, H.; McKillop, A. *Tetrahedron Lett.* **1985**, *26*, 5963–6.

215. Tischler, A. N.; Lanza, T. J. *Tetrahedron Lett.* **1986**, *27*, 1653–6.

216. (a) Sakamoto, T.; Kondo, Y.; Yamanaka, H. *Heterocycles* **1986**, *24*, 31–2. (b) Sakamoto, T.; Kondo, Y.; Iwashita, S.; Yamanaka, H. *Chem. Pharm. Bull.* **1987**, *35*, 1823–8. (c) Sakamoto, T.; Kondo, Y.; Iwashita, S.; Nagano, T.; Yamanaka, H. *Chem. Pharm. Bull.* **1988**, *36*, 1305–8.

217. (a) Sakamoto, T.; Nagano, T.; Kondo, Y.; Yamanaka, H. *Chem. Pharm. Bull.* **1988**, *36*, 2248–52. (b) Sakamoto, T.; Numata, A.; Saitoh, H.; Kondo, Y. *Chem. Pharm. Bull.* **1999**, *47*, 1740–3.

218. Zhu, Z.; Moore, J. S. *J. Org. Chem.* **2000**, *65*, 116–23.

219. Arcadi, A.; Cacchi, S.; Marinelli, F. *Tetrahedron Lett.* **1989**, *30*, 2581–4.

220. Zhong, W.; Gallivan, J. P.; Zhang, Y.; Li, L.; Lester, H. A.; Dougherty, D. A. *Proc. Natl. Acad. Sci. USA* **1998**, *95*, 12088–93.

221. Kondo, Y.; Kojima, S.; Sakamoto, T. *J. Org. Chem.* **1997**, *62*, 6507–11.

222. Yu, M. S.; Lopez de Leon, L.; McGuire, M. A.; Botha, G. *Tetrahedron Lett.* **1998**, *39*, 9347–50.

223. Xu, L.; Lewis, I. R.; Davidsen, S. K.; Summers, J. B. *Tetrahedron Lett.* **1998**, *39*, 5159–62.

224. Shin, K.; Ogasawara, K. *Synlett* **1995**, 859–60.

225. Shin, K.; Ogasawara, K. *Chem. Lett.* **1995**, 289–90.

226. Shin, K.; Ogasawara, K. *Synlett* **1996**, 922–3.

227. Ezquerra, J.; Pedregal, C.; Lamas, C.; Barluenga, J.; Perez, M.; García-Martín, M. A.; González, J. M. *J. Org. Chem.* **1996**, *61*, 5804–12.

228. Soloducho, J. *Tetrahedron Lett.* **1999**, *40*, 2429–30.

229. Blurton, P.; Brickwood, A.; Dhanak, D. *Heterocycles* **1997**, *45*, 2395–403.

230. Botta, M.; Summa, V.; Corelli, F.; Di Pietro, G.; Lombardi, P. *Tetrahedron: Asymmetry* **1996**, *7*, 1263–6.

231. Kondo, Y.; Kojima, S.; Sakamoto, T. *Heterocycles* **1996**, *43*, 2741–6.

232. Yasuhara, A.; Kanamori, Y.; Kaneko, M.; Numata, A.; Kondo, Y.; Sakamoto, T. *J. Chem. Soc., Perkin Trans. 1* **1999**, 529–34.

233. Hiroya, K.; Matsumoto, S.; Sakamoto, T. *Org. Lett.* **2004**, *6*, 2953–6.

234. Amatore, C.; Blart, E.; Genet, J. P.; Jutand, A.; Lemaire-Audoire, S.; Savignac, M. *J. Org. Chem.* **1995**, *60*, 6829–39.

235. Fagnola, M. C.; Candiani, I.; Visentin, G.; Cabri, W.; Zarini, F.; Mongelli, N.; Bedeschi, A. *Tetrahedron Lett.* **1997**, *38*, 2307–10.

236. Collini, M. D.; Ellingboe, J. W. *Tetrahedron Lett.* **1997**, *38*, 7963–6.

237. Zhang, H.-C.; Brumfield, K. K.; Jaroskova, L.; Maryanoff, B. E. *Tetrahedron Lett.* **1998**, *39*, 4449–52.

238. Zhang, H.-C.; Ye, H.; Moretto, A. F.; Brumfield, K. K.; Maryanoff, B. E. *Org. Lett.* **2000**, *2*, 89–92.

239. Rainier, J. D.; Kennedy, A. R.; Chase, E. *Tetrahedron Lett.* **1999**, *40*, 6325–7.

240. Galambos, G.; Szantay, Jr. C.; Tamás, J.; Szántay, C. *Heterocycles* **1993**, *36*, 2241–5.

241. Abbiati, G.; Beccalli, E. M.; Marchesini, A.; Rossi, E. *Synthesis* **2001**, 2477–83.

242. (a) Zhang, H.; Larock, R. C. *J. Org. Chem.* **2002**, *67*, 7048–56. (b) Zhang, H.; Larock, R. C. *J. Org. Chem.* **2002**, *67*, 9318–30.

243. Sumi, S.; Matsumoto, K.; Tokuyama, H.; Fukuyama, T. *Tetrahedron* **2003**, *59*, 8571–87.

244. Haider, N.; Käferböck, J. *Tetrahedron* **2004**, *60*, 6495–507.

245. Liu, J.-J.; Konzelmann, F.; Luk, K.-C. *Tetrahedron Lett.* **2003**, *44*, 2545–8.

246. Chang, K.-J.; Kang, B.-N.; Lee, M.-H.; Jeong, K.-S. *J. Am. Chem. Soc.* **2005**, *127*, 12214–15.

247. Frank, W. C.; Kim, Y. C.; Heck, R. F. *J. Org. Chem.* **1978**, *43*, 2947–9.

248. (a) Yamada, F.; Makita, Y.; Suzuki, T.; Somei, M. *Chem. Pharm. Bull.* **1985**, *33*, 2162–3. (b) Somei, M.; Hasegawa, T.; Kaneko, C. *Heterocycles* **1983**, *20*, 1983–5. (c) Somei, M.; Yamada, F. *Chem. Pharm. Bull.* **1984**, *32*, 5064–5. (d) Yamada, F.; Hasegawa, T.; Wakita, M.; Sugiyama, M.; Somei, M. *Heterocycles* **1986**, *24*, 1223–6. (e) Somei, M.; Ohnishi, H.; Shoken, Y. *Chem. Pharm. Bull.* **1986**, *34*, 677–81. (f) Somei, M.; Makita, Y.; Yamada, F. *Chem. Pharm. Bull.* **1986**, *34*, 948–9. (g) Somei, M.; Saida, Y.; Komura, N. *Chem. Pharm. Bull.* **1986**, *34*, 4116–25. (h) Somei, M.; Funamoto, T.; Ohta, T. *Heterocycles* **1987**, *26*, 1783–4. (i) Somei, M.; Yamada, F.; Ohnishi, H.; Makita, Y.; Kuriki, M. *Heterocycles* **1987**, *26*, 2823–8. (j) Ohta, T.; Shinoda, J.; Somei, M. *Chem. Lett.* **1993**, 797–8.

249. (a) Harrington, P. J.; Hegedus, L. S. *J. Org. Chem.* **1984**, *49*, 2657–62. (b) Harrington, P. J.; Hegedus, L. S.; McDaniel, K. F. *J. Am. Chem. Soc.* **1987**, *109*, 4335–8. (c) Hegedus, L. S.; Toro, J. L.; Miles, W. H.; Harrington, P. J. *J. Org. Chem.* **1987**, *52*, 3319–22.

250. Merlic, C. A.; Semmelhack, M. F. *J. Organomet. Chem.* **1990**, *391*, C23–7.

251. Busacca, C. A.; Johnson, R. E.; Swestock, J. *J. Org. Chem.* **1993**, *58*, 3299–303.

252. Ralbovsky, J. L.; Scola, P. M.; Sugino, E.; Burgos-Garcia, C.; Weinreb, S. M.; Parvez, M. *Heterocycles* **1996**, *43*, 1497–512.

253. Nakagawa, Y.; Irie, K.; Nakamura, Y.; Ohigashi, H.; Hayashi, H. *Biosci. Biotechnol. Biochem.* **1998**, *62*, 1568–73.

254. Sundberg, R. J.; Cherney, R. J. *J. Org. Chem.* **1990**, *55*, 6028–37.

255. Wiedenau, P.; Monse, B.; Blechert, S. *Tetrahedron* **1995**, *51*, 1167–76.

256. Malapel-Andrieu, B.; Mérour, J.-Y. *Tetrahedron Lett.* **1998**, *39*, 39–42.

257. Cacchi, S.; Ciattini, P. G.; Morera, E.; Ortar, G. *Tetrahedron Lett.* **1988**, *29*, 3117–20.

258. Arcadi, A.; Cacchi, S.; Marinelli, F.; Morera, E.; Ortar, G. *Tetrahedron* **1990**, *46*, 7151–64.

259. (a) Akita, Y.; Inoue, A.; Yamamoto, K.; Ohta, A. *Heterocycles* **1985**, *23*, 2327–33. (b) Akita, Y.; Itagaki, Y.; Takizawa, S.; Ohta, A. *Chem. Pharm. Bull.* **1989**, *37*, 1477–80.

260. Rajeswaran, W. G.; Srinivasan, P. C. *Synthesis* **1992**, 835–6.

261. Sezen, B.; Sames, D. *J. Am. Chem. Soc.* **2003**, *125*, 5274–5.

262. Yokoyama, Y.; Matsushima, H.; Takashima, M.; Suzuki, T.; Murakami, Y. *Heterocycles* **1997**, *46*, 133–6.

263. (a) Horwell, D. C.; Nichols, P. D.; Roberts, E. *Tetrahedron Lett.* **1994**, *35*, 939–40. (b) Horwell, D. C.; Nichols, P. D.; Ratcliffe, G. S.; Roberts, E. *J. Org. Chem.* **1994**, *59*, 4418–23.

264. Kalinin, A. V.; Chauder, B. A.; Rakhit, S.; Snieckus, V. *Org. Lett.* **2003**, *5*, 3519–21.

265. Yokoyama, Y.; Kondo, K.; Mitsuhashi, M.; Murakami, Y. *Tetrahedron Lett.* **1996**, *37*, 9309–12.

266. Lee, K. L.; Goh, J. B.; Martin, S. F. *Tetrahedron Lett.* **2001**, *42*, 1635–8.

267. Black, D. St. C.; Keller, P. A.; Kumar, N.; *Tetrahedron* **1992**, *48*, 7601–8.

268. Germain, A. L.; Gilchrist, T. L.; Kemmitt, P. D. *Heterocycles* **1994**, *37*, 697–700.

269. Gilchrist, T. L.; Kemmitt, P. D.; Germain, A. L. *Tetrahedron* **1997**, *53*, 4447–56.

270. Grigg, R.; Sridharan, V.; Stevenson, P.; Sukirthalingam, S.; Worakun, T. *Tetrahedron* **1990**, *46*, 4003–18.

271. (a) Kozikowski, A. P.; Ma, D. *Tetrahedron Lett.* **1991**, *32*, 3317–20. (b) Kozikowski, A. P.; Ma, D.; Brewer, J.; Sun, S.; Costa, E.; Romeo, E.; Guidotti, A. *J. Med. Chem.* **1993**, *36*, 2908–20.

272. Kraus, G. A.; Kim, H. *Synth. Commun.* **1993**, *23*, 55–64.

273. Kannadasan, S.; Srinivasan, P. C. *Synth. Commun.* **2004**, *34*, 1325–35.

274. Desarbre, E.; Mérour, J.-Y. *Heterocycles* **1995**, *41*, 1987–98.

275. Shao, H. W.; Cai, J. C. *Chin. Chem. Lett.* **1996**, *7*, 13–14.

276. Miki, Y.; Shirokishi, H.; Matsushita, K. *Tetrahedron Lett.* **1999**, *40*, 4347–8.

277. (a) Rigby, J. H.; Hughes, R. C.; Heeg, M. J. *J. Am. Chem. Soc.* **1995**, *117*, 7834–5. (b) Rigby, J. H.; Mateo, M. E. *Tetrahedron* **1996**, 52, 10569–82.

278. Harayama, T.; Hori, A.; Abe, H.; Takeuchi, Y. *Heterocycles* **2003**, *60*, 2429–34.

279. (a) Melnyk, P.; Gasche, J.; Thal, C. *Tetrahedron Lett.* **1993**, *34*, 5449–50. (b) Melnyk, P.; Legrand, B.; Gasche, J.; Ducrot, P.; Thal, C. *Tetrahedron* **1995**, *51*, 1941–52.

280. Harayama, T.; Hori, A.; Serban, G.; Morikami, Y.; Matsumoto, T.; Abe, H.; Takeuchi, Y. *Tetrahedron* **2004**, *60*, 10645–9.

281. Kelly, T. R.; Xu, W.; Sundaresan, J. *Tetrahedron Lett.* **1993**, *34*, 6173–6.

282. Grigg, R.; Sridharan, V.; Stevenson, P.; Worakun, T. *J. Chem. Soc., Chem. Commun.* **1986**, 1697–9.

283. (a) Cornec, O.; Joseph, B.; Mérour, J.-Y. *Tetrahedron Lett.* **1995**, *36*, 8587–90. (b) Joseph, B.; Cornec, O.; Mérour, J.-Y. *Tetrahedron* **1998**, *54*, 7765–76.

284. Miki, Y.; Shirokoshi, H.; Asai, M.; Aoki, Y.; Matsukida, H. *Heterocycles* **2003**, *60*, 2095–101.

285. Yoneda, R.; Kimura, T.; Kinomoto, J.; Harusawa, S.; Kurihara, T. *J. Heterocycl. Chem.* **1996**, *33*, 1909–13.

286. Huang, Q.; Larock, R. C. *J. Org. Chem.* **2003**, *68*, 7342–9.

287. Birman, V. B.; Rawal, V. H. *Tetrahedron Lett.* **1998**, *39*, 7219–22.

288. (a) Rawal, V. H.; Michoud, C.; Monestel, R. F. *J. Am. Chem. Soc.* **1993**, *115*, 3030–1. (b) Rawal, V. H.; Michoud, C. *J. Org. Chem.* **1993**, *58*, 5583–4.

289. Rawal, V. H.; Michoud, C. *Tetrahedron Lett.* **1991**, *32*, 1695–8.

290. Rawal, V. H.; Iwasa, S. *J. Org. Chem.* **1994**, *59*, 2685–6.

291. (a) Mori, M.; Chiba, K.; Ban, Y. *Tetrahedron Lett.* **1977**, 1037–40. (b) Mori, M.; Ban, Y. *Tetrahedron Lett.* **1979**, 1133–6. (c) Mori, M.; Kanda, N.; Oda, I.; Ban, Y. *Tetrahedron* **1985**, *41*, 5465–74.

292. Terpko, M. O.; Heck, R. F. *J. Am. Chem. Soc.* **1979**, *101*, 5281–3.

293. For reviews of early work, see (a) Hegedus, L. S. *Angew. Chem. Int. Ed. Engl.* **1988**, *27*, 1113–26. (b) Reference 213c.

294. Larock, R. C.; Babu, S. *Tetrahedron Lett.* **1987**, *28*, 5291–4.

295. Larock, R. C.; Hightower, T. R.; Hasvold, L. A.; Peterson, K. P. *J. Org. Chem.* **1996**, *61*, 3584–5.

296. (a) Larock, R. C.; Yang, H.; Pace, P.; Cacchi, S.; Fabrizi, G. *Tetrahedron Lett.* **1998**, *39*, 1885–8. (b) Larock, R. C.; Pace, P.; Yang, H. *Tetrahedron Lett.* **1998**, *39*, 2515–18. (c) Larock, R. C.; Pace, P.; Yang, H.; Russell, C. E.; Cacchi, S.; Fabrizi, G. *Tetrahedron* **1998**, *54*, 9961–80.

297. Genet, J. P.; Blart, E.; Savignac, M. *Synlett* **1992**, 715–17.

298. Carroll, M. A.; Holmes, A. B. *Chem. Commun.* **1998**, 1395–6.

299. Hegedus, L. S.; Mulhern, T. A.; Mori, A. *J. Org. Chem.* **1985**, *50*, 4282–8.

300. Sundberg, R. J.; Pitts, W. J. *J. Org. Chem.* **1991**, *56*, 3048–54.

301. Sakamoto, T.; Kondo, Y.; Uchiyama, M.; Yamanaka, H. *J. Chem. Soc., Perkin Trans. 1* **1993**, 1941–2.

302. (a) Tietze, L. F.; Grote, T. *J. Org. Chem.* **1994**, *59*, 192–6. (b) Tietze, L. F.; Buhr, W. *Angew. Chem. Int. Ed. Engl.* **1995**, *34*, 1366–8. (c) Tietze, L. F.; Hannemann, R.; Buhr, W.; Lögers, M.; Menningen, P.; Lieb, M.; Starck, D.; Grote, T.; Döring, A.; Schuberth, I. *Angew. Chem. Int. Ed. Engl.* **1996**, *35*, 2674–7.

303. Hoffmann, H. M. R.; Schmidt, B.; Wolff, S. *Tetrahedron* **1989**, *45*, 6113–26.

304. (a) Macor, J. E.; Blank, D. H.; Post, R. J.; Ryan, K. *Tetrahedron Lett.* **1992**, *33*, 8011–14. (b) Macor, J. E.; Ogilvie, R. J.; Wythes, M. J. *Tetrahedron Lett.* **1996**, *37*, 4289–92.

305. Schmitz, W. D.; Denhart, D. J.; Brenner, A. B.; Ditta, J. L.; Mattson, R. J.; Mattson, G. K.; Molski, T. F.; Macor, J. E. *Bioorg. Med. Chem. Lett.* **2005**, *15*, 1619–21.

306. (a) Wensbo, D.; Annby, U.; Gronowitz, S. *Tetrahedron* **1995**, *51*, 10323–42. (b) Wensbo, D.; Gronowitz, S. *Tetrahedron* **1996**, *52*, 14975–88.

307. Hennings, D. D.; Iwasa, S.; Rawal, V. H. *Tetrahedron Lett.* **1997**, *38*, 6379–82.

308. Yang, C.-C.; Sun, P.-J.; Fang, J.-M. *J. Chem. Soc., Chem. Commun.* **1994**, 2629–30.

309. Yun, W.; Mohan, R. *Tetrahedron Lett.* **1996**, *37*, 7189–92.

310. Zhang, H.-C.; Maryanoff, B. E. *J. Org. Chem.* **1997**, *62*, 1804–9.

311. Arumugam, V.; Routledge, A.; Abell, C.; Balasubramanian, S. *Tetrahedron Lett.* **1997**, *38*, 6473–6.

312. Caddick, S.; Kofie, W. *Tetrahedron Lett.* **2002**, *43*, 9347–50.

313. Wu, Q.; Anslyn, E. V. *J. Am. Chem. Soc.* **2004**, *126*, 14682–3.

314. Hegedus, L. S.; Allen, G. F.; Olsen, D. J. *J. Am. Chem. Soc.* **1980**, *102*, 3583–7.

315. Danishefsky, S.; Taniyama, E. *Tetrahedron Lett.* **1983**, *24*, 15–18.

316. (a) For a summary of the author's work, see Grigg, R. *J. Heterocycl. Chem.* **1994**, *31*, 631–9. (b) Grigg, R.; Dorrity, M. J.; Malone, J. F.; Sridharan, V.; Sukirthalingam, S. *Tetrahedron Lett.* **1990**, *31*, 1343–6. (c) Burns, B.; Grigg, R.; Santhakumar, V.; Sridharan, V.; Stevenson, P.; Workun, T. *Tetrahedron Lett.* **1992**, *48*, 7297–320. (d) Grigg, R.; Fretwell, P.; Meerholtz, C.; Sridharan, V. *Tetrahedron* **1994**, *50*, 359–70. (e) Grigg, R.; Sansano, J. M. *Tetrahedron* **1996**, *52*, 13441–54. (f) Grigg, R.; Brown, S.; Sridharan, V.; Uttley, M. D. *Tetrahedron Lett.* **1998**, *39*, 3247–50. (g) Grigg, R.; Sridharan, V.; York, M. *Tetrahedron Lett.* **1998**, *39*, 4139–42. (h) Evans, P.; Grigg, R.; Ramzan, M. I.; Sridharan, V.; York, M. *Tetrahedron Lett.* **1999**, *40*, 3021–4. (i) Grigg, R.; Major, J. P.; Martin, F. M.; Whittaker, M. *Tetrahedron Lett.* **1999**, *40*, 7709–11.

317. Burns, B.; Grigg, R.; Sridharan, V.; Worakun, T. *Tetrahedron Lett.* **1988**, *29*, 4325–8.

318. Burns, B.; Grigg, R.; Ratananukul, P.; Sridharan, V.; Stevenson, P.; Sukirthalingam, S.; Worakun, T. *Tetrahedron Lett.* **1988**, *29*, 5565–8.

319. Grigg, R.; Loganathan, V.; Sridharan, V. *Tetrahedron Lett.* **1996**, *37*, 3399–402.

320. Brown, D.; Grigg, R.; Sridharan, V.; Tambyrajah, V.; Thornton-Pett, M. *Tetrahedron* **1998**, *54*, 2595–606.

321. Grigg, R.; Loganathan, V.; Sukirthalingam, S.; Sridharan, V. *Tetrahedron Lett.* **1990**, *31*, 6573–6.

322. (a) Abelman, M. M.; Oh, T.; Overman, L. E. *J. Org. Chem.* **1987**, *52*, 4130–3. (b) Earley, W. G.; Oh, T.; Overman, L. E. *Tetrahedron Lett.* **1988**, *29*, 3785–8. (c) Madin, A.; Overman, L. E. *Tetrahedron Lett.* **1992**, *33*, 4859–62. (d) Ashimori, A.; Overman, L. E. *J. Org. Chem.* **1992**, *57*, 4571–2. (e) Ashimori, A.; Matsuura, T.; Overman, L. E.; Poon, D. J. *J. Org. Chem.* **1993**, *58*, 6949–51. (f) Overman,

L. E.; Poon, D. J. *Angew. Chem. Int. Ed. Engl.* **1997**, *36*, 518–21. (g) Ashimori, A.; Bachand, B.; Overman, L. E.; Poon, D. J. *J. Am. Chem. Soc.* **1998**, *120*, 6477–87. (h) Ashimori, A.; Bachand, B.; Calter, M. A.; Govek, S. P.; Overman, L. E.; Poon, D. J. *J. Am. Chem. Soc.* **1998**, *120*, 6488–99. (i) Matsuura, T.; Overman, L. E.; Poon, D. J. *J. Am. Chem. Soc.* **1998**, *120*, 6500–3. (j) Overman, L. E.; Paone, D. V.; Stearns, B. A. *J. Am. Chem. Soc.* **1999**, *121*, 7702–3.

323. Newcombe, N. J.; Ya, F.; Vijn, R. J.; Hiemstra, H.; Speckamp, W. N. *J. Chem. Soc., Chem. Commun.* **1994**, 767–8.

324. Iida, H.; Yuasa, Y.; Kibayashi, C. *J. Org. Chem.* **1980**, *45*, 2938–42.

325. Luly, J. R.; Rapoport, H. *J. Org. Chem.* **1984**, *49*, 1671–2.

326. (a) Michael, J. P.; Chang, S.-F.; Wilson, C. *Tetrahedron Lett.* **1993**, *34*, 8365–8. (b) Michael, J. P.; de Koning, C. B.; Petersen, R. L.; Stanbury, T. V. *Tetrahedron Lett.* **2001**, *42*, 7513–16.

327. Kasahara, A.; Izumi, T.; Murakami, S.; Yanai, H.; Takatori, M. *Bull. Chem. Soc. Jpn.* **1986**, *59*, 927–8.

328. Sakamoto, T.; Nagano, T.; Kondo, Y.; Yamanaka, H. *Synthesis* **1990**, 215–18.

329. (a) Latham, E. J.; Stanforth, S. P. *Chem. Commun.* **1996**, 2253–4. (b) Latham, E. J.; Stanforth, S. P. *J. Chem. Soc., Perkin Trans. 1* **1997**, 2059–63.

330. Koerber-Plé, K.; Massiot, G. *Synlett* **1994**, 759–60.

331. (a) Yamazaki, K.; Kondo, Y. *J. Comb. Chem.* **2002**, *4*, 191–2. (b) Yamazaki, K.; Nakamura, Y.; Kondo, Y. *J. Org. Chem.* **2003**, *68*, 6011–19.

332. Baran, P. S.; Guerrero, C. A.; Ambhaikar, N. B.; Hafensteiner, B. D. *Angew. Chem. Int. Ed.* **2005**, *44*, 606–9.

333. Kuethe, J. T.; Wong, A.; Qu, C.; Smitrovich, J.; Davies, I. W.; Hughes, D. L. *J. Org. Chem. Soc.* **2005**, *70*, 2555–67.

334. Hiroi, K.; Hiratsuka, Y.; Watanabe, K.; Abe, I.; Kato, F.; Hiroi, M. *Synlett* **2001**, 263–5.

335. El-Araby, M. E.; Bernacki, R. J.; Makara, G. M.; Pera, P. J.; Anderson, W. K. *Bioorg. Med. Chem.* **2004**, *12*, 2867–79.

336. Chen, C.; Liebermann, D. R.; Larsen, R. D.; Verhoeven, T. R.; Reider, P. J. *J. Org. Chem.* **1997**, *62*, 2676–7.

337. Nazaré, M.; Schneider, C.; Lindenschmidt, A.; Will, D. W. *Angew. Chem. Int. Ed.* **2004**, *43*, 4526–8.

338. Watanabe, T.; Arai, S.; Nishida, A. *Synlett* **2004**, 907–9.

339. Chen, C.; Larsen, R. D. *Org. Syn.* **2000**, *78*, 36–41.

340. (a) Solé, D.; Vallverdú, L.; Peidró, E.; Bonjoch, J. *Chem. Commun.* **2001**, 1888–9. (b) Solé, D.; Vallverdú, L.; Solans, X.; Font-Bardìa, M.; Bonjoch, J. *J. Am. Chem. Soc.* **2003**, *125*, 1587–94.

341. Cheung, W. S.; Patch, R. J.; Player, M. R. *J. Org. Chem.* **2005**, *70*, 3741–4.

342. Witulski, B.; Alayrac, C.; Tevzadze-Saeftel, L. *Angew. Chem. Int. Ed.* **2003**, *42*, 4257–60.

343. (a) Chen, L.-C.; Yang, S.-C.; Wang, H.-M. *Synthesis* **1995**, 385–6. (b) Wang, H.-M.; Chou, H.-L.; Chen, L.-C. *J. Chin. Chem. Soc.* **1995**, *42*, 593–5.

344. Blache, Y.; Sinibaldi-Troin, M.-E.; Voldoire, A.; Chavignon, O.; Gramain, J.-C.; Teulade, J.-C.; Chapat, J.-P. *J. Org. Chem.* **1997**, *62*, 8553–6.

345. Iwaki, T.; Yasuhara, A.; Sakamoto, T. *J. Chem. Soc., Perkin Trans. 1* **1999**, 1505–10.

346. Ames, D. E.; Opalko, A. *Tetrahedron* **1984**, *40*, 1919–25.

347. Lachance, N.; April, M.; Joly, M.-A. *Synthesis* **2005**, 2571–7.

348. Sørensen, U. S.; Pombo-Villar, E. *Helv. Chim. Acta* **2004**, *87*, 82–9.

349. Harris, J. M.; Padwa, A. *Org. Lett.* **2003**, *5*, 4195–7.

350. Ackermann, L.; Kaspar, L. T.; Gschrei, C. J. *Chem. Commun.* **2004**, 2824–5.

351. Bedford, R. B.; Cazin, C. S. J. *Chem. Commun.* **2002**, 2310–11.

352. Liu, Z.; Larock, R. C. *Org. Lett.* **2004**, *6*, 3739–41.

353. Ferreira, I. C. F. R.; Queiroz, M.-J. R. P.; Kirsch, G. *Tetrahedron* **2003**, *59*, 3737–43.

354. (a) Tremont, S. J.; Rahman, H. U. *J. Am. Chem. Soc.* **1984**, *106*, 5759–60. (b) Abdrakhmanov, I. B.; Mustafin, A. G.; Tolstikov, G. A.; Dzhemilev, U. M. *Chem. Heterocycl. Cpds.* **1987**, 420–2.

355. Gowan, M.; Caillé, A. S.; Lau, C. K. *Synlett* **1997**, 1312–14.

356. Sakamoto, T.; Kondo, Y.; Yamanaka, H. *Heterocycles* **1988**, *27*, 453–6.

357. Izumi, T.; Sugano, M.; Konno, T. *J. Heterocycl. Chem.* **1992**, *29*, 899–904.

358. Kasahara, A.; Izumi, T.; Xiao-ping, L. *Chem. Ind.* **1988**, 50–1.

359. Kasahara, A.; Izumi, T.; Kikuchi, T.; Xiao-ping, L. *J. Heterocycl. Chem.* **1987**, *24*, 1555–6.

360. (a) Samizu, K.; Ogasawara, K. *Synlett* **1994**, 499–500. (b) Samizu, K.; Ogasawara, K. *Heterocycles* **1995**, *41*, 1627–9. (c) Sakagami, H.; Ogasawara, K. *Heterocycles* **1999**, *51*, 1131–5.

361. Carlström, A.-S.; Frejd, T. *Acta Chem. Scand.* **1992**, *46*, 163–71.

362. (a) Larock, R. C.; Yum, E. K. *J. Am. Chem. Soc.* **1991**, *113*, 6689–90. (b) Larock, R. C.; Yum, E. K.; Refvik, M. D. *J. Org. Chem.* **1998**, *63*, 7652–62.

363. Jeschke, T.; Wensbo, D.; Annby, U.; Gronowitz, S.; Cohen, L. A. *Tetrahedron Lett.* **1993**, *34*, 6471–4.

364. (a) Ujjainwalla, F.; Warner, D. *Tetrahedron Lett.* **1998**, *39*, 5355–8. (b)Ujjainwalla, F.; Walsh, T. F. *Tetrahedron Lett.* **2001**, *42*, 6441–5.

365. (a) Park, S. S.; Choi, J.-K.; Yum, E. K.; Ha, D.-C. *Tetrahedron Lett.* **1998**, *39*, 627–30. (b) Kang, S. K.; Park, S. S.; Kim, S. S.; Choi, J.-K.; Yum, E. K. *Tetrahedron Lett.* **1999**, *40*, 4379–82.

366. (a) Chen, C.; Lieberman, D. R.; Larsen, R. D.; Reamer, R. A.; Verhoeven, T. R.; Reider, P. J.; Cottrell, I. F.; Houghton, P. G. *Tetrahedron Lett.* **1994**, *35*, 6981–4. (b) Chen, C.; Lieberman, D. R.; Street, L. J.; Guiblin, A. R.; Larsen, R. D.; Verhoeven, T. R. *Synth. Commun.* **1996**, *26*, 1977–84.

367. Larock, R. C.; Doty, M. J.; Han, X. *Tetrahedron Lett.* **1998**, *39*, 5143–6.

368. Maassarani, F.; Pfeffer, M.; Spencer, J.; Wehman, E. *J. Organomet. Chem.* **1994**, *466*, 265–71.

369. Rosso, V. W.; Lust, D. A.; Bernot, P. J.; Grosso, J. A.; Modi, S. P.; Rusowicz, A.; Sedergran, T. C.; Simpson, J. H.; Srivastava, S. K.; Humora, M. J.; Anderson, N. G. *Org. Proc. Res. Dev.* **1997**, *1*, 311–14.

370. (a) Zhang, H.-C.; Brumfield, K. K.; Maryanoff, B. E. *Tetrahedron Lett.* **1997**, *38*, 2439–42. (b) Smith, A. L.; Stevenson, G. I.; Swain, C. J.; Castro, J. L. *Tetrahedron Lett.* **1998**, *39*, 8317–20.

371. For reviews, see (a) Cacchi, S.; Fabrizi, G.; Parisi, L. M. *Heterocycles* **2002**, *58*, 667–82. (b) Zeni, G.; Larock, R. C. *Chem. Rev.* **2004**, *104*, 2285–309.

372. (a) Castle, S. L.; Srikanth, G. S. C. *Org. Lett.* **2003**, *5*, 3611–14. (b) Ma, C.; Yu, S.; He, X.; Liu, X.; Cook, J. M. *Tetrahedron Lett.* **2000**, *41*, 2781–5. (c) Ma, C.; Liu, X.;

Li, X.; Flippen-Anderson, J.; Yu, S.; Cook, J. M. *J. Org. Chem.* **2001**, *66*, 4525–42. (d) Zhou, H.; Liao, X.; Cook, J. M. *Org. Lett.* **2004**, *6*, 249–52.

373. (a) Gathergood, N.; Scammells, P. J. *Org. Lett.* **2003**, *5*, 921–3. (b) Fînaru, A.; Berthault, A.; Besson, T.; Guillaumet, G.; Berteina-Raboin, S. *Tetrahedron Lett.* **2002**, *43*, 787–90. (c) Parmentier, J.-G.; Poissonnet, G.; Goldstein, S. *Heterocycles* **2002**, *56*, 465–76.

374. Hatakeyama, K.; Ohmori, K.; Suzuki, K. *Synlett* **2005**, 1311–15.

375. Hopkins, C. R.; Collar, N. *Tetrahedron Lett.* **2005**, *46*, 1845–8.

376. (a) Chae, J.; Konno, T.; Ishihara, T.; Yamanaka, H. *Chem. Lett.* **2004**, *33*, 314–15. (b) Konno, T.; Chae, J.; Ishihara, T.; Yamanaka, H. *J. Org. Chem.* **2004**, *69*, 8258–65.

377. Shen, M.; Li, G.; Lu, B. Z.; Hossain, A.; Roschangar, F.; Farina, V.; Senanayake, C. H. *Org. Lett.* **2004**, *6*, 4129–32.

378. O'Connor, J. M.; Stallman, B. J.; Clark, W. G.; Shu, A. Y. L.; Spada, R. E.; Stevenson, T. M.; Dieck, H. A. *J. Org. Chem.* **1983**, *48*, 807–9.

379. Larock, R. C.; Berrios-Peña, N.; Narayanan, K. *J. Org. Chem.* **1990**, *55*, 3447–50.

380. Larock, R. C.; Guo, L. *Synlett* **1995**, 465–6.

381. Back, T. G.; Bethell, R. J. *Tetrahedron Lett.* **1998**, *39*, 5463–4.

382. Wang, Y.; Huang, T.-N. *Tetrahedron Lett.* **1998**, *39*, 9605–8.

383. (a) Larock, R. C.; Yum, E. K. *Synlett* **1990**, 529–30. (b) Larock, R. C.; Yum, E. K. *Tetrahedron* **1996**, *52*, 2743–58.

384. Larock, R. C.; Berrios-Peña, N. G.; Fried, C. A. *J. Org. Chem.* **1991**, *56*, 2615–17.

385. (a) Larock, R. C.; Zenner, J. M. *J. Org. Chem.* **1995**, *60*, 482–3. (b) Zenner, J. M.; Larock, R. C. *J. Org. Chem.* **1999**, *64*, 7312–22.

386. Desarbre, E.; Mérour, J.-Y. *Tetrahedron Lett.* **1996**, *37*, 43–6.

387. Larock, R. C.; Liu, C.-L.; Lau, H. H.; Varaprath, S. *Tetrahedron Lett.* **1984**, *25*, 4459–62.

388. (a) Roesch, K. R.; Larock, R. C. *Org. Lett.* **1999**, *1*, 1551–3. (b) Roesch, K. R.; Larock, R. C. *J. Org. Chem.* **2001**, *66*, 412–20.

389. Hegedus, L. S.; Allen, G. F.; Olsen, D. J. *J. Am. Chem. Soc.* **1980**, *102*, 3583–7.

390. Hegedus, L. S.; Winton, P. M.; Varaprath, S. *J. Org. Chem.* **1981**, *46*, 2215–21.

391. Pandey, G. D.; Tiwari, K. P. *Synth. Commun.* **1980**, *10*, 523–7.

392. Mori, M.; Kobayashi, H.; Kimura, M.; Ban, Y. *Heterocycles* **1985**, *23*, 2803–6.

393. (a) Edstrom, E. D.; Yu, T. *J. Org. Chem.* **1995**, *60*, 5382–3. (b) Edstrom, E. D.; Yu, T. *Tetrahedron Lett.* **1994**, *35*, 6985–8.

394. Itahara, T. *Chem. Lett.* **1982**, 1151–2.

395. Belley, M.; Scheigetz, J.; Dubé, P.; Dolman, S. *Synlett* **2001**, 222–5.

396. (a) Kondo, Y.; Sakamoto, T.; Yamanaka, H. *Heterocycles* **1989**, *29*, 1013–16. (b) Kondo, Y.; Shiga, F.; Murata, N.; Sakamoto, T.; Yamanaka, H. *Tetrahedron* **1994**, *50*, 11803–12.

397. Arcadi, A.; Cacchi, S.; Carnicelli, V.; Marinelli, F. *Tetrahedron* **1994**, *50*, 437–52.

398. Battistuzzi, G.; Cacchi, S.; Fabrizi, G.; Marinelli, F.; Paris, L. M. *Org. Lett.* **2002**, *4*, 1355–8.

399. Gabriele, B.; Salerno, G.; Veltri, L.; Costa, M.; Massera, C. *Eur. J. Org. Chem.* **2001**, 4607–13.

400. Herbert, J. M.; McNeil, A. H. *Tetrahedron Lett.* **1998**, *39*, 2421–4.

401. Kumar, K.; Zapf, A.; Michalik, D.; Tillack, A.; Heinrich, T.; Böttcher, H.; Arlt, M.; Beller, M. *Org. Lett.* **2004**, *6*, 7–10.

402. (a) Tollari, S.; Demartin, F.; Cenini, S.; Palmisano, G.; Raimondi, P. *J. Organomet. Chem.* **1997**, *527*, 93–102. (b) Tollari, S.; Cenini, S.; Tunice, C.; Palmisano, G. *Inorg. Chim. Acta* **1998**, *272*, 18–23.

403. Dong, Y.; Busacca, C. A. *J. Org. Chem.* **1997**, *62*, 6464–5.

404. Hirao, K.; Morii, N.; Joh, T.; Takahashi, S. *Tetrahedron Lett.* **1995**, *36*, 6243–6.

405. Iwasaki, M.; Kobayashi, Y.; Li, J.-P.; Matsuzaka, H.; Ishii, Y.; Hidai, M. *J. Org. Chem.* **1991**, *56*, 1922–7.

406. Grigg, R.; Monteith, M.; Sridharan, V.; Terrier, C. *Tetrahedron* **1998**, *54*, 3885–94.

407. Grigg, R.; Kennewell, P.; Teasdale, A. J. *Tetrahedron Lett.* **1992**, *33*, 7789–92.

408. Grigg, R.; Putnikovic, B.; Urch, C. J. *Tetrahedron Lett.* **1996**, *37*, 695–8.

409. Ishikura, M. *Heterocycles* **1995**, *41*, 1385–8.

410. Åkermark, B.; Bäckvall, J. E.; Hegedus, L. S.; Zetterberg, K.; Siirala-Hansén, K.; Sjöberg, K. *J. Organomet. Chem.* **1974**, *72*, 127–38.

411. Hegedus, L. S.; Allen, G. F.; Waterman, E. L. *J. Am. Chem. Soc.* **1976**, *98*, 2674–6.

412. Hegedus, L. S.; Allen, G. F.; Bozell, J. J.; Waterman, E. L. *J. Am. Chem. Soc.* **1978**, *100*, 5800–7.

413. Hegedus, L. S.; Weider, P. R.; Mulhern, T. A.; Asada, H.; D'Andrea, S. *Gazz. Chim. Ital.* **1986**, *116*, 213–19.

414. Hegedus, L. S.; Mulhern, T. A.; Asada, H. *J. Am. Chem. Soc.* **1986**, *108*, 6224–8.

415. Yamaguchi, M.; Arisawa, M.; Hirama, M. *Chem. Commun.* **1998**, 1399–400.

416. Kasahara, A.; Izumi, T.; Murakami, S.; Miyamoto, K.; Hino, T. *J. Heterocycl. Chem.* **1989**, *26*, 1405–13.

417. Hong, C. S.; Seo, J. Y.; Yum, E. K.; Sung, N.-D. *Heterocycles* **2004**, *63*, 631–9.

418. Adams, D. R.; Duncton, M. A. J.; Roffey, J. R. A.; Spencer, J. *Tetrahedron Lett.* **2002**, *43*, 7581–3.

419. Bosch, J.; Roca, T.; Armengol, M.; Fern·ndez-Forner, D. *Tetrahedron* **2001**, *57*, 1041–8.

420. Fang, Y.-Q.; Lautens, M. *Org. Lett.* **2005**, *7*, 3549–52.

421. Thielges, S.; Meddah, E.; Bisseret, P.; Eustache, J. *Tetrahedron Lett.* **2004**, *45*, 907–10.

422. Coleman, C. M.; O'Shea, D. F. *J. Am. Chem. Soc.* **2003**, *125*, 4054–5.

423. Mahanty, J. S.; De, M.; Das, P.; Kundu, N. G. *Tetrahedron* **1997**, *53*, 13397–418.

424. Trost, B. M.; Pedregal, C. *J. Am. Chem. Soc.* **1992**, *114*, 7292–4.

425. Wang, J.; Soundarajan, N.; Liu, N.; Zimmermann, K.; Naidu, B. N. *Tetrahedron Lett.* **2005**, *46*, 907–10.

426. Pearson, S. E.; Nandan, S. *Synthesis* **2005**, 2503–6.

427. (a) Hopkins, C. R.; Collar, N. *Tetrahedron Lett.* **2004**, *45*, 8087–90. (b) Hopkins, C. R.; Collar, N. *Tetrahedron Lett.* **2004**, *45*, 8631–3.

428. Farr, R. N.; Alabaster, R. J.; Chung, J. Y. L.; Craig, B.; Edwards, J. S.; Gibson, A. W.; Ho, G.-J.; Humphrey, G. R.; Johnson, S. A.; Grabowski, E. J. J. *Tetrahedron: Asymmetry* **2003**, *14*, 3503–15.

429. Sendzik, M.; Hui, H. C. *Tetrahedron Lett.* **2003**, *44*, 8697–700.

430. Arcadi, A.; Cacchi, S.; Fabrizi, G.; Marinelli, F.; Parisi, L. M. *Heterocycles* **2004**, *64*, 475–82.

431. van Esseveldt, B. C. J.; van Delft, F. L.; de Gelder, R.; Rutjes, F. P. J. T. *Org. Lett.* **2003**, *5*, 1717–20.

432. Sun, L.-P.; Huang, X.-H.; Dai, W.-M. *Tetrahedron* **2004**, *60*, 10983–92.

433. Tumkevicius, S.; Masevicius, V. *Synlett* **2004**, 2327–30.

434. Dai, W.-M.; Guo, D.-S.; Sun, L.-P.; Huang, X.-H. *Org. Lett*. **2003**, *5*, 2919–22.

435. Kanada, R. M.; Ogasawara, K. *Tetrahedron Lett*. **2001**, *42*, 7311–13.

436. Tatsuta, K.; Mukai, H.; Mitsumoto, K. *J. Antibiot*. **2001**, *54*, 105–8.

437. Yasuhara, A.; Suzuki, N.; Sakamoto, T. *Chem. Pharm. Bull*. **2002**, *50*, 143–5.

438. Pal, M.; Subramanian, V.; Batchu, V. R.; Dager, I. *Synlett* **2004**, 1965–9.

439. Suzuki, N.; Yasaki, S.; Yasuhara, A.; Sakamoto, T. *Chem. Pharm. Bull*. **2003**, *51*, 1170–3.

440. (a) Rodriguez, A. L.; Koradin, C.; Dohle, W.; Knochel, P. *Angew. Chem. Int. Ed*. **2000**, *39*, 2488–90. (b) Koradin, C.; Dohle, W.; Rodriguez, A. L.; Schmid, B.; Knochel, P. *Tetrahedron* **2003**, *59*, 1571–87.

441. Sakai, N.; Annaka, K.; Konakahara, T. *Org. Lett*. **2004**, *6*, 1527–30.

442. Dai, W.-M.; Guo, D.-S.; Sun, L.-P. *Tetrahedron Lett*. **2001**, *42*, 5275–8.

443. Kabalka, G. W.; Wang, L.; Pagni, R. M. *Tetrahedron* **2001**, *57*, 8017–28.

444. (a) Ackermann, L. *Org. Lett*. **2005**, *7*, 439–42. (b) Siebeneicher, H.; Bytschkov, I.; Doye, S. *Angew. Chem. Int. Ed*. **2003**, *42*, 3042–4.

445. Iritani, K.; Matsubara, S.; Utimoto, K. *Tetrahedron Lett*. **1988**, *29*, 1799–802.

446. (a) Arcadi, A.; Cacchi, S.; Marinelli, F. *Tetrahedron Lett*. **1992**, *33*, 3915–18. (b) Cacchi, S.; Carnicelli, V.; Marinelli, F. *J. Organomet. Chem*. **1994**, *475*, 289–96. (c) Cacchi, S.; Fabrizi, G.; Marinelli, F.; Moro, L.; Pace, P. *Synlett* **1997**, 1363–6. (d) Cacchi, S.; Fabrizi, G.; Pace, P. *J. Org. Chem*. **1998**, *63*, 1001–11. (e) Arcadi, A.; Cacchi, S.; Fabrizi, G.; Marinelli, F. *Synlett* **2000**, 394–6. (f) Arcadi, A.; Cacchi, S.; Cassetta, A.; Fabrizi, G.; Parisi, L. M. *Synlett* **2001**, 1605–7. (g) Arcadi, A.; Cacchi, S.; Fabrizi, G.; Parisi, L. M. *Tetrahedron Lett*. **2004**, *45*, 2431–4. (h) Cacchi, S.; Fabrizi, G.; Parisi, L. M. *Synthesis* **2004**, 1889–94. (i) Arcadi, A.; Cacchi, S.; Fabrizi, G.; Marinelli, F.; Parisi, L. M. *J. Org. Chem*. **2005**, *70*, 6213–17.

447. Battistuzzi, G.; Cacchi, S.; Fabrizi, G. *Eur. J. Org. Chem*. **2002**, 2671–81.

448. Saulnier, M. G.; Frennesson, D. B.; Deshpande, M. S.; Vyas, D. M. *Tetrahedron Lett*. **1995**, *36*, 7841–4.

449. (a) Yasuhara, A.; Kaneko, M.; Sakamoto, T. *Heterocycles* **1998**, *48*, 1793–9. (b) Yasuhara, A.; Takeda, Y.; Suzuki, N.; Sakamoto, T. *Chem. Pharm. Bull*. **2002**, *50*, 235–8.

450. Yue, D.; Larock, R. C. *Org. Lett*. **2004**, *6*, 1038–40.

451. Amjad, M.; Knight, D. W. *Tetrahedron Lett*. **2004**, *45*, 539–41.

452. Barluenga, J.; Trincado, M.; Rubio, E.; Gonzàlez, J. M. *Angew. Chem. Int. Ed*. **2003**, *42*, 2406–9.

453. Rainier, J. D.; Kennedy, A. R. *J. Org. Chem*. **2000**, *65*, 6213–16.

454. (a) Takeda, A.; Kamijo, S.; Yamamoto, Y. *J. Am. Chem. Soc*. **2000**, *122*, 5662–3. (b) Kamijo, S.; Sasaki, Y.; Yamamoto, Y. *Tetrahedron Lett*. **2004**, *45*, 35–8.

455. (a) Kamijo, S.; Yamamoto, Y. *Angew. Chem. Int. Ed*. **2002**, *41*, 3230–3. (b) Kamijo, S.; Yamamoto, Y. *J. Org. Chem*. **2003**, *68*, 4764–71.

456. (a) Kamijo, S.; Yamamoto, Y. *J. Am. Chem. Soc*. **2002**, *124*, 11940–5. (b) Onitsuka, K.; Suzuki, S.; Takahashi, S. *Tetrahedron Lett*. **2002**, *43*, 6197–9.

457. (a) Martin, P. *Helv. Chim. Acta* **1984**, *67*, 1647–9. (b) Martin, P. *Tetrahedron Lett*. **1987**, *28*, 1645–6. (c) Martin, P. *Helv. Chim. Acta* **1988**, *71*, 344–7. (d) Martin, P. *Helv. Chim. Acta* **1989**, *72*, 1554–82.

458. (a) Back, T. G.; Bethell, R. J.; Parvez, M.; Taylor, J. A. *J. Org. Chem.* **2001**, *66*, 8599–605. (b) Back, T. G.; Pandyra, A.; Wulff, J. E. *J. Org. Chem.* **2003**, *68*, 3299–302.

459. Witulski, B.; Alayrac, C. *Angew. Chem. Int. Ed.* **2002**, *41*, 3281–4.

460. Lee, C.-Y.; Lin, C.-F.; Lee, J.-L.; Chiu, C.-C.; Lu, W.-D.; Wu, M.-J. *J. Org. Chem.* **2004**, *69*, 2106–10.

461. Serra, S.; Fuganti, C. *Synlett* **2005**, 809–12.

462. Zhang, H.; Larock, R. C. *J. Org. Chem.* **2003**, *68*, 5132–8.

463. (a) Boger, D. L.; Panek, J. S. *Tetrahedron Lett.* **1984**, *25*, 3175–8. (b) Boger, D. L.; Duff, S. R.; Panek, J. S.; Yasuda, M. *J. Org. Chem.* **1985**, *50*. 5782–9. (c) Boger, D. L.; Duff, S. R.; Panek, J. S.; Yasuda, M. *J. Org. Chem.* **1985**, *50*, 5790–5.

464. (a) Nozaki, K.; Takahashi, K.; Nakano, K.; Hiyama, T.; Tang, H.-Z.; Fujiki, M.; Yamaguchi, S.; Tamao, K. *Angew. Chem. Int. Ed.* **2003**, *42*, 2051–3. (b) Kuwahara, A.; Nakano, K.; Nozaki, K. *J. Org. Chem.* **2005**, *70*, 413–19.

465. Willis, M. C.; Brace, G. N.; Holmes, I. P. *Angew. Chem. Int. Ed.* **2005**, *44*, 403–6.

466. (a) Wolfe, J. P.; Wagaw, S.; Buchwald, S. L. *J. Am. Chem. Soc.* **1996**, *118*, 7215–16, and references cited therein. (b) Louie, J.; Driver, M. S.; Hamann, B. C.; Hartwig, J. F. *J. Org. Chem.* **1997**, *62*, 1268–73. (c) Driver, M. S.; Hartwig, J. F. *J. Am. Chem. Soc.* **1996**, *118*, 7217–18, and references cited therein. (d) Hartwig, J. F. *Synlett* **1997**, 329–40. (e) Marcoux, J.-F.; Wagaw, S.; Buchwald, S. L. *J. Org. Chem.* **1997**, *62*, 1568–9. (f) Sadighi, J. P.; Harris, M. C.; Buchwald, S. L. *Tetrahedron Lett.* **1998**, *39*, 5327–30. (g) Wolfe, J. P.; Buchwald, S. L. *J. Org. Chem.* **1997**, *62*, 1264–7. (h) Hartwig, J. F. *Angew. Chem. Int. Ed.* **1998**, *37*, 2046–67. (i) Wolfe, J. P.; Wagaw, S.; Marcoux, J.-F.; Buchwald, S. L. *Acc. Chem. Res.* **1998**, *31*, 805–18. (j) Yang, B. H.; Buchwald, S. L. *J. Organomet. Chem.* **1999**, *576*, 125–46. (k) Hartwig, J. F. *Angew. Chem. Int. Ed.* **1998**, *37*, 2090–2. (l) Belfield, A. J.; Brown, G. R.; Foubister, A. J. *Tetrahedron* **1999**, *55*, 11399–428. (m) Huang, J.; Grasa, G.; Nolan, S. P. *Org. Lett.* **1999**, *1*, 1307–9.

467. (a) Peat, A. J.; Buchwald, S. L. *J. Am. Chem. Soc.* **1996**, *118*, 1028–30. (b) Wolfe, J. P.; Rennels, R. A.; Buchwald, S. L. *Tetrahedron* **1996**, *52*, 7525–46. (c) Wagaw, S.; Rennels, R. A.; Buchwald, S. L. *J. Am. Chem. Soc.* **1997**, *119*, 8451–8. (d) Aoki, K.; Peat, A. J.; Buchwald, S. L. *J. Am. Chem. Soc.* **1998**, *120*, 3068–73. (e) Yang, B. H.; Buchwald, S. L. *Org. Lett.* **1999**, *1*, 35–7.

468. (a) Wagaw, S.; Yang, B. H.; Buchwald, S. L. *J. Am. Chem. Soc.* **1998**, *120*, 6621–2. (b) Wagaw, S.; Yang, B. H.; Buchwald, S. L. *J. Am. Chem. Soc.* **1999**, *121*, 10251–63.

469. Watanabe, M.; Yamamoto, T.; Nishiyama, M. *Angew. Chem. Int. Ed.* **2000**, *39*, 2501–4.

470. Brown, J. A. *Tetrahedron Lett.* **2000**, *41*, 1623–6.

471. Yamazaki, K.; Nakamura, Y.; Kondo, Y. *J. Chem. Soc., Perkin Trans. 1* **2002**, 2137–8.

472. MacNeil, S. L.; Gray, M.; Briggs, L. E.; Li, J. J.; Snieckus, V. *Synlett* **1998**, 419–21.

473. Abouabdellah, A.; Dodd, R. H. *Tetrahedron Lett.* **1998**, *39*, 2119–22.

474. He, F.; Foxman, B. H.; Snider, B. B. *J. Am. Chem. Soc.* **1998**, *120*, 6417–18.

475. Hooper, M. W.; Utsunomiya, M.; Hartwig, J. F. *J. Org. Chem.* **2003**, *68*, 2861–73.

476. Follmann, M.; Graul, F.; Schäfer, T.; Kopec, S.; Hamley, P. *Synlett* **2005**, 1009–11.

477. Old, D. W.; Harris, M. C.; Buchwald, S. L. *Org. Lett.* **2000**, *2*, 1403–6.

478. Movassaghi, M.; Ondrus, A. E. *J. Org. Chem.* **2005**, *70*, 8638–41.

479. Tsang, W. C. P.; Zheng, N.; Buchwald, S. L. *J. Am. Chem. Soc.* **2005**, *127*, 14560–1.

480. (a) Akazome, M.; Kondo, T.; Watanabe, Y. *Chem. Lett.* **1992**, 769–72. (b) Akazome, M.; Kondo, T.; Watanabe, Y. *J. Org. Chem.* **1994**, *59*, 3375–80.

481. (a) Söderberg, B. C.; Shriver, J. A. *J. Org. Chem.* **1997**, *62*, 5838–45. (b) Söderberg, B. C.; Rector, S. R.; O'Neil, S. N. *Tetrahedron Lett.* **1999**, *40*, 3657–60. (c) Söderberg, B. C.; Chisnell, A. C.; O'Neil, S. N.; Shriver, J. A. *J. Org. Chem.* **1999**, *64*, 9731–4. (d) Scott, T. L.; Söderberg, B. C. G. *Tetrahedron Lett.* **2002**, *43*, 1621–4. (e) Dantale, S. W.; Söderberg, B. C. G. *Tetrahedron* **2003**, *59*, 5507–14. (f) Scott, T. L.; Söderberg, B. C. G. *Tetrahedron* **2003**, *59*, 6323–32. (g) For a review, see Söderberg, B. C. G. *Curr. Org. Chem.* **2000**, *4*, 727–64.

482. (a) Tollari, S.; Cenini, S.; Crotti, C.; Gianella, E. *J. Mol. Cat.* **1994**, *87*, 203–14. (b) Tollari, S.; Cenini, S.; Rossi, A.; Palmisano, G. *J. Mol. Cat.* **1998**, *135*, 241–8. (c) Ragaini, F.; Sportiello, P.; Cenini, S. *J. Organomet. Chem.* **1999**, *577*, 283–91.

483. Aoyagi, Y.; Mizusaki, T.; Ohta, A. *Tetrahedron Lett.* **1996**, *37*, 9203–6.

484. (a) Kuethe, J. T.; Wong, A.; Davies, I. W. *Org. Lett.* **2003**, *5*, 3721–3. (b) Kuethe, J. T.; Wong, A.; Davies, I. W. *Org. Lett.* **2003**, *5*, 3975–8. (c) Smitrovich, J. H.; Davies, I. W. *Org. Lett.* **2004**, *6*, 533–5. (d) Kuethe, J. T.; Davies, I. W. *Tetrahedron Lett.* **2004**, *45*, 4009–12.

485. Izumi, T.; Soutome, M.; Miura, T. *J. Heterocycl. Chem.* **1992**, *29*, 1625–9.

486. (a) Uozumi, Y.; Mori, M.; Shibasaki, M. *J. Chem. Soc., Chem. Commun.* **1991**, 81–3. (b) Mori, M.; Uozumi, Y.; Shibasaki, M. *Heterocycles* **1992**, *33*, 819–30.

487. Malapel-Andrieu, B.; Mérour, J.-Y. *Tetrahedron* **1998**, *54*, 11095–110.

488. Trost, B. M.; Molander, G. A. *J. Am. Chem. Soc.* **1981**, *103*, 5969–72.

489. (a) Mann, G.; Hartwig, J. F.; Driver, M. S.; Fernández-Rivas, C. *J. Am. Chem. Soc.* **1998**, *120*, 827–8. (b) Hartwig, J. F.; Kawatsura, M.; Hauck, S. I.; Shaughnessy, K. H.; Alcazar-Roman, L. M. *J. Org. Chem.* **1999**, *64*, 5575–80. (c) Watanabe, M.; Nishiyama, M.; Yamamoto, T.; Koie, Y. *Tetrahedron Lett.* **2000**, *41*, 481–3.

490. Kato, K.; Ono, M.; Akita, H. *Tetrahedron Lett.* **1997**, *38*, 1805–8.

491. Dickens, M. J.; Gilday, J. P.; Mowlem, T. J.; Widdowson, D. A. *Tetrahedron* **1991**, *47*, 8621–34.

492. Itoh, T.; Mase, T. *Org. Lett.* **2004**, *6*, 4587–90.

493. Takayama, H.; Sakai, S. *Chem. Pharm. Bull.* **1989**, *37*, 2256–7.

494. Takayama, H.; Kitajima, M.; Sakai, S. *Heterocycles* **1990**, *30*, 325–7.

495. (a) Anderson, B. A.; Bell, E. C.; Ginah, F. O.; Harn, N. K.; Pagh, L. M.; Wepsiec, J. P. *J. Org. Chem.* **1998**, *63*, 8224–8. (b) Srivastava, R. R.; Collibee, S. E. *Tetrahedron Lett.* **2004**, *45*, 8895–7. (c) Stazi, F.; Palmisano, G.; Turconi, M.; Santagostino, M. *Tetrahedron Lett.* **2005**, *46*, 1815–18.

496. Chae, J.; Buchwald, S. L. *J. Org. Chem.* **2004**, *69*, 3336–9.

497. Bracher, F.; Hildebrand, D.; Ernst, L. *Arch. Pharm.* **1994**, *327*, 121–2.

498. (a) Zhang, D.; Liebeskind, L. S. *J. Org. Chem.* **1996**, *61*, 2594–5. (b) Bailey, W. F.; Jiang, X.-L. *J. Org. Chem.* **1996**, *61*, 2596–7. (c) Yokum, T. S.; Tungaturthi, P. K.; McLaughlin, M. L. *Tetrahedron Lett.* **1997**, *38*, 5111–14.

499. Lemaire-Audoire, S.; Savignac, M.; Genet, J. P.; Bernard, J.-M. *Tetrahedron Lett.* **1995**, *36*, 1267–70.

500. Genet, J. P.; Blart, E.; Savignac, M.; Lemeune, S.; Paris, J.-M. *Tetrahedron Lett.* **1993**, *34*, 4189–92.

501. Takacs, J. M.; Zhu, J. *Tetrahedron Lett.* **1990**, *31*, 1117–20.
502. Reding, M. T.; Fukuyama, T. *Org. Lett.* **1999**, *1*, 973–6.
503. Godleski, S. A.; Villhauer, E. B. *J. Org. Chem.* **1986**, *51*, 486–91.
504. Vangveravong, S.; Nichols, D. E. *J. Org. Chem.* **1995**, *60*, 3409–13.
505. Matsumoto, M.; Watanabe, N.; Kobayashi, H. *Heterocycles* **1987**, *26*, 1479–82.
506. Isomura, K.; Uto, K.; Taniguchi, H. *J. Chem. Soc., Chem. Commun.* **1977**, 664–5.
507. Alper, H.; Mahatantila, C. P. *Heterocycles* **1983**, *20*, 2025–8.
508. Yang, C.-C.; Tai, H.-M.; Sun, P.-J. *J. Chem. Soc., Perkin Trans. 1* **1997**, 2843–50.
509. (a) Genet, J. P.; Grisoni, S. *Tetrahedron Lett.* **1986**, *27*, 4165–8. (b) Genet, J. P.;
       Grisoni, S. *Tetrahedron Lett.* **1988**, *29*, 4543–6. (c) Kardos, N.; Genet, J.-P.
       Tetrahedron: Asym **1994**, *5*, 1525–33.
510. Li, W.-J.; Lin, X.-F.; Wang, J.; Li, G.-L.; Wang, Y.-G. *Synlett* **2005**, 2003–6.
511. Denmark, S. E.; Baird, J. D. *Org. Lett.* **2004**, *6*, 3649–52.
512. Sakai, S.; Yamanaka, E.; Kitajima, M.; Yokota, M.; Aimi, N.; Wongseripatana, S.;
       Ponglux, D. *Tetrahedron Lett.* **1986**, *27*, 4585–8.

Chapter 4

Pyridines

Paul Galatsis

Pyridine was first isolated, like pyrrole, from bone pyrolysates. Its name is derived from the Greek for fire (pyr) and the suffix "idine" used to designate aromatic bases. Pyridine can also be formed from the breakdown of many natural materials in the environment. Pyridine is used as a solvent, in addition to many other uses including products such as pharmaceuticals, vitamins, food flavorings, paints, dyes, rubber products, adhesives, insecticides, and herbicides. Structurally, pyridine derivatives can range in complexity from the relatively simple monosubstituted pyridine, nicotine (**1**), to the highly elaborated sesquiterpene pyridine alkaloids, chuchuhuanines [1] (**2**).

The transformation of benzene to pyridine (Fig. 4.1) by the seemingly straightforward exchange of a CH group for N, allows retention of aromaticity in the six-membered ring, but has a dramatic change on the chemistry of pyridine when compared to that of benzene.

Pyridine is the prototypical electron-poor six-membered ring heterocycle. The aromaticity originally found in the benzene framework is maintained in pyridine *via* overlap with the unhybridized p orbital found on the sp^2 hybridized nitrogen atom that is parallel to the π-system of the carbon framework. The resonance pictures, as well as, the natural atomic charges of pyridine (Fig. 4.2), predict its electron deficient nature.

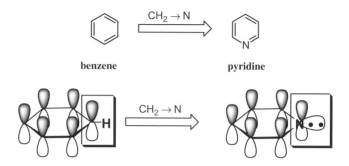

Figure 4.1.

The consequence of this replacement gives pyridine a reduced susceptibility to electrophilic substitution compared to benzene, while being more susceptible to nucleophilic attack. An avenue of chemistry not possible with benzene is the formation of pyridinium salts by donation of the nitrogen lone pair electrons. The resultant salts are still aromatic, however they are much more polarized. This is reflected by the apparent acidity of the corresponding conjugate acid (pK$_a$ 5.2) compared to the acidity of the corresponding conjugate acid of piperidine (pK$_a$ 11.1).

Commercially, pyridine is prepared by the gas phase, high-temperature reaction of crotonaldehyde, formaldehyde, steam, air, and ammonia over a silica-alumina catalyst in 60–70% yield. However, in the laboratory, the challenge is in the preparation of substituted pyridine derivatives in a process that allows one to control regioselectivity and chemoselectivity in the most efficient manner. In this regard the utility of palladium-catalyzed cross-coupling reactions has enabled synthetic chemists by providing the ability to construct highly diversified pyridine derivatives in an efficient fashion [2].

A cursory analysis of the catalytic cycle, as shown in Fig. 4.3, reveals the simplicity of the overall process [2]. An oxidative addition of the organohalide (R$_1$-X) with a derivative of the palladium catalyst (L$_n$Pd^0) affords an intermediate (L$_n$XPdR$_1$) that then undergoes a transmetalation with an organometallic reagent (M-R$_2$). It is the nature of this species that gives rise to the named-reaction variation, as illustrated. Finally, the resultant

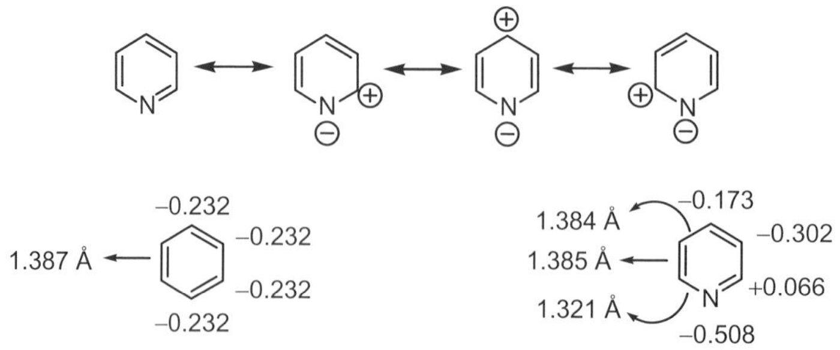

Figure 4.2.

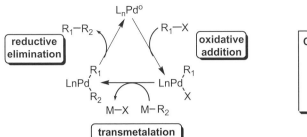

Figure 4.3.

organopalladium species ($L_nR_1PdR_2$) undergoes a reductive elimination to afford the cross-coupled product (R_1-R_2). Phosphine ligands are necessary for nearly all palladium-catalyzed cross-coupling reactions. They have the ability to modulate the catalytic power of the metal. The attributes affected include: (a) solubility of the active species, (b) shielding and steric properties of the catalyst, (c) electron density at the metal center, (d) reactivity of the catalyst in the catalytic cycle, (e) lifetime and turnover number for the catalyst, and (f) enantioselectivity for the reaction with chiral derivatives. These ligands can be classified as monodentate mono-, di-, and triarylphosphines, monodentate trialkylphosphines, bidentate non-chiral phosphines, and chiral phosphines. The nature of substituents (electron withdrawing or electron donating) directly relates to the σ-donating ability of the ligand, thus the nature of the P—M bond. This interaction is also modulated by the steric bulk of the ligand. The greater the steric bulk, as measured by the cone angle, the higher the dissociation rate of the ligand from the palladium and the faster the rate of oxidative addition.

It rapidly becomes apparent that, in order to carry out reactions of this type, certain common reagents need to be readily accessible. The initial organohalide or intermediates suitable for the oxidative addition will require efficient construction or their preparation must be compatible with the subsequent reactions. Additionally, a synthetic approach that maximizes chemical diversity would also be advantageous. For the second step, an appropriate organometallic reagent is required, therefore, as with the oxidative addition step, convenient routes to their access is desirable. With the tremendous gain in popularity of cross-coupling reactions over the past several years, commercial manufacturers have addressed some of these issues. The list of organohalides and organometallic reagents that are commercially available, continues to increase as a result of the increased use and efficacy of this class of reactions. However, in the application of this chemistry to the synthesis of a natural product, the preparation of a polymeric monomer or the elaboration of structure activity relationships (SAR) of a pharmacological species, it may be evident that the chemist will need to prepare these reagents. In the case of pyridine derivatives, one can leverage the unique chemistry of this scaffold to generate these compounds. For the halopyridines, one can take advantage of two general methods, metalation of an appropriately derivatized pyridine or halogen–halogen exchange from a commercially available intermediate.

4.1. Synthesis of halopyridines

4.1.1. Direct metalation followed by quenching with halogens

Direct lithiation of pyridine is a common method of generating the precursors required for the organopalladium cross-coupling reactions. However, if one is not careful, the pyridine ring is susceptible to direct addition of the alkyllithium reagent (Chichibabin reaction) as opposed to deprotonation of the ring.

Most methods employ the concept of complex-induced proximity effect (CIPE) or more simply put, directed ortho metalation (DoM) [3]. This chemistry is characterized by the pre-complexation of the alkyl metal base to a directed metalation group (DMG) that then helps deliver the base to the adjacent (*ortho*) position for deprotonation (Fig. 4.4). Once deprotonation has been completed the DMG assists in the stabilization of the metalated species. A great variety of DMGs have been investigated and include such functionality as halogens, CF_3, OH, OR, $OCONR_2$, $OSONR_2$, SO_2NR_2, SOR, NHCOR, $NHCO_2R$, CO_2H, 2-oxazolino, CONHR, $CONR_2$, COR, N–O.

The DoM reaction of halopyridines with alkyllithium bases takes advantage of the increased kinetic acidity of the ortho hydrogens in aryl halides superimposed upon the relative kinetic acidities of the hydrogens in pyridine (4-position > 3-position > 2-position). Gribble and Saulnier first reported this in the treatment of 2-, 3-, and 4-halopyridines with LDA [4]. The 2-halo derivative **3** afforded 3-lithio species **4**, the 3-halo derivative **5** afforded 4-lithio species **6**, and the 4-halo derivative **7** afforded 3-lithio species **8**.

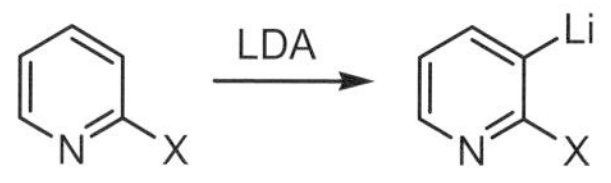

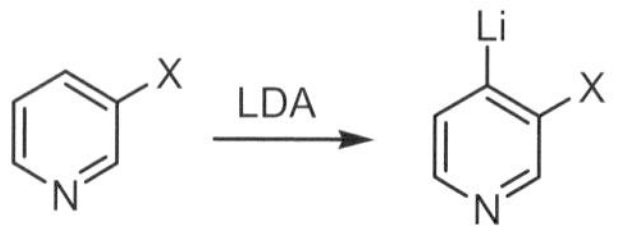

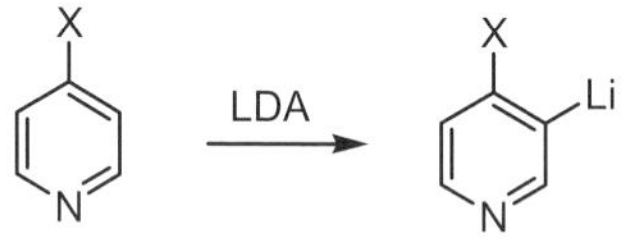

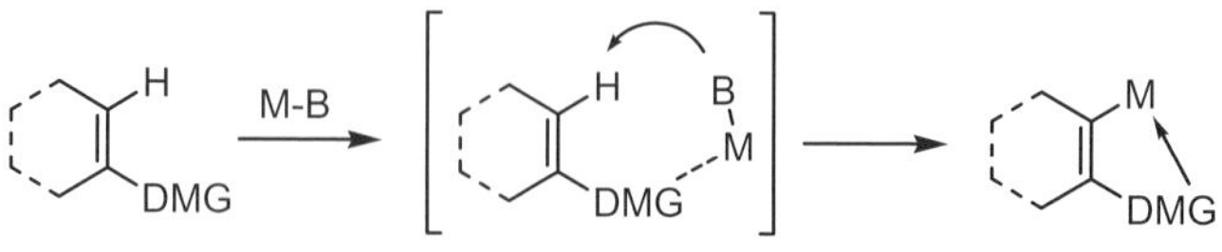

Figure 4.4.

Direct lithiation at the 2-position of the pyridine ring could be accomplished using hexafluoroacetone [5]. The non-bonding electrons of pyridine nitrogen **9** react with the highly polarized carbonyl group of acetone derivative **10** to afford adduct **11** which then delivered the base, LTMP, by prior complexation. The resultant 2-lithio species **12** could then be trapped to produce the desired 2-halopyridines **13**.

9	**10**	**11**	**12**	**13**

E = I   58%
E = Br 50%

One could also accomplish this type of transformation using a Lewis acid complex [6]. Treatment of **9** with BF$_3$-etherate at −78°C with LTMP produced 2-lithio species **15** *via* complex **14**, that could be subsequently trapped with electrophiles.

9   **14**   **15**

In a similar manner, 2-halopyridines **3** could be prepared by exposure to acetyl fluoride. Acylation, by acetyl fluoride, of the nitrogen atom of **9** activated the pyridyl ring for halogenation by appropriate sources of nucleophilic chlorine or bromine.

X = Cl 70%

9   **3**

Direct metalation of the pyridine ring could be accomplished using the complex of *n*-BuLi and *t*-BuOK [7]. While an equilibrium mixture of all 3 possible potassio species (**16**, **17**, and **18**) initially occurred, the polarity of the reaction medium and whether thermodynamic or kinetics conditions were employed influenced the equilibrium, thus giving rise to a preferred product. A polar medium favored formation of 4-potassio species **18**, thermodynamically or kinetically, while 2-potassio species **16** was favored under weakly polar conditions.

9   **16**   **17**   **18**

The alternative complex base *n*-BuLi – LDMAE, obtained from *n*-BuLi and *N,N*-dimethylaminoethanol, has been shown to selectively metalate **9** at the 2-position [8]. Trapping the anion with carbon tetrabromide afforded **19** in excellent yield. Similar results were observed using quinoline as the substrate but with reduced efficiency.

Difficulties in preparing and handling complex bases, along with their exacting experimental conditions have limited their use. Kondo provided a solution to this with a report on the effectiveness of lithium di-*tert*-butyltetramethylpiperidinozincate (TMP-zincate) as a highly chemoselective base [9]. It was determined that pre-complexation of LTMP **20** with di-*tert*-butylzinc was required to generate the active species **21**. The reaction with pyridine at room temperature was found to proceed smoothly to produce the 2-lithio species that could be treated with iodine to afford 2-iodopyridine **22**.

In the case of 2-methoxypyridine **23**, one can produce different substitution patterns by selecting a base with the desired properties [10]. For example, LDA reacted according to the CIPE and afforded 3-lithio species **24**. The properties of the complex base *n*-BuLi – LDMAE were found to be modulated by the addition of sodium amide. This altered its chemistry and resulted in the formation of 6-lithio species **25**.

This complex base, in an un-modulated form, was also shown to provide regioselective 2-lithiation independent of the position of the halogen [11]. Thus, 3-chloropyridine **26** was lithiated to produce **27** and 4-chloropyridine **28** was lithiated to afford **29**.

4.1.2. Metal–halogen exchange

As more and more haloderivatives of pyridine become commercially available, they can be used as starting materials for the preparation of highly derivatized systems. For example, 2,6-dibromopyridine **30** can be desymmeterized by clean monolithiation with *n*-BuLi [12]. The resultant anion **31** can be trapped by an electrophile to afford the desired cross-coupling intermediates **32**.

E = Cl   67%
E = B(OH)$_2$ 53%

Similar results were observed when the unsymmetrical dihalide **33** was exposed to these reaction conditions. Lithiation occurred specifically at the 2-position and when followed by trapping with an electrophile, afforded **34**.

Bromine–magnesium exchange was reported to allow one greater access to more diversified substituted pyridines [13]. Monobromo-substituted pyridines **35** were easily transformed into the corresponding magnesio-derivative that could then be trapped with electrophiles to produce **36**. The versatility of this method was in the regioselectivity with multisubstituted pyridines. For 2,6-dibromopyridine **30**, a single exchange could be accomplished using these reaction conditions to afford **37**. Further regioselectivity was observed for 2,3-dibromopyridine **38** and 2,5-dibromopyridine **40**. In both cases, exchange occurred only at the "meta" position to produce the corresponding 2-bromo-3-substituted pyridine **39** and 2-bromo-5-substituted pyridine **41**, respectively.

This classical method has been improved upon by the introduction of an organomagnesium complex [14]. Reaction of *n*-butyllithium with isopropylmagnesium chloride, in a 2:1 stoichiometry, readily prepared the (*n*-Bu)$_2$(*i*-Pr)MgLi complex. This reagent was shown to efficiently carry out the bromine–magnesium exchange of **42** under non-cryogenic conditions (−10°C). The organomagnesium intermediate was found to be stable for 2 h at −10°C with no loss in yield of the iodine trapped product **43**. No product was obtained when using *i*-PrMgCl. These results stand in contrast to the acidic nature of the protons on the methyl substituent of **42**. Exposure of **42** to *n*-butyllithium at −100°C cleanly formed the lithio-picoline species.

A typical DoM group, such as the pivaloylamino, provided a ready method of introducing a halogen to the pyridine ring as illustrated in the conversion of **44** to **45** [15].

As the number of halogen substituents increases on the pyridine ring, one must be cognizant of the relative stabilities of the intermediates [16]. For 2,3,5-trihalopyridine **46**, lithiation occurred at the more kinetically acidic 4-position giving rise to **47**. Trapping with iodine afforded the expected 4-iodo derivative **48**. Subsequent lithiation did indeed produce 2-lithio species **49**, but its relative instability allowed this species to undergo a "halogen shuffling" procedure to a more stable intermediate. The result of this process gave rise to 2-iodo-4-lithio derivative **50**.

4.1.3. Halogen–halogen exchange

Pyridines bearing a chlorine or bromine at the 2-, 3-, or 4-position **51** can undergo a substitution reaction using trimethylstannyl sodium [17]. The resultant stannylpyridine derivative **52** can participate in an iododestannylation reaction with iodine to give **53**, the result of a net halogen–halogen exchange.

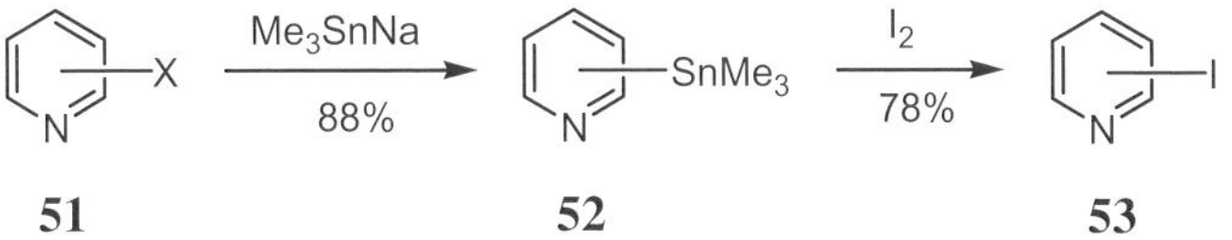

51

X = Cl, Br

52

53

Bromotrimethylsilane or iodotrimethylsilane reacting with 2-halopyridines **3** can participate in an exchange of halides to the corresponding pyridyl bromide or iodide **55**, respectively [18]. The reaction proceeded *via* initial formation of *N*-(trimethylsilyl)pyridyl species **54** followed by an exchange of halogens to afford the product. No reaction was observed with fluoropyridines, presumably due to the reduced nucleophilicity of the pyridyl nitrogen. A similar rationalization was put forth for the lack of reactivity of 2,6-dichloro- and 2,6-dibromopyridine in this reaction. The situation changed when alternate substitution patterns were examined. For 2,3-dichloropyridine **56**, exchange occurred at the 2-position to afford **57**. For 2,4-dichloropyridine **58**, exchange occurred to produce **59**, while 2,5-dichloropyridine **60** exchanged at the 2-position to give **61**.

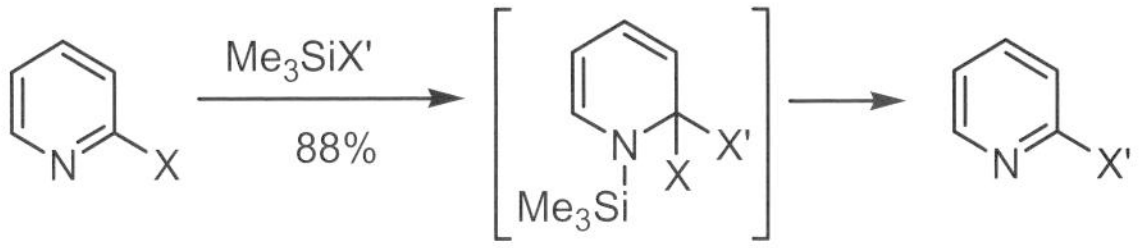

3

54

55

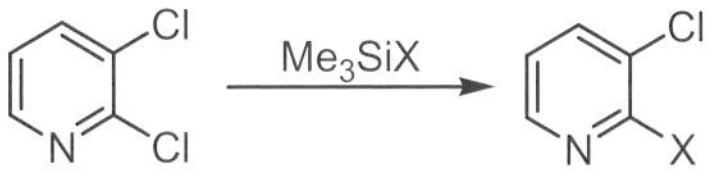

56

57

X = Br 83%
X = I   33%

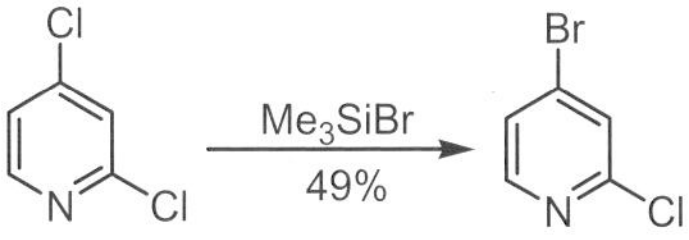

58

59

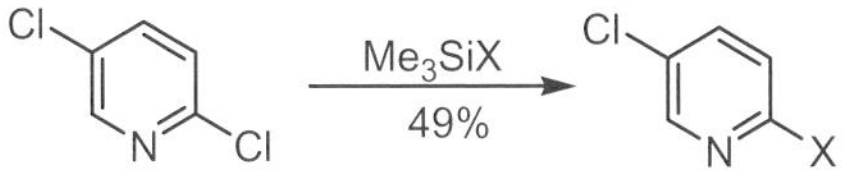

60

61

X = Br 67% (33)
X = I   52%

In a mechanistically similar manner, sodium iodide in the presence of acetyl chloride or under acidic conditions was also capable of exchanging halogens on **51** for iodine to give **53** [19].

51   **53**

The reaction conditions for the halogenation of 3-hydroxypyridine were found to be different depending upon the nature of the halogen [20]. However, it was determined that the alkoxide of **62** was required as the nucleophile in this process. Treating the pyridinolate with sodium hypobromite afforded the 2-bromo derivative **63** in good yield. In order to obtain the corresponding 2-iodo derivative **65**, the reaction conditions required the use of iodine and sodium carbonate to effect the halogenation. The mixed species **64** could be obtained by iodination of the previously prepared bromide **65**.

Similar methodology [21] was used in the preparation of novel heterocyclic systems. Iodination of pyridine **66** afforded **67**, an intermediate used in the preparation of **68**.

While not a halogen–halogen exchange reaction, there are examples of amino groups being exchanged for halogens [22]. It was found to be a general method that pyrylium iodide **70** could react with aminopyridine **69** to generate **71**. Iodide displacement then afforded the corresponding 2-iodopyridine derivative **22**.

An additional example of this type of exchange was illustrated in the diazotization of aminopyridine **69** to the corresponding bromopyridine **19** [23].

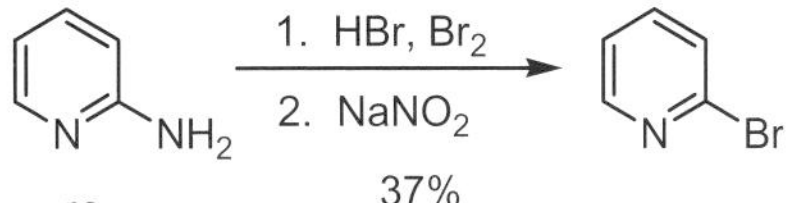

69   1. HBr, Br$_2$ 2. NaNO$_2$ 37%   **19**

4.1.4. Dehydroxy–halogenation

Pyridones or hydroxypyridines can readily be converted to the corresponding halide by a number of different reaction conditions [24]. The use of phosphorous trichloride for transformation was demonstrated in the conversion of **72** to chloride **73**. The exchange of chlorine for iodine, as in **74**, has already been discussed (*vide supra*). The corresponding bromide could be obtained using phosphorous tribromide or phosphorous pentabromide, as in the conversions of **72** to **75** and **76** to **77**.

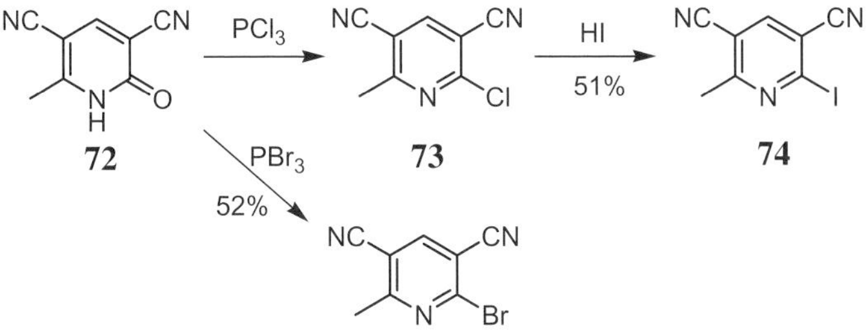

72   PCl$_3$ **73**   HI   51%   **74**   PBr$_3$ 52%   **75**

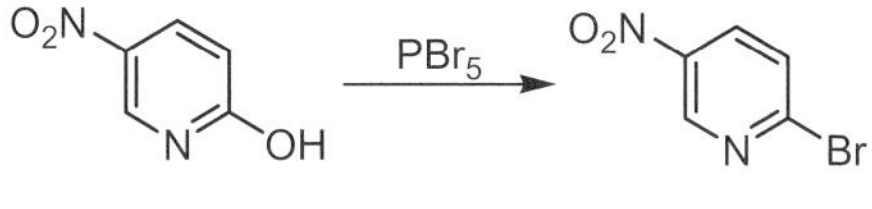

76   PBr$_5$ **77**

Milder conditions also have been reported [25]. For example, treatment of hydroxypyridine **78** with phosphorous pentoxide and tetrabutylammonium bromide (TBAB) carried out the conversion to the corresponding bromide **79**.

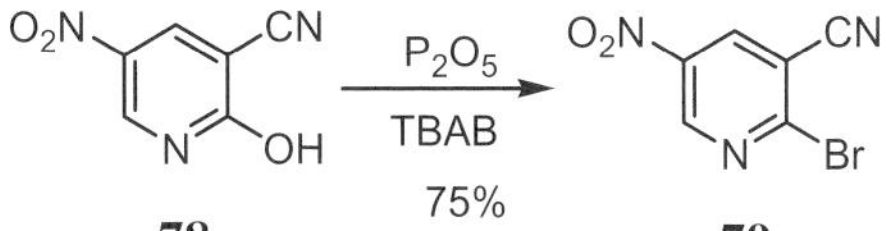

78   P$_2$O$_5$ TBAB   75%   **79**

Additionally, triphenylphosphine was able to facilitate the ability of *N*-halosuccinimides to carry out the desired transformation of **80** to **3** [26].

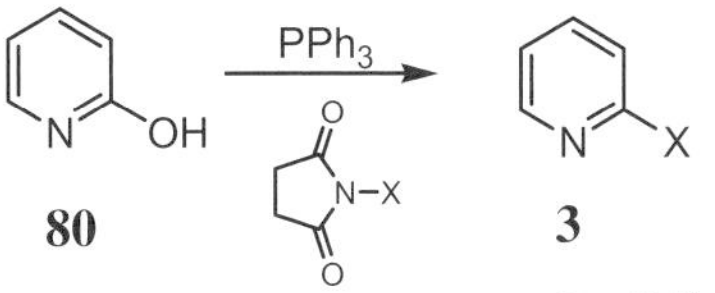

80   PPh$_3$ **3**

X = Cl   43%
X = Br   54%

4.2. Coupling reactions with organometallic reagents

4.2.1. Kumada coupling

Murahashi first reported the Pd-catalyzed variation of the Kumada reaction in 1975. He expanded the original work, reported independently, by Kumada and Tamao and Corriu in 1972. This groundbreaking work on cross-coupling reactions illustrated the ability of Ni(II) complexes to catalyze the reaction of Grignard reagents with alkenyl- or arylhalides [27]. The ensuing 30 years has witnessed an increased utility of metal-catalyzed reactions, in general. The Kumada reaction has the advantage that carbon hybridizations other than sp^2 can be accessed. Once the organobromide has undergone its oxidative addition to the transition-metal catalyst, transmetalation of the Grignard reagent occurs followed by reductive elimination to afford the cross-coupled product. Also, the Kumada reaction has the advantage that it can be conducted over a greater range of reaction temperatures ($-20°C$ to elevated temperatures), but room temperature is the most common due to its ease. While Ni or Pd catalysts can be use to conduct this transformation, this chapter will limit discussion to the Pd-based variations. The main disadvantage resides in the reactivity of the Grignard reagent. For the Kumada reaction to proceed, any additional functional groups present in the reacting partners must be compatible with the nucleophilic or basic nature of the Grignard reagent.

The Kumada reaction has been used for the preparation of polyfunctionalized pyridines. In one example, Grignard reagent **82** was coupled to pyridyl derivative **81** using palladium catalysis to afford **83** in excellent yield [28].

The Grignard reagent of 2-chloro thiophene underwent the Kumada reaction with 2,6-dichloropyridine **84** to produce the monosubstituted pyridine **85**. Mixed heteroarene **86** was afforded when **85** was allowed to react with the Grignard derived from *N*-methyl-2-bromoindole [29].

The bis(organomagnesium) reagent **87** was allowed to undergo a Kumada reaction with the dibromopyridine **40** to afford **88**, an example of a poly(phenylenepyridine) [30].

4.2.2. Negishi coupling

Negishi originally reported that organozinc species could undergo Pd-catalyzed coupling with various organohalides to form the resultant cross-coupled products [31]. Reagents are readily prepared by direct insertion of Zn into alkyl halides. Unlike the Kumada reaction, this method is more compatible with a wide variety of functional groups. The rate of insertion is dependant upon the nature of the substitution pattern. Benzylic and allylic insertion can occur at 0°C, while primary halides require the more forcing conditions of higher concentration and elevated temperatures. Alternatively, reaction partners can be prepared by the reduction of zinc halides to produce more activated zinc (Rieke zinc) that can be paired-up with less reactive organohalides. The organozinc reagents can also be prepared by transmetalation reactions from organolithiums and Grignard reagents. Hydrozincation and carbozincation can also provide access to the requisite organozinc derivatives. This Zn-based reaction offers the advantage of a one-pot reaction sequence, thus avoiding the isolation of intermediate metallic species, while allowing unsymmetrical coupling partners to be generated under mild conditions. Furthermore, the availability of commercial sources of organozinc species can be leveraged to the synthetic chemist's advantage.

The DoM reaction of carbamoyl indole **89** followed by transmetalation with zinc bromide provided the organozinc partner required for the Negishi reaction. Coupling of this intermediate with 3-bromopyridine afforded heterobiaryl **90** [32].

Selective metalation of 2-fluoro-4-iodopyridine **91** followed by trapping with zinc chloride generated **92** that subsequently was reacted with 2,4-dichloropyrimidine to generate pyridinyl-pyrimidine **93** [33]. In addition to the standard reaction conditions, it was found that a microwave-assisted variation also provided the bis(heteroaryl) products in an efficient manner.

Subsequent investigations of this chemistry were focused on preparing analogs of CGP 60474, **94** [34]. This compound was found to be a protein kinase C (PKC) inhibitor. PKC plays a critical role in signal transduction cascades that are involved in cellular proliferation and differentiation. CGP 60474 exhibits selectivity when compared to serine/threonine and tyrosine kinases. Imatinib **95** (Glivec) is a tyrosine kinase inhibitor currently on the market for chronic myeloid leukemia (CML).

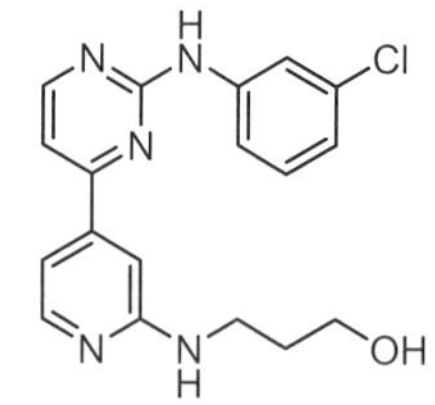

CGP 60474
94

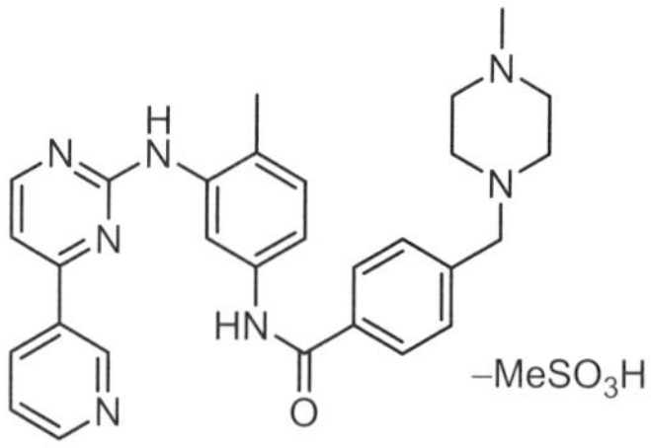

imatinib (Glivec™)
95

Analogs of **94** were prepared in a manner following the precedent set by **93**. However, the yields for **96** did not reproduce those seen in the previous system. This result could be rectified by approaching **96** in a reversed manner. Coupling organozinc reagent **97**, prepared from the protected aniline derivative **98**, with **91** did afford **96** in excellent yield. This intermediate could then be elaborated into derivatives of **94** to study structure-activity relationships of this chemotype.

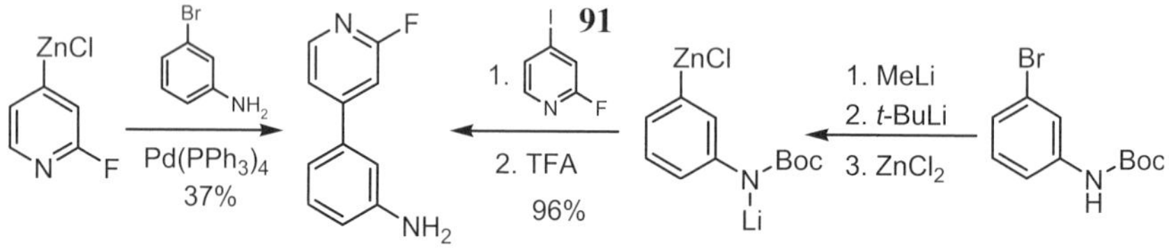

92   **96**   **97**   **98**

Bipyridines are readily prepared by cross-coupling reactions and the Negishi reaction is no exception [35]. Organozinc **99**, prepared in standard fashion from **19**, could be cross-coupled with **102** to generate **100**. Triflate **102** was prepared from aminopyridine **101** *via* diazotization followed by formation of the triflate.

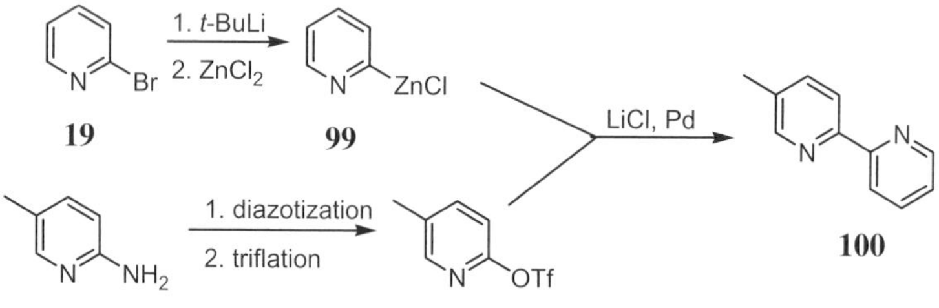

19   **99**   **100**

101   **102**

Analogously, the commercially available pyridylzinc reagent **103**, which can also be prepared directly from **19**, was coupled with substituted halopyridines to generate substituted bipyridyls **104** [36].

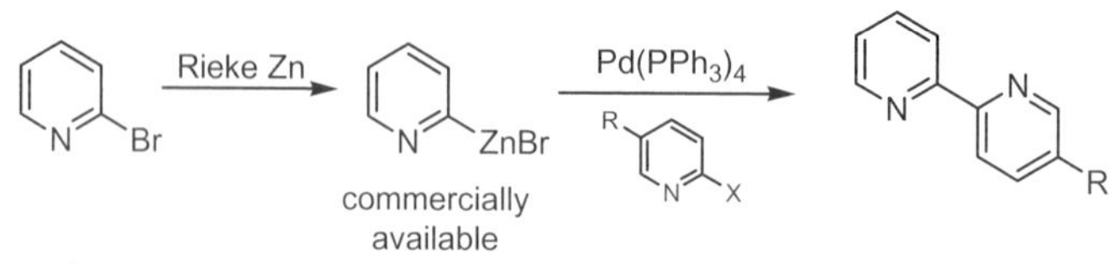

19   **103**   **104**

commercially
available

Corticotropin releasing factor (CRF) ligands (*vide infra*) have been prepared *via* the Negishi reaction [37]. Conversion of iodopyridine **105** to its corresponding organozinc species preceded Pd-catalyzed coupling with **106** to afford **107** in 60% yield. This compound was then able to serve as a key intermediate in the synthesis of CRF analogs.

Converting **108** to the corresponding organozinc intermediate **109** was carried out by transmetalation of the corresponding Grignard reagent. The desired product **110** was then obtained by palladium cross-coupling of **109** with 2-bromopyridine **19** [38].

Highly elaborate helicopodands have been prepared by this method [39]. Metalation of the protected aminopyridine **111** afforded the organozinc reagent **112** that was coupled with dibromohelicene **113** to produce the desired **114**.

Cycloalkylzinc reagents are also capable of being coupled to pyridines [40]. Iodoazetidine **115** was readily transformed to organozinc **116**. Cross-coupling with 2-bromopyridine **19** subsequently afforded **117**.

Polysubstituted bipyridyls are readily prepared by the Negishi reaction [41]. Execution of the standard methods allowed for transformation of **19** into the desired product **118**.

The total synthesis of caerulomycin C **122** employed a Negishi coupling as the key step in the synthetic approach [42]. Dimethoxypyridine **119** was converted to organozinc **120** prior to Pd-catalyzed coupling with 2-bromopyridine **19** to afford **121**, an advanced intermediate used in the synthesis of the natural product.

Metal–halogen exchange of **123** produced organozinc **124** [43]. This compound was then coupled with a chloroquinoline to afford **125**, a key intermediate in the synthesis of camptothecin **126** (*vide infra*).

A three-step, one-pot procedure gave rise to **129** after deprotonation of **127** followed by transmetalation to **128** and palladium-mediated cross-coupling [44].

Non-natural amino acids can be prepared by this method as exemplified by the preparation of **131**. Palladium-mediated coupling of alkylzinc species **130** with **19** afforded **131** [45].

The Pd-catalyzed cross-coupling reactions of **99** provide a brief example of the diversity of functionality and products that are accessible *via* the Negishi reaction [46]. Reaction with 2,6-dibromopyridine **30** provided **132** in good yield, while 2,5-dibromopyridine **40** generated **135**. Partnering with 4-cyano-2-chloropyridine afforded **133** and coupling with an alkynylpyridine provided ready access to **134**.

4.2.3. Suzuki coupling

This reaction consists of a Pd-catalyzed cross-coupling reaction of a (hetero)aryl halide with a (hetero)aryl- or vinylboronic acid derivative to generate unsymmetrical bi-(hetero)aryl compounds. The Suzuki reaction is more advantageous compared to the Kumada and Negishi due to the ready access, through multiple synthetic pathways or commercial sources, of the organoboron reagents. The greater stability of these reagents, in conjunction with the relatively simple experimental reaction conditions, has resulted in a dramatic increase in the use of this variation on the cross-coupling reaction [47].

Electrospray MS studies provided circumstantial evidence for the commonly proposed catalytic cycle of the Suzuki reaction [48]. During the investigation of the reaction of 3-bromopyridine **136** with phenylboronic acids **137** to afford **138**, molecular ion peaks were observed for species consistent with $[(PyrH)Pd(PPh_3)_2Br]^+$ or $[(Pyr)Pd(PPh_3)_2]^+$. Also consistently observed in the spectra was a peak that correlated with [(PyrH) $(R_1R_2C_6H_3)Pd(PPh_3)_2]^+$. Additionally observed, but at much lower levels, was $[(R_1R_2C_6H_3)$ $Pd(PPh_3)_2]^+$ that could have arisen from transmetalation with the catalyst directly. By definition, it is not possible to observe catalytic species so one must question the validity of these observations.

136   **137**   **138**

The choice of base is critical for the Suzuki reaction and in some cases can have a rate accelerating effect on this reaction. In particular, for sterically hindered boronic acids like **139** reacting with halopyridines **140** to afford **141** in non-aqueous solvent [49], it was found that as one increased the basicity not only did the yield of the reaction improve, but also the time required to complete the reaction was dramatically reduced.

139   **140**   **141**

	Yield(%)	Time(h)
Na_2CO_3	26	90
NaOH	40	140
NaOEt	74	4
KOt-Bu	86	4

For the Suzuki reaction, the highest yields can be obtained for bromides, iodides, and triflates reacting with boronic acids. The less reactive aryl chlorides only show reactivity when partnered with electron withdrawing groups. The disconnect in this observation is that chlorides are generally more readily available, least expensive, and in the case of 2- and 4-halopyridines, more stable. The utility of chlorides can be improved using palladium tetrakis(triphenylphosphine) as a catalyst [50]. Anhydrous conditions using potassium phosphate in DMF were inferior when reacting **142** and **143** to generate 3-arylpyridines **144**.

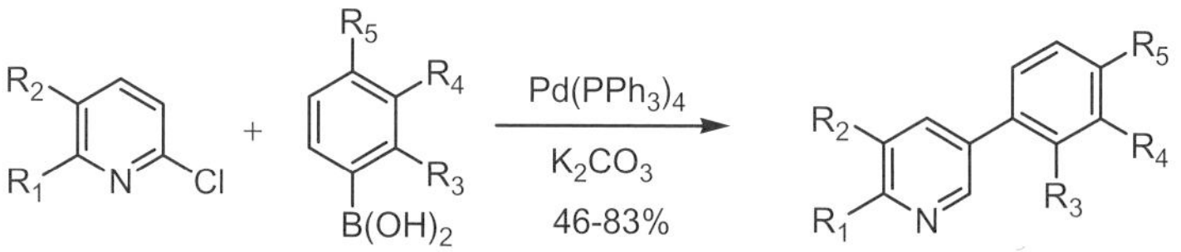

142 **143** **144**

In attempts to improve the efficiency of the Suzuki reaction, various groups have examined ways of making the catalyst more efficient or user-friendly. One example has been to make use of heterogeneous catalysts that have been widely applied to catalytic hydrogenation reactions. The cross-coupling of halopyridines **145** with arylboronic acids **146** that resulted in **147** has been conducted using Pd/C [51]. The addition of phosphine ligands increased the yield 4-fold.

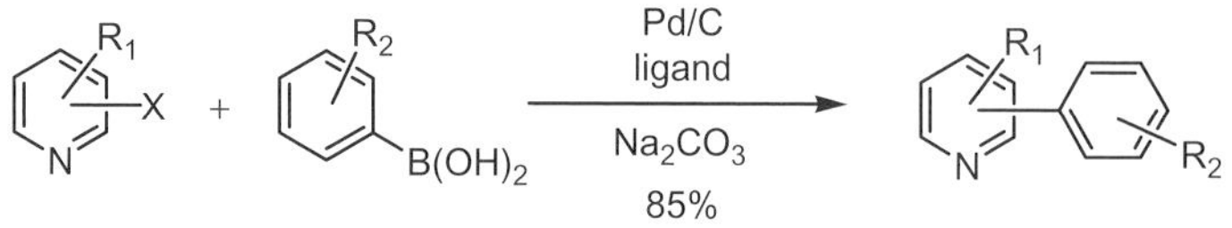

145 **146** **147**

While carefully isolating all products from a reaction of 2-bromopyridine **19** with palladium catalyst, bimetallic complex **148** was isolated and determined to be a pre-catalyst for the Suzuki reaction [52]. Upon examining the utility of this complex, it was found to be able to carry out Suzuki reactions, such as the conversion of **149** and **150** into **151**.

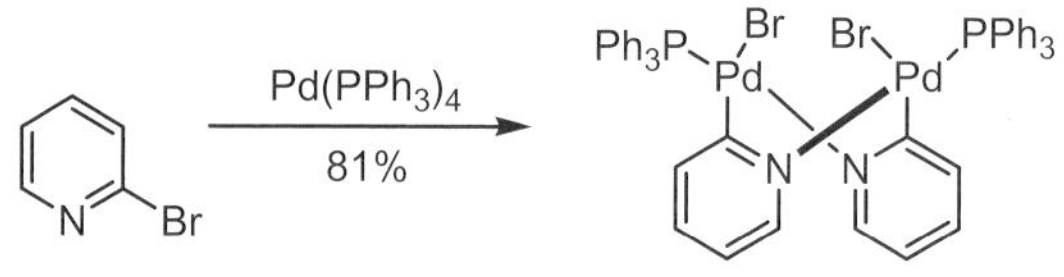

19 **148**

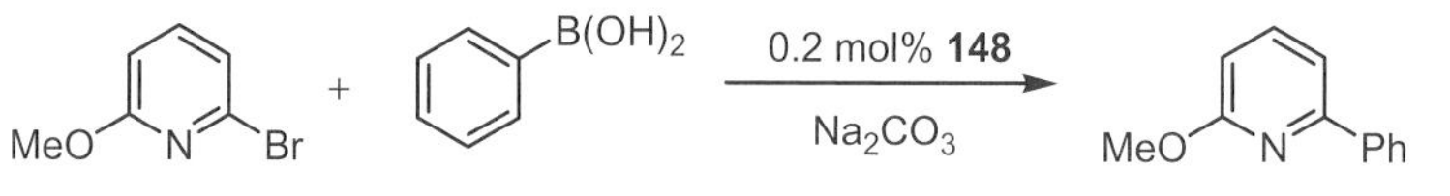

149 **150** **151**

The commercially available palladium catalyst FibreCat has also been applied to the Suzuki reaction [53]. The polymer-supported catalyst with microwave heating gave rise to cross-coupled product in good yield as exemplified in the transformation of **152** and **153** into arylpyridine **154**.

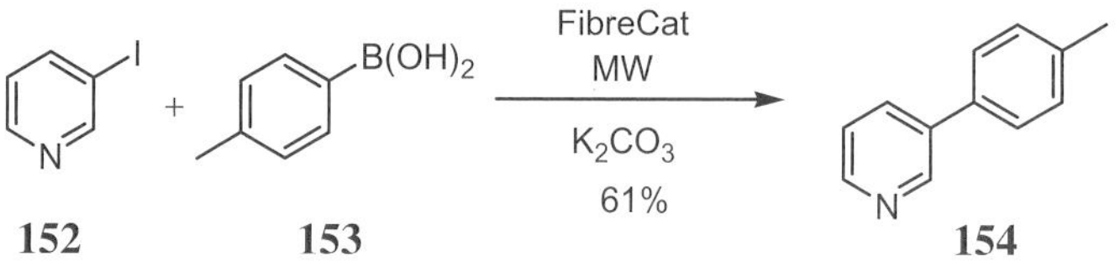

152 **153** **154**

The microencapsulation and cross-linking of polymer chains converted Pd(PPh$_3$)$_4$ to a polymer-incarcerated palladium (PI-Pd) catalyst consisting of a phosphine-free Pd(0) catalyst [54]. The advantage of this catalytic system was that it was recoverable and reusable. Optimized conditions for Suzuki couplings made use of added ligand, as observed for the cross-coupling of **155** with boronic acid **156** to afford **157**.

One additional modification of the catalyst has been recently reported [55]. A yellow-red solution of pH 8–9 was obtained when one equivalent. of PdCl$_2$ was mixed with 2 equivalents of EDTA in water. This aqueous solution was stable at room temperature and in the air. Its utility was demonstrated in the formation of **160** from **158** and **159**.

Liebeskind reported a new method for cross-coupling heteroaromatic thioethers with boronic acids [56]. This required the use of copper(I) thiophene-2-carboxylic acid (CuTC) but allowed the reaction to be conducted under mild and neutral reaction conditions from readily available starting material. As an example, thiomethylpyridine **161** was cross-coupled to boronic acid **162** to afford the biaryl **163**.

Buchwald observed that using solubilizing functionality on the phosphine ligands allowed the Suzuki reaction to be executed under aqueous conditions [57]. This also obviated the need to protect free amino groups as no metal chelation was observed. Thus pyridyl derivative **164** was cross-coupled with boronic acid **165** using the modified ligand **166** to afford **167** in excellent yield.

The nature of the boronic acid has also been investigated [58]. It was observed that organotrifluoroborates **168** had several advantages: easier purification, easier handling, and the ability to tolerate a wider variety of electron withdrawing and electron donating groups. Also, these species are more robust and are less prone to protodeboronation. Thus, **136** could be converted to **169** using reagents such as **168**.

A variation on this theme made use of **170**, prepared from **19**, to improve the nature of the boronic acid species [59]. Indeed, cross-coupling with a variety of aryl halides afforded the desired products **171**.

During a scale-up process it was found that 3-pyridylboroxin **172** was a bench-stable precursor to boronic acids that was easier to purify [60]. During the recrystallization of **165**, prepared from **136**, **172** was isolated. This compound was shown to react, in the Suzuki reaction, to afford **173** in a manner analogous to **165** but with greater ease in handling.

Approaches to adapting the Suzuki reaction to combinatorial chemistry have examined various solid-supported methods [61]. The most straightforward method employed is linking the arylhalide to resin **174** *via* an ester group. With modified resin **175** in-hand, the cross-coupling reaction to generate **176** proceeded.

The pyridyl group has also been immobilized on the resin and was able to survive lithiation conditions necessary to prepare the halopyridine [62]. This approach has also been applied to the Sonogashira, Stille, and Negishi reactions (*vide infra*). In the Suzuki example, the resin-bound pyridine **177** can be converted to the haloderivative **178** using standard lithiation chemistry. With **178** in-hand, cross-coupling and cleavage from the resin afforded **179**.

The Suzuki reaction has been extended to liquid phase organic synthesis (LPOS) by derivatizing carbosilane dendrimers as soluble supports [63]. This occurred without the use of linkers and provided high thermal stability, good accessibility, good solubility in organic solvents, and separation by nanofiltration. Lithiation of pyridyl bromide **136** preceded attachment to the carbosilane dendrimer (G_0-Cl) support. Treatment with methanolic triethylamine quenched any unreacted termini of the support. The supported pyridine **180** was capable of reacting with a variety of cross-coupling partners as exemplified in the formation of **154**.

A pyridylboronic acid derivative **181** has also been attached to a solid-support and does not hinder the Suzuki reaction in that a variety of aryl and heteroaryl halides **182** were able to undergo the reaction in good to excellent yields to afford **183** [64].

As was observed with the Kumada and Negishi reactions, the reactions conditions required to conduct the cross-coupling reaction may not be compatible with a wide variety of functional groups. The mild nature of the Suzuki reaction has been demonstrated by the types of functional groups that can be retained unprotected or *via* methods to modulate the reaction [65]. The coupling of **184** and **185** to generate **186** provided evidence that carboxylic acids and aldehydes are compatible with the Suzuki reaction.

The reaction of **187** with **188** provided evidence that highly substituted bi- and tripyridines **189** could be assembled and that the amino group does not need to be protected in this process [66].

Alternatively, use the ligand 1,1′-bis(di-*tert*-butylphosphino)ferrocene (D-*t*-BPF) allowed retaining the amino group in an unprotected fashion in the Suzuki reaction of **190** with **150** to produce **191** [67].

In the synthetic arena, the Suzuki cross-coupling reaction has seen a great increase in its use and versatility. An initial application was in the synthesis of quaterpyridines [68]. The bis(chlorobipyridine) **192** readily underwent tandem coupling reactions to afford the desired compound **193**.

192   193

The critical and rate-controlling step in the pathway leading to the synthesis of DNA is focused on ribonucleotide reductase (RR). Inhibitors of RR would have great utility as a therapeutic agent against cancer. One such potent inhibitor of ribonuclease diphosphate reductase is 3-AP **196**, and a Suzuki methylation reaction that converted **194** into **195** began the synthesis of this compound [69].

194   195   196

In the development of new chiral nucleophilic catalysts that were DMAP analogs, the chirality was introduced into the framework through the axial chirality inherent in biaryl systems [70]. To this end, bromopyridine **197** was converted to biaryl **198** by cross-coupling with phenylboronic acids.

197   198
                          87-98%

Application of the Suzuki reaction to the synthesis of cytisine **201** required optimization in order for completion [71]. The workers found that isolation of the boronic acid intermediate corresponding to **149** was problematic. The synthesis did proceed uneventfully when the workers opted for using the *in situ* prepared boron-ate complex **199** for the cross-coupling to produce **200**.

149   199

201   200

Hydroboration of the vinyl oxazoline **202** afforded the corresponding boron derivative that was used directly to generate **203**. Hydrolysis then afforded non-natural amino acids as typified by **204** [72].

The 3-amino-2-phenylpiperidine scaffold **205** is an important pharmacophore found in a number of important psychopharmaceutical agents. One retrosynthetic approach to this framework is *via* the corresponding pyridine derivative [73]. It has now been prepared in a single, high-yielding step from inexpensive commercially available starting material and can be scaled to the multi-kilogram level. The free amino group of **190** was protected *in situ* as the imine from benzaldehyde and after work-up afforded the desired cross-coupled product **191**.

Bis-indole alkaloids, such as **208**, possess interesting scaffolds with biological activity. A general route to these compounds that allowed for various heteroaromatic rings to be used as the linker for the indole moieties was desired and the Suzuki reaction of **206** with **207** appeared to be the most direct method for their assembly [74].

Salen ligands have found utility in a number of synthetically useful reactions. Pyridine modified salicylaldehydes **212** have been prepared by the hydrolysis of **211**, the product of a Suzuki reaction of **209** with **210** [75]

Corticotropin releasing factor (CRF), a 41-amino acid peptide originally isolated from hypothalamus, plays a major role in homeostasis by mediating the endocrine, autonomic, immune, and behavioral responses to stress. Imbalances in this system can lead to pathologies such as anxiety, depression, and feeding disorders, therefore ligands mimicking the effects of CRF will be of pharmaceutical benefit [76]. Coupling of pyridyl boronic acid **213** with chloropyridine **214** generated **215**, an advanced intermediate towards **216**, a scaffold found to exhibit CRF activity.

213   214   215   216

Tamoxifen **219** is an important selective estrogen receptor modulator (SERM) that has anti-estrogenic properties of value for the treatment of breast cancer. It does, however, have some undesired side effects. Thus, the development of analogs is of great interest. A pyridyl derivative **218** was assembled *via* the Suzuki reaction of **217** with 3-iodopyridine **152** [77].

217   218   219

Chemistry conceptually similar to that described for tamoxifen analogs **218** has been applied to the synthesis of CDP840 **222**, a phosphodiesterase (PDE) IV inhibitor for the treatment of asthma [78]. In combination with the Liebeskind-variation of the Suzuki reaction, thiopyrimidine ether **220** was cross-coupled to pyridyl-4-boronic acid **210** to afford **221**. Catalytic hydrogenation of the olefin gave rise to **222**.

220   221   222

The Suzuki reaction has been exploited in the development of alternate scaffolds for PDE IV inhibitors [79]. The highly elaborated intermediate **223** was coupled with the pyridyl *N*-oxide **224** to give **225** that possessed the desired biological activity.

223   **224**   **225**

The viability of pyridylboronic acids was investigated as potential intermediates in the preparation of heteroarylpyridines [80]. Boronic acid **227** was prepared from **226** and coupled to 3-bromoquinoline to generate **228**.

226   **227**   **228**

The cytochrome P450 enzyme system constitutes the body's mechanism for handling xenobiotics. Drug–drug interactions are a result of imbalances in this enzyme system and the study of selective antagonists of this system could provide greater insight into these interactions. One such example is nicotinic mimetics used as P450 2A6 antagonists [81]. Pyridylboronic acid **187**, prepared from **158**, could undergo the Suzuki reaction to afford **229**, a compound that displayed antagonistic activity against the P450 enzyme.

158   **187**   **229**

The course of the Suzuki reaction can be redirected if the reaction is conducted under an atmosphere of carbon monoxide. In this modification of the reaction [82], the organo-palladium intermediate is captured by carbon monoxide prior to the transmetalation step. The product from this is the carbonyl-inserted derivative. Application of this variation provided ready access to pyridyl ketones as exemplified in the conversion of **51** with **150** to **230**.

51   **150**   **230**

$X = 2\text{-Br}$ 70% $PdCl_2(PPh_3)_2$
$X = 3\text{-Br}$ 80% $PdCl_2(PCy_3)_2$
$X = 4\text{-Br}$ 81% $PdCl_2(PCy_3)_2$

The nature of the palladium catalyst can be modulated such that one can gain regio-selectivity in the carbonylation reaction [83]. Thus, 2,5-dichloropyridine **60** could be monofunctionalized at the 2-position as in **231**.

4.2.4. Stille coupling

This reaction generates functionalized systems by Pd-catalyzed reaction of sp^2 or sp-hybridized stannanes with organohalides or triflates [84]. The wider tolerance of functional groups has led to the greater scope of this reaction. Its major drawback is the use of potentially toxic tin reagents.

An air and moisture stable, low-coordinate phosphine ligand has been found to be effective in the Stille coupling of pyridine derivatives. This was exemplified in the transformation of **136** to **169** [85].

An improvement in the Stille reaction was found to be the result of a synergistic effect between copper salts and fluoride. Adding copper iodide and cesium fluoride to the standard palladium catalyst resulted in a quantitative yield of **233** from **152** and **232** [86].

Additionally, it was found that regioselectivity could be achieved with copper salts by varying the solvent and the stoichiometry [87]. For the cross-coupling of **235** with **236**, catalytic copper iodide favored formation of **237**, while stoichiometric copper iodide afforded **234**. It was suggested that in DMF, copper favored the palladium species leading to **234**.

A systematic survey of the coupling for **236** and **51** to produce **238** found that other additives could increase the rate and yield of the Stille reaction and the position of the halide did have an effect [88].

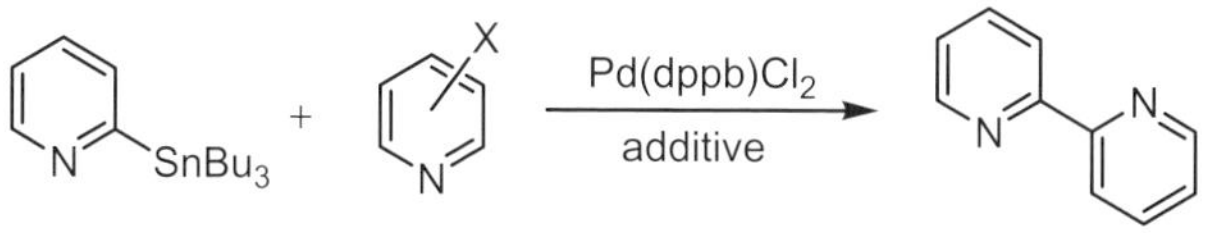

236 **51** **238**

Substitution	Additive	Yield (%)
4-Br	-	47
	Ag_2O	73
	CuO	75
4-Cl	-	13
	CuO	44
3-I	-	39
	CuO	62
3-Br	-	99
	CuO	64

This effect was found to be more general when the synthesis of thieno-1,6-naphthyridines **241** from **239** with **240** could be improved by the use of a CuO additive.

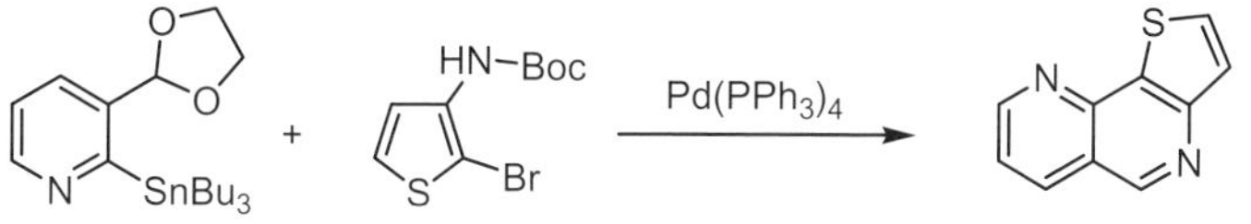

239 **240** **241**

no additive 9%
Ag_2O 24%
CuO 57%

The number of reaction types amenable to the application of combinatorial chemistry continues to increase. The utility of the technology in the preparation of compounds was extended to the Stille reaction [62]. Tin reagents on solid-support **242** were shown to undergo the Stille reaction with **182** to afford **243**. Reversing the nature of the reacting partners, using the solid-supported bromide **178** with aryl stannanes gave the corresponding product **244**.

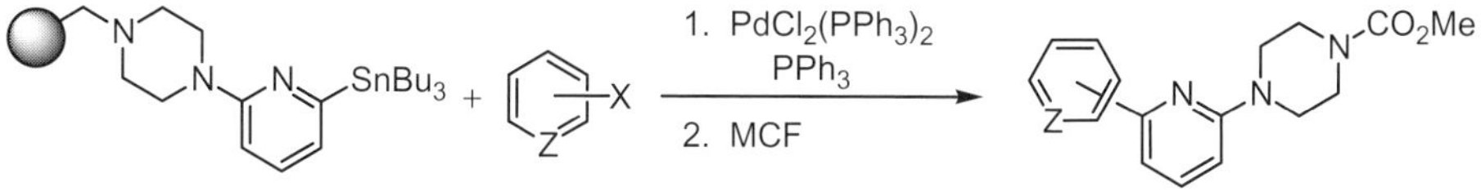

242 **182** **243**

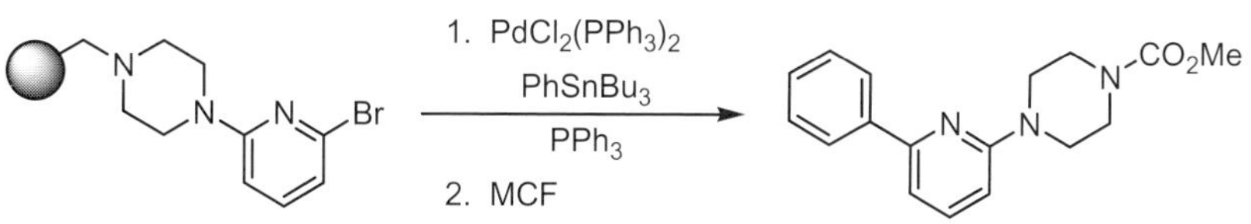

178 **244**

A number of functional groups are compatible with the Stille reaction conditions as illustrated in the preparation of heterobiaryl phosphonates **246** [89]. Stannane **236** could react with aryl phosphonate **245** to produce **246**. These compounds could then be used as Wittig or Wadsworth–Horner–Emmons reaction partners.

The Stille reaction has found utility in the construction of ferrocenyl-derivatives **249** [90]. Formation of these compounds was the result of Pd-catalyzed cross-coupling of **247** with **248**.

Pyridyl stannanes have seen use in the transformation of **250** into **251**. These dinucleating phenanthroline-based ligands have found utility in the synthesis of bimetallic complexes [91].

Bifunctional reagents with respect to the stannane **252** and halide **253** provided rapid entry to polymeric heteroaromatic systems **254** with the potential for use as conductive organic polymers [92].

The first synthesis of septipyridine **257** from the cross-coupling of **255** and **256** provided insight into how to assemble highly elaborate ring systems. This extended the utility of the Stille reaction and pushed the limits of ring system complexity [93].

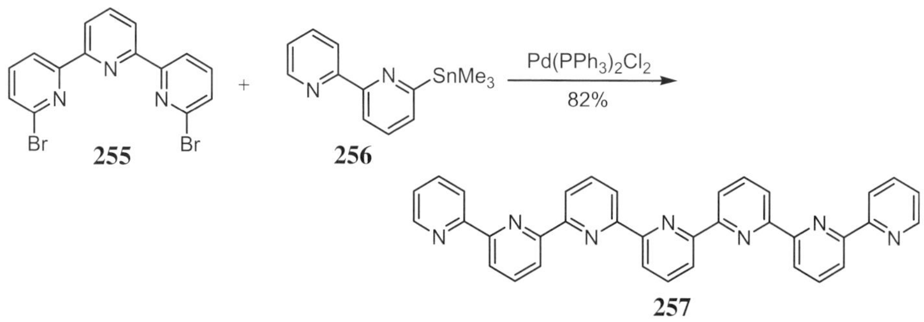

255 + **256** $\xrightarrow{\text{Pd(PPh}_3)_2\text{Cl}_2}$ 82%

257

In addition to the linear array of pyridyl rings, it was also possible to construct cyclic arrays as in cyclosexipyridine **262** [94]. This was accomplished by sequential Stille couplings initiated by reacting **258** with **259** to afford **260**. The remaining bromide of **260** was then coupled with a vinyl stannane derivative that produced **261**. Unmasking of the latent carbonyl functionality preceded elaboration to **262**.

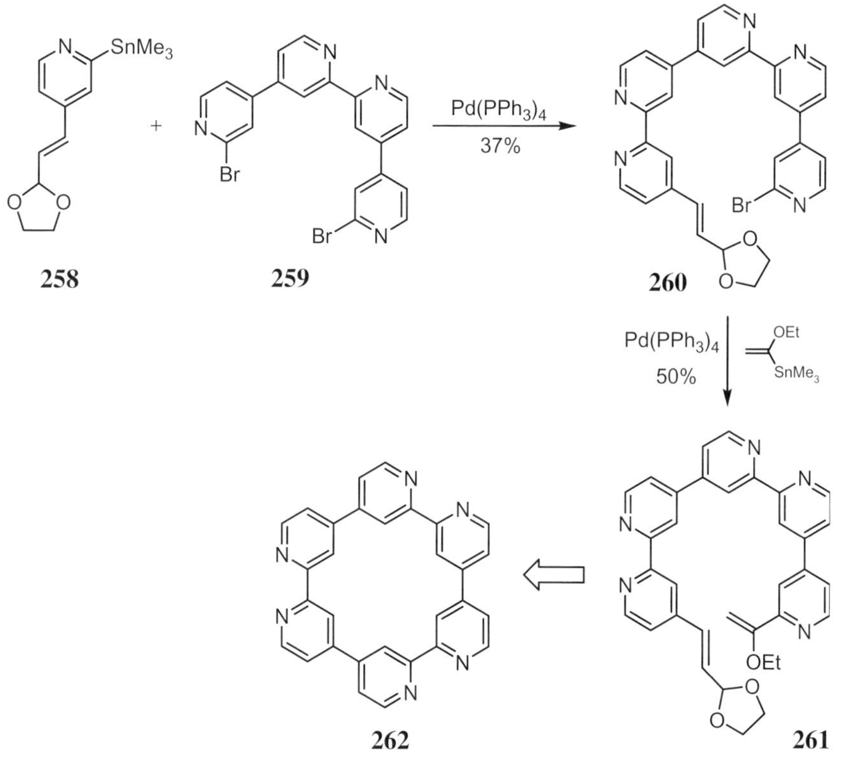

258   **259** $\xrightarrow{\text{Pd(PPh}_3)_4}$ 37%   **260**

Pd(PPh$_3$)$_4$ 50%

261

262

Access to non-symmetrically substituted, highly functionalized poly(bipyridines) would have great utility in the construction of supramolecular frameworks involved in photochemical, electrochemical, or catalytic systems. The Stille reaction provided a departure from the classical methods for the assembly of these skeletons. To this end, the symmetrical bromopyridine **263** participated in cross-coupling with stannane **264** producing bipyridine **265**. The remaining functionality was then exploited to further elaborate these systems [95].

A wide variety of multitopic nitrogen donor oligopyridylimines was made available through Stille cross-coupling reactions [96]. Lithiation of **266** followed by trapping with a tin reagent afforded stannane **267**. Cross-coupling with a variety of arylbromides followed by hydrolysis of the ketal produced **268**. Further reaction with 2,6-diisopropylaniline provided access to the multidentate nitrogen donor ligands for use in the self-assembly of polymetallic complexes.

Multiple reports have described the construction of a wide variety of oligopyridines that are multifunctional with a diversity of utilities [97]. These include a "LEGO" system that has led to the construction of branched frameworks that contain up to 14 pyridine rings. The transformation of **269** and **270** into **271** provides a representative example.

The versatility of the Stille reaction was illustrated in the preparation of all four benzo[4,5]furopyridines [98]. Inter- and intramolecular variations of the reaction were exploited in this work. The alkoxide of **273** underwent a displacement reaction of chloride of **272** to generate **274**. Ring closure to afford **275** was accomplished using the Kelly-variation of the Stille reaction.

For the intermolecular approach, lithiation of **26** and trapping, under standard conditions, produced stannane **276**. This intermediate could be cross-coupled to aryl iodide **273** followed by ring closure to give **277**.

An example of the Stille reaction was used in preparing CD-ring model system **280** for the alkaloid streptonigrin **281** [99]. This was accomplished by the cross-coupling of **278** with **279**.

The nicotinic acetylcholine receptor antagonist (±)-epibatidine **285** is a potent, non-opioid analgesic that can be prepared by a cross-coupling reaction [100]. The advanced intermediate **284** was assembled by the Pd-catalyzed reaction of **282** with **283**.

Derivatives of cephalosporins can serve as selective class C β-lactamase inhibitors and therefore routes to their preparation are of value [101]. Pyridyl analog **287** could be prepared in one of two ways, which differed only in the nature of the reacting partners. β-Lactam stannane **286** could be coupled with pyridyl bromide **19** to afford **287**. Alternatively, pyridyl stannane **236** was reacted with bromide **288** to generate **287**.

Phosphotyrosine plays a central role in the signal transduction processes in multiple cellular systems. Therefore, non-hydrolyzable phosphotyrosine mimetics like **291** have potential use in the pharmacological treatment of disease, e.g. cancer [102]. Pyridyl analog **290** was prepared by the reaction of **289** with **236**.

The nicotine partial agonist cytisine **201** was prepared by employing a Suzuki reaction (*vide supra*). The analog (±)-9-methoxycytisine **295** was generated using a Stille reaction [71]. Iodide **292** was coupled with stannane **293** to generate **294**, an advanced intermediate in the synthesis of **295**.

Imidazo[1,2*a*]pyridines are a "privileged" scaffold in medicinal chemistry, in that it has been found to possess biological activity in a number of disease systems. Stannane **296** was reacted with bromide **136** to afford **297**, an example of such an analog [103].

Endothelin converting enzyme (ECE) inhibitors have been shown to be useful in the treatment of hypertension. One such compound is (+)-(*S*)-WS75624B **301** [104]. Stannane **298** reacting with iodide **299** afforded **300** which was further elaborated into **301**.

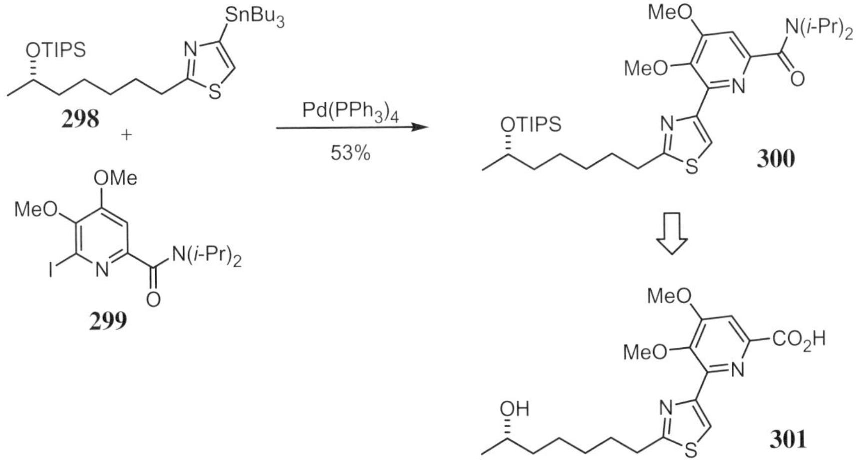

298 + **299** → **300**

Pd(PPh$_3$)$_4$

53%

301

Vascular endothelial growth factor (VEGF) receptor inhibitors possess anti-tumor activity by cutting off the blood supply to growing tumors through anti-angiogenesis. Pyrazine **303** was coupled to stannane **302** to afford **304** [105]. This compound served as an intermediate towards VEGF inhibitors of structure **305**.

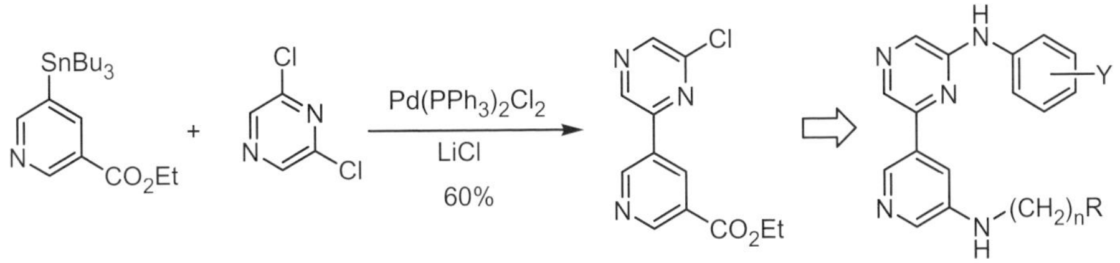

302 **303** **304** **305**

Pd(PPh$_3$)$_2$Cl$_2$

LiCl

60%

Metabatropic glutamate receptor subtype 5 (mGluR5) antagonists are being evaluated for a number of CNS-related indications. Positron emission tomography (PET) ligands provide the ability to visualize the location of the target receptors in the brain. A mGluR5 PET ligand **307** was prepared from precursor **306** by Stille coupling with isotopically labeled methyl iodide in quantitative radiochemical yield [106].

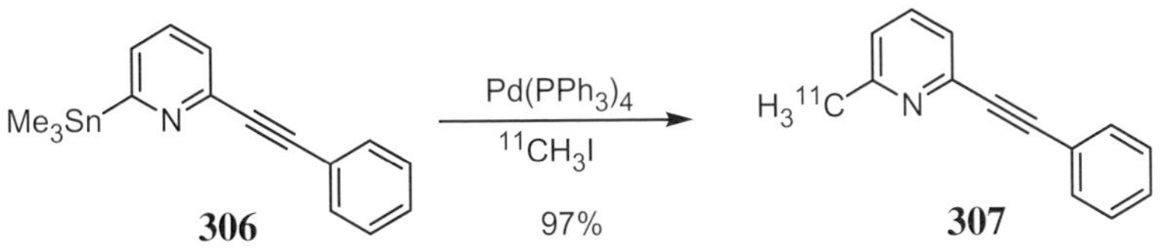

306 **307**

Pd(PPh$_3$)$_4$

^{11}CH$_3$I

97%

4.2.5. Hiyama coupling

In this reaction, the Pd-catalyzed cross-coupling reaction occurs between aryl and alkenyl halides or triflates with organosilyl compounds [107]. The reaction is comparable to the Stille reaction but without the application of potentially toxic tin reagents.

Bromopyridyl derivative **308** was converted to the silyl species **309** by lithium–halogen exchange followed by treatment with ethyltrichlorosilane. This intermediate was not isolated but directly exposed to the cross-coupling reaction conditions to afford the arylated pyridine **310** in good to excellent yields [108].

Pentavalent bis(catechol)silicates **312** were shown to be effective reagents in the Hiyama reaction. It was tolerant to a wide variety of electron donating groups and electron withdrawing groups. Treatment of this reagent with pyridyl triflates **311** under palladium catalysis generated the corresponding cross-coupled products **169** in good yield [109].

Alternate silicates could be prepared and reacted as illustrated in the transformation of **313** into **314** that was cross-coupled to generate **315**.

Incorporation of a halogen into 2-silylpyridines allowed the Hiyama coupling to proceed [110]. It was noted that a significant copper effect existed and that fluoride anion was also necessary for the reaction of **316** with **317** to afford **318**.

4.3. Sonogashira reaction

In this variation of the Pd-catalyzed cross-coupling reaction, which is closely related to the Stephen–Castro reaction, copper acetylides are reacted with (hetero)aryl halides or triflates to produce (hetero)aryl alkynes [111]. The Sonogashira reaction is comparable to the Suzuki or Stille reactions in its scope and functional group tolerability.

Methodological variations of the Sonogashira reaction have been investigated. One such variation is the copper-free version [112]. An example is illustrated in the reaction of **152** with alkyne **319** to produce **320**. The reaction conditions proceeded from room temperature to 50°C and under aerobic conditions.

Alternately, the copper-free conditions can be accomplished using TBAF in conjunction with palladium acetate [113]. Halopyridines **5** could be cross-coupled with alkyne **321**, using these reaction conditions to give rise to **322**.

As was demonstrated with the Suzuki reaction [57], aqueous reaction conditions could be applied to the Sonogashira reaction and have also been investigated, but required changing the nature of the palladium ligand. Alkynylpyridine **324** was prepared by cross-coupling **26** with **323**.

As with other reactions, the Sonogashira can be improved by the use of microwave technology [114]. Treatment of **136** with **325** afforded **326** in 89% yield. This technique not only reduced the time for this reaction, but also improved the yield compared to the standard conditions of 5 h at 80°C.

136   **325**   **326**

Heterogeneous catalyst variations have also been examined and Perlmann's catalyst was found to be superior to the standard Pd/C (11% yield) [115]. This provided for safe handling, high activity, high recovery, and practical access to a variety of compounds, as exemplified by the formation of **328** by coupling **136** with **327**.

136   **327**   84%   **328**

Alternative ligands were examined as a way of modulating catalyst efficiency. Sonogashira reactions were investigated using Tedicyp **329**, a tetradentate phosphine ligand [116]. A wide variety of substituted alkynes **319** could be cross-coupled to halopyridines **51** to afford **330**.

329

51   **319**   15-97%   **330**

As with the Negishi and Suzuki reactions, the same solid-support (*vide supra*) could be adapted to the Sonogashira reaction [62]. The solid-supported bromopyridine **178** could undergo cross-coupling reactions with a variety of alkynes and be released from the solid-support to allow isolation of **331**.

178   2. MCF   **331**

Resin-to-resin transfer reactions (RRTR) involve the preparation of two or more synthetic fragments on separate/different solid-supports, then these fragments are coupled together. This allows for diversification of each fragment separately before the coupling reaction which adds an additional level of diversity. The Sonogashira reaction was investigated as a way to conduct the RRTR coupling process [117]. Improved efficiency was observed when both components were allyl-linked to the resin. Exposure to the palladium catalyst caused the release of the reacting partners from **332** and **333** prior to them undergoing the cross-coupling reaction and ultimately forming **334**.

The synthesis of derivatized, concave pyridines could be accomplished in a straightforward manner [118]. Thus, **335** could be transformed to **336** upon cross-coupling with phenylacetylene. Furthermore, the Sonogashira reaction was also able to generate dimerized variations **338** of these compounds by reacting **337** with **325**.

The Sonogashira reaction has found great utility in the synthesis of compounds for applications as varied as catalysis, recognition phenomenon, supramolecular chemistry, and bioorganic chemistry. The following examples provide a glimpse at the possibilities. One such example is the construction of analogs of Troger's Base **341** from the reaction of **339** with **340** [119].

The synthesis of conjugated molecules with defined length, shape, and topology imparts them with certain molecular, electronic, or photonic properties. One such framework involved the combination of thiophene with bipyridine units. The Sonogashira reaction allowed for the preparation of ditopic and tritopic ligands [120]. One building block for these compounds **343** was generated through several transformations beginning with **342**.

Highly elaborate architectures, such as **346** could be prepared from **344** and **345** using the Sonogashira reaction. These systems could be used as ligands for intramolecular electron-transfer quenching systems [121].

The preparation of nickel porphyrins, difficult by other methods including the Suzuki reaction, can readily be assembled using this methodology [122]. Thus, **347** could be coupled with alkynylpyridine **340** to produce porphyrin **348**. The resultant compounds have found utility in photodynamic therapy, catalysis, or exhibit unique electrical properties.

The Sonogashira reaction played an integral part in the synthesis of steroid hormone derivatives [123]. Ruthenium complex **352** served as a model system to study approaches for the preparation of ^{99m}Tc-estradiol, a radiopharmaceutical for the detection of estrogen receptor-positive breast tumors. Advanced intermediate **351** towards these compounds was generated by the cross-coupling reaction of **349** with **350**.

The Sonogashira reaction has also found utility in the total synthesis arena. This cross-coupling reaction provided access to the intermediate **354** from **353** that was used in the synthesis of (*S*)-(+)-fusarinolic acid **355** [124]. This represented a dramatic improvement in providing a concise and practical synthesis where other approaches had required as many as 10 steps.

A similar refinement of the synthetic route was developed in the approach to niphatesine C **357** (R=Me) and norniphatesine C **357** (R=H) [125], sponge natural product alkaloids with cytotoxic activity. These compounds were readily prepared from the advanced intermediate **356** that was generated by a cross-coupling reaction using **152**.

Tandem Sonogashira reactions beginning with **40** readily assembled **358**. This compound could ultimately be cyclized to **359** which was used as the core scaffold for a key intermediate in the investigation of $\alpha_v\beta_3$ integrin antagonists [126].

SSR182289A **363** is a selective and potent orally-active thrombin inhibitor [127]. The central non-natural amino acid residue **362** was constructed using a Sonogashira reaction. Hydrolysis of the methyl ester followed the cross-coupling reaction of **360** with **361** to produce **362**.

SAR studies of camptothecin **126**, an alkaloid with potent anti-tumor activity, have shown that substitution at the 7-position provided access to compounds with enhanced biological activity. For example, SN-38 **364** has been used in clinical practice, while BNP-1350 **365** is in clinical development. Sonogashira cross-coupling of the advanced intermediate **366** with **325** afforded **367**. This compound served as a common precursor to both **364** and **365** [128].

Non-competitive antagonists of metabotropic glutamate receptors subtype 5 (mGluR5) have been found to affect the brain's reward system. Behaviorally, mGluR5 antagonists inhibit nicotine self-administration and prevent the reinstatement of cocaine self-administration. Methylthiazole ethynyl pyridine (MTEP) **369** has high affinity for the mGluR5 receptor and has been the focus of some of these studies [129]. It could be prepared in straightforward fashion by the cross-coupling reaction of **152** with **368**.

4.4. Heck and intramolecular Heck reactions

Whether the reaction is inter- or intramolecular, the Heck reaction generates vinyl(hetero)arenes or dienes from an alkene and a (hetero)aryl or alkenyl halide [130]. This reaction has great versatility and is applicable to a wide range of aryl and alkene species. Mechanistically, the Heck reaction varies from that depicted in Fig. 4.3. While the oxidative addition of the halogen species occurs, the transmetalation step is replaced by the coordination of the alkene. This is followed by a migratory insertion which essentially substitutes for the cross-coupling step. The product is released not by a reductive elimination, but by a β-hydride elimination sequence (Fig. 4.5).

Imidazolium ionic liquids were found to favor an ionic pathway version of the Heck reaction, in which the bidendate ligand remains complexed throughout the catalytic cycle. These reaction conditions proved to be highly regioselective for the arylation, such that only α-regioselectivity was observed [131]. 3-Chloropyridine **26** was found to arylate

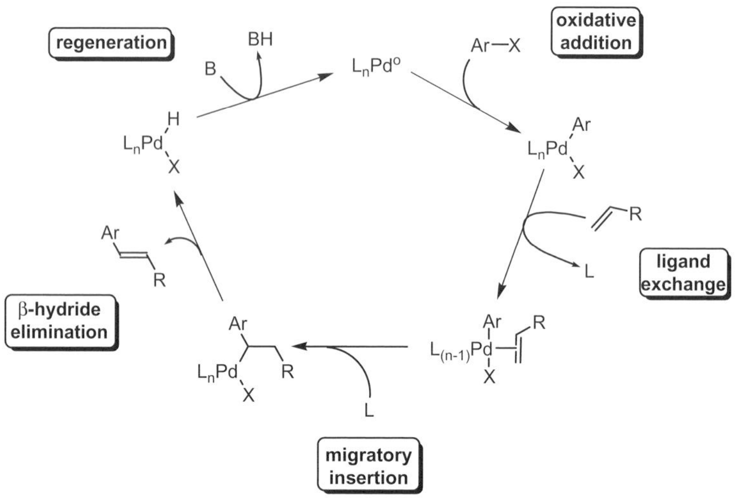

Figure 4.5.

vinyl butyl ether **370** or allyl alcohol **373** to afford the cross-coupled products **371** or **374**, respectively. In the case of **371**, the enol ether could be hydrolyzed to release the latent carbonyl functionality to generate **372**.

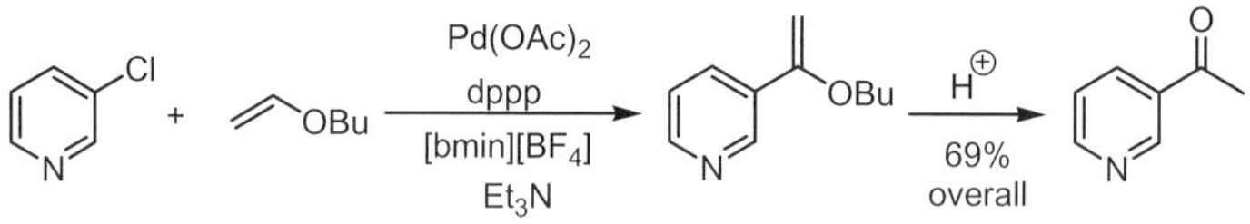

26 **370** **371** **372**

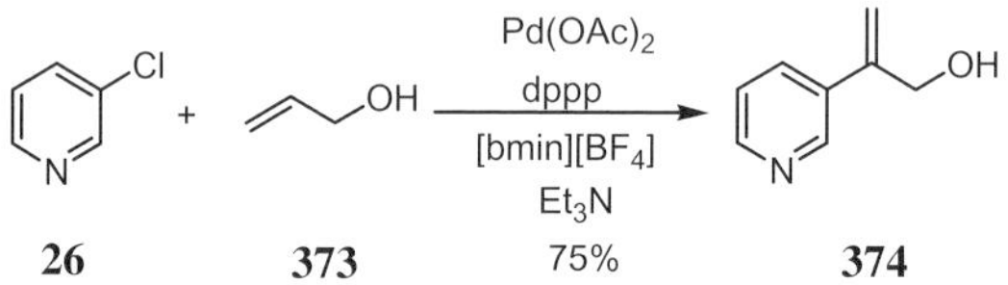

26 **373** **374**

Oxime-derived palladacycles act as efficient catalysts for the Heck reaction [132]. They could be isolated and are found to be thermally stable, not sensitive to air or moisture, and were readily prepared from inexpensive starting material. This was exemplified in the reaction catalyzed by **376** to prepare **377** from **3** and **375**.

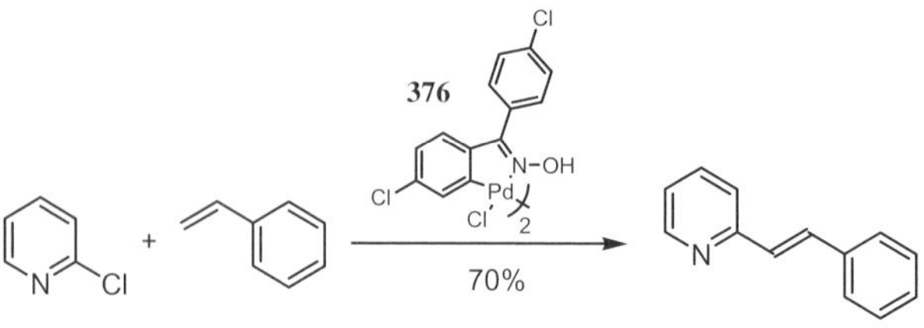

3 **375** **377**

The Heck reaction has been used in a polycondensation format for the generation of orange-red light-emitting hyperbranched and linear polymers [133]. Materials of this sort have utility in full-color flat-panel displays. The new electroluminescent polymer **380** could be obtained by linking bis(vinyl)benzene **379** with the dibromo species **378**.

Conjugated polymers have seen great attention due to their electrical properties. One example is poly(pyridylvinylenes) **384**. These compounds display large red-shifts in their optical absorption upon protonation or alkylation. Thus, there is interest in these materials for LED devices. A representative example of how the Heck reaction was used to assemble these compounds is presented [134]. Palladium-catalyzed vinylation of pyridine **40** afforded **381**. Repeating this chemistry gave rise to the bis(vinyl) pyridine **382**. If excess vinylstannane was employed, **382** could be generated directly from **40** in a single operation. A subsequent Heck reaction with diiodobenzene derivative **383** produced polymer **384**.

The Heck reaction has been used to assemble pyridyl derivatives that have been grafted onto nanoparticles for use in non-linear optical materials [135]. Vinylation of **386** with **385** produced **387**. Hydrolysis of the phosphonate ester then permitted these compounds to be grafted onto titanium and tin oxide supports.

385   **386**   **387**

2-Halopyridines are typically poor substrates for Heck reactions due to the possibility of pyridyl-bridged palladium dimers preventing additional reactions. Improved reaction conditions, of catalyst and the use of DMA as cosolvent, have allowed greater functional group tolerance [136]. This was reduced to practice in the transformation of **388** into **389**.

388   **389**

The 2-indolinone system has been found in a large number of pharmaceutically active compounds. This framework could be constructed by a tandem Heck–Suzuki reaction sequence [137]. Intramolecular Heck reaction of **390** afforded a cyclized intermediate that, *in situ*, reacted with boronic acid **187** to produce **391** in a single pot sequence. The reaction made use of copper thiophene-2-carboxylic acid (CuTC) to generate base-free conditions for the Suzuki reaction.

390   **187**   **391**

An intramolecular variation of the Heck reaction was used in the construction of the monoterpene alkaloid (±)-oxerime **395** [138]. Nucleophilic addition to **392** of an allyl anion equivalent produced **393**, a precursor to an intramolecular Heck cyclization. Palladium-catalyzed ring closure afforded **394**, an intermediate in the synthesis of the natural product **395**.

392   **393**   **394**   **395**

Access to peptide mimetics and enzyme inhibitors was accomplished by the construction of pyridylalanine derivatives [139]. Once incorporated into peptides, the resultant structure was capable of coordinating to cations. The desired amino acid **397** was prepared from **396** by a Heck reaction using an acrylic acid ester followed by hydrogenolysis. This species could then be incorporated into the desired mimetic system **398** that was capable of complexation with europium ions.

A Heck reaction-based approach to the anti-cancer agent 3-AP **196** complements the route utilizing a Suzuki reaction (*vide supra*) [140]. In this example, **190** is cross-coupled to styrene in good yield to produce **399**. The alkene was subsequently elaborated into the sidechain of 3-AP **196**.

L-754,394 **400** was discovered to be a potent and bioavailable inhibitor of HIV protease. The furo[2,3-*b*]pyridine moiety **404** was approached using a Heck reaction to insert a vinyl group that provided a portion of the furan ring [141]. Iodination of 2-hydroxynicotinate **401** generated iodopyridone **402**. This compound could undergo the Pd-catalyzed cross-coupling with vinyltributyltin to afford **403**. This route provided a practical route to this fragment that was amenable to scale-up conditions.

It was possible to combine several of these cross-coupling reactions, in sequence, to ultimately produce highly functionalized pyridine derivatives [142]. A straightforward example began with **136**. Lithiation followed by exchange to organozinc **405** set up the system for a Negishi reaction. The product of this cross-coupling **406** could either undergo a Suzuki reaction to generate **407** or a Heck reaction to produce **408**.

136   **405**   **406**   **407**

408

As with other cross-coupling reactions, it is possible to intercept the palladium intermediates when the reaction is conducted under carbon monoxide pressure [143]. For the example of iodopyridine **152**, pyridyl ketone **409** could be isolated during the reaction with (*Z*)-1-ethoxy-2-(tributylstannyl)ethane.

152   **409**

4.5. Buchwald–Hartwig aminations(C–N bond formation)

The transition-metal catalyzed cross-coupling reaction of (hetero)aryl halides and triflates with primary and secondary amines or (hetero)aryl amines is know as the Buchwald–Hartwig reaction [144]. Mechanistically, this reaction is related to the cross-coupling reactions outlined thus far (Fig. 4.6). The modification arises at the point of transmetalation. This step in the process is substituted with the coordination of the amine reactant. Deprotonation of the amine nitrogen now precedes the reductive elimination step to generate the aryl amine product. This reaction has found utility in the academic setting, for use in natural product total synthesis, and in industry, for the preparation of materials up to the multi-hundred kilogram scale.

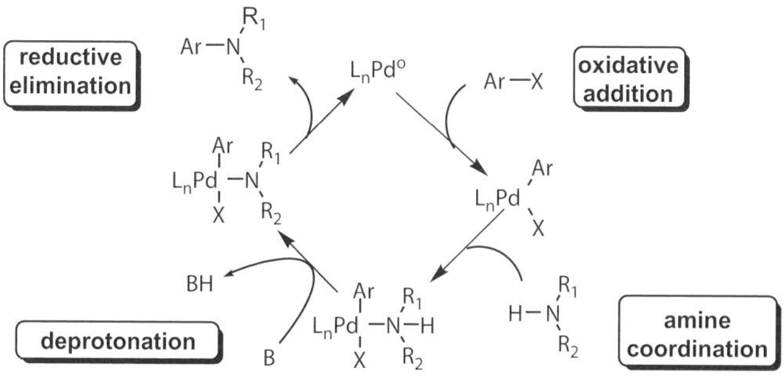

Figure 4.6.

For heteroaromatic systems, this reaction complements nucleophilic aromatic substitutions. The Pd-catalyzed reaction of **19** with **69** afforded **410** in excellent yield [145]. The use of bis-chelating ligands in this chemistry prevented ligand exchange with the pyridine substrate, thereby preventing formation of a bis(pyridyl)palladium species that would terminate the catalytic cycle. As a result of these specific catalytic conditions, this represented the first example of amination of a heteroaromatic halide.

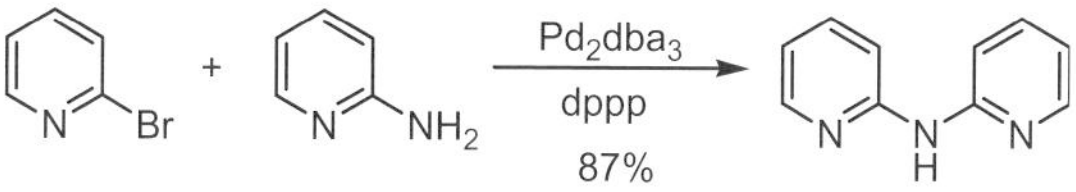

19 **69** **410**

Multiple aminations are possible when polyfunctionalized pyridyl halides **30** are coupled with **69** to generate **411**.

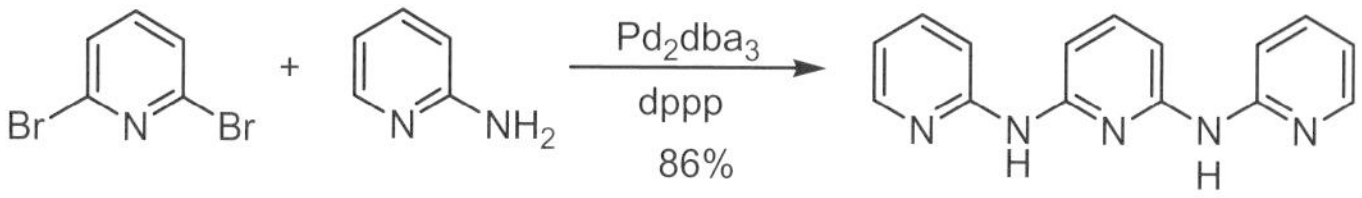

30 **69** **411**

N-heterocyclic carbene ligands, such as IPr **412**, were also shown to be able to facilitate the Buchwald–Hartwig reaction of **26** with **413** to produce **414** [146].

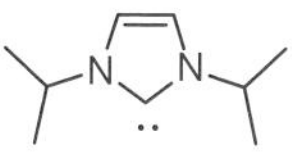

412

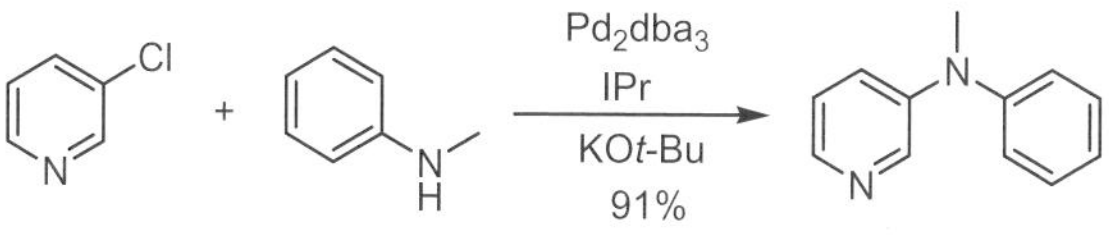

26 **413** **414**

Commercially available air- and water-stable palladium catalysts have been found to be highly efficient in facilitating the Buchwald–Hartwig reaction. The use of more expensive bases, such as cesium carbonate and sodium *tert*-butoxide, can now be substituted with the more common base, potassium hydroxide. In the case of pyridyl derivatives, naphthoquinone(imidazolin-2-ylidene)palladium complexes **415** were found to catalyze the reaction of **26** with **416** to generate **417** [147].

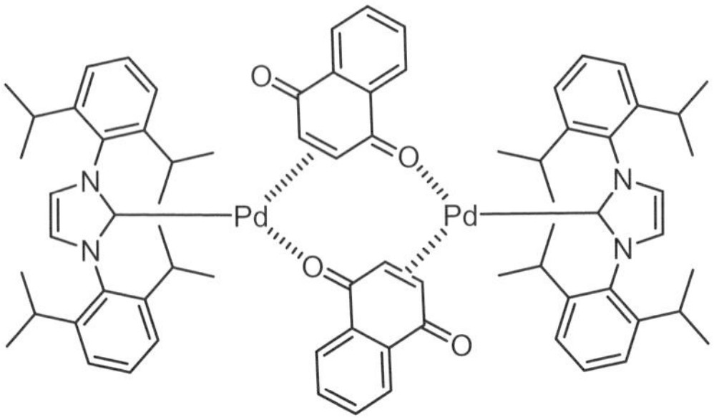

415

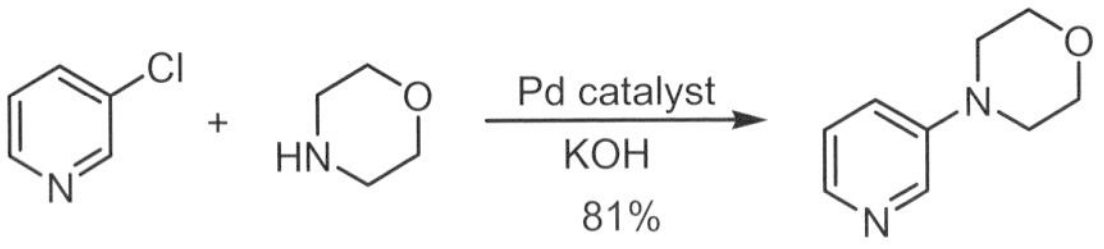

26   **416**   **417**

A different sort of base effect was evaluated in the formation of **419** by amination of **272** with **418**. Different batches of cesium carbonate were examined for particle size and shape, in addition to stoichiometry. Two types were observed based on the supplier and type 2 was found to be optimal for these transformations [148].

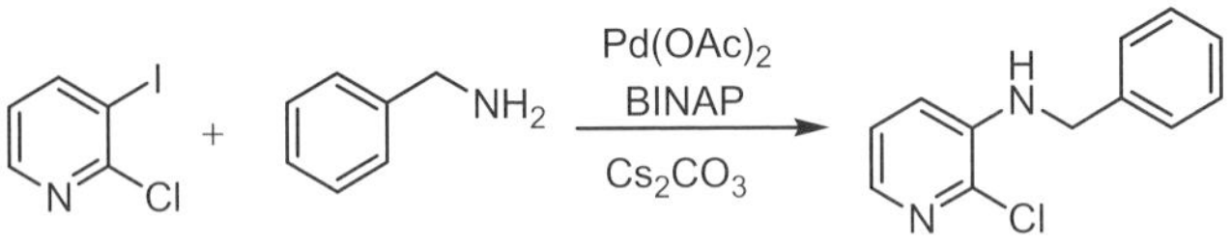

272   **418**   **419**

Proazaphosphatrane ligands, such as **420**, partnered with palladium species generated highly active catalysts for Buchwald–Hartwig reactions. An example of such transformations was illustrated when **26** was able to react with **413** to afford **414** [149].

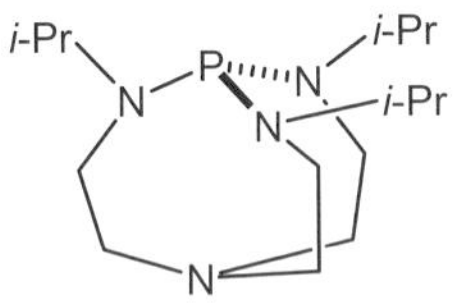

420

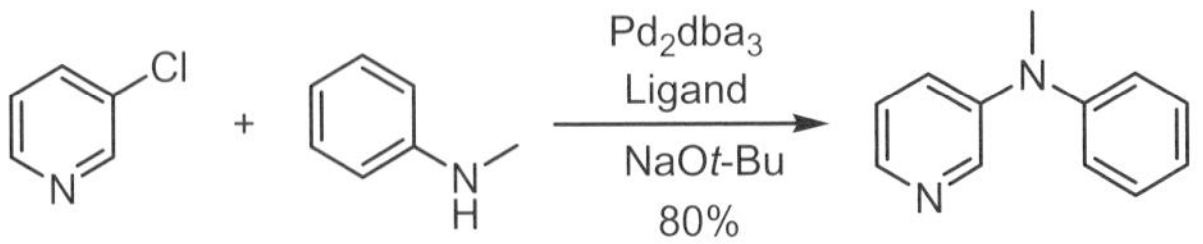

26   **413**   **414**

The application of modern synthetic technology to these Pd-catalyzed reactions has been used to answer the demand of increased efficiency. This was accomplished using temperature-controlled microwave conditions that resulted in decreased reaction times. The reaction of **3** with amine **413** generated **421** in excellent yield exemplified this application [150].

Benzodiazepines are an important class of molecular scaffolds due to their repeated observations in biologically active compounds. An intramolecular version of the Buchwald–Hartwig reaction was applied to the synthesis of pyridobenzodiazepinones. To this end, targets **423** and **425** could be prepared by intramolecular amination of **422** and **424**, respectively [151].

A drug discovery program targeting 7-azabicyclo[2.2.1]heptane systems found utility in bis(imidazol-2-ylidine)palladium complexes in conducting amination reactions [151]. The ligand DiImes-HCl **426** was determined to be optimal in the Buchwald–Hartwig reaction of **427** with **19** to produce **428**.

4.6. Direct C–C bond formation

The direct coupling of aromatic systems has been achieved in the generation of 2-arylpyridines. These frameworks have been of importance in material and medicinal chemistry. The Pd-catalyzed Ullmann reaction afforded 2-nitrophenylpyridine derivatives that served as precursors to 2-substituted-phenylpyridines, which are found to have anti-arrhythmic activity [152]. To this end, 3-iodopyridine **152** could be directly coupled to **429** by palladium-catalysis to afford **430**.

| 152 | 429 | 430 |

While the previous example shows that direct coupling was possible, direct coupling with π-deficient heteroaromatics, like pyridine, has generally proven to be challenging. One way to circumvent the issues related to the pyridine ring nitrogen, was to "protect" it as the *N*-oxide [153]. This had the net effect of blocking side reactions as a result of metal-chelation by this atom. Regioselectivity for the 2-position was observed in the formation of **433** from *N*-oxide **431** partnered with **432**.

| 431 | 432 | 433 |

A general variation on Pd-catalyzed cross-coupling reactions, initially introduced in the discussions of Heck and Suzuki reactions (*vide supra*), involves interrupting the "normal" catalytic cycle and trapping the organopalladium intermediate with carbon monoxide. When the reaction is conducted under an atmosphere of carbon monoxide, this species is capable of causing a ligand exchange. Once coordinated, the carbon monoxide can proceed through carbonyl insertion and reductive elimination to afford carbonylated products (Fig. 4.7).

The simplest of cases is illustrated in the transformation of **434** into **435** [154]. In this case, the carbonyl-inserted intermediate was immediately captured by benzyl alcohol in the reaction medium, thus affording the corresponding benzyl ester of the starting material.

| 434 | 435 |

In principle, any nucleophilic species should be able to react with this carbonylated intermediate. The formation of **436** from **60** illustrated this point with an amino nucleophile giving rise to the amide product **436**.

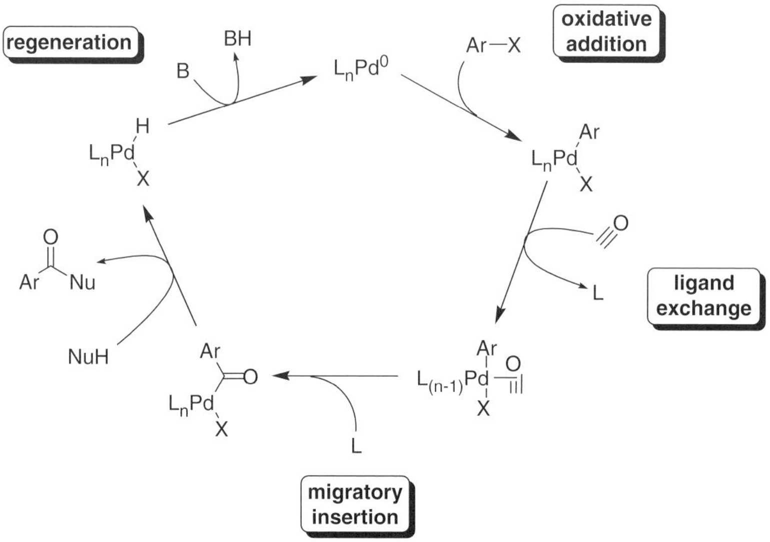

Figure 4.7.

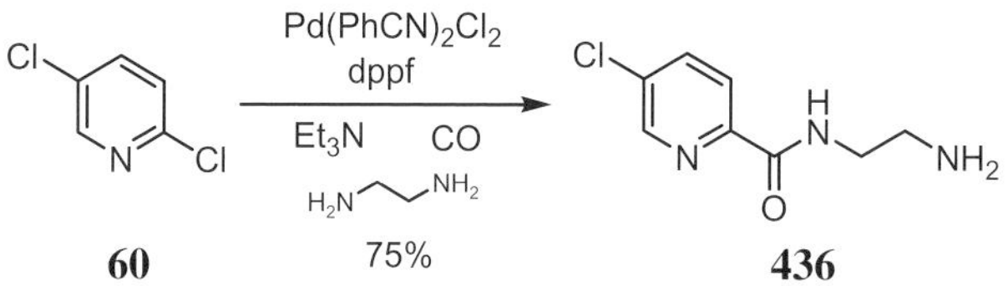

60   75%   **436**

Selection of the appropriate palladium species and corresponding ligand partner permitted regioselective mono- or bis-carbonylation reactions of a difunctional starting material [155]. This was exemplified with dichloropyridine **438**. Using one set of reaction conditions resulted in specific carbonylation at the 2-position to produce **439**. An alternate catalyst/ligand pair resulted in carbonylation at both centers to afford **437**.

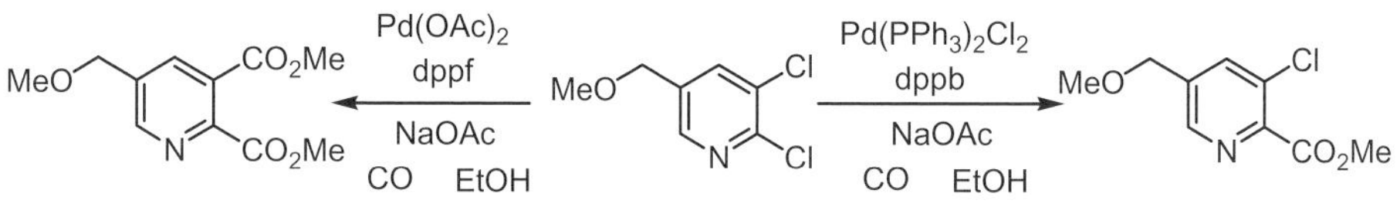

437   90%   **438**   94%   **439**

In a second example of regioselective carbonylation, it was determined that monodentate ligands gave little reaction, while bidentate ligands dramatically increased the rate of the reaction [79]. Thus, in the preparation of an intermediate fragment of the PDE IV inhibitor **225**, the carbonylation of **60** could be effected to afford the desired **440**. This compound could then be carried forward to produce **224**, used in preparing the analog **225**.

60   **440**   **224**   **225**

Multiple analogs of a C2-symmetrical HIV-1 protease inhibitor were prepared using a carbonylation protocol [156]. In this example molybdenum hexacarbonyl was used as a solid source of carbon monoxide for the reaction. Dibromopeptidomimetic **441** could be bis-carbonylated in a high speed format to afford **442**. This one example clearly illustrates the mild and very selective nature of Pd-catalyzed reactions.

441   **442**

4.7. Summary

Figure 4.8 graphically illustrates the result of a simple keyword search of SciFinder using the name reactions and pyridine. The publication frequency shows the dramatic increase in use of these cross-coupling reactions over the past 10 years. In particular, the Sonogashira and the Suzuki reactions have experienced a geometric increase in use.

It becomes apparent from this survey that the only limitation on the use of all these cross-coupling reactions is the imagination of the chemist. The field has matured to the point that most of the common reagents are now commercially available, including the palladium species and ligands to the organometallic compounds including organoboronic acids. For those that cannot be obtained in this manner, there are sufficient protocols to generate the desired compounds in a straightforward and direct manner. While the initial method development for cross-coupling reactions began on substituted aryl systems, all the techniques and technology can be applied to heteroaromatic systems. The examples of leveraging combinatorial chemistry and microwave technology illustrate how versatile these reactions are in synthesis. The diversity of functional group compatibility, combined with this technology, has made these reactions an important tool in chemistry and should be a part of every synthetic chemist's skill set.

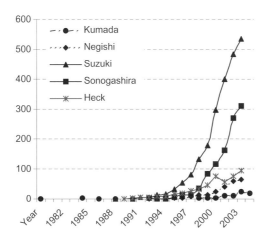

Figure 4.8.

4.8. References

1. Shirota, O.; Sekita, S.; Satake, M.; Morita, H.; Takeya, K.; Itokawa, H. *Heterocycles* **2004**, *63*, 1891–6.
2. (a) Malleron, J-L.; Fiaud, J-C.; Legros, J-Y. *Handbook of Palladium-Catalyzed Organic Reactions* Academic Press, London, **1997**. (b) Tolman, C. A. *Chem. Rev.* **1977**, *77*, 313–48. (c) Wolfe, J. P.; Tomori, H.; Sadighi, J. P.; Yin, J.; Buchwald, S. L. *J. Org. Chem.* **2000**, *65*, 1158–74. (d) Li, J. J. *Prog. Heterocyclic Chem.* **2000**, *12*, 37–56. (e) Hillier, A. C.; Grasa, G. A.; Viciu, M. S.; Lee, H. M.; Yang, C.; Nolan, S. P. *J. Organometal. Chem.* **2002**, *653*, 69–82. (f) Littke, A. F.; Fu, G. C. *Angew. Chem. Int. Ed.* **2002**, *41*, 4176–11. (g) Negishi, E-I. *Handbook of Organopalladium Chemistry for Organic Synthesis*, John Wiley & Sons, NY **2002**. (h) Cardenas, D. J. *Angew. Chem. Int. Ed.* **2003**, *42*, 384–7. (i) Brase, S.; Kirchoff, J. H.; Kibberling, J. *Tetrahedron* **2003**, *59*, 885–939. (j) Chelucci, G.; Orru, G.; Pinna, G. A. *Tetrahedron* **2003**, *59*, 9471–9515. (k) Valentine, D. H., Jr.; Hillhouse, J. H. *Synthesis* **2003**, 2437–60. (l) Schlummer, B.; Scholz, U. *Adv. Synth. Catal.* **2004**, *346*, 1599–626. (m) Nakamura, I.; Yamamoto, Y. *Chem. Rev.* **2004**, *104*, 2127–98. (n) Kappe, C. O. *Angew. Chem. Int. Ed.* **2004**, *43*, 6250–84. (o) Nakamura, I.; Yamamoto, Y. *Chem. Rev.* **2004**, *104*, 2127–98. (p) Schroter, S.; Stock, C.; Bach, T. *Tetrahedron* **2005**, *61*, 2245–67. (q) Handy, C. J.; Manoso, A. S.; McElroy, W. T.; Seganish, W. M.; DeShong, P. *Tetrahedron* **2005**, *61*, 12201–25. (r) Shaughnessy, K. H.; DeVasher, R. B. *Curr. Org. Chem.* **2005**, *9*, 585–604. (s) Christmann, U.; Vilar, R. *Angew. Chem. Int. Ed.* **2005**, *44*, 366–74. (t) Frieman, B. A.; Taft, B. R.; Lee, C-T.; Butler, T.; Lipshutz, B. H. *Synthesis* **2005**, 2989–93. (u) Nicolaou, K. C.; Bulger, P. G.; Sarlah, D. *Angew. Chem. Int. Ed.* **2005**, *44*, 4442–89.
3. (a) Beak, P.; Meyer, A. I. *Acc. Chem. Res.* **1986**, *19*, 356–63. (b) Itami, K.; Kamei, T.; Mitsudo, K.; Nokami, T.; Yoshida, J-I. *J. Org. Chem.* **2001**, *66*, 3970–76. (c) Mongin, F.; Queguiner, G *Tetrahedron* **2001**, *57*, 4059–90. (d) Whisler, M. C.; MacNeil, S.; Snieckus, V.; Beak, P. *Angew. Chem. Int. Ed.* **2004**, *43*, 2206–25.

4. (a) Gribble, G. W.; Saulnier, M. G. *Tetrahedron Lett.* **1980**, *21*, 4137–40. (b) Comins, D. L.; Myoung, Y. C. *J. Org. Chem.* **1990**, *55*, 292–8.

5. Taylor, S. L.; Lee, D. Y.; Martin, J. C. *J. Org. Chem.* **1983**, *48*, 4158–59.

6. (a) Kessar, S. V.; Singh, P.; Singh, K. N.; Dutt, M. *J. Chem. Soc., Chem. Commun.* **1991**, 570–1. (b) Kessar, S. V.; Singh, P. *Chem. Rev.* **1997**, *97*, 721–37.

7. Verbeek, J.; Brandsma, L. *J. Org. Chem.* **1984**, *49*, 3857–9.

8. Gros, P.; Fort, Y.; Caubere, P. *J. Chem. Soc., Perkin Trans. I* **1997**, 3597–3600.

9. Kondo, Y.; Shilai, M.; Uchiyama, M.; Sakamoto, T. *J. Am. Chem. Soc.* **1999**, *121*, 3539–40.

10. Gros, P.; Fort, Y.; Caubere, P. *J. Chem. Soc., Perkin Trans. I* **1998**, 1685–9.

11. Choppin, S.; Gros, P.; Fort, Y. *Eur. J. Org. Chem.* **2001**, 603–6.

12. Bouillon, A.; Lancelot, J-C.; Santos, J. S.D. O.; Collot, V.; Bovy, P. R.; Rault, S. *Tetrahedron* **2003**, *59*, 10043–49.

13. (a) Trecourt, F.; Breton, G.; Bonnet, V.; Mongin, F.; Marsais, F.; Queguiner, G. *Tetrahedron* **2000**, *56*, 1349–60. (b) Ila, H.; Baron, O.; Wagner, A. J.; Knochel, P. *Chem. Commun.* **2006**, 583–93.

14. Kii, S.; Akao, A.; Iida, T.; Mase, T.; Yasuda, N. *Tetrahedron Lett.* **2006**, *47*, 1877–79.

15. Rocca, P.; Marsais, F.; Godard, A.; Queguiner, G. *Tetrahedron* **1993**, *49*, 49–64.

16. Bobbio, C.; Schlosser, M. *Eur. J. Org. Chem.* **2001**, 4533–6.

17. (a) Yamamoto, Y.; Yanagi, A. *Heterocycles* **1981**, 16, 1161–4. (b) Yamamoto, Y.; Yanagi, A. *Chem. Pharm. Bull.* **1982**, 30, 1731–7.

18. Schlosser, M.; Cottet, F. *Eur. J. Org. Chem.* **2002**, 4181–4.

19. (a) Corcoran, R. C.; Bang, S. H. *Tetrahedron Lett.* **1990**, *31*, 6757–8. (b) Hebel, D.; Rozen, S. *J. Org. Chem.* **1991**, *56*, 6298–301. (c) Hebel, D.; Rozen, S. *J. Org. Chem.* **1988**, *53*, 1123–5.

20. Koch, V.; Schnatterer, S. *Synthesis* **1990**, 497–8.

21. Gotchev, D.; Commins, D. L. *Tetrahedron* **2004**, *60*, 11751–8.

22. Eweiss, N. F.; Katritzky, A. R.; Nie, P-L.; Ramsden, C. R. *Synthesis* **1997**, 634–5.

23. Boyer, J. H.; McCane, D. I.; McCarville, W. J.; Tweedie, A. T. *J. Am. Chem. Soc.* **1953**, *75*, 5298–300.

24. (a) Ilavsky, D.; Jehlicka, V.; Kuthan, J. *Coll. Czech. Chem. Commun.* **1979**, *44*, 3288–95. (b) Ager, D. J.; Erickson, R. A.; Froen, D. E.; Prakash, I.; Zhi, B. *Org. Process Res. Develop.* **2004**, *8*, 62–71.

25. Kato, Y.; Okada, S.; Tomimoto, K.; Mase, T. *Tetrahedron Lett.* **2001**, *42*, 4849–951.

26. Sugimoto, O.; Mori, M.; Tanji, K-I. *Tetrahedron Lett.* **1999**, *40*, 7477–8.

27. (a) Tamao, K.; Sumitami, K.; Kumada, M. *J. Am. Chem. Soc.* **1972**, *94*, 4374–6. (b) Tamao, K. *J. Organometal. Chem.* **2002**, *653*, 23–6.

28. Bonnet, V.; Mongin, F.; Trecourt, F.; Queguiner, G.; Knochel P. *Tetrahedron Lett.* **2001**, *42*, 5717–19.

29. Minato, A.; Suzuki, K.; Tamao, K.; Kumada, M. *J. Chem. Soc., Chem. Commun.* **1984**, 511–13.

30. Babudri, F.; Golangiuli, D.; Farinola, G. M.; Naso, F. *Eur. J. Org. Chem.* **2002**, 2785–91.

31. Knochel, P.; Perea, J. J. A.; Jones, P. *Tetrahedron* **1998**, *54*, 8275–319.

32. Anctil, E. J-G.; Snieckus, V. *J. Organomet. Chem.* **2002**, *653*, 150–60.

33. Stanetty, P.; Schnurch, M.; Mihovilovic, M. D. *Synlett* **2003**, 1862–4.

34. Stanetty, P.; Rohrling, J.; Schnurch, M.; Mihovilovic, M. D. *Tetrahedron* **2006**, **62**, 2380–7.

35. Smith, A. P.; Savage, S. A.; Love, J. C.; Fraser, C. L. *Org. Synth.* **2002**, *78*, 51–62.

36. Fang, Y-Q.; Hanan, G. S.; *Synlett* **2003**, 852–4.

37. Arvanitis, A. G.; Arnold, C. R.; Fitzgerald, L. W.; Frietze, W. E.; Olson, R. E.; Gilligan, P. J.; Roberston, D. W. *Bioorg. Med. Chem. Lett.* **2003**, *13*, 289–91.

38. Felding, J.; Kristensen, J.; Bjerrragaard, T.; Sander, L.; Vedso, P.; Begtrup, M. *J. Org. Chem.* **1999**, *64*, 4196–8.

39. Deshayes, K.; Broene, R. D.; Chao, I.; Knobler, C. B.; Diederich, F. *J. Org. Chem.* **1991**, *56*, 6787.

40. Billotte, S. *Synlett* **1998**, 379–80.

41. Savage, S. A.; Smith, A. P.; Fraser, C. L. *J. Org. Chem.* **1998**, *63*, 10048–51.

42. Trecourt, F.; Gervais, B.; Mallet, M.; Queguiner, G. *J. Org. Chem.* **1996**, *61*, 1673–6.

43. Murata, N.; Sugihara, T.; Kondo, Y.; Sakamoto, T. *Synlett* **1997**, 298–300.

44. Gauthier, D. R. Jr.; Szumigala, R. H., Jr.; Dormer, P. G.; Armstrong, J. D. III; Volante, R. P.; Reider, P. J. *Org. Lett.* **2002**, *4*, 375–8.

45. Jackson, R. F.; Wishart, N.; Wood, A.; James, K.; Wythes, M. J. *J. Org. Chem.* **1992**, *57*, 3397–404.

46. Lutzen, A.; Hapke, M. *Eur. J. Org. Chem.* **2002**, 2292–7.

47. (a) Suzuki, A. *J. Organomet. Chem.* **2002**, *653*, 83–90. (b) Kotha, S.; Lahiri, K.; Kashinath, D. *Tetrahedron* **2002**, *58*, 9633–95. (c) Bellina, F.; Carpita, A. Rossi, R. *Synthesis* **2004**, 2419–40. (d) Sasaki, M.; Fuwa, H. *Synlett* **2004**, 1851–74. (e) Tyrell, E.; Brookes, P. *Synthesis* **2004** 469–83. (f) Miura, M. *Angew. Chem. Int. Ed.* **2004**, *43*, 2201–3. (g) Leadbeater, N. E. *Chem. Commun.* **2005**, 2881–902. (h) Baudoin, O. *Eur. J. Org. Chem.* **2005**, 4223–4229. (i) Franzen, R.; Xu, Y. *Can. J. Chem.* **2005**, *83*, 266–72. (j) Suzuki, A. *Chem. Commun.* **2005**, 4759–63.

48. Aliprantis, A. O.; Canary, J. W. *J. Am. Chem. Soc.* **1994**, *116*, 6985–6.

49. Zhang, H.; Chan, K. S. *Tetrahedron Lett.* **1996**, *37*, 1043–4.

50. Lohse, O.; Thevenin, P.; Waldvogel, E. *Synlett* **1999**, 45–8.

51. (a) Tagata, T.; Nishida, M. *J. Org. Chem.* **2003**, *68*, 9412–15. (b) Torii, A. ; Tagata, T.; Nishida, M. *Sci. Tech. Catal.* **2002**, 541–2.

52. Beeby, A.; Bettington, S.; Fairlamb, I. J. S.; Goeta, A. E.; Kapdi, A. R.; Niemela, E. H.; Thompson, A. L. *New J. Chem.* **2004**, *28*, 600–5.

53. Wang, Y.; Sauer, D. R. *Org. Lett.* **2004**, *6*, 2793–6.

54. Okamoto, K.; Akiyama, R.; Kobayashi, S. *Org. Lett.* **2004**, *6*, 1987–90.

55. Korolev, D. N.; Bumagin, N. A. *Tetrahedron Lett.* **2005**, *46*, 5751–4.

56. (a) Liebeskind, L. S.; Srogl, J. *Org. Lett.* **2002**, 4, 979–81. (b) Kusturin, C. L.; Liebeskind, L. S.; Neumann, W. L. *Org. Lett.* **2002**, *4*, 983–5.

57. Buchwald, S. L.; Anderson, K. W. *Angew. Chem. Int. Ed.* **2005**, *44*, 6173–7.

58. Molander, G. A.; Biolatto, B. *J. Org. Chem.* **2003**, *68*, 4302–14.

59. Hodgson, P. B.; Salingue, F. H. *Tetrahedron Lett.* **2004**, *45*, 685–7.

60. Cioffi, C. L.; Spencer, W. T.; Richards, J. J.; Herr, R. J. *J. Org. Chem.* **2004**, *69*, 2210–12.

61. Guiles, J. W.; Johnson, S. G.; Murray, W. V. *J. Org. Chem.* **1996**, *61*, 5169–71.

62. Louerat, F.; Gros, P.; Fort, Y. *Tetrahedron Lett.* **2003**, *44*, 5613–16.

63. Le Notre, J.; Firet, J. J.; Sliedregt, L. A. J. M.; van Steen, B. J.; van Koten, G.; Bebbink, R. J. M. K. *Org. Lett.* 2005, 7, 363–6.

64. Gros, P.; Doudouh, A.; Fort, Y. *Tetrahedron Lett.* **2004**, *45*, 6239–41.

65. Meier, P.; Legraverant, S.; Muller, S.; Schaub, J. *Synthesis* **2003**, 551–4.

66. Thompson, A. E.; Hughes, G.; Batsanov, A. S.; Bryce, M. R.; Parry, P. R.; Tarbit, B. *J. Org. Chem.* **2005**, *70*, 388–90.

67. Itoh, T.; Mase, T. *Tetrahedron Lett.* **2005**, *46*, 3573–7.

68. Zokiewicz, J. A.; Cruskie, M. P., Jr. *Tetrahedron* **1995**, *51*, 11393–400.

69. Niu, C.; Li, J.; Doyle, T. W.; Chen, S-H. *Tetrahedron* **1998**, *54*, 6311–18.

70. Spivey, A. C.; Fekner, T.; Spey, S. E.; Adams, H. *J. Org. Chem.* **1999**, *64*, 9430–43.

71. O'Neill, B. T.; Yohannes, D.; Bundesmann, M. W.; Arnold, E. P. *Org. Lett.* **2000**, *2*, 4201–4.

72. Sabat, M.; Johnson, C. R. *Org. Lett.* **2000**, *2*, 1089–92.

73. Caron, S.; Massett, S. S.; Bogle, D. E.; Castaldi, M. J.; Braish, T. F. *Org. Process Res. Dev.* **2001**, *5*, 254–6.

74. Xiong, W-N.; Yang, C-G.; Jiang, B. *Bioorg. Med. Chem.* **2001**, *9*, 1773–80.

75. Morris, G. A.; Nguyen, S. T. *Tetrahedron Lett.* **2001**, *42*, 2093–6.

76. Arvanitis, A. G.; Arnold, C. R.; Fitzgerald, L. W.; Frietze, W. E.; Olson, R. E.; Gilligan, P. J.; Robertson, D. W. *Bioorg. Med. Chem. Lett.* **2003**, *13*, 289–291.

77. Kamei, T.; Itami, K.; Yoshida, J-I. *Adv. Synth. Catal.* **2004**, *346*, 1824–35.

78. Muraoka, N.; Mineno, M.; Itami, K.; Yoshida, J-I. *J. Org. Chem.* **2005**, *70*, 6933–6.

79. Albaneze-Walker, J.; Murry, J. A.; Soheili, A.; Ceglia, S.; Springfield, S. A.; Bazaral, C.; Dormer, P. G.; Hughes, D. L. *Tetrahedron* **2005**, *61*, 6330–6.

80. Parry, P. R.; Bryce, M. R.; Tarbit, B. *Synthesis* **2003**, 1035–8.

81. Denton, T. T.; Zhang, X.; Cashman, J. R. *J. Med. Chem.* **2005**, *48*, 224–39.

82. Couve-Bonnaire, S.; Carpentier, J-F.; Mortreux, A.; Castanet, Y. *Tetrahedron* **2003**, *59*, 2793–9.

83. Maerten, E.; Hassouna, F.; Couve-Bonnaire, S.; Mortreux, A.; Carpentier, J-F. *Synlett*, **2003**, 1874–6.

84. (a) Pattenden, G.; Sinclair, D. J. *J. Organometal. Chem.* **2002**, *653*, 261–8. (b) Kosugi, M.; Fugami, K. *J. Organometal. Chem.* **2002**, *653*, 50–3. (c) Espinet, P.; Echavarren, A. M. *Angew. Chem. Int. Ed.* **2004**, *43*, 4704–4734. (d) Echavarren, A. M. *Angew. Chem. Int. Ed.* **2005**, *44*, 3962–5.

85. Gajare, A. S.; Jensen, R. S.; Toyota, K.; Yoshifuji, M.; Ozawa, F. *Synlett* **2005**, 144–8.

86. Mee, S. P.H.; Lee, V.; Baldwin, J. E. *Angew. Chem. Int. Ed.* **2004**, *43*, 1132–6.

87. Kim, W-S.; Kim, H-j.; Cho, C-G. *J. Am. Chem. Soc.* **2003**, *125*, 14288–9.

88. Gronowitz, S.; Bjork, P.; Malm, J.; Hornfeldt, A-B. *J. Organomet. Chem.* **1993**, *460*, 127–9.

89. Kennedy, G.; Perboni, A. D. *Tetrahedron Lett.* **1996**, *37*, 7611–14.

90. Liu, C-M.; Zu, Q-H.; Liang, Y-M.; Ma, Y-X. *J. Chem. Res. (S)* **1999**, 636.

91. Lam, F.; Feng, M-Q.; Chan, K. S. *Tetrahedron* **1999**, *55*, 8377–84.

92. Chapman, .M.; Stanforth, S. P.; Berridge, R.; Pozo-Gonzalo, C.; Skabara, P. J. *J. Mater. Chem.* **2002**, *12*, 2292–8.

93. Cardenas, D. J.; Sauvage, J-P. *Synlett* **1996**, 916–18.

94. Kelly, T. R.; Lee, Y-J.; Mears, R. J. *J. Org. Chem.* **1997**, *62*, 2774–81.

95. Puglisi, A.; Banaglia, M.; Roncan, G. *Eur. J. Org. Chem.* **2003**, 1552–8.

96. Champouret, Y. D.M.; Chaggar, R. K.; Dadhiwala, I.; Fawcett, J.; Solan, G. A. *Tetrahedron* **2006**, *62*, 79–89.

97. (a) Pabst, G. R.; Sauer, J. *Tetrahedron* **1999**, *55*, 5067–88. (b) Schubert, U. S.; Eschbaumer, C. *Org. Lett.* **1999**, *1*, 1027–9. (c) Baxter, P. N.; Khoury, R. G.; Lehn, J-M.; Baum, G.; Fenske, D. *Chem. Eur. J.* **2000**, *6*, 4140–8. (d) Fallahpour, R-A. *Synthesis* **2000**,

1138–42. (e) Colasson, B. X.; Dietrich-Buchecker, C.; Sauvage, J-P. *Synlett* **2002**, 271–2. (f) Heller, M.; Schubert, U. S. *Synlett* **2002**, 751–4. (g) Bedel, S.; Ulrich, G.; Picard, C.; Tisnes, P. *Synthesis* **2002**, 1564–70. (h) Heller, M.; Schubert, U. S. *J. Org. Chem.* **2002**, *67*, 8268–72. (i) Puglisi, A.; Benaglia, M.; Roncan, G. *Eur. J. Org. Chem.* **2003**, 1552–8. (j) Zoppellaro, G.; Ivanova, A.; Enkelman, V.; Geies, A.; Baumgarten, M. *Polyhedron* **2003**, *22*, 2099–110. (k) Duan, X-F.; Li, X-H.; Li, F-Y.; Huang, C-H. *Synthetic Commun.* **2004**, *34*, 3227–33. (l) Zong, R.; Thummel, R. P. *J. Am. Chem. Soc.* **2004**, *126*, 10800–1. (m) Martineau, D.; Gros, P.; Fort, Y. *J. Org. Chem.* **2004**, *69*, 7914–18. (n) Darabantu, M.; Boully, L.; Turck, A.; Ple, N. *Tetrahedron* **2005**, *61*, 2897–905. (o) Wu, C-G.; Lin, Y-C.; Wu, C-E.; Huang, P-H.; *Polymer* **2005**, *46*, 3748–57.

 98. Yue, W. S.; Li, J. J. *Org. Lett.* **2002**, *4*, 2201–3.
 99. Crous, R.; Dwyer, C.; Holzapfel, C. W. *Heterocycles* **1999**, *51*, 721–6.
100. Sirisoma, N. S.; Johnson, C. R. *Tetrahedron Lett.* **1998**, *39*, 2059–62.
101. Buynak, J. D.; Doppalapudi, V. R.; Frotan, M.; Kumar, R. *Tetrahedron Lett.* **1999**, *40*, 1281–4.
102. Cockerill, G. S.; Easterfield, H. J.; Percy, J. M.; Pintat, S. *J. Chem. Soc. Perkin Trans. 1* **2000**, 2591–9.
103. Hervet, M.; Thery, I.; Gueiffier, A.; Enguehard-Gueiffier, C. *Helv. Chim. Acta* **2003**, *86*, 3461–9.
104. Stangeland, E. L.; Sammakia, T. *J. Org. Chem.* **2004**, *69*, 2381–5.
105. Kuo, G-H.; Prouty, C.; Wang, A.; Emanuel, S.; DeAngelis, A.; Zhang, Y.; Song, F.; Beall, L.; Connoly, P. J.; Karnachi, P.; Chen, X.; Gruninger, R. H.; Sechler, J.; Fuentes-Pesquera, A.; Middleton, S. A.; Jolliffe, L.; Murray, W. V. *J. Med. Chem.* **2005**, *48*, 4892–909.
106. Yu, M.; Tueckmantel, W.; Wang, X.; Zhu, A.; Kozikowski, A. P.; Brownell, A-L. *Nuclear Med. Biol.* **2005**, *32*, 631–640.
107. (a) Yoshida, J-I. *J. Organomet. Chem.* **2002**, *653*, 105–13. (b) *J. Organomet. Chem.* **2002**, *653*, 69.
108. Hiyama, T. *J. Organomet. Chem.* **2002**, *653*, 58–61.
109. Seganish, W. M.; DeShong, P. *J. Org. Chem.* **2004**, *69*, 1137–43.
110. Pierrat, P.; Gros, P.; Fort, Y. *Org. Lett.* **2005**, *7*, 697–700.
111. (a) Sonogashira, K.; Tohda, Y.; Hagihara, N. *Tetrahedron Lett.* **1975,** *16,* 4467–70. (b) Castro, C. E.; Stephens, R. D. *J. Org. Chem.* **1963,** *28*, 2163. (c) Stephens, R. D.; Castro, C. E. *J. Org. Chem.* **1963**, *28*, 3313–15. (d) Sonogashira, K. *J. Organomet. Chem.* **2002**, *653*, 46–9. (e) Tykwinski, R. R. *Angew. Chem. Int. Ed.* **2003,** *42,* 1566–8. (f) Nagy, A.; Novak, Z.; Kotschy, A. *J. Organometal. Chem.* **2005**, *690*, 4453–61.
112. Liang, B.; Dai, M.; Chen, J.; Yang, Z. *J. Org. Chem.* **2005**, *70*, 391–3.
113. Sorensen, U. S.; Pombo-Villar, E. *Tetrahedron* **2005**, *61*, 2697–703.
114. Erdelyi, M; Gogoll, A. *J. Org. Chem.* **2001**, *66*, 4165–9.
115. Mori, Y.; Seki, M. *J. Org. Chem.* **2003**, *68*, 1571–4.
116. Feuerstein, M.; Doucet, H.; Santelli, M. *Tetrahedron Lett.* **2005**, *46*, 1717–20.
117. Tulla-Puche, J.; Barany, G. *Tetrahedron* **2005**, *61*, 2195–201.
118. (a) Storm, O.; Luning, U. *Eur. J. Org. Chem.* **2002**, 3680–3685. (b) Gil, J. M.; David, S.; Reiff, A. L.; Hegedus, L. S. *Inorg. Chem.* **2005**, *44*, 5858–65.
119. Jensen, J.; Strozyk, M.; Warnmark, K. *Synthesis* **2002**, 2761–5.

120. Goeb, S.; DeNicola, A.; Ziessel, R. *Synthesis* **2005**, 1169–77.

121. Yang, J.; Oh, W. S.; Elder, I. A.; Leventis, N.; Sotiriou-Leventis, C. *Synthetic Commun.* **2003**, *33*, 3317–25.

122. Lin, C-Y.; Chuang, L-C.; Lee, G-H.; Peng, S-M. *J. Organomet. Chem.* **2005**, *690*, 244–8.

123. Arterburn, J. B.; Corona, C.; Rao, K. V.; Carlson, K. E.; Katzenellenbogen, J. A. *J. Org. Chem.* **2003**, *68*, 7063–70.

124. Song, J. J.; Yee, N. K. *J. Org. Chem.* **2001**, *66*, 605–8.

125. (a) Krauss, J.; Bracher, F. *Arch. Pharm. Pharm. Med. Chem.* **2004**, *337*, 371–5. (b) Krauss, J.; Wetzel, I.; Bracher, F. *Nat. Prod. Res.* **2004**, *18*, 397–401.

126. Hartner, F. W.; Hsiao, Y.; Eng, K. K.; Rivera, N. R.; Palucki, M.; Tam, L.; Yasuda, N.; Hughes, D. L.; Weissman, S.; Zewge, D.; King, T.; Tschaen, D.; Volante, R. P. *J. Org. Chem.* **2004**, *69*, 8723–30.

127. Altenburger, J-M.; Lassalle, G. Y.; Matrougui, M.; Galtier, D.; Jetha, J-C.; Bocskei, Z.; Berry, C. N.; Lunven, C.; Lorrain, J.; Herault, J-P.; Schaeffer, P.; O'Conner, S. E.; Herbert, J-M. *Bioorg. Med. Chem.* **2004**, *12*, 1713–30.

128. Luo, Y.; Gao, H.; Li, Y.; Huang, W.; Lu, W.; Zhang, Z. *Tetrahedron* **2006**, *62*, 2465–79.

129. Iso, Y.; Grajkowska, E.; Wroblewski, J. T.; Davis, J.; Goeders, N. E.; Johnson, K. M.; Sanker, S.; Roth, B. L.; Tueckmantel, W.; Kozikowski, A. P. *J. Med. Chem.* **2006**, *49*, 1080–100.

130. (a) Heck, R. F. *J. Am. Chem. Soc.* **1968**, *90*, 5518–26; 5526–31; 5531–4; 5535–8; 5538–42; 5542–6; 5546–8. (b) de Meijere, A.; Meyer, F. E. *Angew. Chem. Int. Ed.* **1994**, *33*, 2379–2411. (c) Dounay, A. B.; Overman, L. E. *Chem. Rev.* **2003**, *103*, 2945–63. (d) Lin, B-L.; Liu, L.; Fu, Y.; Luo, S-W.; Chen, Q.; Guo, Q-X. *Organometallics* **2004**, *23*, 2114–23. (e) Shibasaki, M.; Vogl, E. M.; Ohshima, T. *Adv. Synth. Catal.* **2004**, *346*, 1533–52. (f) Guiry, P. J.; Kiely, D. *Curr. Org. Chem.* **2004**, *8*, 781–94. (g) Alonso, F.; Beletskaya, I. P.; Yus, M. *Tetrahedron* **2005**, *61*, 11711–835. (h) Oestereich, M. *Eur. J. Org. Chem.* **2005**, 783–92.

131. Pei, W.; Mo, J.; Xiao, J. *J. Organometal. Chem.* **2005**, *690*, 3546–51.

132. Alonso, D. A.; Najera, C.; Pacheco, M. C. *Adv. Synth. Catal.* **2002**, *344*, 172–183.

133. Wang, H.; Li, Z.; Jiang, Z.; Liang, Y.; Wang, H.; Quin, J.; Yu, G.; Liu, Y. *J. Polymer Sci.* **2005**, *43*, 493–504.

134. Fu, D-K.; Xu, B.; Swager, T. M. *Tetrahedron* **1997**, *53*, 15487–94.

135. Frantz, R.; Granier, M.; Durand, J-O.; Lanneau, G. F. *Tetrahedron Lett.* **2002**, *43*, 9115–7.

136. Nobert, N.; Hoarau, C.; Celanire, S.; Ribereau, P.; Godard, A.; Queguiner, G.; Marsais, F. *Tetrahedron* **2005**, *61*, 4569–76.

137. Cheung, W. S.; Patch, R. J.; Player, M. R. *J. Org. Chem.* **2005**, *70*, 3741–4.

138. Zhao, J.; Yang, X.; Jia, X.; Luo, S.; Zhai, H. *Tetrahedron* **2003**, *59*, 9379–82.

139. Schmidt, B.; Ehlert, D. K. *Tetrahedron Lett.* **1998**, *39*, 3999–4002.

140. Li, J.; Zheng, L-M.; King, I. Doyle, T. W.; Chen, S-H. *Curr. Med. Chem.* **2001**, *8*, 121–33.

141. (a) Houpis, I.; Choi, W. B.; Reider, P. J.; Molina, A.; Churchill, H.; Lynch, J.; Volante, R. P. *Tetrahedron Lett.* **1994**, *35*, 9355–8. (b) Choi, W-B.; Houpis, I.; Churchill, H. R.O.; Molina, A.; Lynch, J. E.; Volante, R. P.; Reider, P. J.; King, A. O. *Tetrahedron Lett.* **1995**, *36*, 4571–74.

142. (a) Karig, G.; Spencer, J. A.; Gallagher, T. *Org. Lett.* **2001**, *3*, 835–838. (b) Karig, G.; Thasana, N.; Gallagher, T. *Synlett* **2002**, 808–10.

143. Sakamoto, T.; Yasuhara, A.; Kondo, Y.; Yamanaka, H. *Chem. Pharm. Bull.* **1992**, *40*, 1137–9.

144. (a) Guram, A. S.; Rennels, R. A.; Buchwald, S. L. *Angew. Chem. Int. Ed.* **1995**, *34*, 1348–50. (b) Louie, J.; Hartwig, J. F. *Tetrahedron Lett.* **1995**, *36*, 3609–12. (c) Louie, J.; Driver, M. S.; Hamann, B. C.; Hartwig, J. F. *J. Org. Chem.* **1997**, *62*, 1268–73. (d) Hartwig, J. F. *Angew. Chem. Int. Ed.* **1998**, *37*, 2046–67. (e) Hartwig, J. F.; Kawatsura, M.; Hauck, S. I.; Shaughnessy, K. H.; Alcazar-Roman, L. M. *J. Org. Chem.* **1999**, *64*, 5575–80. (f) Ali, M. H.; Buchwald, S. L. *J. Org. Chem.* **2001**, *66*, 2560–5. (g) Schlummer, B.; Scholz, U. *Adv. Synth. Catal.* **2004**, *346*, 1599–626. (h) Loones, K. T. J.; Maes, B. U. W.; Rombouts, G.; Hostyn, S.; Diels, G. *Tetrahedron* **2005**, *61*, 10338–48.

145. Wagaw, S.; Buchwald, S. L. *J. Org. Chem.* **1996**, *61*, 7240–1.

146. Hillier, A. C.; Grasa, G. A.; Vici, H. S.; Lee, H. M.; Yang, C.; Nolan, S. P. *J. Organomet. Chem.* **2002**, *653*, 69–82.

147. Gooβen, L. J.; Paetzold, J.; Briel, O.; Rivass-Nass, A.; Karch, R.; Kayser, B. *Synlett* **2005**, 275–8.

148. Meyers, C.; Maes, B. U.; Loones, K. T. J.; Bal, G.; Lemiere, G. L.F.; Dommisse, R. A. *J. Org. Chem.* **2004**, *69*, 6010–17.

149. (a) Urgaonkar, S.; Verkade, J. G. *J. Org. Chem.* **2004**, *69*, 9135–42. (b) Urgaonkar, S.; Xu, J-H.; Verkade, J. G. *J. Org. Chem.* **2003**, *68*, 8416–23.

150. Maes, B. U. W.; Loones, K. T. J.; Lemiere, G. L.F.; Dommisse, R. A. *Synlett* **2003**, 1822–5.

151. Cheng, J.; Trudell, M. L. *Org. Lett.* **2001**, *3*, 1371–4.

152. Shimizu, N.; Kitamura, T.; Watanabe, K.; Yamaguchi, T.; Shigyo, H.; Ohta, T. *Tetrahedron Lett.* **1993**, *34*, 3421–4.

153. Campeau, L-C.; Rousseaux, S.; Fagnou, K. *J. Am. Chem. Soc.* **2005**, *127*, 18020–1.

154. Beller, M.; Magerlein, W.; Indolese, A. F.; Fischer, C. *Synthesis* **2001**, 1098–109.

155. Bessard, Y.; Roduit, J. P. *Tetrahedron* **1999**, *55*, 393–404.

156. Wannberg, J.; Kaiser, N-F. K.; Vrang, L.; Samuelson, B.; Larhed, M.; Hallberg, A. *J. Comb. Chem.* **2005**, *7*, 611–17.

Chapter 5

Thiophenes and benzo[*b*]thiophenes

Chris Limberakis

Thiophene-containing molecules exist as natural products [1] and synthetic chemotherapeutics [2]. For example, α-terthienyl is isolated from the roots of *Tagetes* sp. (marigold) and has shown potent insecticide activity when treated with ultraviolet (UV) light [3]. On the other hand, tiaprofenic acid (Surgam®), a non-steroidal anti-inflammatory drug (NSAID), is a synthetic thiophene derivative [4]. In addition, clopidogrel bisulfate (Plavix®), an inhibitor of adenosine diphosphate (ADP) induced platelet aggregation, has been an important breakthrough for patients with cardiovascular atherosclerotic disease [5]. More recently, duloxetine hydrochloride (Cymbalta®) [6], a selective serotonin and norepinephrine reuptake inhibitor (SSNRI), has been effective for the treatment of depression [6a], the management of neuropathic pain in diabetics [6b], and for the treatment of stress urinary incontinence [6c]. As for benzo[*b*]thiophenes, they are not very abundant in nature; however, there are some examples as marketed drugs. Arguably one of the most well-known synthetic benzo[*b*]thiophenes is raloxifene hydrochloride (Evista®), a selective estrogen receptor modulator (SERM), that is prescribed for the prevention and treatment of osteoporosis in post-menopausal women [7]. In addition, sertaconazole (Ertaczo®) was recently approved as a topical anti-fungal agent [8].

α-terthienyl   Surgam®   Plavix®   Cymbalta®

Evista®   Ertaczo®

251

1a   **1b**

Thiophene is a useful template for four-carbon homologation *via* reduction [9], as well as a bioisostere of the benzene ring and other heterocycles in medicinal chemistry. Thiophene is a π-electron-excessive heterocycle. It favors electrophilic substitution, which, similar to metalation, takes place preferably at the α-positions due to the electronegativity of the sulfur atom [10]. In comparison to the oxygen atom in furan, the sulfur atom in thiophene has lower electronegativity, so its lone pair electrons are more effectively incorporated into the aromatic system. The aromaticity of thiophene is in between that of benzene and furan. As a consequence, the difference in reactivity of α-halothiophenes and β-halothiophenes is not as pronounced as that of the corresponding halofurans.

Benzothiophene, like thiophene, is a π-electron rich heterocycle. Unlike thiophene, electrophilic attack occurs selectively at the β-position (or 3-position) [10a, 11a]; however, substitution at the α-center (2-position) can also be achieved [11b]. The order of positional reactivity, based on nitration, is $3 > 2 > 6 > 5 > 4 > 7$ [11a, 12a]. Also, halogenation under acidic conditions affords the 3-halobenzothiophene [11c, 13a]. In terms of metalation chemistry, deprotonation occurs preferentially at the 2-position (α) [11d, 13b]; metal–halogen exchange is also favored at the α-center over the β-center [12b].

Although Pd(II), like many transition metals, possesses strong thiophilicity which can lead to poisoning of the catalyst [14], a variety of thiophenes and benzo[*b*]thiophenes have been successful cross-coupling partners [10b, 15].

5.1. Preparation of halothiophenes and halobenzothiophenes

5.1.1. Direct halogenation

Direct halogenation of thiophenes is, undoubtedly, the most commonly used method for synthesizing halothiophenes. Treatment of 3,4-bis[(methoxycarbonyl)methyl]-thiophene (**2**) with N-bromosuccinide NBS resulted in the monobromination product **3**, whereas treatment of **2** with bromine gave rise to the corresponding bisbromination product **4** [16]. Using a direct halogenation with NBS, Gronowitz *et al.* converted 2-methoxythiophene to 2-methoxy-3,5-dibromothiophene (**5**), which was subsequently transformed to 2-methoxy-3-bromothiophene (**6**) *via* regioselective halogen–metal exchange at C(5) followed by quenching with H_2O [17]. 2-Methyl-3-bromo-5-chlorothiophene (**8**), in turn, was obtained from bromination of 2-methyl-5-chlorothiophene (**7**), derived from regioselective chlorination of 2-methylthiophene at C(5) with NCS [18].

2   **3**

Halogenation can be run under milder conditions using more active quaternary ammonium polyhalides such as pyridinium tribromide. The reaction between thiophene and benzyltrimethylammonium tribromide in CH_3COOH–$ZnCl_2$ provided 2,5-dibromothiophene [19]. Chloro- and iodo-substituted thiophene derivatives may be prepared in the same manner. In comparison, bromination of thiophene employing two equivalents of NBS in chloroform gave 2,5-dibromothiophene in 56% yield [20].

The aforementioned direct halogenation methods are also applicable to bisthiophenes [21] and terthiophenes [22]. Taking advantage of the α-activation, tetrabromobisthiophene **10** was converted to the corresponding derivative **11** *via* regioselective reduction [23]. Recognizing that mono-bromination of benzothiophene gave an almost equal amount of 2- and 3-bromothiophene, the preparation of 3-bromo-6-methylbenzothiophene (**13**) began with dibromination of 6-methylbenzothiophene to give dibromide **12**. Selective removal of the C(2)-bromide of **12** using *n*-BuLi, followed by treatment with H_2O, provided **13** [24, 25].

Regioselective bromination of thiophene **14** at the 4-position initially proved problematic [26]. For instance, bromination of **14** with NBS delivered a mixture of **15**, **16a**, and **16b**, while the use of bromine gave a 1.4:1 mixture of **16b** and **16a**. This regioselectivity issue was resolved using a slight excess of bromine in the presence of six equivalents of aluminum chloride and a catalytic amount of iron tribromide to deliver **15** in 79% yield along with the dibrominated thiophene **16a** as the minor product. The authors noted that smooth conversion to **16a** was accomplished using 10 equivalents of bromine in the presence of a catalytic amount of iron tribromide in refluxing carbon tetrachloride.

Iodination of thiophenes may be achieved with N-iodosuccinimide as the example illustrates. In this case, 3,3-dimethyl-2-bisthiophene (**17**) was treated with two equivalents of NIS in dichloromethane to deliver the diiodobisthiophene **18** in excellent yield [27].

3-Iodobenzothiophene **20** was prepared by iodination of 2-trifluoroacetylamino-benzo[*b*]thiophene (**19**), although the same reaction with the regioisomeric 3-trifluoroacetyl-aminobenzo[*b*]thiophene gave only unidentified polymeric material [28]. In another case, thienylpyridine **21** could be either brominated [29], or iodinated to give **22** [30, 31].

Another effective iodination method is the treatment of thienotrifluoroborates and benzothienotrifluoroborates with sodium iodide and an oxidant [32]. Thus, 3-thienotrifluoroborate **23** and 2-benzothienotrifluoroborate **25**, derived from their corresponding boronic acids, were converted to their iodo adducts **24** and **26** in the presence of sodium iodide and chloramine-T in aqueous THF in 83 and 72% yields, respectively. Other oxidants such as *m*-CPBA and hydrogen peroxide delivered the predicted products; however, the yields were significantly lower.

5.1.2. Quenching lithiothiophenes with halogens

2-Lithiothiophene, arising from lithiation of thiophene with *n*-BuLi, was treated with iodine to give 2-iodothiophene, which was allowed to react with sodium malononitrile in the presence of catalytic $PdCl_2(Ph_3P)_2$ to afford thienylmalononitrile **27**. Interestingly, α-metalation of ethyl 3-thiophenecarboxylate was achieved using TMP–zincate **28**, a bulky base. The resulting thienylzincate was treated with I_2 to give 2-iodothiophene **29**. The same sequence with ethyl 2-thiophenecarboxylate furnished 5-iodothiophene **30** [33].

30

Formation of 2-halothiophenes, *via* quenching of the corresponding lithium adduct, is also achieved with 1,2-dihaloethanes [34]. For instance, thienopyran **31** was converted to its 2-lithiothiophne species with *n*-BuLi, and subsequent quenching of this anion with 1,2-diiodoethane delivered iodothiophene **32** in 80% yield [34a]. These authors took the iodination one step further by performing an iodine rearrangement with LDA to deliver the 3-iodothiophene **33** in 50% yield.

31   **32**   **33**

5.2. Oxidative and reductive coupling reactions

A small amount of bisthiophene was isolated when thiophene was treated with Pd(OAc)$_2$ [35]. The oxidative couplings of a thiophene with thiophene, furan, or substituted arenes were achieved in poor to moderate yields using Pd(OAc)$_2$ in HOAc [36–38]. The oxidative couplings of thiophene or benzo[*b*]thiophene with olefins also suffer from inefficiency [39].

2-Bromothiophene-5-carboxaldehyde (**34**) was converted to bisthiophene **35** *via* a Pd-catalyzed reductive homocoupling [40]. The reaction was also applicable to chloro- and iodothiophenes bearing many functional groups. In addition, an Ullmann-type reductive homocoupling of 2-iodothiophene utilizing catalytic Pd/C and three equivalents of Zn provided a practical entry to bisthiophene in 64% yield [41].

34   **35**

5.3. Cross-coupling with organometallic reagents

5.3.1. Kumada coupling

Organomagnesium reagents, which serve as the nucleophiles in the Kumada coupling, are readily synthesized and many of them are commercially available. Even though some Kumada reactions can be run at ambient temperature or lower, many functional groups are not tolerant of Grignard reagents. Nonetheless, in the synthesis of thienylbenzoic acid **36**, the carboxylic acid moiety survived the reaction conditions [42].

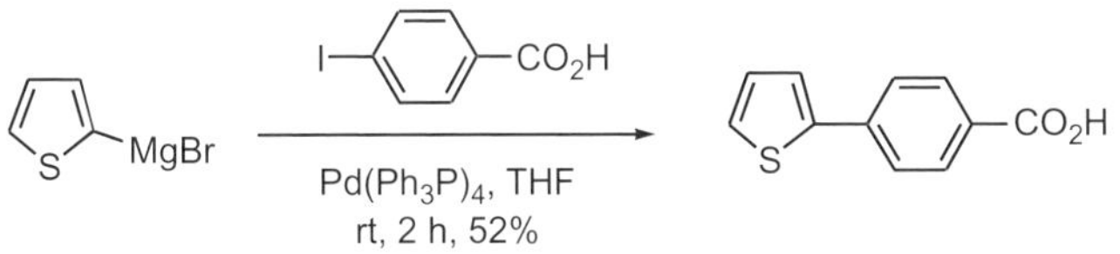

36

In the synthesis of 2-bromobisthiophene (**37**), mono-arylation was achieved using the Kumada coupling of 2-thienylmagnesium bromide with 2,5-dibromothiophene in the presence of Pd(dppf)Cl$_2$, although nickel(0)-catalyzed coupling failed [43]. Unlike the conversion of 2,5-dibromothiophene to **37**, a related thiophene, **38**, was converted to thiophene **39** in high yield in the presence of a nickel catalyst [44].

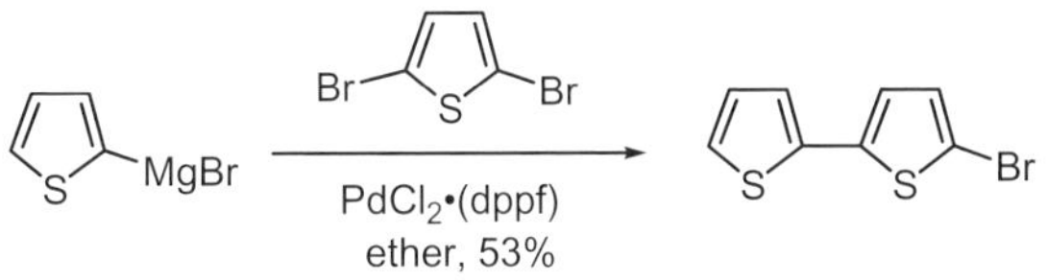

37

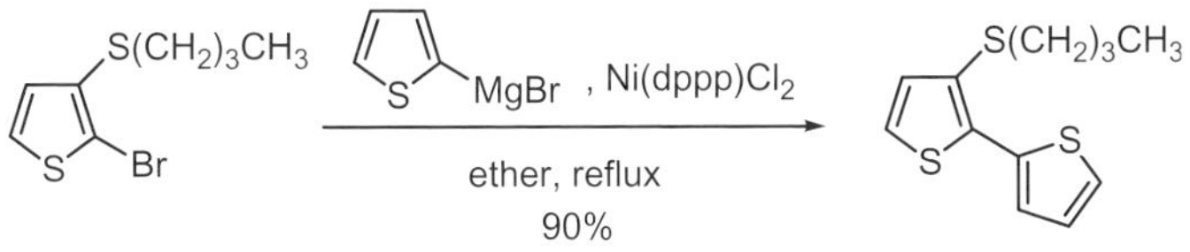

38 **39**

Mono-substitution was also realized when one equivalent of 2-thienylmagnesium bromide was allowed to react with 2,5-dichloropyridine, furnishing thienylpyridine **40** [45]. Furthermore, the Kumada coupling of 5-(2,2′-bithienyl)magnesium bromide (**41**) with 4-bromopyridine proceeded in refluxing ether to give 4-[5-(2-bithien-2′-yl)]pyridine (**42**) [46].

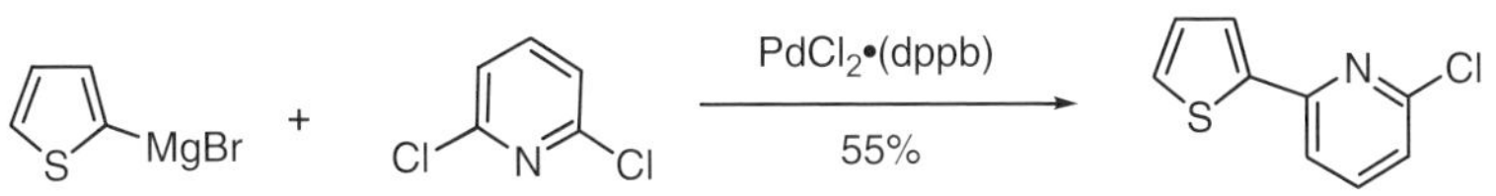

40

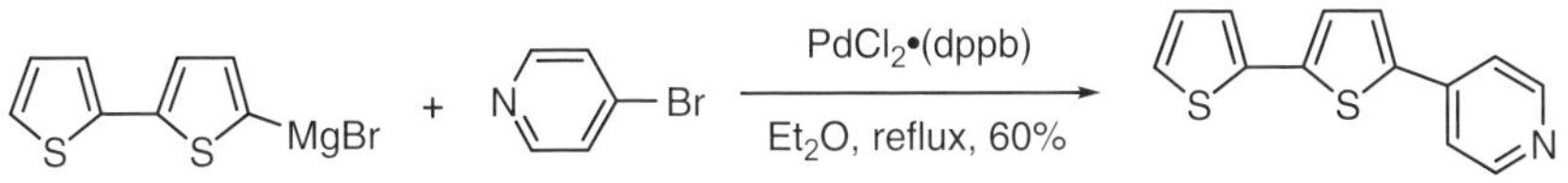

41 **42**

In the direct synthesis of aryl terminal alkynes *via* Pd-catalyzed cross-coupling of aryl halides with ethynylmetals, formation of diarylethynes is one potential side reaction. Indeed, the Kumada coupling of 2-iodo-5-methylthiophene (**43**) with ethynylmagnesium chloride afforded the desired 2-ethynyl-5-methylthiophene (**44**) in only 35% yield, along with 24% of bis(5-methyl-2-thienyl)ethyne (**45**) [47]. The high propensity for H–Mg exchange explained the diarylethyne formation.

43   **44** (35%)   **45** (24%)

5.3.2. Negishi coupling

In comparison to the Kumada coupling, the Negishi coupling tolerates a wider array of functional groups including carbonyl, nitro, and amino; consequently, tedious protection–deprotection sequences are avoided. The requisite organozinc reagents are most conveniently prepared either by reaction of organolithium (organomagnesiun) reagents with zinc chloride or oxidative addition of organohalides with zinc metal.

While the Kumada reaction of 2-iodo-5-methylthiophene (**43**) with ethynylmagnesium chloride gave a substantial amount of diarylacetylene, the Negishi reaction of **43** with ethynylzinc bromide produced ethynylthiophene **46** in 87% yield. Similarly, the Negishi reaction between 2-iodobenzo[*b*]thiophene and ethynylzinc bromide led to 2-ethynyl-benzo[*b*]thiophene (**47**). The limitations of the Negishi reaction for synthesizing arylalkynes reside in the fact that it generally fails with unactivated aryl bromides. However, coupling with many aryl iodides proceeds smoothly [48].

46

47

Halothiophenes were coupled with heteroarylzinc reagents including furylzinc chloride [49] and pyrazolylzinc chloride [50] to deliver heterobiaryls **48** and **51**, respectively.

48

49   **50**   **51**

Gilchrist *et al.* prepared **53** *via* the Negishi coupling of bromoaldehyde **52** and 2-thienylzinc bromide without affecting the enal motif [51, 52]. 3-Thienylzinc bromide **55** was derived from a regioselective *ortho*-lithiation at the C(3) position of 4,4-dimethyl-2-(2-thienyl)oxazoline (**54**) followed by the treatment with ZnBr$_2$. The subsequent Negishi reaction of **55** with iodobenzene afforded tricycle **56** [53]. Additional examples of Negishi couplings of thienylzinc reagents with iodoarenes include the synthesis of arylthiophenes **59a–c** [54], **60** [55], and **61** [56]. Moreover, thienylpyridine **64** was elaborated *via* a Negishi coupling of 2-bromopyridine **62** and thienylzinc bromide **63** [57].

Conventionally, lithiation of thiophene at the C(3) position has been achieved using halogen–metal exchange of 3-bromothiophene with *n*-butyllithium. This method is troublesome, because 3-lithiothiophene has a temperature-dependent stability in polar ethereal solvents and slowly undergoes 2- and 3-positional isomerization as well as decomposition at temperatures higher than –25°C [58, 59]. To overcome this predicament, Rieke metal Zn* was generated from the reduction of $ZnCl_2$ with lithium using naphthalene or biphenyl as an electron carrier in THF. 3-Iodothiophene oxidatively added to Zn* to give the corresponding organozinc reagent **65**, although 3-bromothiophene was inert to Zn*. Subsequently, the Negishi reaction of **65** and 1-iodo-4-nitrobenzene led to arylthiophene **66**.

65   **66**

In recent years an important contribution to the Negishi reaction has been the general method for the coupling of aryl- and heteroaryl chlorides by Fu and coworkers [60]. It has been explained that the low reactivity of arylchlorides is due to the relatively large bond dissociation energy of a sp^2C–Cl relative to the sp^2C–Br and sp^2C–I. Fu and Dai, however, reported that the coupling of the unactivated 3-chlorothiophene with the arylzinc **67** in the presence of 2% $Pd(P(t-Bu)_3)_2$ in a mixture of THF and NMP delivered the 3-arylthiophene **68** in 89% yield.

68

Miyasaka and Rajca used $Pd(P(t-Bu)_3)_2$ to couple 3-bromothiophene to 3-thienylzinc chloride and 3-benzo[b]thienyl chloride to yield thiophene **69** and benzothiophene **70** in 86 and 60% yield, respectively [61]. Unlike the Fu example, the reaction was run at 40°C owing to the greater reactivity of 3-bromothiophene over 3-chlorothiophene.

69

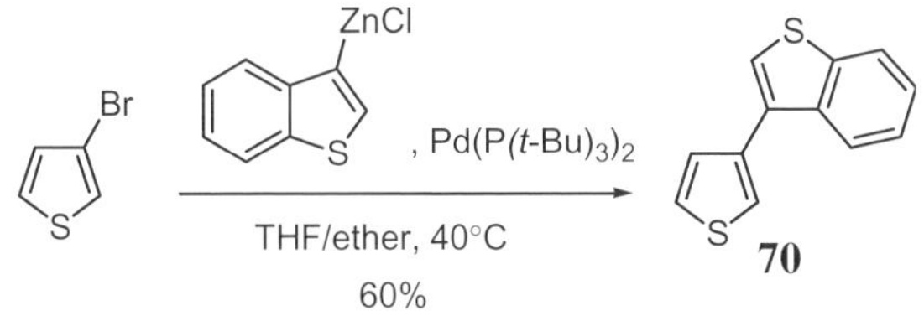

70

The example below illustrates that the Negishi reaction for thiophenes can also be chemoselective. In the case of 5-bromo-5′-iodobithiophene (**71**), thienylzinc reacted at the iodo position in the presence of Pd(dppf)Cl$_2$ to deliver terthiophene **72** in 77% yield [62].

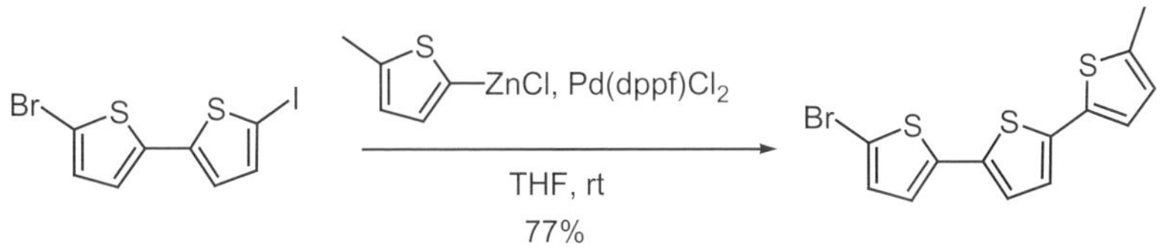

71   **72**

Regioselectivity can also be achieved with the Negishi reaction as illustrated with the conversion of 2,3-dibromothiophene to vinylthiophene **73** [63]. In the event, 2,3-dibromothiophene was allowed to react with 2-propenylzinc chloride in the presence of PdCl$_2$(dppb) to selectively deliver the 2-vinyl-3-bromothiophene in 82% yield [63a].

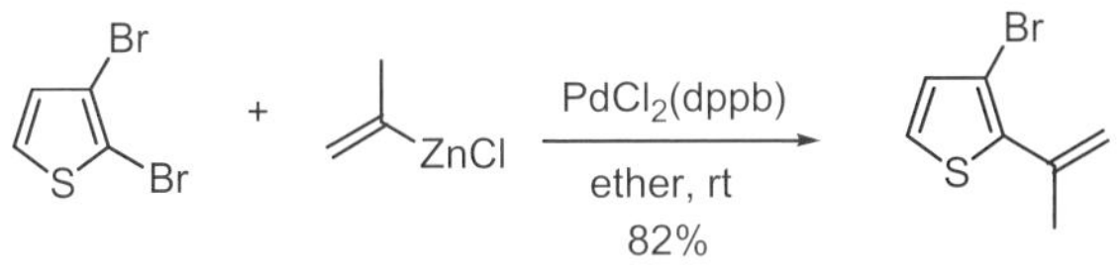

73

5.3.3. Suzuki coupling

5.3.3.1. Preparation of boronic acids, boronates, and trifluoroborates

The Suzuki reaction has been utilized with increasing frequency, particularly as the number of commercially available organoboranes continues to increase. Thienylboronic acid is readily prepared by treatment of a thienyl Grignard reagent or a thienyllithium reagent with a trialkylborate followed by acidic hydrolysis, as exemplified by conversion of bisthiophene **74** to bisthienylboronic acid **75** [64]. Other methods for preparing thienylboronic acid include halogen–metal exchange of a halothiophene followed by quenching with trialkylborate and Pd-catalyzed reaction of a halothiophene with dialkoxyborane [65].

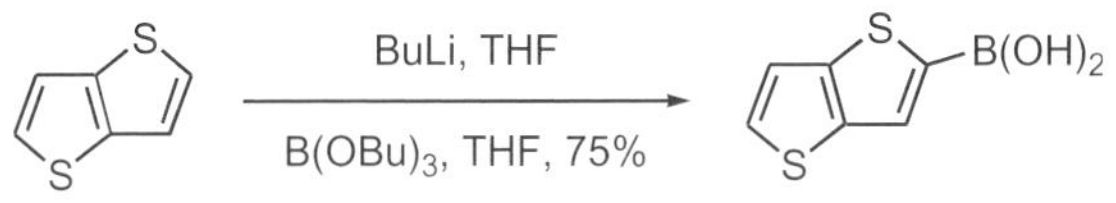

74   **75**

In addition, arylthiophene **76** was obtained by a one-pot Suzuki coupling of *p*-methoxyiodobenzene and 3-bromothiophene *via* an *in situ* boronate formation using one equivalent of bis-pinacolato borane [66]. This method avoided the isolation of boronic acids and is advantageous when base-sensitive groups such as aldehyde, nitriles, and esters are present. However, the cross-coupling yields are low when both aryl halides are electron-poor because of competitive homocoupling during the reaction.

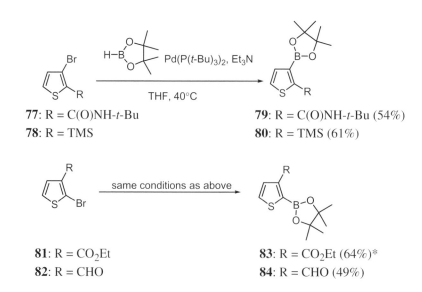

Bromothiophenes can be converted to their boronic esters or boronic acids by using pinacolborane in the presence of Pd(P(*t*-Bu)$_3$)$_2$. As illustrated below, 3-thienylboronates **79** and **80** and 2-thienylboronates **83** and **84** may be formed from their corresponding bromothiophenes in moderate yields. *However, boronate* **83** *was not isolated, instead it was hydrolyzed to the boronic acid; thus, the reported yield reflected isolation of the boronic acid* [67].* In the course of optimizing the formation of the boronic esters, a series of catalysts were surveyed including PdCl$_2$(PPh$_3$)$_2$, PdCl$_2$(dppf), and Pd$_2$dba$_3$. With each of these catalysts the percent conversion ranged from 10–42%, whereas with Pd(P(*t*-Bu)$_3$)$_2$ the conversion was 100%. Interestingly, a mixture of Pd$_2$(dba)$_3$/P(*t*-Bu$_3$) was far superior to Pd$_2$(dba)$_3$ as it gave 100% conversion to the boronate. *Typically, the crude boronic esters were used in the coupling step, since purification led to diminished yields of the boronic ester.* In the example below, boronate **85** was converted to 3-arylthiophene **86** in 59% overall yield.

77: R = C(O)NH-*t*-Bu   **79**: R = C(O)NH-*t*-Bu (54%)
78: R = TMS   **80**: R = TMS (61%)

81: R = CO$_2$Et   **83**: R = CO$_2$Et (64%)*
82: R = CHO   **84**: R = CHO (49%)

85   **86**

Since the discovery of the Suzuki reaction, boronic acids and boronic esters have been used for this coupling; however, in the last few years reports using trifluoroborates have increased [68]. Trifluoroborates offer advantages over boronic acids and boronic esters that include defined structures, ease of large scale preparation, long term stability, and easier reaction workups. The synthesis of trifluoroborates is straightforward for the starting material is the corresponding boronic acid. For example, 3-thienylboronic acid was treated with potassium hydrogen fluoride in aqueous methanol at room temperature to deliver the potassium fluoroborate **23** in 94% yield [69].

5.3.3.2. Couplings with aryl and heteroaryl substrates

Gronowitz and colleagues prepared various thienylpyrimidines using the Suzuki reaction approach [70, 71]. The union of 5-bromo-2,4-di-*t*-butoxypyrimidine (**87**) and 3-methyl-2-thiopheneboronic acid gave thienylpyrimidine **88**, which was then hydrolyzed to the corresponding uracil **89**, a potential antiviral agent. It was also possible to assemble uracil **89** by switching the coupling partners and coupling 2,4-di-*t*-butoxy-5-pyrimidineboronic acid (**90**) with 2-bromo-3-methylthiophene. In addition, the Gronowitz group synthesized a condensed aromatic compound, thieno[2,3-*c*]-1,7-naphthyridine (**95**), enlisting the Suzuki reaction of 2-formylthiophene-3-boronic acid (**93**) and 3-amino-4-iodopyridine (**94**) [72].

87   **88**   **89**

90

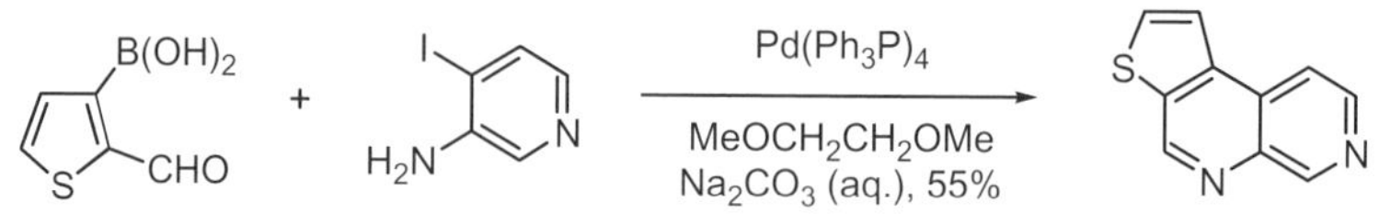

93 **94** **95**

Nettekoven and coworkers exploited the Suzuki reaction to synthesize a key intermediate used in their synthesis 5-thienyl triazolopyridines series [73]. These compounds
have the potential to act as antagonists of adenosine 2a (A2a) receptor, an important
receptor in certain neurodegenerative diseases. The pivotal coupling was accomplished
by using 2-thienylboronic acid and the 5-iodotriazolopyridine **96** in the presence of
Pd(dppf)Cl$_2$ dichloromethane complex to afford the triazolopyridine **97** in 83% yield.
Compound **97** was subsequently acylated with a series of acyl chlorides to deliver the
target triazolopyridines **98**.

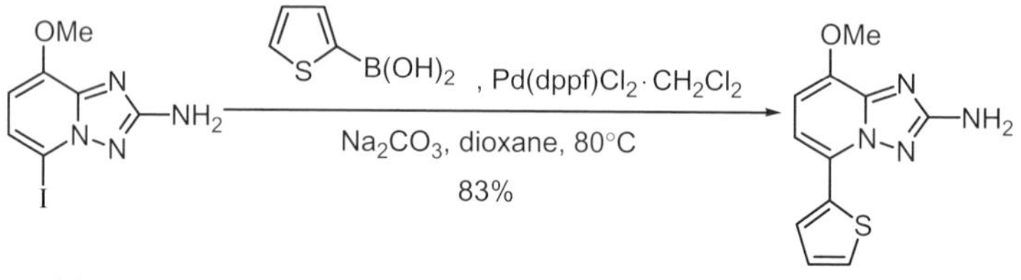

96   **97**

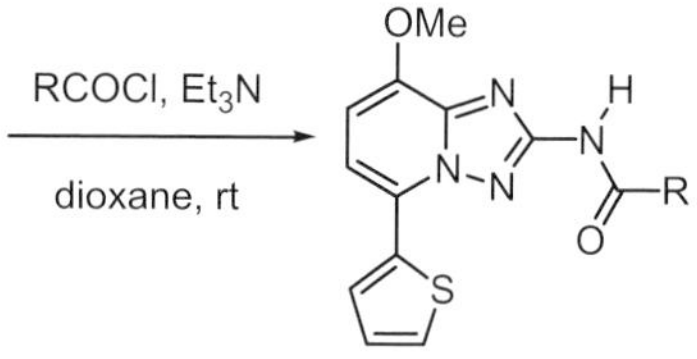

98

Thienylboronic acids are useful building blocks for preparing biaryls and heterobiaryls
employing the Suzuki reaction. In one case, a Suzuki coupling between thiophene-3-
boronic acid and iodocyclopropane **99** was promoted by cesium fluoride to furnish the
adduct **100** with retention of configuration [74]. In another example, the union between
thiophene-3-boronic acid and 5-bromo-2,2-dimethoxy-1,3-indandione (**101**) provided
ninhydrin derivative **102** [75].

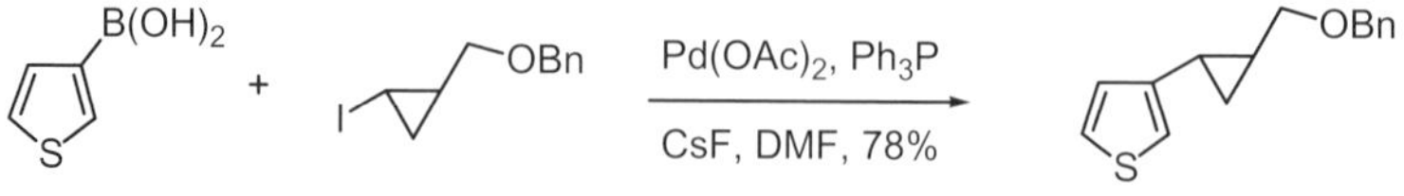

99   **100**

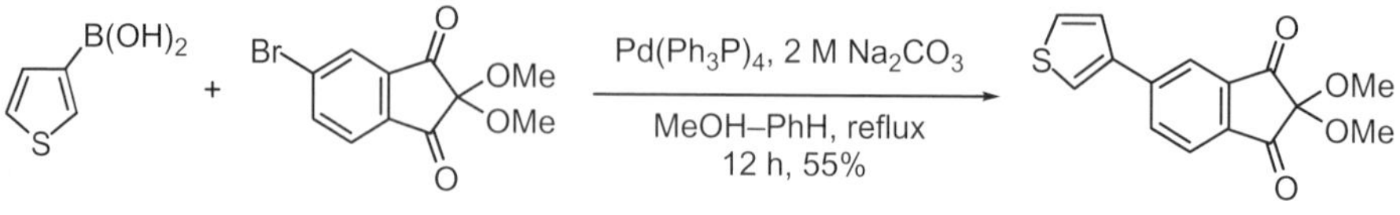

101   **102**

Treatment of thiophene *t*-butyl sulfonamide (**103**) with 2 equivalents of *n*-BuLi formed a dianion in which the second anion resided at the C(5) position. The resulting dianion was quenched with triisopropylborate followed by acidic workup to furnish thienyl-boronic acid **104**, which was then coupled with *p*-bromobenzyl alcohol under basic conditions to afford arylthiophene **105** [76].

103   **104**

105

Introduction of a thiophene substituent onto the pyrazine ring was realized by coupling thiophene-2-boronic acid with bromopyrazine **106** to give thienylpyrazine **107** [77]. In the coupling of indole triflate **108** and thiophene-2-boronic acid, an organic base (triethylamine) gave better results than an inorganic base to provide 3-thienylindole **109** [78].

106   **107**

108   **109**

In their quest to develop new antitumor agents, Mérour and coworkers utilized 2-benzo[*b*]thiophene-2-boronic acid in a pivotal Suzuki reaction [79]. Thus, the benzothiophene boronic acid was coupled very efficiently with triflates **110** and **111** in the presence of palladium tetrakis and aqueous sodium bicarbonate to deliver indoloquinolone **112** and pyridoquinolone **113** in 96 and 89% yield, respectively.

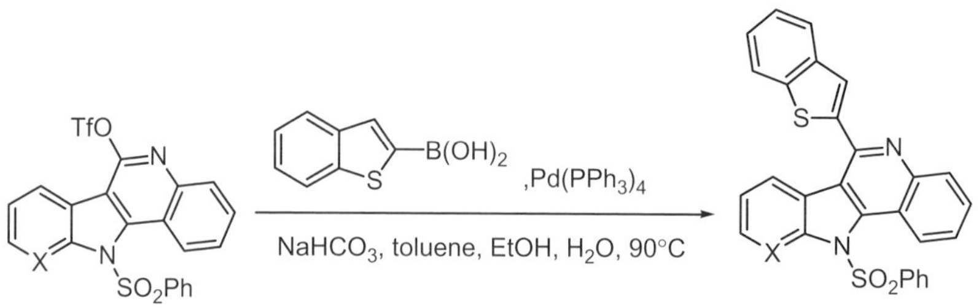

110: X = CH
111: X = N

112: X = CH (96%)
113: X = N (89%)

The scope of the Suzuki reaction was broadened when Coudert and coworkers demonstrated that vinylphosphates of nitrogen containing heterocycles were efficient coupling partners with aryl and heteroaryl boronic acids [80]. The vinylphosphates, synthesized from the corresponding lithium enolates and diphenylchlorophosphate, were considered more stable and economical than their triflate counterparts. One of the heteroarylboronic acids surveyed was 2-benzothiophenylboronic acid. 2-Benzothiophenylboronic acid was coupled to the vinylphosphate of the 4H-benzo[1,4]oxazine **114** and 4H-benzo[1,4]thiazine **115** in the presence of a catalytic amount of Pd(PPh$_3$)$_4$ to afford 4H-benzo[1,4]oxazine **116** and 4H-benzo[1,4]thiazine **117** in 70 and 71% yield, respectively.

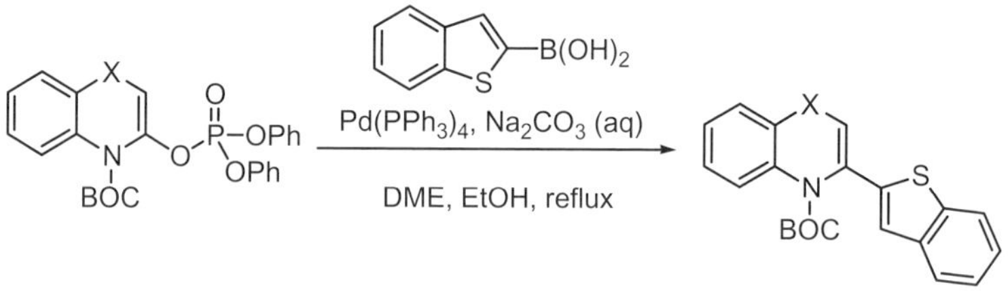

114: X = O
115: X = S

116: X = O (70%)
117: X = S (71%)

The Suzuki reaction of *p*-chlorobenzoyl chloride and thiophene-2-boronic acid was carried out under anhydrous conditions to furnish ketone **118** [81], providing an alternative synthesis of ketones. Behaving like simple aryl halides, iodothiophenes served as coupling partners with phenylboronic acid [82] and thienylboronic acid [83] to deliver biaryls **119** and **120**, respectively.

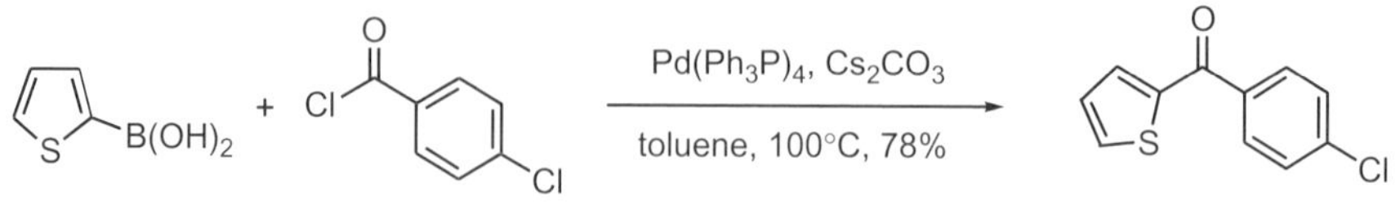

118

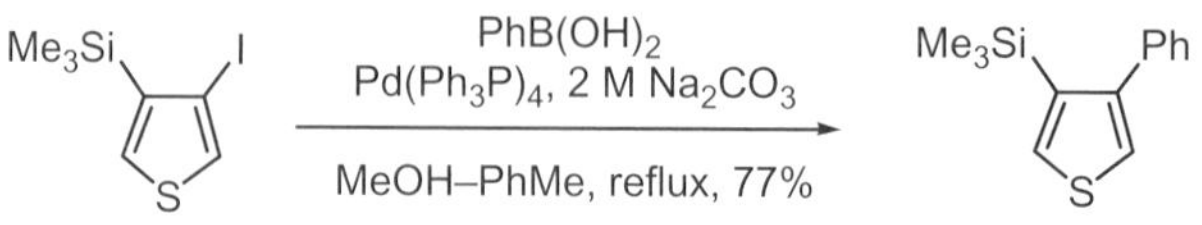

119

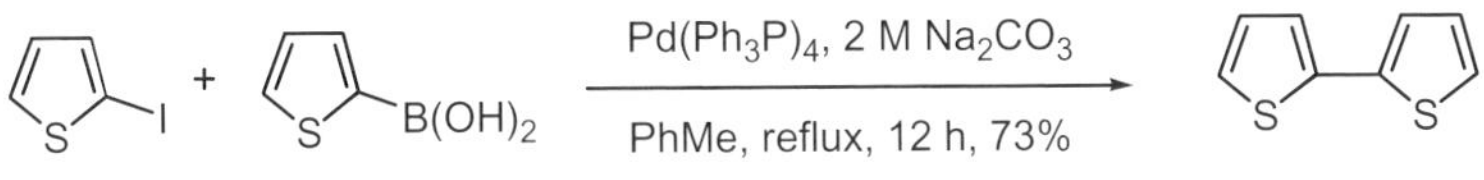

120

Under ligandless conditions, 3-thiophenetrifluoroborate (**23**) was treated with 2-bromopyridine and 3-bromopyridine in the presence of 1% palladium acetate to afford the coupled thiophenes **121a** and **121b** in 60 and 68% yield, respectively [84].

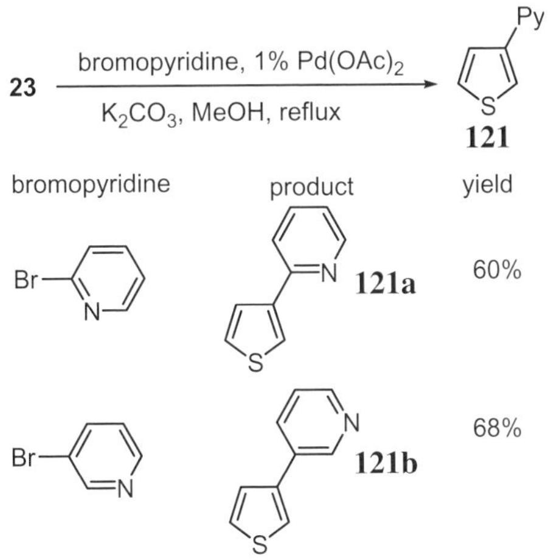

bromopyridine	product	yield

It was discovered, however, that the ligandless conditions were highly dependent on the heteroaryl halide substrate [84]. For example, 5-acetyl-2-bromothiophene reacted with the thiophene in the presence of Pd(OAc)$_2$ to furnish the desired product in 84% yield, but efficient coupling with the more electron-rich 2-bromothiophene required the use of Pd(dppf)Cl$_2$. Other heterocycles were coupled to **23**, but these also required the use of Pd(dppf)Cl$_2$ as shown in the table below.

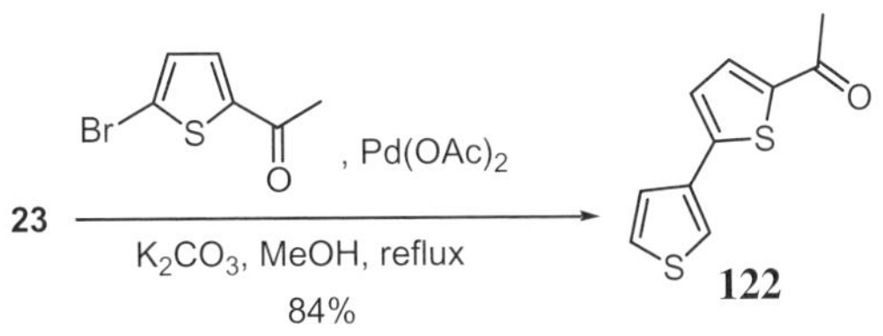

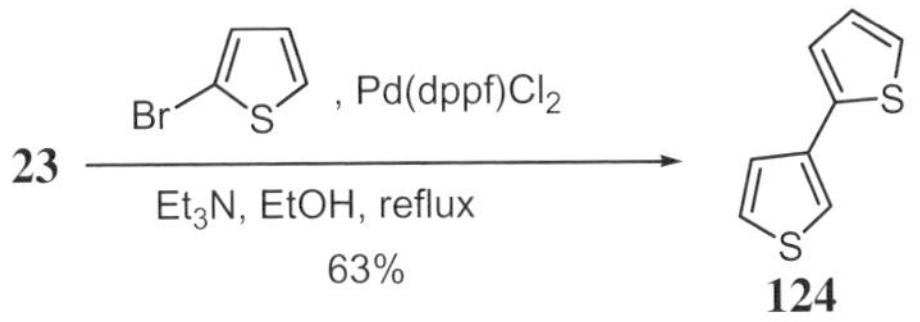

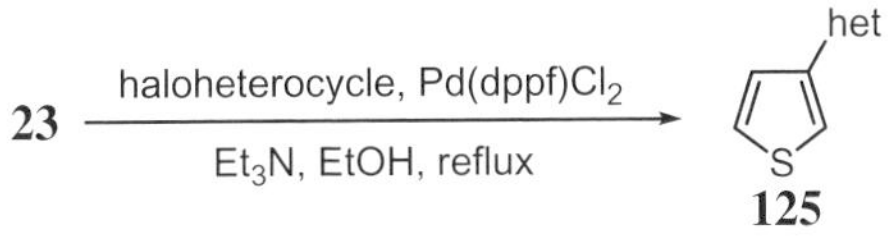

122

124

125

Halide	Product	Yield
	125a	67%
	125b	77%
	125c	83%

Other ligand-promoted couplings have been achieved. For example, Buchwald and Barder reported the coupling of 3-pyridyltrifluoroborate and 5-chlorothiophene-2-carbaldehyde in the presence of palladium acetate and S-Phos to produce the 2-pyridyl thiophene **126** in 75% yield [85]. In addition, Molander and coworkers have also demonstrated a coupling system that works for a diverse array of coupling partners including trifluoroborate thiophenes [86]. For example, **23** reacted with an aryltriflate in the presence of palladium acetate, PCy_3, and cesium carbonate in refluxing aqueous THF to deliver the 3-arylthiophene **127** in 79% yield.

Bussolari and Rehborn made a significant contribution to the Suzuki reaction when they demonstrated that high yielding palladium cross-couplings could be achieved with a series of thiophene-2-carboxaldehydes under mild conditions in water [87]. Typically water is used as a co-solvent in Suzuki reactions since the bases of choice are water soluble inorganic bases. However, Bussolari and Rehborn showed that in the presence of a catalytic amount of palladium acetate and an equimolar amount of tetrabutylammonium bromide, the reaction could be run at room temperature in water to afford arylthiophenes **130a–c** and **132** in yields of 60–82%.

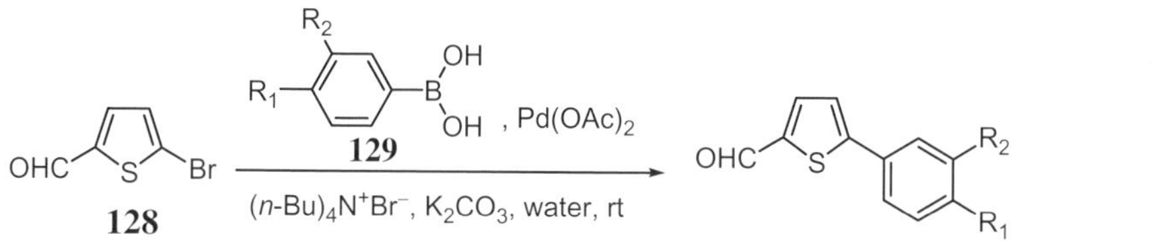

130a: R_1 = H, R_2 = H (67%)
130b: R_1 = H, R_2 = OMe (60%)
130c: R_1 = F, R_2 = H (82%)

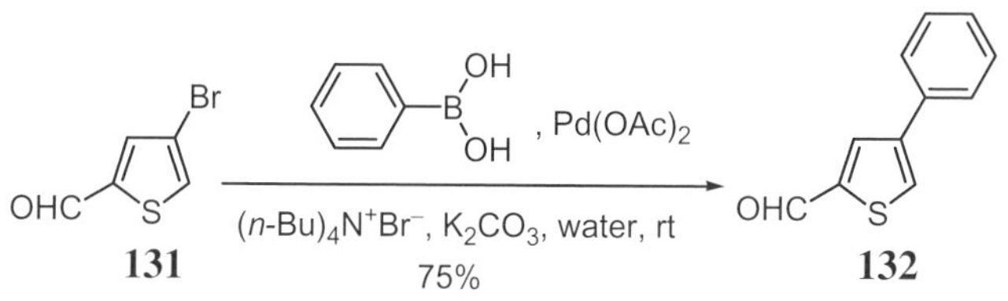

The Suzuki coupling of soluble polyethylene glycol (PEG)-bound bromothiophene **133** and *p*-formylphenylboronic acid provided biaryl **134** [88]. Due to the high solubilizing power of PEG, the reaction was conducted as a liquid-phase synthesis. Treatment of **134** with *o*-pyridinediamine resulted in a two-step-one-pot heterocyclization. After formation of the imine, intermediate, nitrobenzene served as an oxidant in the ring closure step. Finally, transesterification with NaOMe in MeOH resulted in 1*H*-imidazole [4,5-*c*]pyridine **135**.

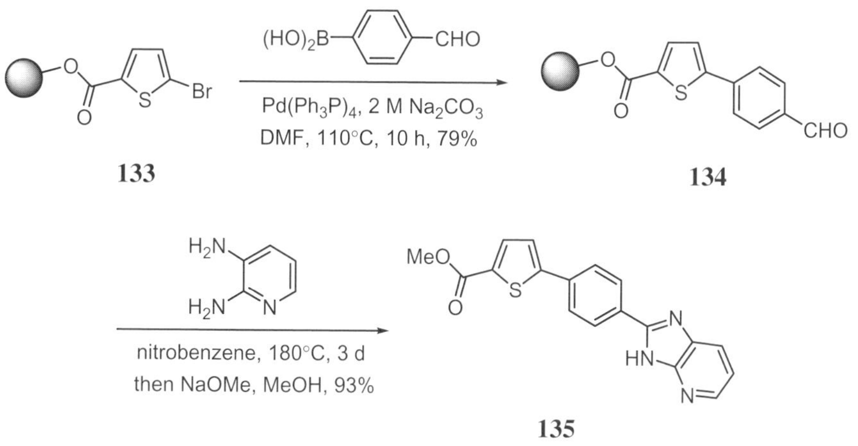

Since the discovery of the Suzuki reaction there have been several improvements in the cross-coupling of thiophenes whether the thienyl moiety is the boronic acid or the halide. One notable example is the Fu modification of the Suzuki reaction, where the catalyst is $Pd_2(dba)_3/P(t-Bu)_3$. For example, 2-chlorothiophene **136** reacted with 2-methylphenyl-boronic acid in the presence of the catalyst system $Pd_2(dba)_3/P(t-Bu)_3$ and potassium fluoride to furnish the 2-arylthiophene **137** in 99% yield [89]. A second example is the use of the catalytic system of $[PdCl(C_3H_5)]_2$ and the tetrapodal phosphorous ligand Tedicyp which requires very low loading [90]. Although the coupling works whether the thiophene is the boronic acid or halide, the turn over numbers TONs is highest when a halothiophene is used. For instance, the coupling between 2-bromothiophene and 2-methylphenylboronic acid in the presence of Tedicyp delivered **138** in 89% yield with a TON of 100,000 [90b]. Although a comparable yield of 87% was obtained when 2-thienylboronic acid coupled to 2-bromotoluene to afford **138**, the TON was only 20 [90b]. In general, Fu found that the Suzuki cross-coupling with 2-bromo and 3-bromothiophene with an arylboronic acid required about 0.1% of the catalyst; however, when the reactivity of the coupling partners was reversed, the catalyst loading increased to 1–10%.

5.3.3.3. Regioselectivity

Regioselectivity has also been achieved with respect to 2,4-dibromothiophenes [91]. Irie and coworkers successfully coupled three isomeric phenylcarboxaldehyde boronic esters **141** at the C(2) position of dibromothiophene **140** in 48–64% yield [91a]. In a related system, the Suzuki coupling between **143** and phenylboronic acid proceeded regioselectively to produce the 2-phenylthiophene **144** in 55% yield [91b].

The same regioselectivity was seen with 2,3-dibromobenzo[*b*]thiophene under standard Suzuki conditions using 2,6-dimethylphenylboronic acid and 2,6-dimethoxyphenylboronic acid as coupling partners in the presence of $Pd(PPh_3)_4$ to deliver 2-aryl-3-bromobenzothiophenes **147a** and **147b** in 95 and 63% yield, respectively [92]. Interestingly, the choice of base can alter the course of this reaction. It is known that the Suzuki reaction is promoted in the presence of strong inorganic bases such as $Ba(OH)_2$, NaOH, and TlOH when sterically hindered boronic acids are used [93]. Thus, the cross-coupling between 2,6-dimethylphenylboronic acid (146a) and 2,6-dimethoxyphenylboronic acid (146b) with 2,3-dibromobenzothiophene (145) in the presence of barium hydroxide afforded the 2,3-diarylbenzo[*b*]thiophenes **147c** and **147d** in good yields [92]. Interestingly, the number of equivalents of the boronic acids for each of the transformations below was identical, 2.6 equivalents.

147a: R = Me (95%)
147b: R = OMe (63%)

147c: R = Me (74%)
147d: R = OMe (65%)

de Lera and coworkers reported some interesting regioselectivity when they compared the Suzuki, Stille, and Sonogashira couplings using 2,3-dibromothiophene as the coupling partner [94]. Under optimal conditions, they discovered that the coupling between vinyl boronic acid **148** and 2,3-dibromothiophene gave a 40.5:1 mixture of **149a:149b**, while the Stille coupling with vinyl stannane **150** afforded a 70:9 mixture of **151a:151b**. Although the Sonogashira coupling between the alkyne and the thiophene gave exclusively the 2-substituted thiophene **153a**, a significant quantity of the dimer **153b** was produced. Thus, based on regioselectivity and yield, the Suzuki coupling was superior to Stille and Sonogashira methods.

5.3.4. Stille coupling

5.3.4.1. Preparation of stannylthiophenes

There are three preferred methods for the preparation of stannylthiophenes although other approaches exist [95]: (a) direct metalation of a thiophene followed by quenching with a stannyl electrophile; (b) halogen–metal exchange of a halothiophene followed by reaction with a stannyl electrophile; and (c) Pd-catalyzed reaction between a halothiophene and hexaalkylditin. Direct metalation of thiophenes followed by reaction with a stannyl electrophile is the most frequently utilized approach to prepare stannylthiophenes. In the absence of *ortho*-directing groups, the metalation occurs at the α positions due to the inductive effect of the C_{sp^2}–S bond [96–99]. Thienylstannane **155** was prepared by treatment of 2-(3-thienyl)-1,3-dioxolane (**154**) with *n*-butyllithium followed by quenching with Me_3SnCl [98–101]. Acidic hydrolysis of dioxolane **155** then produced 2-stannylthiophene-3-carboxaldehyde (**156**). In some cases, addition of TMEDA facilitated the metalation process [102, 103]. In other cases, a bulky base was advantageous if the substrates possessed functional groups susceptible to nucleophilic attack. Thus, direct metalation of (*E*)-3-(2-thienyl)propenoate (**157**) was achieved using LDA, and the resulting lithiothiophene was quenched with trimethyltin chloride to afford stereoisomerically pure (*E*)-3-(5-trimethylstannyl-2-thienyl)propenoate (**158**) [104].

Halogen–metal exchange of a halothiophene followed by quenching with a stannyl electrophile is another approach for preparing stannylthiophenes [105–108]. This protocol is especially useful when regiochemistry is critical, for direct metalation can be less straightforward. In the preparation of 4-stannylthiophene (**160**), addition of *t*-butyl *N*-(4-bromo-3-thienyl)carbamate (**159**) to a cooled solution of butyllithium resulted in a cleaner reaction than when the addition was reversed [105]. Stannane **160** was stable and could be purified by flash chromatography, whereas other regioisomers readily decomposed on silica gel to give the C–Sn bond cleaved product. When 2,4-dibromo-5-methylthiophene (**161**) was treated with one equivalent of *n*-BuLi, a regioselective halogen–metal exchange at the C(2) position was achieved, giving 2-silylthiophene **162** after treatment with trimethylsilyl chloride (TMSCl) [108]. A second halogen–metal exchange at C(4) of **162** was followed by stannylation to give 4-stannylthiophene **163**.

Most functional groups tolerate Pd-catalyzed reactions between a halothiophene and hexaalkylditin for the preparation of stannylthiophenes [109, 110]. In practice, however, this method often suffers from consumption of large quantities of hexaalkylditin because of its disproportionation reaction. When bromothiophene **164** was refluxed with hexamethylditin under Pd(Ph$_3$P)$_4$ catalysis, stannane **165** was obtained in good yield [111].

164 → **165**

Me₃Sn—SnMe₃ / Pd(Ph₃P)₄, toluene, reflux, 68%

5.3.4.2. Alkyl-, vinyl-, and alkynylthiophenes

Methylation of halothiophenes **166** and **168** was accomplished *via* the Stille reaction with tetramethyltin to give methylated thieno[3,2-*b*]pyran **167** [112] and thienyldeoxyuridine **169** [113], respectively. Analogously, the coupling of an allyl chloride, chloromethyl-cephem **170** and 2-tri-*n*-butylstannylthiophene furnished **171**, an intermediate for a C(3) thiophene analog of cephalosporin [114].

The Stille adduct of 2-bromothiophene and 1-ethoxy-2-tributyl-*n*-stannylethene or 1-ethoxy-1-tri-*n*-butylstannylethene is a masked thienyl aldehyde or a masked ketone, respectively [115–118]. Vinyl stannane **173**, derived exclusively as the *E*-isomer from hydrostannation of bis(trimethyl-silyl)propargyl amine (**172**), was coupled with 2-bromothiophene to form (*E*)-cinnamyl amine **174** upon acidic hydrolysis [119, 120]. In another case, stereoisomerically pure phenyl (*E*)-2-tributylstannyl-2-alkenoate **175**, arising from Pd-mediated hydrostannation of phenyl (*E*)-2-alkynoate, was joined with 2-iodothiophene to deliver the stereodefined trisubstituted α,β-unsaturated ester **176** [121, 122].

Liebeskind and associates converted 3,4-diisopropyl squarate (**177**) to stannyl-cyclobutenedione **178** *via* a 1,4-addition-elimination sequence. Compound **178** was then coupled with 2-iodothiophene to afford substituted cyclobutenedione **179** [123]. In another case, 3-lithioquinuclidin-2-ene, generated from the Shapiro reaction of 3-quinu-clidinone (**180**), was quenched with Bu_3SnCl to afford a unique "enamine stannane" **181**. The Stille reaction of stannane **181** and 5-bromo-2-formyl-thiophene then furnished 3-thienylquinuclidine **182** [124].

Vinylphosphates and some vinyl chlorides have been utilized as electrophiles in couplings with stannylthiophenes. The union between vinylphosphate **183** and 2-tributyl-stannylthiophene afforded 3-thienyl-4*H*-1,4-benzoxazine **184** [125], whereas the coupling of 6-chloropyrone **185** and 2-tributylstannylthiophene gave rise to 6-thienyl-4-hydroxy-2*H*-pyan-2-one **186**, offering an opportunity for introducing substitution at the

C(6) position of the pyrone nucleus [126]. Interestingly, allenylstannane, readily prepared from propargyl tosylate by copper-mediated addition of lithium tributylstannane, was coupled with 2-iodothiophene to provide a rapid entry to 2-thienylallene (**187**) [127].

Although the Sonogashira reaction (*vide infra*) is the mildest and most common approach employed to introduce alkyne functionality, it is of limited utility for acetylenes having either electron-withdrawing or electron-donating groups. The Stille reaction of 2-iodothiophene and ethoxy(trimethylstannyl)acetylene, however, allowed formation of ethoxyethynylthiophene (**188**), which was hydrated to ester **189** [128]. Other advantages of using stannanes to prepare arylacetylene include tolerance of more functional groups and better maintenance of stereochemical integrity [129, 130]. In Stille's total synthesis of the naturally occurring thienyl dienynol **193** [130], alkynylstannane **190**, derived from the reaction of ethynylthiophene and (diethylamino)trimethylstannane, was coupled with vinyl iodide **191** to provide isomerically pure dienyne **192**. Subsequent hydrolysis of **192** then gave stereochemically pure (*3E*, *5E*)-8-(2-thienyl)-3,5-octadien-7-yn-1-ol (**193**), a possible natural insecticide isolated from *Crysanthemum macrotum* (Dur.) Ball.

192 → (PPTS, 97%) → **193**

5.3.4.3. Arylthiophenes

A tremendous amount of work has been reported on the synthesis of arylthiophenes and heteroarylthiophenes utilizing the Stille reaction approach. In one case, 2-tributylstannylthiophene was coupled with *p*-acetoxyphenyl iodide to give thienylphenol **194** after hydrolysis [131]. Also, the union of 2-tributylstannylbenzo[*b*]thiophene and *p*-acetyliodobenzene provided arylbenzothiophene **195** using inexpensive Pd/C as a heterogeneous catalyst, CuI as a co-catalyst, and AsPh₃ as a soft ligand [132]. This catalytic system was also applied to 2-tributylstannylthiophene and 4-iodoacetophenone to deliver thiophene **196** in 80% yield [133]. Moreover, Kennedy and Perboni coupled 2-tributylstannylthiophene and 2-chloro-4-bromobenzylphosphonate (**197**) to make heterobiaryl phosphonate **198** [134]. During this reaction, portion-wise addition of the catalyst every two hours was found to be necessary owing to its continuous slow decomposition.

Furthermore, arylthiophenes have been prepared using the Stille coupling of hypervalent iodonium salts [135] or organolead compounds [136, 137] as electrophiles in place of aryl or vinyl halides and triflates. Hypervalent iodonium salts are sufficiently reactive to undergo coupling at room temperature.

Recently, Verkade and coworkers demonstrated that efficient coupling between stannanes and chlorobenzene could be achieved in the presence of $Pd_2(dba)_3$, proazaphosphatrane ligand **202**, and cesium fluoride to provide **203** in good yield [138].

In addition, this coupling system resulted in an efficient coupling between 2-stannylthiophene and 1-bromo-4-methoxy-6-methyl-benzene (**204**) at 50°C to afford **205** in excellent yield.

Li and coworkers discovered that DABCO was an effective ligand when used with palladium acetate in the presence of TBAF in the coupling of stannanes and aryl halides [139].

In the example below, 2-stannylthiophene was allowed to react with *p*-nitroiodobenzene in the presence of the catalytic mixture [*note*: 3 mol% Pd(OAc)$_2$] to afford **206** in quantitative yield. Interestingly, the reaction can be run with 0.0001 mol% of Pd(OAc)$_2$ to deliver **206** in 98% albeit the reaction time is about 2.5 times longer (40 h instead of 16 h).

5.3.4.4. Heteroarylthiophenes

The 2-oxazolinyl group can serve as a masked aldehyde or carboxylic acid and as a director of metalation for both nucleophilic and electrophilic aromatic substitutions. The cross-coupling of aryl Grignard reagents and methylthiooxazoline has been reported with limited examples [140]. However, the Stille reaction of 3-bromothiophene and 2-stannyloxazoline **207**, derived from 4,4-dimethyl-2-oxazoline, did indeed give 2-thienyloxazoline **208** [141]. An asymmetric version of the Stille coupling of chiral 2-bromooxazolines and thiophenestannanes was also described [142a] (albeit in only 20–28% yield), as was the coupling of oxazol-2-yl and 2-oxazolin-2-yltrimethylstannanes with aryl halides [142b].

Some bithienyl and terthienyl derivatives display biological activities including antifungal, nematocidal, and seed germination inhibition. The Stille reaction is regarded as the method of choice for the preparation of these thiophenes. For example, the union of 5-iodo-2-thiophenecarboxaldehyde and 2-tributylstannylthiophene furnished 5-(2′-thienyl)-2-thiophenecarboxaldehyde (**209**) [96]. In addition, terthienyl and bithienyl derivatives have been synthesized using 3,4-dinitro-2,5-dibromothiophene [143] and 3-methylsulfonyl-2-bromothiophene [144, 145] as electrophiles. Furthermore, tetrathiafulvalene (TTF) (210), a π-electron donor in organic conductors, was transformed to the corresponding trimethylstannyl-tetrathiafulvalene 211, which was then coupled with 2,5-dibromothiophene to secure 2,5-thienylbistetrathiafulvalene (**212**) [146].

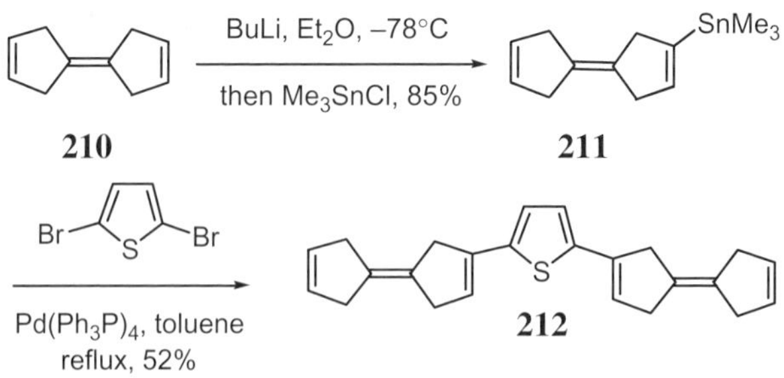

210   211

Pd(Ph$_3$P)$_4$, toluene
reflux, 52%

212

Thienylpyridines are also of great biological interest. Their syntheses *via* the Stille reaction are well precedented. In one case, two equivalents of 2-(trimethylstannyl)pyridine and 2,5-dibromothiophene underwent a Pd-catalyzed cross-coupling reaction to give 2,5-bis(2-pyridyl)thiophene (213) [147, 148]. In a series of papers, Gronowitz's group described the synthesis of pyridine-substituted hydroxythiophenes employing the Stille approach [149, 150]. For instance, 3,5-dibromo-2-methoxythiophene was converted to 3-tri-*n*-methylstannyl-2-methoxythiophene (214) *via* regiochemical halogen–metal exchange at the C(5) position. The subsequent Stille reaction of 214 with 2-bromopyridine produced the desired thienylpyridine 215.

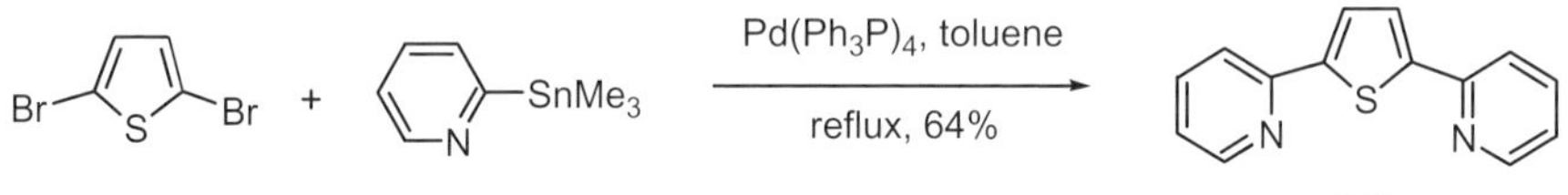

213

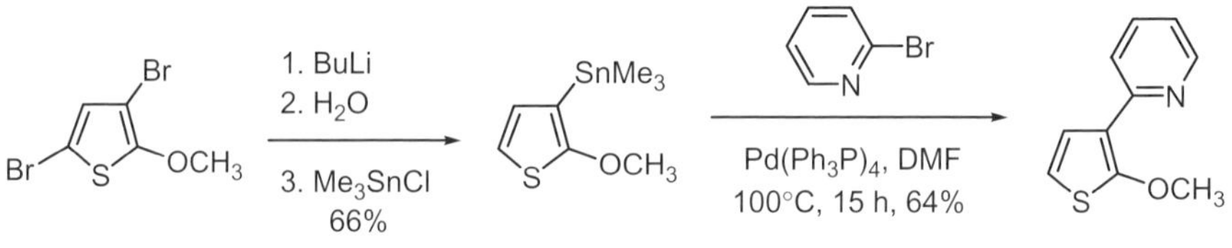

214   215

Gueiffer and coworkers applied the Stille methodology to an imidazo[1,2-a]-pyridine heterocycle to make derivatives of potential biological importance [151]. The imidazo[1,2-a]-pyridinyl stannane 216 was coupled with 2-iodothiophene in the presence of palladium tetrakis to give 217 in 82% yield. The authors noted that the Stille coupling was superior to the Negishi and Suzuki methodologies for the latter two options gave significantly lower yields of the desired coupled products.

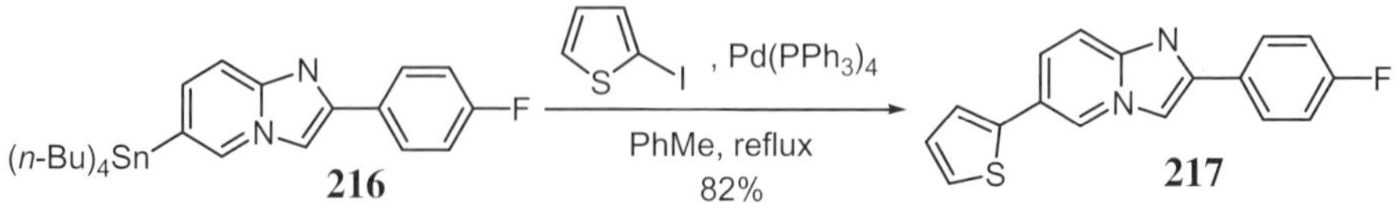

216   217

Li and Yue demonstrated that the Stille coupling of a variety of thienylstannanes with bromoquinoxalines worked very well unlike the Suzuki coupling attempts [152]. For example, thienylstannanes **219** reacted efficiently with the bromoquinoxalines **218** in the presence of Pd(PPh$_3$)$_2$Cl$_2$ and CuI in refluxing tetrahydrofuran to afford heteroarylquinoxalines **220** in 84–92% yield. In addition, they extended this methodology to 2-stannylbenzothiophene and quinxaile **221** to produce the requisite product **222** in 92% yield. It was presumed that the copper iodide accelerates the reaction rate by forming a more reactive organocopper species which promotes transmetalation to palladium.

218a : R = R' = Cl **219a** : R" = CHO   **220a**: R=R'=Cl, R"=CHO (92%)
218b : R = R' = F
                      219b : R" =   **220b**: R=R'=F, R" =   (84%)

                      219c : R" =   **220c**: R=R'=F, R" =   (85%)

221   **92%**   **222**

Baldwin and coworkers expanded the utility of CuI by showing that the Stille coupling can be accelerated in the presence of cesium fluoride; these conditions were particularly effective with sterically hindered or deactivated coupling partners [153]. In addition, they showed that heterocycles benefited from these conditions. For example, 2-stannylthiophene coupled efficiently with 3-iodopyridine in the presence of 10% CuI , two equivalents of cesium fluoride, and Pd(PPh$_3$)$_4$ to deliver the 2-pyridylthiophene **223** in 99% yield. The authors also commented that the reaction could be run at room temperature, albeit the reaction took twice as long.

DMF, 45°C
99%
223

A highly active catalyst for the coupling of 4-chloroquinolone with a variety of aryl and heteroarylstannane was shown to be the air stable palladium–phosphinous species

POPd [154]. Hence, 2-thienylstannane reacted with 4-chloroquinoline **224** in the presence of POPd (**225**) and dicyclohexyl-methylamine (Cy$_2$NMe) to deliver the 4-thienylquinoline **225** in 61% yield. In addition, it was determined that Cy$_2$NMe was superior to sodium acetate, cesium carbonate, or triethylamine.

Like many other palladium cross-coupling methods described in this chapter, one can achieve chemoselectivity with the thiophene ring. For example, cross-coupling between 2,3-dibromothiophene and the furyl stannane **226** in the presence of palladium bistriph-enylphosphine-dichloride resulted in regioselective formation of the 2-furylthiophene **227** in 63% yield [155].

5.3.4.5. Thiophene-containing condensed heteroaromatics

The Stille reaction is the key step in the synthesis of some thiophene-containing condensed heteroaromatics. Enlisting a Stille–Kelly reaction, Iyoda et al. treated dibro-mide **228** with hexamethylditin in the presence of Pd(Ph$_3$P)$_4$ to afford dithienothiophene (**229**) [156].

Gronowitz and coworkers coupled 2-(2-trimethylstannyl-3-thienyl)-1,3-dioxolane (**155**) with tert-butyl N-(ortho-bromothienyl)carbamate (**230**) to give the Stille adduct, which underwent acid-catalyzed deprotection and cyclization to deliver dithienopyridine **231** [157]. The uncharacteristically low yield of the Stille coupling was presumably due to the steric bulk of **155** which decreased the rate of transmetalation. In contrast, *t*-butyl *N*-(2-trimethylstannyl-3-thienyl)carbamate (**232**) was coupled with 3-iodo-2-formylpyridine to produce thieno[3,2-[*b*][2,8]naphthyridine (**233**) in good yield [158].

1. Pd(Ph$_3$P)$_4$, DMF
100–120°C, 24 h

2. 2 N HCl, 43%

155   +   **230**   →   **231**

PdCl$_2$•(dppb), CuO

DMF, 100°C, 2 h, 78%

232   →   **233**

5.3.4.6. Cu(I) thiophene-2-carboxylate (CuTC)

Since its introduction by Allred and Liebsekind in 1996 [159], copper thiophene-2-carboxylate (CuTC) has emerged as a mild and useful reagent for mediating the cross-coupling of organostannanes with vinyl iodides at room temperature. The CuTC is especially effective for substrates that are not stable at high temperature. In Paterson's total synthesis of elaiolide, he enlisted a CuTC-promoted Stille cyclodimerization of vinyl iodide **234** to afford the 16-membered macrocycle **235** under very mild conditions [160].

NMP, 15 min., 80%

PMBO   OH   O

234   →   **235**

5.3.5 Hiyama coupling

Palladium-catalyzed cross-coupling reactions of organosilicon compounds and organic halides display higher stereoselectivity and chemoselectivity than other organometallic reagents, and organosilicon substrates are readily available. However, because the C–Si bond is not as polarized as other carbon–metal bonds, introduction of a fluorine atom into the silyl group of organosilanes is necessary to accelerate Pd-mediated cross-coupling reactions. In the presence of catalytic η^3-allylpalladium chloride dimer and two equivalents of KF, the cross-coupling of ethyl(2-thienyl)difluorosilane and methyl 3-iodo-2-thiophenecarboxylate gave adduct **236** under relatively forcing conditions [161].

236

5.4. Sonogashira reaction

The Sonogashira reactions of both α-halothiophenes [162] and β-halothiophenes [163] proceed smoothly even for fairly complicated molecules as illustrated by the transformation of brotizolam (**237**) to alkyne **238** [164]. Interestingly, 3,4-bis(trimethylsilyl)thiophene (**240**), derived from the intermolecular cyclization of 4-phenylthiazole (**239**) and bis(trimethylsilyl)acetylene, underwent consecutive iodination and Sonogashira reaction to make 3,4-bisalkynylthiophenes [165]. Consequently, a regiospecific mono-*ipso*-iodination of **240** gave iodothiophene **241**, which was coupled with phenylacetylene to afford alkynylthiophene **242**. A second iodination and a Sonogashira reaction then provided the unsymmetrically substituted 3,4-bisalkynylthiophene **243**.

237   **238**

239   **240**   **241**

242   **243**

Using NaOH as the base, diarylacetylenes have been synthesized from either 2-methyl-3-butyn-2-ol [166] or trimethylsilylacetylene [167]. In both cases, NaOH unmasked the protecting groups after the first coupling reaction, revealing the additional terminal alkynyl functionality. Therefore, coupling the adduct **244**, derived from 2-iodothiophene and 2-methyl-3-butyn-2-ol, with 2-iodobenzothiophene provided diarylacetylene **245** [166]. Analogously, dithienylacetylene (**246**) was obtained when 2-iodothiophene and trimethylsilylacetylene were subjected to the same conditions [167].

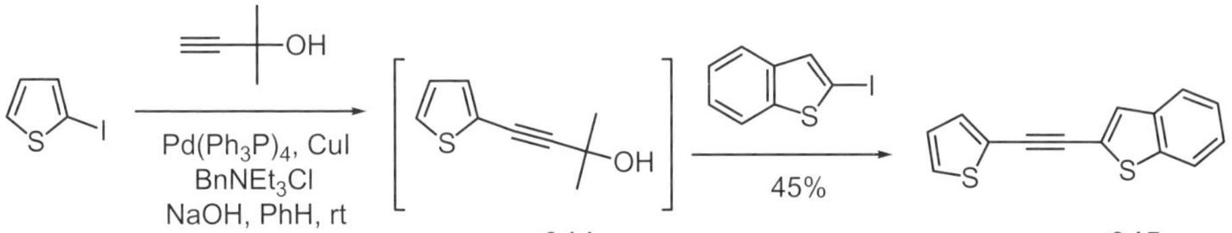

244

245

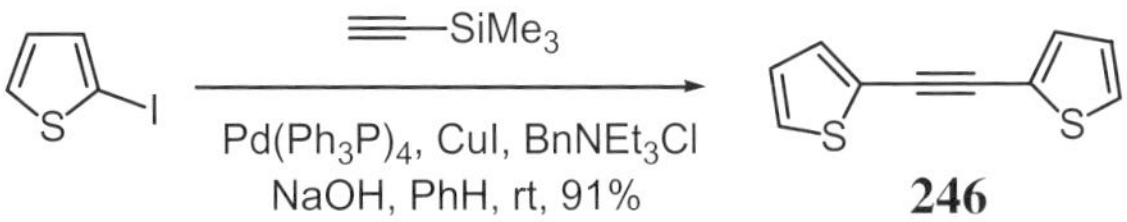

246

Advantage has been taken of the aforementioned observations in the synthesis of a terthiophene natural product, arctic acid (**250**) [168]. Palladium-catalyzed carbonylation of bromobisthiophene **37**, obtained from the Kumada coupling of 2-thienylmagnesium bromide and 2,5-dibromothiophene, gave bithiophene ester **247**, which was converted to iodide **248** by reaction with iodine and yellow mercuric oxide. Subsequent propynylation of **248** was then realized using the Sonogashira reaction with prop-1-yne to give bisthienyl alkyne **249**, which was subsequently hydrolyzed to **250**, a natural product isolated from the root of *Arctium lappa*.

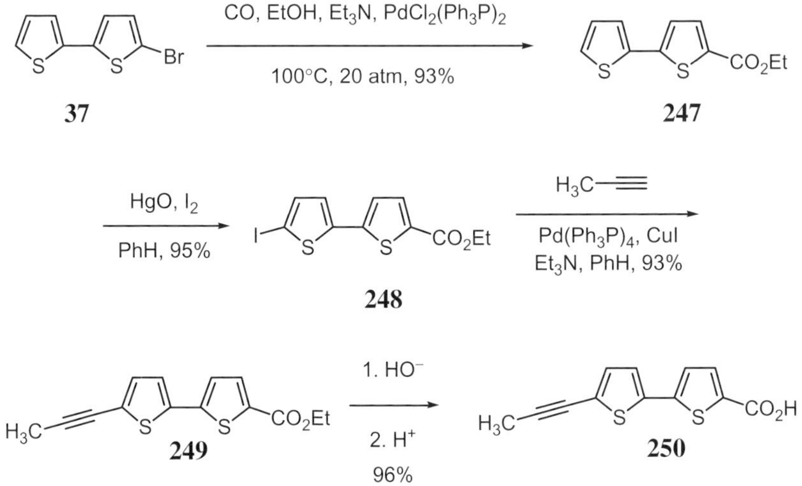

37

247

248

249

250

In addition to using the ligand Tedicyp in the Suzuki reaction, Santelli and coworkers have introduced this ligand to the Sonogashira reaction with a variety of heterocycles including 2- and 3-bromothiophene [169]. In the representative examples below, 2-bromothiophene was allowed to react with phenylacetylene (**251a**) and but-3-yn-1-ol (**251b**) in the presence of [Pd(C$_3$H$_5$)Cl]$_2$/Tedicyp and CuI to afford the acetylenes **252a** and **252b** in 95 and 52% yield, respectively. The authors noted that the reaction between 2-bromothiophene and the acetylenes did not proceed without the phosphorous ligand. Similar results were reported with 3-bromothiophene.

Microwave irradiation has also proven to be an effective reaction technique to increase reaction rates for the Sonogashira reactions. For example, Erdélyi and Gogoll reported the first homogeneous Sonogashira reaction and applied to 2-iodothiophene, 3-iodothiophene, and 3-bromothiophene. 2-Iodothiophene and 3-iodothiophene reacted with trimethylacetylene in the presence of Pd(PPh$_3$)$_2$Cl$_2$, copper iodide and diethylamine to afford 2-thienoacetylene **253** and 3-thienoacetylene **254** in 86 and 88% yield, respectively [170]. These yields were consistent with conducting the Sonogashira reaction under conventional thermal heating. The situation with the less reactive 3-bromothiophene, however, was different, and the microwave conditions demonstrated their superiority to standard heating. Under microwave conditions the corresponding product **255** was delivered in 81% while the literature yield under thermal conditions was only 28%.

5.5. Heck and intramolecular Heck reactions

Following the standard Heck protocol, 2-iodo-thiophene **256** reacted with methylacrylate in the presence of PdCl$_2$(PPh$_3$)$_2$ to deliver the α,β-unsaturated ester **257** in 66% yield [171].

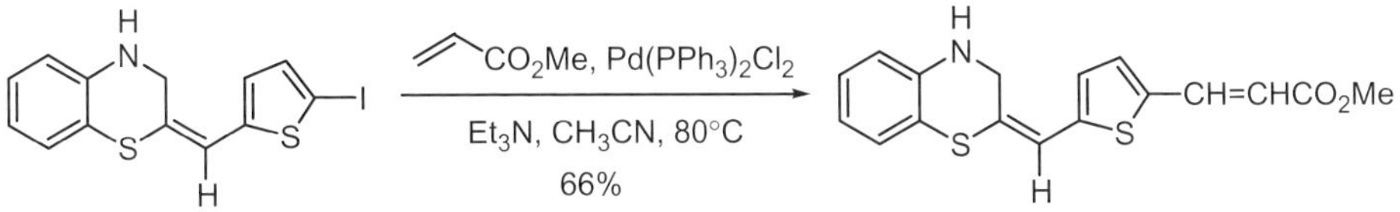

256   **257**

Both 2-bromothiophene and 3-bromothiophene have been coupled with allyl alcohols to make thienylated α,β-unsaturated ketones [172]. Iodothiophenes were more reactive than the corresponding bromides, whereas the chlorothiophenes were unreactive. As expected, 2-bromothiophene was two to three times more reactive than 3-bromothiophene. In addition to the expected Heck adduct **259**, the reaction of 2-bromothiophene **258** with 1-methylprop-2-en-1-ol also resulted in the regioisomer **260** [173].

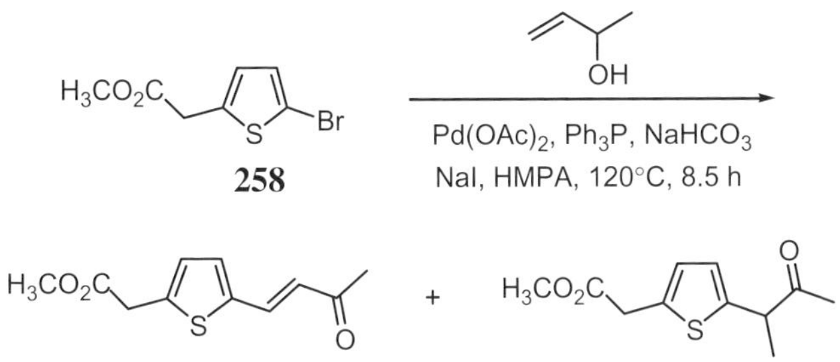

258

259   +   **260**

Fu and coworkers have utilized their very successful catalytic mixture of Pd$_2$(dba)$_3$/P(t-Bu)$_3$/Cy$_2$NMe to effect the Heck coupling between the 3-chlorothiophene **261** and 2,3-dihydrofuran at room temperature to furnish 2-thienyl-2,3-dihydrofuran **262** in 87% yield [174]. The coupling proceeded as expected, but olefin isomerization occurred.

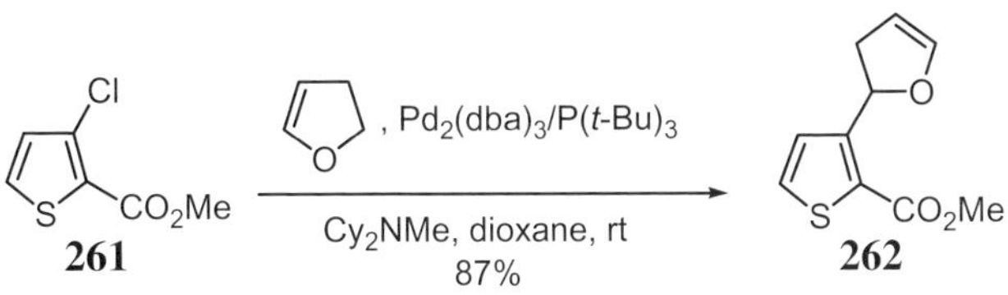

261   **262**

Santelli and coworkers reported an efficient coupling between 2- and 3-bromothiophene with acrolein ethylene acetal in the presence of PdCl$_2$(C$_3$H$_5$)$_2$ and the tetrapodal phosphine ligand Tedicyp [175]. The development of Tedicyp was an important contribution, since without this ligand, couplings with acrolein acetals had led to mixtures of products.

The two Heck reaction between the bromothiophenes and acrolein acetal delivered the corresponding products **263** and **264** in 80 and 70% yield [175b].

Ohta's group investigated the heteroaryl Heck reaction of thiophenes and benzothiophenes with aryl halides [176] and chloropyrazines [177]. Addition of the electrophiles invariably took place at C(2) as exemplified by the formation of arylbenzothiophene **265** from the reaction of benzothiophene and p-bromobenzaldehyde [176].

Lemaire and coworkers developed a method for the synthesis of 2-arylbenzo[b]thiophenes, biologically important compounds, by using a catalytic system free of phosphine [178]. The method involved using $Pd(OAc)_2$ in combination with tetra-butylammonium bromide (Jeffrey ligand free Heck conditions) [178a–c] or dicyclohexyl-18-crown-6 (DCH-18-C-6) in the presence of potassium carbonate [178b–d]. For example, 3-cyanothiophene (**266**) was coupled with 3-chloro-1-brombenzene using both additives to deliver compound **267** in good yield. Although yields were similar, the reaction run in the presence of the crown ether delivered a 30:1 ratio of **267** to **268** [178c] while tetrabutylammonium bromide gave a 5:1 mixture [178c].

Additive	% Yield of **267**	**267:268**
n-Bu₄NBr	72	5:1
DCH-18-C-6	76	30:1

This methodology was extended to a variety of 3-substituted benzothiophenes bearing electron-withdrawing and electron-donating substituents on the thiophene ring and a series of aryl- and heteroaryl bromides as shown in the table below [178c]. For entries 1 and 2, tetrabutylammonium bromide was used because the crown either caused decomposition of the benzo[*b*]thiophenecarboxaldehyde. In contrast, DCH-18-6 (entry 6) was shown to be superior to the tetrabutylammonium bromide (entry 5) with coupling of **269e** with 1-bromo-2-methylbenzene since there was a significant increase in the yield of **270e**.

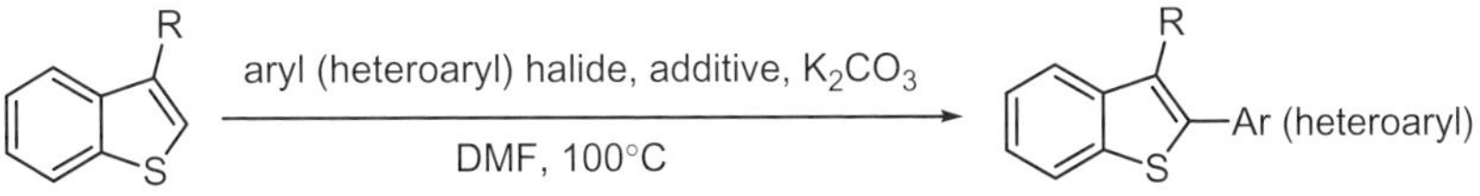

Entry	R	aryl(het)bromide	additive	Product	% yield
1	CHO **269a**	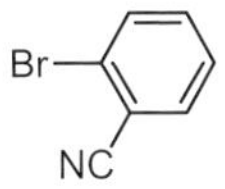 Br— (NC)	*n*-Bu$_4$NBr	**270a**	54
2	CHO **269b**	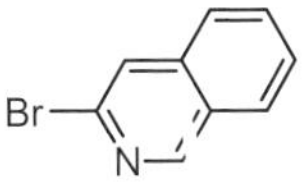 Br— (isoquinoline)	*n*-Bu$_4$NBr	**270b**	52
3	OCH$_2$CF$_3$ **269c**	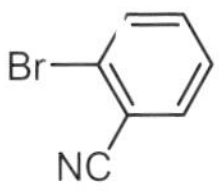 Br— (NC)	DCH-18-C-6	**270c**	76
4	OCH$_2$CF$_3$ **269d**	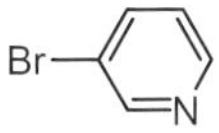 Br— (pyridine)	DCH-18-C-6	**270d**	59
5	OMe **269e**	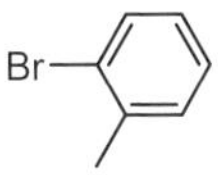 Br—	*n*-Bu$_4$NBr	**270e**	42
6	OMe **269e**	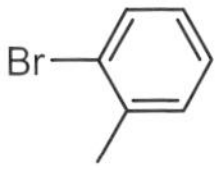 Br—	DCH-18-C-6	**270e**	75

This methodology was also applied to 2-aryl-3-substituted benzothiophenes **271** and **273** bearing either a nitrogen at the 3-position or a phenolic oxygen to give benzothiophenes **272** and **274** in good yields [178d].

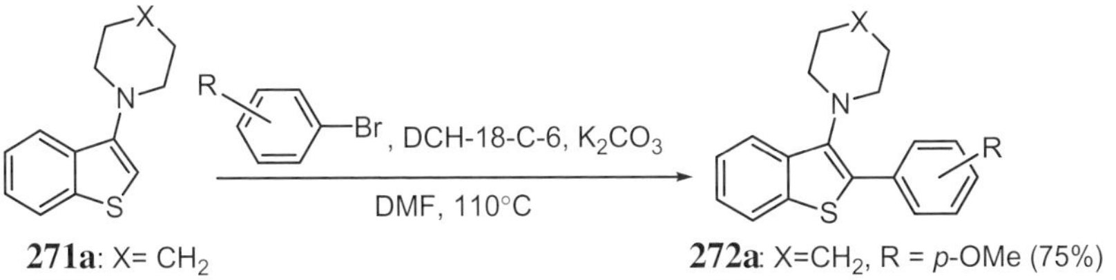

271a: X= CH$_2$
271b: X=O
271c: X = NCH$_3$

272a: X=CH$_2$, R = *p*-OMe (75%)
272b: X = O, R = *m*-CN (65%)
272c: X = NCH$_3$, R = *m*-CN (52%)

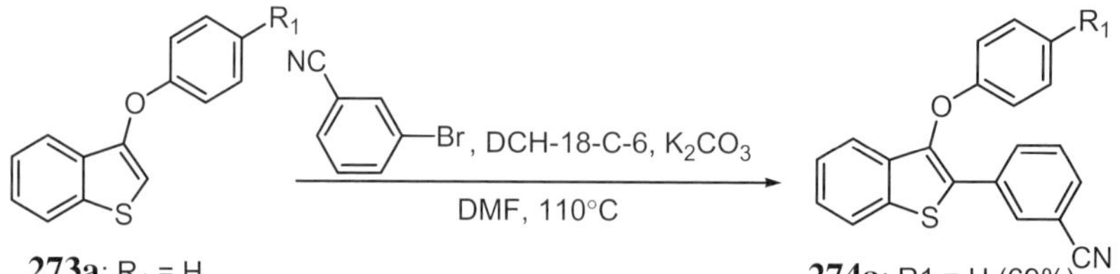

273a: R$_1$ = H
273b: R$_1$ = F
273c: R$_1$ = CN

274a: R1 = H (69%)
274b: R1 = F (67%)
274c: R1 = CN (73%)

As expected, thiophenes and aryl halides coupled in the presence of tetrabutylammonium bromide as illustrated with 2-thienylnitrile with iodobenzene to afford the arylation product **275** [178a].

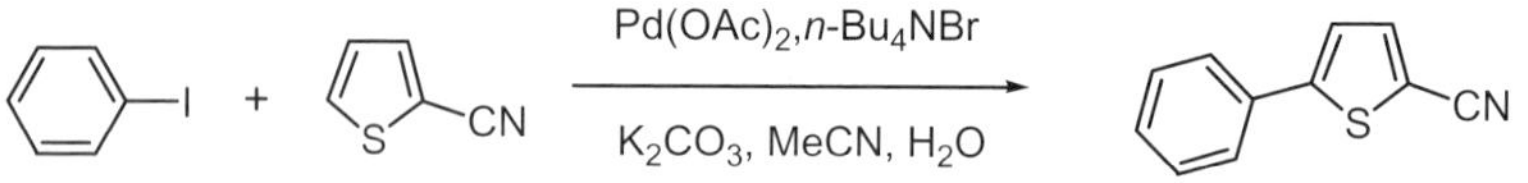

While the intramolecular Heck reaction has been widely used to synthesize indoles and benzofurans, not many applications have been found in the preparation of benzothiophenes because of the thiophilicity of the Pd(II) species. Pleixats and coworkers treated iodophenylsulfide **276**, obtained from o-iodoaniline and crotyl bromide in two steps, with Pd(Ph$_3$P)$_4$ and Et$_3$N in refluxing acetonitrile to form the intramolecular Heck cyclization product **277** [175]. The mechanism is akin to that of the Mori–Ban indole synthesis (see page 27). In another case, the intramolecular Heck cyclization of enamidone **278** with a pendant thienylbromide moiety furnished the 6-endo-trig product, indolizine **279**, in 63% yield, along with the debrominated enamidone **280** in 37% yield [179].

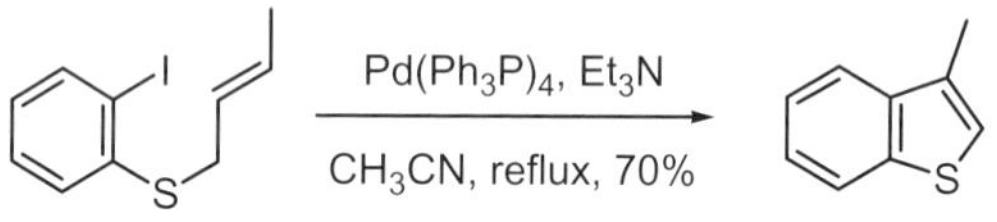

276   **277**

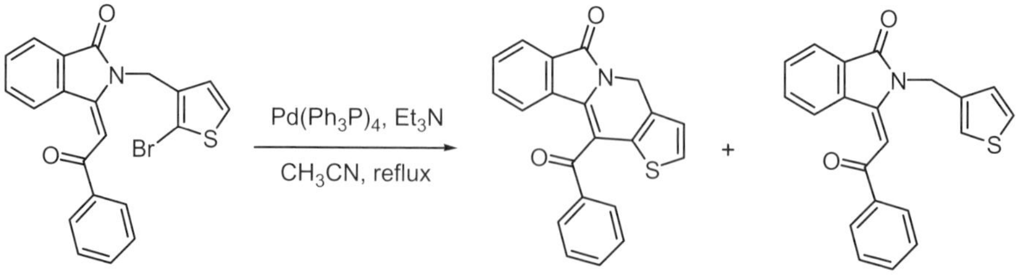

278   **279**   **280**

5.6. Carbonylation reactions

Halothiophenes take part in Pd-catalyzed alkoxycarbonylations in the presence of CO, alcohol, and base. In order to avoid the inconvenience of pressurized carbon monoxide, alkyl formate may be used as a safe surrogate [180]. In one of the many examples, 2-iodothiophene was carbonylated to the corresponding methyl ester using methylformate in place of CO.

Lin and Yamamoto described a Pd-catalyzed carbonylation of benzyl alcohols [181]. Thus, under the agency of palladium catalysis and promotion by HI, 3-thiophenemethanol was carbonylated to give 3-thiopheneacetic acid as a major product along with methylthiophene as a minor one.

The Hiyama group discovered that transmetalation of pentacoordinate silicate occurs in Pd-catalyzed reactions. They also successfully conducted Pd-catalyzed alkoxycarbonylation of organofluorosilanes with organic halides under the promotion of fluoride ion [182]. One salient feature of such a silicon-based carbonylative reaction is its remarkable functional group accommodation, even allowing a carbonyl group on either coupling partner. For example, a three-component carbonylation of 2-thienyldifluorosilane and *m*-iodobenzaldehyde was carried out in the presence of CO (1 atm), $(\eta^3\text{-}C_3H_5Pd)_2$, and KF in DMI to form ketone **282**. In another three-component carbonylative cross-coupling reaction, the hypervalent iodonium salt **283,** a halide surrogate, was joined with *p*-methoxyphenylboronic acid at room temperature to afford ketone **284** [183].

282

283   **284**

5.7. Buchwald–Hartwig aminations

Historically, α-halothiophenes have been considered as poor substrates for the Pd-catalyzed amination because of the strong thiophilicity of Pd(II). However, in recent years many reports have been published showing that the halo- and aminothiophenes and halo-aminobenzo[b]thiophenes react under the Buchwald–Hartwig arylamination conditions. In addition a detailed study by Hartwig and coworkers examined the scope and mechanism of Pd-catalyzed aminations in five-membered ring heterocycles [184].

The first example was published by Watanabe and coworkers. They successfully aminated both α- and β-halothiophenes [185]. In a strategy employing the sterically hindered, electron-rich phosphine ligand P(t-Bu)$_3$, NaOt-Bu as the base, and Pd(OAc)$_2$ as the catalyst, 2,5-dibromothiophene was bisaminated with diphenylamine to afford 2,5-bis(diphenylamino)-thiophene (285). This method is also applicable to 3-bromothiophene (69% yield), whereas the monoamination of 3,4-dibromothiophene was low-yielding (12%).

285

Since this first account of arylamination of a thiophene using the Buchwald–Hartwig conditions, others have addressed the coupling of arylamines, heteroarylamines, and aliphatic amines to thiophene and benzothiophene templates [186]. For example, Luker and coworkers showed that the coupling between butylamine and 3-halo or 3-triflatethiophene-2-carboxylic acid methyl ester thiophenes 286 in the presence of Pd$_2$(dba)$_3$ and cesium carbonate in toluene at 110°C delivered the 3-aminothiophenes 287 in good to excellent yield [186a]. This method was, however, restricted to the examples where the halogen or triflate was in conjugation with an electron withdrawing group.

286a: X = Br   **287a**: X = Br (94%, 100%)
286b: X = Cl   **287b**: X = Cl (68%)
286c: X = OTf   **287c**: X = OTf (96%)

Kirsch and coworkers have investigated the coupling between a number of anilines bearing electron-donating and electron-withdrawing groups with trisubstituted bromothiophenes. They discovered that reactions involving electron-rich anilines tended to produce higher yields and required a lower catalyst/ligand loading as compared to the electron-deficient anilines [186c]. For example, the reaction between 2-methoxyaniline, an electron rich aniline, and bromothiophene 288 in the presence of 3 mol% Pd(OAc)$_2$, 2,2′-bis(diphenylphosphino)-1,1′-binapthyl (BINAP, 4 mol%), and cesium carbonate in toluene at 115°C gave a 65% yield of 289a. However, the reaction between 288 and 2-acetylaniline, an electron-deficient aniline, afforded a modest 45% yield of 289b and required 10 mol% of Pd(dba)$_2$ and 10 mol% of BINAP.

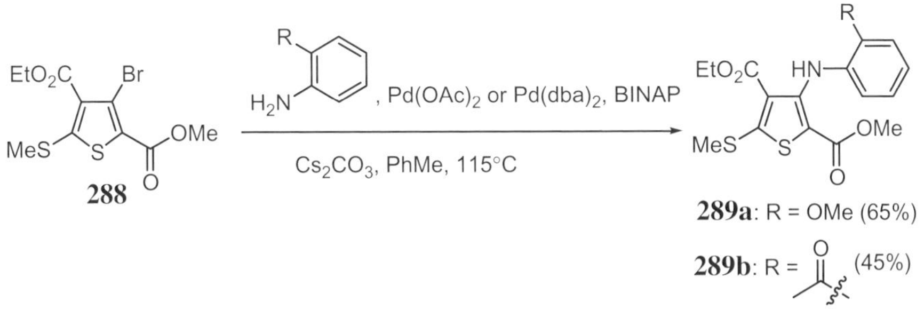

288

289a: R = OMe (65%)

289b: R = (45%)

Kirsch and coworkers extended this methodology to include benzo[*b*]thiophenes [187]. They presented examples where 6-aminobenzo[*b*]thiophenes were coupled with aryl bromides as well as the converse between 6-bromobenzo[*b*]thiophenes and anilines. For example, benzothiophene **290** and 3,4-dimethoxyaniline in the presence of Pd(OAc)$_2$, and racemic-BINAP to deliver **291** in 70% yield. In addition, the 6-aminobenzothiophene **292** reacted with 4-fluoro-1-bromobenzene using the same catalytic system to furnish the 6-aminobenzothiophene **293** in excellent yield [187a].

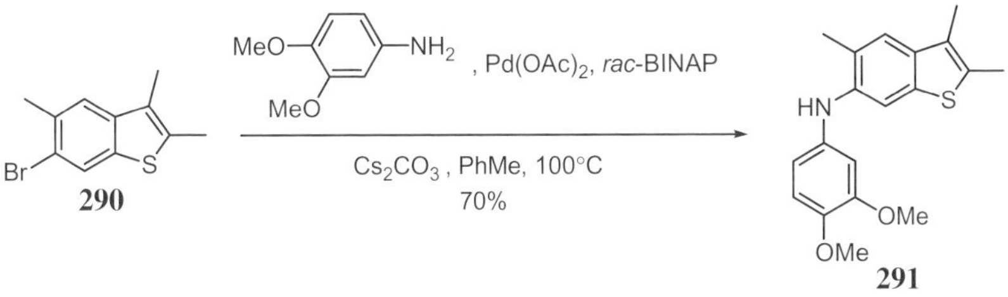

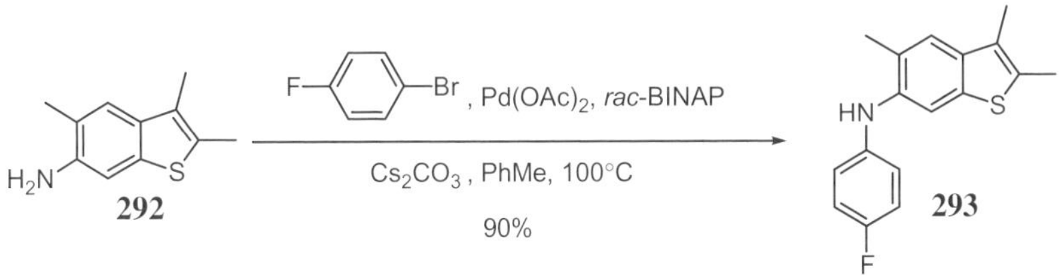

290   **291**

292   **293**

Palladium-catalyzed amination of the benzo[*b*]thiophene nucleus has also been applied to the synthesis of 3-[(4-pyridinyl)amino]benzo[*b*]thiophenes **295** which are selective serotonin re-uptake inhibitors [187b]. The synthesis of several of the inhibitors is illustrated.

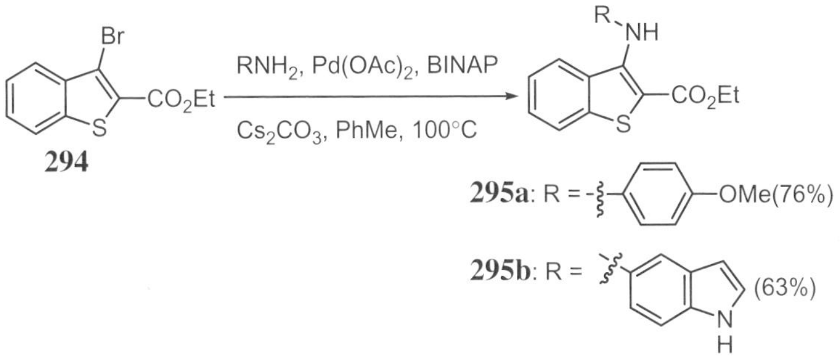

294

295a: R = -⧙⟩—OMe(76%)

295b: R = (63%)

5.8. Miscellaneous

5.8.1. Palladium-catalyzed C—P bond formation

The formation of an sp^3-hybridized C—P bond is readily achievable using the Michaelis–Arbuzov reaction. Such an approach is not applicable to form heteroaryl C—P bonds in which the carbon atoms are sp^2 hybridized, whereas palladium catalysis does provide a useful method for C_{sp^2}—P bond formation. The first report on Pd-catalyzed C—P bond formation was revealed by Hirao *et al.* [188–190]. Xu's group further expanded the scope of these reactions [191, 192]. They coupled 2-bromothiophene with *n*-butyl benzenephosphite to form *n*-butyl arylphosphinate **296** [191]. In addition, the coupling of 2-bromothiophene and an alkylarylphosphinate was also successful [192]. For the mechanism, see page 20.

296

5.8.2. Palladium-catalyzed cycloisomerization

In contrast to the prevalence of furan preparation *via* Pd-catalyzed heteroannulation methods, a scarcity of literature precedents are found for thiophene syntheses *via* such an approach, again possibly due to the thiophilicity of the Pd(II)-intermediates. Nevertheless, Gabriele's group synthesized substituted thiophenes from (Z)-2-en-4-yne-1-thiols utilizing Pd-catalyzed cycloisomerization [193]. For example, using PdI$_2$ as the catalyst and KI as the solubilizing agent, the Pd(II)-catalyzed cycloisomerization of **297** gave rise to thiophene **300**. Presumably, coordination of the triple bond with the Pd(II) species gives **298**, which is followed by nucleophilic attack by the SH group, forming **299**. Protonolysis of **299** is then followed by aromatization to furnish **300**.

297   **298**   **299**   **300**

To summarize, electrophilic substitutions and metalations of thiophenes take place preferably at the α-positions due to the electronegativity of the sulfur atom. This is the consequence of the more effective incorporation of lone pair electrons on the sulfur into the aromatic system. In contrast, electrophilic substitution of benzo[*b*]thiophenes occurs at the β-position; however, the preference is not strong and substitution can result in other positions. Like thiophene, metalation and metal–halogen exchange are favored at the α-position. Also, oxidative couplings of thiophenes and benzothiophenes are not very synthetically useful due to low yields and the consumption of stoichiometric amounts of Pd(OAc)$_2$; however reductive couplings appear to be more viable. Moreover, the palladium cross-coupling of thiophenes and benzo[*b*]thiophenes are well documented in the literature including the Kumada, Negishi, Hiyama, Suzuki, Stille couplings with the latter two reactions being most prevalent. In addition, the Heck and Buchwald–Hartwig couplings are also finding much popularity. Because of these positive results in palladium chemistry, the thiophene and benzo[*b*]thiophene moieties have become an integral part of medicinal chemistry and material science.

5.9. References

1. Russel, R. K. in *Comprehensive Heterocyclic Chemistry II: Thiophenes and Their Benzo Derivatives*. Bird, C. W. Ed.; Elsevier Science Ltd.: Oxford, **1996**; Vol. 2; pp. 680–682.

2 (a) Press, J. B. in *The Chemistry of Heterocyclic compounds: Thiophene and Its Derivatives*. Gronowitz, S. Ed; Wiley and Sons: New York, **1985**; Vol. 44, pt 1; pp. 353–456. (b) Russel, R. K. in *Comprehensive Heterocyclic Chemistry II: Thiophenes and Their Benzo Derivatives*. Bird, C. W. Ed.; Elsevier Science Ltd.: Oxford, **1996**; Vol. 2; pp. 682–92.

3. Nivsarkar, M.; Cherian, B.; Padh, H. *Curr. Sci.* **2001**, *81*, 667–72.

4. (a) Plosker, G. L.; Wagstaff, A. J. *Drugs* **1995**, *50*, 1050–75. (b) Sorkin, E. M.; Brogden, R. N. *Drugs* **1985**, *29*, 208–35.

5. (a) Li, J. J; Johnson, D. S.; Sliskovic, D. R.; Roth, B. D. in *Contemporary Drug Synthesis*; John Wiley & Sons, Inc.: Hoboken (NJ), **2004**, pp. 1–10. (b) Rodgers, J. E.; Steinhubl, S. R. *Expert Rev. Cardiovasc. Ther.* **2003**, *4*, 507–22.

6. (a) Bymaster, F. P.; Beedle, E. E.; Findlay, J.; Gallagher, P. T.; Krushinski, J. H.; Mitchell, S.; Robertson, D. W.; Thompson, D. C.; Wallace, L.; Wong, D. T. *Bioorg. Med. Chem. Lett.* **2003**, *13*, 4477–80. (b) Goldstein, D. J.; Lu, Y.; Detke, M. J.; Lee, T. C.; Iyengar, S. *Pain* **2005**, *116*, 109–18. (c) van Kerrebroeck, P. *BJU Int.* **2004**, *94 (S1)*, 31–7.

7. (a) Miller, C. P. *Curr. Pharm. Des.* **2002**, *8*, 2089–111. (b) Bradley, D. A.; Godfrey A. G.; Schmid. C. R. *Tetrahedron Lett.* **1999**, *40*, 5155–59.

8. Carillo-Muñoz, A. J.; Giusiano, G.; Ezkurra, P. A.; Quindós, G. *Expert. Rev. Anti. Infect. Ther.* **2005**, *3*, 333–42.

9. Yoshida, Z.-i.; Yamada, Y.; Tamaru, Y. *Chem. Lett.* **1977**, 423–4.

10. (a) Rajappa, S.; Natekar, M. in *Comprehensive Heterocyclic Chemistry II*. Bird C. W. Ed.; Elsevier Science Ltd: Oxford, **1996**, vol. 2, pp. 501–6. (b) *Ibid.*, pp. 594–601.

11. (a) Gupta, R. R.; Kumar, M.; Gupta, V. in *Heterocyclic Chemistry II*; Springer–Verlag: Berlin, **1999**; p. 327–8. (b) *Ibid.*, p. 327. (c) *Ibid.*, pp. 328–329. (d) *Ibid.*, pp. 334.

12. (a) Jouele, J. A.; Mills, K.; Smith, G. F. in *Heterocyclic Chemistry*. 3rd edition; Chapman & Hall: London, **1995**, pp. 350–1. (b) *Ibid.*, p. 353.

13. (a) Katritzky, A. R.; Pozharskii, A. F. in *Handbook of Heterocyclic Chemistry*; 2nd edition; Elsevier Science Ltd: Oxford, **2000**; p. 310. (b) *Ibid.*, p. 320.

14. Smith, G. V.; Notheisz, F.; Zsigmond, A. G.; Bartok, M. *Stud. Surf. Sci. Catal.* **1993**, *75* (*New Frontiers in Catalysis, Pt. C*), 2463–6.

15. Schröter, S.; Stock, C.; Bach, T. *Tetrahedron* **2005**, *61*, 2245–67.

16. Fazio, A.; Gabriele, B.; Salerno, G.; Destri, S. *Tetrahedron* **1999**, *55*, 485–502.
17. Zhang, Y.; Hörnfeldt, A.-B.; Gronowitz, S. *J. Heterocycl. Chem.* **1995**, *32*, 435–44.
18. Lucas, L.; van Esch, J.; Kellogg, R. M.; Feringa, B. L. *Tetrahedron Lett.* **1999**, *40*, 1775–8.
19. Okamoto, T.; Kakinami, T.; Fujimoto, H.; Kajigaeshi, S. *Bull. Chem. Soc. Jpn.* **1991**, *64*, 2566–8.
20. Mitchell, R. H.; Chen, Y.; Zhang, J. *Org. Prep. Proc. Int.* **1997**, *29*, 715–9.
21. Rossi, R.; Carpita, A.; Lezzi, A. *Tetrahedron* **1984**, *40*, 2773–9.
22. Hucke, A.; Cava, M. P. *Tetrahedron* **1998**, *63*, 7413–7.
23. Otsubo, T.; Kono, Y.; Hozo, N.; Miyamoto, H.; Aso, Y.; Ogura, F.; Tanaka, T.; Sawada, M. *Bull. Chem. Soc. Jpn.* **1993**, *66*, 2033–41.
24. Clark, P. D.; Clarke, K.; Scrowston, R. M.; Sutton, T. M. *J. Chem. Res. Synop.* **1978**, 10.
25. Cross, P. E.; Dickinson, R. P.; Parry, M. J.; Randall, M. J. *J. Med. Chem.* **1986**, *29*, 1637–43.
26. Kiryanov, A. A.; Seed, A. J.; Sampson, P. *Tetrahedron Lett.* **2001**, *42*, 8797–800.
27. Sotgiu, G.; Zambianchi, M.; Barbarella, G.; Botta, C. *Tetrahedron* **2002**, *58*, 2245–51.
28. Prats, M.; Gálvez, C. *Heterocycles* **1992**, *34*, 149–56.
29. Abbotto, A.; Bradmante, S.; Facchetti, A.; Pagani, G. *J. Org. Chem.* **1997**, *62*, 5755–65.
30. Nakajima, R.; Iida, H.; Hara, T. *Bull. Chem. Soc. Jpn.* **1990**, *63*, 636–7.
31. Takahashi, K.; Tarutani, S. *Heterocycles* **1996**, *43*, 1927–35.
32. Kabalka, G. W.; Mereddy, A. R. *Tetrahedron Lett.* **2004**, *45*, 343–5.
33. Kondo, Y.; Shilai, M.; Uchiyama, M.; Sakamoto, T. *J. Am. Chem. Soc.* **1999**, *121*, 3539–40.
34. (a) Torrado, A.; Lamas, C.; Agejas, J.; Jiménez, A.; Díaz, N.; Gilmore, J.; Boot, J.; Findlay, J.; Hayhurst, L.; Wallace, L.; Broadmore, R.; Tomlinson, R. *Bioorg. Med. Chem.* **2004**, *12*, 5277–95. (b) Yannopoulos, C. G.; Xu, P.; Ni, F.; Chan, L.; Pereira, O. Z.; Reddy, T. J.; Das, S. K.; Poisson, C.; Nygun-Ba, N.; Turcotte, N.; Proulx, M.; Halab, L.; Wang, W.; Bédard, J.; Morin, N.; Hamel, M.; Nicolas, O.; Bilimoria, D.; L'Heureux, L.; Bethell, R.; Dionne, G. *Bioorg. Med. Chem. Lett.* **2004**, *14*, 5333–7.
35. Eberson, L.; Gomez-Gozalez, L. *Acta Chem. Scand.* **1973**, *27*, 1249–54.
36. Kozhevnikov, I. V. *React. Kinet. Catal. Lett.* **1976**, *4*, 451–8.
37. Itahara, T.; Hashimoto, M.; Yumisashi, H. *Synthesis* **1984**, 255–6.
38. Itahara, T. *J. Org. Chem.* **1985**, *50*, 5272–5.
39. Fujiwara, Y.; Maruyama, O.; Yoshidomi, M.; Taniguchi, H. *J. Org. Chem.* **1981**, *46*, 851–5.
40. Hassan, J.; Lavenot, L.; Gozzi, C.; Lemaire, M. *Tetrahedron Lett.* **1999**, *40*, 857–8.
41. Venkatraman, S.; Li, C.-J. *Org. Lett.* **1999**, *1*, 1133–5.
42. Amatore, C.; Jutand, A.; Negri, S.; Fauvarque, J. F. *J. Organomet. Chem.* **1990**, *390*, 389–98.
43. (a) Minato, A.; Tamao, K.; Hayashi, T.; Suzuki, K.; Kumada, M. *Tetrahedron Lett.* **1980**, *21*, 845–8. (b) Strässler, C.; Davis, N. E.; Kool, E. T. *Helv. Chim. Acta* **1999**, *82*, 2160–71.
44. Iarossi, D.; Mucci, A.; Schenetti, L.; Sodini, V. *J. Heterocyclic Chem.* **1999**, *36*, 241–7.
45. Abbotto, A.; Bradmante, S.; Facchetti, A.; Pagani, G. *J. Org. Chem.* **1997**, *62*, 5755–65.

46. Minato, A.; Suzuki, K.; Tamao, K.; Kumada, M. *J. Chem. Soc., Chem. Commun.* **1984**, 511–3.
47. Gilchrist, T. L.; Healy, M. A. M. *Tetrahedron Lett.* **1990**, *31*, 5807–10.
48. Gilchrist, T. L.; Healy, M. A. M. *Tetrahedron* **1993**, *49*, 2543–56.
49. Minato, A.; Suzuki, K.; Tamao, K.; Kumada, M. *J. Chem. Soc., Chem. Commun.* **1984**, 511–13.
50. Felding, J.; Kristensen, J.; Bjerregaard, T.; Sander, L.; Vedsø, Bergtrup, M. *J. Org. Chem.* **1999**, *64*, 4196–8.
51. Ennis, D. S.; Gilchrist, T. L. *Tetrahedron Lett.* **1990**, *46*, 2623–32.
52. Gronowitz, S. in *Organic Sulphur—Structure, Mechanism, and Synthesis*; Sterling, C. J. M. Ed.; Butterworths: London, **1975**, pp. 203–28.
53. Moses, P.; Gronowitz, S. *Arkiv. Kemi.* **1961**, *18*, 119.
54. Takahashi, K.; Sakai, T. *Chem. Lett.* **1993**, 157–60.
55. Betzemeier, B.; Knochel, P. *Angew. Chem., Int. Ed. Eng.* **1997**, *36*, 2623–4.
56. Ribereau, P.; Pasteur, P. *Bull. Soc. Chim. Fr.* **1969**, 2076–9.
57. Brandão, M. A.; de Oliveira, A. B.; Snieckus, V. *Tetrahedron Lett.* **1993**, *34*, 2437–40
58. Wu, X.; Rieke, R. D. *J. Org. Chem.* **1995**, *60*, 6658–9.
59. Rieke, R. D.; Kim, S.-H.; Wu, X. *J. Org. Chem.* **1997**, *62*, 6921–7.
60. Dai, C.; Fu, G. C. *J. Am. Chem. Soc.* **2001**, *123*, 2719–24.
61. Miyasaka, M.; Rajca, A. *Synlett* **2004**, 177–81.
62. Boas, U.; Dhanabalan, A.; Greve, D.; Meiijer, E. W. *Synlett* **2001**, 634–6.
63. (a) Tamao, K.; Nakamura, K.; Ishii, H.; Yamaguchi, S.; Shiro, M. *J. Am. Chem. Soc.* **1996**, *118*, 12469–70. Also see: (b) Negishi, E.; Luo, F. T.; Frisbee, R.; Matsushita, H. *Heterocycles* **1982**, *18*, 117. (c) Minato, A.; Suzuki, K.; Tamao, K.; Kumada, M. *J Chem. Soc., Chem. Comm.* **1984**, 511. (d) Minato, A.; Suzuki, K.; Tamao, K. *J. Am. Chem. Soc.* **1987**, *109*, 1257–8.
64. Prim, D.; Kirsch, G. *J. Chem. Soc., Perkin Trans. 1* **1994**, 2603–6.
65. (a) Murata, M.; Oyama, T.; Watanabe, S.; Masuda, Y. *J. Org. Chem.* **2000**, *65*, 164–8; (b) Murata, M.; Watanabe, S.; Masuda, Y. *J. Org. Chem.* **1997**, *62*, 6458–9.
66. Giroux, A.; Han, Y.; Prasit, P. *Tetrahedron Lett.* **1997**, *38*, 3841–4.
67. Christophersen, C.; Begtrup, M.; Ebdrup, S.; Petersen, H.; Vedsø, P. *J. Org. Chem.* **2003**, *68*, 9513–16.
68. Molander, G. A.; Figueroa, R. *Aldrichimica Acta* **2005**, *38*, 49–56.
69. Molander, G. A.; Biolatto, B. *Org. Lett.* **2002**, *4*, 1867–70.
70. Wellmar, U.; Hörnfeldt, A.-B.; Gronowitz, S. *J. Heterocycl. Chem.* **1995**, *32*, 1159–63.
71. Peters, D.; Hörnfeldt, A.-B.; Gronowitz, S. *J. Heterocycl. Chem.* **1990**, *27*, 2165–73.
72. Malm, J.; Rehn, B.; Hörnfeldt, A.-B.; Gronowitz, S. *J. Heterocycl. Chem.* **1994**, *31*, 11–15.
73. Nettekoven, M.; Püllmann, B.; Schmitt, S. *Synthesis* **2003**, 1649–52.
74. Charette, A. B.; Giroux, A. *J. Org. Chem.* **1996**, *61*, 8718–9.
75. Hark, R. R.; Hauze, D. B.; Petrovskaia, O.; Joullié, M. M.; Jaouhari, R.; McComiskey, P. *Tetrahedron Lett.* **1994**, *35*, 7719–22.
76. Kevin, N. J.; Rivero, R. A.; Greenlee, W. J.; Chang, R. S. L.; Chen, T. B. *Biorg. Med. Chem. Lett.* **1994**, *4*, 189–94.
77. Jones, K.; Keenan, M.; Hibbert, F. *Synlett* **1996**, 509–10.

78. Malapel-Andrieu, B.; Mérour, J.-Y. *Tetrahedron* **1998**, *54*, 11079–94.

79. Mouaddib, A.; Joseph, B.; Hasnaoui, A.; Mérour, J.-Y. *Synthesis* **2000**, 549–56.

80. Lepifre, F.; Buon, C.; Rabot, R.; Bouyssou, P.; Coudert, G. *Tetrahedron Lett.* **1999**, *40*, 6373–6.

81. Haddach, M.; McCathy, J. R. *Tetrahedron Lett.* **1999**, *40*, 3109–12.

82. Ye, X.-S.; Wong, H. N. C. *J. Org. Chem.* **1997**, *62*, 1940–54.

83. Andersen, N. G.; Maddaford, S. P.; Keay, B. A. *J. Org. Chem.* **1996**, *61*, 9556–9.

84. Molander, G. A.; Biolatto *J. Org. Chem.* **2003**, *68*, 4302–14.

85. Barder, T. E.; Buchwald, S. L. *Org. Lett.* **2004**, *6*, 2649–52.

86. Molander, G. A.; Petrillo, D. E.; Landzberg, N. K.; Rohanna, J. C.; Biolatto, B. *Synlett* **2005**, 1763–6.

87. Bussolari, J. C.; Rehborn, D. C. *Org. Lett.* **1999**, *1*, 965–7.

88. Blettner, C.; König, W. A.; Rühter, G.; Stenzel, W.; Schotten, T. *Synlett* **1999**, 307–10.

89. Littke, A. F.; Dai, C.; Fu, G. C. *J. Am. Chem. Soc.* **2000**, *122*, 4020–8.

90. (a) Feuerstein, M; Doucet, H.; Santelli, M. *Tetrahedron Lett.* **2001**, *42*, 5659–62. (b) Kondolff, I.; Doucet, H.; Santelli, M. *Synlett* **2005**, 2057–61.

91. (a) Yamamoto, S.; Matsuda, K.; Irie, M. *Chem Eur. J.* **2003**, *9*, 4878–86. (b) Kodani, T.; Matsuda, K.; Yamada, T.; Kobatake, S.; Irie, M. *J. Am. Chem. Soc.* **2000**, *122*, 9631–7. (c) Bussolari, J. C.; Rehborn, D. C. *Organic Lett.* **1999**, *1*, 965–7. (d) Dallemagne, P.; Khanh, L. P.; Alsaïdi, A.; Varlet, I.; Collot, V.; Paillet, M.; Bureau, R.; Rault, S. *Bioorg. Med. Chem. Lett.* **2003**, *13*, 1161–7. (e) Gallant, M.; Belley, M.; Carrière; Chateauneuf, A.; Denis, D.; Lachance, N.; Lamontagne, S.; Metters, M. M.; Sawyer, N.; Slipetz, D.; Truchon, J. F.; Labelle, M. *Bioorg. Med. Chem. Lett.* **2003**, *13*, 3813–16.

92. Heynderickx, A.; Samat, A.; Guglielmetti, R. *Synthesis* **2002**, 213–16.

93. Miyaura, N.; Suzuki, A. *Chem. Rev.* **1995**, *95*, 2457–83.

94. Pereira, R; Iglesias, B; de Lera, A. R. *Tetrahedron* **2001**, *57*, 7871–81.

95. Davies, A. G. *Organotin Chemistry*; VCH, Weinheim, FRG, **1997**, 329 pp.

96. Crisp, G. T. *Synth. Comm.* **1989**, *19*, 307–16.

97. Zhang, Y.; Hörnfeldt, A-B.; Gronowitz, S. *Synthesis* **1989**, *2*, 130–1.

98. Prim, D.; Kirsch, G. *J. Chem. Soc., Perkin Trans. 1* **1994**, 2603–6.

99. Zhang, Y.; Hörnfeldt, A.-B.; Gronowitz, S. *J. Heterocycl. Chem.* **1995**, *32*, 771–7.

100. Gronowitz, S.; Hörnfeldt, A.-B.; Yang, Y. *Chem. Scrip.* **1988**, *28*, 275–9.

101. Malm, J.; Hörnfeldt, A.-B.; Gronowitz, S. *Heterocycles* **1993**, *35*, 245–62.

102. Tamao, K.; Yamaguchi, S.; Shiozaki, M.; Nakagawa, Y.; Ito, Y. *J. Am. Chem. Soc.* **1992**, *114*, 5867–9.

103. Hucke, A.; Cava, M. P. *J. Org. Chem.* **1998**, *63*, 7413–17.

104. Rossi, R.; Carpita, A.; Ciofalo, M.; Lippolis, V. *Tetrahedron* **1991**, *47*, 8443–60.

105. Björk, P.; Aekermann, T.; Hörnfeldt, A-B.; Gronowitz, S. *J. Heterocycl. Chem.* **1995**, *32*, 751–4.

106. Otsubo, T.; Kono, Y.; Hozo, N.; Miyamoto, H.; Aso, Y.; Ogura, F.; Tanaka, T.; Sawada, M. *Bull. Chem. Soc. Jpn.* **1993**, *66*, 2033–41.

107. Gronowitz, S.; Yang, Y.; Hörnfeldt, A-B. *Acta Chem. Scand.* **1992**, *46*, 654–60.

108. Yoshida, S.; Kubo, H.; Saika, T.; Katsumura, S. *Chem. Lett.* **1996**, 139–40.

109. Kosugi, M.; Shimizu, K.; Ohtani, A.; Migita, T. *Chem. Lett.* **1981**, 829–30.

110. Kosugi, M.; Ohta, T.; Migita, T. *Bull. Chem. Soc. Jpn.* **1983**, *56*, 3855–6.

111. Rivero, R. A.; Kevin, N. J.; Allen, E. E. *Biorg. Med. Chem. Lett.* **1993**, *3*, 1119–24.

112. Sanfilippo, P. J.; McNally, J. J.; Press, J. B.; Fitzpatrick, L. J.; Urbanski, M. J.; Katz, L. B.; Giardino, E.; Falotico, R.; Salata, J.; Moore, Jr., J. B.; Miller, W. *J. Med. Chem.* **1992**, *35*, 4425–33.

113. Wigerinck, P.; Kerremans, L.; Claes, P.; Snoeck, R.; Maudgal, P.; De Clercq, E.; Herdewijn, P. *J. Med. Chem.* **1993**, *36*, 538–43.

114. Park, H.; Lee, J. Y.; Lee, Y. S.; Park, J. O.; Koh, S. B.; Ham, W-H. *J. Antibiotics* **1994**, *47*, 606–8.

115. Sakamoto, T.; Kondo, Y.; Yasuhara, A.; Yamanaka, H. *Heterocycles* **1990**, *31*, 219–21.

116. Sakamoto, T.; Kondo, Y.; Yasuhara, A.; Yamanaka, H. *Tetrahedron* **1991**, *47*, 1877–86.

117. Crisp, G.; Glink, P. T. *Tetrahedron* **1994**, *50*, 3213–34.

118. Johannes, H.-H.; Grahn, W.; Reisner, A.; Jones, P. G. *Tetrahedron Lett.* **1995**, *36*, 7225–8.

119. Corriu, R. J. P.; Bolin, G.; Moreau, J. J. E. *Tetrahedron Lett.* **1991**, *32*, 4121–4.

120. Corriu, R. J. P.; Geng, B.; Moreau, J. J. E. *J. Org. Chem.* **1993**, *58*, 1443–8.

121. Hollingworth, G. J.; Sweeney, J. B. *Tetrahedron Lett.* **1992**, *33*, 7049–52.

122. Bellina, F.; Carpita, A.; De Santis, M.; Rossi, R. *Tetrahedron* **1994**, *50*, 12029–46.

123. Liebeskind, L. S.; Fengl, R. W. *J. Org. Chem.* **1990**, *55*, 5359–64.

124. Nodvall, G.; Sundquist, S.; Nilvebrant, L.; Hacksell, U. *Biorg. Med. Chem. Lett.* **1994**, *4*, 2837–40.

125. Buon, C.; Bouyssou, P.; Coudert, G. *Tetrahedron Lett.* **1999**, *40*, 701–2.

126. May, P. D.; Larsen, S. D. *Synlett* **1997**, 895–6.

127. Aidhen, I. S.; Braslau, R. *Synth. Comm.* **1994**, *24*, 789–97.

128. Sakamoto, T.; Yasuhara, A.; Kondo, Y.; Yamanaka, H. *Chem. Pharm. Bull.* **1994**, *42*, 2032–5.

129. Ye, X.-S.; Wong, H. N. C. *Chem. Commun.* **1996**, 339–40.

130. Stille, J. K.; Simpson, J. H. *J. Am. Chem. Soc.* **1987**, *109*, 2138–52.

131. Bailey, T. R. *Tetrahedron Lett.* **1986**, *27*, 4407–10.

132. Roth, G. P.; Farina, V. *Tetrahedron Lett.* **1995**, *36*, 2191–4.

133. Liebeskind, L. S.; Cabrera-Peña, E. *Organic Syn.* **2004**, *Coll. Vol 10*, 9–12.

134. Kennedy, G.; Perboni, A. D. *Tetrahedron Lett.* **1996**, *37*, 7611–14.

135. Kang, S.-K.; Lee, H.-W.; Jang, S.-B.; Kim, T.-H.; Kim, J.-S. *Synth. Commun.* **1996**, *26*, 4311–18.

136. Kang, S.-K.; Ryu, H.-C.; Choi. S.-C. *Chem. Commun.* **1998**, *12*, 1317–18.

137. Kang, S.-K.; Lim, K.-H.; Ho, P.-S.; Yoon, S.-K.; Son, H.-J. *Synth. Commun.* **1998**, *28*, 1481–9.

138. Su, W.; Urgaonkar, S.; McLaughlin, P. A.; Verkade, J. G. *J. Am. Chem. Soc.* **2004**, *126*, 16433–9.

139. Li, J.-H.; Liang, Y.; Wang, D.-P.; Liu, W.-J.; Xie, Y.-X; Yin, D.-L. *J. Org., Chem.* **2005**, *70*, 2832–4.

140. Pridgen, L. N.; Killmer, L. B. *J. Org. Chem.* **1981**, *46*, 5402–4.

141. Dondoni, A.; Fogagnolo, M.; Fantin, G.; Medici, A.; Pedrini, P. *Tetrahedron Lett.* **1986**, *27*, 5269–70.

142. (a) Meyers, A. I.; Novachek, K. A. *Tetrahedron Lett.* **1996**, *37*, 1747–8. (b) Dondoni, A.; Fantin, G.; Fogagnolo, M.; Medici, A.; Pedrini, P. *Synthesis* **1987**, 693–6.

143. Kitamura, C.; Tanaka, S.; Yamashita, Y. *J. Chem. Soc., Chem. Commun.* **1994**, 1585–6.

144. Folli, U.; Iarossi, D.; Montorsi, M.; Mucci, A.; Schenetti, L. *J. Chem. Soc., Perkin Trans. 1* **1995**, 537–40.

145. Barbarella, G.; Zambianchi, M.; Sotgiu, G.; Bongini, A. *Tetrahedron* **1997**, *53*, 9401–6.

146. Iyoda, M.; Kuwatani, Y.; Ueno, N.; Oda, M. *J. Chem. Soc., Perkin Trans. 1* **1992**, 158–9.

147. Takahashi, K.; Nihira, T. *Bull. Chem. Soc. Jpn.* **1992**, *65*, 1855–9.

148. Takahashi, K.; Nihira, T.; Akiyama, K.; Ikegami, Y.; Fukuyo, E. *J. Chem. Soc., Chem. Commun.* **1992**, 620–2.

149. Gronowitz, S.; Björk, P.; Malm, J.; Hörnfeldt, A-B. *J. Organomet. Chem.* **1993**, *460*, 127–9.

150. Zhang, Y.; Hörnfeldt, A.-B.; Gronowitz, S. *J. Heterocycl. Chem.* **1995**, *32*, 435–44.

151. Hervet, M.; Théry, I.; Gueiffier, A.; Enguehard-Gueiffier, C. *Helv. Chim. Acta* **2003**, *86*, 3461–9.

152. Li, J. J.; Yue, W. S. *Tetrahedron Lett.* **1999**, *40*, 4507–10.

153. Mee, S. P. H.; Lee, V.; Baldwin, J. E. *Chem. Eur. J.* **2005**, *11*, 3294–308.

154. Wolf, C.; Lerebours, R. *J. Org. Chem.* **2003**, *68*, 7077–84.

155. Yamamura, K.; Kusuhara, N.; Kondou, A.; Hashimoto, M. *Tetrahedron* **2002**, *58*, 7653–61.

156. Iyoda, M.; Miura, M.; Sasaki, S.; Kabir, S. M. H.; Kuwatani, Y.; Yoshida, M. *Tetrahedron Lett.* **1997**, *38*, 4581–2.

157. Zhang, Y.; Hörnfeldt, A.-B.; Gronowitz, S. *Synthesis* **1989**, *2*, 130–1.

158. Malm, J.; Hörnfeldt, A.-B.; Gronowitz, S. *Tetrahedron Lett.* **1992**, *33*, 2199–202.

159. Allred, G. D.; Liebeskind, L. S. *J. Am. Chem. Soc.* **1996**, *118*, 2748–9.

160. Paterson, I.; Lombart, H.-G.; Allerton, C. *Org. Lett.* **1999**, *1*, 19–22.

161. Hatanaka, Y.; Fukushima, S.; Hiyama, T. *Heterocycles* **1990**, *30*, 303–6.

162. Nguefack, J.-F.; Bolitt, V.; Sinou, D. *Tetrahedron Lett.* **1996**, *37*, 5527–30.

163. John, J. A.; Tour, J. M. *Tetrahedron* **1997**, *53*, 15515–34.

164. Walser, A.; Flynn, T.; Mason, C.; Crowley, H.; Maresca, C.; O'Donnell, M. *J. Med. Chem.* **1991**, *34*, 1440–6.

165. Ye, X.-S.; Wong, H. N. C. *Chem. Commun.* **1996**, 339–40.

166. Rossi, R.; Carpita, A.; Lezzi, A. *Tetrahedron* **1984**, *40*, 2773–9.

167. D'Auria, M. *Synth. Commun.* **1992**, *22*, 2393–9.

168. Carpita, A.; Rossi, R. *Gazz. Chim. Ital.* **1985**, *115*, 575–83.

169. Feuerstein, M.; Doucet, H.; Santelli, M. *Tetrahedron Lett.* **2005**, *46*, 1717–20.

170. Erdélyi, M.; Gogoll, A. *J. Org. Chem.* **2001**, *66*, 4165–9.

171. Kundu, N. G.; Nandi, B. *J .Org. Chem.* **2001**, *66*, 4563–75.

172. Tamaru, Y.; Yamada, Y.; Yoshida, Z.-I. *Tetrahedron* **1979**, *35*, 329–40.

173. Arnau, N.; Moreno-Mañãs, M.; Pleixats, R. *Tetrahedron* **1993**, *49*, 11019–28.

174. Littke, A. F.; Fu. G. C. *J. Am. Chem. Soc.* **2001**, *123*, 6989–7000.

175. (a) Berthiol, F.; Feuerstein, M.; Doucet, H.; Santelli, M. *Tetrahedron Lett.* **2002**, *43*, 5625–8. (b) Lemhardi, M.; Doucet, H.; Santelli, M. *Tetrahedron* **2004**, *60*, 11533–40.

176. Ohta, A.; Akita, Y.; Ohkuwa, T.; Chiba, M.; Fukunaka, R.; Miyafuji, A.; Nakata, T.; Tani, N. Aoyagi, Y. *Heterocycles* **1990**, *31*, 1951–7.

177. Aoyagi, Y.; Inoue, A.; Koizumi, I.; Hashimoto, R.; Tokunaga, K.; Gohma, K.; Komatsu, J.; Sekine, K.; Miyafuji, A.; Konoh, J. Honma, R. Akita, Y.; Ohta, A. *Heterocycles* **1992**, *33*, 257–72.

178. (a) Lavenot, L.; Gozzi, C.; Ilg, K.; Orlova, I.; Penalva, V.; Lemaire, M. *J. Organomet. Chem.* **1998**, *567*, 49–55. (b) Chabert, J. F. D.; Gozzi, C.; Lemaire, M. *Tetrahedron Lett.* **2002**, *43*, 1829–33. (c) Chabert, J. F. D.; Joucla, L.; David, E.; Lemaire, M. *Tetrahedron* **2004**, *60*, 3221–30. (d) David, E.; Perrin, J.; Pellet-Rostaing, S.; Chabert, J. F. D.; Lemaire, M. *J. Org. Chem.* **2005**, *70*, 3569–73.

179. Pigeon, P.; Decroix, B. *Tetrahedron Lett.* **1996**, *37*, 7707–10.

180. Carpentier, J.-F.; Castanet, Y.; Brocard, J.; Mortreux, A.; Petit, F. *Tetrahedron Lett.* **1991**, *32*, 4705–8.

181. Lin, Y.-S.; Yamamoto, A. *Tetrahedron Lett.* **1997**, *38*, 3747–50.

182. Hatanaka, Y.; Fukushima, S.; Hiyama, T. *Tetrahedron* **1992**, *48*, 2113–26.

183. Kang, S.-K.; Lim, K.-H.; Ho, P.-S.; Yoon, S.-K.; Son, H.-J. *Synth. Commun.* **1998**, *28*, 1481–9.

184. Hooper, M. W.; Utsonomiya, M.; Hartwig, J. F. *J. Org. Chem.* **2003**, *68*, 2861–73.

185. Watanabe, M.; Yamamoto, T.; Nishiyama, M. *Chem. Commun.* **2000**, 133–4.

186. (a) Luker, T. J.; Beaton, H. G.; Whiting, M.; Mete, A.; Cheshire, D. R. *Tetrahedron Lett.* **2000**, *41*, 7731–5. (b) Ogawa, K.; Radke, K. R.; Rothstein, S. D.; Rasmussen, S. C. *J. Org. Chem.* **2001**, *66*, 9067–70. (c) Hesse, S.; Queiroz, M.-J. R. P.; Kirsch, G. *Synthesis* **2005**, 2373–8.

187. (a) Ferreira, I. C. F. R.; Queiroz, M.-J. R. P.; Kirsch, G. *Tetrahedron* **2003**, *59*, 975–81. (b) Queiroz, M.-J. R. P.; Begouin, A.; Ferreira, I. C. F. R.; Kirsch, G.; Calhelha, R. C.; Barbosa, S.; Estevinho, L. M. *Eur. J. Org. Chem.* **2004**, 3679–85.

188. Hirao, T.; Masunaga, T.; Ohshiro, Y.; Agawa, T. *Tetrahedron Lett.* **1980**, *21*, 3595–8.

189. Hirao, T.; Masunaga, T.; Ohshiro, Y.; Agawa, T. *Synthesis* **1981**, 56–7.

190. Hirao, T.; Masunaga, T.; Yamada, Y.; Ohshiro, Y.; Agawa, T. *Bull. Chem. Soc. Jpn.* **1982**, *55*, 909–13.

191. Xu, Y.; Li, Z.; Xia, J.; Guo, H.; Huang, Y. *Synthesis* **1983**, 377–8.

192. Xu, Y.; Zhang, J. *Synthesis* **1984**, 778–80.

193. Gabriele, B.; Salerno, G.; Fazio, A. *Org. Lett.* **2000**, *2*, 351–2.

Chapter 6

Furans and benzo[*b*]furans

Kevin M. Shea

Furan-containing molecules are found in both natural products and pharmaceuticals [1]. At one time, furfural was produced in great quantities from corncobs. Perillene, a secondary plant metabolite, is an example of naturally occurring furan. Furan is frequently used as a bioisostere of a benzene ring in medicinal chemistry. For example, Ranitidine (Zantac) marketed by Glaxo for treatment of ulcers and gastric acid reflux was one of the first "blockbuster drugs" with annual sales of over 1 billion dollars.

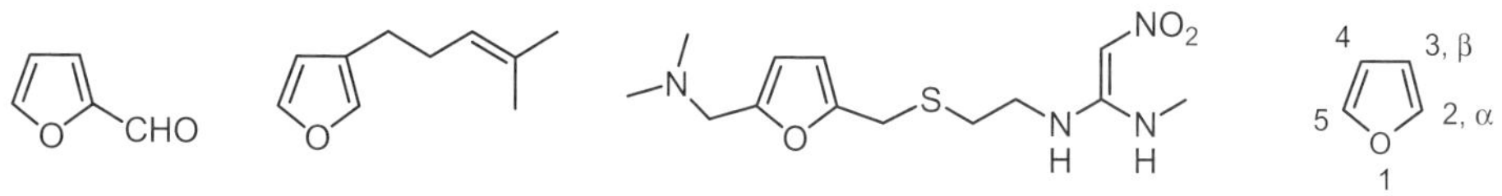

furfural   perillene   Ranitidine   furan

Furan is a π-electron-excessive heteroarene. Electrophilic substitution and metalation take place regioselectively at C(2) due to the electronegativity of the oxygen atom. In comparison to the sulfur atom in thiophene, the oxygen in furan has higher electronegativity. Therefore, the lone pair of electrons on the oxygen are less effectively incorporated into the aromatic system, which contributes to the pronounced difference of reactivity of the α- and β-positions [2]. In the context of palladium chemistry, furan and indole are the two classes of heterocycles that have attracted most attention. The synthesis of benzofurans and indoles using palladium chemistry has been a very active and prolific field. Contrary to all-carbon aryl halides, regioselective coupling reactions are routinely accomplished at C(2) for a 2,3-dihalofuran derivative. However, unlike 2-chloropyridines (an electron-deficient heterocycle), 2-chlorofurans are not sufficiently activated to add oxidatively to Pd(0).

6.1. Synthesis of halofurans and halobenzo[*b*]furans

6.1.1. Halofurans

Commercially available 3-bromofuran is prepared from furfural *via* the four-step sequence shown below [3, 4]. 3-Iodofuran can be synthesized in two steps by sequential deiodination of tetraiodofuran [5].

Simply treating furan in DMF with Br$_2$ at 20–30°C results in 70% of 2-bromofuran and 8% of 2,5-dibromofuran. Further treatment of 2-bromofuran with another equivalent of Br$_2$ at 30–40°C produces 2,5-dibromofuran in 48% yield [6, 7]. Alternatively, deprotonation of furan at the C(2) position using EtLi followed by quenching with Br$_2$ provides 2-bromofuran [8]. Similarly, quenching 3-lithiofuran, derived from halogen–metal exchange between 3-bromofuran and EtLi, with hexachloroethane yields 3-chlorofuran [9].

The strategies mentioned above, or slight modifications, are also applicable to the synthesis of a variety of substituted halofurans. For example, trisubstituted bromofuran 2 is available upon treatment of disubstituted furan 1 with bromine in DMF [10]. Bromination of 2-phenylfuran with bromine, sodium acetate, and acetic acid furnishes 5-bromo-2-phenylfuran in 77% yield [11]. NBS is another useful bromine source that leads to substitution in the 2 or 5 positions [12, 13]. However, when these positions are already occupied, NBS reacts to provide 3- or 4-bromo derivatives. A regioselective example involves treatment of disubstituted furan 3 with NBS to furnish exclusively trisubstituted bromide 4 [14].

Iodonation of furans can be accomplished by reaction of a lithiated furan with iodine. For example, 2-iodo-5-*n*-butylfuran is available from 2-*n*-butylfuran upon treatment with *n*-butyllithium followed by I$_2$ [15].

6.1.2. Halobenzofurans

2-Iodobenzofuran is readily prepared by lithiation of benzofuran followed by iodine quench [16]. Bromination of benzofuran with bromine under basic conditions yields 2,3-dibromobenzofuran, which, when subjected to a regioselective halogen/metal exchange at the C(2) position with *n*-BuLi followed by quenching with methanol, furnishes 3-bromobenzofuran [17]. 2,3-Dibromobenzofuran can also be easily prepared using a related 3-step sequence that proceeds in 75% overall yield from benzofuran [18]. Benzofurans already substituted in the 3-position yield 2-bromo derivatives upon treatment with bromine and potassium acetate [19].

A common strategy for the synthesis of 3-halobenzofurans involves a 2-step procedure of halogenation, then elimination. Benzofuran **5** reacts with bromine to yield tribromide **6** which furnishes dibromobenzofuran **7** upon treatment with potassium hydroxide [20]. Chlorination can be achieved *via* an analogous reaction sequence [21]. 2,3,5-Tribromobenzofuran can be prepared using a similar strategy in three steps and 52% overall yield from 5-bromobenzofuran (**5**) [18].

Nitrobenzofurans are readily converted to bromobenzofurans by phosphorus tribro-mide. For example, 2-nitrobenzofuran **8** is transformed into 2-bromobenzofuran **9** in 74% yield [22].

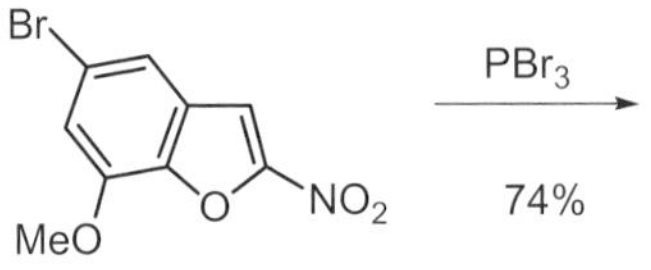

Literature is replete with synthetic methods to prepare 5-bromobenzofurans. One of the more practical syntheses [23, 24] commences with etherification of 4-bromophenol with bromoacetaldehyde diethyl acetal using either NaH in DMF or KOH in DMSO. Treatment of the resulting aryloxyacetaldehyde acetal with polyphosphoric acid (PPA) affords 5-bromobenzofuran in good yield *via* intramolecular cyclocondensation. However, cyclization of *m*-aryloxyacetaldehyde acetal **10** results in a mixture of two regioisomers, 6-bromobenzofuran (**11**) and 4-bromobenzofuran (**12**). Finally, 7-bro-mobenzofuran (**14**) can be prepared similarly using the intramolecular cyclocondensation of aryloxyacetaldehyde acetal **13** generated from etherification of 2-bromophenol with bromoacetaldehyde diethyl acetal.

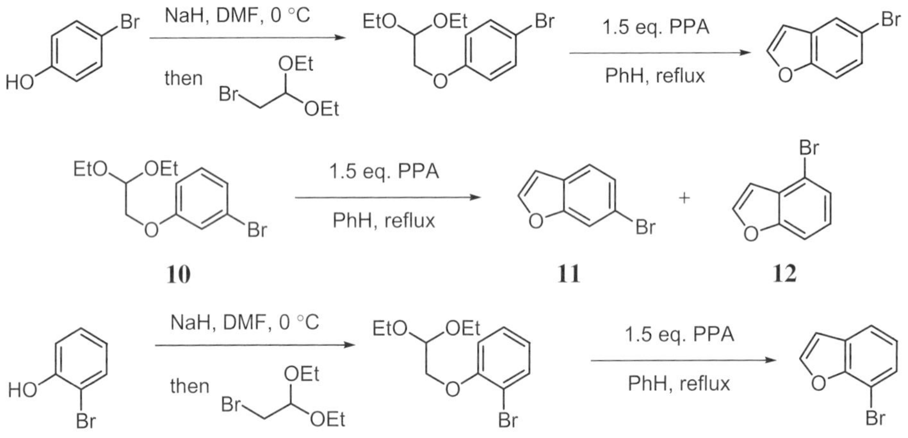

6.2. Oxidative coupling/cyclization

When furan or substituted furans are subjected to classic oxidative coupling conditions [Pd(OAc)$_2$ in refluxing HOAc], 2,2′-bifuran and 2,3′-bifuran are the major and minor products, respectively [25, 26]. Similar results are observed for the arylation of furans using Pd(OAc)$_2$ [27]. The oxidative couplings of furan or benzo[*b*]furan with olefins also proceed analogously [28]. These types of coupling reactions, however, have limited syn-thetic utility since they all consume at least one equivalent of palladium acetate.

One strategy for the synthesis of 2,2′-bifuran using a catalytic amount of palladium is illustrated below. Homocoupling of 2-tributylstannylfuran using 10 mol% of palladium acetate and two equivalents of copper(II) chloride as the reoxidant provides the target bifuran in 91% yield [29].

Tsuji's group coupled 2-methylfuran with ethyl acrylate to afford adduct **15** *via* a Pd-catalyzed Fujiwara–Moritani/oxidative Heck reaction using *tert*-butyl peroxybenzoate to reoxidize Pd(0) to Pd(II) [30]. The palladation of 2-methylfuran takes place at the electron-rich C(5) in a fashion akin to electrophilic aromatic substitution. The perbenzoate acts as a hydrogen acceptor.

An interesting application of the Fujiwara–Moritani/oxidative Heck reaction for the synthesis of benzofurans was recently reported by the Stoltz lab [31]. A variety of allyl phenyl ethers (all containing electron-rich aryl components) react with 10 mol% palladium acetate, 20 mol% ethyl nicotinate, 20 mol% sodium acetate, and one equivalent of benzoquinone at 100°C to provide benzofurans in 52–79% yield (e.g. **16**→**17**). The mechanism of this transformation begins with arene palladation of Pd(II) followed by olefin insertion, β-hydrogen elimination, and olefin isomerization to the thermodynamically favored benzofuran product. The resulting Pd(0) species is then oxidized to Pd(II) thus regenerating the active catalyst.

The phenolic oxygen on 2-allyl-4-bromophenol (**18**) readily undergoes oxypalladation using a catalytic amount of $PdCl_2$ and three equivalents of $Cu(OAc)_2$, to yield the corresponding benzofuran **19**. This process, akin to the Wacker oxidation, is catalytic in terms of palladium, and $Cu(OAc)_2$ serves as the oxidant [32]. Benzofuran **21**, a key intermediate in Kishi's total synthesis of aklavinone [33], was synthesized *via* the oxidative cyclization of phenol **20** using stoichiometric amounts of a Pd(II) salt.

20   **21**

6.3. Coupling reactions with organometallic reagents

Furan derivatives participate in the standard variety of cross-coupling reactions, as well as some more exotic transformations. Importantly, regioselective cross-couplings are well-known for dihalofurans and di- and trihalobenzofurans. The 2-position is the most electron-poor carbon in both furans and benzofurans. Thus, 2,3-, and 2,4-dihalofurans react *via* cross-coupling reactions exclusively at the 2-position. Similar selectivity is observed for di- and trihalofurans containing a halogen in the 2-position. Several recent reviews highlight applications of cross-coupling reactions involving furans [1, 34, 35].

6.3.1. Kumada coupling

The Kumada coupling is used sparingly for the functionalization of furans. Rossi and associates described one example in which 2-bromofuran reacts with 2-thienylmagnesium bromide to assemble thienylfuran **22** [36]. The reaction proceeds readily at room temperature in high overall yield.

22

6.3.2. Negishi coupling

The Negishi coupling is a popular method for the preparation of furan and benzofuran derivatives. The heteroaromatic component can be used as either the organozinc reagent or the halogen-containing coupling partner. Reactions in the 2-position are the most common, although functionalization at the 3-position is possible.

Treating 2-furyllithium (derived from deprotonation of furan with *n*-BuLi) with $ZnCl_2$ gives 2-furylzinc chloride, which then couples with 1,3-dibromobenzene to furnish bis-furylbenzene **23** [37, 38]. In addition, 2-furylzinc chloride couples with 4-iodobenzoic acid to yield adduct **24** without interference from the carboxylic acid [39]. Furylzinc halides also couple efficiently with vinyl bromides [40].

23

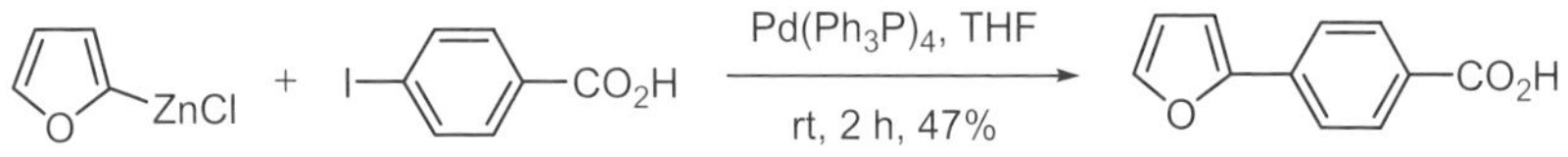

24

Snieckus *et al.* prepared the furan amide organozinc reagent **27** *via ortho*-lithiation of furyl amide **25** followed by reaction of the resulting 2-lithiofuran species **26** with $ZnBr_2$ [41]. The subsequent Negishi reaction of **27** and 2-bromotoluene generates phenylfuran **28**. Gilchrist's group carried out an *ortho*-lithiation using an oxazoline directing group [42]. Thus, directed *ortho*-lithiation of 2-furyloxazoline **29** with *s*-BuLi followed by treatment with $ZnBr_2$ provides 3-furylzinc bromide **30**, which was subsequently joined with aryl-, benzyl-, and vinyl halides to give the expected adducts such as **31**.

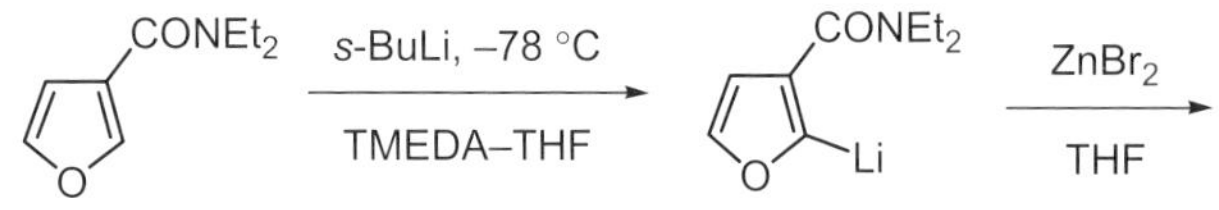

25   **26**

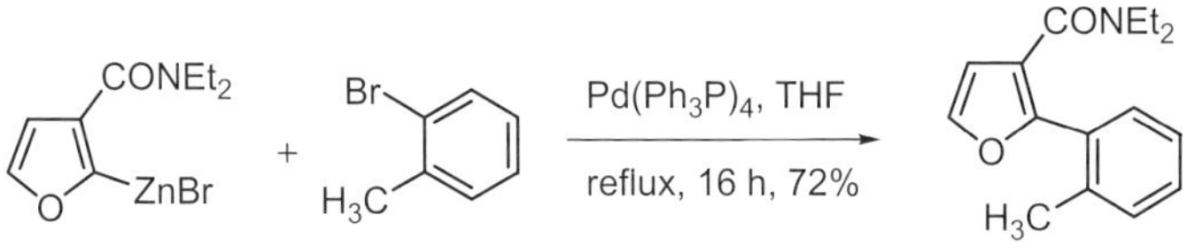

27   **28**

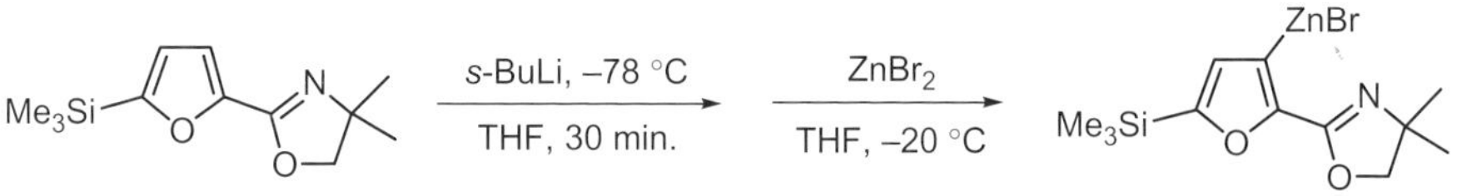

29   **30**

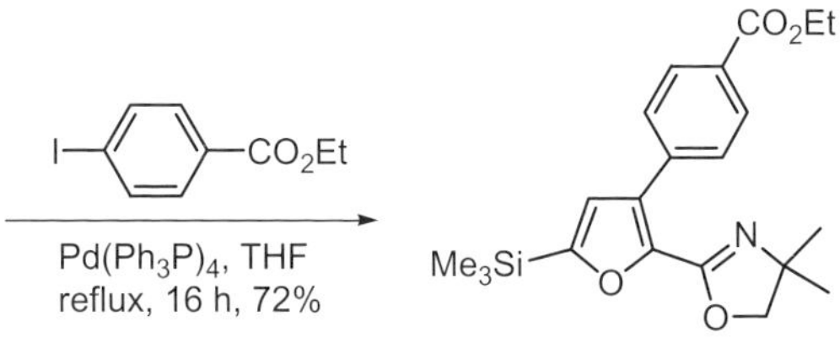

31

These couplings exemplify the functional group tolerance of the Negishi reaction conditions. Carboxylic acids, esters, amides, and even free anilines are compatible with the reaction conditions, as illustrated by the synthesis of furylaniline **32** [43].

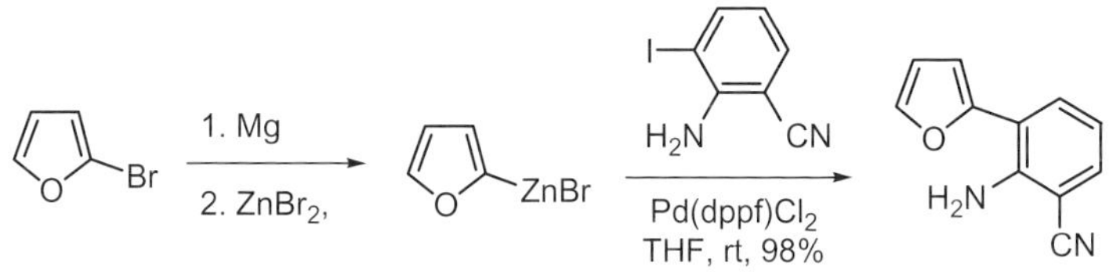

32

Both vinyl- and aryl triflates have been cross-coupled with 2-furylzinc chloride [44–46]. Since vinyl triflates are easily obtained from the corresponding ketones, they are useful substrates in Pd-catalyzed reactions. In the following example, a Negishi coupling of 2-furylzinc chloride and indol-5-yl triflate (**33**) provide an expeditious entry to 2-(5′-indolyl)furan (**34**). Protection of the NH in the indole ring is not required. A similar reaction was successful with pyridyl- and quinolinyl triflates.

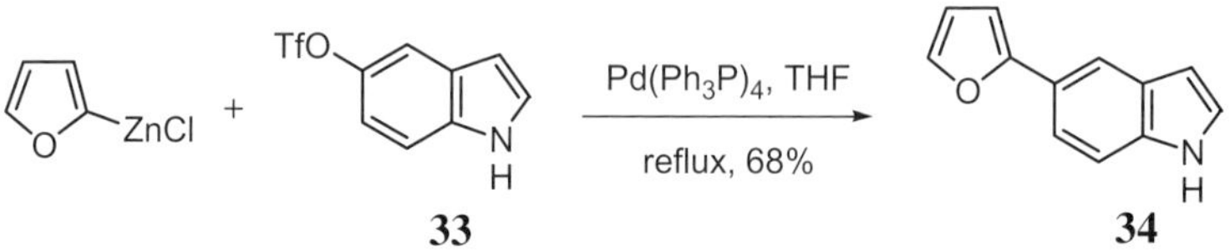

33   **34**

Furylzinc halides also undergo coupling with vinyl and aryl tellurides in a variation of the Negishi reaction. In one example, Z-vinylic telluride **35** reacts with 2-furylzinc chloride to furnish 2-vinylfuran **36** in 75% yield [47].

35   75%   **36**

Like ordinary aryl halides, furyl- and benzofuryl halides participate in Negishi couplings as nucleophiles. The electrophilic coupling partners range from aryl- [48], alkynyl- (**37→38**) [49, 50], vinyl- (**39→40**) [51, 52], to alkylzinc reagents [12, 53, 54]. For example, with the aid of ultrasonic activation, organozinc reagent **42** was generated from the protected β-iodoalanine derivative **41** and was then effectively coupled *in situ* with 2-furoyl chloride to give enantiomerically pure protected 4-oxo-α-amino acid **43** [53].

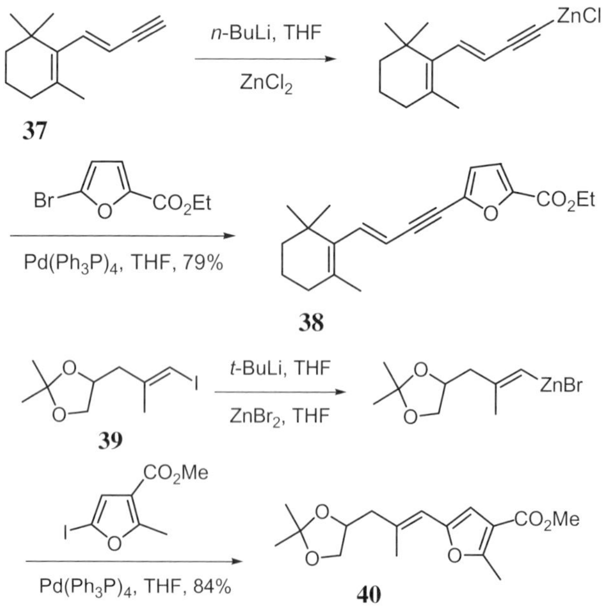

37

38

39

40

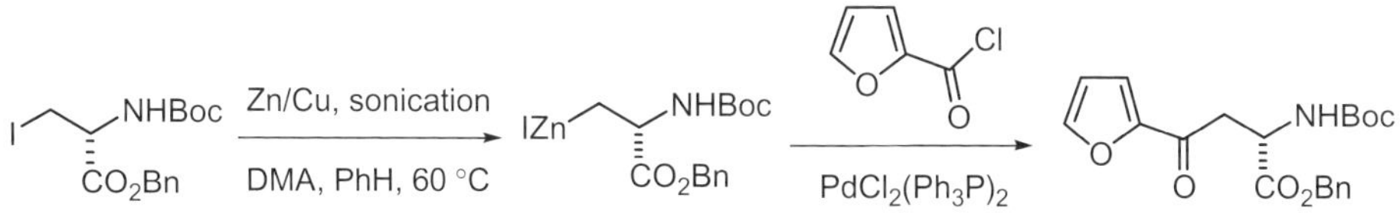

41 → **42** → **43**

It is also possible to use the Negishi coupling for the polymerization of dibromofurans. First, 3-octylfuran is converted to 2,5-dibromo-3-octylfuran (**44**) upon addition of bromine. Treatment of this material with Rieke zinc followed by tetrakis provides a 40% yield of poly(3-octylfuran) (**45**) [55].

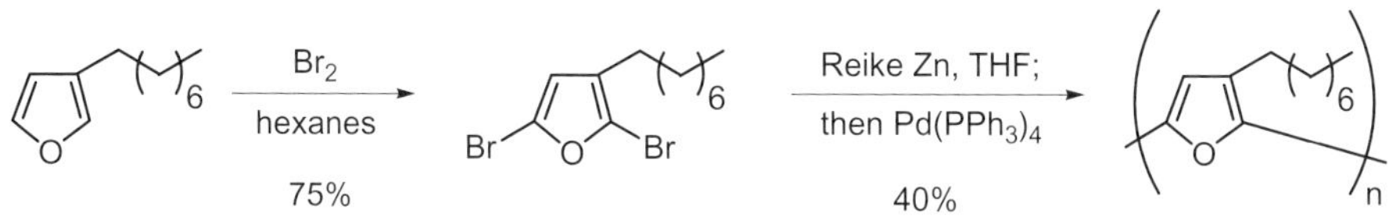

44 → **45**

Unlike normal all-carbon aryl halides, regioselective Negishi couplings of dihalofurans and trihalobenzofurans are synthetically useful. Bach and coworkers have studied both systems extensively and use a variety of cross-coupling reactions to produce functionalized furans and benzofurans. In one example of a selective Negishi reaction, 2,3-dibromofuran **46** couples with 2-furylzinc chloride at C(2) to afford bisfuran **47** [56, 57]. A related compound, bromide **48**, was prepared by a regioselective Sonogashira coupling of **46** (see Section 6.4) and then subjected to Negishi conditions with methylzinc chloride to provide trisubstituted furan **49**.

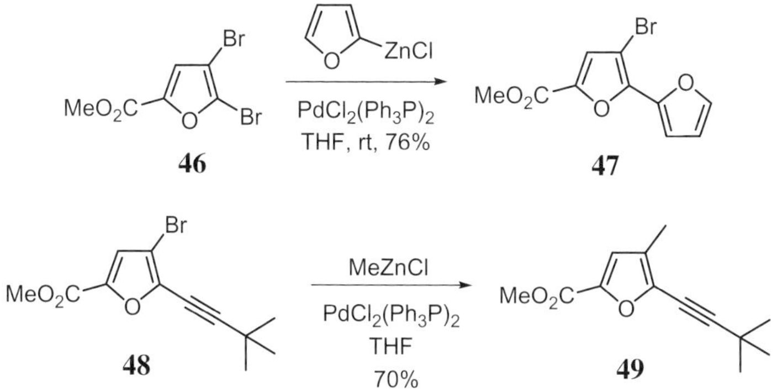

46 → **47**

48 → **49**

The Bach group can also perform selective reactions starting with 2,3,5-tribromobenzofuran (**50**). Dibromide **51** is available *via* a Negishi coupling of **50** with the appropriate arylzinc chloride. A nickel-catalyzed Kumada coupling selectively functionalizes the 5-position to yield bromide **52**. The final substituent is installed by another Negishi coupling, this time with methylzinc chloride to afford trisubstituted furan **53** [18, 58, 59].

6.3.3. Suzuki coupling

The Suzuki reaction is an excellent method for the functionalization of furans and ben-zofurans. The aromatic groups have served as both the boron component [60] and the halide partner in a large number of reactions. A variety of furan and benzofuran substi-tution patterns are accessible *via* Suzuki couplings [1, 34].

6.3.3.1. Furylboronic acids

Furylboronic acid, behaving like a simple arylboronic acid, has been coupled with a myriad of oragnohalides and triflates. The coupling between 2-furylboronic acid and bromopyridine **54** produces furylpyridine **55** [61], while reaction with an aryl iodide smoothly provides a 2-aryl substituted furan [62]. Triflates like furanone **56** readily couple with 2-furylboronic acid to provide monosubstituted furan products (e.g. **57**) [63, 64].

A variety of 5-aryl- and 5-heteroaryl-2-furaldehydes are available in one step starting with 5-(diethoxymethyl)-2-furylboronic acid (**58**). For example, thiophene substituted furaldehyde **60** is generated by Suzuki coupling of **58** and bromothiophene **59** followed by an acidic workup [65].

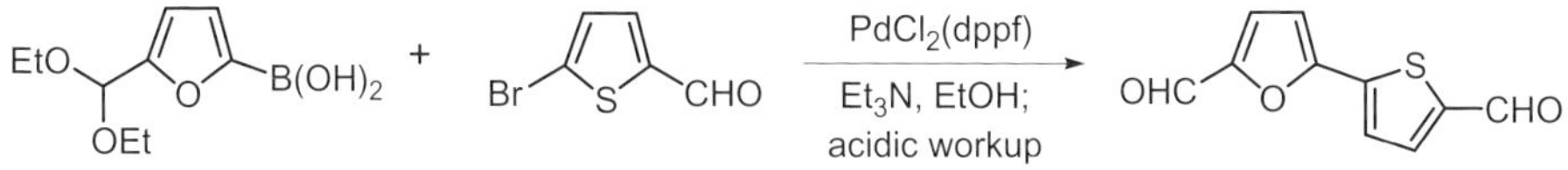

58   **59**   90%   **60**

Adapting Gribble's method for synthesizing indol-3-yl triflate [66], Mérour *et al.* converted 2-formyl-1-(phenylsulfonyl)-1*H*-indole (**61**) to indol-2-yl triflate **62** in two steps. Triflate **62** was subsequently coupled with benzofuryl-2-boronic acid to furnish 2-benzofurylindole **63** [67, 68].

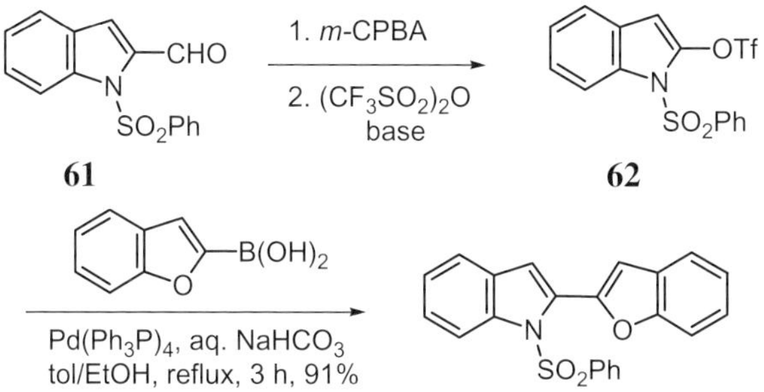

61   **62**

63

Suzuki couplings of 3-furylboronic acids are less common than the 2-substituted derivatives; however, this is a useful strategy for the formation of a carbon–carbon bond at the 3-position. Several 3-arylfurans are available *via* the coupling of aryl bromides with 3-furylboronic acids. For example, bromoaniline **64** reacts to yield furan **66** in the presence of Pd catalyst **65** [69]. In a different report, 2-bromoacetaniline (**67**) couples with 2-formyl-3-furylboronic acid (**68**) [70]. The resulting Suzuki coupling adduct undergoes a spontaneous cyclization, forming tricyclic furo[2,3-*c*]quinoline **69**.

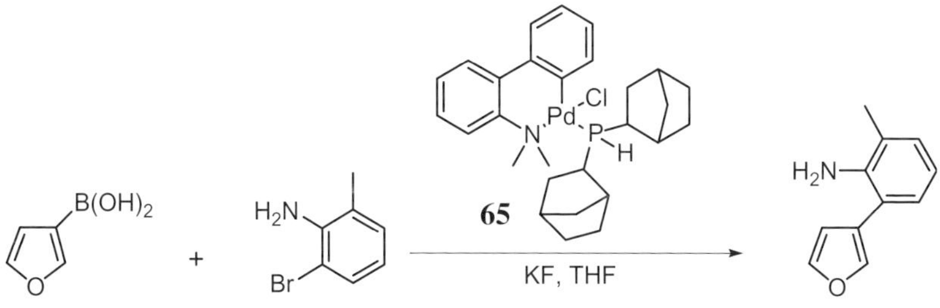

64   **65**   67%   **66**

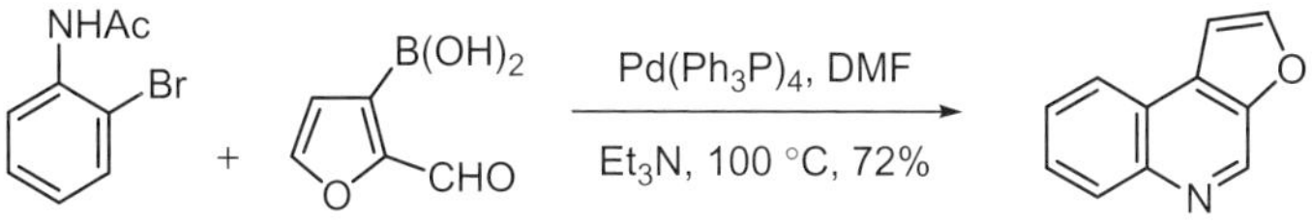

67   **68**   **69**

Wong and coworkers have developed an excellent procedure for the synthesis of *tris*(4-trimethylsilylfuran-3-yl)boroxine (**70**) and its use in Suzuki couplings. A recent example from their lab highlights the use of **70** in a reaction with highly substituted allyl bromide **71**. The product of this reaction, disubstituted furan **72**, is a synthetic intermediate in a model study towards the synthesis of furanoeudesmanes [71].

6.3.3.2. Furans as electrophiles

Like simple aryl halides, furyl halides take part in Suzuki couplings as electrophiles [72, 73]. Young and Martin coupled 2-bromofuran with 5-indolylboronic acid to prepare 5-substituted indole **73** [74]. Terashima's group cross-coupled 3-bromofuran with diethyl-(4-isoquinolyl)borane (**74**) to make 4-substituted isoquinoline **75** [75]; 2- and 3-substituted isoquinolines were synthesized in the same fashion [76].

Vachal and Toth recently described a general strategy for the synthesis of 2,5-diaryl-furans that involves a Stille coupling (see Section 6.3.4) of 2-tributylstannylfuran with phenyl iodide, bromination, and a Suzuki coupling with an aryl boronic acid. In one case, diarylfuran **76** is produced in 65% overall yield after the three-step sequence [11, 77]. A series of furan[3,2-*b*]pyrroles can be prepared using a multistep sequence involving two Suzuki couplings between bromofurans and arylboronic acids [78]. Several examples of disubstituted furans with potential medicinal applications have been prepared using Suzuki reactions of halofurans with arylboronic acids [12, 79, 80].

Halofurans, like other aryl halides, participate in B-alkyl Suzuki reactions. Kim's group successfully coupled iodofuran **77** with alkyl 9-BBN derivative **78** to provide disubstituted furan **79** in 95% yield [15].

Unlike simple aryl halides, a regioselective Suzuki coupling of 2,4-dibromofuran may be achieved at C(2) [81]. The coupling between 2,4-dibromofuran and pyrimidylboronic acid **80** provides furylpyrimidine **81**, which can then be hydrolyzed to 5-substituted uracil **82**. Selectivity is also possible for reactions of 4,5-dibromo-2-furaldehyde. Schreiber and coworkers recently reported the synthesis of a variety of bis-functionalized furaldehydes using two sequential Suzuki couplings. Starting with resin-bound alkene **83**, they first perform a hydroboration with 9-BBN followed by a regioselective Suzuki coupling at the 5-position of 4,5-dibromo-2-furaldehyde to yield bromofuran **84**. A second Suzuki reaction with seven different arylboronic acids furnishes the trisubstituted furan products **85** [10].

Furostifoline (90), a furo[3,2-a]carbazole, was isolated from *Murraya euchrestifolia*. Timári's total synthesis of 90 commences with alkylation of bromocresol 86 with bromoacetaldehyde diethyl acetal and P_4O_{10}-promoted cyclization to furnish 5-bromo-7-methylbenzofuran (87) [82]. The Suzuki coupling of boronic acid 88, derived from 87, with *o*-bromonitrobenzene yields biaryl 89. Nitrene generation, achieved *via* deoxygenation of nitro compound 89 using triethyl phosphite, is followed by cyclization to furostifoline (90).

The Pd-catalyzed three-component cross-coupling reaction among aryl metal reagents, carbon monoxide, and aryl electrophiles is a straightforward and convenient route for the synthesis of unsymmetrical biaryl ketones. The reaction of electron-deficient electrophiles generally suffers from a side reaction that gives the direct coupling product without monoxide insertion. Miyaura developed an efficient Pd-catalyzed carbonylative three-component cross-coupling reaction of an arylboronic acid with aryl electrophiles including a bromofuran substrate [83]. Using $Pd(Ph_3P)_4$ as catalyst, the unsymmetrical biaryl ketone 92 is synthesized from 2-bromofuran 91. It is notable that $PdCl_2(dppf)$, the catalyst of choice for other aryl halides, gives exclusive direct coupling product without insertion of CO.

6.3.4. Stille coupling

Along with the Suzuki reaction, the Stille reaction is one of the most popular Pd-catalyzed transformations involving furans [1, 34]. Furans participate as both the stannane and halogen component, although the former is more popular. As previously highlighted, reaction at the 2-position is more common and convenient, although functionalization at the 3-position is well-known.

6.3.4.1. Furan motif as a nucleophile (stannane)

The commercial availability of 2-trialkylstannylfurans and 2-tributylstannylbenzofuran make them, by far, the most widely studied stannylfuran derivatives in the Stille reaction.

Furylstannanes that are not commercially available may be prepared by a number of methods, one of which involves direct metalation of a furan followed by quenching with tin chloride (93→94) [84]. The second method for furylstannane preparation uses halogen–metal exchange to generate the lithiofuran species, which is then quenched by tin chloride (95→96) [85]. The third method, more suitable to base-sensitive substrates, is Pd-catalyzed coupling of a halofuran or halobenzofuran with hexaalkylditin (97→98) [86].

Interestingly, the alkyne-oxazole Diels–Alder cycloaddition strategy provides a unique entry to some furyl stannanes [87]. Thus, thermolysis of bis(tributylstannyl)acetylene (99) and 4-phenyloxazole (100) lead to a separable mixture of 3,4-bis(tributylstannyl)furan (101, 19% yield) and 3-tributylstannylfuran (102, 23% yield).

The furyl- and benzofurylstannanes can be coupled with a variety of electrophiles including vinyl triflates [88, 89], vinyl halides [90], aryl halides [91, 92], and heteroaryl halides, specifically isoxazoles [93], thiophenes [94–96], 7-azaindoles [97], purines [98], and pyridinium cations [99].

Stille couplings of stannylfurans using novel conditions and catalyst systems have been reported by several groups. Bannwarth successfully coupled aryl bromides with 2-tributylstannylfuran in supercritical carbon dioxide with perfluorotagged and untagged palladium complexes [100]. Verkade employed bulky proazaphosphatrane ligands to couple vinyl and aryl chlorides with 2-tributylstannylfuran in high yields [101]. Li demonstrated the efficacy of a palladium acetate/DABCO catalyst system for the reaction of aryl bromides with 2-tributylstannylfuran in near quantitative yields [102].

The groups of Liebeskind and Guillaumet independently reported Stille couplings of 2-tributylstannylfuran with heteroaromatic thioethers [103, 104]. Both groups demonstrated that these transformations are Pd-catalyzed copper-promoted reactions. Using catalytic tetrakis (Pd(PPh$_3$)$_4$) and copper bromide/dimethyl sulfide complex in refluxing dimethoxyethane, the Guillaumet group can couple 2-tributylstannylfuran with methylthiobenzothiazole (**103** R = Me) to afford furan **104** in 81% yield. The Liebeskind lab reacts the same substrates with catalytic PdCl(PPh$_3$)$_2$(CH$_2$Ph) and copper(I) 3-methylsalicylate (CuMeSal) to furnish **104** in 36% yield. However, they see a dramatic improvement in efficiency when reacting phenylthiobenzothiazole (**103** R = Ph) with 2-tributylstannylfuran in the presence of tetrakis and CuMeSal. These conditions afford the desired product in 96% yield.

Grigg and coworkers have used 2-tributylstannylfuran to terminate a variety of interesting tandem cyclization-anion capture reactions. In one example, aryl iodide **105** reacts *via* a 5-*exo-trig* process to provide dihydroindolone **106** [105–107].

In their synthetic studies towards lophotoxin and pukalide, Paterson and coworkers explored both intermolecular and intramolecular Stille coupling reactions [108]. The intermolecular approach between vinyl iodide **107** and furylstannane **108** is more successful, giving adduct **109** in 67% yield. The intramolecular version provides the macrocyclized 14-membered lactone in only 15% yield. Pattenden reported a related investigation towards the synthesis of lophotoxin and observed comparable inter- and intramolecular reactivity in similar Stille reactions [109].

In the total synthesis of moracin M (**114**), a phytoalexin isolated from infected white mulberry, Widdowson *et al.* prepared 2-stannylated benzofuran **111** from benzofuran **110** *via* direct metalation and treatment with Me₃SnCl [110]. Stannane **111** was then coupled with aryl iodide **112** to afford adduct **113**, which was desilylated to moracin M (**114**).

Meyers and Novachek described the Stille coupling of 2-bromooxazoline with a furylstannane to produce furyloxazoline **115** [111]. Liebeskind and Wang conducted a benzannulation of a furylcyclobutenone that was prepared *in situ* using a Stille coupling of 4-chloro-2-cyclobutenone **116** with 2-tributylstannylfuran. Under the reaction conditions, the Stille product furnishes benzofuranol **117** *via* a dienyl ketene intermediate [112].

Selective couplings between dihaloheterocycles and 2-tributylstannylfuran have proven useful in several syntheses. The group of de Lera successfully coupled 2,3-dibromothiophene to yield exclusively 3-bromo-2-furanylthiophene (**118**) [113]. Cho demonstrated that 3,5-dibromo-2-pyrone reacts with complete selectivity to furnish 5-furanyl product **119** [114]. Finally, Gundersen prepared a variety of 9-aryl- and 9-benzyl-6-(2-furyl)purines. For example, purines **121** are available *via* selective Stille coupling of 2-tributylstannylfuran and dichloropurine **120** [115].

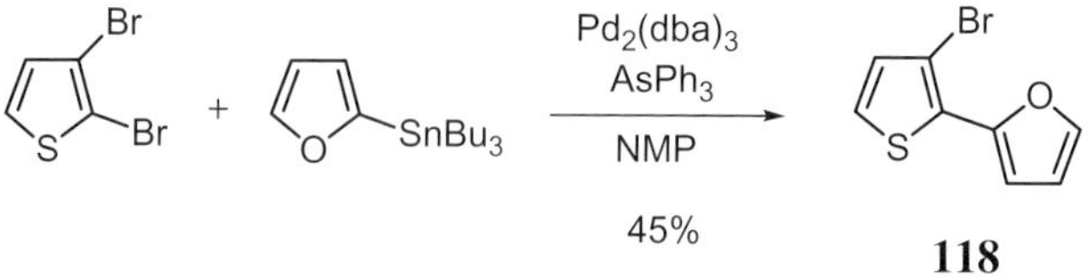

118

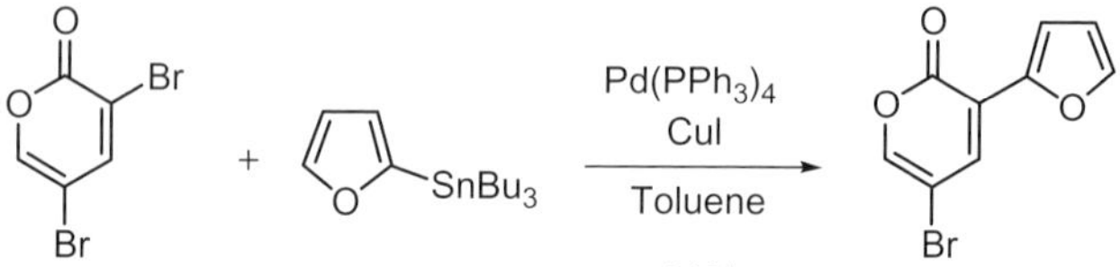

119

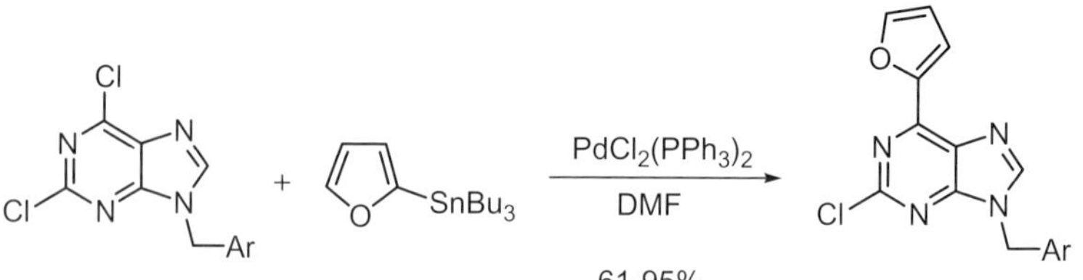

120   **121**

6.3.4.2. Furan motif as an electrophile (halide or triflate)

Like most aryl halides, furyl halides and furyl triflates couple with a variety of organostannanes including alkenyl, aryl, and heteroaryl stannanes in the presence of catalytic palladium. Carbamoylstannane **122** is prepared by treating lithiated piperidine with carbon monoxide and tributyltin chloride sequentially. The Stille reaction of **122** and 3-bromofuran then gives rise to amide **123** [116]. In another example, lithiation of 4,4-dimethyl-2-oxazoline followed by quenching with Me₃SnCl results in 2-(tributylstannyl)-4,4-dimethyl-2-oxazoline (**124**) in 70–80% yield [117]. Subsequent Stille reaction of **124** with 3-bromofuran affords 2-(3′-furyl)-4,4-dimethyl-2-oxazoline (**125**).

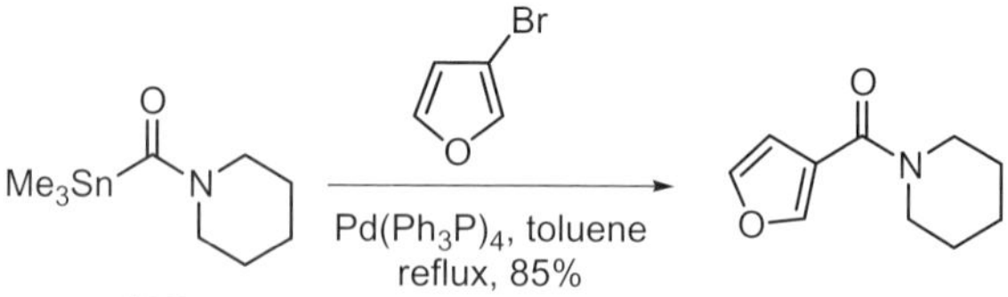

122   **123**

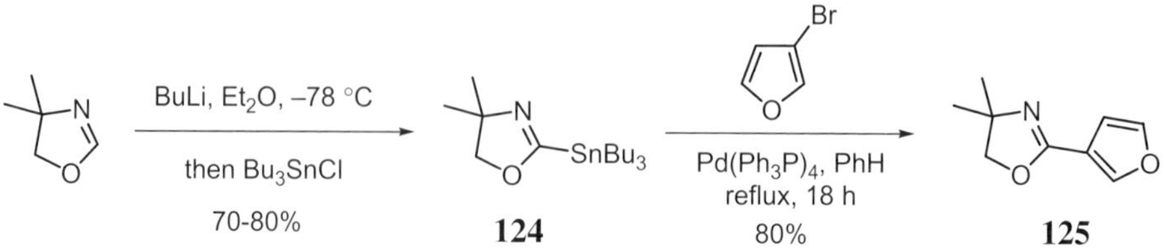

124   **125**

Hudkins *et al.* prepared a new reagent, 1-carboxy-2-(tributylstannyl)indole (**126**), using CO_2 to introduce an amine-protecting group [118]. The Stille coupling of **126** with 2-bromobenzofuran gives the 2-substituted indole **127** after concomitant removal of the carboxylic acid protecting group. BOC, SEM, or phenylsulfonyl groups proved much more difficult to cleave. Snieckus and associates conducted Stille cross-coupling reactions on a solid support *via* an ester linker to make styryl, biaryl, and heterobiaryl compounds [119]. In this report, bromofuran **128** was successfully coupled with both vinyl- and arylstannanes; the product of the reaction with vinyl tributyltin, furan **129**, is shown below.

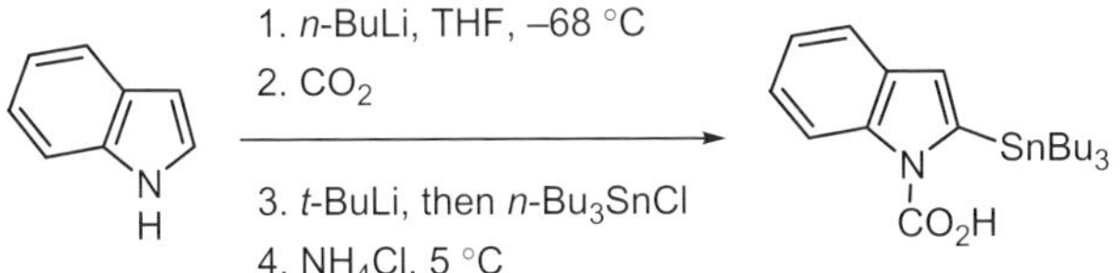

126

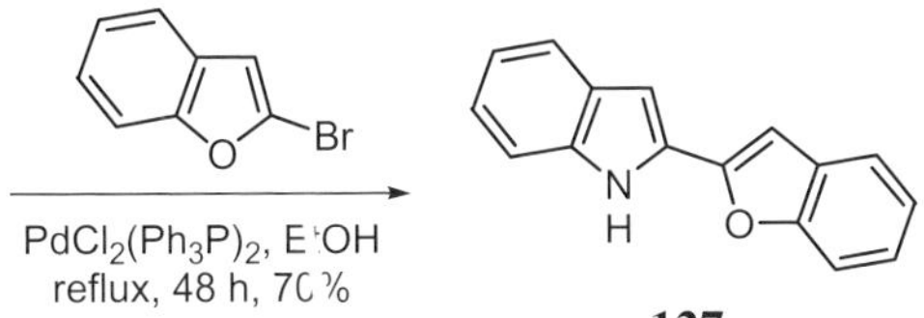

127

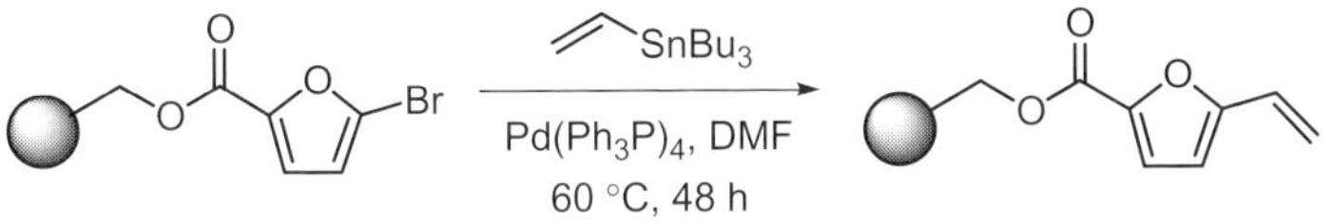

128 **129**

Unlike all-carbon aryl halide substrates, regioselective Stille coupling of 2,3-dihalofurans has been observed. Taking advantage of C(2) activation, allyl furan **130** is available upon coupling of 2,3-dibromofuran **46** with allyl tributyltin [56, 57]. Likewise, dibromobenzofurans react exclusively in the 2-position; coupling of benzofuran **131** with arylstannane **132** furnishes 2-arylbenzofuran **133** in excellent yield [120].

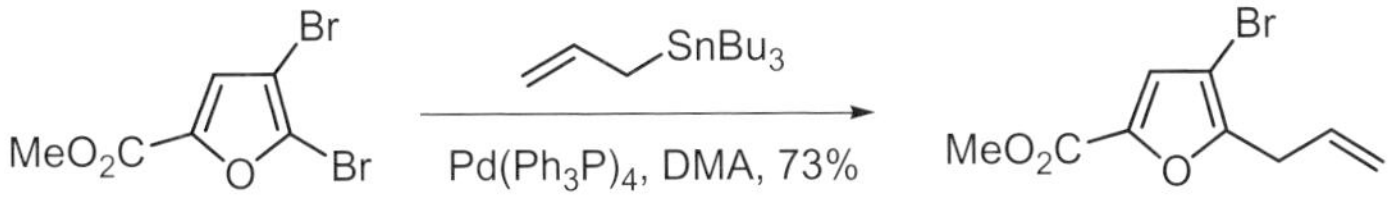

46 **130**

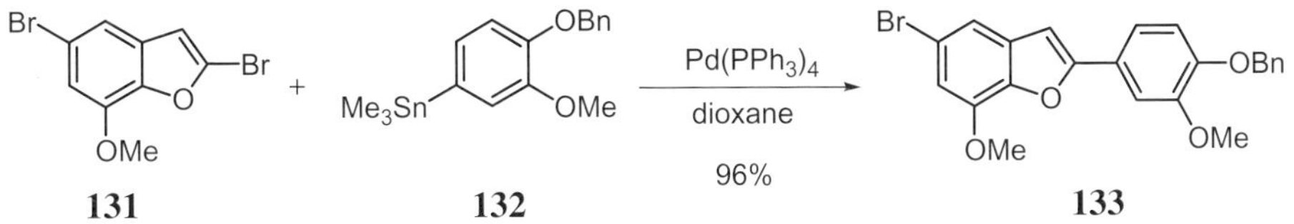

131 **132** **133**

6.3.5. Stille–Kelly reaction

The Stille–Kelly reaction is rarely used in conjunction with furans. Sessler and cowork-ers applied this strategy successfully as part of their synthetic pathway for the synthesis of furan-containing porphyrin derivatives. In the presence of hexabutylditin, Pd-catalyzed homocoupling of bromofuran **134** takes place to give bifuran diester **135** [121].

$$\text{Br}_2, \text{DMF} \quad 74\% \qquad \text{Bu}_3\text{Sn–SnBu}_3 \quad \text{Pd(Ph}_3\text{P)}_4, \text{tol} \quad \text{reflux, 54\%}$$

134   **135**

6.3.6. Hiyama coupling

The Hiyama coupling offers an alternative when selectivity and/or availability of other reagents are problematic. Hiyama *et al.* coupled alkyltrifluorosilane **137** with 2-bromo-furan **136** to give the corresponding cross-coupled product **138** in moderate yield in the presence of catalytic Pd(Ph$_3$P)$_4$ and three equivalents of TBAF [122, 123]. In this case, more than one equivalent of fluoride ion was needed to form a pentacoordinated silicate. Alkyltrifluorosilane **137** was prepared by hydrosilylation of the corresponding terminal olefin with trichlorosilane followed by fluorination with CuF$_2$.

$$\text{Pd(Ph}_3\text{P)}_4, \text{TBAF} \quad \text{THF, 36\%}$$

136   **137**   **138**

A variant of the Hiyama coupling for the preparation of aryl amides was reported by Cunico. Reaction of carbamoylsilane **139** with 3-bromofuran in the presence of bis(tri-*t*-butylphosphine)palladium provides amide **140** in 61% yield [124].

$$\text{Pd(P}t\text{-Bu}_3)_2 \quad \text{toluene}$$

139   61%   **140**

6.4. Sonogashira reaction

The most common use of the Sonogashira reaction in relation to the chemistry of furans is to synthesize benzofurans from orthohalophenols. These heteroannulation reac-tions are described in detail in Section 6.6.2. In general, the Sonogashira reaction works well to provide alkynyl substituted furans and benzofurans. Reactions of 2-halofurans smoothly provide access to a variety of 2-alkynyl furans. As part of a two-step strategy for the synthesis of hetarylacetylenes, the Sonogashira reaction was used to prepare alkynol **142** from the combination of 2-bromofuran with 3-methyl-1-butyn-3-ol (**141**). Exposure of alcohol **142** to catalytic potassium hydroxide initiates a "retro-Favorski" reaction to yield 2-alkynylfuran [125]. Another example of the use of the Sonogashira reaction for the synthesis of a monosubstituted furan involves reaction of 2-iodofuran with 1-ethynylcyclohexene to furnish alkyne **144** in 81% yield [126].

Several groups have successfully used 5-bromo-2-furaldehyde in Sonogashira reactions for the synthesis of 5-alkynyl-2-furaldehydes. Müller's group coupled 5-bromo-2-furaldehyde with phenylacetylene to form disubstituted furan **145** in excellent yield [127]. Lewis and coworkers coupled two hydroxyureas with 5-bromo-2-furaldehyde to provide alkyne products **146** in poor to good yields [128]. It is possible to efficiently prepare other disubstituted furans *via* a Sonogashira coupling. Namely, 2,5-dialkynylfurans are available beginning with 2,5-dibromofuran [129].

Although more synthetically challenging, Sonogashira reactions at the furan 3-position are possible. Yasuhara successfully coupled 3-bromofuran **147** with alkynylcarbamate **148** to furnish alkynylfuran **149** in 79% yield [130]. However, trimethylsilylacetylene did not provide any of the corresponding coupling product in a similar reaction with 3-bromofuran **147**. Another failure of the Sonogashira reaction, this time with 3-bromobenzofuran (**150**), was recorded by Yamanaka and coworkers [131]. Only resinous materials were obtained from **150** even though the same reaction worked well (58–96% yields) for 3-iodoben-zothiophene.

Due to activation of the furan α-positions, regioselective Sonogashira reactions can be achieved at C(2) rather than C(3) [56, 57]. For example, dibromofuran **151** reacts smoothly with *t*-butylacetylene to provide trisubstituted furan **152**. Similar selectivity is observed for Sonogashira reactions of 2,3-dibromo and 2,3,5-tribromobenzofurans [18, 58].

6.5. Heck, intramolecular Heck, and heteroaryl Heck reactions

Halofurans and halobenzofurans readily participate in Heck reactions with alkenes, and it is possible to synthesize benzofurans *via* intramolecular Heck reactions of aryl vinyl ethers. The heteroaromatic behavior of furans enable them to undergo heteroaryl Heck reactions to provide substituted furans.

6.5.1. Intermolecular Heck reaction

Although the Heck reactions of heteroaryl halides are now commonplace [132–135], few examples are found using organohalide substrates possessing a carboxylic acid moiety [136]. However, 4,5-dibromo-2-furancarboxylic acid (**153**) undergoes a Heck reaction with ethyl acrylate to afford diacrylate **154** [137].

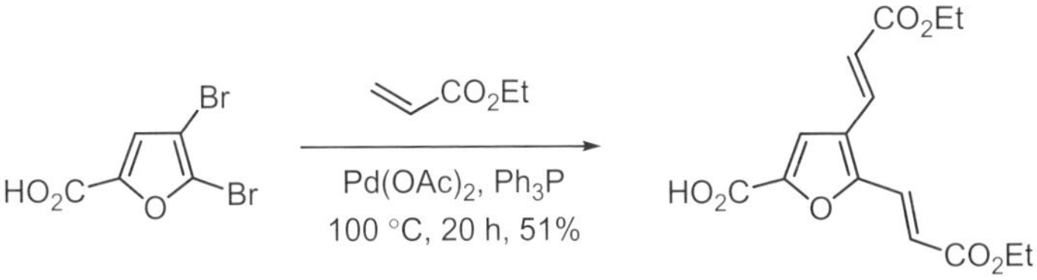

153 → **154**

The intermolecular Heck reaction of 3-bromofuran and tosylallylamine **155** gives adduct **156** under the classical Heck conditions [138]. Subsequent Rh-catalyzed hydroformylation with ring closure occurs regioselectively to furnish the hydroxypyrrolidine, which dehydrates upon exposure to catalytic HCl to afford dihydropyrrole **157**.

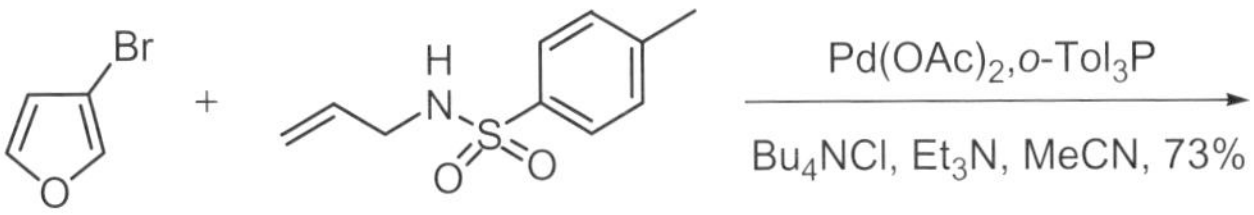

155

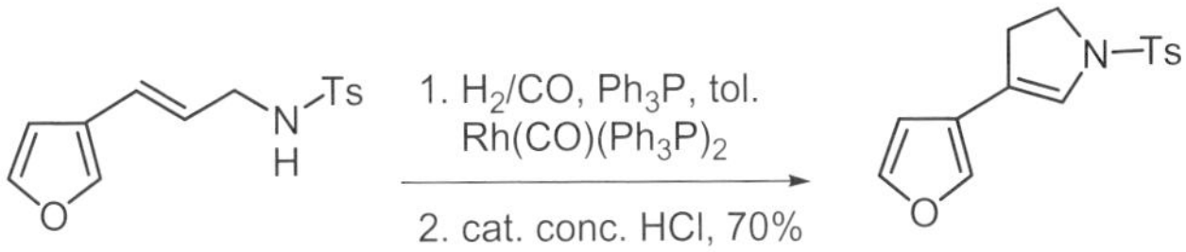

156 **157**

6.5.2. Intramolecular Heck reaction

Two distinct types of intramolecular Heck reactions are important in the chemistry of furans. The most straightforward are transformations involving halofurans reacting intramolecularly with alkenes. For example, Muratake *et al.* exploited the intramolecular Heck cyclization to establish the tricyclic core structure *en route* to the synthesis of a furan analog of duocarmycin SA, a potent cytotoxic antibiotic [139]. Under Jeffery's phase-transfer catalysis conditions, substrate **158** is converted to tricyclic derivatives **159** and **160** as an inseparable mixture (*ca.* 4:1) of two double bond isomers.

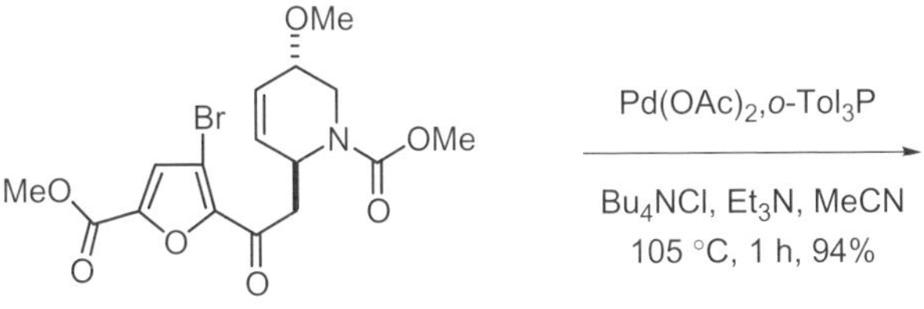

158

                    159   160

The second type of intramolecular Heck reactions involves conversion of an aryl halide into a furan. Similar to the Pd-catalyzed pyrrole and thiophene annulations, an intramolecular Heck reaction of substrate **161** furnishes benzofuran **162** [140]. Such an approach has become a popular means of synthesizing fused furans.

        161   38%   162

An intramolecular Heck cyclization strategy was developed for the construction of indole and benzofuran rings on solid support [141], enabling rapid generation of small-molecular libraries by simultaneous parallel or combinatorial synthesis. The S_N2 displacement of resin-bound γ-bromocrotonyl amide **164** with *o*-iodophenol **163** affords the cyclization precursor **165**. A subsequent intramolecular Heck reaction using Jeffery's "ligand-free" conditions furnishes, after double bond tautomerization, the resin-bound benzofurans, which are then cleaved with 30% TFA in CH_2Cl_2 to deliver the desired benzofuran derivatives **166** in excellent yields and purity.

        163   164   165

                                        166

Rawal's group developed an intramolecular aryl Heck cyclization method to synthesize benzofurans, indoles, and benzopyrans [142]. The rate of cyclization is significantly accelerated in the presence of bases, presumably because the phenolate anion formed under the reaction conditions is much more reactive as a soft nucleophile than phenol. In the presence of a catalytic amount of Herrmann's dimeric palladacyclic catalyst (**168**) [143], and three equivalents of Cs_2CO_3 in DMA, vinyl iodide **167** is transformed into "*ortho*" and "*para*" benzofuran **169** and **170**. In the mechanism proposed by Rawal, oxidative addition of phenolate **171** to Pd(0) is followed by nucleophilic attack of the ambident phenolate anion on σ-palladium intermediate **172** to afford aryl–vinyl palladium species **173** after rearomatization of the presumed cyclohexadienone intermediate. Reductive elimination of palladium followed by isomerization of the exocyclic double bond furnishes **169**.

6.5.3. Heteroaryl Heck reaction

Intermolecular heteroaryl Heck reactions occur readily at the more electron-rich 2-position in furans and benzofurans, while intramolecular reactions have been used to functionalize both the 2- and 3-positions. Ohta's group coupled aryl bromides such as 2-bromonitrobenzene with benzofuran [144]. They later described the heteroaryl Heck reactions of chloropyrazines with both furan and benzofuran [145].

Grigg's group synthesized unique bicyclic β-lactam **175** *via* an intramolecular Heck reaction from **174** [146, 147]. The 7-membered ring was formed *via* an unusual insertion at *C(3)* of the furan.

6.6. Heteroannulation

In addition to catalyzing reactions of furans and benzofurans, palladium also plays an important role in a wide variety of reactions that produce the furan and benzofuran ring structures from acyclic and monocyclic starting materials, respectively [1, 148, 149]. The six major classes of furan and benzofuran precursors are highlighted below.

6.6.1. Propargyl carbonates

In 1985, Tsuji's group carried out a Pd-catalyzed reaction of propargyl carbonate with methyl acetoacetate as a soft carbonucleophile under *neutral conditions* to afford 4,5-dihydrofuran **176** [150–152]. The resulting unstable **176** readily isomerizes to furan **177** under acidic conditions. In addition, they reported formation of disubstituted furan **179** *via* a Pd-catalyzed heteroannulation of hydroxy propargylic carbonate **178** [153]. Presumably, an allenylpalladium complex (*cf.* **181**) is the key intermediate.

The heteroannulation reactions of propargyl carbonates mentioned above proceed *without addition of a base* as explained by the mechanism detailed below. First, oxidative addition of propargyl carbonate to Pd(0) yields allenylpalladium(II) carbonate **180**. Decarboxylation of **180** releases one molecule of CO_2 and a methoxide anion, along with allenylpalladium intermediate **181**. At this point, the self-generated methoxide anion serves as a base to deprotonate methyl acetoacetate to form the corresponding enolate, which attacks the *sp* carbon of **181** to give the palladium carbene complex **182**. Isomerization of **182** leads to the π-allylpalladium complex **183**, which then undergoes an *O*-alkylation with the carbonyl oxygen and gives rise to the *exo*-methylenefuran **176** following elimination of Pd(0).

An extension of Tsuji's methodology has recently been reported by Liang in which he is able to perform a three-component coupling for the preparation of tetrasubstituted furans [154]. A combination of methyl or ethyl acetoacetate with propargyl bromide or carbonate and an aryl iodide enables the formation of tetrasubstituted furans **184**. The proposed mechanism for this transformation is similar to that of Tsuji in that an allenyl-palladium intermediate is postulated. However, a three-component reaction is possible because the initial oxidative addition in this reaction is with the aryl iodide. Interaction of this Pd(II) intermediate with the propargyl bromide or carbonate leads to the allenyl-palladium species and, eventually, the product **184**.

The final example of a propargyl carbonate reaction was reported by Yoshida and Ihara and enables the formation of benzofurans [155]. Interestingly, their substrate incorporates both a propargyl carbonate and an alkynol, the subject of the following section. Propargyl carbonate **185** reacts with 2-methyl-1,3-cyclopentadione (**186**) to provide benzofuran **187** in 87% yield. In analogy to the previous examples in this section, an allenyl-palladium species was postulated as the key intermediate in the reaction mechanism. Yoshida and Ihara also report a rather surprising result when using dimethyl malonate as the nucleophile. Instead of the expected C-alkylation product, they obtain 83% yield of the O-alkylation product **188**.

185   **186**   **187**

185   **188**

6.6.2. Alkynols

The most popular Pd-catalyzed method for the production of furans and benzofurans involves reactions of alkynols. Acyclic alkynols are converted into furans, while benzene substituted alkynols are transformed into benzofurans. The use of this strategy is widespread for the synthesis of benzofurans; however, it is occasionally used for the syntheses of furans. For example, intramolecular alkoxylation of alkyne **189** proceeds *via* an alkenylpalladium complex and subsequent carbonylation to form furan **190** [156, 157]. In addition, 3-hydroxyalkyl-benzo[*b*]furans were prepared by Bishop *et al. via* a Pd-catalyzed heteroannulation of silyl-protected alkynols with 2-iodophenol in a fashion akin to the Larock indole synthesis [158]. In a related series of experiments, Qing demonstrated that alkynes **191** could be efficiently converted into furans **192** [159].

189   **190**

191   **192**

An impressive alternate approach to the synthesis of furans beginning with alkynols was developed by Balme [160, 161] and subsequently applied in a total synthesis by Morimoto [162]. Balme discovered that a three-component coupling reaction between a propargylic alkoxide, a conjugate addition acceptor, and an unsaturated halide yields a variety of di- and trisubstituted furans. In one example, propargyl alcohol, diethyl ethoxymethylene malonate (**193**), and iodobenzene combine to furnish disubstituted furan **194** in

53% yield [160]. The mechanism of this reaction begins with the formation of the propargylic alkoxide and subsequent conjugate addition to Michael acceptor **193** to yield enolate **195**. The palladium(II) oxidative addition product from reaction of iodobenzene and the palladium catalyst then associates with the alkyne π-electrons to provide **196**. This activated alkyne undergoes a 5-*exo* cyclization to yield alkenylpalladium species **197**. Reductive elimination furnishes **198** which decarboxylates upon addition of potassium *t*-butoxide to provide **199**. Elimination of ethoxide followed by isomerization completes the reaction and yields furan **194**.

Morimoto successfully used this strategy as a key step in his synthesis of nordehydrocacalohastine derivates **203** [162]. Trisubstituted furan **202** is available upon the three-component coupling of enynol **201**, Michael acceptor **193**, and *o*-iodotoluene. The desired product, **203**, can then be prepared in five to six steps from **202**.

202   **203**

Benzene-containing alkynols are frequently used to prepare a variety of benzofuran products. 2-Phenylbenzofuran **206** is readily available from reaction of *o*-bromophenol and phenylacetylene [163]. The mechanism of this reaction involves a spontaneous cyclization *via* the intramolecular alkoxylation of alkyne **204** (the coupling adduct of *o*-bromophenol and phenyl acetylene) to yield benzofurylpalladium complex **205** and, finally, 2-phenylbenzofuran **206** [164].

204   **205**   **206**

The intermediacy of benzofurylpalladium complex **205** was confirmed by trapping it with various electrophiles, including allyl halides [165], aryl halides [166], vinyl triflates, propargyl carbonates (giving rise to 3-allenylbenzofurans) [167], and carbonylating reagents [168–170]. Moreover, Cacchi *et al.* took advantage of the benzofurylpalladium intermediate and synthesized 2,3-disubstituted benzofurans from propargylic *o*-(alkyl) phenyl ethers [171, 172]. Scammells's synthesis of benzofuran **208** serves as an example [168]. A sequential Pd-catalyzed annulation and alkoxylcarbonylation of alkynyl phenol **207** gives **208**, an intermediate in the synthesis of XH-14, a potent antagonist of the A_1 adenosine receptor isolated from the plant *Salvia miltiorrhiza*.

207   **208**

A solid-phase synthesis of 2-substituted benzofurans was accomplished *via* Pd-catalyzed heteroannulation of acetylenes [173]. The starting carboxylic acid **209** was directly linked to the hydroxy resin TentaGel™S-OH using Mitsunobu conditions to give the resin-bound ester, which was then deacetylated by mild alkaline hydrolysis to generate phenol **210**. Treating **210** with terminal alkynes in the presence of $PdCl_2(Ph_3P)_2$, CuI, and tetramethylguanidine (TMG) in DMF smoothly produced the heteroannulation products as resin-bound benzofurans **211**. Cleavage was subsequently performed with base to deliver 2-substituted benzofuran carboxylic acids **212** after neutralization.

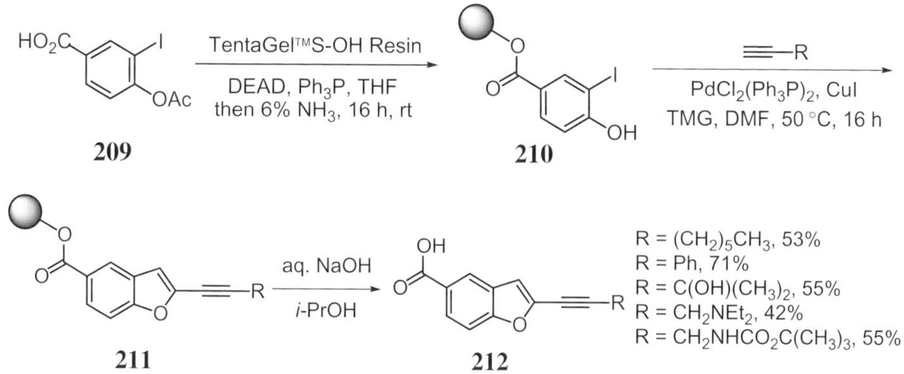

R = (CH₂)₅CH₃, 53%
R = Ph, 71%
R = C(OH)(CH₃)₂, 55%
R = CH₂NEt₂, 42%
R = CH₂NHCO₂C(CH₃)₃, 55%

6.6.3. Dibromoalkenols

Bisseret recently reported the conversion of dibromoalkenols, a new class of benzofuran precursors, into 2-phosphonylated and 2-arylbenzofurans [174]. For example, reaction of dibromalkenol **213** with phenyl boronic anhydride provides 2-phenylbenzofuran in 80% yield. The proposed mechanism for this transformation starts with a Suzuki coupling of the more reactive *E* vinyl bromide and the boronic anhydride followed by a second Pd-catalyzed coupling between the *Z* vinyl bromide and the alcohol oxygen.

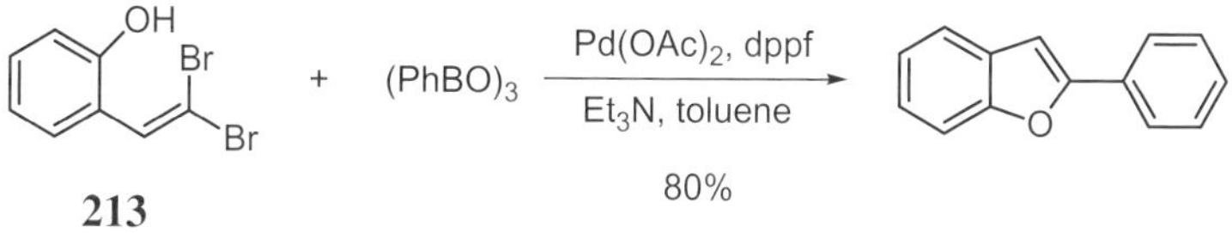

6.6.4. Alkynones

Palladium-catalyzed isomerization of ynones to furans has been an active area of research over the last twenty years. Huang *et al.* described a Pd-catalyzed rearrangement of α,β-acetylenic ketones to furans in moderate yield [175]. For example, Pd(dba)₂ promotes the isomerization of alkyne **214** to a putative allenyl ketone intermediate **215**, which subsequently cyclizes to the corresponding furan **216**.

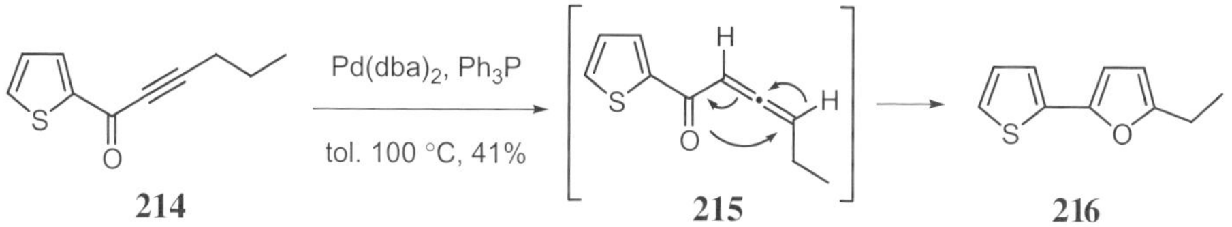

Ling also described the conversion of a variety of α,β-acetylenic ketones into 2,5-disubstituted furans [176]. Numerous palladium catalysts were investigated for the conversion of alkynones **217** into furans **218** and 3,3′-bifurans **219**. Optimization of the catalyst system enabled production of either **218** or **219**.

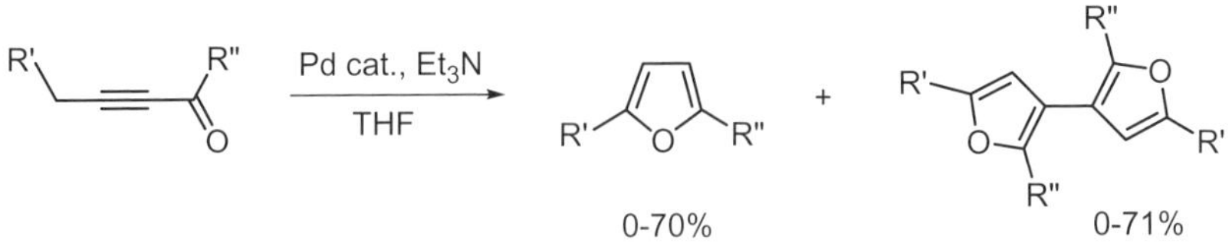

0-70%   0-71%
217   218   219

Müller reported an intriguing one-pot three-component reaction for the preparation of 3-halofurans [177]. The key α,β-acetylenic ketone 221, an intermediate in this transformation, is prepared *via* a Sonogashira reaction of a propargyl ether and an acid chloride. An illustrative example involves combination of benzoyl chloride with protected propargyl alcohol to provide alkynone 221 after the Sonogashira coupling. Addition of PTSA and NaCl initiates deprotection of the alcohol, conjugate addition of the chloride, and cyclization of the enone alcohol to yield the disubstituted furan product 220. This versatile strategy works well with several different acid chlorides. Furthermore, use of NaI in place of NaCl yields iodofuran products that can be further functionalized by Suzuki couplings with aryl boronic acids.

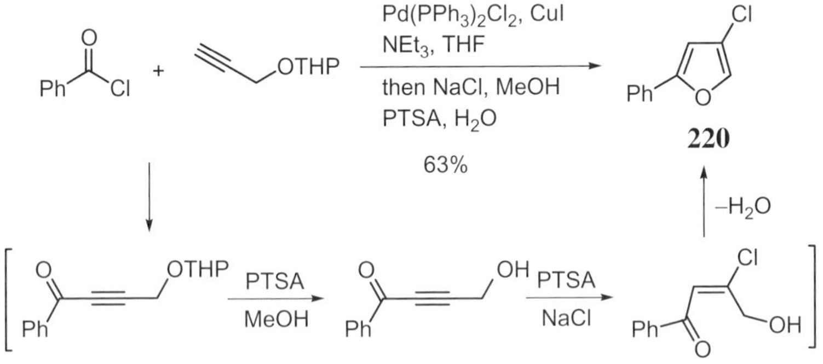

Utimoto *et al.* synthesized substituted furans using a Pd-catalyzed rearrangement of easily accessible β,γ-acetylenic ketones [178]. One plausible pathway is illustrated here using the transformation of β,γ-acetylenic ketone 222 to 2,5-disubstituted furan 223. Enolization of 222 is followed by an intramolecular oxypalladation of the resulting enol 224 to form furylpalladium(II) species 225, which is subsequently treated with acid to give furan 223. Intercepting the furylpalladium(II) species 225 with an electrophile should result in a carbodepalladation in place of protodepalladation. This hypothesis was realized in a tandem intramolecular alkoxylation of β,γ-acetylenic ketone 222 to afford trisubstituted furan 226 when allyl chloride was added to the original recipe [178]. 2,2-Dimethyloxirane was used as a proton scavenger, ensuring exclusive formation of 3-allylated 2,5-disubstituted furan 226 without contamination by protonated furans.

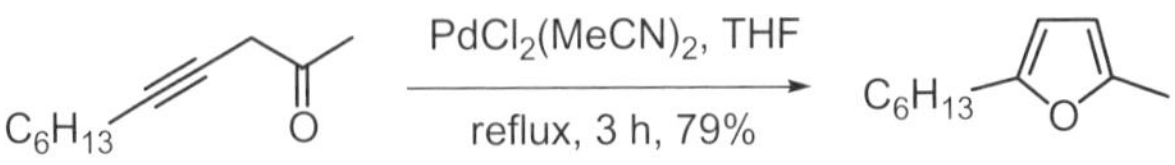

222   223

A similar substituted furan synthesis takes place *via* a Pd-catalyzed tandem carbonyla-tion–arylation using an α,β-acetylenic ketone, carbon monoxide, and bromothiophene [179].

Although less common than conversions of α,β-, and of β,γ-acetylenic ketones, γ,δ-alkynones can also lead to formation of furans. The γ,δ-acetylenic ketone **227** under-goes a Pd-catalyzed cyclization to furnish 2,5-disubstituted furan **228** (no yield provided) [180].

6.6.5. Enones

Enones are converted into furans with much less frequency than alkynones. One exam-ple of this transformation entails conversion of β-iodo-β,γ-enone **229** into 2,5-disubstituted furan **230** *via* a Pd-catalyzed cyclization using Herrmann's palladacycle catalyst **168** [181].

6.6.6. Allenones

Allenones have also been transformed into substituted furans *via* Pd-catalyzed het-eroannulation. Ma and Zhang converted 1,2-dienyl ketones and organic halides to sub-stituted furans under the agency of $Pd(Ph_3P)_4$ and Ag_2CO_3. Addition of 10 mol% of Ag_2CO_3 was crucial for the reaction of the organic halide. For example, allenone **231** and 5-bromopyrimidine were converted to pyrimidylfuran **232** in excellent yield [182, 183].

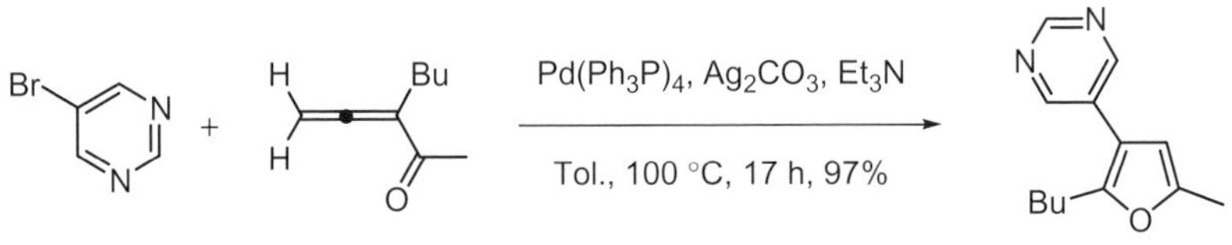

6.7. Carbonylation and C–N and C–O bond formation

Palladium-catalyzed alkoxylcarbonylation of furan and benzofuran were achieved in the presence of $Hg(O_2CCF_3)_2$ in ethanol with low efficiency [184]. Other heterocycles including thiophene and pyrrole were also carbonylated to give the corresponding esters in low yields using the same method.

Applying Buchwald's Pd-catalyzed amination methodology, Thomas and coworkers prepared a range of bicyclic piperazines [185]. While Pd-catalyzed amination of 5-bromobenzofuran leads to 5-benzofurylpiperizine **233** in 65% yield after deprotection, the corresponding reaction of 7-bromobenzofuran only gives 7-benzofurylpiperizine **234** in 20% yield. They speculated that either steric hindrance of the oxidative addition inter-mediate or the interaction between the oxygen lone pair and the metal center are respon-sible for the low yield. The debrominated benzofuran is the major by-product.

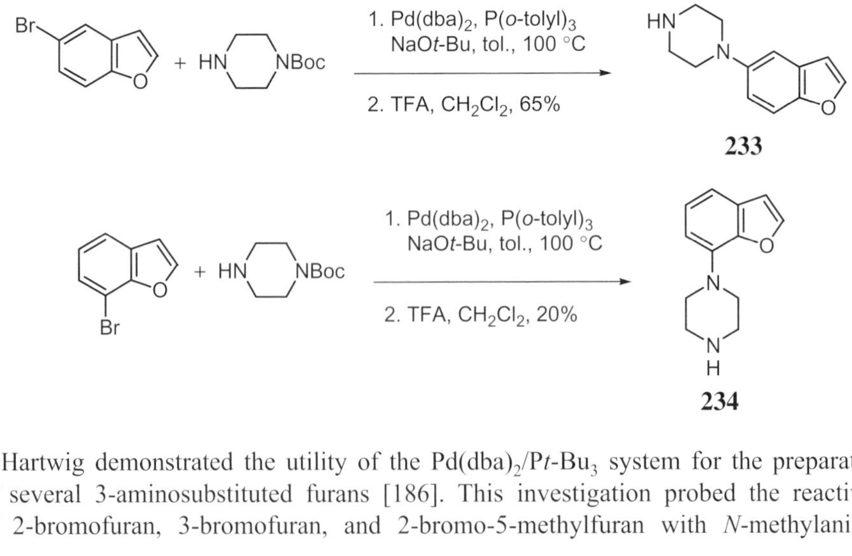

Hartwig demonstrated the utility of the $Pd(dba)_2/Pt$-Bu_3 system for the preparation of several 3-aminosubstituted furans [186]. This investigation probed the reactivity of 2-bromofuran, 3-bromofuran, and 2-bromo-5-methylfuran with *N*-methylaniline and diphenylamine. The desired products are obtained in poor to good yield; the best example involves coupling of 3-bromofuran with *N*-methylaniline to furnish 3-aminofuran **235** in 85% yield.

Benzofurans are available *via O*-arylation of enolates as reported by Willis [187]. In the simplest example, tricyclic benzofuran **237** was prepared from bromoketone **236** upon treatment with cesium carbonate and DPEphos. In total, eight other benzofurans were prepared in good to excellent yield using this methodology.

In conclusion, due to the activation effects stemming from the electronegativity of the oxygen atom on the α-positions of furans and benzofurans, regioselective coupling can be attained for Pd-catalyzed reactions. Regioselective cross-couplings (Suzuki, Stille, and Sonogashira reactions) of α,β-dihalofurans can be achieved at the α-positions. Unlike 2-chloropyridines, 2-chlorofurans are not sufficiently activated to oxidatively add to Pd(0).

The most unique feature of furan synthesis using palladium chemistry is heteroannulation. Enones, ynones, and ynols all have been annulated into furans and benzofurans. More importantly, trapping the reactive Pd(II) intermediates at different stages with electrophiles offers unique opportunities to synthesize substituted furans and benzofurans.

6.8. References

1. For a review of furans, see König, B. in *Science of Synthesis*; Maas, G., Regitz, M. Eds.; Houben-Weyl Methods of Molecular Transformations, Category 2; Georg Thieme Verlag: New York, **2001**; Vol. 9, 183–286.
2. Balaban, A. T.; Oniciu, D. C.; Katritzky, A. R. *Chem. Rev.* **2004**, *104*, 2777–812.
3. Zaluski, M. C.; Robba, M.; Bonhomme, M. *Bull. Soc. Chim. Fr.* **1970**, 1838–43.
4. Chadwick, D. J.; Chambers, J. Hargreaves, H. E.; Meakins, G. D.; Snowden, R. C. *J. Chem. Soc., Perkin Trans. 1* **1973**, 2327–32.
5. Gorzynski, M.; Rewicki, D. *Liebigs Ann. Chem.* **1986**, 625–37.
6. Keegstra, M. A.; Klomp, A. J. A.; Brandsma, L. *Synth. Commun.* **1990**, *20*, 3371–4.
7. Sornay, R.; Meunier, J.-M.; Fourari, P. *Bull. Soc. Chim. Fr.* **1971**, 990–6.
8. Verkruijsse, H. D.; Keegstra, M. A.; Brandsma, L. *Synth. Commun.* **1989**, *19*, 1047–50.

9. Gronowitz, S.; Hörnfeldt, A. B.; Pettersson, K. *Synth. Commun.* **1973**, *3*, 213–18.

10. Burke, M. D.; Berger, E. M.; Schreiber, S. L. *J. Am. Chem. Soc.* **2004**, *126*, 14095–104.

11. Vachal, P.; Toth, L. M. *Tetrahedron Lett.* **2004**, *45*, 7157–61.

12. Ismail, M. A.; Brun, R.; Wenzler, T.; Tanious, F. A.; Wilson, W. D.; Boykin, D. W. *J. Med. Chem.* **2004**, *47*, 3658–64.

13. Khatuya, H. *Tetrahedron Lett.* **2001**, *42*, 2643–4.

14. Milkiewicz, K. L.; Parks, D. J.; Lu, T. *Tetrahedron Lett.* **2003**, *44*, 4257–60.

15. Kim, G.; Jung, S.; Lee, E.; Kim, N. *J. Org. Chem.* **2003**, *68*, 5395–8.

16. Mann, I. S.; Widdowson, D. A.; Clough, J. M. *Tetrahedron* **1991**, *47*, 7981–90.

17. Benincori, T.; Brenna, E.; Sannicolo, F.; Trimarco, L.; Antognazza, P.; Casarotti, E.; Demartin, F.; Pilati, T. *J. Org. Chem.* **1996**, *61*, 6244–51.

18. Bach, T.; Bartels, M. *Synthesis* **2003**, 925–39.

19. Malamas, M. S.; Sredy, J.; Moxham, C.; Katz, A.; Xu, W.; McDevitt, R.; Adebayo, F. O.; Sawicki, D. R.; Seestaller, L.; Sullivan, D.; Taylor, J. R. *J. Med. Chem.* **2000**, *43*, 1293–310.

20. Hamilton, C. J.; Saravanamuthu, A.; Fairlamb, A. H.; Eggleston, I. M. *Bioorg. Med. Chem.* **2003**, *11*, 3683–93.

21. Plé, P. A.; Green, T. P.; Hennequin, L. F.; Curwen, J.; Fennell, M.; Allen, J.; Lambert-van der Brempt, C.; Costello, G. *J. Med. Chem.* **2004**, *47*, 871–87.

22. Lin, S.-Y.; Chen, C.-L.; Lee, Y.-J. *J. Org. Chem.* **2003**, *68*, 2968–71.

23. Tasker, A. S.; Sorensen, B. K.; Jae, H.-S.; Winn, M.; von Geldern, T. W.; et al. *J. Med. Chem.* **1997**, *40*, 322–30.

24. Barker, P.; Finke, P.; Thompson, K. *Synth. Commun.* **1989**, *19*, 257–65.

25. Kozhevnikov, I. V. *React. Kinet. Catal. Lett.* **1976**, *4*, 451–8.

26. Itahara, T.; Hashimoto, M.; Yumisashi, H. *Synthesis* **1984**, 255–6.

27. Itahara, T. *J. Org. Chem.* **1985**, *50*, 5272–5.

28. Fujiwara, Y.; Maruyama, O.; Yoshidomi, M.; Taniguchi, H. *J. Org. Chem.* **1981**, *46*, 851–5.

29. Parrish, J. P.; Flander, V. L.; Floyd, R. J.; Jung, K. W. *Tetrahedron Lett.* **2001**, *42*, 7729–31.

30. Tsuji, J.; Nagashima, H. *Tetrahedron* **1984**, *40*, 2699–702.

31. Zhang, H.; Ferreira, E. M.; Stoltz, B. M. *Angew. Chem. Int. Ed.* **2004**, *43*, 6144–8.

32. Roshchin, A. I.; Kel'chevski, S. M.; Bumagin, N. A. *J. Organomet. Chem.* **1998**, *560*, 163–7.

33. Pearlman, B. A.; MaNamara, J. M.; Hasan, I.; Hatakeyama, S.; Sekizaki, H.; Kishi, Y. *J. Am. Chem. Soc.* **1981**, *103*, 4248–51.

34. Schröter, S.; Stock, C.; Bach, T. *Tetrahedron* **2005**, *61*, 2245–67.

35. Chinchilla, R.; Nájera, C.; Yus, M. *Chem. Rev.* **2004**, *104*, 2667–722.

36. Carpita, A.; Rossi, R.; Veracini, C. A. *Tetrahedron* **1985**, *41*, 1919–30.

37. Klingstedt, T.; Frejd, T. *Organometallics* **1983**, *2*, 598–600.

38. Pelter, A.; Rowland, M.; Jenkins, I. H. *Tetrahedron Lett.* **1987**, *28*, 5213–6.

39. Amatore, C.; Jutand, A.; Negri, S.; Fauvarque, J. F. *J. Organomet. Chem.* **1990**, *390*, 389–98.

40. Villiers, P.; Vicart, N.; Ramondenc, Y.; Plé, G. *Eur. J. Org. Chem.* **2001**, 561–74.

41. Brandão, M. A. F.; de Oliveira, A. B.; Snieckus, V. *Tetrahedron Lett.* **1993**, *34*, 2437–40.

42. Ennis, D. S.; Gilchrist, T. L. *Tetrahedron* **1990**, *46*, 2623–32.
43. Campbell, J. B.; Firor, J. W.; Davenport, T. W. *Synth. Commun.* **1989**, *19*, 2265–72.
44. Arcadi, A.; Burini, A.; Cacchi, S.; Delmastro, M.; Marinelli, F.; Pietrani, B. *Synlett* **1990**, 47–8.
45. Sakamoto, T.; Kondo, Y.; Watanabe, R.; Yamanaka, H. *Chem. Pharm. Bull.* **1986**, *34*, 2719–24.
46. Sakamoto, T.; Katoh, E.; Kondo, Y.; Yamanaka, H. *Heterocycles* **1988**, *27*, 1353–6.
47. Zeni, G.; Alves, D.; Braga, A. L.; Stefani, H. A.; Nogueira, C. W. *Tetrahedron Lett.* **2004**, *45*, 4823–6.
48. Negishi, E.-I.; Takahashi, T.; King, A. O. *Org. Synth.* **1988**, *66*, 67–74.
49. Chandraratna, R. A. S.; Gillett, S. J.; Song, T. K.; Attard, J.; Vuligonda, S.; Garst, M. E.; Arefieg, T.; Gill, D. W.; Wheeler, L. *Bioorg. Med. Chem. Lett.* **1995**, *5*, 523–7.
50. Negishi, E.-I.; Xu, C.; Tan, Z.; Kotora, M. *Heterocycles* **1997**, *46*, 209–14.
51. Paterson, I.; Gardner, M.; Banks, B. J. *Tetrahedron* **1989**, *45*, 5283–92.
52. Sauvêtre, R.; Gillet, J.-P.; Normant, J.-F. *Synthesis* **1986**, 538–43.
53. Jackson, R. F. W.; Wishart, N.; Wood, A.; James, K.; Wythes, J. *J. Org. Chem.* **1992**, *57*, 3397–404.
54. Tsubuki, M.; Takahashi, K.; Honda, T. *J. Org. Chem.* **2003**, *68*, 10183–6.
55. Politis, J. K.; Nemes, J. C.; Curtis, M. D. *J. Am. Chem. Soc.* **2001**, *123*, 2537–47.
56. Bach, T.; Krüger, L. *Synlett* **1998**, 1185–6.
57. Bach, T.; Krüger, L. *Eur. J. Org. Chem.* **1999**, 2045–57.
58. Bach, T.; Bartels, M. *Synlett* **2001**, 1284–6.
59. Bach, T.; Bartels, M. *Tetrahedron Lett.* **2002**, *43*, 9125–7.
60. For reviews of furyl- and benzofuranylboronic acids, see: Tyrrell, E.; Brookes, P. *Synthesis* **2004**, 469–83 and Ref. 35.
61. Thompson, W. J.; Gaudino, J. *J. Org. Chem.* **1984**, *49*, 5237–43.
62. Macklin, T. K.; Snieckus, V. *Org. Lett.* **2005**, *7*, 2519–22.
63. Wu, J.; Zhu, Q.; Wang, L.; Fathi, R.; Yang, Z. *J. Org. Chem.* **2003**, *68*, 670–3.
64. Kamino, T.; Kuramochi, K.; Kobayashi, S. *Tetrahedron Lett.* **2003**, *44*, 7349–51.
65. McClure, M. S.; Roschangar, F.; Hodson, S. J.; Millar, A.; Osterhout, M. H. *Synthesis* **2001**, 1681–5.
66. Gribble, G. W.; Conway, S. C. *Synth. Commun.* **1992**, *22*, 2129–41.
67. Bourlot, A. S.; Desarbre, E.; Mérour, J.-Y. *Synthesis* **1994**, 411–6.
68. Benoit, J.; Malapel, B.; Mérour, J.-Y. *Synth. Commun.* **1996**, *26*, 3289–95.
69. Liu, B.; Moffett, K. K.; Joseph, R. W.; Dorsey, B. D. *Tetrahedron Lett.* **2005**, *46*, 1779–82.
70. Yang, Y. *Synth. Commun.* **1989**, *19*, 1001–8.
71. Yick, C.-Y.; Tsang, T.-K.; Wong, H. N. C. *Tetrahedron* **2003**, *59*, 325–33.
72. Moody, C. J.; Doyle, K. J.; Elliott, M. C.; Mowlem, T. J. *J. Chem. Soc., Perkin Trans. 1* **1977**, 2413–9.
73. Song, A. Z.; Wong, H. N. C. *J. Org. Chem.* **1994**, *59*, 33–41.
74. Young, Y.; Martin, A. R. *Heterocycles* **1992**, *34*, 1395–8.
75. Ishikura, M.; Oda, I.; Terashima, M. *Heterocycles* **1987**, *26*, 1603–10.
76. Ishikura, M.; Oda, I.; Terashima, M. *Heterocycles* **1985**, *23*, 2375–86.
77. For a different example of this strategy, see Batista-Parra, A.; Venkitachalam, S.; Wilson, W. D.; Boykin, D. W. *Heterocycles* **2003**, *60*, 1367–76.
78. Milkiewicz, K. L.; Parks, D. J.; Lu, T. *Tetrahedron Lett.* **2003**, *44*, 4257–60.

79. Lee, S.; Yi, K. Y.; Hwang, S. K.; Lee, B. H.; Yoo, S.; Lee, K. *J. Med. Chem.* **2005**, *48*, 2882–91.

80. Jiang, W.; Sui, Z.; Macielag, M. J.; Walsh, S. P.; Fiordeliso, J. J.; Lanter, J. C.; Guan, J.; Qiu, Y.; Kraft, P.; Bhattacharjee, S.; Craig, E.; Haynes-Johnson, D.; John, T. M.; Clancy, J. *J. Med. Chem.* **2003**, *46*, 441–4.

81. Wellmar, U.; Hörnfeldt, A.-B.; Gronowitz, S. *J. Heterocycl. Chem.* **1995**, *32*, 1159–63.

82. Soós, T.; Timári, G.; Hajós, G. *Tetrahedron Lett.* **1999**, *40*, 8607–9.

83. Ishiyama, T.; Kizaki, H.; Hayashi, T.; Suzuki, A.; Miyaura, N. *J. Org. Chem.* **1998**, *63*, 4726–31.

84. Gronowitz, S.; Timari, G. *J. Heterocycl. Chem.* **1990**, *27*, 1159–60.

85. Katsumura, S.; Fujiwara, S.; Isoe, S. *Tetrahedron Lett.* **1988**, *29*, 1173–6.

86. Arcadi, A.; Cacchi, S.; Fabrizi, G.; Marinelli, F.; Moro, L. *Synlett* **1999**, 1432–4.

87. (a) Yang, Y.; Wong, H. N. C. *J. Chem. Soc., Chem. Commun.* **1992**, 656–8; (b) Yang, Y.; Wong, H. N. C. *Tetrahedron* **1994**, *50*, 9583–608.

88. Chacun-Lefèvre L.; Joseph, B.; Mérour, J.-Y. *Tetrahedron* **2000**, *56*, 4491–9.

89. Langer, P.; Eckardt, T.; Schneider, T.; Göbel, C.; Herbst-Irmer, R. *J. Org. Chem.* **2001**, *66*, 2222–6.

90. DeBoos, G.; Fullbrook, J. J.; Owton, W. M.; Percy, J. M.; Thomas, A. C. *Synlett* **2000**, 963–6.

91. Bailey, T. R. *Tetrahedron Lett.* **1986**, *27*, 4407–10.

92. Sasabe, M.; Houda, Y.; Takagi, H.; Sugane, T.; Bo, X.; Yamamura, K. *J. Chem. Soc., Perkin Trans. 1* **2000**, 3786–90.

93. Labadie, S. S. *Synth. Commun.* **1994**, *24*, 709–19.

94. Hucke, A.; Cava, M. P. *J. Org. Chem.* **1998**, *63*, 7413–7.

95. Miyata, Y.; Nishinaga, T.; Komatsu, K. *J. Org. Chem.* **2005**, *70*, 1147–53.

96. Yamamura, K.; Kusuhara, N.; Kondou, A.; Hashimoto, M. *Tetrahedron* **2002**, *58*, 7653–61.

97. Chi, S. M.; Choi, J.-K.; Yum, E. K.; Chi, D. Y. *Tetrahedron Lett.* **2000**, *41*, 919–22.

98. Brill, W. D.-D.; Riva-Toniolo, C. *Tetrahedron Lett.* **2001**, *42*, 6515–8.

99. García-Cuadrado, D.; Cuadro, A. M.; Alvarez-Builla, J.; Vaquero, J. J. *Synlett* **2002**, 1904–6.

100. Osswald, T.; Schneider, S.; Wang, S.; Bannwarth, W. *Tetrahedron Lett.* **2001**, *42*, 2965–7.

101. Su, W.; Urgaonkar, S.; McLaughlin, P. A.; Verkade, J. G. *J. Am. Chem. Soc.* **2004**, *126*, 16433–9.

102. Li, J.-H.; Liang, Y.; Wang, E.-P.; Liu, W.-J.; Xie, Y.-X.; Yin, D.-L. *J. Org. Chem.* **2005**, *70*, 2832–4.

103. Egi, M.; Liebeskind, L. S. *Org. Lett.* **2003**, *5*, 801–2.

104. Alphonse, F.-A.; Suzenet, F.; Keromnes, A.; Lebret, B.; Guillaumet, G. *Org. Lett.* **2003**, *5*, 803–5.

105. Fretwell, P.; Grigg, R.; Sansano, J. M.; Sridharan, V.; Sukirthalingam, S.; Wilson, D.; Redpath, J. *Tetrahedron* **2000**, *56*, 7525–39.

106. Fielding, M. R.; Grigg, R.; Urch, C. J. *J. Chem. Soc., Chem. Commun.* **2000**, 2239–40.

107. Böhmer, J.; Grigg, R.; Marchbank, J. D. *J. Chem. Soc., Chem. Commun.* **2002**, 768–9.

108. Paterson, I.; Brown, R. E.; Urch, C. J. *Tetrahedron Lett.* **1999**, *40*, 5807–10.

109. Cases, M.; de Turiso, F. G.-L.; Hadjisoteriou, M. S.; Pattenden, G. *Org. Biomol. Chem.* **2005**, *3*, 2786–804.

110. Mann, I. S.; Widdowson, D. A.; Clough, J. M. *Tetrahedron* **1991**, *47*, 7981–90.

111. Meyers, A. I.; Novachek, K. A. *Tetrahedron Lett.* **1996**, *37*, 1747–8.

112. Libeskind, L. S.; Wang, J. *J. Org. Chem.* **1993**, *58*, 3550–6.

113. Pereira, R.; Iglesias, B.; de Lera, A. R. *Tetrahedron* **2001**, *57*, 7871–81.

114. Kim, W.-S.; Kim, H.-J.; Cho, C.-G. *Tetrahedron Lett.* **2002**, *43*, 9015–7.

115. Bakkestuen, A. K.; Gundersen, L.-L.; Utenova, B. T. *J. Med. Chem.* **2005**, *48*, 2710–23.

116. Lindsay, C. M.; Widdowson, D. A. *J. Chem. Soc., Perkin Trans. 1* **1988**, 569–73.

117. Dodoni, A.; Fantin, G.; Fagagnolo, M.; Medici, A.; Pedrini, P. *Synthesis* **1987**, 693–6.

118. Hudkins, R. L.; Diebold, J. L.; Marsh, F. D. *J. Org. Chem.* **1995**, *60*, 6218–20.

119. Chamoin, S.; Houldworth, S.; Snieckus, V. *Tetrahedron Lett.* **1998**, *39*, 4175–8.

120. Lin, S.-Y.; Chen, C.-L.; Lee, Y.-J. *J. Org. Chem.* **2003**, *68*, 2968–71.

121. Sessler, J. L.; Hoehner, M. C.; Gebauer, A.; Andrievsky, A.; Lynch, V. *J. Org. Chem.* **1997**, *62*, 9251–60.

122. Matsuhashi, H.; Kuroboshi, M.; Hatanaka, Y.; Hiyama, T. *Tetrahedron Lett.* **1994**, *35*, 6507–10.

123. Matsuhashi, H.; Asai, S.; Hirabayashi, K.; Hatanaka, Y.; Mori, A.; Hiyama, T. *Bull. Chem. Soc. Jpn.* **1997**, *70*, 437–44.

124. Cunico, R. F.; Maity, B. C. *Org. Lett.* **2002**, *4*, 4357–9.

125. Mal'kina, A. G.; Brandsma, L.; Vasilevsky, B. A. *Synthesis* **1996**, 589–90.

126. Siebeneicher, H.; Doye, S. *Eur. J. Org. Chem.* **2002**, 1213–20.

127. Karpov, A. S.; Rominger, F.; Müller, T. J. J. *J. Org. Chem.* **2003**, *68*, 1503–11.

128. Lewis, T. A.; Bayless, L.; Eckman, J. B.; Ellis, J. L.; Grewal, G.; Libertine, L.; Nicolas, J. M.; Scannell, R. T.; Wels, B. F.; Wenberg, K.; Wypij, D. M. *Bioorg. Med. Chem. Lett.* **2004**, *14*, 2265–8.

129. Garcia, J.; Lopez, M.; Romeu, J. *Synlett* **1999**, 429–31.

130. Yasuhara, A.; Suzuki, N.; Sakamoto, T. *Chem. Pharm. Bull.* **2002**, *50*, 143–5.

131. Sakamoto, T.; Kondo, Y.; Watanabe, R.; Yamanaka, H. *Chem. Pharm. Bull.* **1988**, *36*, 2248–52.

132. Lemhadri, M.; Doucet, H.; Santelli, M. *Tetrahedron* **2004**, *60*, 11533–40.

133. Lightfoot, A. P.; Maw, G.; Thirsk, C.; Twiddle, S. J. R.; Witing, A. *Tetrahedron Lett.* **2003**, *44*, 7645–8.

134. Berthio, F.; Feuerstein, M.; Doucet, H.; Santelli, M. *Tetrahedron Lett.* **2002**, *43*, 5625–8.

135. Karabelas, K.; Hallberg, A. *J. Org. Chem.* **1986**, *51*, 5286–90.

136. Crisp, G.; O'Donoghue, A. I. *Synth. Commun.* **1989**, *19*, 1745–58.

137. Karminski-Zamola, G.; Dogan, J.; Bajic, M. *Heterocycles* **1994**, *38*, 759–67.

138. Busacca, C. A.; Dong, Y. *Tetrahedron Lett.* **1996**, *37*, 3947–50.

139. Muratake, H.; Okabe, K.; Takahashi, M.; Tonegawa, M.; Natsume, M. *Chem. Pharm. Bull.* **1997**, *45*, 799–806.

140. Henke, B. R.; Aquino, C. J.; Birkemo, L. S.; Croom, D. K.; Dougherty, Jr., R. W.; Ervin, G. N. *J. Med. Chem.* **1997**, *40*, 2706–25.

141. Zhang, H.-C.; Maryanoff, B. E. *J. Org. Chem.* **1997**, *62*, 1804–9.

142. Hennings, D. D.; Iwasa, S.; Rawal, V. H. *Tetrahedron Lett.* **1999**, *38*, 6379–82.

143. Herrmann, W. A.; Brossmer, C.; Öfele, K.; Reisinger, C. P.; Priermeier, T.; Beller, M.; Fischer, H. *Angew. Chem., Int. Ed. Engl.* **1995**, *34*, 1844–8.

144. Ohta, A.; Akita, Y.; Ohkuwa, T.; Chiba, M.; Fukunaka, R.; Miyafuji, A.; Nakata, T.; Tani, N. *Heterocycles* **1990**, *31*, 1951–7.

145. Aoyagi, Y.; Inoue, A.; Koizumi, I.; Hashimoto, R.; Tokunaga, K.; Gohma, K.; Komatsu, J.; Sekine, K.; Miyafuji, A.; Konoh, J.; Honma, R.; Akita, Y.; Ohta, A. *Heterocycles* **1992**, *34*, 257–72.

146. Burwood, M.; Davies, B.; Diaz, I.; Grigg, R.; Molina, P.; Sridharan, V.; Hughes, M. *Tetrahedron Lett.* **1995**, *36*, 9053–6.

147. Grigg, R.; Fretwell, P.; Meerholtz, C.; Sridharan, V. *Tetrahedron Lett.* **1994**, *50*, 359–70.

148. Kirsch, G.; Hesse, S.; Comel, A. *Curr. Org. Synth.* **2004**, 47–63.

149. Brown, R. C. D. *Angew. Chem. Int. Ed.* **2005**, *44*, 850–2.

150. Tsuji, J.; Watanabe, H.; Minami, I.; Shimizu, I. *J. Am. Chem. Soc.* **1985**, *107*, 2196–8.

151. Minami, I.; Yuhara, M.; Watanabe, H.; Tsuji, J. *J. Organomet. Chem.* **1987**, *334*, 225–42.

152. Greeves, N.; Torode, J. S. *Synthesis* **1993**, *62*, 1009–13.

153. Tsuji, J.; Mandai, T. in *Metal-catalyzed Cross-coupling Reactions*, Diederich, F.; Stang, P. J. Eds.; Wiley-VCH: Weinhein, Germany, **1998**, 485–6.

154. Duan, X.; Liu, X.; Guo, L.; Liao, M.; Liu, W.-M.; Liang, Y. *J. Org. Chem.* **2005**, *70*, 6980–83.

155. Yoshida, M.; Morishita, Y.; Fujita, M.; Ihara, M. *Tetrahedron Lett.* **2004**, *45*, 1861–4.

156. Gabriele, B.; Salerno, G.; De Pascalli, F.; Scianò, G. T.; Costa, M.; Chiusoli, G. P. *Tetrahedron Lett.* **1997**, *38*, 6877–80.

157. Gabriele, B.; Salerno, G.; Lauria, E. *J. Org. Chem.* **1999**, *64*, 7687–99.

158. Bishop, B. C.; Cottrell, I. F.; Hands, D. *Synthesis* **1997**, 1315–20.

159. Qing, F.-L.; Gao, W.-Z.; Ying, J. *J. Org. Chem.* **2000**, *65*, 2003–6.

160. Garcon, S.; Vassiliou, S.; Cavicchioli, M.; Hartmann, B.; Monteiro, N.; Balme, G. *J. Org. Chem.* **2001**, *66*, 4069–73.

161. Bottex, M.; Cavicchioli, M.; Hartmann, B.; Monteiro, N.; Balme, G. *J. Org. Chem.* **2001**, *66*, 175–9.

162. Doe, M.; Shibue, T.; Haraguchi, H.; Morimoto, Y. *Org. Lett.* **2005**, *7*, 1765–8.

163. Villemin, D.; Goussu, D. *Heterocycles* **1989**, *29*, 1255–61.

164. For recent examples of the use of this strategy for the synthesis of benzofurans, see: (a) Sanz, R.; Castroviejo, M. P.; Fernández, Y.; Fañanás, F. J. *J. Org. Chem.* **2005**, *70*, 6548-51. (b) Dai, W.-M.; Lai, K. W. *Tetrahedron Lett.* **2001**, *43*, 9377-80. (c) Kabalka, G. W.; Wang, L.; Pagni, R. M. *Tetrahedron* **2001**, *57*, 8017–28.

165. Monteiro, N.; Balme, G. *Synlett* **1998**, 746–7.

166. Arcadi, A.; Cacchi, S.; Rosario, M. D.; Fabrizi, G.; Marinelli, F. *J. Org. Chem.* **1996**, *61*, 9280–8.

167. Monteiro, N.; Arnold, A.; Balme, G. *Synlett* **1998**, 1111–3.

168. Lütjens, H.; Scammells, P. J. *Synlett* **1999**, 1079–81.

169. Arcadi, A.; Cacchi, S.; Giuseppe, S. D.; Fabrizi, G.; Marinelli, F. *Org. Lett.* **2002**, *4*, 2409–12.

170. Liao, Y.; Smith, J.; Fathi, R.; Yang, Z. *Org. Lett.* **2005**, *7*, 2707–9.

171. Cacchi, S.; Fabrizi, G.; Moro, L. *Tetrahedron Lett.* **1998**, *39*, 5101–4.

172. Cacchi, S.; Fabrizi, G.; Moro, L. *Synlett* **1998**, 741–5.

173. Fancelli, D.; Fagnola, M. C.; Severino, D.; Bedeschi, A. *Tetrahedron Lett.* **1997**, *38*, 2311–4.

174. Thielges, S.; Meddah, E.; Bisseret, P.; Eustache, J. *Tetrahedron Lett.* **2004**, *45*, 907–10.

175. Sheng, H.; Lin, S.; Huang, Y. Z. *Tetrahedron Lett.* **1986**, *27*, 4893–4.

176. Jeevanandam, A.; Narkunan, K.; Ling, Y.-C. *J. Org. Chem.* **2001**, *66*, 6014–20.

177. Karpov, A. S.; Merkul, E.; Oeser, T.; Müller, T. J. J. *J. Chem. Soc., Chem. Commun.* **2005**, 2581–3.

178. Fukuda, Y.; Shiragami, H.; Utimoto, K.; Nozaki, H. *J. Org. Chem.* **1991**, *56*, 5816–9.

179. Okura, K.; Furuune, M.; Miura, M.; Nomura, M. *J. Org. Chem.* **1992**, *57*, 4754–6.

180. Picquet, M.; Bruneau, C.; Dixneuf, P. H. *Tetrahedron* **1999**, *55*, 3937–48.

181. Luo, F.-T.; Jeevanandam, A.; Bajji, C. *Tetrahedron Lett.* **1999**, *27*, 4893–4.

182. Ma, S.; Zhang, J. *J. Chem. Soc., Chem. Commun.* **2000**, 117–8.

183. For several recent examples, see: (a) Ma, S.; Gu, Z.; Yu, Z. *J. Org. Chem.* **2005**, *70*, 6291-4. (b) Ma, S.; Zhang, J.; Lu, L. *Chem. Eur. J.* **2003**, *9*, 2447–56.

184. Jaouhari, R.; Dixneuf, P. H.; Lécolier, S. *Tetrahedron Lett.* **1986**, *27*, 6315–8.

185. Kerrigan, F.; Martin, C.; Thomas, G. H. *Tetrahedron Lett.* **1998**, *39*, 2219–22.

186. Hooper, M. W.; Utsunomiya, M.; Hartwig, J. F. *J. Org. Chem.* **2003**, *68*, 2861–73.

187. Willis, M. C.; Taylor, D.; Gillmore, A. T. *Org. Lett.* **2004**, *6*, 4755–57.

Chapter 7

Thiazoles and benzothiazoles

Richard J. Mullins and Adam M. Azman

Thiazoles play a prominent role in nature. For example, the thiazolium ring present in vitamin B1 serves as an electron sink and its coenzyme form is important for the decarboxylation of α-keto-acids. Furthermore, thiazoles are useful building blocks in pharmaceutical agents as exemplified by 2-(4-chlorophenyl)thiazole-4-acetic acid, a synthetic anti-inflammatory agent.

| Vitamin B₁ | 2-(4-chlorophenyl)-
thiazole-4-acetic acid | thiazole |

Thiazole is a π-electron-excessive heterocycle. The electronegativity of the N-atom at the 3-position makes C(2) partially electropositive and therefore susceptible to nucleophilic attack. In contrast, electrophilic substitution of thiazoles preferentially takes place at the electron-rich C(5) position. More relevant to palladium chemistry, 2-halothiazoles and 2-halobenzothiazoles are prone to undergo oxidative addition to Pd(0) and the resulting σ-heteroaryl palladium complexes participate in various coupling reactions. Even 2-chlorothiazole and 2-chlorobenzothiazole are viable substrates for Pd-catalyzed reactions.

7.1. Synthesis of halothiazoles

Two of the most frequently used approaches for halothiazole synthesis are direct halogenation of thiazoles and the Sandmeyer reaction of aminothiazoles. The third method, an exchange between a stannylthiazole and a halogen, is not practical in the context of palladium chemistry simply because the stannylthiazole can be used directly in a Stille coupling.

7.1.1. Direct halogenation

Simple thiazole cannot be directly halogenated under standard conditions, but 2-methylthiazole can be brominated at the 5-position. If there are two substituents on the thiazole ring, the last vacant position then may be readily halogenated [1, 2].

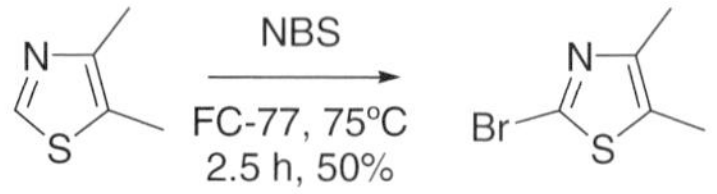

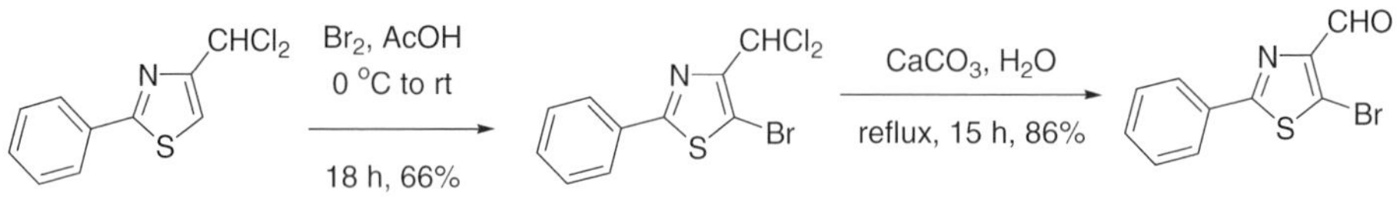

Dehydroxy-halogenation of thiazolidinedione (1) with phosphorus oxybromide led to 2,4-dibromothiazole (2) [3], whereas the same reaction conducted in DMF resulted in 2,4-dibromo-5-formylthiazole (3) via the Vilsmeier reaction [4].

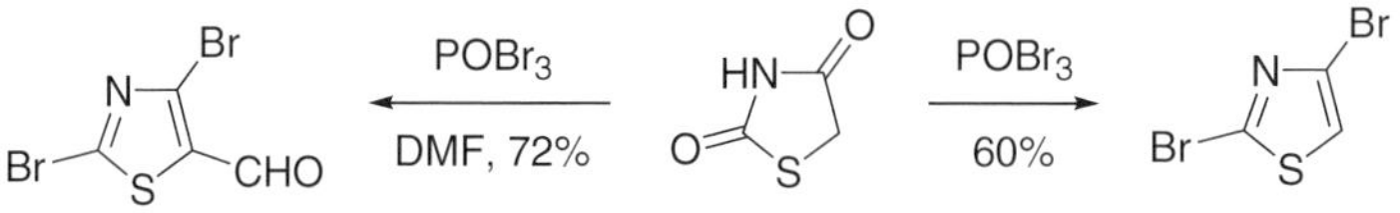

One drawback of the above reaction conditions involves the generation of toxic HBr upon quenching of the reaction. Thus, a more facile bromination procedure was developed in which phosphorus pentoxide was used in concert with a quaternary ammonium bromide to convert 2-hydroxybenzothiazole (4) into 2-bromobenzothiazole (5) [5].

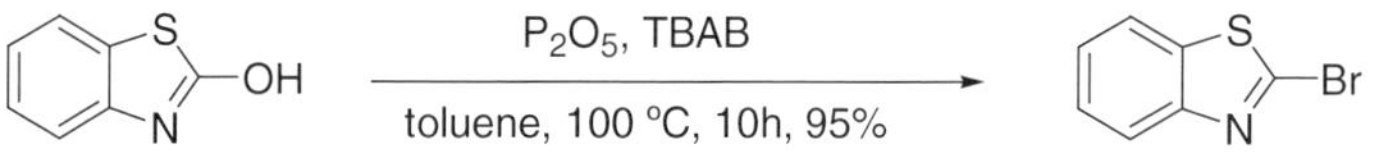

Commercially available 2,5-dibromothiazole can be alternatively generated by bromination of thiazole or 2-bromothiazole [6]. 2,4-Dibromothiazole and 2,5-dibromothiazole are among the most useful building blocks in the synthesis of thiazole-containing molecules. Regioselective bromination was achieved at C(4) when 2-amino-6-trifluoromethoxybenzothiazole (6) was treated with bromine in acetic acid to afford 4-bromobenzothiazole 7 [7].

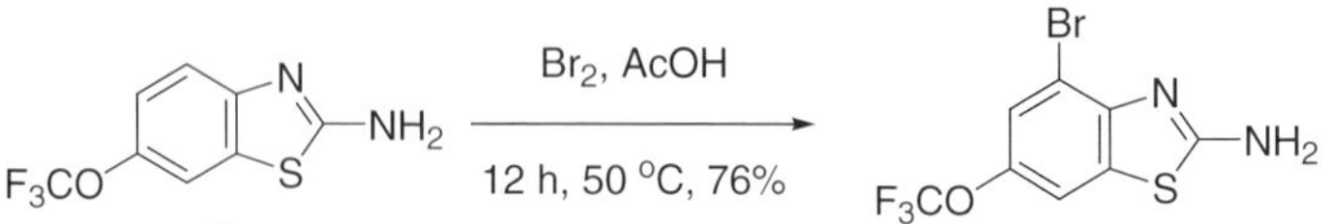

7.1.2. Sandmeyer reaction

The Sandmeyer reaction converts aminothiazoles, often commercially available, to halothiazoles via the intermediacy of diazonium salts. For instance, 2-aminothiazole was transformed into 2-iodothiazole [8, 9]. The reaction tolerates a number of functional groups including nitro and ester groups as shown below [10].

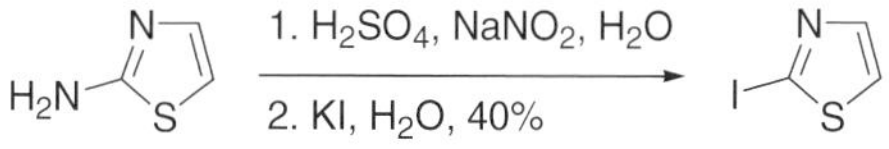

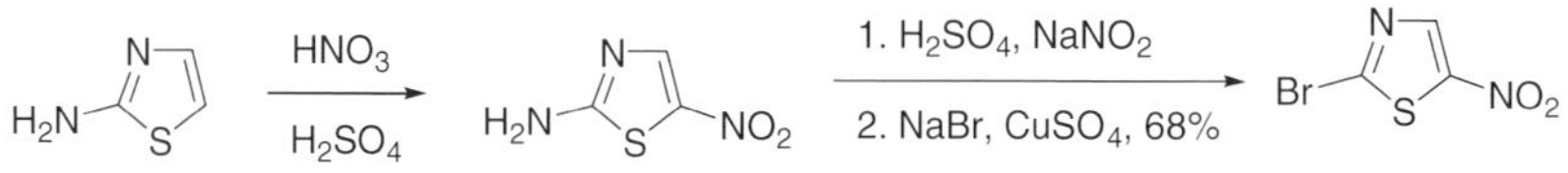

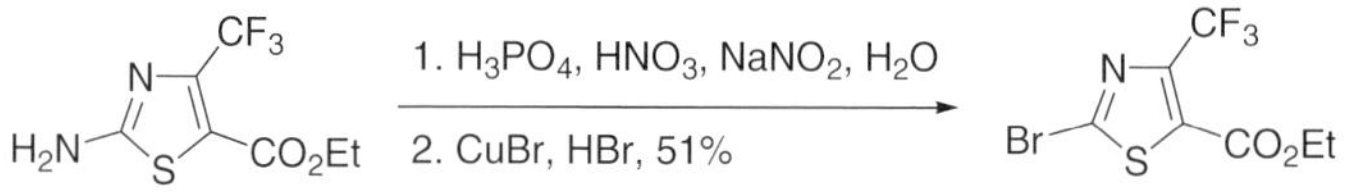

Some variants of the classic Sandmeyer conditions have been used for halothiazole synthesis. For example, sodium nitrite can be replaced with isoamyl nitrite or *tert*-butyl nitrite as illustrated by the transformation of 2-aminobenzothiazole **8** to 2-bromobenzothiazole **9** [11, 12].

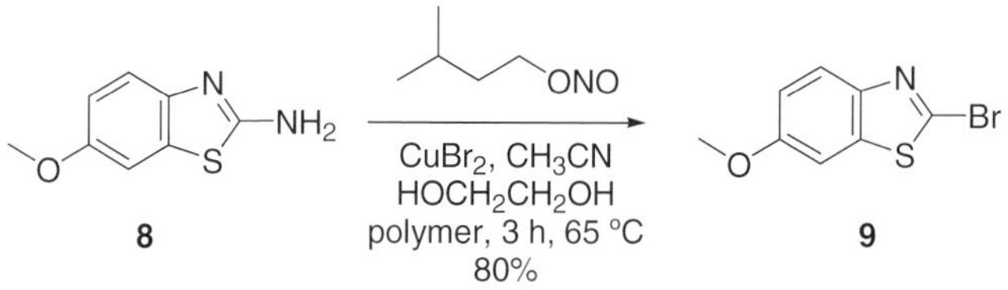

7.2. Coupling reactions with organometallic reagents

7.2.1. Negishi coupling

7.2.1.1. Thiazole as Negishi nucleophile

2-Thiazolylzinc halides have been widely employed as nucleophiles in Negishi couplings [13–15]. Deprotonation of thiazole followed by treatment with $ZnCl_2$ results in efficient preparation of 2-thiazolylzinc chloride. Subsequent Negishi coupling of thiazolylzinc chloride with 5-iodo-2′-deoxyuridine **10** elaborated the 5-substituted pyrimidine nucleoside **11** with the assistance of one equivalent of $ZnCl_2$ [15].

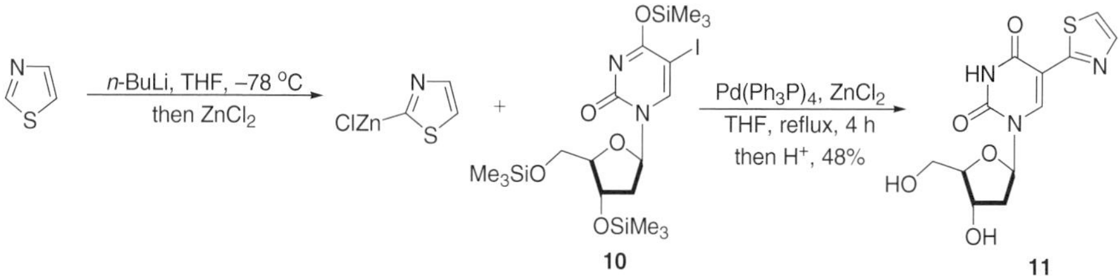

2-Thiazolylzinc bromide, formed *in situ* by the quenching of lithiothiazole with ZnBr$_2$, was coupled with 2-iodopyridine **12** to give thiazolylpyridine **13**. Hydrolysis of **13** then led to thiazolylpyridine acid **14**, an inhibitor of endothelin conversion enzyme-1 (ECE-1) [16].

Pd(II) intermediates obtained *in situ* from the intramolecular arylation of alkynes are viable coupling partners with organozinc reagents. For instance, treatment of 4-(*o*-iodophenyl)-1-butyne (**15**) with a catalytic amount of Pd(Ph$_3$P)$_4$ resulted in the *cis* insertion intermediate **16** [17]. Coupling of the reactive species **16** with 2-benzothiazolylzinc chloride, derived from treating 2-benzothiazolyllithium with one equivalent of anhydrous ZnCl$_2$, gave (*Z*)-1-indanylidene-substituted benzothiazole **17**, along with by-product **18**. Such a tandem reaction provides a quick entry to (*Z*)-1-indanylidene-substituted heteroaryls, which are otherwise not easily synthesized.

Palladium-catalyzed reactions with aryl triflates have been extensively used in organic synthesis. The electron-withdrawing ability of the CF$_3$SO$_2$ group is thought to be essential for a rapid insertion of Pd(0) to the C−O bond of the aryl triflate. Knochel's group reported an alternative using aryl nonaflates (Nf = SO$_2$(CF$_2$)$_3$CF$_3$) that are readily prepared and stable to flash chromatography [18, 19]. For instance, reaction of 4-iodophenol with commercially available FSO$_2$(CF$_2$)$_3$CF$_3$ in the presence of Et$_3$N in Et$_2$O afforded aryl nonaflate **19** in excellent yield. The Negishi reaction of bifunctional **19** with 2-thiazolylzinc bromide, derived from oxidative addition of 2-bromothiazole to Zn(0), led to arylthiazole **20** in the presence of Pd(dba)$_2$ and tri-*o*-furylphosphine (TFP). The resulting arylthiazole **20** with the pendant nonaflate functional group was then subjected to another Negishi coupling reaction with 3-trifluoromethylphenylzinc bromide to afford 4-(thiazol-2-yl)-3′-(trifluoromethyl)biphenyl (**21**).

The Negishi coupling, as well as other important transition metal-catalyzed coupling reactions, often relies on the presence of phosphine ligands to accelerate the coupling process. Unfortunately, the presence of these ligands complicates the separation and purification of the desired compound. Further destructive chemical transformation via oxidation of the phosphine, is often required to enable these separation processes. The expensive preparation of chiral phosphines suggests the necessity for the ability to recover these ligands in usable form, while accomplishing the goal of improved separation ability. Lipshutz and coworkers have demonstrated the precipitation of phosphine ligands from crude organometallic coupling reactions [20]. Specifically, this protocol was showcased in the Negishi reaction of metallated thiazole **22** and aryl bromide **23**. Following precipitation of the triphenylphosphine–copper complex, the product **24** was obtained in high yield and sufficient purity. It is notable that the more sophisticated phosphine ligands can be recovered in pure, reusable form by treatment of the phosphine–copper complex with a dithiol.

7.2.1.2. Thiazole as Negishi electrophile

Halothiazoles have found considerable utility as electrophiles in the Negishi coupling reaction. In an ECE-1-related project, the total synthesis of WS75624 B (**28**) commenced with regioselective halogen–metal exchange at C(2), and the resulting 2-lithio-4-bromothiazole was treated with thionolactone **25**. The intermediate sulfide was trapped with methyl iodide to give mixed ketal **26**. The Negishi reaction of **26** and the organozinc reagent derived from dimethoxypyridine provided adduct **27**, which was then further manipulated to accomplish the total synthesis of **28**, a non-peptide inhibitor of ECE [21a]. Recently, Bach and Heuser reported consecutive and regioselective Negishi and Suzuki couplings using 2,4-dibromothiazole [21b].

This reaction protocol has also found utility in materials synthesis. Specifically, the synthesis of 2,5-dithiazolyl-3,4-diphenylsilole **32** was accomplished via an elegant one pot procedure [22]. Following the intramolecular reductive cyclization of **29** and quenching of excess reducing agent Li-naphthalenide by chlorotriphenyl silane, production of the 2,5-dizinc silole **31** was accomplished via transmetalation with **30**. Negishi coupling was then accomplished in the same pot between **31** and 2-bromothiazole to yield **32**.

The Snieckus group has demonstrated an interesting variant of the Negishi cross coupling protocol with halothiazoles as an alternative to the classical Friedel–Crafts acylation methodology [23]. *Ortho*-metalation of *O*-vinyl carbamate **33** with *s*-BuLi was followed by transmetalation with zinc bromide. Upon addition of a palladium-catalyst, the cross coupling reaction with 2-bromothiazole was effective in producing **35**. Surprising in this study was the fact that Pd(PPh₃)₄ showed low reactivity for aryl bromides and triflates, necessitating the use of PdCl₂(dppf) to effect sufficient reactivity. Hydrolysis of **35** would then be expected to provide a methyl ketone, which would be the product of a Friedel–Crafts reaction at C(2) of the thiazole nucleus. Thus, this reaction manifold offers advantages over traditional Friedel–Crafts chemistry in that C(2) of the thiazole is usually not susceptible to attack by electrophilic species.

The difluorovinylzinc reagent **38** was generated by treating the corresponding vinyllithium **37**, derived from deprotonation of terminal fluoroolefin **36**, with ZnCl$_2$. Subsequent coupling of **38** with 2-iodobenzothiazole gave adduct **39** stereoselectively [24].

An alternative strategy for preparing the organozinc species was developed by the Knochel group [25]. Utilizing mesitylmagnesium bromide, a low temperature iodine/magnesium exchange reaction occurs with **40**, producing the organometallic **41**. Mesitylmagnesium bromide was chosen over phenylmagnesium bromide to avoid side reaction with the aryl iodide produced via the halogen–metal exchange. Transmetalation with ZnBr$_2$ thus provided an aryl organozinc reagent which underwent coupling with 2-bromothiazole to give **42**. Of significance in this report is the fact that the mild iodine/magnesium exchange reaction avoids side reactivity with the sensitive nitro functional group. Additionally, this method provides access to some nitro-containing biphenyls which are not available via electrophilic nitration.

Besides halothiazoles, thiomethyl groups have also found utility as leaving groups in the Negishi coupling reaction of thiazoles [26]. For instance, reaction of **43** with benzyl-zinc bromide under the normal Negishi conditions resulted in **44**.

7.2.2. Suzuki coupling

The synthesis of thiazole boronic acids is quite complicated. In fact, attempts to pre-pare 2-thiazoleboronic acid were unsuccessful [27]. As a consequence, the Suzuki reac-tion involving thiazoles is only possible when the thiazole component serves as the electrophilic coupling partner. Halothiazoles have shown the greatest utility in this regard. For instance, 2,5-di-(2-thienyl)thiazole (**45**) was prepared by the union of 2,5-dibromothiazole and easily accessible 2-thiopheneboronic acid [27]. Unfortunately, the yield was poor and analogous reactions of 2,5-dibromothiazole with 3-thiopheneboronic acid and 2-selenopheneboronic acid both failed.

Snieckus *et al.* enlisted a combination of directed *ortho*-lithiation and Suzuki coupling to assemble some unsymmetrical heterobiaryls [28]. Carboxamidophenylboronic acid **47** was derived from sequential metalation of amide **46** and treatment with B(OMe)$_3$ followed by acidic workup. Hetero cross-coupling of **47** with 2-bromothiazole occurred smoothly to furnish phenylthiazole **48**. Similarly, a hetero cross-coupling between 2-bromothiazole and 3-formyl-4-methoxyphenylboronic acid produced a heterobiaryl as an intermediate of an orally bioavailable NK$_1$ receptor antagonist [29].

An alternate strategy was utilized in the synthesis of a potential thyroid receptor ligand [30]. Electrophilic bromination was followed by the Miyaura boronic ester synthesis to yield **49** [31]. Suzuki coupling between **49** and 2-bromothiazole was then accomplished to provide **50**, a precursor to the derivative **51**. Unfortunately, the presence of the thiazole at C(3), in **51** resulted in a decrease in binding affinity and selectivity as compared to the lead compound.

In addition to arylthiazoles, heteroarylthiazoles also have been synthesized using halothiazoles and heteroarylboronic acids. Suzuki coupling of 2-bromothiazole and 5-indolylboronic acid led to 5-substituted indole **52** [32]. The Suzuki coupling of 2,4-dibromothiazole with 2,4-di-*t*-butoxy-5-pyrimidineboronic acid (**53**) resulted in selective heteroarylation at the 2-position to give pyrimidylthiazole **54** although the yield was low [33–35].

2-Bromobenzothiazole has also been used as the electrophile in the Suzuki coupling reactions with a variety of structurally diverse boronic acids [36]. Among these examples, derivative **55**, a positron emission tomography (PET) probe precursor was produced for the *in vivo* imaging of Alzheimer's disease.

Laurenti and Santelli have recently introduced the tetraphosphine ligand (Tedicyp) **57** for use in palladium coupling reactions [37]. The catalyst formed from the combination of **57** and [PdCl(C$_3$H$_5$)$_2$]$_2$ has been shown to be remarkably efficient for allylic substitution and Heck vinylation reactions [38]. Additionally, efficient yields have been achieved in the Suzuki coupling reaction between arylboronic acids and other heteroarylbromides [39]. Specifically, the coupling reaction between 2-bromothiazole and benzeneboronic acid to produce **56** was achieved using a substrate/catalyst ratio of 10,000:1.

Molander has published a series of papers demonstrating the utility of potassium alkyl, alkenyl-, alkynyl-, and aryltrifluoroborates in palladium coupling reactions. The crystallinity and air-stability of these trifluoroborate salts make the use of these an interesting alternative to the use of boronic acids or esters. Good yields have been obtained in several related palladium coupling processes, which are most easily classified as Suzuki couplings. The broad applicability of this process is demonstrated by the production of **58** [40], **59** [41], and **60** [42].

7.2.3. Stille coupling

7.2.3.1. Synthesis of stannylthiazoles

Two frequently used methods for preparing stannylthiazoles involve either direct metalation or halogen–metal exchange followed by treatment with alkyltin chloride. Dondoni *et al.* described a preparation of 2-, 4-, and 5-trimethylstannylthiazoles in 1986 [43]. For instance, 2-trimethylstannylthiazole was readily obtained by quenching 2-lithiothiazole, derived from direct metalation of thiazole, with trimethyltin chloride.

1. *n*-BuLi, Et$_2$O, –78 °C

2. Me$_3$SnCl, 96%

On the other hand, the preparation of 5-trimethylstannylthiazole began with selective deprotonation of 2-trimethylsilylthiazole at C(5) followed by treatment with Me$_3$SnCl, giving rise to 2-trimethylsilyl-5-trimethylstannylthiazole (**63**). Deprotection of **63** under acidic conditions then gave 5-trimethylstannylthiazole [43]. In this case, the trimethylsilyl group serves as a protecting group, preventing deprotonation from occurring at the normally more acidic C(2) position of the thiazole ring.

n-BuLi, Et$_2$O, –78 °C

then Me$_3$SnCl, 90%

THF, H$^+$

rt, 93%

63

Similarly, 2-(tributylstannyl)benzothiazole was prepared by deprotonation of benzothiazole with *n*-BuLi followed by the addition of tributyltin chloride [44, 45].

n-BuLi, THF

Bu$_3$SnCl, 80%

The second method to prepare stannylthiazoles involves a halogen–metal exchange followed by quenching with trimethyl- or tributyltin chloride. Gronowitz *et al.* carried out the halogen–metal exchange of 2-bromothiazole and treated the resulting 2-lithiothiazole with a solution of Bu$_3$SnCl in Et$_2$O to furnish 2-tributylstannylthiazole [46].

n-BuLi, Et$_2$O, –70 °C

Bu$_3$SnCl, –70 °C, 4 h

49%

It has been demonstrated in the literature that the halogen–metal exchange of 2,4-dibromothiazole occurs predominantly at the C(2) position. 2-Stannyl-4-bromothiazole can be prepared using this strategy [47].

1. *n*-BuLi, Et$_2$O, –78 °C

2. Me$_3$SnCl, 89%

In order to prepare 4-trimethylstannylthiazoles selectively, C(2) was once more protected with the easily removable trimethylsilyl group. Regioselective halogen–metal exchange of 2,4-dibromothiazole occurred at C(2) and the resulting 2-lithio-4-bromothiazole was trapped with trimethylsilyl chloride to give 2-trimethylsilyl-4-bromothiazole (**61**). Subsequent halogen–metal exchange of **61** followed by treatment with Me_3SnCl led to **62**, which was then hydrolyzed to 4-trimethylstannylthiazole [43].

The preparation of stannylthiazoles via ditin chemistry has not been widely utilized. In one case, the synthesis of 4-trimethylstannylthiazole **64** started with selective halogen–metal exchange at C(2) by treating 2,4-dibromothiazole with *n*-BuLi [48]. Trapping the resulting 2-lithio-4-bromothiazole with propanal and subsequent Jones oxidation secured 4-bromothiazole **63**. The Pd-catalyzed reaction of **63** with hexamethyldistannane in the presence of $PdCl_2(Ph_3P)_2$ provided 4-tributylstannylthiazole **64**.

A novel homolytic substitution yielded 2-(tributylstannyl)benzothiazole [49]. Thus, 2-(alkylsulfonyl)benzothiazole **65** was allowed to react with 2 equivalents of tributyltin hydride in the presence of catalytic azobisisobutyronitrile (AIBN) in refluxing benzene, affording 2-(tributylstannyl)benzothiazole along with tributylstannylsulfinate **66**.

7.2.3.2. Thiazole as Stille nucleophile

The Stille coupling of stannylthiazole with an assortment of electrophiles has been achieved. The electrophilic partners span a wide spectrum of halides including allyl chlorides. Chloromethylcephem **67** was coupled with 2-tri-*n*-butylstannylthiazole to give **68**, which was then manipulated into a C(3) thiazole analog of cephalosporin [50].

67 → **68**

With respect to the coupling reactions of stannylthiazoles with aryl halides, the union of 4-chlorobromobenzene and 2-tributylstannylthiazole constructed arylthiazole **69** [51]. The Stille reaction of 4-bromo-2-nitrobenzylphosphonate (**70**) and 2-tributylstannylthiazole led to heterobiaryl phosphonate **71**, which may be utilized as a substrate in a Wadsworth–Horner–Emmons reaction or a bioisosteric analog of a carboxylic acid [52]. The phosphonate did not interfere with the reaction. In addition, the coupling of 5-bromo-2,2-dimethoxy-1,3-indandione (**72**) and 2-tributylstannylbenzothiazole resulted in adduct **73**, which was then hydrolyzed to 5-(2′-benzothiazolyl)ninhydrin [53].

69

70 **71**

72 **73**

An attempt to generalize the ditin chemistry presented above for the generation of a stannylthiazole proved unsuccessful. Even simple bromothiazole was recalcitrant towards such a method. However, treatment of **74** with hexamethyldistannane in the presence of $PdCl_2(Ph_3P)_2$ gave dimerized product **75** and debrominated product **76** as the two major products [54]. The isolation of the dimer **75** and nuclear magnetic resonance (NMR) evidence indicated that the desired 2-stannylthiazole from **74** was generated during the course of the reaction. Thus, trapping the stannane *in situ* could be accomplished when pyridyl triflate **77** was refluxed in the presence of $PdCl_2(Ph_3P)_2$ with slow addition of a dioxane solution of **78**, furnishing dimethyl sulfomycinamate (**79**), the methanolysis product of the antibiotic sulfomycin I [54].

An additional Stille reaction involves treatment of a 2-(tributylstannyl)benzothiazole with bromobenzene under the influence of $PdCl_2(Ph_3P)_2$ leading to 2-phenylbenzothiazole (**80**) [44, 45].

In another case, 2-tributylstannylthiazole was coupled with 2′-deoxyuridine iodide **81** to furnish 5-thiazolyl-2′-deoxyuridine **82** [55].

Stannylthiazoles have also been coupled with an array of heterocyclic halides. Reaction of 2.4 equivalents of 2-trimethylstannylthiazole with 2,5-dibromothiophene gave 2,5-di-(2′-thiazolyl)-thiophene (**83**) [56]. Similarly, Dondoni's group coupled 2,5-dibromothiazole with 2 equivalents of 2-trimethylstannylthiazole to assemble trithiazole **84** [57].

6-Substituted thiazolylindole **87** was formed via the Stille coupling of 6-iodoindole **85** and 5-tributylstannylthiazole **86** [58]. Stannane **86** was not stable at temperatures higher than 60°C and decomposition occurred presumably by amidic hydrogen transfer on the 5-position. Simply blocking NH with a methyl group greatly improved the yield.

Synthesis of a thiazolylpyridine via the Stille coupling has been reported [59, 60]. Also described was a Stille coupling of a bromoquinolizinium salt **88** and 2-tributylstannylth-iazole [61]. The reaction was conducted at room temperature to give a substituted quino-lizinium salt **89**.

In a similar manner, Stille coupling of 2-tributylstannylthiazole and 5-bromopyrimi-dine **90** afforded 5-(2′-thiazolyl)-2,4-bis(trimethylsilyl)pyrimidine (**91**), which was hydrolyzed with acid to the corresponding uracil **92** [46].

The Liebeskind group cross-coupled 4-chloro-2-cyclobutenone **93** with 2-tributylstan-nyl-benzothiazole to synthesize α-pyridone-based azaheteroaromatics [62]. The adduct **94** underwent a thermal rearrangement to afford a transient vinylketene **95**, which then intramolecularly cyclized onto the C–N double bond of benzothiazole, giving rise to thia-zolo[3,2-*a*]pyridin-5-one **96**. In another case, 2-acetyl-4-trimethylstannylthizaole (**97**) was coupled with an acid chloride **98** to form the desired ketone **99** [63].

Nicolaou *et al.* took advantage of the Stille coupling to install the thiazole motif of epothilones and to conduct analog synthesis using isosteres of thiazole [64]. Thus, the union of vinyl iodide **100** and stannane **101** led efficiently to adduct **102** as the precursor to epothilone E. All the functional groups including hydroxyl groups, ketone, ester, and olefins were preserved. This is a good example of functional group tolerance in the Stille coupling.

Kelly's total synthesis of micrococcinic acid (**113**) is probably the best example to showcase the utility of the Stille reaction in the synthesis of thiazole-containing molecules [48]. The first Stille coupling was carried out between bromothiazole **103** and trimethylstannylpyridine **104** to form thiazolylpyridine **105**. Similarly, the second coupling between trimethylstannylthiazole **107** and bromothiazole **103** afforded dithiazole **108**. The ethoxy group in **105** and the trimethysilyl unit in **108** were modified to give bromides **106** and **109**, respectively. Subsequently, the Stille–Kelly reaction featuring a 1:1 mixture of **106** and **109** with (Me_3Sn–$SnMe_3$) and $PdCl_2(Ph_3P)_2$ furnished the desired cross-coupled **110**. Conventional functional group transformations led to triflate **111**, which was subjected to the fourth Stille coupling with stannane **112** to secure the pentacycle which was converted to micrococcinic acid (**113**) via acidic cleavage of the two amides. Kelly's total synthesis of micrococcinic acid (**113**) outlined here is a strong testimony to the utility of the Stille coupling reaction even for very complex molecules.

7.2.3.3. Thiazole as Stille electrophile

The Stille reaction has also found application utilizing halo- and other substituted thiazoles as electrophilic coupling partners with aryl- and alkenylstannanes. In the search for more potent inhibitors of the norepinephrine transporter, a variety of heterocyclic groups were introduced via the Stille coupling between alkenyl stannane **114** and heteroaryl halides [65]. The introduction of the thiazole nucleus was accomplished using 2-iodothiazole in the presence of $Pd_2(dba)_3$, Ph_3As, and CuI, producing **115** in an acceptable yield. Unfortunately, derivatives which contained a nitrogen in the appended heterocycle showed relatively little inhibition of reuptake at monoamine transporters ([³H]DA uptake Ki 5.09 nm for **115**). It was proposed that this and other similar heteroaromatics are too basic, thus making the binding disfavored in the lipophilic pocket of the norepinephrine transporter.

In preparing a library of β-lactamase inhibitors, 2-bromothiazole was utilized in the Stille coupling reaction with **116** and **117** [66]. Under identical conditions, the potential inhibitors were separately produced in good yields. Following the coupling reaction, the thioethers were subsequently oxidized to the corresponding sulfones **118** and **119**.

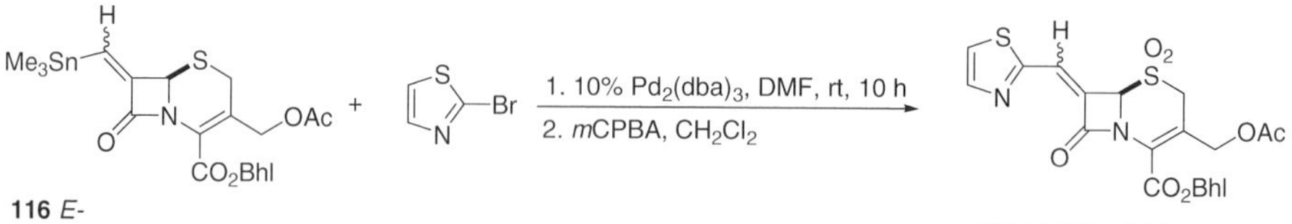

116 *E-*
117 *Z-*

118 *E-* 68% yield
119 *Z-* 48% yield

In addition, Undheim *et al.* coupled 2-bromothiazole with 2-methylthio-5-stannylpyrimidine (**120**) to assemble the pyrimidinethiazole **121** [67]. Stannane **120** was prepared by Pd-catalyzed coupling between 2-methylthio-5-bromopyrimidine and hexamethyldistannane in the presence of fluoride ion at ambient temperature.

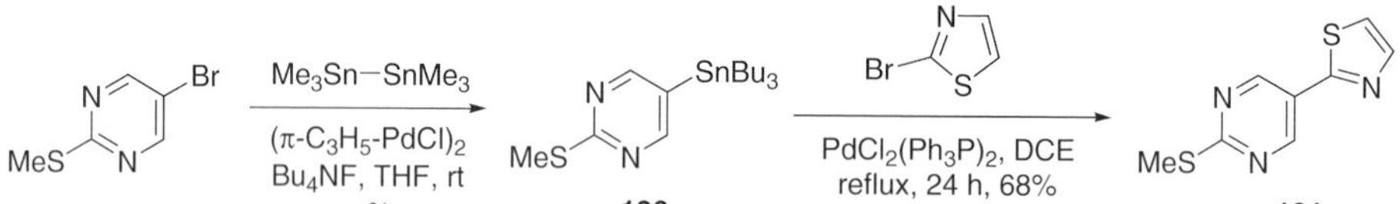

120

121

In 2000, Liebeskind and Srogl reported a Pd-catalyzed boronic acid-thioether cross-coupling protocol which was mediated by copper(I)-carboxylate [68]. Recognizing their increased accessibility, this protocol was successfully applied to the Stille-type coupling reactions of organostannanes with heteroaromatic thioethers [69]. Specifically, the Pd-catalyzed reaction of **122** and stannane **123** was promoted by copper(I)-3-methylsalicylate to provide **124** in excellent yield.

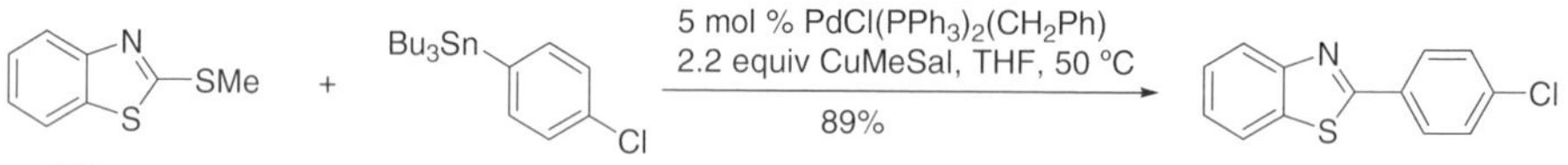

122

123

124

Using 2-furylstannane as the Stille coupling partner, the reaction was found to be rather sluggish under the conditions described above. This decrease in yield was attributed to the decomposition of the Palladium-catalyst over prolonged reaction times. Thus, a switch to the more stable $Pd(PPh_3)_4$, as well as replacement of the *S*-methyl ether with a more labile *S*-phenyl ether (as in **126**), resulted in improved reactivity [69].

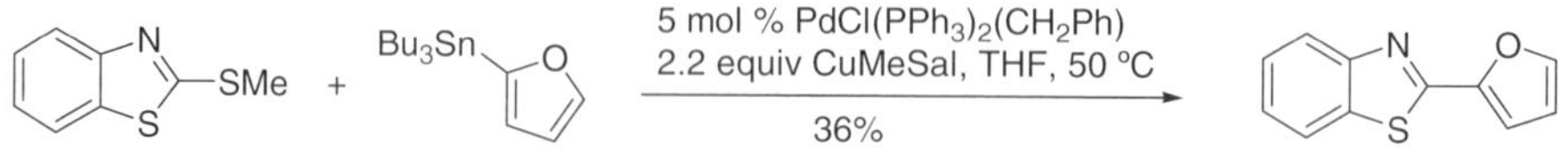

122

125

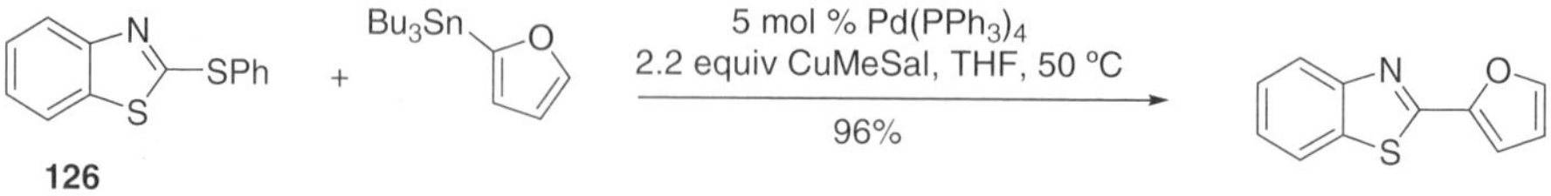

126

125

Around the same time that Liebeskind reported the above, Guillaumet and coworkers achieved similar results [70]. The only difference in the two works was in their noted improvement in the reactions utilizing CuBr–DMS as the copper cofactor. For example, the use of the *S*-methyl ether substituent (in **122**) in coupling with 2-stannylfuran produced **125** in improved yield as compared to the related Liebeskind example.

7.3. Sonogashira reaction

In 1987, Yamanaka's group described a Pd-catalyzed reaction of halothiazoles with terminal acetylenes [71]. While the yield for the Sonogashira reaction of 2-bromo-4-phenylthiazole (**126**) with trimethylacetylene to afford **127** was moderate (36% after desilylation), the coupling of 4-bromothiazole and 5-bromo-4-methylthiazole with phenylacetylene gave the desired internal acetylenes **128** and **129** in 71 and 65% yield, respectively.

2-Iodobenzothiazole was coupled with trimethylsilylacetylene to give adduct **130** which was readily desilylated to furnish 2-ethynyl-1,3-benzothiazole (**131**) [72].

Dimethyl propargyl alcohol **132** serves as a mask for the corresponding terminal acetylene. Therefore, basic cleavage of **132** unveiled the terminal acetylene, which was coupled *in situ* with 2-bromobenzothiazole in the presence of a phase-transfer catalyst to afford the unsymmetrical diarylbutadiyne **133** [73].

The Sonogashira reaction of 2-substituted-5-acetyl-4-thiazolyl triflate **134** and phenyl-acetylene led to 3-alkynylthiazole **135**, which subsequently underwent a *6-endo-dig* annulation in the presence of ammonia to produce pyrido[3,4-*c*]thiazole **136** [74].

Recognizing the potential advantages of a conformationally rigid alkyne spacer in the side chain, Höfle and Karama utilized a Sonogashira reaction for epothilone analog synthesis [75]. Specifically, the coupling reaction between 4-bromo-2-methylthiazole and alkyne **137** was carried out over 6 hours at elevated temperatures. In a mere four steps, involving a dicyclohexyl carbodiimide (DCC) coupling, ring-closing metathesis, protecting group removal and epoxidation, the efficient synthesis of analog **139** was completed. Unfortunately, the alkyne analog **139** showed only moderate cytotoxicity (IC$_{50}$ ≥ 500 ng/mL) against the standard mouse fibroblast cell line L929.

137   **138**   **139**

Researchers at Merck employed the Sonogashira coupling reaction of **140** and 2-bromothiazole for the synthesis of a potential non-nucleoside HIV-1 reverse transcriptase inhibitor [76]. Initial studies on this coupling reaction were slowed by the presence of acetylenic homocoupling when run in the presence of cuprous iodide. This side reaction, commonly observed in the Sonogashira coupling has been attributed to oxidative homocoupling of the copper acetylide, prior to the desired Pd-catalyzed coupling. Removal of the copper salt, however, provides a solution to the dimerization problem. While **141** was produced in a meager 37% yield, it was found to be one of the more potent inhibitors of HIV-1 RT (IC_{50} = 10.9 nM). Additionally, **141** proved to be an efficient anti-HIV-1-agent in a cell culture (CIC_{95} = 12 nM).

140   **141**

In a reaction which demonstrates the functional group tolerance of the Sonogashira reaction, Cristalli and coworkers synthesized **143** as one of the number of derivatives of adenosine-5'-N-ethyluronamide [77]. This, along with several other derivatives was screened to assess potency in binding at the A_{2a} over the A_1 adenosine receptor. Compound **143** was found to be virtually non-selective for the A_{2a} receptor (2.1:1), but possessed substantial vasodilating activity.

142   **143**

The Sonogashira reaction has been combined with an alkyne hydroamination for the synthesis of several 2-arylethylamine compounds [78]. Specifically, compounds **144** and **145** were produced in this manner. Following coupling under standard Sonogashira conditions, a regioselective Cp_2TiMe-catalyzed hydroamination produced a ketimine, which was reduced to provide the 2-arylethylamine. A high level of diversity can be obtained utilizing this process, as three commercial building blocks (aryl halides, terminal alkynes, and primary amines) are coupled in a straightforward and efficient process.

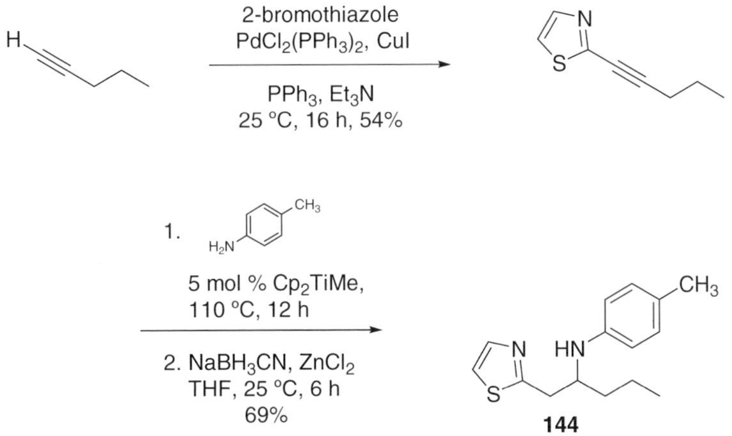

144

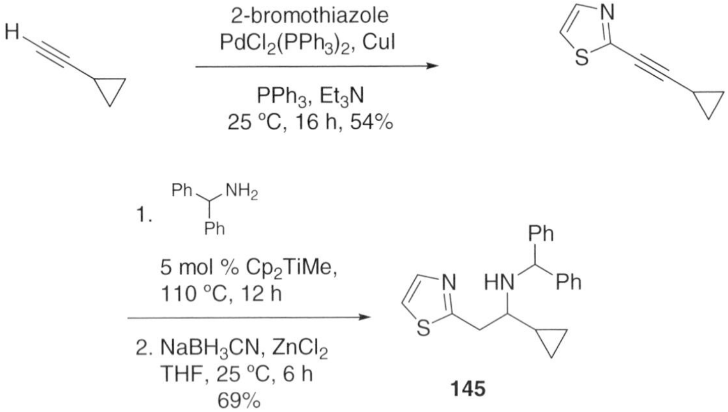

145

Müller and coworkers have recently developed a coupling-isomerization reaction, initially identified as a side reaction which occurred under standard Sonogashira conditions [79]. As demonstrated below, the coupling reaction is followed by a shuffling of oxidation states via an alkyne-allene isomerization [80]. The product, α,β-unsaturated ketone **146**, is reminiscent of a product which would be obtained from a Heck reaction. The utility of this reaction was further demonstrated when diamine **147** was added to the reaction pot. Following a conjugate addition reaction and imine formation, compound **148** resulted from the three-component, one-pot reaction sequence enabled by the coupling-isomerization reaction.

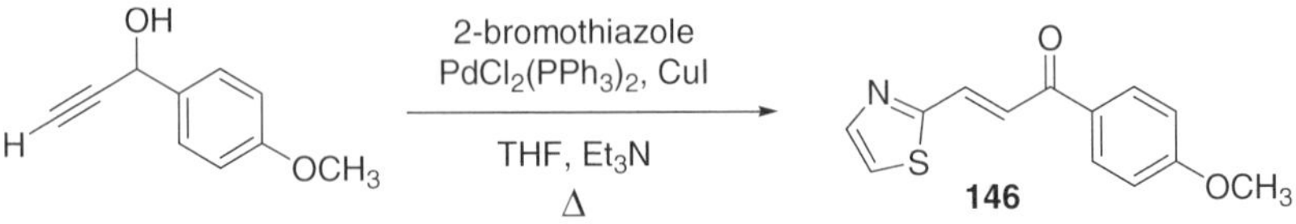

146

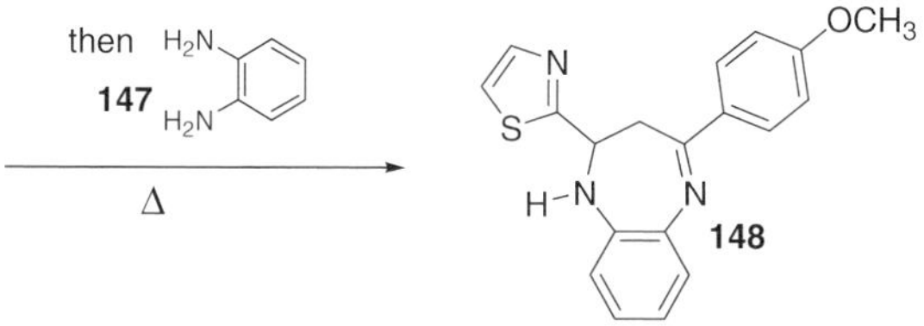

In a related one-pot procedure, the power of this methodology was further demonstrated when α,β-unsaturated ketone **149** was intercepted by enamine **150**. Following Michael addition and imine hydrolysis, 1,5-diketone **151** was produced [81].

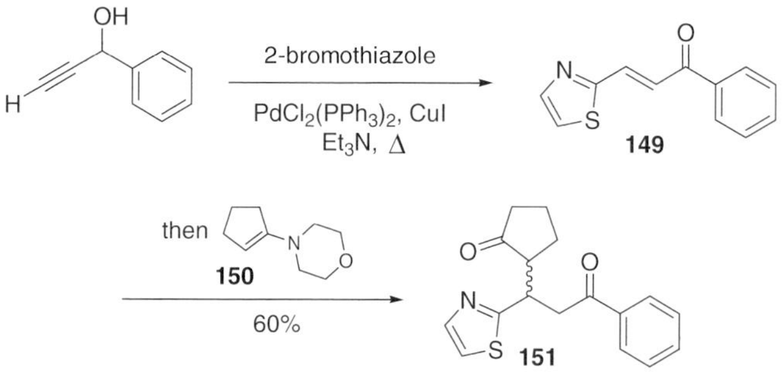

7.4. Heck and heteroaryl Heck reactions

Heck *et al.* reported that Pd-catalyzed reaction of 2-bromothiazole with methyl acrylate using tri-(*o*-tolyl)phosphine as the ligand failed to give significant amounts of the adduct [82]. During the investigation on 2-halo-1,3-azoles conducted by Yamanaka's group, they discovered that both a Sonogashira reaction with acetylenes and a Heck reaction with terminal olefins gave a large amount of resinous substance, probably due to ring-cleavage caused by palladation at the 2-position of the substrates [83]. Similarly, the Heck reaction of 4-bromothiazole with terminal olefins gave the adducts in only 8–19% yields. The results for the Heck reactions of 5-bromothiazoles were idiosyncratic as well. Very different results were observed for different substrates. As illustrated below, 5-bromothiazole **152** was subjected to the Heck reaction conditions with ethyl acrylate to afford **153** in 61% yield, whereas the same reaction using styrene led to almost exclusively the reduction product **154** along with only 1% of the Heck adduct.

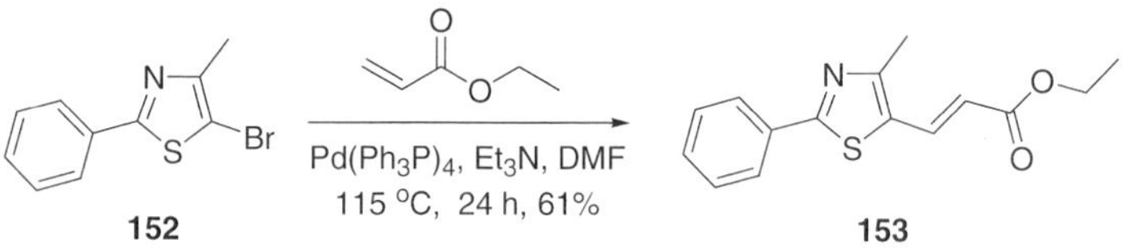

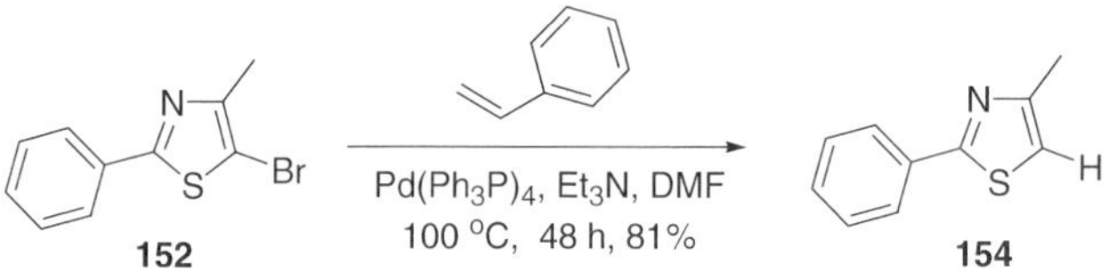

In contrast, thiazoles and benzothiazoles are suitable recipients for the "heteroaryl Heck" reaction. Treatment of 2-chloro-3,6-diisobutylpyrazine (155) with thiazole led to regioselective addition at C(5), giving rise to adduct 156 [84]. A similar reaction between 2-chloro-3,6-diethylpyrazine (157) and benzo[b]thiazole took place at C(2) exclusively to afford pyrazinylbenzothiazole 158 [84].

An analogous result was observed for the reaction of iodobenzene and 2-methylthiazole to furnish 5-phenyl-2-methylthiazole (163) [85]. The mechanism for the heteroaryl Heck reaction may be exemplified by the formation of 163. Oxidative addition of iodobenzene to Pd(0) gives intermediate 160, which subsequently inserts onto thiazole regioselectively at the electron-rich C(5) position to form 161. The regioselectivity of the Heck reaction is analogous to that seen with electrophilic substitution on thiazole, which is known to be C(5) > C(4) > C(2) [86]. The base present deprotonates the insertion adduct 161, giving rise to aryl(thiazolyl)palladium(II) intermediate 162, which then undergoes a reductive elimination to afford 5-phenylthiazole (163) and regenerates Pd(0) for next catalytic cycle.

This reaction motif has been nicely extended to the preparation of a number of 2,5-disubstituted thiazoles. Optimization of the reaction conditions demonstrated the need for very bulky phosphine ligands, such as P(t-Bu)$_3$ and P(biphenyl-2-yl)(t-Bu)$_2$ [87]. Ultimately, the reaction has proven to be quite general, efficiently producing **164**, **165**, and **166**.

Remarkably, this reaction motif has been successfully applied for the synthesis of sterically hindered 2,3,5-trisubstituted thiazoles, as well. The authors proposed that the carbamoyl group at C(5) serves as a blocking group, allowing coupling to occur at C(4) [87]. After the carbamoyl group is cleaved under the reaction conditions, further arylation occurs at C(5). This process is quite efficient in the production of **167**.

In many cases, selectivity for monoarylation, under conditions described above, is rather unreliable. Kondo and coworkers have described a solution to this problem, employing the arylation coupling on solid support [88]. As expected, the coupling reaction of immobilized iodobenzoate was followed by methanolysis to produce exclusively the 5-arylated thiazole **168**. Addition of CuI caused a complete reversal in regioselectivity, directing the arylation to the 2-position of the thiazole nucleus, as in **169**. While intriguing, this behavior is not unique to the solid-phase version of this reaction. The immobilized product **169** could then participate in a second coupling/methanolysis sequence, resulting in **171**. Utilizing this method, unsymmetrical 2,5-diarylated thiazoles can be obtained.

An additional limitation to the heteroaryl Heck process deals with the high tempera-tures required for sufficient reactivity. The high temperatures typically necessary for this process can be circumvented using the conditions developed by Maleczka and coworkers [89]. The addition of polymethylhydroxysilane and cesium fluoride to the reaction mix-ture allows coupling to proceed in yields similar to those previously obtained. However, these reactions are complete after stirring for 17 hours at room temperature. Specifically, arylation of benzothiazole occurs to provide **172**.

7.5. Carbonylation

The Pd-catalyzed alkoxycarbonylation of 5-bromothiazole using $PdCl_2(Ph_3P)_2$ as the catalyst in the presence of CO, EtOH, and Et_3N led to smooth formation of 5-ethoxycar-bonylthiazole (**173**) [90]. Et_3N served as an HBr scavenger, facilitating the transforma-tion. This is one of the first examples of the Pd-catalyzed alkoxycarbonylation of a haloheterocycle, providing a variety of heterocyclic esters.

7.6. C−N bond formation

An extension of Buchwald's Pd-catalyzed amination led to amination of 2-chloroben-
zothiazole with piperidine **174** to form **175** [91]. The marked reactivity enhancement of
the chloride is attributed to the polarization at C(2). The methodology is also suitable for
2-chlorobenzoxazole and 2-chlorobenzoimidazole.

Several examples of C−N bond formation have emerged from the Hartwig laboratories [92]. Notable is the use of $Pd(O_2CCF_3)_2$ with K_3PO_4 as the base of choice for the amination reaction between morpholine and 2-bromothiazole. It had been found that
2-bromothiazole decomposed using the stronger base, NaO^tBu.

7.7. Site selective coupling reactions

The ease of halogen displacement from the 2-position has enabled a large number of site
selective coupling reactions of a thiazole nucleus containing multiple halogens. Research in
this area has been pioneered in the Bach laboratories [93]. Their impressive synthesis of cystothiazole E (**180**) demonstrates the versatility of this protocol [94]. Following a Negishi coupling at the 2-position of the 2,4-dibromothiazole, hydrogenation of the alkene provided **177**.
A second Negishi coupling using the organozinc species derived from **177** occurred once
again, selectively, at the 2-position of **176**. This impressive synthetic sequence finally culminated in the Suzuki reaction between **178** and the highly functionalized boronic acid **179**.

Another member of the family, cystothiazole A (**185**) was synthesized in a similar manner by Panek and coworkers, taking advantage of site selective coupling reactions [95]. Once again, a regioselective Negishi coupling provided **177**. Following conversion to the 4-stannane derivative **181**, a regioselective Stille reaction was utilized with the bistriflate **182**. A second Stille coupling with **184** was employed at the 4-position of **183** to complete the synthesis of cystothiazole A (**185**).

In a comprehensive investigation of the regioselectivity of palladium coupling reactions on thiazole **186**, Hodgetts demonstrated that complete site selectivity can be obtained in Suzuki, Stille, Sonogashira, and Negishi reactions [96]. As shown in the following table, coupling took place exclusively at the 2-position under conditions which are standard for each reaction type. In all but the Sonogashira, yields were excellent. No byproducts associated with substitution at the 5-position were observed.

Reaction Type	Substrate (R-M)	Time (h)	Yield (%)
Suzuki	PhB(OH)$_2$	4	81
Stille	CH$_2$CHSnBu$_3$	12	91
Sonogashira	PhCCH	4	19
Negishi	(2-Pyridyl)ZnBr	12	74

The product of the Suzuki coupling **187** was then submitted to each of the different coupling reactions under standard conditions to produce trisubstituted thiazoles [96]. Notably, these reactions were noticeably slower than the first substitution, which is to be expected given the ease of substitution at C(2) as compared to C(5). Additionally, a larger excess of organometallic was necessary to drive the reactions to completion.

Reaction Type	Substrate (R-M)	Time (h)	Yield (%)
Suzuki	PhB(OH)$_2$	16	87
Stille	CH$_2$CHSnBu$_3$	24	72
Sonogashira	PhCCH	24	56
Negishi	(2-Pyridyl)ZnBr	24	70

The utility of this process was further demonstrated when the one-pot Suzuki coupling procedure was demonstrated to produce **188** in excellent yield [96]. Procedurally, after the first coupling had occurred, phenyl boronic acid was added along with additional Pd(PPh$_3$)$_4$.

Finally, the ester group at C(4) provides a handle for further functionalization. It can be removed via a hydrolysis/decarboxylation sequence to provide the disubstituted thiazole nucleus [96]. On the other hand, hydrolysis of **188** was followed by a Hunsdiecker reaction to produce bromide **189**.

With the bromide **189** secured, each major coupling reaction efficiently occurs at the C(4) position, as shown in the following table [96]. This methodology, being compatible with several organometallic processes, offers great flexibility for the synthesis of di- and trisubstituted thiazole derivatives.

Reaction Type	Substrate (R-M)	Time (h)	Yield (%)
Suzuki	PhB(OH)$_2$	2	94
Stille	CH$_2$CHSnBu$_3$	24	81
Sonogashira	PhCCH	24	61
Negishi	(2-Pyridyl)ZnBr	24	75

The power of this methodology has been demonstrated by the facile synthesis of a thia-zolo[4,5-c]quinolinone, a class of molecule which was anticipated to have high affinity for the GABA receptor [97]. In a single reaction flask, beginning with **186**, consecutive, site-selective Suzuki couplings were followed by acyl substitution to provide **190**.

To summarize, among the Pd-catalyzed cross-coupling reactions involving thiazoles, the Stille coupling has once again proven to be the most robust among all organometallics, displaying better tolerance of a wide variety of functional groups. In an academic research laboratory or a drug discovery setting, it is the method of choice for preparing substituted thiazoles. However, an alternative route, such as Negishi coupling should be pursued first in a large-scale preparation simply because of stannane toxicity.

For the Heck reactions involving halothiazoles, 2-bromo- and 4-bromothiazoles tend to give resinous products, whereas some 5-bromothiazoles may form the desired Heck adduct with appropriate olefins. With respect to the heteroaryl Heck reaction using a thia-zole as a coupling partner, the addition occurs regioselectively at the electron-rich C(5) position, whereas it occurs at C(2) for benzothiazole.

7.8. References

1. Al Hariri, M.; Galley, O.; Pautel, F.; Fillion, H. *Eur. J. Org. Chem.* **1998**, *4*, 593–4.
2. Ceulemans, E.; Dyall, L. K.; Dehaen, W. *Tetrahedron* **1999**, *55*, 1977–88.
3. Reynaud, P.; Robba, M.; Moreau, R. C. *Bull. Soc. Chim. Fr.* **1962**, 1735–8.
4. Kerdesky, F. A. J.; Seif, L. S. *Synth. Commun.* **1995**, *25*, 2639–715.
5. Kato, Y.; Okada, S.; Tomimoto, K.; Mase, T. *Tetrahedron Lett.* **2001**, *42*, 4849–51.
6. Klein, P. *Helv. Chim. Acta* **1954**, *37*, 2057–67.
7. Mignani, S.; Audiu, F.; Le Belvec, J.; Nemecek, C.; Barreau, M.; Jimonet, P.; Gueremy, C. *Synth. Commun.* **1992**, *22*, 2769–80.
8. Neenan, T.; Whitesides, G. M. *J. Org. Chem.* **1988**, *53*, 2489–96.

9. Gellis, A.; Vanelle, P.; Maldonado, J.; Crozet, M. P. *Tetrahedron Lett.* **1997**, *38*, 2085–6.

10. Lee, L. F.; Schleppnik, F. M.; Howe, R. K. *J. Heterocycl. Chem.* **1985**, *22*, 1621–30.

11. Suzuki, N.; Nomoto, T.; Toya, Y.; Yoda, B.; Saeki, A. *Chem. Express* **1992**, *7*, 717–20.

12. Suzuki, N.; Nomoto, T.; Toya, Y.; Kanamori, N.; Yoda, B.; Saeki, A. *Biosci. Biotechnol. Biochem.* **1993**, *57*, 1561–2.

13. Vincent, P.; Beaucourt, J.; Pichat, L. *Tetrahedron Lett.* **1984**, *25*, 201–2.

14. Bell, A.; Roberts, D.; Ruddoch, K. *Tetrahedron Lett.* **1988**, *29*, 5013–6.

15. Wigerinck, P.; Snoeck, R.; Claes, P.; De Clercq, E.; Herdewijn, P. *J. Med. Chem.* **1991**, *34*, 2383–9.

16. Massa, M. A.; Patt, W. C.; Ahn, K.; Sisneros, A. M.; Herman, S. B.; Doherty, A. *Bioorg. Med. Chem. Lett.* **1998**, *8*, 2117–22.

17. Luo, F.-T.; Wang, R.-T. *Heterocycles* **1990**, *30*, 1543–8.

18. Rottländer, M.; Knochel, P. *J. Org. Chem.* **1998**, *63*, 203–8.

19. Prasad, A. S. B.; Stevenson, T. M.; Citineni, J. R.; Nyzam, V.; Knochel, P. *Tetrahedron* **1997**, *53*, 7237–54.

20. Lipshutz, B. H.; Frieman, B.; Birkedal, H. *Org. Lett.* **2004**, *6*, 2305–08.

21. (a) Huang, S.-T.; Gordon, D. M. *Tetrahedron Lett.* **1998**, *39*, 9335–8. (b) Bach, T.; Heuser, S. *Tetrahedron Lett.* **2000**, *41*, 1707–10.

22. Yamaguchi, S.; Endo, T.; Uchida, M.; Izumizawa, T.; Furukawa, K.; Tamao, K. *Chem. Eur. J.* **2000**, *6*, 1683–92.

23. Superchi, S.; Sotomayor, N.; Miao, G.; Joseph, B.; Snieckus, V. *Tetrahedron Lett.* **1996**, *37*, 6057–60.

24. Gillet, J.-P.; Sayvetre, R.; Normant, J.-F. *Tetrahedron Lett.* **1985**, *26*, 3999–4002.

25. Sapoutzis, I.; Dube, H.; Knochel, P. *Adv. Synth. Catal.* **2004**, *346*, 709–12.

26. Angiolelli, M. E.; Casalnuovo, A. L.; Selby, T. P. *Synlett*, **2000**, 905–7.

27. Gronowitz, S.; Peters, D. *Heterocycles* **1990**, *30*, 645–8.

28. Sharp, M. J.; Snieckus, V. *Tetrahedron Lett.* **1985**, *27*, 5997–6000.

29. Ward, P.; Armour, D. R.; Bays, D. E.; Evans, B.; Giblin, G. M. P. *et al. J. Med. Chem.* **1995**, *38*, 4985–92.

30. Hangeland, J. J.; Doweyko, A. M.; Dejneka, T.; Friends, T. J.; Devasthale, P.; Mellström, K.; Sandberg, J.; Grynfarb, M.; Sack, J. S.; Einspahr, H.; Färnegårdh, M.; Husman, B.; Ljunggren, J.; Koehler, K.; Sheppard, C.; Malm, J.; Ryono, D. E. *Bioorg. Med. Chem. Lett.* **2004**, *14*, 3549–53.

31. Ishiyama, T.; Murata, M.; Miyaura, N. *J. Org. Chem.* **1995**, *60*, 7508–10.

32. Young, Y.; Martin, A. R. *Heterocycles* **1992**, *34*, 1395–8.

33. Peters, D.; Hörnfeldt, A.-B.; Gronowitz, S. *J. Heterocycl. Chem.* **1990**, *27*, 2165–73.

34. Peters, D.; Hörnfeldt, A.-B.; Gronowitz, S. *J. Heterocycl. Chem.* **1991**, *28*, 529–31.

35. Wellmar, U.; Hörnfeldt, A.-B.; Gronowitz, S. *J. Heterocycl. Chem.* **1995**, *32*, 1159–63.

36. Majo, V. J.; Prabhakaran, J.; Mann, J. J.; Kumar, J. S. D. *Tetrahedron Lett.* **2003**, *44*, 8535–7.

37. Laurenti, D.; Santelli, M. *Org. Prep. Proc. Int.* **1999**, *31*, 245–94.

38. (a) Laurenti, D.; Feuerstein, M.; Pepé, G.; Doucet, H.; Santelli, M. *J. Org. Chem.* **2001**, *66*, 1633–37; (b) Feuerstein, M.; Laurenti, D.; Doucet, H.; Santelli, M. *Chem. Comm.* **2001**, 43–4; (c) Feuerstein, M.; Doucet, H.; Santelli, M. *J. Org. Chem.* **2001**, *66*, 5923–5.

39. Feuerstein, M.; Doucet, H.; Santelli, M. *J. Organomet. Chem.* **2003**, *687*, 327–36.
40. Molander, G. A.; Katona, B. W.; Machrouhi, F. *J. Org. Chem.* **2002**, *67*, 8416–23.
41. Molander, G.; Bernardi, C. R. *J. Org. Chem.* **2002**, *67*, 8424–9.
42. Molander, G. A.; Biolatto, B. *J. Org. Chem.* **2003**, *68*, 4302–14.
43. Dondoni, A.; Mastellari, A. R.; Medici, A.; Negrini, E. *Synthesis* **1986**, 757–60.
44. Kosugi, M.; Koshiba, M.; Atoh, A.; Sano, H.; Migita, T. *Bull. Chem. Soc. Jpn.* **1986**, *59*, 677–9.
45. Molloy, K. C.; Waterfield, P. C.; Mahon, M. F. *J. Organomet. Chem.* **1989**, *365*, 61–73.
46. Peters, D.; Hörnfeldt, A.-B.; Gronowitz, S. *J. Heterocycl. Chem.* **1990**, *27*, 2165–73.
47. Kelly, T. R.; Lang, F. *Tetrahedron Lett.* **1995**, *36*, 9293–6.
48. Kelly, T. R.; Jagoe, C. T.; Gu, Z. *Tetrahedron Lett.* **1991**, *32*, 4263–6.
49. Watanabe, Y.; Ueno, Y.; Araki, T.; Endo, T.; Okawara, M. *Tetrahedron Lett.* **1991**, *27*, 215–8.
50. Park, H.; Lee, J. Y.; Lee, Y. S.; Park, J. O.; Koh, S. B.; Ham, W.-H. *J. Antibiot.* **1994**, *47*, 606–8.
51. Bailey, T. R. *Tetrahedron Lett.* **1987**, *27*, 4407–10.
52. Kennedy, G.; Perboni, A. D. *Tetrahedron Lett.* **1996**, *37*, 7611–4.
53. Hark, R. R.; Hauze, D. B.; Petrovskaia, O.; Joullié, M. M.; Jaouhari, R.; McComisky, P. *Tetrahedron Lett.* **1994**, *35*, 7719–22.
54. Kelly, T. R.; Lang, F. *J. Org. Chem.* **1996**, *61*, 4623–33.
55. Gutierrez, A. J.; Terhorst, T. J.; Matteucci, M. D.; Froehler, B. C. *J. Am. Chem. Soc.* **1994**, *116*, 5540–4.
56. Gronowitz, S.; Peters, D. *Heterocycles* **1990**, *30*, 645–58.
57. Dondoni, A.; Fogagnolo, M.; Medici, A.; Negrini, E. *Synthesis* **1987**, 185–6.
58. Benhida, R.; Lecubin, F.; Fourrey, J.-L.; Castellasnos, L. R.; Quintro, L. *Tetrahedron Lett.* **1999**, *40*, 5701–3.
59. Malm, J.; Hörnfeldt, A-B.; Gronowitz, S. *Tetrahedron Lett.* **1992**, *33*, 2199–202.
60. Gronowitz, S.; Björk, P.; Malm, J.; Hörnfeldt, A.-B. *J. Organomet. Chem.* **1993**, *460*, 127–9.
61. Barchín, B. M.; Valenciano, J.; Cuadro, A. M.; Alvarez-Builla, J.; Vaquero, J. *Org. Lett.* **1999**, *1*, 545–7.
62. Birchler, A. G.; Liu, F.; Liebeskind, L. S. *J. Org. Chem.* **1994**, *59*, 7737–45.
63. Dondoni, A.; Fantin, G.; Fogagnolo, M.; Mastellari, A.; Medici, A. *Gazz. Chim. Ital.* **1988**, *118*, 211–32.
64. Nicolaou, K. C.; He, Y.; Roschangar, F.; King, N. P.; Vourloumis, D.; Li, T. *Angew. Chem., Int. Ed.* **1998**, *37*, 84–7.
65. Zhou, J.; Kläβ, T.; Zhang, A.; Johnson, K. M.; Wang, C. Z.; Ye, Y.; Kozikowski, A. P. *Bioorg. Med. Chem. Lett.* **2003**, *13*, 3565–9.
66. Buynak, J. D.; Doppalapudi, V. R.; Frotan, M.; Kumar, R.; Chambers, A. *Tetrahedron* **2000**, *56*, 5709–18.
67. Sandosham, J.; Undheim, K. *Acta Chem. Scand.* **1989**, *43*, 684–9.
68. Liebeskind, L. S.; Srogl, J. *J. Am. Chem. Soc.* **2000**, *122*, 11260–1.
69. Egi, M.; Liebeskind, L. S. *Org. Lett.* **2003**, *5*, 801–2.
70. Alphonse, F.-A.; Suzenet, F.; Keromnes, A.; Lebret, B.; Guillaumet, G. *Org. Lett.* **2003**, 803–5.
71. Sakamoto, T.; Nagata, H.; Kondo, Y.; Shiraiwa, M.; Yamanaka, H. *Chem. Pharm. Bull.* **1987**, *35*, 823–8.

72. Schlegel, J.; Maas, G. *Synthesis* **1999**, 100–6.
73. Nye, S. A.; Potts, K. T. *Synthesis* **1988**, 375–7.
74. Arcadi, A.; Attanasi, O. A.; Guidi, B.; Rossi, E.; Santeusanio, S. *Chem. Lett.* **1999**, 59–60.
75. Karama, U.; Höfle, G. *Eur. J. Org. Chem.* **2003**, 1042–9.
76. Tucker, T. J.; Lyle, T. A.; Wiscount, C. M.; Britcher, S. F.; Young, S. D.; Sanders, W. M.; Lumma, W. C.; Goldman, M. E.; O'Brien, J. A.; Ball, R. G.; Homnick, C. F.; Schleif, W. A.; Emini, E. A.; Huff, J. R.; Anderson, P. S. *J. Med. Chem.* **1994**, *37*, 2437–44.
77. Cristalli, G.; Camaioni, E.; Vittori, S.; Volpinin, R.; Borea, P. A.; Conti, A.; Dionisotti, S.; Ongini, E.; Monopoli, A. *J. Med. Chem.* **1995**, *38*, 1462–72.
78. Siebeneicher, H.; Doye, S. *Eur. J. Org. Chem.* **2002**, 1213–20.
79. Müller, T. J. J.; Ansorge, M.; Aktah, D. *Angew. Chem., Int. Ed.* **2000**, *39*, 1253–6.
80. Braun, R. U.; Müller, T. J. J. *Tetrahedron* **2004**, *60*, 9463–9.
81. Yehia, N. A. M.; Polborn, K.; Müller, T. J. J. *Tetrahedron Lett.* **2002**, *43*, 6907–10.
82. Frank, W. C.; Kim, Y.; Heck, R. F. *J. Org. Chem.* **1978**, *43*, 2947–9.
83. Sakamoto, T.; Nagata, H.; Kondo, Y.; Shiraiwa, M.; Yamanaka, H. *Chem. Pharm. Bull.* **1987**, *35*, 823–8.
84. Aoyagi, Y.; Inoue, A.; Koizumi, I.; Hashimoto, R.; Tokunaga, K.; Gohma, K.; Komatsu, J.; Sekine, K.; Miyafuji, A.; Konoh, J. Honma, R. Akita, Y.; Ohta, A. *Heterocycles* **1992**, *33(1)*, 257–72.
85. Pivsa-Art, S.; Satoh, T.; Kawamura, Y.; Miura, M.; Nomura, M. *Bull. Chem. Soc. Jap.* **1998**, *71*, 467–73.
86. Potts, K. T. *Comprehensive Heterocyclic Chemistry*: Pergamon Press, Oxford, **1984**, Vols 5 and 6.
87. Yokooji, A.; Okazawa, T.; Satoh, T.; Miura, M.; Nomura, M. *Tetrahedron*, **2003**, *59*, 5685–9.
88. Kondo, Y.; Komine, T.; Sakamoto, T. *Org. Lett.* **2000**, *2*, 3111–3.
89. Gallagher, W. P.; Maleczka, R. E. *J. Org. Chem.* **2003**, *68*, 6775–9.
90. Head, R. A.; Ibbotson, A. *Tetrahedron Lett.* **1984**, *25*, 5939–42.
91. Hong, Y.; Tanoury, G. J.; Wilkinson, H. S.; Bakale, R. P.; Wald, S. A.; Senanayake, C. H. *Tetrahedron Lett.* **1997**, *38*, 5607–10.
92. Hooper, M. W.; Utsunomiya, M.; Hartwig, J. F. *J. Org. Chem.* **2003**, *68*, 2861–73.
93. Schröter, Sve.; Stock, C.; Bach, T. *Tetrahedron* **2005**, *61*, 2245–67.
94. Bach, T.; Heuser, S. *Angew. Chem. Int. Ed.* **2001**, *40*, 3184–5.
95. Shao, J.; Panek, J. S. *Org. Lett.* **2004**, *6*, 3083–5.
96. Hodgetts, K. J.; Kershaw, M. T. *Org. Lett.* **2002**, *4*, 1363–5.
97. Hodgetts, K. J.; Kershaw, M. T. *Org. Lett.* **2003**, *5*, 2911–4.

Chapter 8

Oxazoles and benzoxazoles

Marudai Balasubramanian

8.1. Introduction

Like thiazole, oxazole is a π-electron-excessive heterocycle. The electronegativity of the *N*-atom attracts electrons so that C(2) is partially electropositive and therefore susceptible to nucleophilic attack. However, electrophilic substitution of oxazoles takes place at the electron-rich position C(5) preferentially. More relevant to palladium chemistry, 2-halooxazoles or 2-halobenzoxazoles are prone to oxidative addition to Pd(0). Even 2-chlorooxazole and 2-chlorobenzoxazole are viable substrates for pd-catalyzed reactions.

During the last decade, several 2,4-disubstituted oxazole-containing natural products have been isolated and found to be biologically active and their total synthesis attracted much synthetic effort. The prominent examples are hennoxazole A (antiviral), leucascandrolide A (cytotoxic and antifungal), phorboxazole A & B, disorazoles A1, C1, D1, virginiamycin M2 (antibiotic), and diazonamide A. Hennoxazole A was isolated from the sponge *polyfibrospongia sp*. It shows activity against herpes simplex virus type 1 (IC$_{50}$ = 0.6 µg/mL) and is a peripheral analgesic. Diazonamide A, in turn, is a secondary metabolite of the colonial ascidian, *Diazona chinensis*, a marine species collected from ceilings of small caves in the Philippines. It is potent *in vitro* against HCT-116 human colon carcinoma and B-16 murine melanoma cancer cell lines (IC$_{50}$ < 15 ng/mL). On the other hand, zoxazoleamine, a sedative and muscle relaxant, is an example of synthetic pharmaceutical agents.

Directly linked bisoxazole core as a unique feature of hennoxazole A's structure is only found in the polycyclic marine alkaloids diazonamides A–B and cyanobacterium derived muscoride A and hennoxazole A. The marine natural product hennoxazole A was synthesized by a convergent synthesis [1].

(−)-Hennoxazole A

Diazonamide A is also a marine natural product. An indolyl bisoxazole as a potential intermediate for the synthesis of diazonamide A has been reported [2–4, 5a–c].

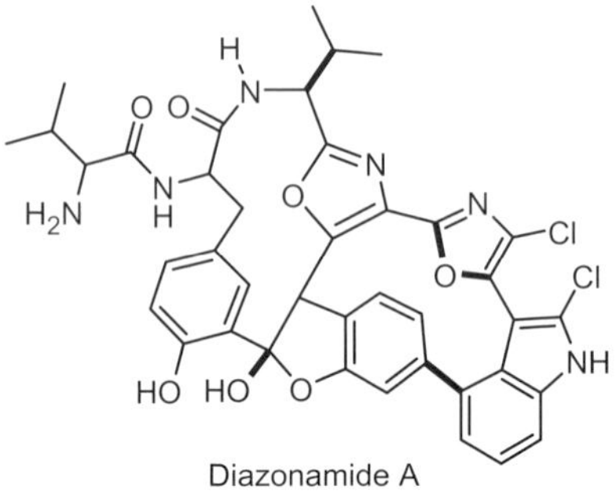

Diazonamide A

Leucascandrolide A was isolated from the sponge *Leucascandra caveolata*. The natural product displays strong *in vitro* cytotoxicity against KB and P388 cancer cell lines and is also a potent antifungal, inhibiting the growth of *Candida albicans*. A few synthetic routes for leucascandrolide A were described [6, 7].

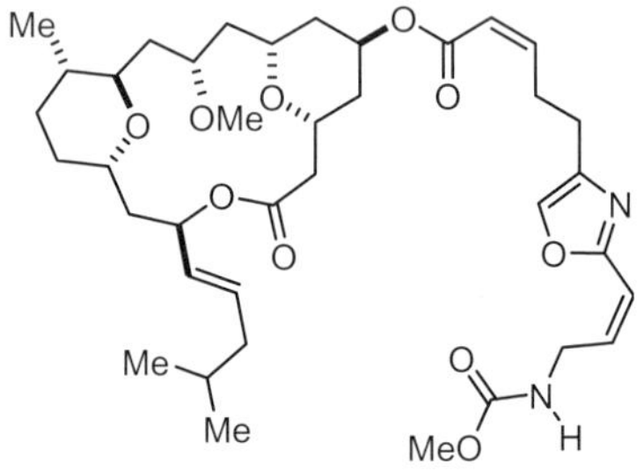

Leucascandrolide A

Phorboxazole A and B are marine natural products isolated from recently discovered species of Indian Ocean sponge near Muiron Island, western Australia. The phorboxazoles are among the most cytostatic natural products known, inhibiting the growth of tumor cells at nanomolar concentrations.

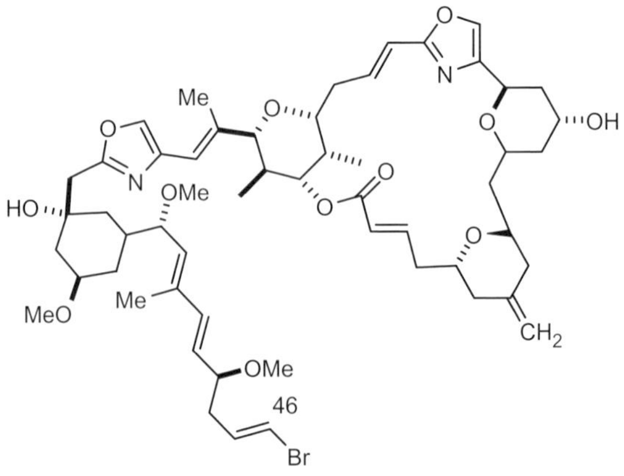

Phorboxazole A

Synthesis of phorboxazole A has been accomplished [8, 9]. The C(46) terminus of phorboxazole A was modified to incorporate a biotin-terminated linker *via* a direct Sonogashira reaction with tris-polyethyleneglycol vinyl iodide-biotin ester using catalytic $PdCl_2(PPh_3)_2$, CuI, and NEt_3 in THF. This process demonstrated the utility of mild C–C bond formation in the context of the phorboxazole architecture and provided a potential affinity probe [10].

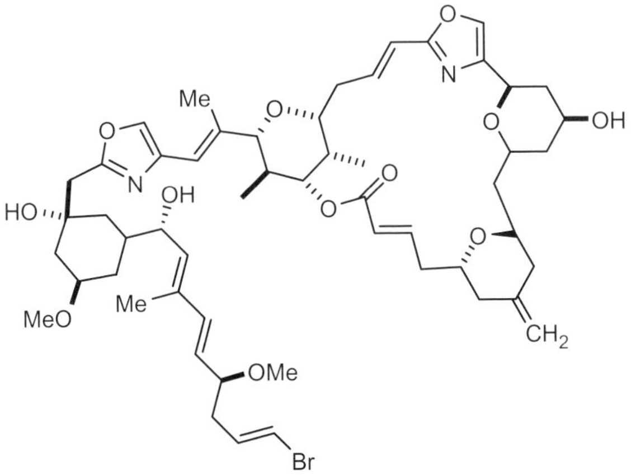

Phorboxazole B

The synthesis of phorboxazole B has been accomplished in 27 linear steps [11] in an overall yield of 12.6%. The key fragment couplings include a metalated oxazole alkylation and an oxazole-stabilized Wittig olefination.

The disorazole is comprised of a family of 29 closely related macrolides, which were isolated from the myxobacterium *Sorangium cellulosum*. Disorazole A1, the major component of the crude extraction residue, possesses antifungal activity, decay of microtubules in subnanomolar concentration, initiates cell cycle arrest in G2/M phase, and competes *in vitro* with vinsblastine for the tublin binding site. The retrosynthetic disconnections of disorazole A1 and a few stereoselective synthetic routes are described [12].

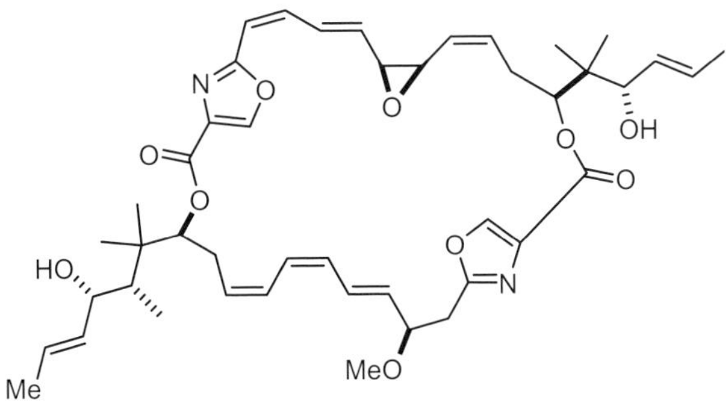

Disorazole A1

The stereoselective synthesis of the monomeric subunit of the macrolide dimer disorazole A1 and C1 [12–16] has been accomplished by a convergent coupling using the Sonogashira [12, 13] and the Stille reactions [14].

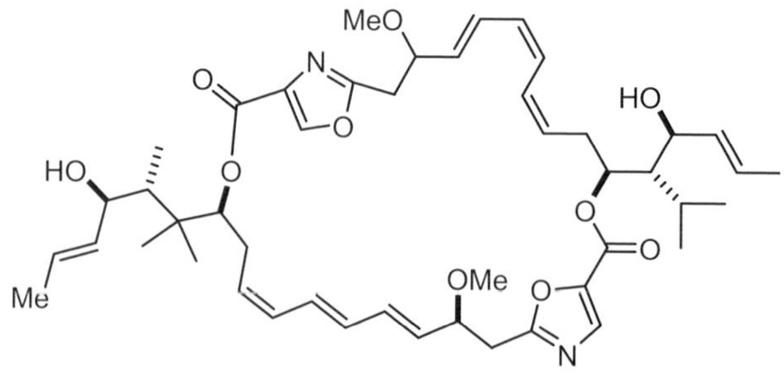

Disorazole C1

The synthesis of disorazole D1 involving the Sonogashira cross-coupling reaction is outlined [16].

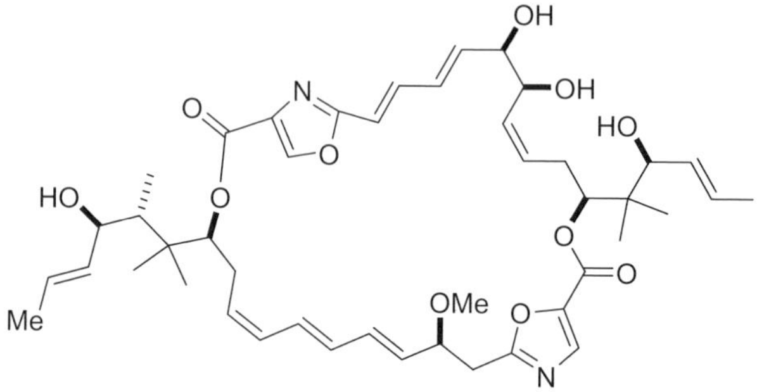

Disorazole D1

Rhizoxin D is a 16-membered macrolide isolated from the fungus *Rhizopus chinensis Rh-2* and it showed remarkable antimitotic properties. Rhizoxin D was synthesized from four subunits and the final step involved the Stille coupling of the oxazole fragment with an iodolactone [17].

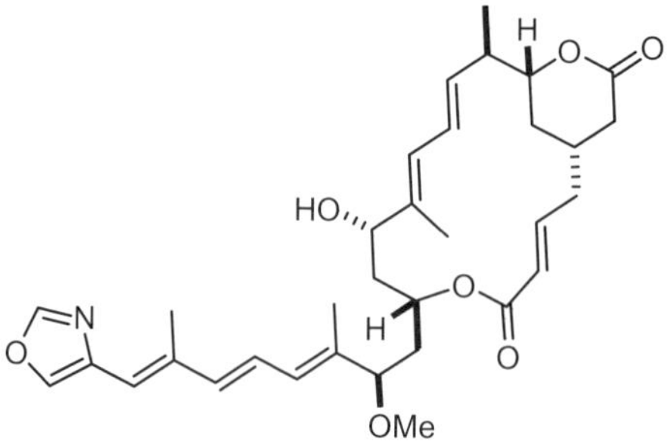

Rhizoxin D

Virginiamycin is a macrolactone with antibacterial activity, and pimprinin was isolated from *Streptomyces pimprina*.

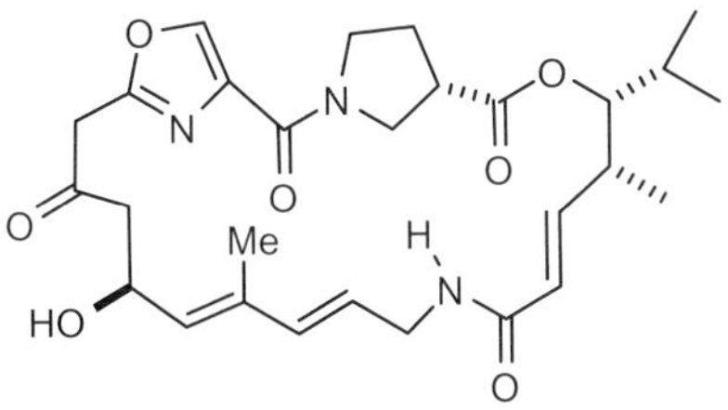

(−)-Virginiamycin

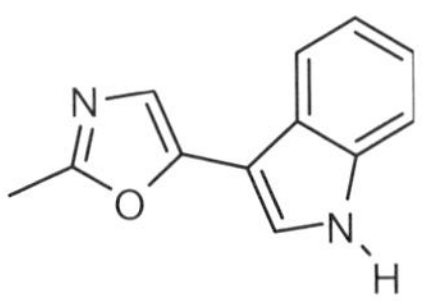

Pimprinin

A few pharmaceuticals derived from oxazole are in use. The anti-inflammatory and analgesic actions of 2-diethylamino-4,5-diphenyloxazole are known.

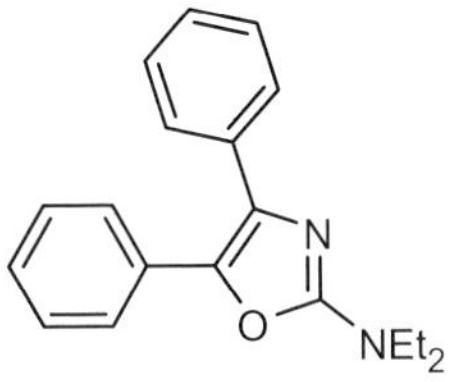

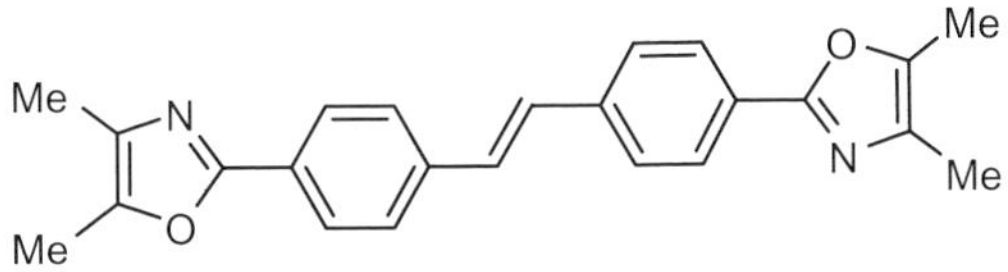

4,4'-Bisoxazol-2-ylstilbene

Aryl-substituted oxazoles are strongly fluorescent. In solution they are suitable as luminous substances for liquid scintillation counters and optical brighteners. 4,4'-Bisoxazol-2-ylstilbene is added to washing agents. During the washing process, it is absorbed by the fibers so that the clothes appear to be "whiter than white" as a result of its blue fluorescence.

8.2. Synthesis of halooxazoles and halobenzoxazoles

8.2.1. Direct halogenation

Condensation of 2-bromoethylamine hydrobromide with benzoyl chloride in benzene in the presence of five equivalents of Et_3N gave 2-phenyl-4,5-dihydrooxazole (**1**) in 67% yield [18]. Treatment of **1** with three equivalents of NBS in boiling CCl_4 in the presence of AIBN led to 5-bromo-2-phenyloxazole (**2**). Presumably, sequential bromination and dehydrobromination of **1** led to 2-phenyloxazole, which underwent further bromination to afford **2**.

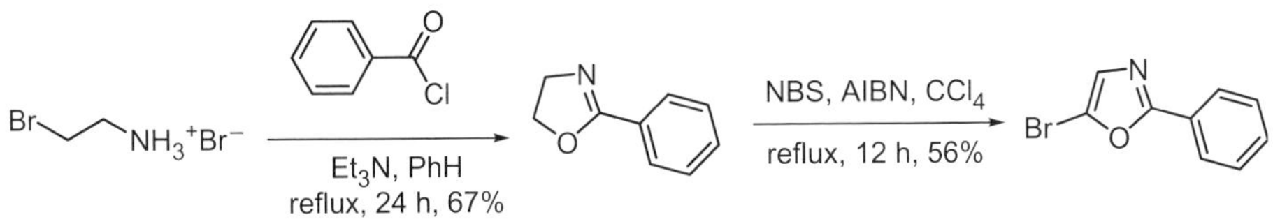

8.2.2. Metalation and halogen quench

Barrett and Kohrt deprotonated oxazole **3** using *n*-BuLi and then quenched the resulting oxazol-2-yllithium with iodine to prepare the desired 2-iodooxazole **4** [19].

3   **4**

The regiochemistry for trapping lithiooxazole depends upon the oxazole substituents as well as the nature of the electrophile. Hodges *et al.* observed that the major product of reaction between lithiated oxazole (**5** + **6**) and benzaldehyde was the C(4)-substituted oxazole **7**, resulting from reaction of the dominant acyclic valence bond tautomer **5** *via* the initial aldol adduct **6** followed by proton transfer and recyclization [20].

5   **6**   **7**

Vedejs and Luchetta developed a method for the iodination of oxazoles at C(4) *via* 2-lithiooxazoles by exploiting the aforementioned equilibrium between cyclic **5** and acyclic **6** valence bond tautomers of 2-lithiooxazole [21]. When 5-(*p*-tolyl)oxazole (**8**) was treated with lithium hexamethyldisilazide (LiHMDS) in THF followed by treatment with 1,2-diiodoethane as the electrophile, 2-iodooxazole **9** was obtained exclusively. On the other hand, when 50 volume percentage of DMPU was added *prior to* the addition of the base, 4-iodooxazole **10** was isolated as the predominant product (73%) with ca. 2% of **9** and ca. 5% of the 2,4-diiodooxazole derivative.

8   **9**

10

8.2.3. Sandmeyer reaction

Aminooxazole **11**, readily obtained by reaction of *N*-Boc-L-Val-OH with amino-malononitrile *p*-toluenesulfonate and EDC in pyridine [22], was converted directly to bromooxazole **12** by *in situ* bromination *via* a nitrosamine intermediate.

11

12

While halogenation and Sandmeyer reactions are suitable for preparation of oxazolyl halides, benzoxazolyl halides with halogen on the benzene ring moiety may be synthesized *via* other approaches. For instance, 5-halobenzoxazoles **13** were prepared by treating 4-halo-2-aminophenols with trimethylorthoformate and concentrated aqueous HCl [23].

13

8.3. Coupling reactions with organometallic reagents

There are limited precedents for elaborating the oxazole nucleus using Pd-catalyzed cross-coupling reactions, partially because of the ring-opening tendency of oxazol-2-yllithium. Nonetheless, the rapid progress in palladium chemistry compounded with the biological importance of oxazole-containing entities will certainly spur more interest in this area.

Condensation of aryl halides with various active methylene compounds is readily promoted by catalytic action of palladium to give the corresponding arene derivatives containing a functionalized ethyl group [24]. Yamanaka *et al.* extended this chemistry to haloazoles including oxazoles, thiazoles, and imidazoles [25]. Thus, in the presence of $Pd(Ph_3P)_4$, 2-chlorooxazole **14** was refluxed with phenylsulfonylacetonitrile and NaH to form 4,5-diphenyl-α-phenylsulfonyl-2-oxazoloacetonitrile (**15**), which existed predominantly as its enamine tautomer. In a similar fashion, 4-bromooxazole and 5-bromooxazole also were condensed with phenylsulfonylacetonitrile under the same conditions.

14   **15**

8.3.1. Negishi coupling

Organozinc reagents exhibit greater functional group compatibility than the corresponding organolithium and Grignard reagents. Therefore, Negishi coupling has found wide applications in organic synthesis. By using a sequence of regiocontrolled halogenation and Pd-catalyzed coupling reactions, the synthesis of various substituted oxazoles from ethyl 2-chlorooxazole-4-carboxylate was accomplished. The methodology was applied to the synthesis of a series of 2,4- and 2,5-disubstituted, and 2,3,5-trisubstituted oxazoles. 2-Pyridyloxazole **16** was prepared from 2-chlorooxazole-4-ethylcarboxylate *via* the Negishi coupling with 2-pyridylzinc bromide [26].

An improved and scalable method for the Pd-catalyzed cross-coupling reaction of oxazole-2-ylzinc derivatives with aryl bromides was reported [27]. Zinc chloride ($ZnCl_2$) was used in the coupling reactions and the products were obtained in excellent yields, tolerating an array of functional groups on the aryl ring [27].

Fluorovinylzinc **17**, generated by treating the corresponding vinyllithium with $ZnCl_2$, was coupled with 2-iodobenzoxazole to produce fluorovinylbenzoxazole **18** stereoselectively [28, 29].

2-Oxazolylzinc chloride, prepared from the corresponding 2-lithiooxazole, underwent a Negishi reaction with 4-iodooxazole **19** in the presence of Pd(dba)$_2$ and trifuranylphosphine (TFP) to give bisoxazole **20** [21].

In like fashion, the cross-coupling of iodoindoline **21** with 2-oxazolylzinc chloride led to an oxazole-containing partial ergot alkaloid **22**, a potent 5-HT$_{1A}$ agonist.

2-Oxazolylzinc chloride reagents can be prepared from transmetalation of 2-oxazolyl-lithium with ZnCl$_2$ [30]. A strong covalent bond with carbon and relatively low oxophilicity suggested that the equilibrium for zinc derivatives should sufficiently favor the metalated oxazol-2-ylzinc species to yield a productive reaction. Indeed, Anderson *et al.* successfully carried out the Negishi coupling of oxazol-2-ylzinc chloride reagents with aryl halides and acid chlorides as well as organotriflates [31a,b, 32]. Therefore, deprotonation of benzoxazole using *n*-BuLi followed by transmetalation with ZnCl$_2$ gave 2-chlorozincbenzoxazole (**23**), which was cross-coupled with 1-naphthalene triflate to provide the expected naphthalenylbenzoxazole (**24**). The Pd(0) here was derived from the treatment of PdCl$_2$(Ph$_3$P)$_2$ with *n*-BuLi.

An improved and scalable method for 2-arylbenzoxazole **25** was reported by Reeder *et al. via* the Negishi reaction of **23** with aryl bromides [27].

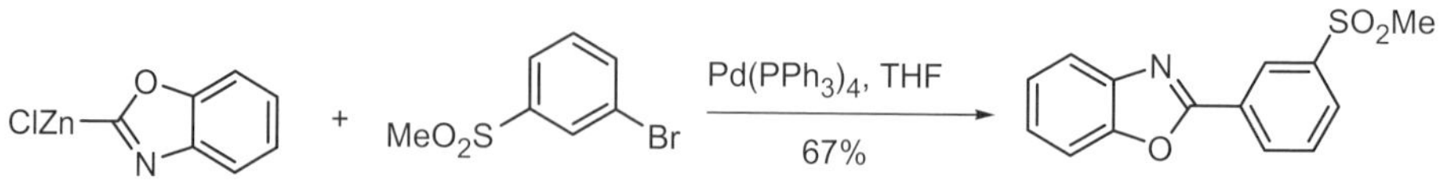

23   **25**

8.3.2. Suzuki coupling

The Suzuki reaction represents the Pd-catalyzed cross-coupling reaction between organoboron reagents and aryl and heteroaryl halides. Vachal and Toth have developed a versatile approach to the synthesis of 2,5-diaryoxazoles using a Suzuki coupling [33]. Thus, 2,5-diaryloxazoles were prepared from 2-bromo-5-aryloxazole and electron rich and electron deficient boronic acids [33].

The readily available ethyl 2-chlorooxazole-4-carboxylate proved to be a versatile scaffold for the synthesis of 2,4-disubstituted oxazoles and 2,4,5-trisubstituted oxazoles. 2,4-Disubstituted oxazole **26** was prepared from 2-chlorooxazole-4-ethylcarboxylate *via* the Suzuki coupling with phenylboronic acid [26]. The carboxylic functionality at C(4) of **26** could then be exploited by a variety of synthetic transformations.

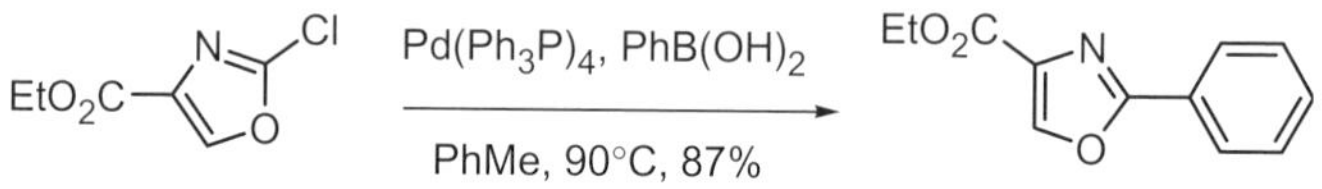

26

Maekawa *et al.* reported that a series of ω-oxazolylalkanoic acid derivatives as antidiabetic agents, promoting the glucose-dependent secretion of insulin. ω-(5-Oxazolyl)alkanoates **27**, the key intermediates for the synthesis of target compounds, were prepared *via* Suzuki coupling of substituted 2-chlorooxazole with phenylboronic acid [34].

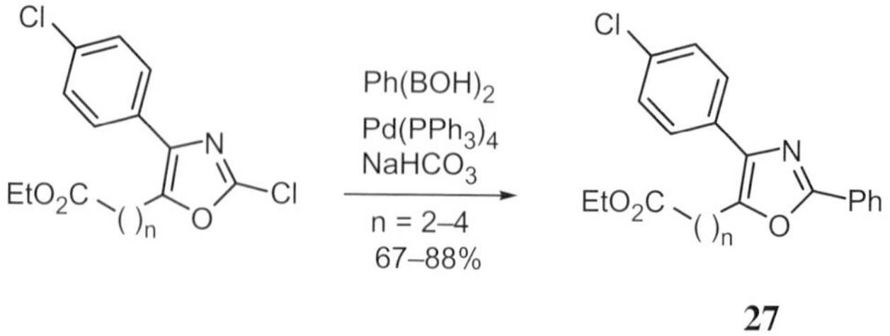

27

A Pd-catalyzed base-free coupling of arylboronic acids with π-deficient heteroaromatic thioethers was mediated by copper(I) thiophene-2-carboxylate CuTC [35]. Benzoxazole with a thioglycolamide pendant (**28**) was coupled with arylboronic acid in the presence of Pd$_2$(dba)$_3$/tris(2-furyl)phosphine (TFP) to give 2-arylbenzoxazole **29**. The modifiable leaving group of this system suggested possible applications ranging from solid support-based reagents to pendant substrate recognition [35].

Attempts to prepare oxazolylboronic acids have failed probably due to the equilibrium between the cyclic and acyclic valence bond tautomers of the lithiooxazoles. A somewhat relevant Suzuki coupling involved the Pd-catalyzed cross-coupling of 6-bromo-2-phenyloxazolo[4,5-*b*]pyridine (**30**) with phenylboronic acid to provide 6-phenyl-2-phenyloxazolo[4,5-*b*]pyridine (**31**) [36].

A series of 2-aryloxazoloquinolones **32** having high affinity for the GABA receptor were prepared *via* a Pd-catalyzed intramolecular diaryl coupling of ethyl 5-bromo-2-phenyl-oxazole-4-carboxylate with 2-aminophenylboronic acid followed by cyclization [37]. This one-pot intramolecular Suzuki reaction was followed by cyclization to afford a tricyclic compound **32**.

8.3.3. Suzuki–Miyaura coupling

Application of the resin "capture-release" methodology to macrocyclization *via* an intramolecular Suzuki–Miyaura coupling was reported recently [38]. Aryl boronic acid was trapped by an ammonium hydroxide-form Dowex® Ion Exchanger resin (D–OH⁻), leading to polymer-ionically bound borates **33**, which further cyclized into macroheterocycles **34** under the Suzuki–Miyaura coupling conditions [38].

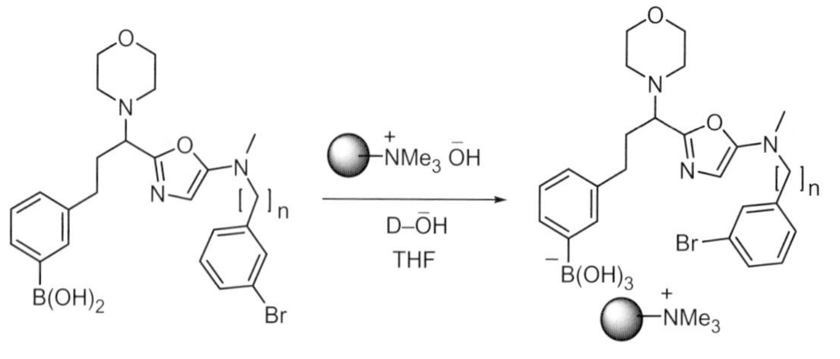

33

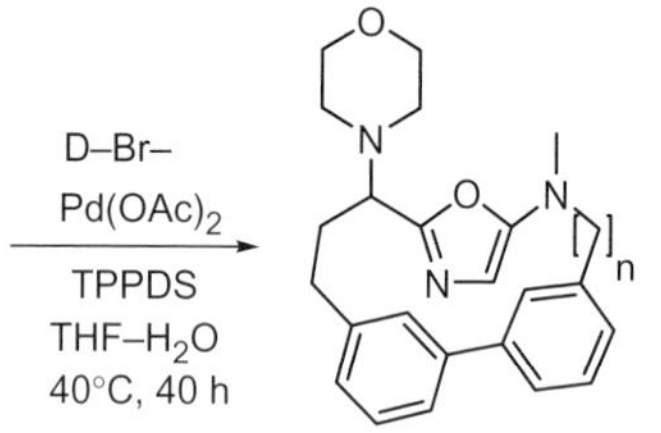

n = 1, 22%; n = 2, 16%; n = 3, 20%

34

8.3.4. Stille coupling

The Stille coupling reaction is the most versatile method among all Pd-catalyzed cross-coupling reactions with organometallic reagents. By lithiation of 4-methyloxazole with *n*-BuLi and subsequent quenching with trimethyltin chloride, Dondoni *et al.* prepared 2-trimethylstannyl-4-methyloxazole [39], which was then coupled with aryl- and heteroaryl-halides to provide the expected 2-aryloxazole. Thus, 2-trimethylstannyl-4-methyloxazole was coupled with 3-bromo-pyridine to afford oxazolylpyridine **35**.

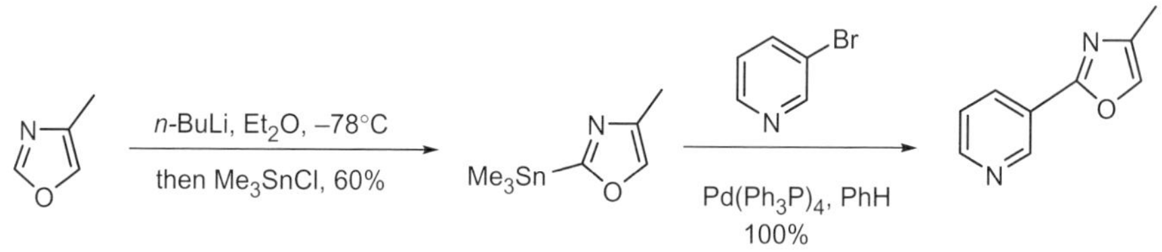

35

Smith III *et al.* reported an effective synthesis of a variety of 2,4-orthogonally func-tionalized oxazoles with shorter reaction times and modest to excellent overall yields (48–90%). 2-Chloromethyl-4-vinyloxazole (**36**) was obtained in 78% yield *via* the Stille coupling of 2-chloromethyl-4-triflate with vinyltributyltin [40].

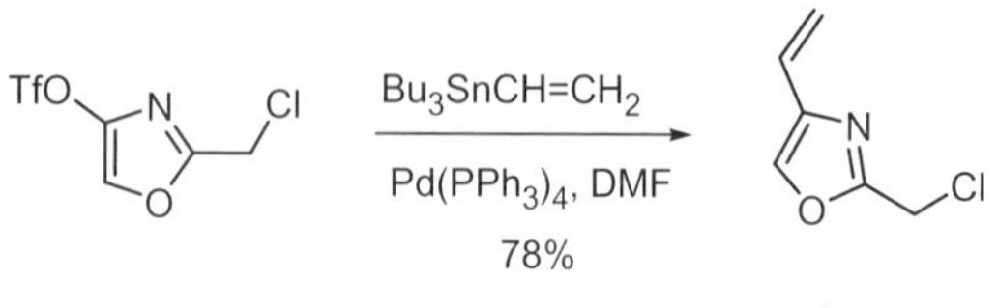

36

2-Vinyloxazole-4-ethylcarboxylate (**37**) was prepared from 2-chlorooxazole-4-ethyl-carboxylate and vinyltrimethylstannane [26]. The Suzuki, Stille, and Negishi coupling reactions were successfully used to install substitutents at the C(2) oxazole position. The carboxylic and vinyl functionalities could then be further exploited by a variety of synthetic transformations.

37

4-Bromo- and 4-bromoalkyl-2,5-diphenyloxazole **38** were subjected to the Stille coupling with a range of commercially available tributyltin reagents. Tri-2-furylphosphine/Pd$_2$(dba)$_3$ was used as an effective catalyst. Copper(II) oxide enhanced the Stille coupling reactions of 2,5-diphenyl-4-tributylstannanyloxazole with various electrophiles [41]. Such method offered an efficient synthetic route to prepare resins from oxazole-containing monomers such as 2,5-diphenyl-4-vinyloxazole.

38   **39**

40

Kelly and Lang successfully carried out the total synthesis of dimethyl sulfomycinate *via* a Stille coupling of oxazolyl triflate **42** with an array of organostannanes [42, 43]. Thus, 2-aryl-4-oxalone **41** was transformed into the corresponding triflate **42**, which was then coupled with 2-trimethylstannylpyridine under the agency of Pd(Ph$_3$P)$_4$ and LiCl to provide adduct **43** [42]. The couplings of triflate **42** with phenyl-, vinyl-, and phenylethynyl trimethyltin all proceeded in excellent yields. Unfortunately, application of such method to more delicate systems in the natural product failed, and hence the oxazole moiety was installed from acyclic precursors.

41

42   **43**

Analogously, Barrett and Kohrt transformed 2-phenyl-4-oxalone into triflate **44**, which was then converted to the corresponding stannane **45** using Pd-catalyzed coupling with hexa-methyldistannane. Subsequent coupling of **45** with 2-iodooxazole provided bisoxazole **46** [19].

44

45   **46**

The Stille couplings also have been exploited in the synthesis of the aromatic macrocyclic core of diazonamide A [22, 43]. Pattenden's group utilized the Pd-catalyzed coupling between the 3-stannyl substituted indole **47** and the 3-bromooxazole **48** to provide a particularly expeditious route to the ring system **49** [43]. In addition, Harran's group secured the connection between bromooxazole **12** and vinylstannane **50** using a Stille coupling [22].

47   **48**   **49**

12   **50**   **51**

In synthetic studies toward the synthesis of phorboxazole described by Schaus and Panek, although their initial attempts of carboalumination of terminal alkynes followed by Pd-catalyzed cross-coupling of the resulting vinylmetallic intermediates **52** with **44** were successful, the strategy failed to install the C(27)–C(29) olefin of phorboxazole. Nonetheless, employing a Stille coupling of oxazole-4-triflate **44** and a vinyl stannane, they constructed the C(26)–C(31) subunit of phorboxazole [44]. The trimethylstannane was found to be advantageous over the corresponding tributylstannane presumably due to the ease of transmetalation.

Palladium-catalyzed cross-coupling of terminal alkynes with 4-trifloyloxazole was studied for the construction of the C(26)–C(31) subunit of phorboxazole A [44]. The Stille coupling of triflate oxazole (**54**) with vinyl stannane (**55**) afforded the key intermediate for phorboxazole A [8, 9].

The Stille coupling of 4-bromomethyloxazole **56** with vinyltributyltin in the presence of Pd$_2$(dba)$_3$ and tri(2-furyl)phosphine produced 4-allyloxazole **57** in 82% yield [7]. Allyloxazole **57** is a key intermediate in the synthesis of leucascandrolide A.

Synthesis of conjugated ester **59** required for the synthesis of inthomycin involved the Stille coupling of 2-bromomethyloxazole and organostannanes **58**. Bromobis (triphenylphosphine)-*N*-succinimidepalladium (II) was used as a novel catalyst for the Stille cross-coupling reactions of allylic and benzylic halides with conjugated organo-stannanes [45].

58   **59**

Rhizoxin D was synthesized from four subunits [17]. The Stille coupling of oxazole stannane **60** and iodolactone **61** was achieved in the presence of palladium(II) chloride bis(acetonitrile) [17]. The final segment D towards the synthesis of rhizoxin D involved the critical bond formation of C(19)–C(20), which was also achieved *via* a Stille coupling.

60   **61**

On the other hand, the stereoselective synthesis of the masked half of the disorazole A1 involved the Stille coupling of oxazole-dibromoolefine **62** with diene **63** and the expected key intermediate, bromotriene **64**, was assembled in excellent yield [15].

62   **63**   **64**

Retrosynthetic analysis of disorazole C1 suggested that the cyclodimerization of acid **67** would be a viable synthetic route. This was accomplished by addition of trimethylstannane **65** to a solution of dieneyl iodide **66** *via* a Stille coupling in the presence of a catalytic amount of Pd(CH$_3$CN)Cl$_2$ to give triene **67** in 76% yield [14].

65   **66**

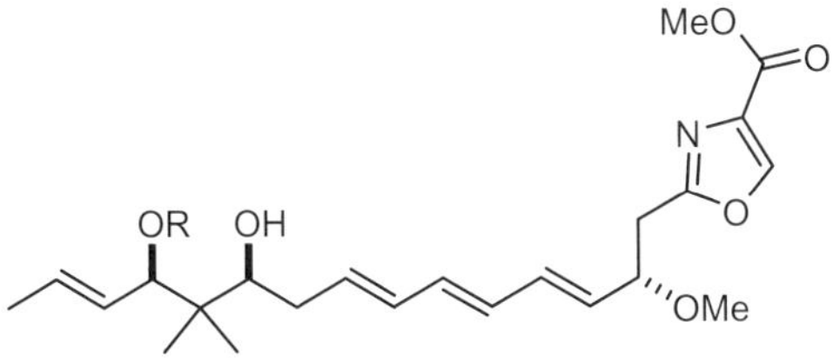

67

The synthesis of a novel oxazole building block, 4-bromomethyl-2-chlorooxazole, and its Pd-catalyzed cross-coupling reactions to prepare a range of 2,4-disubstituted oxazoles, were described [46]. Selectivity for the 4-bromomethyl position was observed with the Stille coupling effected in good to excellent yields or the Suzuki coupling in moderate yields to provide a range of 4-substituted-2-chloroxazoles **68** [46]. Subsequent coupling at the 2-chloro-position was achieved through either the Stille or the Suzuki reactions to provide vinyloxazole **69** in excellent yields (**69**) [46].

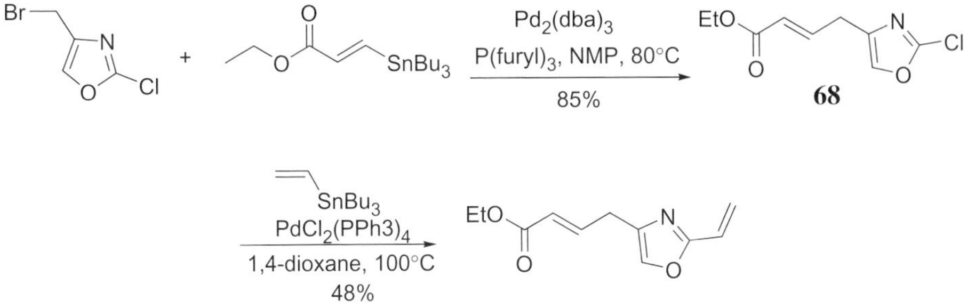

68

69

2-Benzoxazolyllithium undergoes a ring-opening tautomerization. However, intercepting the cyclic lithium species with R$_3$SnCl furnishes the desired 2-trialkylstannylbenzoxazole (**70**). Interestingly, trapping the oxazolyllithium system with Me$_3$SiCl predominantly resulted in the acyclic product [47]. The Stille coupling of 2-(tributyl-stannyl)benzoxazole with bromobenzene under the influence of PdCl$_2$(Ph$_3$P)$_2$ led to 2-phenylbenzooxazole (**71**) [48].

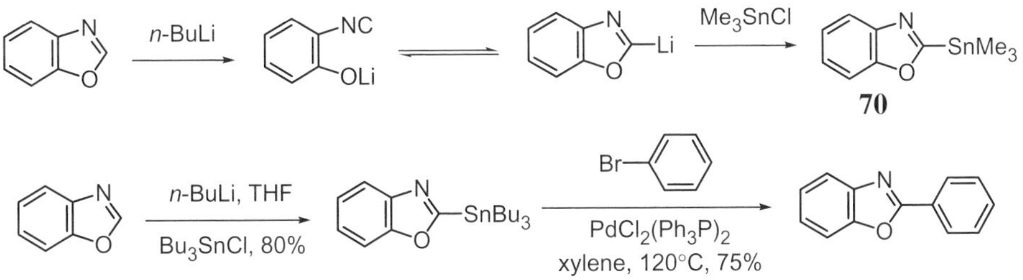

70

71

Similarly, the Stille coupling of 2-trimethylstannylbenzoxazole with acid chlorides such as pivaloyl chloride and benzoylchloride provided corresponding heterocyclic ketones [49].

8.3.5. Sonogashira coupling

The Sonogashira reaction is of considerable value in heterocyclic synthesis. Heteroaryl halides like bromooxazoles are viable substrates for the Pd-catalyzed cross-coupling reactions with terminal acetylene in the presence of Pd/Cu catalyst. In 1987, Yamanaka's group described the Pd-catalyzed reactions of halothiazoles with terminal acetylenes [50]. Submission of 4-bromo- (**72**) and 5-bromo-4-methyloxazoles (**73**) to the Sonogashira reaction conditions with phenylacetylene led to the expected acetylenes (**74** and **75**).

The synthetic utility of phenyliodooxazole **76** was demonstrated through a Sonogashira coupling with alkylacetylene and silanes. 4-Alkynyl- and silylated-oxazoles **77** were prepared in moderate yields [51].

The Sonogashira coupling of triflate of oxazole **44** with alkyne in the presence of catalytic amount of Pd(0) and CuI readily afforded 2,4-disubstituted oxazole **78**. The coupling process was clean and devoid of side reactions such as alkyne homocoupling [52]. The application of this methodology was explored towards the C(1)–C(11) fragment of a natural product leucascandrolide A.

Further retrosynthetic analysis of leucascandrolide A led to several synthetic strategies and one of the key steps was the C(1)–C(11) bond making using intermediates **81** and cyclic lactones.

Oxazole triflate **79** was treated with alkynamide **80** under the Sonogashira coupling conditions to give the corresponding side chain precursor **81** [6].

An efficient and convergent synthesis of the C(1)–C(11) side chain **83** of leucascandrolide A has been achieved. The key bond connection was made through the use of a Sonogashira cross-coupling [53, 54]. Oxazolyl triflate **82** was treated with alkynamide **80** using the Sonogashira coupling reactions to provide the side chain precursor **83** (part of the synthesis of *leucascandrolide A*) [53, 54].

Retrosynthetic disconnections of disorazole A1 and C1 envisioned the masked fragment **86,** which could be synthesized *via* the Sonogashira coupling C(10)–C(11) of vinyliodide **84** and enyne **85** in the presence of $PdCl_2(PPh_3)_2$ [12].

Alternatively, the coupling of half masked alkyne **87** and dienyliodide **88** gave key intermediate **89** for the synthesis of disorazole C1 [13].

The Sonogashira coupling was used for the assembly of the protected C(1)–C(11) enyne **92** towards the synthesis of disorazole D1. Best yields were achieved using $Pd_2Cl_2(PPh_3)_2$ and by the addition of alkyne **91** after premixing the catalyst, copper salt, and vinyl iodide **90** in degassed DMF [16].

8.4. Heck and heteroaryl Heck reactions

Yamanaka's group reported the Heck reactions of 4-bromo- and 5-bromo-4-methyloxazoles **72** and **73** with ethyl acrylate and acrylonitrile, giving the coupled products **93** and **94**, respectively [50].

73   **94**

Oxazoles and benzoxazoles are viable participants in the heteroaryl Heck reactions. In their important work published in 1992, Ohta and colleagues demonstrated that oxazoles and benzoxazoles, along with other π-sufficient aromatic heterocycles such as furans, benzofurans, thiophenes, benzothiophenes, pyrroles, thiazole, and imidazoles are acceptable recipient partners for the heteroaryl Heck reactions of chloropyrazines [55]. Therefore, treatment of 2-chloro-3,6-diethylpyrazine (**95**) with oxazole led to regioselective addition at C(5), giving rise to oxazolylpyrazine **96**. Similar results were obtained for the heteroaryl Heck reaction of iodobenzene or bromobenzene with oxazole and benzoxazole [56].

95   **96**

The Hantzch–Panek condensation of ethylpyruvate with acrylamide followed by treatment with trifluoroaceticanhydride (TFAA) provided 2-vinyloxazole-4-ethylcarboxylate **37** in good yield [57]. The Stille coupling of 2-chlorooxazole-4-ethylcarboxylate with vinyltrimethylstannane also provided 2-vinyloxazole **37** [26]. The reactivity of **37** towards cycloaddition reactions with enone and coupling reactions with aryl halides were explored [57]. Heck reaction of vinyloxazole with p-iodoanisole gave the cinnamyl oxazole **97** in 71% yield [57].

37   **97**

Palladium-catalyzed direct arylation of 2-phenyl-5-oxazolecarboxanilide (**98**) with bromobenzene in the presence of Pd(OAc)$_2$ proceeded to give 2,4,5-triphenyloxazole (**99**) (33%) along with considerable amount of 2,4-diphenyloxazole (20%) [58]. In this reaction carbamoyl group acted as a directing group for the arylation at the 3-position, after which it cleaved and the C(2) and C(5) were arylated to give 2,3,5-triphenyloxazole [58].

98   **99**

2,5-Disubstituted oxazole **101** was prepared from *N*-propargylamides **100** and 4-iodoaceto-phenone, *via* the sequential Pd-catalyzed coupling/base-catalyzed cyclization process using Pd$_2$(dba)$_3$ as the catalyst and P(2-furyl)$_3$ as the ligand [59].

100   **101**

Coupling reaction between 2-chloro-3,6-diisobutylpyrazine (**102**) and benz[*b*]oxazole took place at C(2) exclusively to afford pyrazinylbenzoxazole **103** [55].

102   **103**

In the search of new non-peptide glycoprotein GPIIb/GPIIIa antagonists, heterocyclic scaffolds such as oxazolepiperidine have been explored. Such framework has both acidic and basic functionalities, which may confer the pharmacological properties to the desired molecule. The Heck reaction of bromooxazolopyridines **104** with methyl acrylate in the presence of palladium acetate afforded methylpropenoate–oxazolopyridines **105** in 70 and 88% yields, respectively [60].

104

105

8.5. Carbonylation

A unique Pd(II)-promoted *ortho*-esterification of 2,5-diphenyloxazole has been described [61]. When 2,5-diphenyloxazole was heated with 2.5 equivalents of Pd(OAc)$_2$ in acetic acid and CCl$_4$, regioselective palladation took place, giving rise to arylpalladium(II) σ-complex **106** in almost quantitative yield. The regioselectivity observed reflects the strong coordination ability of the oxazole nitrogen atom to the palladium atom.

Complex **106** was then dissolved in MeOH–THF (1:1) and the solution was stirred at 0°C to produce 2-(5′-phenyloxazol-2-yl)-benzoate **107**. A drawback of this method is the use of 2.5 equivalents of Pd(OAc)$_2$, making it less useful for preparative purposes.

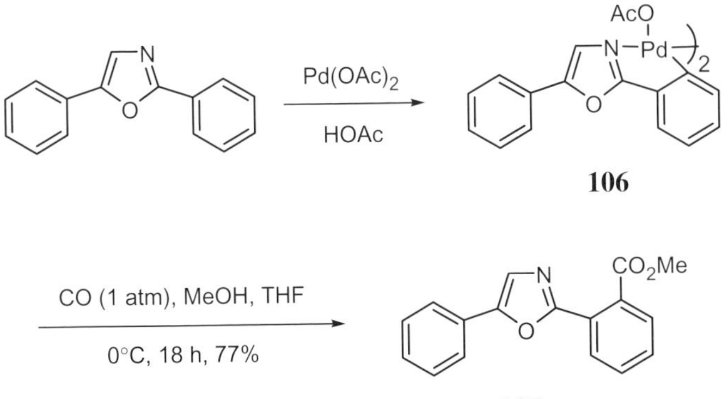

106

107

Prop-2-ynylamides **108** and **109** have been carbonylated under oxidative conditions to give oxazolines **110** and bisoxazolines **111** bearing an (alkoxycarbonyl)methylene chain at the 5-position in good yields. The cyclization–alkoxycarbonylation process was carried out in alcoholic media at 50–70°C and under 24 bar pressure of 3:1 CO/air in the presence of catalytic amounts of 10% Pd/C or PdI$_2$ in conjunction with KI [62].

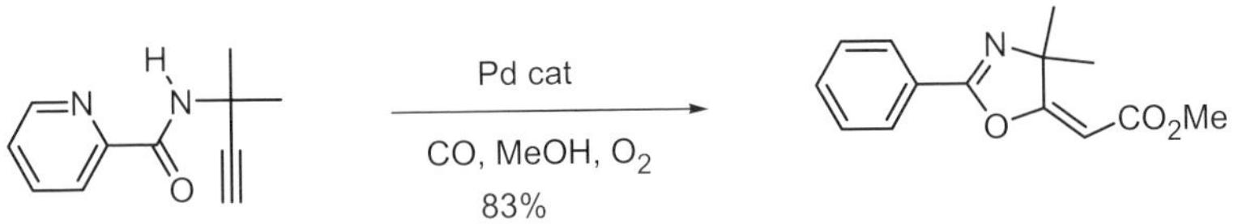

108

110

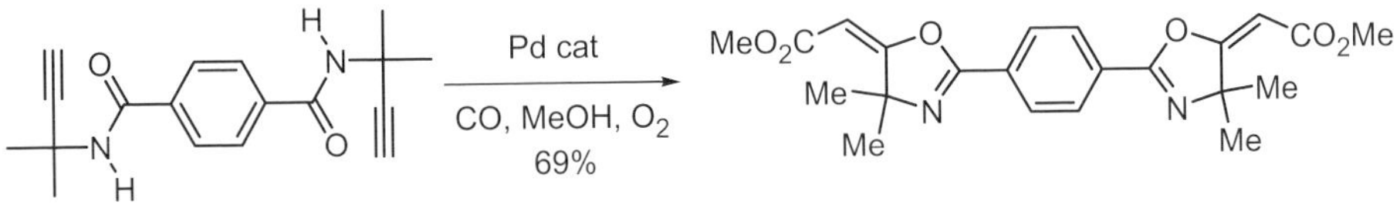

109

111

2-Aryl- and 2-heteroaryl-benzoxazoles (**112, 113**) were prepared by Pd-catalyzed three-component condensation of aryl halides with *o*-aminophenols and carbon monoxide followed by dehydrative cyclization [63, 64]. A variant of such methodology using *o*-fluorophenylamines in place of *o*-aminophenols was used to synthesize arylbenzoxazoles.

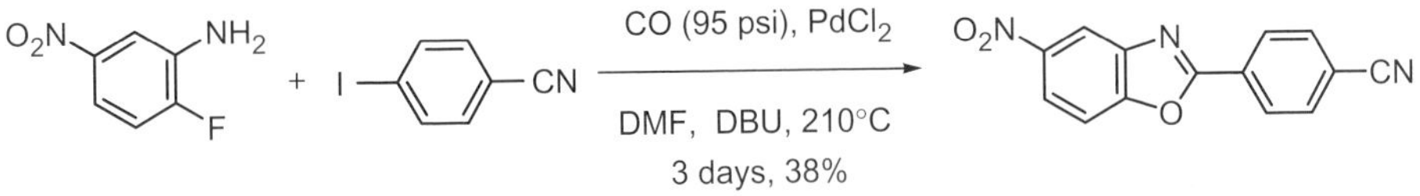

112

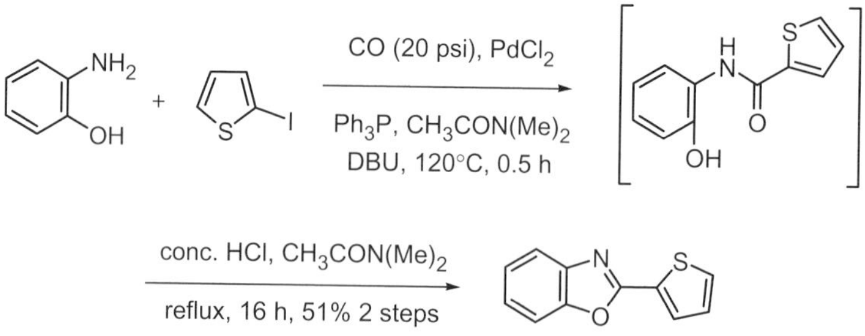

113

8.6. Palladium-catalyzed amination

An extension of Buchwald's catalytic amination using palladium led to amination of 2-chlorobenzoxazole [65]. The marked reactivity enhancement of the chloride is attributed to the polarization at C(2). Palladium-catalyzed amination of electron rich 2-chlorobenzoxazole with morpholine gave 2-morpholinylbenzoxazole (**114**) in substantial yield in the presence of $Pd_2(dba)_3$/BINAP [65]. 2-Morpholinylbenzoxazole (**114**) also was obtained from 2-chlorobenzoxazole in 51% yield with $Pd(O_2CCF_3)_2$/$P(t$-$Bu)_3$ [66].

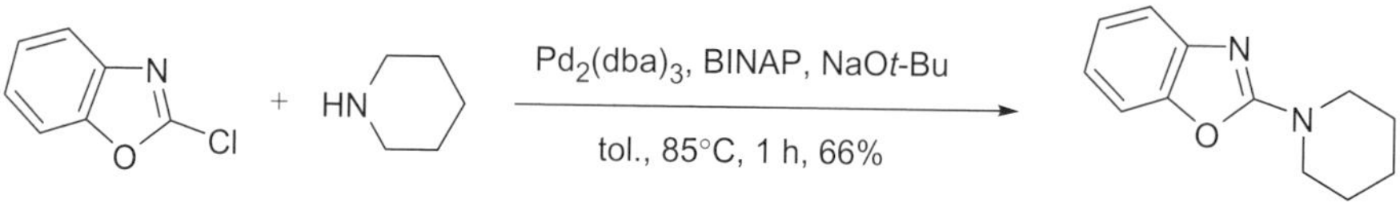

114

Recent studies on the C–C cleavage reaction of diynes proceeded smoothly with $Pd(NO_3)_3$ as the best among all palladium catalysts tested [67]. Substituted benzoxazoles **115** and **116** along with alkanones were derived from C–C bond cleavage of diynes through hydroamination [68].

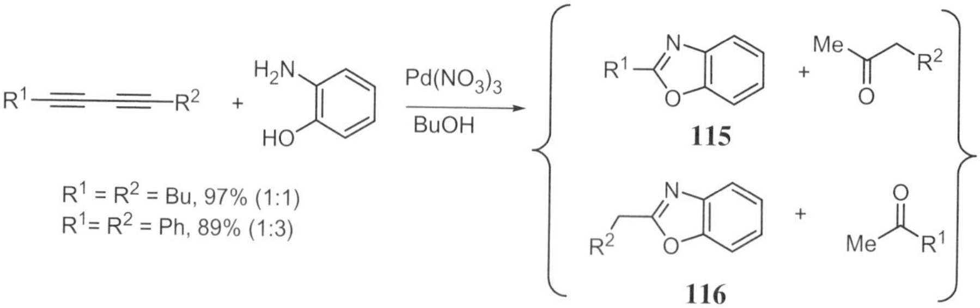

$R^1 = R^2 = Bu$, 97% (1:1)
$R^1 = R^2 = Ph$, 89% (1:3)

115

116

8.7. Carbopalladation of nitriles

Insertion of an internal alkyne into an aryl palladium intermediate and subsequent cyclization onto nitrile present at *ortho* position offers a novel synthetic route for heterocyclic compounds. Naphthoxazole **118** was prepared *via* the Pd-catalyzed annulation of 2-acetoxy-2-(2-iodophenyl)acetonitrile **117** with diphenylacetylene in moderate yield [69].

8.8. Quaterfuran and Quinquifuran

For more than a decade fluors have been sought for use as wave shifters in fiber light-guides [70]. Benzoxazole terminated quater- and quinquifurans were both stable and fast exhibiting a green fluorescence with decay times of about 2.4 ns. In general furan compounds, quaterfuran, and quinquifuran were prepared in high yields by means of the Ullmann reaction or by Pd-catalyzed unsymmetrical coupling. 2-Chlorozincfuran underwent the Negishi coupling with bromofuranylbenzoxazole **119** using PdCl$_2$–dppb to give bifurylbenzoxazole **120**, which was subsequently iodinated to iodobifurylbenzoxazole (**120**) [70]. An Ullmann coupling of **120** with copper bronze then gave quaterfuran (**121**) in good yield [70].

Dilithiation of terfuran with BuLi followed by exchange with ZnCl$_2$ produced 2,5″-di(chlorozinc)terfuran (**122**) [70]. The Negishi coupling of **122** with 2-(5-bromo-2-furyl)-5-t-butylbenzoxazole (**123**) in the presence of PdCl$_2$–dppb produced quinquifuran (**124**) [70].

123

$$\xrightarrow[\text{11\%}]{\text{PdCl}_2, \text{ dppb}}$$

124   Quinquifuran

8.9. Summary

In recent years, several natural products containing 2,4-disubstituted oxazoles have been isolated and their synthetic routes were investigated. Several key intermediates required for the synthesis of target compounds were achieved *via* the Stille, Suzuki, Heck, and Sonogashira reactions. In addition, several 2,5-diaryloxzoles were prepared from 2-halo-5-aryloxazoles *via* the Suzuki coupling. These coupling reactions enabled manipulation and elaboration of the functional groups further in the 2,4-disubstituted oxazoles system to more useful synthetic intermediates. Oxazol-2-ylzinc chloride reagents can be prepared from the transmetalation of oxazol-2-yllithium with $ZnCl_2$, whereas oxazol-2-ylboronic acid is unprecedented. Stannyloxazoles are readily prepared and the Stille reaction is still the method of choice for synthesizing oxazole-containing molecules if the toxicity is not of great concern. 4-Bromo- and 5-bromo-4-methyloxazoles are viable substrates for both the Sonogashira and Heck reactions. The heteroaryl Heck reaction using oxazole as a recipient partner takes place predominantly at the electron-rich position C(5), whereas it occurs at C(2) for benzoxazole.

8.10. References

1. Yokokawa, F.; Asano, T.; Shioiri, T. *Org. Lett.* **2000**, *2*, 4169–71.
2. Bagley, M. C.; Hind, S. L.; Moody, C. J. *Tetrahedron Lett.* **2000**, *41*, 6897–900.
3. Bagley, M. C; Moody, C. J.; Pepper, A. G. *Tetrahedron Lett.* **2000**, *41*, 6901–4.
4. Nicolaou, K. C.; Snyder, S. A.; Giuseppone, N.; Huang, X.; Bella, M.; Reddy, M. V.; Bheema Rao, P.; Koumbis, A. E.; Giannakakou, P.; O'Brate, A. *J. Am. Chem. Soc.* **2004**, *126*, 10174–82.
5. (a) Nicolaou, K. C.; Snyder, S. A.; Giuseppone, N.; Huang, X.; Simonsen, K. B.; Koumbis, A. E.; Bigot, A. *J. Am. Chem. Soc.* **2004**, *126*, 10162–73. (b) Vedejs, E.; Barda, D. A. *Org. Lett.* **2000**, *2* (8), 1033–5. (c) Zajac, M. A.; Vedejs, E. *Org. Lett.* **2004**, *6* (2), 237–40.
6. Paterson, I.; Tudge, M. *Tetrahedron* **2003**, *59*, 6833–49.

7. Hornberger, K. R.; Hamblett, C. L.; Leighton, J. L. *J. Am. Chem. Soc.* **2000**, *122*, 12894–5.

8. Smith III, A. B.; Verhoest, P. R.; Minbiole, K. P.; Schelhaas, M. *J. Am. Chem. Soc.* **2001**, *123*, 4834–6.

9. Smith III, A. B.; Minbiole, K. P.; Verhoest, P. R.; Schelhaas, M. *J. Am. Chem. Soc.* **2001**, *123*, 10942–53.

10. Hansen, T. M.; Engler, M. M.; Forsyth, C. J. *Bioorg. Med. Chem. Lett.* **2003**, *13*, 2127–30.

11. Evans, D. A.; Fitch, D.; Smith, T. E.; Cee, V. J. *J. Am. Chem. Soc.* **2000**, *122*, 10033–46.

12. Hartung, I. V.; Niess, B.; Haustedt, L. O.; Hoffmann, H. M. R. *Org. Lett.* **2002**, *4*, 3239–42.

13. Hillier, M. C., Price, A. T.; Meyers, A. I. *J. Org. Chem.* **2001**, *66*, 6037–45.

14. Hillier, M. C.; Park, D. H.; Price, A. T.; Ng, R.; Meyers, A. I. *Tetrahedron Lett.* **2000**, *41*, 2821–4.

15. Hartung, I. V.; Eggert, U.; Haustedt, L. O.; Niess, B.; Scheäfer, P. M.; Hoffmann, H. M. R. *Synthesis* **2003**, *12*, 1844–50.

16. Haustedt, L. O.; Panicker, B.; Kleinert, M.; Hartung, I. V.; Eggert, U.; Niess, B.; Hoffmann, H. M. R. *Tetrahedron* **2003**, *59*, 6967–77.

17. White, J. D.; Blakemore, P. R.; Green, N. J.; Bryon Hauser, E.; Holoboski, M. A; Keown, L. E.; Nylund Kolz, C. S.; Phillips, B. W. *J. Org. Chem.* **2002**, *67*, 7750–60.

18. Kashima, C.; Arao, H. *Synthesis* **1989**, 873–4.

19. Barrett, A. G. M.; Kohrt, J. T. *Synlett* **1995**, 415–16.

20. Hodges, J. C.; Patt, W. C.; Connolly, C. J. *J. Org. Chem.* **1991**, *56*, 449–52.

21. Vedejs, E.; Luchetta, L. M. *J. Org. Chem.* **1999**, *64*, 1011–14.

22. Jeong, S.; Chen, X.; Harran, P. G. *J. Org. Chem.* **1998**, *63*, 8640–1.

23. Kunz, K. R.; Taylor, E. W.; Hutton, H. M.; Blackburn, B. J. *Org. Prep. Proced. Int.* **1990**, *22*, 613–18.

24. Sakamoto, T.; Katoh, E.; Kondo, Y.; Yamanaka, H. *Chem. Pharm. Bull.* **1990**, *38*, 1513–17.

25. Sakamoto, T.; Kondo, Y.; Suginome, T.; Ohba, S.; Yamanaka, H. *Synthesis* **1992**, 552–4.

26. Hodgetts, K. J.; Kershaw, M. T. *Org. Lett.* **2002**, *4*, 2905–7.

27. Reeder, M. R.; Gleaves, H. E.; Hoover, S. A.; Imbordino, R. J.; Pangborn, J. J. *Org. Process Res. Dev.* **2003**, *7*, 696–9.

28. Gillet, J.-P.; SayvÍtre, R.; Normant, J.-F. *Tetrahedron Lett.* **1985**, *26*, 3999–4002.

29. Gillet, J.-P.; SauvÍtre, R.; Normant, J.-F. *Synthesis* **1986**, 538–43.

30. Miller, R. D.; Lee, V. Y.; Moylan, C. R. *Chem. Mater.* **1994**, *6*, 1023–32.

31. (a) Harn, N. K.; Gramer, C. J.; Anderson, B. A. *Tetrahedron Lett.* **1995**, *36, 9453–6. (b) Anderson, B. A.; Harn, N. K. *Synthesis* **1996**, 583–5.

32. Anderson, B. A.; Becke, L. M.; Booher, R. N.; Flaugh, M. F.; Harn, N. K.; Kress, T. J.; Varie, D. L.; Wepsiec, J. P. *J. Org. Chem.* **1997**, *62*, 8634–9.

33. Vachal, P.; Toth, M. L. *Tetrahedron Lett.* **2004**, *45*, 7157–61.

34. Maekawa, T.; Sakai, N.; Tawada, H.; Murase, K.; Hazama, M.; Sugiyama, Y.; Momose, Y. *Chem. Pharm. Bull.* **2003**, *51*, 565–73.

35. Liebeskind, L. S.; Srogl, J. *Org. Lett.* **2002**, *4*, 979–81.

36. Viaud, M.-C.; Jamoneau, P.; Savelon, L.; Guillaumet, G. *Heterocycles* **1995**, *41*, 2799–810.

37. Hodgetts, K. J.; Kershaw, M. T. *Org. Lett.* **2003**, *5*, 2911–14.
38. Lobrégat, V.; Alcaraz, G.; Bienaymé, H.; Vaultier, M. *Chem. Commun.* **2001**, 817–18.
39. Dondoni, A.; Fantin, G.; Fogagnolo, M.; Medici, A.; Pedrini, P. *Synthesis* **1987**, *8*, 693–6.
40. Smith III, A. B.; Minbiole, K. P.; Freeze, S. *Synlett* **2001**, *11*, 1739.
41. Clapham B.; Sutherland, A. J. *J. Org. Chem.* **2001**, *66*, 9033–7.
42. Kelly, T. R.; Lang, F. *J. Org. Chem.* **1996**, *61*, 4623–33.
43. Boto, A.; Ling, M.; Meek, G.; Pattenden, G. *Tetrahedron Lett.* **1998**, *39*, 8167–70.
44. Schaus, J. V.; Panek, J. S. *Org. Lett.* **2000**, *2*, 469–71.
45. Crawforth, C. M.; Burling, S.; Fairlamb, I. J. S.; Taylor, R. J. K; Whitwood, A. C. *Chem. Commun.* **2003**, 2194–5.
46. Young, G. L.; Smith, S. A.; Taylor, R. J. K. *Tetrahedron Lett.* **2004**, *45*, 3797–801.
47. Jutzi, P.; Gilge, U. *J. Organomet. Chem.* **1983**, *246*, 159–62.
48. Kosugi, M.; Koshiba, M.; Atoh, A.; Sano, H.; Migita, T. *Bull. Chem. Soc. Jpn.* **1986**, *59*, 677–9.
49. Jutzi, P.; Gilge, U. *J. Heterocycl. Chem.* **1983**, *20*, 1011–14.
50. Sakamoto, T.; Nagata, H.; Kondo, Y.; Shiraiwa, M.; Yamanaka, H. *Chem. Pharm. Bull.* **1987**, *35*, 823–8.
51. Ducept, P. C.; Marsden, S. P. *Synlett* **2000**, *5*, 692–4.
52. Langille, N. F.; Dakin, L. A.; Panek, J. S. *Org. Lett.* **2002**, *4*, 2485–88.
53. Dakin, L. A.; Langille, N. F.; Panek, J. S. *J. Org. Chem.* **2002**, *67*, 6812–15.
54. Paterson, I.; Tudge, M. *Angew. Chem. Int. Ed. Engl.* **2003**, *42*, 343–7.
55. Aoyagi, Y.; Inoue, A.; Koizumi, I.; Hashimoto, R.; Tokunaga, K.; Gohma, K.; Komatsu, J.; Sekine, K.; Miyafuji, A.; Konoh, J.; Honma, R.; Akita, Y.; Ohta, A. *Heterocycles* **1992**, *34*, 257–72.
56. Pivsa-Art, S.; Satoh, T.; Kawamura, Y.; Miura, M.; Nomura, M. *Bull Chem. Soc. Jpn.* **1998**, *71*, 467–73.
57. Ahmed, F.; Donaldson, W. A. *Synth. Commun.* **2003**, *33*, 2685–93.
58. Yokooji, A.; Okazawa, T.; Satoh, T.; Miura, M.; Nomura, M. *Tetrahedron* **2003**, *59*, 5685–9.
59. Arcadi, A.; Cacchi, S.; Cascia, L.; Fabrizi, G.; Marinelli, F. *Org. Lett.* **2001**, *3*, 2501–4.
60. Grumel, V.; Mérour, J.-Y.; Guillaumet, G. *Heterocycles* **2001**, *55*, 1329–45.
61. Sakakibara, T.; Kume, T.; Ohyabu, T.; Hase, T. *Chem. Pharm. Bull.* **1989**, *37*, 1694–7.
62. Bacchi, A.; Costa, M.; Gabriele, B.; Pelizzi, G.; Salerno, G. *J. Org. Chem.* **2002**, *67*, 4450–7.
63. Perry, R. J.; Wilson, B. D.; Miller, R. J. *J. Org. Chem.* **1992**, *57*, 2883–7.
64. Perry, R. J.; Wilson, B. D. *J. Org. Chem.* **1992**, *57*, 6351–4.
65. Hong, Y.; Tanoury, G. J.; Wilkinson, H. S.; Bakale, R. P.; Wald, S. A.; Senanayake, C. H. *Tetrahedron Lett.* **1997**, *38*, 5607–10.
66. Hooper, M. W.; Utsunomiya, M.; Hartwig, J. F. *J Org. Chem.* **2003**, *68*, 2861–73.
67. Shimada, T.; Yamamoto, Y. *J. Am. Chem. Soc.* **2002**, *124*, 12670.
68. Shimada, T.; Yamamoto, Y. *J. Am. Chem. Soc.* **2003**, *125*, 6646–7.
69. Tian, Q.; Pletnev, A. A.; Larcock, R. C. *J. Org. Chem.* **2003**, *68*, 339–47.
70. Kauffman, J. M.; Moyna, G. *J. Heterocyclic Chem.* **2002**, *39*, 981–8.

Chapter 9

Imidazoles

Richard J. Mullins and Adam M. Azman

The imidazole ring is present in a number of biologically important molecules as exemplified by the amino acid histidine. It can serve as a general base (pK_a = 7.1) or a ligand for various metals (e.g. Zn, etc.) in biological systems. Furthermore, the chemistry of imidazole is prevalent in protein and DNA biomolecules in the form of histidine or adenine/guanine, respectively.

histidine   imidazole

Imidazole is a π-electron-excessive heterocycle. Electrophilic substitution normally occurs at C(4) or C(5), whereas nucleophilic substitution takes place at C(2). The order of reactivity for electrophilic substitution for azoles is:

imidazole > thiazole > oxazole

9.1. Synthesis of haloimidazoles

The C(2) position on 1-alkyl or 1-arylimidazole possesses the most acidic hydrogen, and a carbanion is readily generated at C(2) upon treatment with a strong base, such as *n*-butyllithium. The resulting carbanion can be treated by a halogen electrophile to furnish 2-haloimidazoles. Kirk selectively deprotonated 1-tritylimidazole with *n*-butyllithium and quenched the resulting 2-lithioimidazole species with NIS, NBS, or *tert*-butyl hypochlorite, respectively, to make the corresponding 2-iodo-, 2-bromo-, or 2-chloroimidazole [1]. An extension of Kirk's method to 1,4-disubstituted imidazole **1** provided an entry to the corresponding 2-iodoimidazole **2** [2].

407

Electrophilic iodination of 2-substituted imidazoles using NIS in DMF led to 4,5-diiodoimidazoles [3, 4].

Treating imidazole with bromine in a NaOH solution afforded the complete ring bromination product, 2,4,5-tribromoimidazole. 4-Bromoimidazole was then produced in 62% yield upon selective reductive debromination [5]. The same procedure converted 2-methylimidazole and 1-methylimidazole to 4(5)-iodo-2-methylimidazole [6] and 4(5)-bromo-1-methylimidazole [7], respectively.

By carefully monitoring the pH of the reaction, regioselective deiodination was achieved. Treating imidazole with iodine at pH 12 furnished 2,5-diiodoimidazole. Selective reductive deiodination secured 2-iodoimidazole, which upon bromination afforded 4,5-dibromo-2-iodoimidazole [6].

An alternative to the above procedure involves the use of aqueous potassium dichloroiodate as an iodating agent. Interestingly, the order of addition of the iodinating reagent is important for efficient reactivity. For example, the addition of an aqueous solution of imidazole to the KICl$_2$ resulted in the production of **3** in high yield. On the other hand, the order had to be reversed in order to iodinate the 2-substituted imidazoles to produce **4** and **5** [8].

The use of I$_2$ in combination with HIO$_4$ has also been reported for the synthesis of **6** and **7** [9]. In either case, barely a trace of the monoiodinated product was detected.

Having appropriate substituents on the imidazole ring, Haseltine and Wang were able to carry out a regioselective bromination of 1-benzyl-5-methylimidazole in excellent yield [10].

A variant of the Sandmeyer reaction transformed 5-aminoimidazole **8** into the corresponding 5-iodoimidazole **9**. Here the diazotization was accomplished using isoamyl nitrite [11, 12].

While dealing with imidazoles, an important characteristic is their annular tautomerism. A tautomeric equilibrium for many imidazoles is rapidly achieved at room temperature. In some tautomeric pairs, though, one tautomer often predominates over the other. For instance, 4(5)-bromoimidazole favors the 4-bromo tautomer in a 30:1 ratio, whereas 4(5)-nitroimidazole exists predominantly as the 4-nitro tautomer (700:1) [13]. 4(5)-Methoxyimidazole has a ratio of 2.5:1 for the 4- and 5-methoxy tautomers.

9.2. Homocoupling reaction

Homocoupling to produce 2,2′-biimidazoles has been very well developed [14]. As one example of this process, the homocoupling of iodide **10** to yield **11** *via* this method was described by Sessler and coworkers.

Unfortunately, homocoupling to provide 4,4′-biimidazoles has not been studied as extensively. Pyne and Cliff revealed that the Palladium(0)-catalyzed homocoupling of 4-iodo-1-(triphenylmethyl)imidazole (**12a**) or its 2-methyl analog **12b** delivered the 4, 4′-biimidazoles **13a** and **13b**, respectively [6].

12a, R = H
12b, R = Me

13a, 69%
13b, 63%

9.3. Coupling reactions with organometallic reagents

9.3.1. Negishi coupling

There are several ways to prepare imidazol-2-ylzinc reagents. First, direct metalation followed by $ZnCl_2$ quench is suitable for preparation of imidazol-2-ylzinc chlorides. Treating 1-dimethylaminosulfonylimidazole with *n*-BuLi and quenching the resulting 2-lithioimidazole species with $ZnCl_2$ generated the organozinc reagent *in situ*. The organozinc reagent was then cross-coupled with 2-bromopyridine to give the unsymmetrical pyridylimidazole [15].

In a similar manner, consecutive Negishi reactions were carried out in one pot between **14** and 2,6-dibromopyridine. The application of this reaction resulted in a shorter and more simple synthesis of the ligand **15** [16].

This method of imidazolylzinc chloride synthesis has also found application in the synthesis of a sulfonyl chloride to be used in the preparation of some potential factor Xa inhibitors [17]. Thus, following imidazole metalation and quenching with $ZnCl_2$, a chemoselective coupling reaction occurred to provide **16**. Treatment with sulfur dioxide provided the sulfinic acid, which upon treatment with thionyl chloride, provided the requisite sulfonyl chloride **17**.

Direct metalation was also employed by Li and coworkers in the synthesis of non-peptide, small molecule antagonists for the Interleukin-8 receptor [18]. Standard conditions were utilized to produce organozinc reagent **18**. The coupling between **18** and **19** was then conducted utilizing a palladate complex (prepared from the reaction of $PdCl_2(PPh_3)_4$ and *n*-BuLi) to produce **20**. While the reaction did not proceed in a remarkable yield, the power of the Negishi methodology is demonstrated in its ability to form C—C bonds in this sterically hindered environment. Unfortunately, the presence of the imidazolyl substituent resulted in no increased binding activity relative to the lead compound.

Finally, this reaction motif has been used in the synthesis of analogs of the antipsychotic sertindole for structure-activity relationship (SAR) studies [19]. Specifically, the substitution was effected at the 5-position of the indole nucleus to give analog **21**.

Second, halogen–metal exchange followed by quenching with $ZnCl_2$ works for all three possible haloimidazoles. Bromide **23** was derived from direct bromination at C(4) from 2-(4-fluorophenyl)-5*H*-pyrrolo[1,2-*a*]imidazole (**22**) using bromine. Halogen–metal exchange of **23** with *n*-BuLi followed by quenching with $ZnCl_2$ furnished organozinc reagent **24** *in situ*, which then was coupled with vinyl iodide **25** in the presence of $Pd(Ph_3P)_4$ to form tetrahydropyridine derivative **26** [20]. Adduct **26** was an intermediate to a potent inhibitor of cyclooxygenase and 5-lipoxygenase enzymes. In addition, imidazol-4-ylzinc reagent **27** was easily generated by treating 1-trityl-4-iodoimidazole with ethyl Grignard reagent followed by the addition of ZnCl$_2$ [21]. The corresponding imidazol-4-ylzinc bromide was also made in the same fashion using ZnBr$_2$ [22]. The Negishi reaction of **27** with triflate **28** in the presence of LiCl led to quinolinylimidazole **29**.

Negishi coupling between an organozinc reagent prepared in this manner has also been demonstrated using 2-bromopyridine as the electrophile. Specifically, this study was aimed at the preparation of potential MAP kinase inhibitors [23]. Thus, following halogen–metal exchange of **30** and transmetalation, the Negishi coupling proceeded in an excellent yield to give **31**.

The mild nature of the transmetalation conditions was demonstrated by the Knochel group in the preparation of **33** [24]. The ester group on **32** was unaffected during the halogen–metal exchange process. Transmetalation to give the organozinc halide was followed by the coupling reaction with iodobenzene. Compound **33** was isolated in pure form in a respectable 66% yield.

32   **33**

Third, direct oxidative addition of organohalides to Zn(0) offers another synthetic strategy of preparing imidazol-2-ylzinc reagents. This method is advantageous over the usual transmetalations using organolithium or Grignard reagents because of the better tolerance of functional groups. Treating 1-methyl-2-iodoimidazole with readily available zinc dust, the Knochel group produced the expected imidazol-2-zinc iodide [25]. The subsequent Negishi reaction was carried out with vinyl-, aryl-, and heteroaryl halides.

In addition to its use as the nucleophile in the Negishi reacton, imidazolyl iodides have found some usage as electrophiles under these conditions. Evans and Bach prepared organozinc reagent **35** from quantitative metalation of iodide **34** by using an activated Zn/Cu couple in the presence of an ester functionality. Treating 2-iodoimidazole **36** with three equivalents of **35** in the presence of $PdCl_2(Ph_3P)_2$ afforded adduct **37**, which was then transformed into diphthine (**38**) in five additional steps [26].

34   **35**

37

9.3.2. Suzuki coupling

Imidazolylboronic acids are not well-known. 5-Imidazolylboronic acid was prepared *in situ* by Ohta's group using selective lithiation at C(5) from 1,2-protected imidazole.

However, the subsequent Suzuki coupling with 3-bromoindole gave the expected indolylimidazole in only 7% yield [27]. This observation indicated that more investigation into imidazolylboronic acids is necessary to better understand their synthesis and behavior.

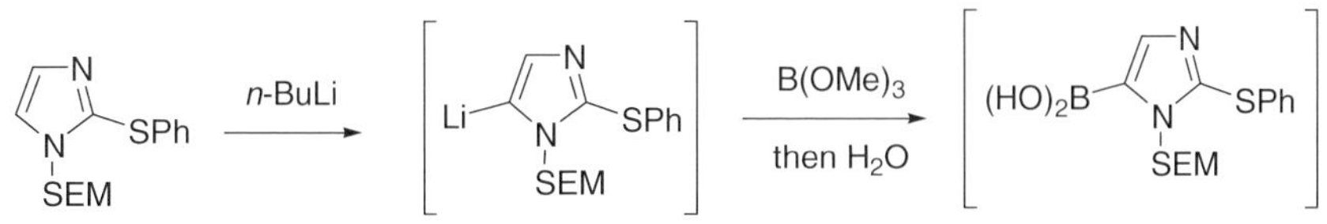

Nonetheless, to make imidazole-containing molecules, haloimidazoles may serve as electrophilic coupling partners for the Suzuki reaction. As described in Section 9.1, regioselective bromination at the C(4) position could be achieved for a 1,5-dialkylimidazole using NBS in CH_3CN. The Suzuki coupling of 1-benzyl-4-bromo-5-methylimidazole with phenylboronic acid assembled the unsymmetrical heterobiaryl in 93% yield, whereas the corresponding Stille reaction with phenyltrimethyltin proceeded in only moderate yield (51%) [10].

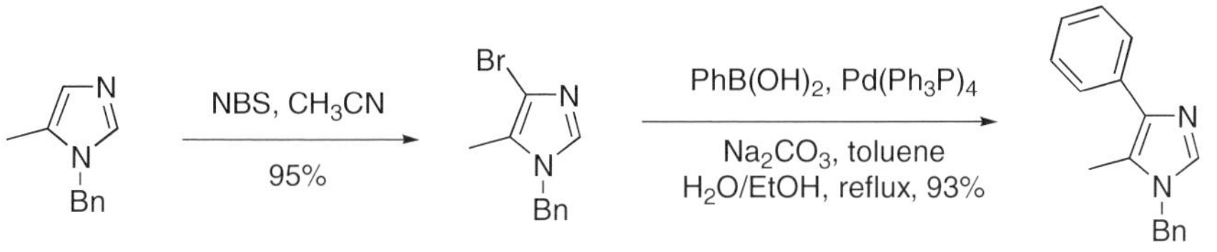

In similar studies, the 4,5-disubstituted-1-methyl imidazole **41** has been prepared. In an effort to synthesize molecules with this substitution pattern, Shapiro and Gomz-Lor utilized a carboxylate at C(2) to serve as a blocking group [28]. As shown below, regioselective halogen–metal exchange allowed for the introduction of the carboxylate at C(2). A second halogen–metal exchange at C(5), resulted in formylation with DMF to give **39**. Finally, the blocking carboxylate group could be removed under strongly acidic conditions. The 4,5-disubstituted imidazole **41** was thus produced in high yield upon coupling with boronate ester **40**.

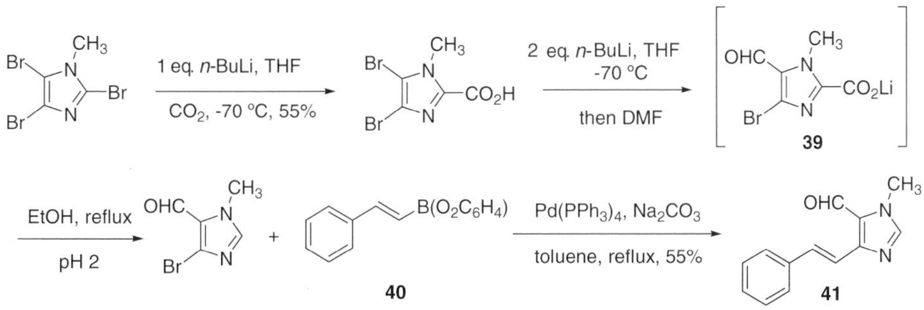

There have been several similar investigations involving the use of imidazolyl halides as the electrophile in the Suzuki reaction. Predominantly, these reactions have been

utilized for synthesis in search of biologically active compounds containing the imidazole nucleus. The synthesis of **42** [23], **44** [29], **45** [30], and **46** [31] serve as examples
in this regard.

Perhaps, the most elegant demonstration of the Suzuki coupling in imidazole synthesis
comes from efforts for the synthesis of the natural product nortopsentin C. Nortopsentin C (**47**), along with nortopsentins A and B, was isolated from the marine sponge
Spongosorites ruetzleri. They all possess a characteristic 2,4-bisindolylimidazole skeleton and exhibit cytotoxic and antifungal activities. Successive and regioselective diarylation *via* Suzuki coupling reactions using halogenated imidazole to make nortopsentins
have been disclosed in the total synthesis of **47** by Ohta's group [32, 33]. The *N*-protected
2,4,5-triiodoimidazole (**48**) was coupled with one equivalent of the 3-indolylboronic acid
49 to install indolylimidazole **50**. The Suzuki reaction occurred regioselectively at C(2)
of the imidazole ring. Subsequently, a regioselective halogen–metal exchange reaction
took place predominantly at C(5) to access **51**. A second Suzuki coupling reaction at C(4)
of **51** with 6-bromo-3-indolylboronic acid **52** resulted in the assembly of the entire skeleton. Two consecutive deprotection reactions removed all three silyl groups to deliver the
natural product (**47**). This synthesis elegantly exemplifies the chemoselectivity of the
Suzuki reaction between an arylbromide and an aryliodide. The desired regioselectivity
was achieved *via* manipulation of the substrate functionalities.

nortopsentin C (**47**)

9.3.3. Stille coupling

In contrast to the scarcity of precedents in the imidazolylboronic acids, abundant reports exist on imidazolylstannanes. In the Stille reaction, the imidazole nucleus may serve as either the nucleophile as an imidazolylstannane or the electrophile as an imidazole halide. One advantage of the Stille reaction is that it often tolerates delicate functionalities in both partners.

One method for preparing imidazolylstannanes is direct metalation followed by treatment with R₃SnCl [34]. 1-Methyl-2-tributylstannylimidazole, derived in such manner, was coupled with 3-bromobenzylphosphonate to furnish heterobiaryl phosphonate **54** [35]. Under the same reaction conditions, 4-bromobenzylphosphonate led to the adduct in 69% yield, whereas only 24% yield was obtained for 2-bromobenzylphosphonate. The low yield encountered for the *ortho* derivative may be attributed to the steric factors to which the Stille reaction has been reported to be sensitive [36]. Heterobiaryl phosphonates such as **54** are not only substrates for the Wadsworth–Horner–Emmons reaction, but also bioisosteric analogs of the carboxylic acid group.

Analogously, 5-tributylstannylimidazole **56** was easily obtained from the regioselective deprotonation of 1,2-disubstituted imidazole **55** at C(5), followed by the treatment with tributyltin chloride [37]. In the presence of 2.6 equivalents of LiCl, the Stille reaction of **56** with aryl triflate **57** afforded the desired 1,2,5-trisubstituted imidazole **58** with 2,6-di-*tert*-butyl-4-methylphenol (BHT) as a radical scavenger. Reversal of the nucleophile and electrophile of the Stille reaction also provided satisfactory results. For example, the coupling reaction of 5-bromoimidazole **60**, derived from imidazole **59** *via* a regioselective bromination at C(5), and vinylstannane **61** produced adduct **62** [37].

Steglich's group coupled *N*-methyl bromoindolylmaleimide **63** with 5-tributylstannyl-1-methyl-1*H*-imidazole to afford **64** and then converted **64** to didemnimide C (**65**), an alkaloid isolated from the Caribbean ascidian *Didemnum conchyliatum* [38]. 5-Tributylstannyl-1-methyl-1*H*-imidazole, in turn, was prepared according to Undheim's direct metalation method from 1-methyl-1*H*-imidazole [39].

Achab conducted a Stille coupling using 4-iodoimidazole **66** as the electrophilic coupling partner in his synthesis of topsentin. The connection between an indole and an imidazole was realized by the Stille reaction of **66** with indolylstannane **67** to furnish adduct **68**. Global deprotection then transformed **68** into topsentin (**69**), a marine bis-indole alkaloid [40]. In another case, Hibino *et al.* carried out the Stille reaction of stan-nylimidazole **70** with 3-iodoindole **71** to assemble indolylimidazole **72**, an important intermediate for the total synthesis of grossularines-1 and -2 [41]. Stannane **70**, in turn, was synthesized from the corresponding 5-bromoimidazole *via* halogen–metal exchange followed by treatment with Me$_3$SnCl. This strategy has also been applied toward the syn-thesis of some structural analogs of the marine cytotoxic agents, grossularines 1 and 2, and eudistomin U [42, 43].

Interestingly, with only a slight difference in the imidazole *N*-protection, an anomalous Stille reaction occurred for the stannane **74** [44]. When the protecting group of the 5-stannylimidazole was changed from methyl into 1-ethoxymethyl, the reaction between 3-iodoindole **73** and 1-ethoxymethyl-5-trimethylstannyl-2-methylthioimidazole (**74**) gave not only the normal *ipso* product **75** but also the *cine* product **76**.

Lovely and coworkers have shown the Stille to efficiently provide 4-vinylimidazoles [45]. As is typical, substitution took place at C(4) of the imidazole. The resulting vinyl imidazoles were found to be suitable substrates in a Diels–Alder reaction, as demonstrated in the production of **78** from coupling product **77**. Notably, the expected isomerization back to the imidazole nucleus could be avoided using the lower boiling chlorinated solvents such as chloroform and methylene chloride.

Using the Stille reaction similar to the one described immediately above, a new synthetic route to imidazo[4,5-*c*]pyridines has been delineated [46]. Following the coupling reaction to provide alkenyl imidazole **79b**, an electrocyclic reaction was employed to provide **80** after aromaticity was attained by the elimination of methanol. The flexibility of this process is demonstrated by the fact that nearly identical yields are possible regardless of whether the Stille coupling reaction is done on compound **79a** or the aldehyde compound that immediately precedes it in the synthetic sequence.

Additional literature precedents for the Stille reactions in which imidazolylstannanes are used are listed below [47–51].

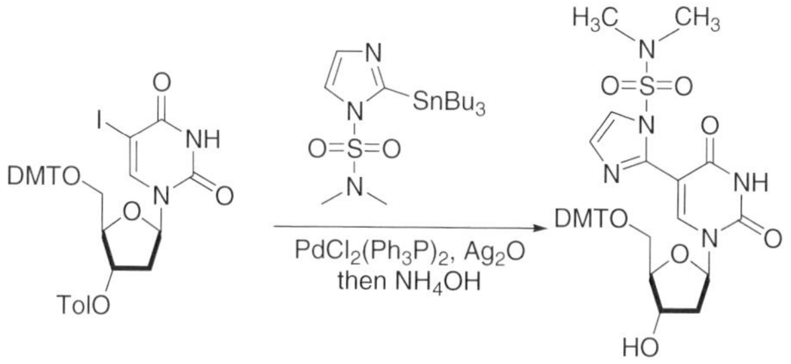

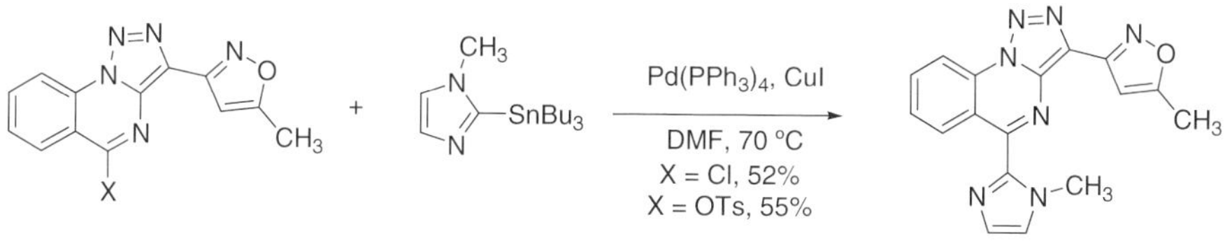

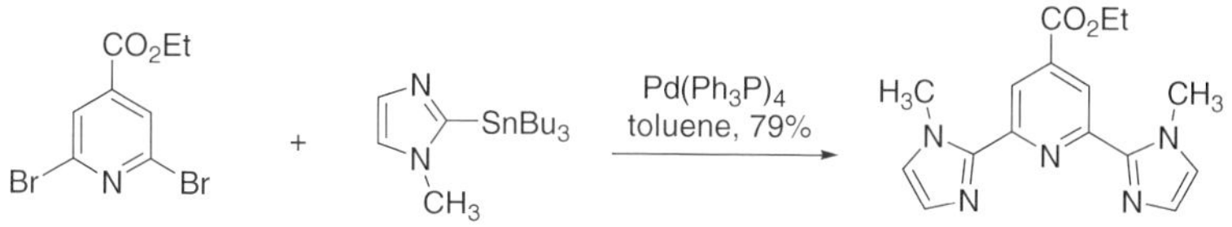

Most literature precedents assemble the heterobiaryls at the C(4) position. In one instance, however, the substitution took place at the C(5) position. As shown below, treatment of 5-iodoimidazole derivative **81** with (E)-1-tributylstannylprop-1-en-3-ol (**82**) in the presence of PdCl$_2$(PhCN)$_2$ led to adduct **83** in 68% yield [52].

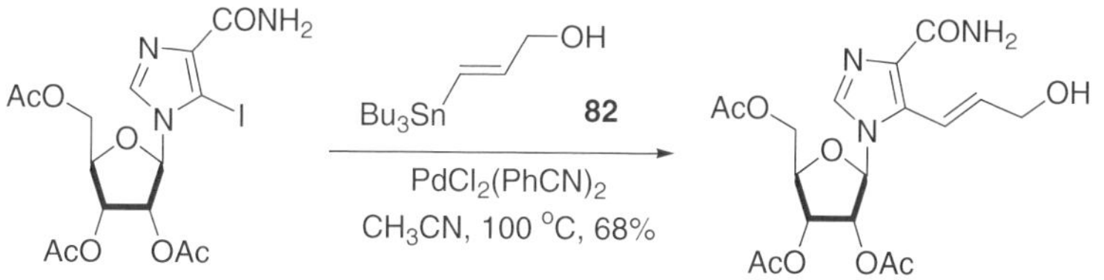

N-protected 4-iodoimidazoles have been employed to synthesize an assortment of imidazole–heteroaryls [53, 54]. For example, coupling between stannyl furan and imidazole **84** proceeded efficiently to produce **85** [31].

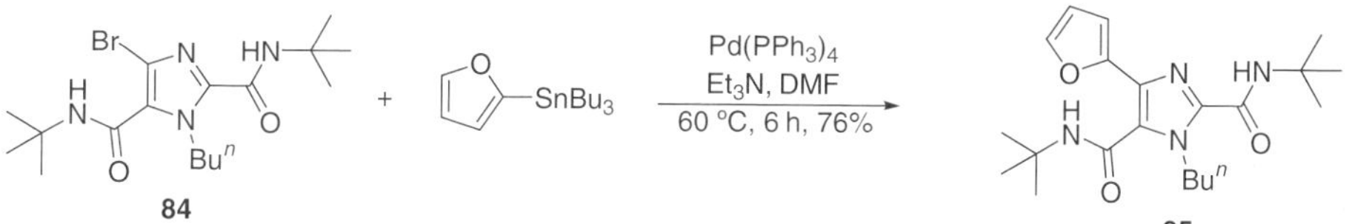

By taking advantage of the Stille reaction, **90**, a naturally occurring imidazole alkaloid was synthesized [55–58]. Pyne and Cliff protected 4-iodoimidazole with chloromethylethyl ether, forming two tautomers **86** and **87** in a 1:9 ratio. After isolation, the Stille coupling of **87** with vinylstannane **88** [(*E*) : (*Z*) = 88 : 22] afforded isomerically pure alkene **89**, which was then manipulated in five additional steps into (1*R*, 2*S*, 3*R*)-2-acetyl-4(5)-(1,2,3,4-tetrahydroxybutyl)imidazole (THI, **90**), a constituent of Caramel Color III.

9.4. Sonogashira reaction

Regardless of the position of the iodide [C(2), C(4), or C(5)] on the imidazole ring, the Sonogashira reaction of an iodo-*N*-methylimidazole and simple acetylenes gave unsatisfactory results during initial investigations [59]. However, when diiodoimidazole **91**, derived from iodination of 1,2-dimethylimidazole, was submitted to the classic Sonogashira reaction conditions with trimethylsilylacetylene, diyne **92** was isolated in 68% yield. The same reaction using propargyl alcohol did not work as well (17%, 34% yields) when performed on substrate **91** [4].

In contrast, an excellent yield was obtained for the reaction between 5-iodoimidazole and propargyl alcohol *in the absence of CuI* to install the 5-alkynyl derivative [12, 60]. The adduct was then deprotected to 5-alkynyl-1-β-D-ribofuranosylimidazole-4-carboxamide **94**, an antileukemic agent. It is noteworthy that if the Sonogashira reaction was conducted in the presence of CuI, the yield dropped to 19%, most likely due to alkyne homocoupling.

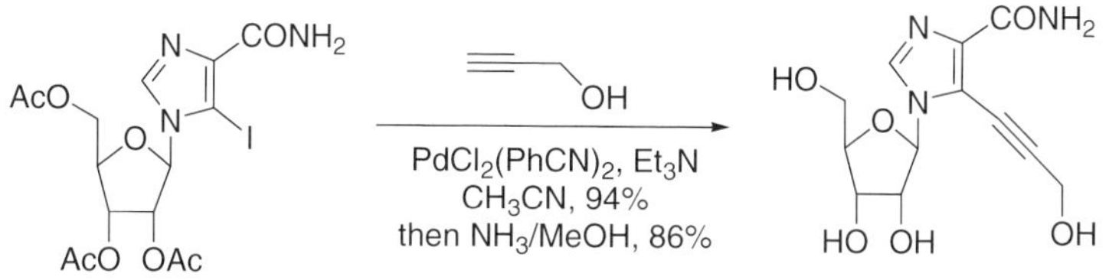

93   **94**

This process has been demonstrated to have significant flexibility, if not wide applicability in the preparation of **95** [23], **96** [61, 62], and **97** [31].

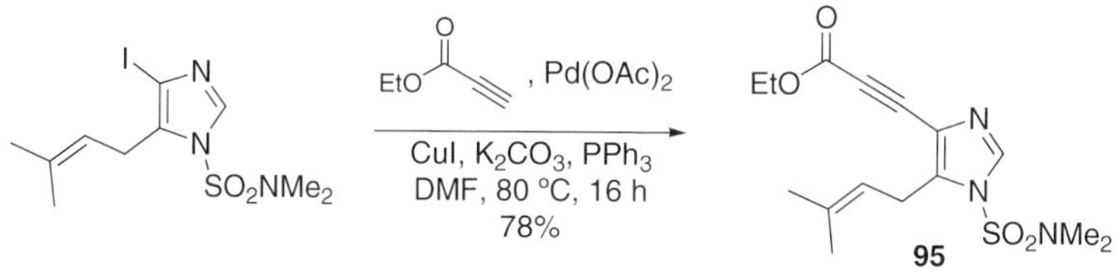

95

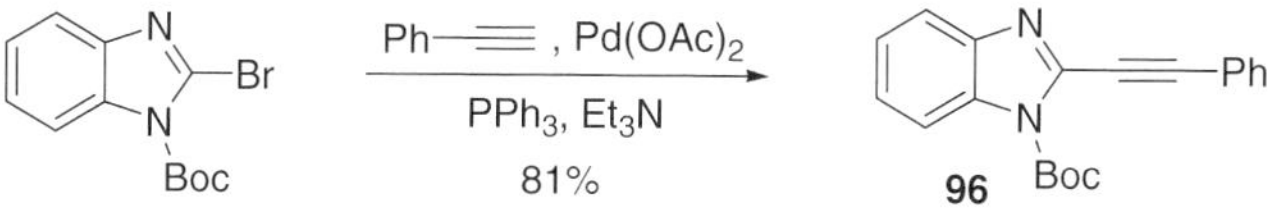

96

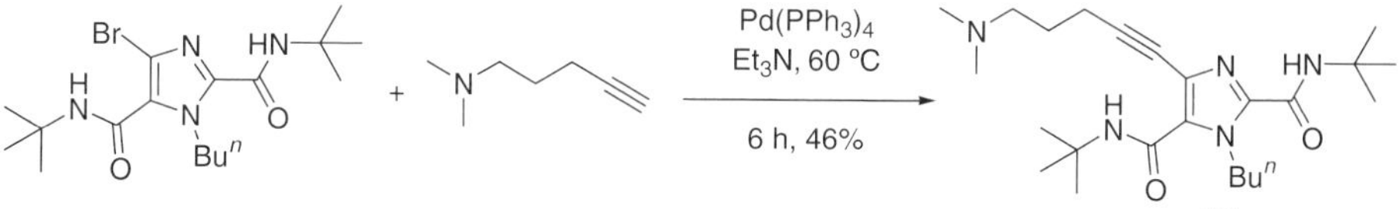

97

The natural product keramidine (**98**) was isolated from *Agelas* sp. in 1984 and was found to be an antagonist on serotonergic receptors of the rabbit aorta [63]. It is one of the number of structurally diverse pyrrole–imidazole alkaloids, the common skeleton of which was first observed in oroidin. Keramidine (**98**) poses a challenge for synthesis as a result of the Z-alkene substituted at the 5-position of the imidazole ring. Synthesis of this structure on gram scale should enable further syntheses of other structurally related oroidin alkaloids. The key step in the efficient six-step synthesis involved the Sonogashira reaction of alkyne **99** with the protected 4-iodoimidazole **100**. Protecting group cleavage *via* methylation at N(3) of **101** resulted in **102**. Metalation at C(2) of the imidazole nucleus was followed by quenching with tosyl azide to provide **103**. The synthesis of keramidine was thus competed following Boc removal, amide bond formation with the activated trichloromethyl ketone **104** and double hydrogenation using Lindlar's catalyst to prepare the amine functionality at C(2) of the imidazole, as well as establish the Z-olefin geometry in the natural product.

Regioselective functionalization of histidine at the C(2) position of the imidazole nucleus has been a synthetic challenge. Evans and Bach observed an anomalous phenomenon when diiodide **105** was subjected to standard Sonogashira reaction conditions [26]. Upon coupling in the presence of phosphine ligands, adduct **106** was isolated in 52% yield. This pivotal reaction thus solved the dual problems of the dehalogenation at the C(4) atom and the bond construction at the C(2) atom in a single operation. More interestingly, when the same reaction was effected in the presence of Pd(dba)$_2$, phenylacetylene served as *a reducing agent* to afford the C(2) monoiodoimidazole by a selective H–I exchange at the C(4) atom.

9.5. Heck and heteroaryl Heck reactions

Generally, the intermolecular Heck reaction between 2-iodo-, 4-iodo-, and 5-iodo-1-methylimidazoles and olefins suffers from low yields (<25%). Therefore, these transformations are of limited synthetic utility [41]. In one case, variable yields for adduct **108**

(15–58%) were observed for the Heck reaction of 5-bromo-1-methyl-2-phenylthio-1*H*-imidazole (**107**) and a large excess of methyl acrylate [64].

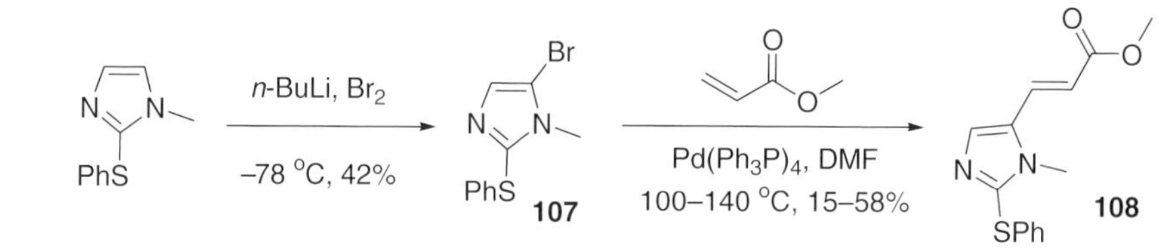

More recently, however, the Heck reaction utilizing other imidazolyl bromides has suggested the validity of this method in preparing substituted imidazole derivatives. In conjunction with their search for biologically active materials, the Avery group has utilized the Heck reaction to provide substituted imidazole carboxamides [31]. Thus, the Heck reaction between imidazole **109** and 4-methoxystyrene produced **110** in 43% yield. While this yield is not outstanding, when the identical conditions were applied to coupling between **109** and methyl acrylate, a gratifying 85% yield of **111** was obtained. Importantly, substitution in these reactions occurred in a sterically hindered environment, suggesting more potential for this process than had been indicated by previous studies.

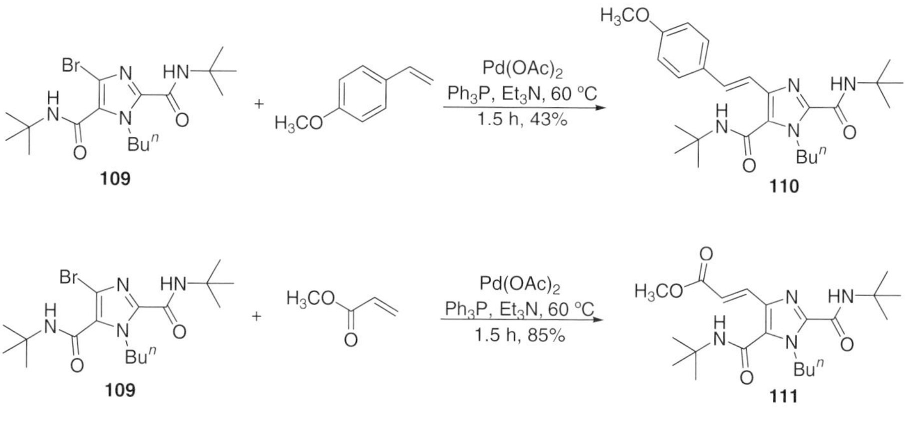

The application of the Heck reaction to imidazole containing natural products was demonstrated in the first total synthesis of the antifungal alkaloid, fungerin (**113**) [65]. The critical step in this synthesis was the coupling reaction at the 4-position of the imidazole nucleus. Contrasting some of the previous results in the literature, two examples shown below demonstrate coupling between *N*-protected-4-iodoimidazoles **112** and **114**, and methyl acrylate to give **113** and **115**, respectively. It is possible that these results shed light on the failures of the original attempts at employing the Heck reaction in related systems. While the *N*-methyl derivative **112** underwent reaction in 52% yield to provide the natural product **113**, the switch to a more electron withdrawing protecting group at the 1-position allowed for increased yield in the coupling step. Compound **115** was also converted to the natural product, fungerin (**113**) in two further steps, involving deprotection and methylation of the imidazole nitrogen.

Ohta's group thoroughly studied the heteroaryl Heck reactions of chloropyrazines and π-electron-rich heteroaryls [64, 66, 67]. The substitution occurred at the electron-rich C(5) position of the imidazole ring for the heteroaryl Heck reaction of 2-chloro-3,6-dimethylpyrazine and N-methylimidazole.

Miura's group carried out a heteroaryl Heck reaction of bromobenzene and 1-methylimidazole and isolated both mono-arylation (53%) and bis-arylation products [68]. In accord with Ohta's observation, the first arylation took place at the electron-rich C(5) and the second arylation occurred at the more electron-poor C(2).

In the cases above, the 1-position of the imidazole is necessarily protected, making this method somewhat limited in its scope. Recently, an attractive strategy has been delineated by the Sames group to circumvent this problem [69]. In their studies, the free NH-imidazole could be selective arylated at the 4-position to provide 4-phenylimidazole in high yield. This exciting result is enabled by the use of the MgO as a base which is presumed to bind with the nitrogen, completely inhibiting N-arylation. Interestingly, by adding a small amount of copper iodide to the reaction, the reaction selectivity switches with C(2) being the favored site of arylation. These results have been extended to include the C-selective arylation of several other free (NH)-azoles.

A very general method for arylation of an imidazo[1,2-*a*]pyrimidine was elaborated by Li and coworkers [70]. Recognizing that the heteroaryl Heck reaction occurs at the most electron-rich position of the heteroaromatic ring, it was suspected that this reaction should be quite regioselective for the imidazole nucleus in this system. This proved to be the case as arylation with several bromides was found to proceed with complete regioselectivity for the 3-position of the heterocycle. The yields in these reactions are generally good, though the reaction is less efficient when an electron donating substituent is attached to the aryl bromide. This is consistent with the similar findings in furan arylation studies [71].

An intramolecular heteroaryl Heck was the pivotal step in the synthesis of 5-butyl-1-methyl-1*H*-imidazo[4,5-*c*]quinolin-4(5*H*)-one (**116**), a potent antiasthmatic agent [72]. The optimum yield was obtained under Jeffery's "ligand-free" conditions, echoing Ohta's observation for the intermolecular version. Once again, the C_{aryl}–C_{aryl} bond was constructed at the C(5) position of the imidazole ring. Another intramolecular heteroaryl Heck cyclization of pyrrole and imidazole derivatives was also reported to assemble annulated isoindoles [73].

9.6. Tsuji–Trost reaction

The Tsuji–Trost reaction is the Pd(0)-catalyzed allylation of a nucleophile [74–77]. The NH group in imidazole can take part as a nucleophile in the Tsuji–Trost reaction, whose applications are found in both nucleoside and carbohydrate chemistry. Starting from cyclopentadiene and paraformaldehyde, cyclopentenyl allylic acetate **117** was prepared in diastereomerically-enriched form *via* a Prins reaction [78]. Treating **117** with imidazole under Pd(0) catalysis provided the *N*-alkylated imidazole **118**.

Extending the aforementioned methodology from imidazole to adenine, the Tsuji–Trost reaction between the sodium salt of adenine and allylic acetate **119** gave **120** as a 82:18 mixture of *cis:trans* isomers. Carbocyclic nucleoside **120** was advantageous over normal nucleosides as a drug candidate because it was not susceptible to degradation *in vivo* by nucleosidases and phosphorylases [78].

In the carbohydrate chemistry arena, the Tsuji–Trost reaction has been applied to construct *N*-glycosidic bonds [79]. In the presence of $Pd_2(dba)_3$, the reaction of 2,3-unsaturated hexopyranoside **121** and imidazole afforded *N*-glycopyranoside **122** regiospecifically at the anomeric center with the retention of configuration. In terms of the stereochemistry, the oxidative addition of allylic substrate **121** to Pd(0) formed the π-allyl complex with the inversion of configuration, then nucleophilic attack by imidazole proceeded with another inversion of the configuration. Therefore, the overall stereochemical outcome is the retention of configuration.

Moreover, Pd(0)-catalyzed allylations of imidazole with cyclopentadiene monoepoxide led to imidazole-substituted cyclopentenol in moderate yield [80].

Palladium(0)-catalyzed allylations of 4(5)-nitroimidazole, 2-methyl-4(5)-nitroimida-zole, 4(5)-bromoimidazole, and 4(5)-methoxyimidazole resulted in complicated mix-tures, which did not necessarily reflect the tautomeric ratios of the starting material [7]. For example, poor regioselectivity for the products (**123** and **124**) was observed in the Tsuji–Trost reaction of 4(5)-bromoimidazole with cinnamyl carbonate. However, the same reaction with 4(5)-nitroimidazole and 2-methyl-4(5)-nitroimidazole led predomi-nantly to the 1-allylation products. In addition, removal of the *N*-imidazole allyl groups can be selectively effected under mild conditions by Pd-catalyzed π-allyl chemistry [81].

123   **124** (**123**:**134** = 17:83)

9.7. Phosphonylation

Phosphorus esters are important isosteres for carboxylic acid derivatives possessing similar structural and electronic properties. Although formation of a C_{sp^3}–P bond is relatively straightforward, making a C_{sp^2}–P bond is more cumbersome. Recently, Pd(0)-catalyzed coupling reactions have opened up a new venue for C_{sp^2}–P bond creation.

In order to synthesize biologically relevant phosphonylimidazole **127**, bromoimidazole **126** was derived from radical-initiated bromination of methyl 1-*p*-methoxybenzyl-2-thiomethyl-5-imidazolylcarboxylate (**125**) [82]. The thiomethyl group served to block the C(2) position, which would otherwise undergo preferential halogenation under these con-ditions. As expected, a variety of Arbusov–Michaelis reaction conditions failed even under forcing conditions. On the other hand, Pd-catalyzed phosphorylation of **126** with diethyl phosphite led to methyl-4-diethylphosphonyl-1-*p*-methoxybenzyl-2-thiomethyl-5-imidazolylcarboxylate (**127**). After further manipulations, the desired phosphonic acid-linked aminoimidazoles, which resembled intermediates formed during purine biosynthesis, were accessed.

125   **126**   **127**

In conclusion, the imidazolylzinc and imidazolyltin reagents are undoubtedly easier to prepare than the corresponding imidazolylboronic acids. As a consequence, if the imidazole fragment is to serve as a nucleophile in a Pd-catalyzed cross-coupling reaction, the Negishi and Stille reactions are better choices than the Suzuki coupling. However, if the imidazole fragment is to serve as an electrophile, the choice is not that crucial. In other words, most imidazole-containing molecules may be assembled by judiciously choosing appropriate coupling partners.

9.8. References

1. Kirk, K. L. *J. Org. Chem.* **1978**, *43*, 4381–3.
2. Bond, R. F.; Bredenkamp, M. W.; Holzafel, C. W. *Synth. Commun.* **1989**, *19,* 2551–66.
3. Bell, A. S.; Campbell, S. F.; Morris, D. S.; Roberts, D. A.; Stefaniak, M. H. *J. Med. Chem.* **1989**, *32*, 1552–8.
4. Kim, G.; Kang, S.; Ryu, Y.; Keum, G.; Seo, M. J. *Synth. Commun.* **1999**, *29*, 507–12.
5. Stensiö, K. E.; Wahlberg, K.; Wahren, R. *Acta Chem. Scand., Sect. B.* **1973**, *27*, 2179–83.
6. Cliff, M. D.; Pyne, S. G. *Synthesis* **1994**, 681–2.
7. Katritzky, A. R.; Slawinski, J. J.; Brunner, F.; Gorun, S. *J. Chem. Soc., Perkin Trans. 1,* **1989**, 1139–45.
8. Garden, S. J.; Torres, J. T.; de Souza Melo, S. C.; Lima, A. S.; Pinto, A. C.; Lima, E. L. S. *Tetrahedron Lett.* **2001**, *42*, 2089–92.
9. Hayakawa, Y.; Kimoto, H.; Cohen, L. A.; Kirk, K. L. *J. Org. Chem.* **1998**, *63*, 9448–54.
10. Wang, D.; Haseltine, J. *J. Heterocycl. Chem.* **1994**, *31*, 1637–9.
11. Matsuda, A.; Minakawa, N.; Sasaki, T.; Ueda, T. *Chem. Pharm. Bull.* **1988**, *36*, 2730–3.
12. Minakawa, N.; Takeda, T.; Sasaki, T.; Matsuda, A.; Ueda, T. *J. Med. Chem.* **1991**, *34*, 778–86.
13. Arnau, N.; Arredondo, Y.; Moreno–Mañas, M.; Pleixats, R.; Villarroya, M. *J. Heterocycl. Chem.* **1995**, *32*, 1325–34.
14. Allen, W. E.; Fowler, C. J.; Lynch, V. M.; Sessler, J. L. *Chem. Eur. J.* **2001**, *7*, 721–9.
15. Bell, A. S.; Roberts, D. A.; Ruddock, K. S. *Tetrahedron Lett.* **1988**, *29*, 5013–6.
16. Stupka, G.; Gremaud, L.; Bernardinelli, G.; Williams, A. F. *Dalton Trans.* **2004**, 407–12.
17. Choi-Sledeski, Y. M.; McGarry, D. G.; Green, D. G.; Mason, H. J.; Becker, M. R.; Davis, R. S.; Ewing, W. R.; Dankulich, W. P.; Manetta, V. E.; Morris, R. L.; Spada, A. P.; Cheney, D. L.; Brown, K. D.; Colussi, D. J.; Chu, V.; Heran, C. L.; Morgan, S. R.; Bentlye, R. G.; Leadley, R. J.; Maignan, S.; Guilloteau, J.-P.; Dunwiddie, C. T.; Pauls, H. W. *J. Med. Chem.* **1999**, *42*, 3572–87.
18. Li, J. J.; Carson, K. G.; Trivedi, B. K.; Yue, W. S.; Ye, Q.; Glynn, R. A.; Miller, S. R.; Connor, D. T.; Roth, B. D.; Luly, J. R.; Low, J. E.; Heilig, D. J.; Yang, W.; Qin, S.; Hunt, S. *Bioorg. Med. Chem. Lett.* **2003**, *11*, 3777–90.
19. Balle, T.; Perregaard, J.; Ramirez, M. T.; Larsen, A. K.; Søby, K. K.; Liljefors, T.; Andersen, K. *J. Med. Chem.* **2003**, *46*, 265–83.
20. Heys, J. R.; Villani, A. J.; Mastrocola, A. R. *J. Labeled Compds. Radiopharm.* **1996**, *38,* 761–9.

21. Jetter, M. C.; Reitz, A. B. *Synthesis* **1998**, 829–31.
22. Turner, R. M.; Ley, S. V.; Lindell, S. D. *Synlett* **1993**, 748–50.
23. Dobler, M. R. *Tetrahedron Lett.* **2003**, *44*, 7115–17.
24. Abarbri, M.; Thibonnet, J.; Bérillon, L.; Dehmel, F.; Rottländer, M.; Knochel, P. *J. Org. Chem.* **2000**, *65*, 4618–34.
25. Prasad, A. S. B.; Stevenson, T. M.; Citineni, J. R.; Zyzam, V.; Knochel, P. *Tetrahedron* **1997**, *53*, 7237–54.
26. Evans, D. A.; Bach, T. *Angew. Chem., Int. Ed. Engl.* **1993**, *32,* 1326–7.
27. Kawasaki, I.; Yamashita, M.; Ohta, S. *J. Chem. Soc., Chem. Commun.* **1994**, *18*, 2085–6.
28. Shapiro, G.; Gomz-Lor, B. *J. Org. Chem.* **1994**, *59*, 5524–6.
29. Wu, Z.; Frale, M. E.; Bilodeau, M. T.; Kaufman, M. L.; Tasber, E. S.; Balitza, A. E.; Hartman, G. D.; Coll, K. E.; Rickert, K.; Shipman, J.; Shi, B.; Sepp-Lorenzino, L.; Thomas, K. A. *Bioorg. Med. Chem. Lett.* **2004**, *14*, 909–12.
30. Blass, B. E.; Huang, C. T.; Kawamoto, R. M.; Li, M.; Liu, S.; Portlock, D. E.; Rennells, W. M.; Simmons, M. *Bioorg. Med. Chem. Lett.* **2000**, *10*, 1543–5.
31. Chittiboyina, A. G.; Reddy, R.; Watkins, E. B.; Avery, M. A. *Tetrahedron Lett.* **2004**, *45*, 1869–72.
32. Kawasaki, I.; Yamashita, M.; Ohta, S. *Chem. Pharm. Bull.* **1996**, *44*, 1831–9 and references cited therein.
33. Kawasaki, I.; Katsuma, H.; Nakayama, Y.; Yamashita, M.; Ohta, S. *Heterocycles* **1998**, *48,* 748–50.
34. Molloy, K. C.; Waterfield, P. C.; Mahon, M. F. *J. Organomet. Chem.* **1989**, *365,* 61–73.
35. Kennedy, G.; Perboni, A. D. *Tetrahedron Lett.* **1996**, *37*, 7611–14.
36. Saá, J. M.; Martel, J. M.; García-Rosa, G. *J. Org. Chem.* **1992**, *57*, 678–85.
37. Keenan, R. M.; Weinstock, J.J.; Finkelstein, J. A.; Franz, R. G.; Gaitanopoulos, D. E.; Girard, G. R.; Hill, D. T.; Morgan, T. M.; Samanen, J. M.; Hempel, J.; Eggleaton, D. S.; Aiyar, N.; Griffin, E.; Ohlstein, E. H.; Stack,, E.J.; Weidley, E. F.; Edwards, R. *J. Med. Chem.* **1992**, *35*, 3858–72.
38. Terpin, A.; Winklhofer, C.; Schumann, S.; Steglich, W. *Tetrahedron* **1998**, *54*, 1745–52.
39. Gaare, K.; Repstad, T.; Benneche, T.; Undheim, K. *Acta Chem. Scand.* **1993**, *47*, 57–62.
40. Achab, S. *Tetrahedron Lett.* **1996**, *37*, 5503–6.
41. Choshi, T.; Yamada, S.; Sugino, E.; Kuwada, T.; Hibino, S. *J. Org. Chem.* **1995**, *60*, 5899–904.
42. Achab, S.; Guyot, M.; Potier, P. *Tetrahedron Lett.* **1995**, *36*, 2615–18.
43. Achab, S.; Diker, K.; Potier, P. *Tetrahedron Lett.* **2001**, *42*, 8825–8.
44. Choshi, T.; Yamada, Nobuhiro, J.; Mihara, Y.; S.; Sugino, E.; Hibino, S. *Heterocycles* **1998**, *48*, 11–14.
45. Lovely, C. J.; Du, H.; Dias, H. V. R. *Org. Lett.* **2001**, *3*, 1319–22.
46. Yashioka, H.; Choshi, T.; Sugino, E.; Hibino, S. *Heterocycles*, **1995**, *41*, 161–74.
47. Kosugi, M.; Koshiba, M.; Atoh, A.; Sano, H.; Migita, T. *Bull. Chem. Soc. Jpn.* **1986**, *59*, 677–9.
48. Beard, R. L.; Colon, D. F.; Klein, E. S.; Vorse, K. A.; Chandraratna, R. A. S. *Biorg. Med. Chem. Lett.* **1995,** *5*, 2729–34.

49. Gutierrez, A. J.; Terhorst, T. J.; Matteucci, M. D.; Froehler, B. C. *J. Am. Chem. Soc.* **1994**, *116*, 5540–4.

50. Jones, P.; Chambers, M. *Tetrahedron* **2002**, *58*, 9973–81.

51. Vermonden, T.; Branowska, D.; Marcelis, A. T. M.; Sundhölter, E. J. R. *Tetrahedron* **2003**, *59*, 5039–45.

52. Minakawa, N.; Sasaki, T.; Matsuda, A. *Biorg. Med. Chem. Lett.* **1993**, *3*, 183–6.

53. Shi, G.; Cao, Z.; Zhang, X. *J. Org. Chem.* **1995**, *60*, 6608–11.

54. Hutchinson, J. H.; Cook, J. J.; Brashear, K. M.; Breslin, M. J.; Glass, J. D.; Gould, R. J.; Halczenko, R. J.; Holahan, M. A.; Lynch, R. J.; Sitko, G. R.; Stranieri, M. T.; Harman, G. D. *J. Med. Chem.* **1996**, *39*, 4583–91.

55. Cliff, M. D.; Pyne, S. G. *J. Org. Chem.* **1995**, *60*, 2378–83.

56. Cliff, M. D.; Pyne, S. G. *Tetrahedron Lett.* **1995**, *36*, 5969–72.

57. Cliff, M. D.; Pyne, S. G. *Tetrahedron* **1996**, *52*, 13703–12.

58. Cliff, M. D.; Pyne, S. G. *J. Org. Chem.* **1997**, *62*, 1023–32.

59. Sakamoto, T.; Nagata, H.; Kondo, Y.; Shiraiwa, M.; Yamanaka, H. *Chem. Pharm. Bull.* **1987**, *35*, 823–8.

60. Matsuda, A.; Minakawa, N.; Ueda, T. *Nucleic Acids Symp. Ser.* **1988**, *20*, 13–14.

61. Nadipuram, A. K.; David, W. M.; Kumar, D.; Kerwin, S. M. *Org. Lett.* **2002**, *4*, 4543–6.

62. Kerwin, S. M.; Nadipuram, A. *Synlett* **2004**, 1404–8.

63. Lindel, T.; Hochgürtel, M. *Tetrahedron Lett.* **1998**, *39*, 2541–4.

64. Yamashita, M.; Oda, M.; Hayashi, K.; Kawasaki, I.; Ohta, S. *Heterocycles* **1998**, *48*, 2543–50.

65. Benhida, R.; Lezama, R.; Fourrey, J.-L. *Tetrahedron Lett.* **1998**, *39*, 5963–4.

66. Ohta, A.; Akita, Y.; Ohkuwa, T.; Chiba, M.; Fukunaka, R.; Miyafuji, A.; Nakata, T.; Tani, N. *Heterocycles* **1990**, *31*, 1951–7.

67. Aoyagi, Y.; Inoue, A.; Koizumi, I.; Hashimoto, R.; Tokunaga, K.; Gohma, K.; Komatsu, J.; Sekine, K.; Miyafuji, A.; Konoh, J.; Honma, R.; Akita, Y.; Ohta, A. *Heterocycles* **1992**, *34*, 257–72.

68. Pivsa-Art, S.; Satph T.; Yawamura, Y.; Miura, M.; Nomura, M. *Bull. Chem. Soc. Jpn.* **1998**, *71*, 467–73.

69. (a) Sezen, B.; Sames, D. *J. Am. Chem. Soc.* **2003**, *125*, 5274–5; (b) Sezen, B.; Sames, D. *J. Am. Chem. Soc.* **2003**, *125*, 10580–5.

70. Li, W.; Neslos, D. P.; Jensen, M. S.; Hoerrner, R. S.; Javadi, G. J.; Cai, D.; Larsen, R. D. *Org. Lett.* **2003**, *5*, 4835–7.

71. (a) Orig. Ref 43 (b) Glover, B.; Harvey, K. A.; Liu, B.; Sharp, M. J.; Tymoschenko, M. F. *Org. Lett.* **2003**, *5*, 301–4.

72. Kuroda, T.; Suzuki, F. *Tetrahedron Lett.* **1991**, *32*, 6915–8.

73. Huang, J.; Du, M. *Youji Huaxue* **1994**, *14*, 604–8.

74. Tsuji, J.; Takahashi, H.; Morikawa, M. *Tetrahedron Lett.* **1965**, 4387–90.

75. Tsuji, J. *Acc. Chem. Res.* **1969**, *2*, 144–52.

76. Godleski, S. A. in *Comprehensive Organic Synthesis*, Trost, B. M. and Fleming, I. Eds., Vol. 4, Chapter 3.3, Pergamon: Oxford, **1991**.

77. Moreno-Mañas, M.; Pleixats, R. in *Advances in Heterocyclic Chemistry* Katritzky, A. R. Ed.; Academic Press: New York, **1996**, *66*, 73–129.

78. Saville-Stones, E. A.; Lindell, S. D.; Jennings, N. S.; Head, J. C.; Ford, M. J. *J. Chem. Soc., Perkin Trans. 1* **1991**, 2603–4.

79. Bolitt, V.; Chaguir, B.; Sinou, D. *Tetrahedron Lett.* **1992**, *33*, 2481–4.
80. Arnau, N.; Cortes, J.; Moreno-Mañas, M.; Pleixats, R.; Villarroya, M. *J. Heterocycl. Chem.* **1997**, *34*, 233–9.
81. Kimbonguila, A. M.; Boucida, S.; Guibe, F.; Loffet, A. *Tetrahedron* **1997**, *53*, 12525–38.
82. Lin, J.; Thompson, C. M. *J. Heterocycl. Chem.* **1994**, *31*, 1701–5.

Chapter 10

Pyrazines and quinoxalines

Marudai Balasubramanian

10.1. Pyrazines

Minuscule quantities of naturally occurring pyrazines are found in some foodstuffs and are largely responsible for their flavor and aroma. For example, 3-methylpropyl-2-methoxypyrazine occurs in green peas and wine, and a seasoned wine connoisseur can identify a parts per trillion quantity. In addition, 2-methyl-6-vinylpyrazine exists in coffee.

Pyrazine is an electron-deficient, 6π-electron heteroaromatic compound, wherein the inductive effects of the nitrogen atoms induce a partially positive charge on the carbon atoms. Consequently, oxidative addition of chloropyrazine takes place more readily than chlorobenzene although the C–Cl bond lengths for both chloropyrazine and chlorobenzene are virtually the same—1.733 [1] and 1.745 Å [2], respectively. Simple chloropyrazines and their *N*-oxides undergo a wide range of C–C bond forming reactions under standard palladium-catalyzed reaction conditions. It is worth noting that oxidative addition of chlorobenzene to Pd(0) occurs, using sterically hindered, electron-rich phosphine ligands as reported by Reetz *et al.* [3, 4]. This enhanced reactivity may be ascribed to the observation that oxidative addition of an aryl chloride is greatly promoted by electron-rich palladium complexes.

Dragmacidins (A–F) represent an emerging class of bioactive marine natural products from deepwater sponges including *dragmacidon, halicortex, spongosorities, hexadella*, and the tunicate *didemnum candidum*. Four dragmacidins A–D, contained a piperazine linker and displayed modest antifungal, antiviral, and cytotoxic activities [5]. Structurally more complex aminoimidazole and guanidine-containing pyrazinones dragmacidins D–F, also possess a wide range of interesting biological properties [5].

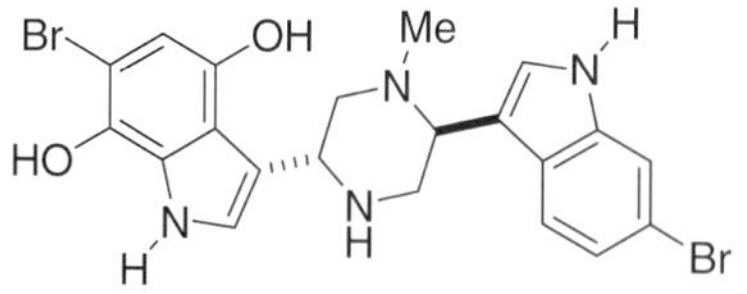

Dragmacidin

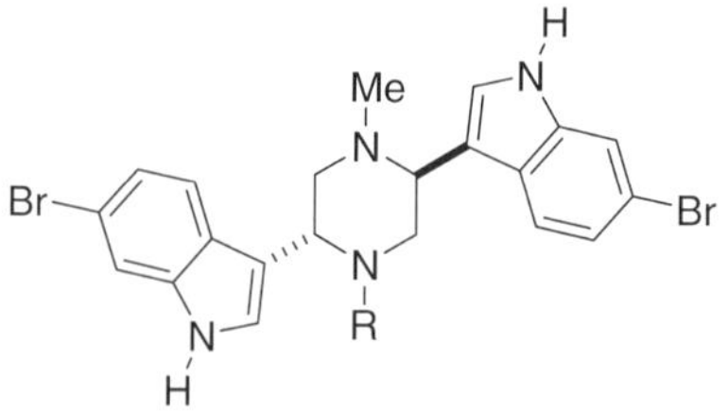

R = H,   Dragmacidin A
R = Me, Dragmacidin B

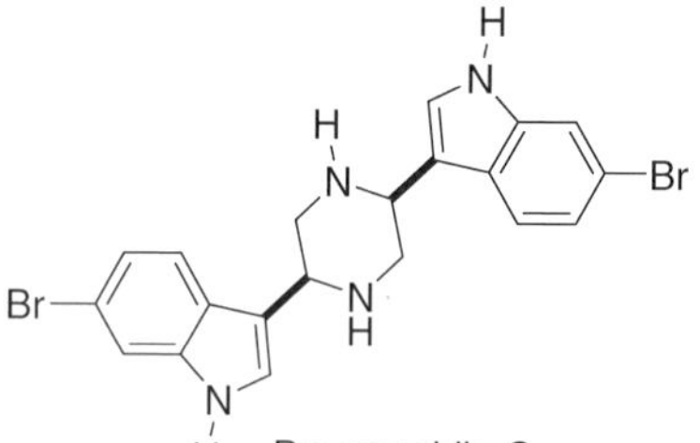

Dragmacidin C

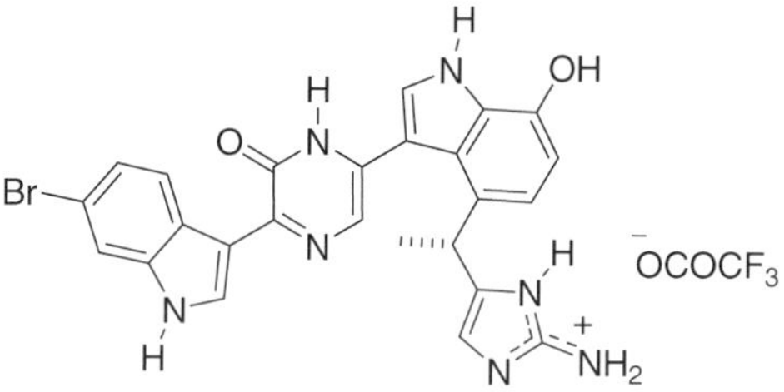

Dragmacidin D

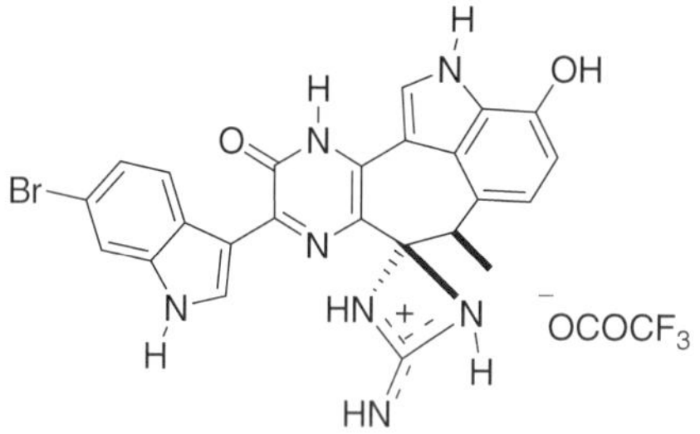

Dragmacidin E

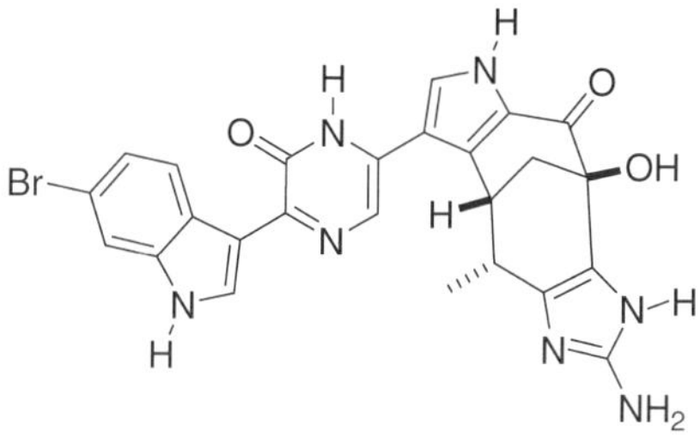

Dragmacidin F

Dragmacidin D is a potent inhibitor of serinethreonine protein phosphate (PP). Preliminary evidence suggests that the dragmacidin D is a selective inhibitor of PP1 versus PP2A. It is also a nonsteroidal anti-inflammatory agent since it inhibits resiniferitoxin-induced inflammation in mouse ear models. Dragmacidin D was found to selectively inhibit neutral nitric oxide synthase (iNOS) in the presence of inducible NOS. Endogenous nitric oxide (NO), produced *via* NOS-mediated metabolism of *L*-arginine, is known to play a role in a variety of regulatory functions such as the control of blood pressure, antibacterial activity, gastric motility, and neurotransmission [5]. Stoltz and Jiang research groups have reported the total synthesis of dragmacidin D–F [5–7]. The key steps involved the regioselective introduction of two indole units using the Suzuki and the Stille cross-coupling reactions.

In 1999, Durán *et al.* isolated two pyrazine alkaloids from *Botryllus leachi*: Botryllazines A–B [8]. These compounds exhibited cytotoxicity against human tumor cells. Regioselective metalation of pyrazines and cross-coupling reactions provided an easy access to botryllazines A and B from chloropyrazines with good yields [8].

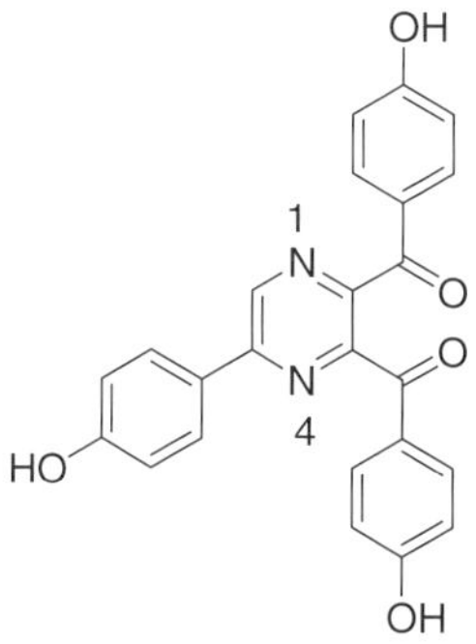

Botryllazine A

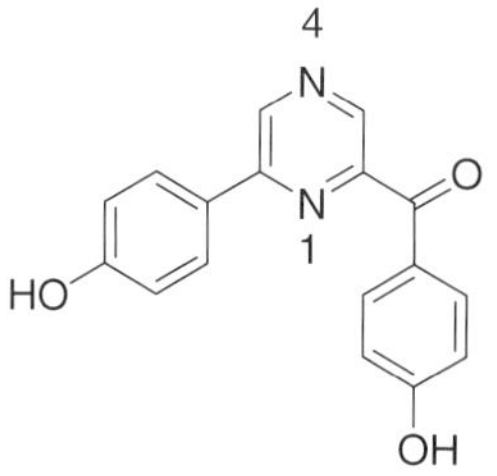

Botryllazine B

Aequorea victoria is widely distributed in marine organisms, e.g. coelenterates, fish, squids, and shrimps. Coelenterazine is a light-producing compound originally found in jellyfish. Benthocyanin A is a powerful radical scavenger from the mycelium of *Streptomyces prunicolor*. A new route to phenazine *via* the Pd(II)-catalyzed intramolecular amination of aryl bromide is described [9].

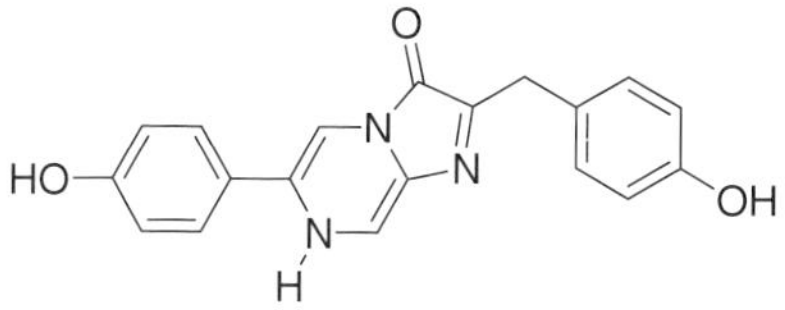

Coelenterazine

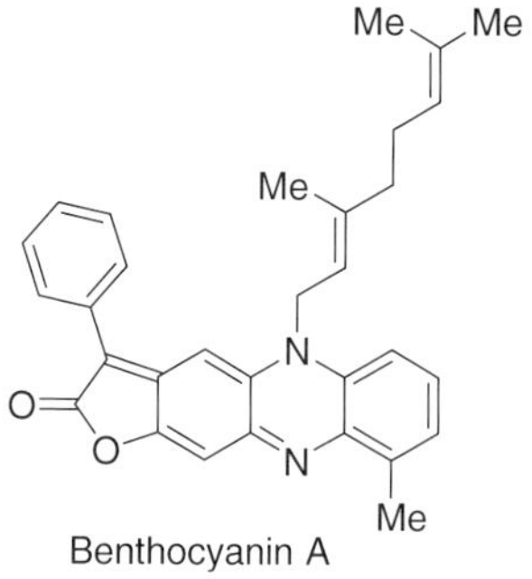

Benthocyanin A

10.2. Coupling reactions with organometallic reagents

10.2.1. Kumada reaction

The nickel-catalyzed cross-coupling reaction of fluoropyrazine and phenylmagnesium bromide was achieved in high yield using commercially available ligands such as dppe, dppp, and dppf to afford 2-phenylpyrazine **1** [10]. The conditions were suitable for the reaction of electron-poor fluorosubstrates and to a lesser extent, for reaction of fluorobenzenes.

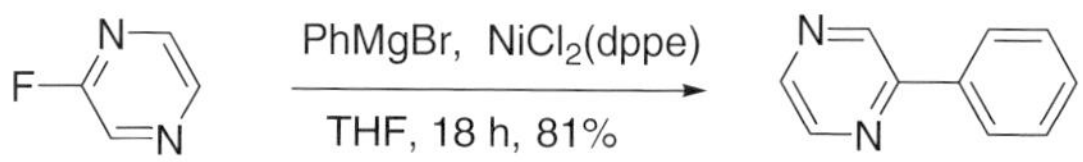

1

Substituted 2-phenylpyrazines **2** were prepared from 2-chloropyrazine and arylmagnesium chloride *via* Pd-catalyzed reactions [11]. These coupling reactions were carried out under mild conditions, often below 0°C, and extended to further functionalize halopy-ridines, haloquinolines, and halodiazines.

2

With palladium catalysis, both 2,6-dichloropyrazine **3** and chloropyrazine-*N*-oxide **4** were methylated using trimethylaluminum to give adducts **5** and **6**, respectively [12]. Trimethylaluminum was found to be a useful reagent for the alkylation of substituted pyrazines and their *N*-oxides [12].

3   **5**

4   **6**

In contrast to the aforementioned alkylations with heteroaryls, Pd-catalyzed alkylation reactions of chloropyrazines with alkyl organometallic reagents bearing β-sp^3-hydrides have sometimes proven to be quite troublesome. In addition to coupling, concomitant dehalogenation may also take place [13, 14]. When 2-chloro-3,6-dimethylpyrazine (**7**) was heated with triethylaluminum in the presence of K_2CO_3 and Pd(Ph$_3$P)$_4$, the desired ethylation product **8** was obtained in only 23% yield along with 49% of the dechlorinated product, 2,5-dimethylpyrazine (**9**) [13, 14]. Similar results were also observed when the alkylmetal was diethylzinc. Formation of dechlorinated products with this substrate was unavoidable but could be alleviated with organoboron reagents. The Suzuki coupling of **7** with triethylborane gave **8** in 74% yield along with only 5% of **9** [13, 14].

	7	**8**	**9**
AlEt$_3$, Pd(Ph$_3$P)$_4$, 1,4-dioxane		23%	49%
ZnEt$_2$, Pd(Ph$_3$P)$_4$, K_2CO_3, DMF		25%	49%
BEt$_3$, Pd(Ph$_3$P)$_4$, K_2CO_3, DMF		74%	5%

Similarly, ethylation reactions using triethylaluminum, diethylzinc, and triethylborane of the corresponding 1- and 4-oxide (10 and 11) have been described. Pentylation and octylation of 2-chloropyrazines and their 4-oxides were also feasible using pentylstannane and octylstannane; however, pentylation and octylation of 2-chloropyrazine 1-oxide failed [15].

10   **11**

What is a proposed mechanism for formation of the dechlorinated product 9? Presumably, oxidative addition of 7 to Pd(0) generates 12, which undergoes a transmetalation process to give the ethylpalladium(II) species 13. Intermediate 13 has two competing destinations: (a) reductive elimination to deliver the desired adduct 8 and (b) β-hydride elimination to furnish palladium(II) hydride 14, which subsequently undergoes a reductive elimination, giving rise to the dechlorinated product 9 [15].

10.2.2. Negishi coupling

Organozinc reagents exhibit greater functional group compatibility than organolithium and Grignard reagents. (±)-USB-165 is a potent neuronal nicotinic acetylcholine receptor (nAChR) ligand, which displays functional selectivity between nAChR subtypes. Using USB-165 as a lead structure, pyridines and pyridazines were synthesized using the Negishi coupling. Cross-coupling of vinyl triflate 15 with 2-pyrazinylzincbromide 16 using Pd(PPh$_3$)$_4$ gave coupled product 17 [16].

Palladium-catalyzed coupling of arylalkyl zinc bromide with 2-thiomethylpyrazine
(18) provided a regioselective 2-benzylpyrazine (19) in high yield [17]. Many benzylzinc
reagents and methylthio-N-heterocycles are straightforward to prepare, and Negishi coupling reactions are a useful method for the regioselective introduction of heterocyclic
substituents.

Contrasting results were observed in the Pd-catalyzed alkynylation with alkylzincs
versus the Sonogashira alkynylation. For example, reaction of iodopyrazine with ethyl-3-
bromozinc-propynate (20) provided the coupled product alkynylpyrazine 21 [18].
However, the corresponding Sonogashira reaction of iodopyrazine with propynoate provided 21 in only 37% yield [18].

Benzylfluoropyrazines have been used as building blocks to synthesize a wide range
of arylbenzylfluoropyrazines, which exhibit mesomeric properties. Alkyl or arylbenzylfluoropyrazines 23 were conveniently prepared from iodofluorobenzylpyrazines (22) via
arylation by Suzuki method [19] or alkylation under the Negishi conditions [17].

10.2.3. Stille coupling

The Stille reaction is a Pd-catalyzed cross-coupling between an organostannane and
electrophile, and regarded as one of the versatile methods in organic synthesis.
Bioisosters of (−)ferruginine (ethyl pyrazinylazabicyclooctenecarboxylic acid, 25) as
ligand for nicotinic acetylcholine receptor having pyrazine motif were synthesized via the
Stille coupling of bicyclic triflate 24 and 2-tributylstannylpyrazine [20].

A Pd-catalyzed copper-mediated coupling of pyrazinylthioether **26** with an arylstannane produced 2-(4-chlorophenyl)pyrazine (**27**) in excellent yield [21]. Copper(I) 3-methylsali-cylate was superior to copper(I) thiophene-2-carboxylate. No coupling occurred between **26** and 4-chlorophenyltri-*n*-butylstannane in the presence of 5% Pd(PPh$_3$)$_4$; however, with copper(I) 3-methylsalicylate, **27** was obtained in 93% yield [21].

The coupling reaction of benzofuranylthioethers (**28**) and tri-*n*-butylpyrazinylstannane was carried out in the presence of Pd$_2$(dba)$_3$–TFP and CuI diphenylphosphinate (CuDPP) as a unique mediator for these Pd-catalyzed coupling reactions [22]. A mild synthetic method for biheteroaryl ketone **29** was developed under neutral reaction conditions.

A convenient synthetic method was described to introduce reactive functionalities as well as heterocyclic moieties at the C(2) position of 4-(*N,N*-dimethylamino)pyridine (4-DMAP) *via* an unprecedented direct lithiation with BuLi–LiDMAE reagent [23]. Stille coupling of 2-chloropyrazine with 2-tributyltin-4-(*N,N*-dimethylamino)pyridine in the presence of PdCl$_2$(PPh$_3$)$_4$ afforded 2-pyrazinyl-(4-*N,N*-dimethylamino)pyridine (**30**) [23]. New useful 4-DMAP-containing synthons as polyhetero-cycles have been efficiently prepared.

Acylpyrazines, constituents in foodstuffs and pheromones, have been prepared by the Minisci-type radical reactions of pyrazines even though these methods often suffer from poor regioselectivity of aromatic substitution. Alternatively, synthesis of acetylpyrazines and propionylpyrazines **33** was achieved *via* a Stille coupling of bromopyrazines **31** with tributyl-(1-ethoxyalkenyl)tin (**32**), followed by acidic hydrolysis of the ether derivative [24].

Coupling of β-fluoro-β-trifluoromethyl-α-phenylvinylstannane (**34**) with 2-iodopy-razine in the presence of Pd(PPPh$_3$)$_4$ produced tetrasubstituted alkene **35** in 65% yield [25]. Most of these coupling reactions proceeded with retention of configuration of the double bond.

34   **35**

A new strategy for the synthesis of DUB-165, a nicotinic acetylcholine receptor (nAChRs), and its bioisosteres series containing pyridine and pyrazine moiety have been described [26]. The key step in the synthesis of bioisoster **37** was the Stille cross-coupling of tributylstannylpyrazine with the vinyl triflate of the N-protected 9-azabicyclononene **36** [26].

36   **37**

A number of (Z)-9-(heteroarylmethylene)-7-azatricyclo[4.3.1.0^{3,7}]decanes were syn-thesized to evaluate their activities at dopamine, serotonin, and norepinephrine receptors. The Stille coupling of vinylstannane **38** with 2-pyrazinyl bromide in the presence of Pd$_2$(dba)$_3$ and AsPh$_3$ gave the pyrazine derivative **39** in moderate yield [27]. The yields were significantly improved with AsPh$_3$ ligand.

38   **39**

α-Tetralones were synthesized to test their affinity and efficacy at the benzodiazepine site of GABA$_A$ receptors containing a α-subunit [28]. The Stille coupling of substituted bromotetralone **40** with 2-tributyltinpyrazine in the presence of Pd(PPh$_3$)$_4$ furnished pyrazinyltetralone (**41**) [28].

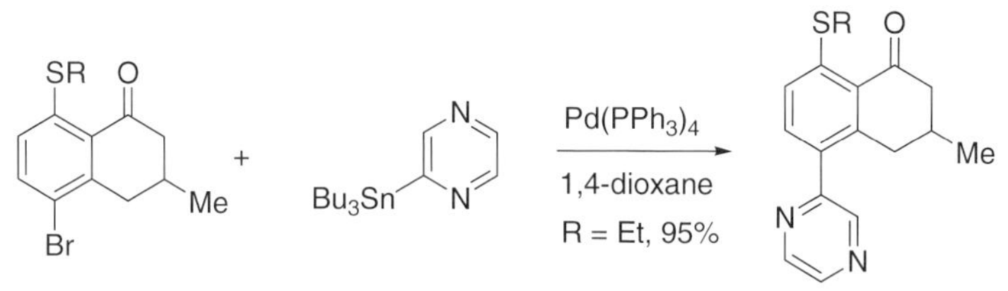

40   **41**

The Stille coupling of 2-chloro-6-tributyltinylpyrazine with 4-methoxybenzoylchloride produced a key intermediate **43** for the synthesis of pyrazine alkaloids botryllazine A–B [29].

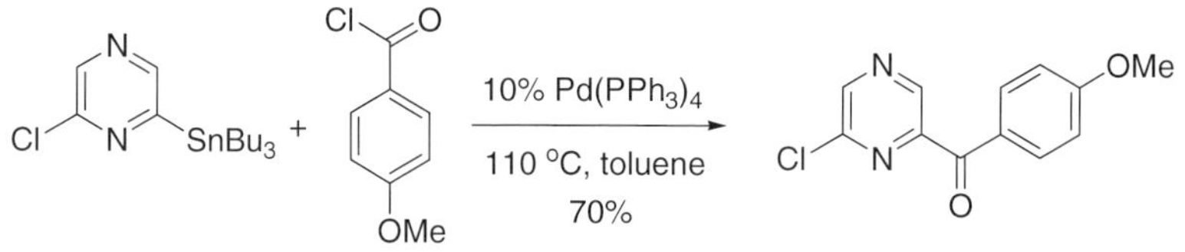

42   **43**

The Stille reaction of 2-chloro-3,6-diisopropylpyrazine (**44**) and 2-chloro-3,6-diiso-propylpyrazine 4-oxide (**45**) with tetra(*p*-methoxyphenyl)stannane (readily prepared *in situ* from the corresponding Grignard reagent and SnCl$_4$) led to the corresponding arylation products **46** and **47**, respectively [15]. Additional Stille coupling reactions of halopyrazines and their *N*-oxides have been carried out with tetraphenyltin and aryl-, heteroaryl-, allyl-, and alkylstannanes [30, 31].

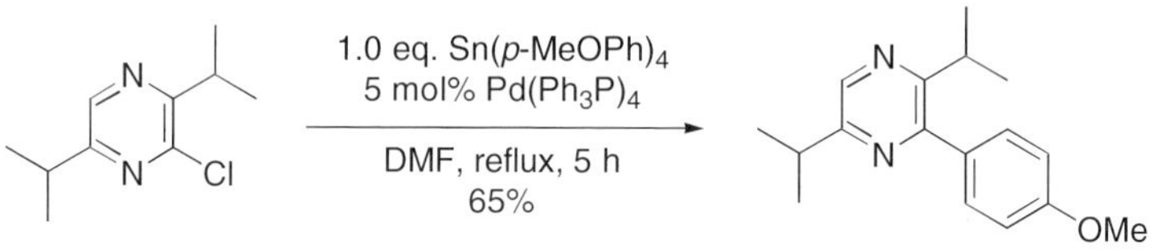

44   **46**

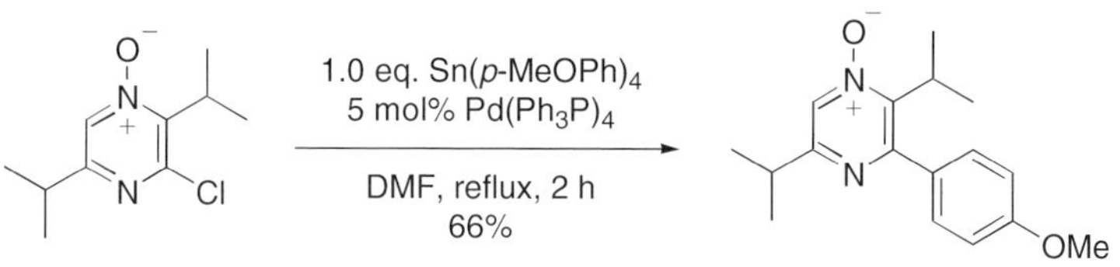

45   **47**

Interestingly, the Stille coupling of quaternary pyridylstannane **48** with 2-chloropyrazine afforded adduct **49** [32]. *N*-Methylated 3-(tributylstannyl)pyridine **48** was easily prepared by refluxing 3-(tributylstannyl)pyridine with methyl tosylate in EtOAc. In contrast, only a 29% yield of the coupling adduct was isolated from the Stille reaction of 3-(tributyl-stannyl)pyridine *N*-oxide and 2-chloropyrazine.

48   **49**

The Stille coupling of tetraethyltin and chloropyrazine **50** led to ethylpyrazine **51**, an important intermediate for preparing quinuclidinylpyrazine derivatives as muscarinic agonists [33]. In this particular case, the reductive elimination took place faster than β-hydride elimination.

50   **51**

10.2.4. Suzuki coupling

Although pyrazinylboron reagents are unknown, the Suzuki couplings have been conducted using halopyrazines and other boronic acids. As one of the early examples, 6-bromo-3-aminopyrazinoate **52** was coupled with 2-furylboronic acid to furnish furylpyrazine **53** [34]. Among three palladium catalysts examined, Pd(dppf)(OAc)$_2$ was found to be the best choice. In the same manner, 3-furyl and 4-pyridyl substituents were also introduced on to **52** in good yields. Moreover, bromopyrazine **54** and 2-thiopheneboronic acids were coupled to deliver thienylpyrazine **55** [35]. Also reported was a simpler version of a similar Suzuki coupling between 2-chloropyrazine and phenylboronic acid [36, 37].

52   **53**

54   **55**

5-Arylated indoles are important intermediates for the synthesis of various agonists and antagonists of the central nervous system neurotransmitter serotonin (5-hydroxy-tryptamine, 5HT). Suzuki coupling provides a general and regioselective methodology for the synthesis of 5-arylated indoles. 5-Indolylboronic acid **56**, easily obtained from commercially available 5-bromoindole, was coupled efficiently with 2-chloropyrazine to furnish 5-(2-pyrazinyl)indole (**57**) [38].

56   **57**

The Pd-catalyzed Suzuki coupling of dichloropyrazines **58** with arylboronic acid has been reported. The reaction proceeded smoothly with bis(triphenylphosphine)palladium(II) dichloride under anaerobic conditions to give the corresponding diarylpiperazine **59** in moderate to good yields [39].

58   **59**

The Suzuki coupling of *p*-phenylalanine boronic acid (**60**) with 2-chloropyrazine under microwave irradiation afforded 4-pyrazinylphenylalanine (**61**) as a free amino acid in high yield [40]. The amino group and carboxyl group were well tolerated under microwave irradiation and offered a straightforward approach to the synthesis of unprotected 4-arylphenylalanines [40].

60   **61**

The Suzuki coupling of 2-chloropyrazine with *o*-pivaloylaminophenyl boronic acid (**62**) led to the formation of substituted phenylpyrazine **63** which was further transformed into a tricyclic system, pyrazinoindazole *via* removal of the pivaloyl group followed by diazotization [41].

62   **63**

Coupling of 2-bromo-5-aminopyrazine (**64**) with 2-methoxy-4-pyridyl boronic acid (**65**) in the presence of Pd(PPh$_3$)$_4$ afforded the corresponding pyridylpyrazine **66** [42]. More electron-deficient heterocycles gave the lowest yields of coupled products.

Iodofluorobenzylpyrazines **67** were arylated *via* the Suzuki [19] or Negishi [17] coupling reactions to provide corresponding benzylfluoropyrazines **68** [19].

Synthesis of pyrazine alkaloid, botryllazine A, involved a combination of the Stille and Suzuki coupling reactions. The Suzuki coupling of 6-substituted 2-chloropyrazine **69** with 4-methoxyboronic acid produced 2,6-diarylpyrazine **70** [29].

α-(Trifluoromethyl)ethenyl boronic acid (**72**) was conveniently prepared from the reaction of readily available 2-bromotrifluoropropene with alkyl borate and magnesium in one-pot. The Suzuki coupling of 2-amino-3-methoxy-5-bromopyrazine (**71**) with boronic acid **72** provided styrene derivative, **73** in excellent yield [43].

A one step approach to heteroaryl benzoic acids from readily accessible heteroaryl halides and 4-carboxybenzene boronic acid was described by Gong and Pauls. Pyrazinylbenzoic acid (**74**) was prepared *via* the Suzuki coupling of 4-carboxybenzene boronic acid with 2-chloropyrazine in the presence of Pd(PPh$_3$)$_4$ in high yield [44].

2,6-Bis-[1-(ethoxycarbonyl)-ethylamino]3,5-diaryl-1,4-pyrazine (**75**) is a powerful inhibitor of 2,2′-azobis(2-amidinopropane)dihydrochloride (AAPH)-induced linoleate peroxidation. Its key intermediate, 2,6-diamino-3, 5-diaryl-1,4-pyrazine (**76**) was prepared *via* the Suzuki coupling of 2,6-diamino-3, 5-dibromopyrazine and arylboronic acids [45].

75

76

A key intermediate required for the synthesis of colenterazine was reported by Adamczyk *et al*. [46]. Treatment of 3-benzyl-5-bromo-2-pyrazinamine (**77**) with commercially available 4-methoxyphenylboronic acid in the presence of dppb and Pd(PhCN)$_2$Cl$_2$ afforded 3-benzyl-5-(4-methoxyphenyl)-2-pyrazinamine (**78**) in 88% yield [46].

77   **78**

Organotrifluoroborates in Pd-coupled reactions proved to be more reactive than the corresponding boronic acids or esters. In general they are more robust, easily purified, easier to handle, and less prone to protodeboronation. The coupling of aryl- and electron-rich heteroaryltrifluoroborates with aryl and activated heteroaryl bromides proceeds readily under ligandless conditions. When deactivated aryl- and heteroaryltrifluoroborates are coupled with aryl and heteroaryl bromides and chlorides, a low loading (0.5–2.5%) of PdCl$_2$(dppf)–CH$_2$Cl$_2$ efficiently catalyzes the coupling reactions. 2-Chloropyrazine was treated with 1-naphthyltrifluoroborate (**79**) and 3-thiophenetrifluoroborate (**80**) to give corresponding cross-coupled products **81** (85%) and **82** (83%) yields, respectively [47].

79   **81**

Pyrazinetriflate **83** was prepared from corresponding hydroxypyrazine by reacting with trifluoromethanesulfonic acid anhydride and N,N-diisopropylethylamine. The Suzuki coupling of pyrazinetriflate **83** with arylboronic acids using Pd(PPh$_3$)$_4$ gave tosyl-amidepyrazines **84** [48].

A key intermediate required for the synthesis of coelenterazine was described, where aryl 2-aminopyrazines (**85, 86, 87**) were effectively prepared *via* Suzuki coupling start-ing from the corresponding 2-amino-5-bromopyrazines (**88, 89, 90**) and t-butyl-dimethylsilyloxy-4-phenyl boronic acid (**91**) [49, 50].

Bisindolylpyrazines were synthesized and evaluated for cytotoxic activity against diverse human tumor cells [51]. The Suzuki cross-coupling of 2-amino-3-indolyl-5-bromopyrazine (**92**) with 5-indolylboronic acid **93** furnished 3,5-(bisindolyl)-2-aminopyrazine (**94**) [51].

2-(4-Methoxyphenyl)pyrazine (**95**) was obtained from 2-chloropyrazine *via* a Suzuki coupling [52].

A direct approach for selective construction of properly substituted bis(indole) pyrazine **100**, the skeleton of a marine alkaloid dragmacidin D, has been reported. The Suzuki and the Stille cross-coupling reactions are key steps involved for regioselective introduction of two indole units [6].

Another synthesis of dragmacidin D is described using the Suzuki methodology [5, 7]. Coupling of borylated indoles with readily available chloropyrazine proceeded smoothly under standard Suzuki conditions. The selective bond formation constructed in the dragmacidin F framework involved the fusion of the indole **102** and alkoxypyrazine **101** subunits while leaving the indolyl bromide at C(6) intact [5]. In the critical halogen-selective Suzuki coupling reaction, indolylboronic acid ester **104** was treated with dibromide **103** in the presence of Pd(PPh$_3$)$_4$.

Fluoropyrazines are used as building blocks to synthesize various alkylaryl and diarylpyrazines, which are aza-analogs of terphenyl. They are expected to provide rod-like molecules with potential liquid crystal applications. Cross-coupling reactions of 2-fluoro-6-tributylstannyl-pyrazine **108** and aryl iodides or bromides under the Stille conditions afforded bis(diarylpyrazine) **110** in moderate yields [53, 54].

10.2.5. Sonogashira coupling

Akita and Ohta disclosed one of the earliest Sonogashira reactions of chloropyrazines and their *N*-oxides [55, 56]. The union of 2-chloro-3,6-dimethylpyrazine and phenylacetylene led to 2,5-dimethyl-3-phenylethynylpyrazine (**111**). Subsequent Lindlar reduction of **111** yielded (*Z*)-2,5-dimethyl-3-styrylpyrazine (**112**), a natural product isolated from mandibular gland secretion of the Argentine ants, *Iridomyrmex humilis*.

111   **112**

Initially, **113** was obtained as the Sonogashira adduct of 2-chloro-3,6-diisobutylpyrazine and trimethylsilylacetylene. Interestingly, **113** underwent an additional Sonogashira coupling with 2-chloropyrazine to afford unsymmetrical 1,2-bispyrazinylacetylene (**114**) in excellent yield [57]. Here, desilylation occurred *in situ*, and the resulting terminal alkyne was then coupled with 2-chloropyrazine.

113   **114**

Under standard Sonogashira reaction conditions, [Pd(Ph$_3$P)$_2$Cl$_2$, CuI, Et$_2$NH], many alkynylpyrazines have been synthesized [58–62]. Taylor and Ray coupled 2-amino-3-cyano-5-bromopyrazine (**115**) with *t*-butyl-4-ethynylbenzoate (**116**) to give **117** [63]. Alkynylpyrazine **117** is an intermediate in the synthesis of methotrexate analogs as chemotherapeutic agents.

115   **117**

Having pyrazinylacetylenes in hand, one could convert the alkyne functionality into the corresponding ketone *via* hydration [64]. Thus, the coupling of iodide **118** and acetylene **119** produced pyrazinylalkyne **120**. Subsequent exposure of **120** to aqueous sodium sulfide and aqueous hydrochloric acid in methanol led to ketone **121**. Such a maneuver provides additional opportunities for further manipulation of the alkynes derived from the Sonogashira coupling reactions.

The standard Sonogashira reaction conditions were not successful for the coupling reaction of 3-chloropyrazine 1-oxide **122** and 1-hexyne [65]. In contrast, treatment of **122** and 1-hexyne with Pd(Ph$_3$P)$_4$ and KOAc produced 3-(1-hexynyl)pyrazine-1-oxide (**123**), together with the codimeric product, (*E*)-enyne **124** [65]. Presumably, the codimerization product **124** resulted from the *cis* addition of 1-hexyne to adduct **123**.

The Sonogashira reaction of 2-chloropyrazine 1-oxide gave only recovered starting material. Pentylation and octylation of 2-chloropyrazine 1-oxide also failed [15]. Possible explanations for these results are either catalyst agglomeration or metal formation from pyrazinylpalladium complex.

Quéguiner's group has accomplished lithiation of a *sym*-disubstituted pyrazine, 2,6-dimethoxypyrazine (**125**), with lithium 2,2,6,6-tetramethylpiperidine (LTMP). The resulting lithiated intermediate was quenched with I$_2$ to give 3-iodo-2,6-dimethoxypyrazine (**126**) and 3,5-diiodo-2,6-dimethoxypyrazine (**127**) [66]. Iodide **128** was then coupled with phenylacetylene to provide adduct **129**.

Quéguiner's metalation/cross-coupling strategy was applied to the total synthesis of arglecin (**132**), an antiarythmic pyrazine natural product extracted from cultures of *strep-tomyces* and *lavandurae*. Starting from commercially available 2,6-dichloropyrazine, iodopyrazine **130** was prepared in four steps comprising Cl–I exchange, S_NAr with sodium methoxide, metalation followed by quenching with isobutanal and dehydration. The Sonogashira reaction of **130** with propargyl alcohol provided **131**, which was subsequently transformed into arglecin **132** in five steps [67].

Cypridina luciferin, an imidazopyrazinone, is responsible for the bioluminescence of the crustacean *Cypridina (vargula) hilgendorfii* and deep-sea fishes. *Cypridina luciferin* shows a typical luciferin-luciferase reaction to emit a blue light of 465 nm from an oxyluciferin–luciferase complex. Regioselective introduction of a C(3) unit was achieved with *N*-Boc-propargylamine under Sonogashira coupling conditions to yield an alkynyl compound. A key intermediate in the synthesis of *Cypridina luciferin* is described *via* the Suzuki coupling of aminopyrazine derivative **134** with *N*-tosylindolyl-3-boronic acid (**135**) to afford indolylaminopyrazine **136** in 80% yield [68].

Various synthetic routes for 2-fluoro-9-oxime ketolides were reported. The 6-O-propargyl **137** was derivatized with various heteroaryl bromides using a Sonogashira coupling reaction to give ketolide **138** in excellent yield [69].

137   **138**

The Sonogashira coupling of 2-iodopyrazine with methyl-5-hexynoate (**139**) afforded alkyne (**140**) in 81% yield [70]. Subsequent hydrostannation of **140** and iodine displacement of the tributyltin group of **141**, produced vinyl iodide **142** in 76% yield [70]. Compound **142** was further coupled with 3,4-dimethoxyphenylboronic acid **143** under the Suzuki conditions to afford alkenyldiarylmethane (ADAM) **144** [70].

139   **140**

141   **142**

144, 52%

The Pd-catalyzed coupling of substituted indoline **145** with diaryl-2,3-dicyanopyrazine **146** was carried out in the presence of Pd(PPh$_3$)$_4$, CuI, NEt$_3$ to provide a novel spiropyran **147** [71]. This procedure is useful for combining two functional dye compounds that have totally different functionalities.

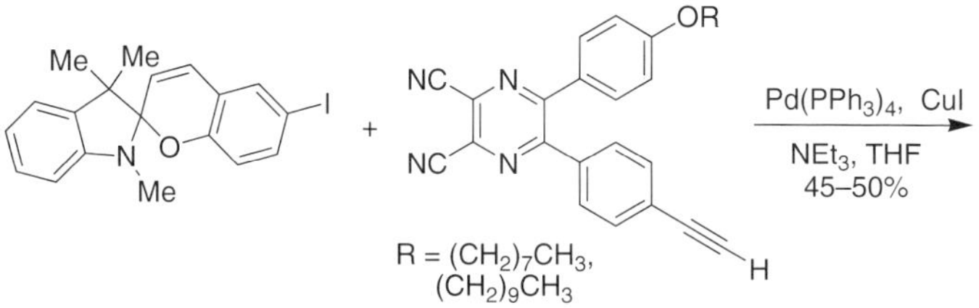

145 + **146**

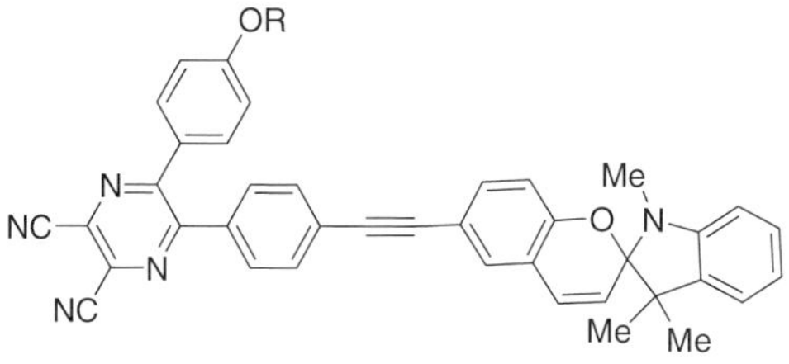

$$\xrightarrow[\substack{\text{NEt}_3, \text{ THF} \\ 45-50\%}]{\text{Pd(PPh}_3)_4, \text{ CuI}}$$

R = (CH$_2$)$_7$CH$_3$,
(CH$_2$)$_9$CH$_3$

147

10.3. Palladium-catalyzed amination

The scope of the Pd-catalyzed *N*-arylation of heteroarylamines has been mostly limited to aminopyridines and analogs and no examples with functional groups on the aryl halides or the heteroarylamine have been reported. 2-Chloropyrazine was coupled with 2-amino-4-methylthiazole (**148**) in the presence of Pd$_2$(dba)$_3$ to provide **149**. Similarly, 2-aminopyrazine was treated with *m*-bromotoluene to afford **150** in 98% yield [72].

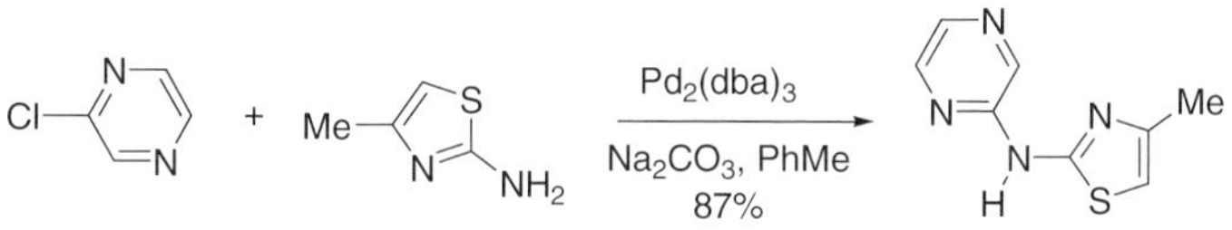

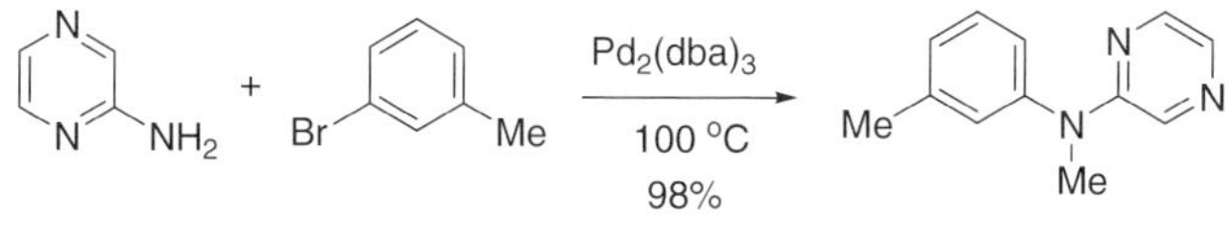

High-speed Pd-catalyzed amination of 2-chloropyrazine with *N*-methylaniline under temperature-controlled microwave heating produced 2-(*N*-methyl-*N*-phenyl)pyrazine in high yield (**151**) [73].

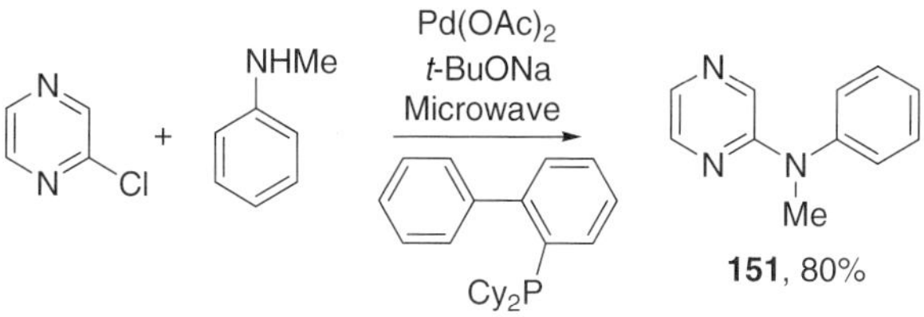

Selective amination of 2,3-dichloropyridine with 2-aminopyrazine produced 2-pyrazinyl-(3-chloro-2-pyridyl)amine (**152**) in 91% yield [74]. The use of mild amination conditions resulted in maximum selectivity with tolerance of base sensitive functional groups.

152

10.4. Heck reaction

Akita and Ohta revealed one of the early Heck reactions of halopyrazines [75]. They reacted 2-chloro-3,6-dimethylpyrazine with styrene in the presence of Pd(Ph₃P)₄ and KOAc using *N,N*-dimethylacetamide (DMA) as solvent to make (*E*)-2,5-dimethyl-3-styrylpyrazine (**153**). This methodology was later extended to 2-chloropyrazine-*N*-oxides although the yields were modest (28–38%) [76].

153

Bearing a structural resemblance to the *N*-linked nucleosides, *C*-nucleosides possess different physicochemical and biochemical properties. By a stereospecific Heck cross-coupling reaction between ribofuranoidglycal **154** and an iodopyrazine **155**, the Townsend group synthesized a novel 2′-deoxy-β-D-ribofuranosylpyrazine *C*-nucleoside **156** [77, 78]. While both hydroxyl groups in **154** were protected as their TBDMS ethers, the Heck reaction between **154** and **155** was followed by desilylation to produce **156** with exclusively the β-configuration. Protection of the two amino groups in **155** was not needed possibly due to the weak basicity of the amino nitrogen on the electron-deficient pyrazine ring. Finally, **156** was stereospecifically reduced using sodium triacetoxyboro-hydride to deliver 2′-deoxy-β-D-ribofuranosylpyrazine *C*-nucleoside **157**.

Ohta *et al.* thoroughly studied the heteroaryl Heck reactions of halopyrazines with both π-electron-rich and π-electron-deficient heteroaryls. These Pd-catalyzed direct couplings are advantageous over coupling with organometallics because the conversion of the heteroaryls to the corresponding organometallic reagents is obviated. A heterobiaryl, 2-(pyrazin-2-yl)indole **159**, was obtained from the heteroaryl Heck reaction of 2-chloro-3,6-diisobutylpyrazine (**158**) and indole [79]. Ohta *et al.* also coupled other 2-chloro-3,6-dialkylpyrazines with indole in moderate to good yields under the same conditions. In these cases, the couplings took place regioselectively at the C(2) position of indole. However, when the coupling reaction of **158** was conducted with 1-tosylindole, 2% of the coupling occurred at the C(3) position of 1-tosylindole [80]. Moreover, when the halopyrazines were 2-chloro-3,6-diethylpyrazine and 2-chloro-3,6-dimethylpyrazine, the ratio of the C(3)/C(2) coupling increased to 7 and 12%, respectively.

158   **159**

Furthermore, Ohta's group successfully conducted heteroaryl Heck reactions of chloropyrazines with many π-electron-rich heteroaryls including furan, thiophene, benzo[*b*]furan, and benzo[*b*] thiophene [81, 82]. In reactions of chloropyrazines with furan, thiophene, and pyrrole, disubstituted heterocycles were also isolated, albeit in low yields. Along with the disubstituted furan **161**, the mono-arylation product **160** was isolated when 2-chloro-3,6-diethylpyrazine and furan were refluxed in the presence of Pd(Ph₃P)₄ and KOAc. In the case of 2-chloro-3,6-dimethylpyrazine and thiophene, monothienylpyrazine **162** was the sole product. When 2-chloro-3,6-diisobutylpyrazine was used as substrate, 9% of the disubstituted thiophene was detected. Analogous to the couplings with furan and thiophene, the heteroaryl Heck reactions of chloro-3,6-diethylpyrazine with benzo[*b*]furan and benzo[*b*]thiophene produced **163** and **164**, respectively.

160, 63%   **161**, 16%

162

163, X = O, 54%
164, X = S, 81%

Although the heteroaryl Heck reactions of chloropyrazines with pyrrole itself were low-yielding for both mono- and bis-arylation products, higher yields were obtained for N-phenylsulfonylpyrrole. Bulkier alkyl substituents on the pyrazine ring promoted the formation of C(3)-substituted pyrroles. The C(3)-substituted pyrrole **166** was the major product (62%) for the coupling of **158** and N-phenylsulfonylpyrrole, while C(2)-substituted pyrrole **165** was a minor product (15%).

158   **165**, 15%   **166**, 62%

Similar results were observed for the heteroaryl Heck reactions of chloropyrazines with π-electron-rich heteroaryls including oxazole, thiazole, benz[*b*]oxazole, benz[*b*]thiazole, and N-methylimidazole.

167, X = O, 72%
168, X = S, 61%

169, X = O, 65%
170, X = S, 59%

158   **171**, 40%

10.5. Carbonylation reactions

To synthesize pyrazinecarboxylic esters and pyrazinecarboxamides, chloropyrazines were subjected to Pd-catalyzed carbonylation reactions in either alcohols or dialkylamines [83, 84].

2-Chloro-3,6-dimethylpyrazine was smoothly carbonylated in methanol containing a catalytic amount of $Pd(dba)_2$ and Ph_3P to give 2-methoxycarbonyl-3,6-dimethylpyrazine (**172**). Somewhat lower yields were observed in the preparation of pyrazinecarboxylic diesters under the same conditions.

CO (40 kg/cm^{-2})/MeOH
$Et_3N/Pd(dba)_2/Ph_3P$
150 °C, 16 h, >85%

172

In comparison to the ease of alkoxycarbonylation of halopyrazines, the aminocarbonylation was dramatically influenced by both the phosphorus ligands and carbon monoxide pressure. Pyrazinecarboxamide was prepared *via* aminocarbonylation using diethylamine as solvent. For example, 2-chloropyrazine was converted into 2-(diethylaminocarbonyl) pyrazine (**173**) in 85% yield along with 8% of the nucleophilic substitution product **174**. Intriguingly, when butylamine was used, the corresponding S_NAr product similar to **174** was the major product (72% yield), whereas the amide was isolated as a minor product (28%).

CO (40 kg/cm^{-2})/Et_2NH
$Pd(OAc)_2/Ph_3P$
120 °C, 16 h

173, 85%   +   **174**, 8%

Palladium-catalyzed carbonylation of 2-chloropyrazine 1-oxide failed, whereas that of 3-chloropyrazine 1-oxide (**175**) proceeded without deoxygenation of the *N*-oxide function to give 3-methoxycarbonylpyrazine 1-oxide (**176**). This observation was in accord with the failure of Stille reactions of 2-chloropyrazine 1-oxide [15, 38].

CO (40 kg/cm^{-2})/MeOH
$Et_3N/Pd(dba)_2/Ph_3P$
120 °C, 16 h, 70%

175   **176**

Efficient Pd-catalyzed carbonylation of 2-chloropyrazine afforded pyrazine carboxylic acid esters (**177**) in 82% yield in the presence of dppf and dppb ligands [85].

CO, $PdCl_2(PhCN)_2$, dppf
130 °C, 25 bar CO
82%

177

10.6. Cyanation of pyrazines

Palladium-catalyzed coupling reactions between chloropyrazines and organometallic reagents without β-sp³-hydrides are straightforward and generally proceed in good yields. In 1981, Ohta *et al.* introduced a cyano group by refluxing chloropyrazine **158** with KCN in DMF in the presence of a catalytic amount of Pd(Ph₃P)₄ to give cyanopyrazine **178** [86]. Analogously, cyanation of 2-amino-5-bromopyrazine with KCN in 18-crown-6 was facilitated by Pd(Ph₃P)₄ and CuI [87].

2-Chloro-5,6-diarylpyrazine **179** was treated with KCN in the presence of Pd(PPh₃)₄ to produce 2-cyano-5,6-diarylpyrazine **180** in 81% yield [88].

10.7. Deoxygenation of heteroamine-*N*-oxide

Deoxygenation of heteroaromatic *N*-oxides **181** to the corresponding amine **182** was achieved under mild conditions. Polymethylhydrosilaoxane (PMHS) in the presence of either tetrakis(triphenylphosphine)palladium(0) [Pd(PPh₃)₄], titanium (IV) isopropoxide [Ti(*i*-PrO₄)₄], or palladium on carbon (Pd/C) [89] were employed in this transformation.

Dechlorination of chloropyrazines and their *N*-oxides was accomplished using either Pd(Ph₃P)₄ and HCO₂Na in DMF [90–92] or hydrogenation with Pd(Ph₃P)₄ and KOAc in DMF [75]. The *N*-oxide functionality was not reduced under these conditions.

10.8. Quinoxalines

10.8.1. Coupling reactions with organometallic reagents

10.8.1.1. Stille coupling

The Pd-catalyzed cross-coupling reactions on the benzene ring of quinoxalines with organometallic reagents are rather straightforward, and usually proceed in good yields. The Stille reaction has been utilized to synthesize aryl-substituted quinoxalines and related heteroarenes as novel herbicides [93]. To prepare the two-regioisomeric aryldifluoromethoxyquinoxalines **185** and **186**, a 2:5 mixture of bromoquinoxalines **183** and **184** was refluxed in toluene with *meta*-trifluoromethylphenylstannane in the presence of Pd(Ph$_3$P)$_4$. In parallel, two regioisomeric aryltrifluoromethylquinoxalines **189** and **190** were isolated from the Stille reaction of bromoquinoxalines **187** and **188** (**187**/**188** = 3:1) under the same conditions.

R = OCHF$_2$, **183**	**184**	55%	**185**
R = CF$_3$, **187**	**188**	82%	**189**

The Stille coupling of bromoquinoxaline-*N*-oxide **191** and *meta*-trifluoromethylphenylstannane produced aryldifluoromethoxyquinoxalines **192**. Later, **192** was deoxygenated by catalytic hydrogenation to give pure **185** [93].

The Stille reaction of bromoquinoxaline **193** and vinylstannane delivered vinylquinoxaline **194**. In addition, **194** was further manipulated to a 5-piperidino-ethylquinoxaline-2,3-dione **195** as an AMPA receptor antagonist [94]. Palladium-catalyzed nucleophilic substitution on the benzene ring has also been described [95]. Thus, transformation of 5,8-diiodoquinoxalines to quinoxaline-5,8-dimalononitriles with sodium malononitrile was promoted by $PdCl_2$–$(Ph_3P)_2$.

191   **194**

195

The Pd-mediated cross-coupling reactions on the pyrazine ring of quinoxaline with organometallic reagents are more cumbersome. In one example, the Stille reaction of 2-chloroquinoxaline **196** with benzylstannane **197**, produced only 32% of the coupling product **198** [96]. Compound **197** was in turn obtained from the S_N2 displacement of corresponding biphenylbromide with tri-*n*-butyllithiostannane.

196   **198**

In another case, the Stille reaction of the simple 2-iodoquinoxaline **199** and 1,3-dithiole-2-thione stannanes **200** gave the adduct **201** only 0–10% yield, although the use of copper thiophene-2-carboxylate (CuTC) gave 80% yield of **201** [97].

199   **200**   **201**

Indolylquinoxaline **203** was obtained from the union of 3-bromoquinoxalin-2-ylamine **202** and 1-phenylsulfonyl-2-tri-*n*-butylstannyl-1*H*-indole [98]. A wide variety of heterocyclic stannanes bearing various functional groups underwent Stille coupling with **202** under the same conditions to give the corresponding adducts in 72–98% yields.

Synthesis and SAR studies have been done for a series of 2-amino-3-heteroarylquinoxalines. The Stille reaction of furylstannane with 2-amino-3-bromo-quinoxaline **202** gave 2-furylquinoxaline **204** [99].

2,3-Dipyrrolylquinoxalines with extended chromophores are efficient fluorimetric sensors for pyrophosphate. The Stille coupling of 5,8-dibromo-2,3-di(pyrrol-2-yl) quinoxaline with aryltributylstannane produced 5,8-diaryl substituted quinoxalines **205** as possible sensors [100].

10.8.1.2. Sonogashira reaction

The Sonogashira reaction is of considerable value in heterocyclic synthesis. It has been conducted on the pyrazine ring of quinoxaline and the resulting alkynyl- and dialkynylquinoxalines were subsequently used to synthesize condensed quinoxalines [101–104]. Ames *et al.* prepared unsymmetrical diynes from 2,3-dichloroquinoxalines. Thus, condensation of 2-chloroquinoxaline with an excess of phenylacetylene furnished 2-phenylethynylquinoxaline (**206**). Displacement of the chloride with the amine also occurred when the condensation was carried out in the presence of diethylamine. Treatment of **206** with a large excess of aqueous dimethylamine led to ketone **207** that exists predominantly in the intramolecularly hydrogen-bonded enol form **208**.

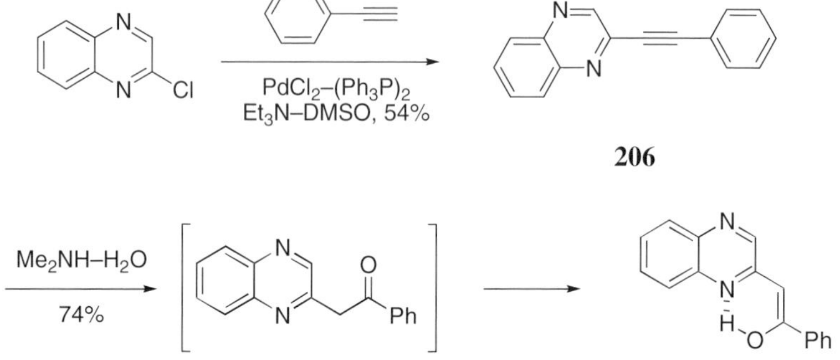

206

207   **208**

Condensed quinoxalines have been synthesized using 2-chloro-3-alkynylquinoxalines as common intermediates. 2-Chloro-3-phenylethynylquinoxaline (**209**) was prepared from 2,3-dichloroquinoxaline. Action of methylamine on **209** gave 1,2-disubstituted pyrrolo[2,3-*b*]quinoxaline **210** [101]. Similarly, 2-phenylthieno[2,3-*b*]quinoxaline (**211**) was synthesized by treating **209** with ethanolic sodium sulfide [101].

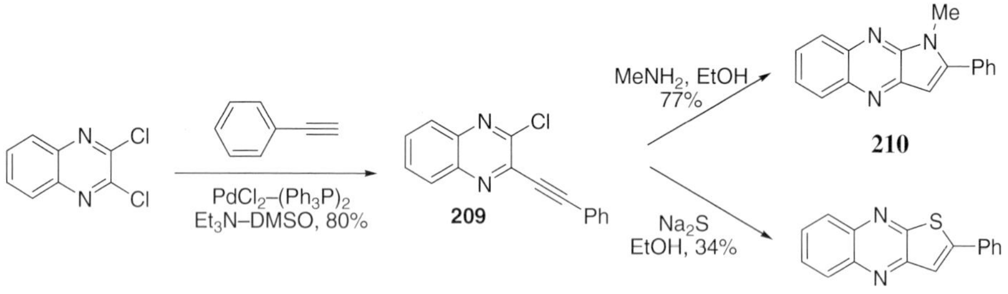

210

209

211

Likewise, when **209** was refluxed with aqueous KOH in dioxane, the corresponding 2-phenylfurano[2,3-*b*]quinoxaline was produced in 67% yield [101]. The Sonogashira reactions of chloro- and dichloroquinoxalines and trimethylsilylacetylene [105, 106] or but-3-yn-2-ol [107] have also been documented.

2-Chloroquinoxaline and 2,6-dichloroquinoxaline were coupled with trimethylsilylacetylene to give ethynylquinoxaline and further desilylated to **212** [108].

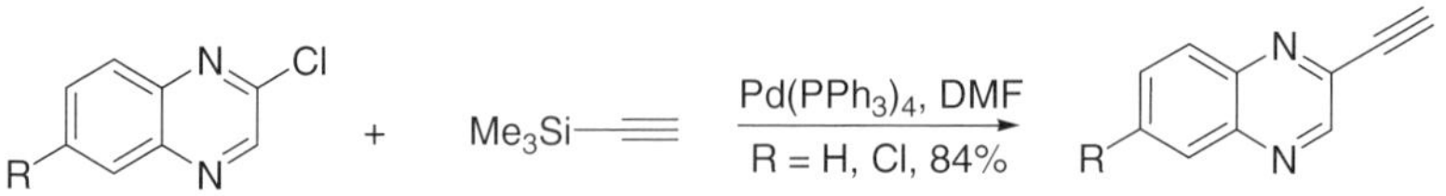

212

The cross-coupling of 3-chloro-7-methoxy-1-methylquinoxaline-2-one (**213**) and 2-chloro-5-methoxycarbonyl-3-methylquinoxaline (**214**) with phenylacetylene in the presence of Pd(PPh₃)₄ gave the corresponding alkynes, 7-methoxy-1-methyl-3-(4-methoxy-carbonyl)-phenylethynylquinoxaline-2-one (**215**), and 5-methoxycarbonyl-3-methyl-2-phenylethynyl-quinoxaline (**216**), respectively [109].

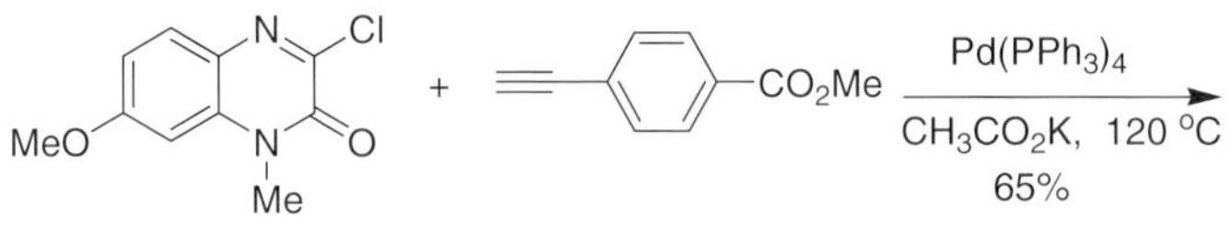

213

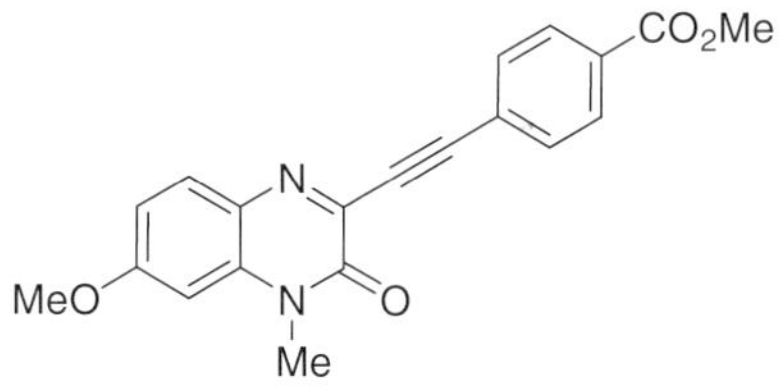

215

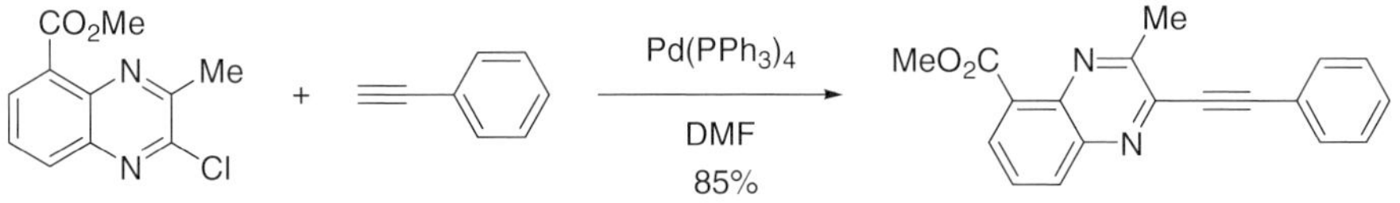

214   **216**

The Pd-catalyzed coupling of 2-halo-, and 2,6-dihaloquinoxalines (**217**) with propargyl alcohol was reported [110]. This process worked well for the simple alkyl-, and a range of aryl-alkynes, resulting in the corresponding 2-substituted quinoxalines.

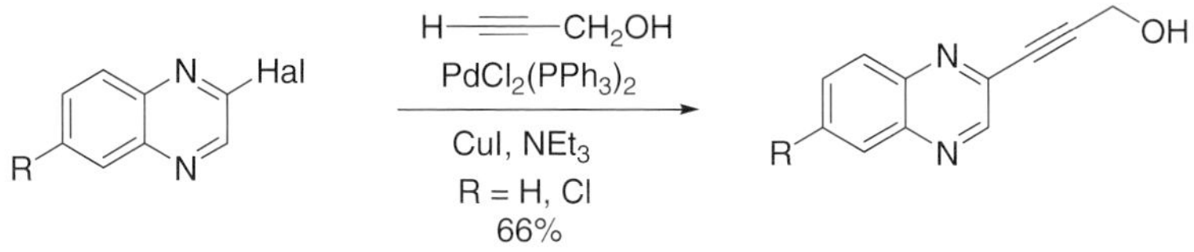

217   **218**

A new series of rigid, protected terminal dialkyne, 2,3-diphenyl-5,8-bis(ethynyl)quinox-aline (**220**) was prepared *via* the cross-coupling of trimethylsilyl protected alkynyl ligand and dihaloquinoxaline (**219**) [111].

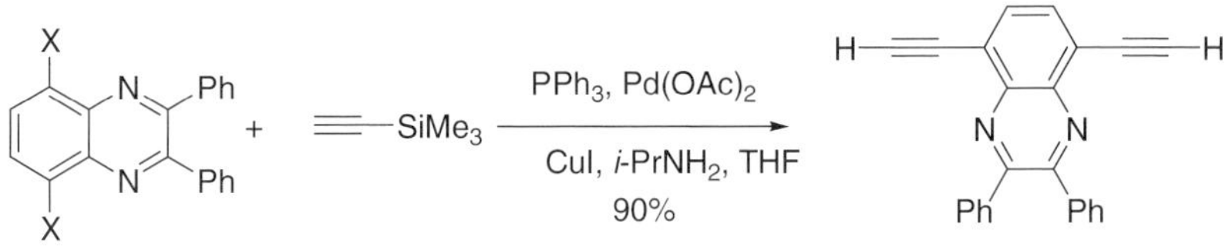

219   **220**

Palladium-catalyzed coupling of 2-chloroquinoxaline with racemic but-3-yn-2-ol furnished pyrazinylalkynol **221** [112].

221

Palladium-catalyzed C−P bond formation on the benzene ring of quinoxaline has been reported. Phosphoric acid ester **223** was prepared from 7-bromoquinoxaline **222** and diethylphosphite *via* a Heck-type reaction [113].

222   **223**

10.8.2. Intramolecular Heck reaction

Quinoxalines with their unique 1,4-diazine moiety are not ideal substrates for the Heck reaction because they are not only more labile than the corresponding naphthalenes, but also are strong chelating agents and behave as either monodentate or bidentate ligands. Aminoquinoxalines are especially strong chelating agents, and catalytic efficiency is difficult to achieve for such substrates. Li synthesized a variety of 3-substituted pyrrolo[2,3-*b*] quinoxalines *via* an intramolecular Heck reaction of allyl-3-haloquinoxalin-2-ylamines under Jeffery's "ligand-free" conditions. As an example, 3-methylpyrrolo[2,3-*b*]quinoxaline (**225**) was prepared from allyl-3-chloroquinoxalin-2-ylamine (**224**) [114]. Apparently, the external double bond from the initial cyclization process underwent a facile rearrangement to deliver the thermodynamically more stable pyrrole ring. The enhanced reactivity and yield are presumably due to the coordination and thereby solvation of the palladium intermediates by chloride ions present in the reaction mixture. Once the "locked" palladium catalyst is released from the substrates, the catalytic cycle continues smoothly [114].

224   **225**

10.8.3. Carbonylation reactions

Nitrene can be generated by Pd-catalyzed deoxygenation of organic nitro compounds with CO. Carbonylation of nitroenamine **226**, derived from 2-nitroaniline and α-substituted aldehyde, was carried out using Pd(dba)$_2$ and dppb. A mixture of 1,2-dihydroquinoxaline **227** and 3,4-dihydroquinoxaline **228** was obtained *via* sequential reactions of carbon monoxide insertion, heterocyclization, and *N*-alkylation followed by dealkylation [115].

| **226** | **227** (71%) | **228** (11%) |

10.8.4. Hydrogenation

The reaction of 2-methylquinoxaline in MeOH in the presence of 1mol% either RR[BDPBzP]Ir [COD]OTf or RR[BDPBzP]Rh [NBD]OTf under hydrogen pressure of 290 psi gave (2S) 2-methyl-1,2,3,4-tetrahydroquinoxaline (**229**) in 40–93%, respectively. The Rh precursor was more reactive than the Ir analog, which gave a higher enantiomeric excess. The higher stereoselectivity of Ir versus Rh is most likely kinetic in nature, i.e. the greater kinetic inertness of Ir compounds as compared to Rh analogs may be crucial for the stabilization of the intermediates which underwent the enantiotype hydride transfer [116].

229

10.8.5. Amination

Benthocyanin A is a powerful radical scavenger from the mycelium of *Streptomyces prunicolor*. A new route to phenazine is described *via* the Pd(II)-catalyzed intramolecular amination of aryl bromide [9]. Treatment of 2-bromo-3-nitro-methylbenzoate (**230**) and 6-aminobenzofuran derivatives **231** with Pd(OAc)$_2$ gave nitrodiphenylamine **232** in 99%. Catalytic hydrogenation of **232** followed by bromination gave the phenylenediamine derivative **233** in 91% yield. Cyclization of the diamine **233** provided the key intermediate phenazine **234**.

| **230** | **231** | **232** |

233   234

The cyclization o-phenylenediamine with 2-butene-1,4-diol catalyzed by palladium acetate coordinated with (R)-BINAP as the chiral ligand leads to optically active 1,2,3,4-tetrahydro-2-vinylquinoxaline with 19ee and 58% yield [117].

A general method for the solid phase synthesis of various arylaminoheterocycles has been described [118]. Dichloroquinoxalines can be captured onto solid support and further elaborated by aromatic substitutions with amines at elevated temperatures or by anilines *via* Pd-catalyzed cross-coupling reactions [118].

10.8.6. Cyanation of pyridopyrazine

Cyanation of chloropyridopyrazine **235** with Zn(CN)$_2$ in the presence of catalytic amounts of Zn powder, Pd$_2$(dba)$_3$, and dppf afforded 5-cyanopyridopyrazine **236** in 59% yield [119].

235   236

10.8.7. Pyrrolopyrazine

Pyrrolopyrazine is a core structure that has drawn some attention from the pharmaceutical industry, both as a surrogate structure for indole or azaindole and as a novel kinase inhibitor [120]. The synthesis of 6,7-diphenylpyrrolopyrazines (**238**) was achieved *via* a Pd-catalyzed heteroannulation of **237**, utilizing both conventional and microwave heating conditions. The reaction proceeded under microwave irradiation conditions to yield the desired products in modest yield [121].

237 + Ph—≡—Ph → **238**

PdCl₂dppf, LiCl
Na₂CO₃, DMF
MW (100W)
41%

10.8.8. Lumazines

A series of lumazines having a propionyl, β-alkoxy, or β-hydroxypropionyl group at C(6) were recently isolated from the metabolites of the swimming plychaete, *Odontosyllis undecimdonta*. This marine creature is commonly called fire worm since it luminesces during spawning at sunset in the middle of autumn at Toyama Bay, Japan [122]. The synthesis of 1,3-dimethyl-6-propionylpteridine-2,4-dione (**241**) was achieved *via* a cross-coupling of 6-bromolumazine (**239**) with 1-ethoxyprop-1-eneyltin (**240**) in the presence of a palladium catalyst and copper iodide [122].

239 → **241**

Bu₃Sn **240**
Pd(PPh₃)₄ MeCN, CuI
R = H, 77%; R = Me, 96%

To summarize, both chloropyrazines and chloroquinoxalines are sufficiently activated to serve as viable substrates for palladium chemistry under standard conditions. In contrast to chlorobenzene, the inductive effect of the two nitrogen atoms polarizes the C−N bonds. Therefore, oxidative additions of both chloropyrazines and chloroquinoxalines to Pd(0) occur readily. One exception is 2-chloropyrazine-*N*-oxide, which does not behave as a simple chloropyrazine. All Pd-catalyzed reactions with 2-chloropyrazine-*N*-oxide failed, presumably because the nitrogen atom no longer possesses the electronegativity required for activation.

Although the boranes of pyrazines and quinoxalines are yet to be made, their halides are legitimate electrophilic coupling partners. Halopyrazines and haloquinoxalines undergo Kumada, Negishi, Suzuki, Sonogashira, and Heck reactions. The heteroaryl Heck reactions of chloropyrazines are very well studied, whereas those of haloquinoxalines are yet to be seen. The heteroaryl Heck reactions are expected to provide good methods to prepare heterobiaryls without the necessity of making heteroaryl organometallic reagents.

10.9. References

1. Noorduin, L.; Swen, S. *Cryst. Struct. Commun.* **1976**, *5*, 153–5.
2. Andre, D.; Fourme, R.; Renaud, M. *Acta Crystallogr., Sect. B.* **1971**, *27*, 2371–80.
3. Reetz, M. T.; Lohmer, G.; Schwickardi, R. *Angew. Chem., Int. Ed. Engl.* **1998**, *37*, 481–3.

4. (a) Littke, A. F.; Fu, G. C. *Angew. Chem., Int. Ed. Engl.* **1998**, *37*, 3387–8.
   (b) Littke, A. F.; Fu, G. C. *J. Org. Chem.* **1999**, *64*, 10–1. (c) Buchwald, S. L.;
   Wolfe, J. P. *Angew. Chem., Int. Ed. Enl.* **1999**, *38*, 2413–16.

5. Garg, N. K.; Sarpong, R.; Stoltz, B. M. *J. Am. Chem. Soc.* **2002**, *124*, 13179–84.

6. Yang, C.-G.; Liu, G.; Jiang, B. *J. Org. Chem.* **2002**, 67, 9392–6.

7. Garg, N. K.; Caspi, D. D.; Stoltz, B. M. *J. Am. Chem. Soc.* **2004**, *126*, 9552–3.

8. Durán, R.; Zubía, E.; Ortega, M. J.; Naranjo, S.; Salvá, J. *Tetrahedron* **1999**, *55*,
   13225–32.

9. Emoto, T.; Kubosaki, N.; Yamagiwa, Y.; Kamikawa, T. *Tetrahedron Lett.* **2000**, *41*,
   355–8.

10. Mongin, F.; Mojovic, L.; Guillamet, B.; Trécourt, F.; Quéguiner, G. *J. Org. Chem.*
    2002, *67*, 8991–4.

11. Bonnet, V.; Mongin, F.; Trécourt, F.; Quéguiner, G.; Knochel, P. *Tetrahedron* **2002**,
    58, 4429–48.

12. (a) Ohta, A.; Inoue, A.; Watanabe, T. *Heterocycles* **1984**, *22*, 2317–21. (b) Ohta, A.;
    Inoue, A.; Ohtsuka, K.; Watanabe, T. *Heterocycles* **1985**, *23*, 133–7.

13. Ohta, A.; Ohta, M.; Igarashi, Y.; Saeki, K.; Yuasa, K.; Mori, T. *Heterocycles* **1987**,
    26, 2449–54.

14. Ohta, A.; Itoh, R.; Kaneko, Y.; Koike, H.; Yuasa, K. *Heterocycles* **1989**, *29*, 939–45.

15. Watanabe, T.; Hayashi, K.; Sakurada, J.; Ohki, M.; Takamatsu, N.; Hirohata, H.;
    Takeuchi, K.; Yuasa, K.; Ohta, A. *Heterocycles* **1989**, *29*, 123–31.

16. Sharples, C. G. V.; Karig, G.; Simpson, G. L.; Spencer, J. A.; Wright, E.; Millar, N. S.;
    Wonnacott, S.; Gallagher, T. *J. Med. Chem.* **2002**, *45*, 3235–45.

17. Angiolelli, M. E.; Casalnuovo, A. L.; Selby, T. P. *Synlett* **2000**, *6*, 905–7.

18. Negishi, E.; Qian, M.; Zeng, F.; Anastasia, L.; Babinski, D. *Org. Lett.* **2003**, *5*,
    1597–600.

19. Toudic, F.; Plé, N.; Turck, A.; Quéguiner, G. *Tetrahedron* **2002**, *58*, 283–93.

20. Gundisch, D.; Harms, K.; Schwarz, S.; Seitz, G.; Stubbs, M. T.; Wegge, T. *Bioorg.
    Med. Chem.* **2001**, *9*, 2683–91.

21. Egi, M.; Liebeskind, L. S. *Org. Lett.* **2003**, *5*, 801–2.

22. Wittenberg, R.; Srogl, J.; Egi, M.; Liebeskind, L. *Org. Lett.* **2003**, *5*, 3033–5.

23. Cuperly, D.; Gros, P.; Fort, Y. *J. Org. Chem.* **2002**, *67*, 238–41.

24. Sato, N.; Narita, N. *Synthesis* **2001**, *10*, 1551–5.

25. Jeong, I. H.; Kim, M. S.; Park, Y. S. *Bull. Korean Chem. Soc.* **2002**, *23*, 1823–6.

26. Gohlke, H.; Gundisch, D.; Schwarz, S.; Seitz, G.; Tilotta, M. C.; Wegge, T. *J. Med.
    Chem.* **2002**, *45*, 1064–72.

27. Zhou, J.; Kläβ, T.; Zhang, A.; Johnson, K. M.; Wang, C. Z.; Ye, Y.; Kozikowski, A. P.
    Bioorg. Med. Chem. Lett. **2003**, *13*, 3565–69.

28. Szekeres, H. J.; Atack, J. R.; Chambers, M. S.; Cook, S. M.; Macaulay, A. J.:
    Pillai, G. V.; MacLeod, A. M. *Bioorg. Med. Chem. Lett.* 2004, *14*, 2871–75.

29. Buron, F.; Plé, N.; Turck, A.; Quéguiner, G. *J. Org. Chem.* **2005**, *70*, 2616–21.

30. Ohta, A.; Ohta, M.; Watanabe, T. *Heterocycles* **1986**, *24*, 785–92.

31. Nakamura, H.; Takeuchi, D.; Murai, A. *Synlett* **1995**, 1227–8.

32. (a) Zoltewicz, J. A.; Cruskie, M. P. Jr. *J. Org. Chem.* **1995**, *60*, 3487–93.
    (b) Zoltewicz, J. A.; Maier, N. M.; Fabian, W. M. F. *J. Org. Chem.* **1997**, *62*, 3215–9.

33. Street, L. J.; Baker, R.; Book, T.; Reeve, A. J.; Saunders, J.; Willson, T.; Marwood, R. S.;
    Patel, S.; Freedman, S. B. *J. Med. Chem.* **1992**, *35*, 295–305.

34. Thompson, W. J.; Jones, J. H.; Lyle, P. A.; Thies, J. E. *J. Org. Chem.* **1988**, *53*, 2052–5.
35. Jones, K.; Keenan, M.; Hibbert, F. *Synlett* **1996**, 509–10.
36. Mitchell, M. B.; Wallbank, P. J. *Tetrahedron Lett.* **1991**, *32*, 2273–6.
37. Ali, N. M.; McKillop, A.; Mitchell, M. B.; Rebelo, R. A.; Wallbank, P. J. *Tetrahedron* **1992**, *48*, 8117–26.
38. Yang, Y.; Martin, A. R. *Heterocycles* **1992**, *34*, 1395–8.
39. Schultheiss, N.; Bosch, E. *Heterocycles* **2003**, *60*, 1891–97.
40. Gong, Y.; He, W. *Org. Lett.* **2002**, *4*, 3803–5.
41. Tapolcsányi, P.; Krajsovszky, G.; Andó, R.; Lipcsey, P.; Horváth, G.; Mátyus, P.; Riedl, Z.; Hajós, G.; Maes, B. U. W.; Lemière, G. L. F. *Tetrahedron* **2002**, *58*, 10137–43.
42. Parry, P. R.; Wang, C.; Batsanov, A. S.; Bryce, M. R.;Tarbit, B. *J. Org. Chem.* **2002**, *67*, 7541–43.
43. Jiang, B.; Wang, Q.-F.; Yang, C.-G.; Xu, M. *Tetrahedron Lett.* **2001**, 4083–5.
44. Gong, Y.; Pauls, H. W. *Synlett* **2000**, *6*, 829–31.
45. Cavalier, J.-F., Burton, M.; Tollenaere, C. D; Dussart, F.; Marchand, C.; Rees, J.-F.; Marchand, -B. J. *Synthesis* **2001**, *5*, 768–72.
46. Adamczyk, M.; Johnson, D. D.; Mattingly, P. G.; Pan, Y.; Reddy, R. E. *Org. Prep. Proced. Int.* **2001**, *33*, 477–85.
47. Molander, G. A.; Biolatto, B. *J. Org. Chem.* **2003**, *68*, 4302–14.
48. Kuse, M.; Kondo, N.; Ohyabu, Y.; Isobe, M. *Tetrahedron* **2004**, *60*, 835–40.
49. Jeanjot, P.; Bruyneel, F.; Arrault, A.; Gharbi, S.; Cavalier, J.-F.; Abels, A.; Marchand, C.; Touillaux, R.; Rees, J.-F.; Marchand-Brynaert, J. *Synthesis* **2003**, *4*, 513–22.
50. Adamczyk, M.; Akireddy, S. R.; Johnson, D. D.; Mattingly, P. G.; Pan, Y.; Reddy, R. E. *Tetrahedron* **2003**, *59*, 8129–42.
51. Jiang, B.; Yang, C.-G, Xiong, W.-N.; Wang, J. *Bioorg. Med Chem.* **2001**, *9*, 1149–54.
52. Nakamura, T.; Sato, M.; Kakinuma, H.; Miyata, N.; Taniguchi, K.; Bando, K.; Koda, A.; Kameo, K. *J. Med. Chem.* **2003**, *46*, 5416.
53. Glendening, M. E.; Goodby, J. W.; Hird, M.; Toyne, K. J. *J. Chem. Soc., Perkin Trans. 2*, **1999**, 481–91.
54. Toudic, F.; Heynderickx, A.; Plé, N.; Truck, A.; Quéguiner, G. *Tetrahedron* **2003**, *59*, 6375–84.
55. Akita, Y.; Ohta, A. *Heterocycles* **1982**, *19*, 329–31.
56. Akita, Y.; Inoue, A.; Ohta, A. *Chem. Pharm. Bull.* **1986**, *34*, 1447–58.
57. Akita, Y.; Kanekawa, H.; Kawasaki, T.; Shiratori, I.; Ohta, A. *J. Heterocycl. Chem.* **1988**, *25*, 975–7.
58. Sakamoto, T.; Shiraiwa, M.; Kondo, Y.; Yamanaka, H. *Synthesis* **1983**, 312–4.
59. Yamanaka, H.; Mizugaki, M.; Sakamoto, T.; Sagi, M.; Nakagawa, Y.; Takayama, H.; Ishibashi, M.; Miyazaki, H. *Chem. Pharm. Bull.* **1983**, *31*, 4549–53.
60. Taylor, E. C.; Ray, P. S. *J. Org. Chem.* **1987**, *52*, 3997–4000.
61. Taylor, E. C.; Dötzer, R. *J. Org. Chem.* **1991**, *56*, 1816–22.
62. Nakamura, H.; Wu, C.; Takeuchi, D.; Murai, A. *Tetrahedron Lett.* **1998**, *39*, 301–4.
63. Taylor, E. C.; Ray, P. S. *J. Org. Chem.* **1988**, *53*, 35–8.
64. Chapdelaine, M. J.; Warwick, P. J.; Shaw, A. *J. Org. Chem.* **1989**, *54*, 1218–21.
65. Sato, N.; Hayakawa, A.; Takeuchi, R. *J. Heterocycl. Chem.* **1990**, *27*, 503–6.
66. Turck, A.; Trohay, D.; Mojovic, L.; Plé, N.; Quéguiner, G. *J. Organomet. Chem.* **1991**, *412*, 301–10.

67. Turck, A.; Plé, N.; Dognon, D.; Harmoy, C.; Quéguiner, G. *J. Heterocycl. Chem.* **1994**, *31*, 1449–53.
68. Nakamura, H.; Aizawa, M.; Takeuchi, D.; Murai, A.; Shimoura, O. *Tetrahedron Lett.* **2000**, *41*, 2185–8.
69. Beebe, X.; Yang, F.; Bui, M. H.; Mitten, M. J.; Ma, Z.; Nilius, A. M.; Djuric, S. W. *Bioorg. Med. Chem. Lett.* **2004**, *14*, 2417–21.
70. Xu, G.; Hartman, L.; Wargo, H.; Turpin, J. A.; Buckheit, R. W.; Cushman, M. *Bioorg. Med. Chem.* **2002**, *10*, 283–90.
71. Lee, B. H.; Jaung, J. Y.; Jeong, S. H. *Bull. Korean Chem. Soc.* **2002**, *23*, 1045–6.
72. Yin, J.; Zhao, M. M; Huffman, M. A.; Mcnamara, J. M. *Org. Lett.* **2002**, *4*, 3481–4.
73. Maes, B. U. W.; Loones, K. T. J.; Lemière, G. L. F.; Dommisse, R. A. *Synlett* **2003**, *12*, 1822–5.
74. Jonckers, T. H. M.; Maes, B. U. W.; Lemière, G. L. F.; Dommisse, R. *Tetrahedron* **2001**, *57*, 7027–34.
75. Akita, Y.; Inoue, A.; Mori, Y.; Ohta, A. *Heterocycles* **1986**, *24*, 2093–7.
76. Akita, Y.; Noguchi, T.; Sugimoto, M.; Ohta, A. *J. Heterocycl. Chem.* **1986**, *23*, 1481–5.
77. Chen, J. J.; Walker, J. A. II; Liu, W.; Wise, D. S.; Townsend, L. B. *Tetrahedron Lett.* **1995**, *36*, 8363–6.
78. Walker, J. A. II; Chen, J. J.; Hinkley, J. M.; Wise, D. S.; Townsend, L. B. *Nucleosides Nucleotides* **1997**, *16*, 1999–2012.
79. Akita, Y.; Inoue, A.; Yamamoto, K.; Ohta, A.; Kurihara, T.; Shimizu, M. *Heterocycles* **1985**, *23*, 2327–33.
80. Akita, Y.; Itagaki, Y.; Takizawa, S.; Ohta, A. *Chem. Pharm. Bull.* **1989**, *37*, 1477–80.
81. Ohta, A.; Akita, Y.; Ohkuwa, T.; Chiba, M.; Fukunaka, R.; Miyafuji, A.; Nakata, T.; Tani, N.; Aoyagi, Y. *Heterocycles* **1990**, *31*, 1951–7.
82. Aoyagi, Y.; Inoue, A.; Koizumi, I.; Hashimoto, R.; Tokunaga, K.; Gohma, K.; Komatsu, J.; Sekine, K.; Miyafuji, A.; Konoh, J.; Honma, R.; Akita, Y.; Ohta, A. *Heterocycles* **1992**, *33*, 257–72.
83. Takeuchi, R.; Suzuki, K.; Sato, N. *Synthesis* **1990**, 923–4.
84. Takeuchi, R.; Suzuki, K.; Sato, N. *J. Mol. Cat.* **1991**, *66*, 277–8.
85. Beller, M.; Mägerlein, W.; Indolese, A. F.; Fischer, C. *Synthesis* **2001**, *7*, 1098–109.
86. Akita, Y.; Shimazaki, M.; Ohta, A. *Synthesis* **1981**, 974–5.
87. Sato, N.; Suzuki, M. *J. Heterocycl. Chem.* **1987**, *24*, 1371–2.
88. Matsuda, T.; Aoki, T.; Ohgiya, T.; Koshi, T.; Ohkuchi, M.; Shigyo, H. *Bioorg. Med. Chem. Lett.* **2001**, *11*, 2369–72.
89. Chandrasekhar, S.; Reddy, C. R.; Jagadeeshwar Rao, R.; Madhusundana, R. J. *Synlett* **2002**, *2*, 349–51.
90. Helquist, P. *Tetrahedron Lett.* **1978**, 1913–14.
91. Akita, Y.; Ohta, A. *Heterocycles* **1981**, *16*, 1325–8.
92. Ohta, A.; Akita, Y.; Takizawa, M.; Kurihara, S.; Masano, S.; Watanabe, T. *Chem. Pharm. Bull.* **1978**, *26*, 2046–53.
93. *Synthesis and Chemistry of Agrochemicals IV*, Selby, T. P.; Denes, R.; Kilama, J. J.; Smith, B. K. Eds.; ACS Symposium Series 584; Chapter 16, American Chemical Society; Washington, DC, **1995**, pp 171–85.
94. Auberson, Y. P.; Bischoff, S.; Moretti, R.; Schmutz, M.; Veenstra, S. J. *Bioorg. Med. Chem. Lett.* **1998**, *8*, 65–70.

95. Tsubata, Y.; Suzuki, T.; Miyashi, T.; Yamashita, Y. *J. Org. Chem.* **1992**, *57*, 6749–55.
96. Kim, K. S.; Qian, L.; Dickinson, K. E. J.; Delaney, C. L.; Bird, J. E.; Waldron, T. L.; Moreland, S. *Bioorg. Med. Chem. Lett.* **1993**, *3*, 2667–70.
97. Dinsmore, A.; Garner, C. D.; Joule, J. A. *Tetrahedron* **1998**, *54*, 3291–302.
98. Li, J. J.; Yue, W. S. *Tetrahedron Lett.* **1999**, *40*, 4507–10.
99. Li, J. J.; Carson, K. G.; Trivedi, B. K.; Yue, W. S.; Ye, Q.; Glynn, R. A.; Miller, S. R.; Connor, D. T., Roth, B. D.; Luly, J. R.; Low, J. E.; Heilig, D. J.; Yng, W.; Qin, S.; Hunt, S. *Bioorg. Med. Chem.* **2003**, *11*, 3777–90.
100. Aldakov, D.; Anzenbacher, P. *Chem. Commun.* **2003**, 1394–5.
101. Ames, D. E.; Brohi, M. I. *J. Chem. Soc., Perkin Trans. 1* **1980**, *7*, 1384–9.
102. Ames, D. E.; Bull D.; Takundwa, C. *Synthesis* **1981**, 364–5.
103. Ames, D. E.; Mitchell, J. C.; Takundwa, C. C. *J. Chem. Res., Synop.* **1985**, 144–5.
104. Ames, D. E.; Mitchell, J. C.; Takundwa, C. C. *J. Chem. Res., Miniprint* **1985**, 1683–96.
105. Dinsmore, A.; Birks, J. H.; Garner, C. D.; Joule, J. A. *J. Chem. Soc., Perkin Trans. 1* **1997**, 801–7.
106. Kim, C.-S.; Russell, K. C. *J. Org. Chem.* **1998**, *63*, 8229–34.
107. Bradshaw, B.; Dinsmore, A.; Garner, C. D.; Joule, J. A. *J. Chem. Soc., Chem. Commun.* **1998**, 417–8.
108. Armengol, M.; Joule, J. A. *J. Chem. Soc., Perkin Trans. 1* **2001**, 154–8.
109. Katoh, A.; Yoshida, T.; Ohkanda, J. *Heterocycles* **2000**, *52*, 911–20.
110. Armengol, M.; Joule, J. A. *J. Chem. Soc., Perkin Trans. 1* **2001**, 978–84.
111. Khan, M. S.; Al-Suti, M. K.; Al-Mandhary, M. R. A.; Ahrens, B.; Bjernemose, J. K.; Mahon, M. F.; Male, L.; Ratithby, P. R.; Friend, R. H.; Köhler, A.; Wilson, J. S. *Dalton Trans.* **2003**, 65–73.
112. Bradshaw, B.; Dinsmore, D.; Collision, D.; Garner, C.D.; Joule, J.A. *J. Chem. Soc. Perkin. Trans. 1* **2001**, 3232–8.
113. Acklin, P.; Allgeier, H.; Auberson, Y. P.; Bischoff, S.; Ofner, S.; Sauer, D.; Schmutz, M. *Bioorg. Med. Chem. Lett.* **1998**, *8*, 493–8.
114. Li, J. J. *J. Org. Chem.* **1999**, *64*, 8425–7.
115. Söderberg, B. C. G.; Wallace, J. M.; Tamariz, J. *Org. Lett.* **2002**, *4*, 1339–42.
116. Bianchini, C.; Barbaro, P.; Scapacci, G. *J. Organometallic Chem.* **2001**, *621*, 26–33.
117. Yang S. C.; Shue Y. J.; Liu, P. C. *Organometallics* **2002**, *21*, 2013–16.
118. Ding, S.; Gray, N.S.; Wu, X.; Ding, Q.; Schultz, P.G. *J. Am. Chem. Soc.* **2002**, *124*, 1594–6.
119. Mederski, W. W. K. R.; Kux, D.; Knoth, M.; Schwarzkopf-Hofmann, M. *J. Heterocycles* **2003**, *60*, 925–32.
120. Kunick, C.; Lauenroth, K.; Leost, M.; Meijer, L.; Lemcke, T. *Bioorg. Med. Chem. Lett.* **2004**, *14*, 413–16.
121. Hopkins, C. R.; Collar, N. *Tetrahedron Lett.* **2005**, *46*, 1845–8.
122. Satu, N.; Fukuya, S. *J. Chem. Soc., Perkin Trans. 1* **2000**, 89–95.

Chapter 11

Pyrimidines

Michael Palucki

The pyrimidine entity is one of the most prominent structures found in nucleic acid chemistry. Pyrimidine derivatives including uracil, thymine, cytosine, adenine, and guanine are fundamental building blocks for deoxyribonucleic acid (DNA) and ribonucleic acid (RNA). Vitamin B1 (thiamine) is a well-known example of a naturally occurring pyrimidine that is encountered in our daily lives. Synthetic pyrimidine-containing compounds also occupy a prominent place in the pharmaceutical arena. Pyrimethamine and Trimethoprim are two representative pyrimidine-containing chemotherapeutics. Pyrimethamine is a dihydrofolate reductase inhibitor; effective for toxoplasmosis in combination with a sulfonamide; whereas Trimethoprim is an antimalarial drug, widely used as a general systemic antibacterial agent in combination with sulfamethoxazole.

Vitamin B₁   Pyrimethamine   Trimethoprim   pyrimidine

Due to the electronegativity of the two nitrogen atoms, the pyrimidine ring is a deactivated, π-electron-deficient heterocycle. Its chemical behavior is comparable to that of 1,3-dinitrobenzene or 3-nitropyridine. One or more electron-donating substituents on the pyrimidine ring are often required for effecting electrophilic substitution reactions. In contrast, nucleophilic displacement is significantly more facile on a pyrimidine ring compared to a pyridine ring. This trend also extends to palladium chemistry: Pd(0) undergoes oxidative addition more readily with 4-chloropyrimidine than with 2-chloropyridine.

Remarkable differences in the inherent reactivity for each C-position on pyrimidinyl halides and triflates have been observed. The C(4) and C(6) positions of a halopyrimidine are more prone to S_NAr processes than the C(2) position. The order of S_NAr displacement for halopyrimidines is:

$$C(4) \quad > \quad C(2) \quad >> \quad C(5)$$

This trend is also observed in palladium chemistry where the general order for oxidative addition often correlates with that of nucleophilic substitution. Not only are 2-, 4-, and 6-chloropyrimidines viable substrates for palladium-catalyzed reactions, but 4- and 6-chloropyrimidines react more readily than 2-chloropyrimidines.

Undheim and Benneche reviewed the Pd-catalyzed reactions of pyrimidines, among other π-deficient azaheterocycles including pyridines, quinolines, and pyrazines, in 1990 [1] and 1995 [2]. A review by Kalinin also contains some early examples in which C–C formation on the pyrimidine ring is accomplished using Pd-catalyzed reactions [3]. In this chapter, we will systematically survey the palladium chemistry involving pyrimidines.

11.1. Synthesis of pyrimidinyl halides and triflates

Pyrimidinyl halides are not only precursors for Pd-catalyzed reactions, but are also found in a number of pharmaceutically active agents. One of the most frequently employed approaches for halopyrimidine synthesis is direct halogenation. When pyrimidinium hydrochloride and 2-aminopyrimidine were treated with bromine, 5-bromopyrimidine and 2-amino-5-bromopyrimidine were obtained, respectively, *via* an addition–elimination process instead of an aromatic electrophilic substitution [4, 5]. Analogously, 2-chloro-5-bromopyrimidine (**1**) was generated from direct halogenation of 2-hydroxypyrimidine [6]. Treating **1** with HI then gave 2-iodo-5-bromopyrimidine (**2**). In the preparation of 5-bromo-4,6-dimethoxypyrimidine (**4**), N-bromosuccinimide was found to be superior to bromine for the bromination of 4,6-dimethoxypyrimidine (**3**) [7].

5-Iodopyrimidine **7** was prepared by iodination of 2,4-diaminopyrimidine **6**, which was derived from commercially available 2-amino-4-chloro-6-methylpyrimidine (**5**) *via* an S_NAr reaction with ammonia [8]. Similarly, iodination of 6-chloro-2,4-dimethoxypyrim-idine (8) with *N*-iodosuccinimide in trifluoroacetic acid led to dihalopyrimidine 9 [9].

Another reliable method of halopyrimidine synthesis is "dehydroxy-halogenation". Refluxing pyrimidinones 10 and 13 with phosphorus oxychloride was followed by treating the resulting chloropyrimidines with hydroiodic acid to afford iodopyrimidine 11 and 14, respectively [10, 11]. 4-Chloropyrimidinone 13, on the other hand, was prepared by direct halogenation of pyrimidone 12.

Although dehydroxy-halogenation of pyrimidones works well using phosphorus oxy-chloride and phosphorus oxybromide, both reagents are moisture-sensitive and phosphorus oxybromide is relatively expensive. Extending the well-known halogenation of alcohols using triphenylphosphine and *N*-halosuccinimide, Sugimoto *et al.* halogenated several π-deficient hydroxyheterocycles including hydroxypyridine, hydroxyquinoline, hydroxyquinoxaline, and hydroxypyrimidine [12]. Thus, treating hydroxypyrimidine 15 with triphenylphosphine and *N*-halosuccinimide resulted in halopyrimidine 16 under mild conditions.

The preparation of 6-membered ring heterocyclic triflates including pyrimidinyl triflate is a challenging task [13]. Simple 2-pyrimidinyl triflates and 2-methylthio-4-pyrimidinyl triflates with hydrogen at the electron-deficient 4/6-positions are unstable. However, when there is an additional substituent on the pyrimidine ring, the corresponding pyrimidinyl triflate becomes sufficiently stable, allowing isolation *via* flash chromatography. Therefore, treating 6-methyluracil (17) with NaH followed by triflic anhydride gave rise to bis-triflate 18. 2-Methylthio-4-pyrimidinyl triflate (19) and 2-pyrimidinyl triflate (20) were also prepared using triflic anhydride from respective pyrimidones [14]. In these cases, triflic anhydride was found to be more convenient than *N*-phenylmethanesulfonimide (PhNTf$_2$) due to comparative ease of purification of the resulting triflates.

11.2. Coupling reactions with organometallic reagents

11.2.1. Negishi coupling

Pyrimidinylzinc chloride 22 was generated *in situ* by halogen–metal exchange of 5-bromo-4,6-dimethoxypyrimidine (21) with *n*-BuLi followed by treatment with ZnCl$_2$ [15]. The subsequent Negishi coupling of 22 with 3,4-dinitrobromobenzene gave phenylpyrimidine 23.

The Reformatsky reagent, a classic organozinc reagent, has proven useful for the Negishi reaction with organohalides including halopyrimidines. Reformatsky reagent 25 was prepared from metalation of ethyl bromoacetate with fresh zinc metal. While 2-iodopyrimidine 24 readily coupled with Reformatsky reagent 25 to afford ethyl 4,6-dimethyl-2-pyrimidine acetate (**26**), the reaction of 2-bromopyrimidine with **25** was much less efficient and 2-chloropyrimidine was virtually inert under such reaction conditions [16]. Ethyl 2, 6-dimethyl-4-pyrimidineacetate (**28**) was synthesized in a similar fashion from 4-iodo-2,6-dimethylpyrimidine (**27**) and **25**. However, the desired coupling product was not observed for the same reaction of 5-iodo-2,6-dimethylpyrimidine.

Researchers at Pfizer were interested in preparing 3-aryl azetidines in efforts to iden-tify new drug candidates for the treatment of asthma. A general approach was developed in which the organozinc species of the azetidine **29** was prepared *via* Zn insertion into the C–I bond followed by Pd-catalyzed cross-coupling with aryl halides. Using this process, 5-bromopyridine was coupled to produce **31** in 46% yield [17].

The 1-(-4-fluorophenyl)indole is a common pharmacophore found in antipsychotic drugs. In a one-pot procedure, halogen–metal exchange of bromide **32** with *n*-BuLi at ˜ −78°C in THF followed by transmetalation with ZnCl₂ cleanly provides the aryl zinc species **33**. In the same pot, addition of an aryl halide, Pd(Ph₃P)₄ in DMF and heating to 80°C provided the desired Negishi cross-coupled product **34**. Using this method, 2-bro-mopyridine and 4-bromopyridine were successfully coupled to **33** in 66 and 65% isolated yield, respectively [18].

a: Ar = 2-pyrimidine: 66% yield
b: Ar = 4-pyrimidine: 65% yield

Oxidative cyclization of substituted enyne carboxylic acid with I_2 provides the corresponding 5-substituted iodopyrones. Zinc insertion into the C–I bond provides an organozinc species for Negishi coupling reactions. Indeed, it was found that iodopyrone **36** can be converted to the organozinc intermediate and subsequently cross coupled to 4-bromopyrimidine in 72% isolated yield [19].

Ferrocene derived ligands have found widespread use in asymmetric catalysis [20]. Metalation of the cyclopentadiene ring with *t*-BuLi followed by transmetalation to zinc provide a Negishi cross-coupling partner that can be employed to access structurally distinct ferrocene molecules. Using this methodology, the parent ferrocene **38** can be coupled to 5-bromopyridine in 44% isolated yield [21]. Similarly, *ortho*-metalation of the more substituted ferrocene **41** followed by transmetalation with $ZnBr_2$ at low temperature, provide organozinc intermediate **42** which can be used in a Negishi cross-coupling reaction. Cross-coupling to 2-iodopyrimidine provides the desired product **43** in very good yield [22].

The marked increase of stability of organozinc reagents compared to lithium or magnesium organometallics allows the Negishi reactions to be carried out at high temperatures. Organozinc reagent **45**, derived from pyridazine **44** by *ortho*-lithiation and treatment with $ZnCl_2$, was coupled with 5-bromopyrimidine to form pyrimidinylpyradazine **46** at 65°C [23]. Sonication was found to shorten the reaction time significantly and improve the yield. In a similar fashion, *ortho*-lithiation of 2-methylthio-4-chloropyrimidine (**47**) using LTMP and subsequent treatment with $ZnCl_2$ led to organozinc reagent **48**, which was then cross coupled with iodobenzene to furnish 5-arylpyrimidine **49**.

In efforts to prepare substituted methylthio-*N*-heterocycles, researchers at Dupont discovered a new type of Negishi coupling reaction in which benzylic zinc reagents were found to undergo a Pd-catalyzed cross-coupling reaction at the C—S site. 2-(Methylthio) pyrimidine **50** was found to be a particularly reactive substrate for this reaction. Indeed, a 71% isolated yield using 1 mol% Pd(PPh₃)₄ was obtained with benzyl zinc bromide and 2-(methylthio)pyrimidine in THF. This new cross-coupling method provides a convenient alternative to the use of pyrimidine halides [24].

Pyrimidine triflates are also suitable substrates for Negishi reactions. The coupling of 2,4-pyrimidinyl triflate (**52**) with 2.5 equivalents of *p*-anisylzinc bromide gave 2,4-dianisylpyrimidine **53** [23]. The Negishi reaction of 2,4-pyrimidinyl bis-triflate **54** and *p*-anisylzinc bromide occurred predominantly at C(4). Interestingly, since *p*-anisylzinc bromide was generated by treating *p*-bromoanisole with *n*-BuLi followed by reaction with ZnBr₂, some butylzinc bromide was present in the reaction mixture. As a consequence, the second Negishi reaction with butylzinc bromide proceeded without detectable β-hydride elimination to produce 2-butylpyrimidine **55** [25].

Although alkyl halides are prone to β-hydride elimination, alkylzinc or alkylboron reagents can take part in Negishi or Suzuki coupling to install alkyl substituents onto pyrimidine rings. Indeed, alkylpyrimidine **57** was synthesized from pyrimidinyl triflate **56** using *n*-butylzinc chloride. Furthermore, vinylpyrimidine **59** and thienylpyrimidine **60** were prepared also from the corresponding triflates, respectively [26].

Alkynyl zinc reagents can be prepared from vinyl dichlorides *via* halogen–metal exchange with *n*-BuLi, followed by transmetalation to Zn. Alkynyl zinc reagents prepared using this method have been shown to effectively undergo Negishi cross-coupling with a variety of aryl iodides including 2-iodopyrimidine [27].

11.2.2. Suzuki coupling

Like simple carbocyclic arylboronic acids, pyrimidineboronic acids couple with organohalides and organotriflates under palladium catalysis. However, simple 5-pyrimidineboronic acid **65** is not trivial to make due to competing nucleophilic addition of the anionic intermediate to the azomethine bond. It is preferable to reverse the coupling partners—using 5-halopyrimidine to couple with other easily accessible boronic acids to prepare the same Suzuki adduct. Nonetheless, the Gronowitz group successfully prepared **65** from 5-bromopyrimidine **64** *via* a halogen–metal exchange using *n*-BuLi at an extremely low temperature followed by treatment with tributylborate and basic hydrolysis [28].

For the halogen–metal exchange reaction of bulkier halopyrimidines, steric hindrance retards the nucleophilic attack at the azomethine bond. As a consequence, halogen–metal exchange of 5-bromo-2,4-di-*t*-butoxypyrimidine (**68**) with *n*-BuLi could be carried out at –75°C [28]. The resulting lithiated pyrimidine was then treated with *n*-butylborate followed by basic hydrolysis and acidification to provide 2,4-di-*t*-butoxy-5-pyrimidineboronic acid (**69**). 5-Bromopyrimidine **68** was prepared from 5-bromouracil in two steps consisting of a dehydroxy-halogenation with phosphorus oxychloride and an S_NAr displacement with sodium *t*-butoxide.

The Suzuki coupling of **69** was utilized to prepare 5-substituted uracils as potential antiviral agents [29, 30]. Adduct **70**, derived from **69** and 2-bromo-3-methylthiophene, was transformed to the corresponding uracil **71** *via* acidic hydrolysis. Conveniently, reversal of the coupling partners also resulted in formation of adduct **70**, assembled from the Suzuki coupling of 5-bromo-2,4-di-*t*-butoxypyrimidine (**68**) and 3-methyl-2-thiopheneboronic acid.

Similarly, 5-bromo-2-methoxypyrimidine **75** can be reacted with *n*-BuLi at −78°C to effect the halogen–metal exchange followed by sequential quenching with triisopropylborate in THF and then water to give the 2-methoxy-5-pyrimidine boronic acid **76** in 61% isolated yield. The product boronic acid can be cross coupled to a variety of aryl halides [31].

2-Chloropyrimidine was coupled with diethyl (3-pyridyl)borane in the presence of Pd(Ph$_3$P)$_4$, Bu$_4$NBr, and KOH to afford 3-(2′-pyrimidinyl)pyridine [32]. Likewise, the Suzuki coupling of 2-bromopyrimidine with diethyl (4-pyridyl)borane (**79**) led to 4-(2′-pyrimidinyl)pyridine (**80**) in 50% yield, whereas 2-chloropyrimidine produced **80** in only 20% yield under the same conditions [33]. Diethyl (4-pyridyl)borane (**79**), on the other hand, was readily accessible from sequential treatment of 4-bromopyridine with *n*-BuLi and diethylmethoxyborane.

In attempts to prepare oligo- and poly(aryl/heteroaryl) systems, the use of 2,5-dibromopyrimidine **81** was required in a sequential fashion. Treating **81** with three equivalents of benzene boronic acid in the presence of catalytic Pd(PPh$_3$)$_4$ in refluxing THF, gave only two isolable products. The mono-Suzuki product **83** resulting from reaction at the 2-position was obtained in 43% yield and the bis-Suzuki product **84** was obtained in 32% yield. The results of this reaction establish that the bromine at the 2-position is more reactive than the bromine at the 5-position [34].

The coupling of 2- and 5-pyrimidinylbromides with 5-indolylboronic acid afforded 5-pyrimidinylindoles [35]. 5-Indolylboronic acid was readily prepared from commercially available 5-bromoindole by a one-pot process involving treating 5-lithio-1-potassioindole with tributylborate followed by acidic hydrolysis. Meanwhile, it was also discovered that Pd(dppb)Cl$_2$ possessed greater effectiveness than Pd(Ph$_3$P)$_4$ for the coupling of 2-chloropyrimidine **85** and arylboronic acid **86** to afford 2-arylpyrimidine **87** [36, 37].

Conjugated oligomers with alternating phenylene–pyrimidine moieties were prepared *via* sequential Suzuki coupling reactions starting with 2-bromo-5-pyrimidine **88**. Selective coupling of **88** at the iodo-substituted carbon was carried out under standard Suzuki conditions to give the desired product **89** in high yield. Anhydrous conditions coupled with the use of the bulky electron rich tri-*t*-butyl phosphine was found to be the best conditions for the second Suzuki reaction [38].

The 3,4-diarylisoxazole is a subunit utilized in pharmaceutically active agents. Indeed, one therapeutic compound containing this subunit is Valdecoxib [39]. Not surprisingly, a convenient approach to 3,4-diarylisoxazoles using the Suzuki cross-coupling reaction was developed. Using 5-bromopyrimidine **64**, Suzuki cross coupling to boronic acid **91** was accomplished in 60% yield using the standard Pd(PPh$_3$)$_4$/Na$_2$CO$_3$ system [40].

Preparation of 6-aryl-2,4-diaminopyrimidines **94** was accomplished *via* Suzuki reactions of a variety of aryl boronic acids with 6-chloro-2,4-diaminopyrimidine **93**. Using Pd(PPh$_3$)$_4$ as catalyst, DME as solvent, and 2M Na$_2$CO$_3$ as base in a biphasic system, moderate to excellent yields were obtained for a variety of aryl boronic acids [41].

In efforts to prepare heteroaryl benzoic acids, a convenient approach using the Suzuki coupling of carboxybenzene boronic acids with heteroaryl halides was developed. Thus, 4-carboxybenzene boronic acid **96** was coupled to 2-amino-5-bromopyrimidine **95** in 76% isolated yield using Pd(PPh$_3$)$_4$ as catalyst, and Na$_2$CO$_3$ as base in MeCN/H$_2$O solvent. Using the same protocol but with two equivalents of 2,6-dichloropyrimidine, regioselective Suzuki coupling at the C(6) carbon can be accomplished in 81% yield [42].

For the Suzuki coupling of 5-bromopyrimidine and protected *p*-boronophenylalanine **52** to assemble **53**, Pd(dppf)Cl$_2$ was found to be an effective catalyst [43], whereas using 2-bromopyrimidine as the substrate gave the corresponding adduct in only 20% yield.

Solid support synthesis of libraries of low molecular weight molecules with potential biological activity is a common approach to identifying leads in medicinal chemistry. It is not surprising, then, that given the importance of the pyrimidine moiety in biological systems, a solid support synthesis of a library utilizing the pyrimidine as a core diversity element was recently reported. Resin supported chloropyrimidine **104** was coupled with a variety of boronic acids using $Pd_2(dba)_3/P(t\text{-}Bu)_3$ as catalyst system and KF as base in moderate yield [44].

The Suzuki reaction of halo-pyrimidines with organoboron derivatives is not limited to $Csp^2\text{-}Csp^2$ coupling, but can also be accomplished in a $Csp^3\text{-}Csp^2$ fashion. Hydroboration of terminal alkene **107** with 9-BBN followed by Pd-catalyzed Suzuki cross coupling with 5-bromopyrimidine provides the desired alkyl substituted product **108**. The classical Suzuki conditions proved insufficient, whereas the combination of $Pd(OAc)_2$ and a biphenyl monophosphine ligand developed by Buchwald provided the desired product in 55% yield [45].

Potassium trifluoroaryl borates have emerged as a convenient alternative to aryl boronic acids in Suzuki cross-coupling reactions. This is primarily due to their robustness, ease of preparation, and observation that they are less prone to protodeboronation relative to aryl boronic acids. Suzuki cross coupling of potassium trifluorophenyl borate **109** with 5-bromopyridine **64** using $Pd(OAc)_2$, K_2CO_3 in refluxing MeOH provides the desire product **110** in 92% isolated yield [46]. Similarly, alkenyl [47] and alkynyl trifluoroborates [48] such as potassium *trans*-styryl trifluoroborate **111** and potassium (1-hexyn-1-yl)- trifluoroborate **113** have been coupled to **64** and 2,4,6-trichloropyrimidine **114** in good yield.

The electron-deficient nature of the pyrimidine ring system provides the opportunity for Suzuki cross coupling of aryl boronic acids with pyrimidine derivatives containing non-traditional leaving groups. Indeed, as shown in the Negishi portion of this chapter, pyrimidine methyl sulfides can undergo Pd-catalyzed cross-coupling reactions, similar to that of pyrimidine halides. Suzuki cross coupling of the pyrimidine methyl sulfide **116** with boronic acid **117** provides the desired product **118** in 86% isolated yield [49].

11.2.3. Stille coupling

11.2.3.1. The pyrimidine motif as a nucleophile

In 1989, Undheim and colleagues prepared 5-tributylstannylpyrimidine **121** *via* lithiation of 2-methylthio-5-bromopyrimidine **119** and subsequent reaction with tributyltin chloride [50]. The same technique has been utilized to synthesize other 5-stannylpyrimidine derivatives such as 2-*tert*-butyldimethoxy-5-stannylpyrimidine [51]. The best yield for preparing 2-methylthio-4-tributylstannylpyrimidine (**123**) was obtained using 2-methylthio-4-iodopyrimidine (**122**) as the precursor. However, treating **122** with tributylstannyl lithium gave **123** in only 52% yield [52]. In addition, 4-stannylated pyrimidines were also prepared from the corresponding pyrimidinyl halides *via* a halogen–metal exchange followed by quenching with trialkyltin chloride.

122 → **123**

Another important approach for stannylpyrimidine synthesis is the S_NAr reaction of a pyrimidinyl halide with a stannyl anion [52]. Treating 2-chloro-5-bromopyrimidine with Bu_3SnLi led to 2-chloro-5-tributylstannylpyrimidine (**124**) in 53% yield. Similarly, treatment of 2-chloropyrimidine with either Bu_3SnLi or Me_3SnLi gave 2-tributylstannylpyrimidine or 2-trimethylstannylpyrimidine in 84 or 46% yield, respectively. Furthermore, a regioselective substitution was achieved for the S_NAr reaction of 2,4-dibromopyrimidine with Me_3SnNa to afford 4-bromo-2-trimethylstannylpyrimidine (**125**) in 78% yield, whereas using Me_3SnLi resulted in, surprisingly, extensive polymerization. An attempt to conduct the stannylation of 2,5-dichloropyrimidine with Bu_3SnLi was also unsuccessful.

124

125

Both of the two aforementioned methods must be carried out at low temperatures. On the other hand, the Pd-catalyzed coupling reaction between 2-methylthio-5-bromopyrimidine and hexamethyldistannane in the presence of fluoride ion can be run at ambient temperature to prepare 2-methylthio-5-tributylstannylpyrimidine (**121**) [53]. The Stille reaction of **121** and 5-bromofurfural then afforded adduct **126**. Although Sn–Sn bonds are known to be thermally stable, the weakening of the Sn–Sn bond can be achieved through the formation of complexes such as $[(Bu_3SnSn_3X)^-Bu_4N^+]$ (X = F or Cl). During the formation of **121**, it was found that bis(π-allylpalladium chloride) was the best catalyst in terms of reaction rate and yield. Furthermore, bromides of thiophene, pyridine, and thiazole were coupled with **121**, and electron-withdrawing groups on the heteroaryl halides appeared to activate these electrophiles.

121

126

Thermal decarboxylation of pyrimidylcarboxylic organotin esters is another means to prepare the corresponding stannylpyrimidines [54]. This method obviates the intermediacy of lithiated pyrimidine species that would undergo undesired reactions at higher temperatures. The decarboxylation occurs at the activated positions. Therefore, thermal decarboxylation of tributyltin carboxylate **128**, derived from refluxing carboxylic acid **127** with bis(tributyltin) oxide, provided 4-stannylpyrimidine **129**. Addition of certain Pd(II) complexes such as bis(acetonitrile)palladium(II) dichloride improved the yields, whereas AIBN and illumination failed to significantly affect the yield.

In the literature, most Stille couplings involving pyrimidines were conducted using pyrimidinyl halides and other organostannanes. There are limited applications of stannylpyrimidines in the Stille coupling. However, if the coupling partner has delicate functional groups, choosing a stannylpyrimidine as the nucleophile does have its merits as showcased by the coupling of **121** with 5-bromofurfural to make **60**, as well as the union of 5-tributylstannylpyrimidine with 4-bromo-2-nitrobenzylphosphonate (**130**) to provide heterobiaryl phosphonate **131** [55].

In attempts to identify molecules that target the Melanacortin receptor, chemists at Procter and Gamble developed the 2,4,5-trisubstituted tetrahydropyrans as peptidomimetic scaffolds. These molecules were prepared from Pd-catalyzed cross coupling of functionalized 5-bromodihydropyrans and the appropriate cross-coupling partner. Indeed, the Stille cross coupling of 5-trimethylstannylpyrimidine with the 5-bromodihydropyran derivative **132** using Pd(PPh$_3$)$_4$ in DMF with KF as base was accomplished in good yields [56].

132 **133**

Enol triflates have emerged as attractive alternatives to vinyl halides in the Stille coupling partner due to their ease of preparation from readily available carbonyl compounds. The addition of LiCl has been found to be beneficial to Stille enol triflate coupling reactions. Thus, it was not surprising that coupling 5-tributylstannylpyrimidine to enol triflate **134** proceeded in good yield in the presence of LiCl. In this case, the addition of CuI as cocatalysts was found to also be beneficial to the reaction outcome [57].

134 **135**

Subtle differences in the structures of the coupling partner can have a significant effect on the outcome of a Stille reaction. For example, Stille cross coupling of iodoindole **136a** and the pyrimidine tin derivative **137** using the combination of Pd$_2$dba$_3$/PPh$_3$/CuI affords the desired product in 75% yield. By comparison, Stille cross coupling using the same conditions but with **136b** affords the desired product in only 40% yield [58].

136 **137** **138**
a: R = Ts: 75%
b: R = Ac: 40%

One of the most appealing features of Pd-catalyzed cross-coupling reactions is their ability to tolerate a wide variety of functional groups. This feature allows for these reactions to be implemented towards the end of a total synthesis. This is exemplified in the preparation of Epothiolone B analogs [59]. Stille reaction of tributylstannylpyrimidine derivative **140** with the highly functionalized vinyl iodide **139** provides the desired Epothiolone B analog **141**.

139 + **140** → **141**

Thioesters are readily accessible and have recently been shown to undergo Pd-catalyzed cross coupling with boronic acids to prepare ketone products. This chemistry has expanded to Stille cross coupling. The key to the reaction is the addition of stoichiometric amounts of Cu^I-diphenylphosphinate [60]. Stille coupling of thioester **142** with 2-tributylstannylpyrimidine affords the desired product **143** in 61% isolate yield.

142 + → **143**

11.2.3.2. The pyrimidine motif as an electrophile

In the more prolific aspect of the Stille couplings involving a pyrimidine fragment, pyrimidinyl halides or triflates have been coupled with a variety of stannanes. When there is only one reactive halide on the pyrimidine ring, the reaction outcome is straightforward with no regiochemical concern. The simpler stannanes are vinyl stannanes [61–63]. More complicated variants include stannylquinones [64] and 1-(trialkylsilyloxy)vinyltin [65] as illustrated by the synthesis of **144**.

144

When there is more than one halide on the pyrimidine ring, multiple regioisomeric products can be obtained. The 4(6)-position in pyrimidine is more reactive than the 2-position and regiospecific coupling can be achieved. The reaction of 2,4-dichloropyrimidine and styrylstannane first proceeded regiospecifically at C(4), giving rise to **145**, which was subsequently coupled with phenylstannane under more forcing conditions to afford disubstituted pyrimidine **146** [62].

145   **146**

Only bromo- and iodopyrimidine provide sufficient activation for the Stille coupling of a 5-halopyrimidines to take place. Undheim *et al.* discovered that sequential substitution of 5-bromo-2,4-dichloropyrimidine occurred according to the order of reactivity, 4-Cl > 5-Br > 2-Cl. The regio- and chemoselectivity were elegantly demonstrated by stepwise introduction of three different substituents onto 5-bromo-2,4-dichloropyrimidine **147** [54].

In addition, some Stille adducts have been further manipulated to form condensed heteroaromatic ring systems. The coupling of 4-acetylamino-5-bromopyrimidine **151** and (*E*)-1-ethoxy-2-(tributylstannyl)ethene resulted in (*E*)-4-acetylamino-5-(2-ethoxyethenyl) pyrimidine **152**, which then cyclized under acidic conditions to furnish pyrrolo[2,3-*d*] pyrimidine **153**. Pyrrolo[3,2-*d*]pyrimidines were also synthesized in a similar fashion by using 5-acetylamino-4-bromopyrimidine [66].

Halopyrimidines also couple with stannanes of heterocycles such as furans [67], azaindoles [68], pyridines [69–72], thiazoles, pyrroles [72], and thiophenes [73]. A representative example is the coupling of 3-tributylstannyl-7-azaindole **154** with 5-bromopyrimidine to furnish heterobiaryl **155** after acidic hydrolysis [68]. Moreover, a selective substitution at the 5-position was achieved when 4-chloro-5-iodopyrimidine **156** was allowed to react with 2-thienylstannane to provide thienylpyrimidine **157** [73].

Recent advances in functionalization of 2-chloropyridine *via* C(6) lithiation followed by tributylstannation has provided access to 6-tributylstannyl-2-chloropyridine **158**, which in turn can be coupled to halopyrimidines under standard Stille conditions [74]. Subsequent refinement of this procedure has provided a one-pot protocol for the lithiation/electrophilic stannation/Stille coupling to provide heteroaryl substituted pyrimidines, wherein the tributylstannane intermediate is not isolated [75].

The synthesis of pyrimidinyl thioethers using palladium catalysis poses a synthetic challenge due to the potential multiple coordination of either pyrimidinyl halides or the product to the catalyst, resulting in inhibition of catalysis. Nonetheless, the Pd-catalyzed coupling of organotin sulfides with halopyrimidines was successfully carried out using the Stille coupling [76]. Both 5-bromopyrimidine and activated 2-chloropyrimidine were coupled with phenyltributyltin sulfide to give the expected phenylsulfide derivatives **160** and **161**, respectively.

Pyrimidine thioethers may also be synthesized *via* direct Pd-catalyzed C–S bond formation between halopyrimidines and thiolate anions. For very unreactive thiol nucleophiles such as 2-thiopyrimidine **162**, both a strong base and a palladium catalyst are essential. Without a palladium catalyst or replacing *t*-BuONa with K_2CO_3, the reaction failed to furnish the desired pyrimidine thioether **163** [77].

While the use of heteroaromatic halides and triflates in cross-coupling reactions have grown over the years, these substrates often have limited availability and/or stability. Consequently, research efforts have focused on expanding cross-coupling reactions from traditional aromatic C−X bonds wherein X = halide or triflate, to more stable and perhaps more accessible electrophilic leaving groups. To this end, several reports have emerged describing cross-coupling reactions with heteroaromatic thioethers. Both methyl thioethers [78] and aryl thioethers [79] have been shown to be effective leaving groups.

11.2.4. Organozirconium, organoaluminum, and organoindium reagents

Many examples exist for Pd-catalyzed cross couplings of alkenylzirconocenes with simple carbocyclic aryl or alkenyl halides, whereas few precedents are seen for the coupling of alkenylzirconocenes with heteroaryl halides. Undheim and coworkers reported a Pd-catalyzed cross coupling of 2,4-dichloropyrimidine with alkenylzirconocene [80]. Hydrozirconation of hexyne readily took place at room temperature with zirconocene chloride hydride in benzene. The resulting hexenylzirconocene chloride (**168**) was then coupled with 2,4-dichloropyrimidine at the more electrophilic 4-position, giving rise to 2-chloro-4-[(*E*)-1-hexenyl]pyrimidine (**169**).

Palladium-catalyzed alkylations are generally rare because of the ease with which β-hydride elimination occurs. However, alkylation of 2,4-dichloropyrimidine with trimethylaluminum in the presence of Pd(Ph₃P)₄ was achieved without detectable β-hydride elimination. The preference for coupling at the 4-position was maintained for the formation of 2-chloro-4-methylpyrimidine (**170**), which then underwent an additional cross coupling with triisobutylaluminum to afford 2,4-dialkylpyrimidine **171** [81].

Sodium tetraalkynyl aluminate salts can be prepared from sodium aluminum hydride **172** and four equivalents of a terminal alkyne. The sodium tetraalkynyl aluminate salts are air stable and easily handled. The fairly reactive alkynyl substituent can be transferred in a cross-coupling mode to aryl bromides such as 5-bromopyridine. All four alkynes are transferrable from one aluminate salt [82].

Although traditional neutral alkyl aluminum reagents have been employed in cross-coupling reactions, the pyrophoric nature of the reagents have limited their overall utility. Addition of chelating ligands can stabilize the aluminum reagent, rendering it less pyrophoric, however, its reactivity is also tempered. Organoindium reagents have recently been shown to be effective in Pd-catalyzed cross coupling, and methylation of 5-bromopyrimidine is an example. The organoindium dimer **174** is fairly air stable and easy to handle [83].

11.3. Sonogashira reaction

Activated and deactivated positions in halopyridines exhibit marked difference in reactivity in palladium chemistry, whereas little difference in reactivity was observed among 2-, 4-, and 5-positions of halopyrimidines for their Sonogashira reactions [84]. While 2-iodo-4,6-dimethylpyrimidine was the most suitable substrate for preparing internal alkyne **176**, the reaction of either the corresponding bromide or chloride was less efficient [85]. Good to excellent yields were obtained for the preparation of alkynylpyrimidines from most terminal alkynes with the exception of propargyl alcohols. Later reports showed that at elevated temperature (100°C), the Sonogashira reaction of both 2- and 4-chloropyrim-idines with trimethylsilylacetylene proceeded to give, after basic hydrolysis, the corre-sponding ethynylpyrimidines in 64 and 61% yields, respectively [86].

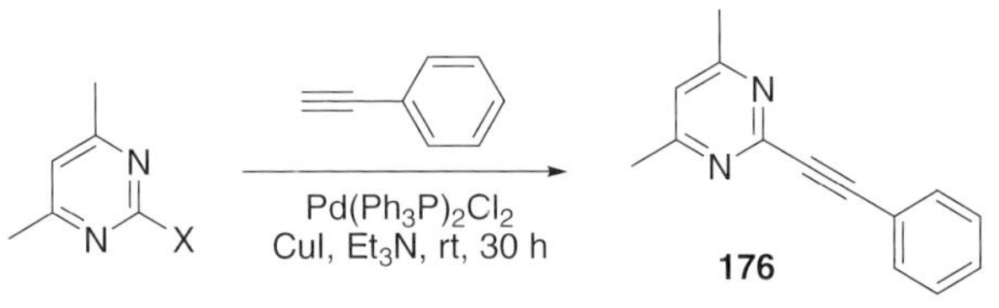

X = Cl, 5%
X = Br, 37%
X = I, 95%

176

Due to its mild reaction conditions and tolerance of many functional groups, the Sonogashira reaction has been utilized extensively in the coupling of halopyrimidines with a variety of terminal alkynes. Halopyrimidine substrates including 2-iodo [87], 4-iodo-[88], 5-bromo- [89], and 5-iodopyrimidines [90] have been successfully coupled with ter-minal alkynes.

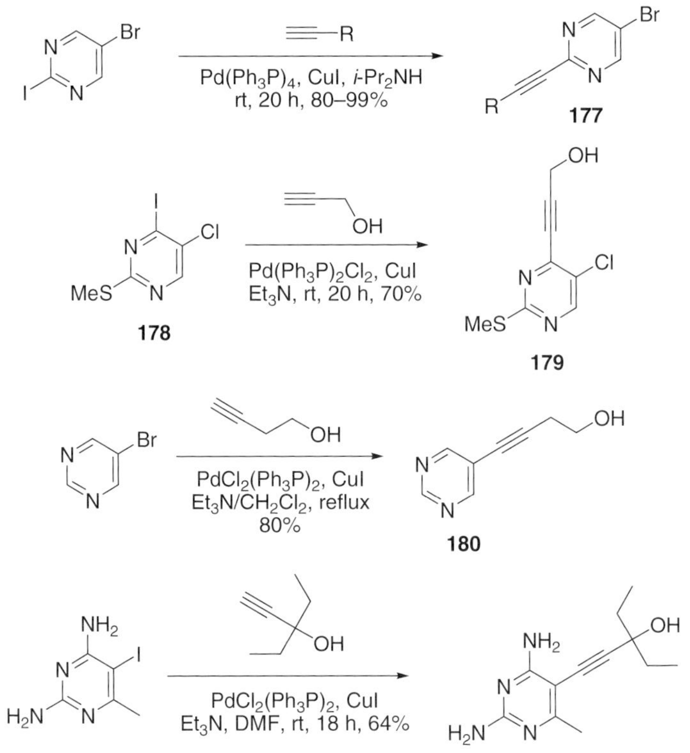

177

178

179

180

181

182

The Sonogashira reaction of 5-bromopyrimidine with *N,N*-dimethylpropargylamine gave aminoalkyne **183** [91], whereas the union of 6-methyl-2-phenyl-4-iodopyrimidine and 3,3,3-triethoxy-1-propyne afforded ester **185** after acidic hydrolysis [92].

Kim and Russell synthesized 5,6-diethynyl-2,4-dimethoxypyrimidine (**189**) starting from iodination of 5-chloro-2,4-dimethoxypyrimidine [93]. Very careful experimentation resulted in optimal conditions for the Sonogashira reaction of dihalopyrimidine **187** with trimethylsilylacetylene to provide bis-alkyne **188**. The temperature appeared to be crucial. Only mono-substitution for the iodine was observed at lower temperature, whereas Bergman cyclization seemed to occur at temperatures higher than 120°C. Subsequent desilylation of **188** then delivered diethynylpyrimidine **189**.

The Sonogashira products of halopyrimidines with pendant functional groups are good precursors for synthesizing condensed heteroaromatic compounds. Yamanaka's group prepared four different condensed heteroaromatics from further manipulations of such Sonogashira products. Those four condensed heteroaromatics are 5-oxo-7-pyrido[4,3-*d*] pyrimidine **192** [94], furo[2,3-*d*]pyrimidine **195** [95, 96], thieno[2,3-*d*]pyrimidine **198** [95, 96], and pyrrolo[2,3-*d*]pyrimidine **202** [97–100].

First, 4-chloropyrimidine **190** was treated with phenylacetylene to give alkynylpyrimidine **191** [94]. The fact that the Sonogashira reaction proceeded readily at room temperature may be ascribed to the electron-withdrawing effect of the neighboring ethoxycarbonyl group on **190**. When heated with ammonia in EtOH, alkynylpyrimidine ester **191** cyclized efficiently to produce pyridyl lactam **192**.

In addition, 2,6-dimethyl-5-iodo-4(3*H*)-pyrimidone (**193**) was allowed to react with phenylacetylene to give internal alkyne **194**, which underwent a spontaneous cyclization to furnish 2,4-dimethyl-6-phenylfuro[2,3-*d*]pyrimidine (**195**) [95, 96].

Dehydroxy-halogenation of **193** using POCl₃ led to dihalopyrimidine **196**, which was subsequently coupled with phenylacetylene to give 4-chloro-5-alkynylpyrimidine **197** [95, 96]. Subsequent treatment of **197** with sodium hydrosulfide in refluxing ethanol gave 2,4-dimethyl-6-phenylthieno[2,3-*d*]pyrimidine (**198**).

Yamanaka and associates developed a method for the synthesis of 2-butylindole from the Sonogashira adduct of ethyl 2-bromophenylcarbamate and 1-hexyne [97, 98]. Extension of that method to pyridines led to the synthesis of pyrrolopyridines [99]. However, the method was not applicable to the synthesis of pyrrolo[2,3-*d*]pyrimidines. They then developed an alternative route involving an initial S$_N$Ar displacement at the 4-position of 4,5-dihalopyrimidine followed by a Sonogashira coupling at the 5-position [100]. Thus, 5-iodopyrimidine **200** was obtained from an S$_N$Ar displacement at the 4-position of a 4-chloro-5-iodo-2-methylthiopyrimidine (**199**). The subsequent Sonogashira reaction of **200** with trimethylacetylene at 80°C resulted in adduct **201**, which spontaneously cyclized to pyrrolo[2,3-*d*]pyrimidine **202**.

201   **202**

Several Sonogashira adducts of heteroaromatics including some pyridines (see Section 4.3) and pyrimidines underwent an unexpected isomerization [101]. This observed isomerization appeared to be idiosyncratic, and substrate-dependent. The normal Sonogashira adduct **204** was obtained when 2-methylthio-5-iodo-6-methylpyrimidine (**203**) was reacted with but-3-yn-ol, whereas chalcone **205**, derived from isomerization of the normal Sonogashira adduct, was the major product when the reaction was carried out with 1-phenylprop-2-yn-1-ol.

203   **204**

203   **205**

The preparation of oxygen heterocycles that contain a pendant pyrimidine ring has been accomplished *via* one-pot or through process Sonogashira coupling followed by cyclization to prepare furopyridines and trisubstituted furans. Sonogashira coupling of 5-bromopyrimidine with 2-ethynyl-3-pyridinol **206** affords the desired furopyridine **207** in 45% yield [102]. Likewise, Sonogashira coupling of 5-bromopyrimidine with 2-propyne-1,3-diketone **208** provides the trisubstitute furan in 98% isolated yield [103].

206   **207**

208   **209**

Achieving a one-pot double cross-coupling reaction on a dihaloaromatic substrate can often be difficult due to the deactivating nature of the first cross-coupled substituent.

The electronic nature of pyrimidines, however, provides a more favorable basis for accomplishing a double cross-coupling reaction. A good example is the double Sonogashira reaction of 2,4-dibromo-6-n-pentylpyrimidine with propyne **210** [104]. The desire double Sonogashira product **211** was obtained in quantitative yield.

The presence of different halides on dihalopyrimidines provides the opportunity to achieve a double Sonogashira cross coupling with different terminal alkynes. This has been demonstrated sequentially as shown with 6-chloro-5-iodo-2,4-dimethoxy pyrimidine **212**. In this case, the first coupling was achieved in 85% yield at the 5-iodocarbon with a TIPS protected acetylene **213**, and the second was achieved at the 6-chlorocarbon with 5-hexyn-1-ol in 65% yield, including removal of the TIPS group [105]. A one-pot double Sonogashira reaction with different alkynes was efficiently demonstrated on 5-bromo-2-iodopyrimidine. In this case, a substituted phenyl acetylene was coupled first followed by a TMS protected acetylene. The overall one-pot yield, including removal of the TMS group to give **215** was 74% [106].

Heteroannulation of alkynes is an efficient method for the preparation of complex heterocycles. Preparation of the preannulation substrate is often accomplished *via* Sonogashira coupling. Indeed, this protocol has been utilized effectively for the preparation of quinazolinones [107] and benzothioazoles [108]. The reactions have been carried out on substituted iodopyrimidines to provide the desired complex heterocycles.

11.4. Heck reaction

Analogous to simple carbocyclic aryl halides, 5-halopyrimidines readily take part in Pd-catalyzed olefinations under standard Heck conditions. In a simple case, Yamanaka *et al.* synthesized ethyl 2,4-dimethyl-5-pyrimidineacrylate (**223**) *via* the Heck reaction of 5-iodo-2,4-dimethylpyrimidine and ethyl acrylate [109].

The Heck reaction of 2- or 4(6)-halopyrimidines was less straightforward. Initial attempts with the Heck reaction of 4-iodopyrimidines without a substituent at the 5-position were plagued by homocoupling, giving rise to bis-pyrimidine **224** as the major product [109]. Yamanaka and colleagues later discovered that the homocoupling could be eliminated if the reaction was carried out *in the absence of* Ph_3P [110–112]. The same was true for the elimination of Ph_3P from the coupling reaction of 2-iodopyrimidine. Many palladium catalysts such as Pd/C, were found to be effective for these olefinations. If there was a substituent such as I, Br, Cl, EtO, Et on the 5-position of the 4-iodopyrim-idines, Ph_3P was still effective as the ligand [113]. For instance, selective olefination of 2-isopropyl-4-iodo-5-bromo-6-methylpyrimidine (**227**) with styrene was achieved to furnish styrylpyrimidine **228**.

Studies of the Heck reaction of substituted halopyrimidines with methyl vinyl ketone revealed an interesting observation [114–116]: although the reactions of bromopyrimidines gave the usual olefinic substituted product, the reaction of 5-iodopyrimidine **229** with methyl vinyl ketone led to the *addition* of pyrimidine to the double bond, giving rise to substituted 5-(3-oxobutyl)-pyrimidine **230**. Two plausible pathways involve either a radical or an anion intermediate. Both possible mechanisms start with insertion of the initially formed σ-complex **231** to methyl vinyl ketone, leading to intermediate **232**. In pathway a, **232** undergoes a homolytic fission to afford **233**, which abstracts a hydrogen from either triethylamine or excess methyl vinyl ketone, giving rise to **230**. In pathway b, a heterolytic fission occurs, leading to carbanion **234**, which is then protonated by the tertiary ammonium salt generated in the catalytic cycle to give **230**.

Walker and associates described a heteroaryl Heck reaction of 2,4-dimethoxy-5-iodopyrimidine with thiophene [117]. They found that it was advantageous to carry out the thienylation in the presence of water as opposed to anhydrous conditions. Thienylation of less reactive 2,4-dimethoxy-5-bromopyrimidine gave the product in a lower yield (38%).

Researchers at Schering Plough were interested in preparing a variety substituted Azetidinones for testing as Cholesterol Absorption Inhibitors. In that sense, the Heck reaction is known to be an effective reaction that provides a useful diversity element and was employed as such. Heck reaction of 5-bromopyridine with chiral the terminal olefin provided the desired product [118].

11.5. The carbonylation reaction

Palladium-catalyzed alkoxycarbonylation enables synthesis of a variety of heterocyclic esters that are otherwise not easily prepared. 5-Bromopyrimidine was transformed into 5-ethoxycarbonylpyrimidine in quantitative yield employing the Pd-catalyzed alkoxycarbonylation. The alkoxycarbonylation of 2-chloro-4,6-dimethoxypyrimidine, in turn, led to benzyl 4,6-dimethoxypyrimidine-2-carboxylate (**240**), whereas alkoxycarbonylation of 2-(chloromethyl)-4,6-dimethoxypyrimidine provided pyrimidinyl-2-acetate **242** [119]. 4,6-Dimethoxypyrimidines **240** and **242** are both important intermediates for the preparation of antihypertensive and antithrombotic drugs.

241 → CO (15 bar), EtOH, Pd(OAc)₂ / dppf, CH₃CO₂Na, 145 °C, 2 h, 82% → **242**

11.6. Heteroannulation

To make tryptophan analogs, Gronowitz and coworkers conducted a pyrrole annulation from an aminoiodopyrimidine utilizing the Larock indole synthesis conditions (see Section 1.10) [120]. They prepared heterocondensed pyrrole **244** by treating 4-amino-5-iodopyrimidine **243** with trimethylsilyl propargyl alcohol under the influence of a palladium catalyst. The regiochemical outcome was governed by steric effects.

243 → HO—≡—SiMe₃ / Pd(OAc)₂, Ph₃P, Et₃N, n-Bu₄NCl / DMF, 100 °C, 19 h, 51% → **244**

In conclusion, the palladium chemistry of pyrimidines has its own characteristics when compared to carbocyclic arenes and other nitrogen-containing heterocycles such as pyridine and imidazole. One salient feature of halopyrimidines is that the C(4) and C(6) positions are more activated than C(2). As a result, 2-, 4-, and 6-chloropyrimidines are viable substrates for Pd-catalyzed reactions and 4- and 6-chloropyrimidines react more readily than 2-chloropyrimidines. For the Sonogashira reaction, though, there is little difference in the reactivity among 2-, 4-, and 5-positions of substituted halopyrimidines. Not only is the Sonogashira reaction a reliable method to make alkynylpyrimidines, the Sonogashira adducts are also good substrates for synthesizing condensed heteroaromatics, as is the Pd-catalyzed heteroannulation strategy.

In choosing a particular cross-coupling reaction, the Stille coupling is the method of choice for two reasons. First, the Stille coupling conditions tolerate a wide variety of functional groups. Second, many pyrimidinylstannanes can be made using sodium or lithium stannanes without going through the intermediacy of an organolithium or organomagnesium species. On the other hand, although both Negishi and Suzuki reactions are less common than the Stille reaction, alkyl groups can be transferred from either alkylzinc or alkylboron reagents under somewhat forcing conditions, giving rise to alkylpyrimidines.

With regard to the Heck reaction, 5-halopyrimidines are the best substrates. The Heck reactions of both 4-halopyrimidines and 2-halopyrimidines are difficult unless a 5-substituent is present. Nonetheless, the Pd-catalyzed alkenylation of both 2- and 4-halopyrimidines proceeds smoothly in the absence of Ph₃P.

11.7. References

1. Undheim, K.; Benneche, T. *Heterocycles* **1990**, *30*, 1155–93.
2. Undheim, K.; Benneche, T. in *Adv. Heterocycl. Chem.* **1995**, *62*, 305–418.
3. Kalinin, K. N. *Synthesis* **1991**, 413–32.
4. Pews, R. G. *Heterocycles* **1990**, *31*, 109–14.
5. Sato, N.; Takeuchi, R. *Synthesis* **1990**, 659–60.
6. Falck-Pedersoen, M. L.; Benneche, T.; Undheim, K. *Acta Chem. Scand., Sec. B* **1989**, *B43*, 251–8.
7. Caton, P. L.; Grant, M. S.; Pain, D. L.; Slack, R. *J. Chem. Soc.* **1965**, 5467–73.
8. Jones, M. L.; Baccanari, D. P.; Tansik, R. L.; Boytos, C. M.; Rudolph, S. K.; Kuyper, L. F. *J. Heterocycl. Chem.* **1999**, *36*, 145–8.
9. Bhatt, R. S.; Kundu, N. G.; Chwang, T. L.; Heidelberger, C. *J. Heterocycl. Chem.* **1981**, *18*, 771–4.
10. Edo, K.; Sakamoto, T.; Yamanaka, H. *Heterocycles* **1979**, *12*, 383–6.
11. Solberg, J.; Undheim, K. *Acta Chem. Scand., Sec. B* **1986**, *B40*, 381–6.
12. Sugimoto, O.; Mori, M.; Tanji, K.-i. *Tetrahedron Lett.* **1999**, *40*, 7477–8.
13. Sandosham, J.; Undheim, K.; Rise, F. *Heterocycles* **1993**, *35*, 235–44.
14. Sandosham, J.; Undheim, K. *Heterocycles* **1994**, *37*, 501–14.
15. D'Alarcao, M.; Bakthavachalam, V.; Leonard, N. J. *J. Org. Chem.* **1985**, *50*, 2456–61.
16. Yamanaka, H.; An-Naka, M.; Kondo, Y.; Sakamoto, T. *Chem. Pharm. Bull.* **1985**, *33*, 4309–13.
17. Billotte, S. *Synlett*, **1998**, 379–80.
18. Balle, T.; Andersen, K.; Vedso, P. *Synthesis* **2002**, 1509–12.
19. Bellina, F.; Biagetti, M.; Carpita, A.; Rossi, R. *Tet. Lett.* **2001**, *42*, 2859–63.
20. Togni, A.; Hayashi, T. *Ferrocenes: Homogeneous Catalysis, Organic Synthesis, Material Science*, VCH: Weinheim, 1995.
21. Horikoshi, R.; Nambu, C.; Mochida, T. *Inorg. Chem.* **2003**, *42*, 6868–75.
22. Kloetzing, R. J.; Lotz, M.; Knochel, P. *Tetrahedron Asymm.* **2003**, *12*, 255–64.
23. Turck, A.; Plé, N.; Leprétre-Gaquére, A.; Quéguiner, G. *Heterocycles* **1998**, *49*, 205–14.
24. Angiolelli, M. E.; Casalnuovo, A. L.; Selby, T. P. *Synlett* **2000**, 905–7.
25. Sandosham, J.; Undheim, K. *Heterocycles* **1993**, *35*, 235–44.
26. Sandosham, J.; Undheim, K. *Heterocycles* **1994**, *37*, 501–14.
27. Rodriguez, D.; Castedo, L.; Saa, C. *Synlett* **2004**, *5*, 783–6.
28. Gronowitz, S.; Hörnfeldt, A.-B.; Musil, T. *Chem. Scr.* **1986**, *26*, 305–9.
29. Peters, D.; Hörnfeldt, A.-B.; Gronowitz, S. *J. Heterocycl. Chem.* **1990**, *27*, 2165–73.
30. Wellmar, U.; Hörnfeldt, A.-B.; Gronowitz, S. *J. Heterocycl. Chem.* **1995**, *32*, 1159–63.
31. Saygili, N.; Batsanov, A. S.; Bryce, M. R. *Org. Biomol. Chem.* **2004**, *2*, 852–7.
32. Ishikura, M.; Kamada, M.; Terashima, M. *Synthesis* **1984**, 936–8.
33. Ishikura, M.; Ohta, T.; Terashima, M. *Chem. Pharm. Bull.* **1985**, *33*, 4755–63.
34. Hughes, G.; Wang, C.; Batsanov, A. S.; Fern, M.; Frank, S.; Bryce, M. R.; Perepichka, I. F.; Monkman, A. P.; Lyons, B. P. *Org. Biomol. Chem.* **2003**, *1*, 3069–77.
35. Yang, Y.; Martin, A. R. *Heterocycles* **1992**, *34*, 1395–8.
36. Mitchell, M. B.; Walbank, P. J. *Tetrahedron Lett.* **1991**, *32*, 2273–6.

37. Ali, N. M.; McKillop, A.; Mitchell, M. B.; Rebelo, R. A.; Walbank, P. J. *Tetrahedron* **1992**, *48*, 8117–26.

38. Wong, K.-T.; Hung, T. S.; Lin, Y.; Wi, C.-C.; Lee, G.-H.; Peng, S.-M.; Chio, C. H.; Su, Y. O. *Org. Lett.* **2002**, *4*, 513–6.

39. Talley, J. J.; Brown, J. D.; Carter, J. S.; Graneto, M. J.; Koboldt, C. M.; Masferrer, J. L.; Perkins, W. E.; Rogers, R. S.; Shaffer, A. F.; Zhang, Y. Y.; Zweifel, B. S.; Seibert, K. *J. Med. Chem.* **2000**, *43*, 775–7.

40. Dileep Kumar, J. S.; Ho, M. M.; Leung, J. M.; Toyokuni, T. *Adv. Synth. Catal.* **2002**, *344*, 1146–51.

41. Cooke, G.; de Cremiers, H. A.; Rotello, V. M.; Tarbit, B.; Vanderstraeten, P. E. *Tetrahedron* **2001**, *57*, 2787–89.

42. Gong, Y.; Pauls, H. W. *Synlett* **2000**, 829–31.

43. Satoh, Y.; Gude, C.; Chan, K.; Firooznia, F. *Tetrahedron Lett.* **1997**, *38*, 7645–8.

44. Wade, J. V.; Krueger, C. A. *J. Comb. Chem.* **2003**, *5*, 267–72.

45. Iglesias, B.; Alveraz, R.; de Lera, A. R. *Tetrahedron* **2000**, *57*, 3125–30.

46. Molander, G.; Biolatto, B. *J. Org. Chem.* **2003**, *68*, 4302–14.

47. Molander, G. A.; Bernardi, C. R. *J. Org. Chem.* **2002**, *67*, 8424–9.

48. Molander, G. A.; Katona, B. W.; Machrouhi, F. *J. Org. Chem.* **2002**, *67*, 8416–23.

49. Liebeskind, L. S.; Srogl, J. *Organic Lett.* **2002**, *4*, 979–81.

50. Sandosham, J.; Benneche, T.; Moeller, B. S.; Undheim, K. *Acta Chem. Scand., Ser. B* **1988**, *B42*, 455–61.

51. Arukwe, J.; Benneche, T.; Undheim, K. *J. Chem. Soc., Perkin Trans. 1* **1989**, 255–9.

52. Sandosham, J.; Undheim, K. *Tetrahedron* **1994**, *50*, 275–84.

53. Sandosham, J.; Undheim, K. *Acta Chem. Scand.* **1989**, *43*, 684–9.

54. Majeed, A. J.; Antonsen, O.; Benneche, T.; Undheim, K. *Tetrahedron* **1989**, *45*, 993–1006.

55. Kennedy, G.; Perboni, A. D. *Tetrahedron Lett.* **1996**, *37*, 7611–14.

56. Kulesza, A.; Ebetino, F. H.; Mishra, R. K.; Cross-Doersen, D.; Mazur, A. W. *Organic Lett.* **2003**, *5*, 1163–6.

57. Gohlke, H.; Gundisch, D.; Schwarz, S.; Seitzh, G.; Tilotta, M. C.; Wegge, T *J. Med. Chem.* **2002**, *45*, 1064–73.

58. Ahaider, A.; Fernandez, D.; Danelon, G.; Cuevas, C.; Manzanares, I.; Albericio, F.; Joule, J. A.; Alvarez, M. *J. Org. Chem.* **2003**, *68*, 10020–29.

59. Nicolaou, K. C.; Sasmal, P. K.; Rassias, G.; Reddy, M. V.; Altmann, K.-H.; Wartmann, M.; O'Brate, A.; Giannakakou, P. *Angew. Chem. Int. Ed* **2003**, *42*, 3515–20.

60. Wittenberg, R.; Srogl, J.; Masahiro, E.; Liebeskind, L. S. *Organic Lett.* **2003**, *5*, 3033–5.

61. Solberg, J.; Undheim, K. *Acta Chem. Scand.* **1987**, *B41*, 712–6.

62. Sandosham, J.; Undheim, K. *Acta Chem. Scand.* **1989**, *43*, 62–8.

63. Benneche, T. *Acta Chem. Scand.* **1990**, *44*, 927–31.

64. Liebeskind, L. S.; Riesinger, S. W. *J. Org. Chem.* **1993**, *58*, 408–13.

65. Verlhac, J.-B.; Pereyre, M.; Shin, H. *Organometallics*, **1991**, *10*, 3007–9.

66. Sakamoto, T.; Satoh, C.; Kondo, Y.; Yamanaka, H. *Chem. Pharm. Bull.* **1993**, *41*, 81–6.

67. Sandosham, J.; Undheim, K. *Acta Chem. Scand.* **1994**, *48*, 279–82.

68. Alvarez, M.; Fernández, D.; Joule, J. A. *Synthesis* **1999**, 615–20.

69. Gros, P.; Fort, Y. *Synthesis* **1999**, 754–6.

70. Schubert, U. S.; Eschbaumer, C. *Org. Lett.* **1999**, *1*, 1027–9.
71. Gronowitz, S.; Hörnfeldt, A.-B.; Musil, T. *Chem. Scr.* **1986**, *26*, 305–9.
72. Peters, D.; Hörnfeldt, A.-B.; Gronowitz, S. *J. Heterocycl. Chem.* **1990**, *27*, 2165–73.
73. Kondo, Y.; Watanabe, R.; Sakamoto, T.; Yamanaka, H. *Chem. Pharm. Bull.* **1989**, *37*, 2933–6.
74. Choppin, S.; Gros, P.; Fort, Y. *Organic Lett.* **2000**, *2*, 803–5.
75. Mathieu, J.; Gros, P.; Fort, Y. *Tetrahedron Lett.* **2001**, *42*, 1879–81.
76. Chen, J.; Crisp, G. T. *Synth. Commun.* **1992**, *22*, 683–6.
77. Harr, M. S.; Presley, A. L.; Thoraresen, A. *Synlett* **1999**, 1579–81.
78. Alphonse, F.-A.; Suzenet, F.; Keromnes, A.; Lebret, B.; Guillaumet, G. *Organic Lett.* **2003**, *5*, 803–5.
79. Liebeskind, L. S.; Masahiro, E. *Organic Lett.* **2003**, *5*, 801–2.
80. Mangalagiu, I.; Benneche, T.; Undheim, K. *Acta Chem. Scand.* **1996**, *50*, 914–7.
81. Lu, Q.; Mangalagiu, I.; Benneche, T.; Undheim, K. *Acta Chem. Scand.* **1997**, *51*, 302–6.
82. Gelman, D.; Tsvcelikhovsky, D.; Molander, G. A.; Blum, J. *J. Org. Chem.* **2002**, *67*, 6287–90.
83. Jaber, N.; Schumann, H.; Blum, J. *J. Heterocyclic Chem.* **2003**, *40*, 565–7.
84. Edo, K.; Yamanaka, H.; Sakamoto, T. *Heterocycles* **1978**, *9*, 271–4.
85. Edo, K.; Sakamoto, T.; Yamanaka, H. *Chem. Pharm. Bull.* **1978**, *26*, 3843–50.
86. Sakamoto, T.; Shirawa, M.; Kondo, Y.; Yamanaka, H. *Synthesis* **1983**, 312–4.
87. Shibata, T.; Yonekubo, S.; Soai, K. *Angew. Chem. Int. Ed.* **1999**, *38*, 659–61.
88. Solberg, J.; Undheim, K. *Acta Chem. Scand.* **1986**, *B40*, 381–6.
89. Tilley, J. W.; Levitan, P.; Lind, J.; Welton, A. F.; Crowley, H. J.; Tobias, L. D.; O'Donnell, M. *J. Med. Chem.* **1987**, *30*, 185–93.
90. Jones, M. L.; Baccanari, D. P.; Tansik, R. L.; Boytos, C. M.; Rudolph, S. K.; Kuyper, L. F. *J. Heterocycl. Chem.* **1999**, *36*, 145–8.
91. Bleicher, L. S.; Cosford, N. D. P.; Herbaut, A.; McCallum, J. S.; McDonald, I. A. *J. Org. Chem.* **1998**, *63*, 1109–18.
92. Sakamoto, T.; Shiga, F.; Yasuhara, A.; Uchiyama, D.; Kondo, Y.; Yamanaka, H. *Synthesis* **1992**, 746–8.
93. Kim, C.-S.; Russell, K. C. *J. Org. Chem.* **1998**, *63*, 8229–34.
94. Sakamoto, T.; Kondo, Y.; Yamanaka, H. *Chem. Pharm. Bull.* **1982**, *30*, 2410–16.
95. Sakamoto, T.; Kondo, Y.; Yamanaka, H. *Chem. Pharm. Bull.* **1982**, *30*, 2417–20.
96. Sakamoto, T.; Kondo, Y.; Watanabe, R.; Yamanaka, H. *Chem. Pharm. Bull.* **1986**, *34*, 2719–24.
97. Sakamoto, T.; Kondo, Y.; Yamanaka, H. *Heterocycles* **1986**, *24*, 31–2.
98. Sakamoto, T.; Kondo, Y.; Iwashita, S.; Yamanaka, H. *Chem. Pharm. Bull.* **1987**, *35*, 1823–8.
99. Sakamoto, T.; Kondo, Y.; Iwashita, S.; Nagano, T.; Yamanaka, H. *Chem. Pharm. Bull.* **1988**, *36*, 1305–8.
100. Kondo, Y.; Watanabe, R.; Sakamoto, T.; Yamanaka, H. *Chem. Pharm. Bull.* **1989**, *37*, 2933–6.
101. Minn, K. *Synlett* **1991**, 115–6.
102. Arcadi, A.; Cacchi, S.; Di Giuseppe, S.; Fabrizi, G.; Marinelli, F. *Synlett* **2002**, *3*, 454–7.

103. Arcadi, A.; Cacchi, S.; Fabrizi, G.; Marinelli, F.; Parisi, L. M. *Tetrahedron* **2003**, *59*, 4661–71.

104. Kim, J. T.; Gevorgyan, V. *Organic Lett.* **2002**, *4*, 4697–99.

105. Choy, N.; Blanco, B.; Wen, J.; Krishan, A.; Russel, K. C. *Organic Lett.* **2000**, *2*, 3761–4.

106. Wong, K.-T.; Che Hsu, C. *Organic Lett.* **2001**, *3*, 173–5.

107. Kundu, N. G.; Chaudhuri, G. *Tetrahedron* **2001**, *57*, 6833–42.

108. Kundu, N. G.; Nandi, B. *J. Org. Chem.* **2001**, *66*, 4563–75.

109. Edo, K.; Sakamoto, T.; Yamanaka, H. *Chem. Pharm. Bull.* **1979**, *27*, 193–7.

110. Sakamoto, T.; Arakida, H.; Edo, K.; Yamanaka, H. *Heterocycles* **1981**, *16*, 965–8.

111. Sakamoto, T.; Kondo, Y.; Yamanaka, H. *Chem. Pharm. Bull.* **1982**, *30*, 2417–20.

112. Sakamoto, T.; Arakida, H.; Edo, K.; Yamanaka, H. *Chem. Pharm. Bull.* **1982**, *30*, 3647–56.

113. Wada, A.; Yamamoto, J.; Hase, T.; Nagai, S.; Kanatomo, S. *Synthesis* **1986**, 555–6.

114. Wada, A.; Yasuda, H.; Kanatomo, S. *Synthesis* **1988**, 771–5.

115. Wada, A.; Ohki, K.; Nagai, S.; Kanatomo, S. *J. Heterocycl. Chem.* **1991**, *28*, 509–12.

116. Basnak, I.; Takatori, S.; Walker, R. T. *Tetrahedron Lett.* **1997**, *38*, 4869–72.

117. Head, R. A.; Ibbotson, A. *Tetrahedron Lett.* **1984**, *25*, 5939–42.

118. Bessard, Y.; Crettaz, R. *Tetrahedron* **1999**, *55*, 405–12.

119. Rosenblum, S. B.; Huynh, T.; Afonso, A.; Davies Jr. H. R. *Tetrahedron* **2000**, *56*, 5735–42.

120. Wensbo, D.; Eriksson, A.; Jeschke, T.; Annby, U.; Gronowitz, S. *Tetrahedron Lett.* **1993**, *34*, 2823–6.

Chapter 12

Quinolines

Nadia M. Ahmad

Quinolines play an indispensable role in medicinal chemistry. Quinine (**1**) is one of the oldest medicines used to fight malaria, whereas one of the latest quinoline-containing drugs is montelukast (**2**, Singulair®), an anti-asthmatic drug. Thus quinoline chemistry has always attracted the attention of medicinal chemists. The unique characteristics of quinoline chemistry, stem from the stereoelectronic effects that the nitrogen atom has exerted on the quinoline molecule. Quinoline (**3**) is a π-electron-deficient heterocycle. Due to the electronegativity of the nitrogen atom, the α and γ positions bear a partial positive charge, making these C(2) and C(4) positions prone to nucleophilic attacks. A similar trend occurs in the context of palladium chemistry. Halogens at the α and γ positions of quinoline are more susceptible to oxidative addition to Pd(0) in comparison to simple carbocyclic aryl halides. As a consequence, even 2-chloro- and 4-chloro-quinolines undergo palladium-catalyzed reactions under standard conditions, a phenomenon not frequently observed in carbocyclic chloroaryl compounds.

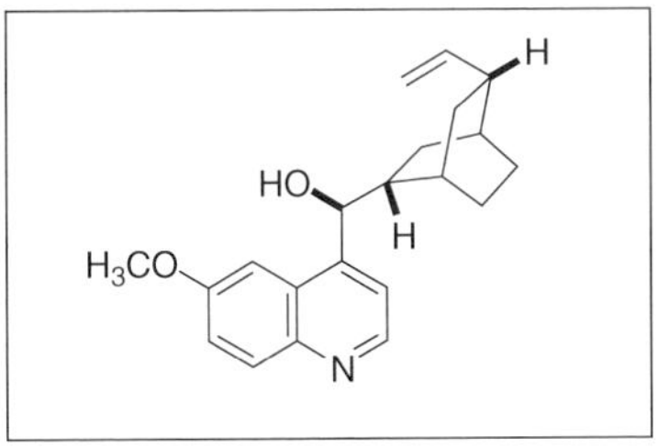

1, quinine

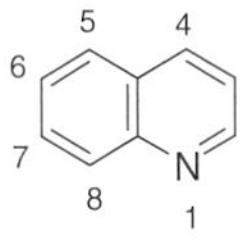

3, quinoline

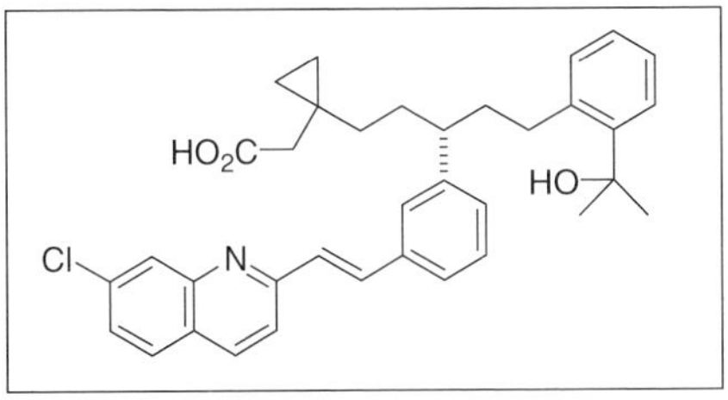

2, montelukast

12.1. Synthesis of quinoline electrophiles

12.1.1. Halogenation of quinolones

A popular method to prepare haloquinolines is the halogenation of quinolones using oxyphosphorus halides, most notably POCl₃. The carbonyl can be located either at the C(2) or the C(4) positions. As depicted in Scheme 1, the C(2) position of quinolone **4** was chlorinated with POCl₃ to give 6-bromo-2-chloroquinoline (**5**) [1]. The subsequent S$_N$Ar displacement of the chlorine substituent on **5** with sodium methoxide led to 6-bromo-2-methoxyquinoline. Analogously, chlorination at the C(4) position of quinolones is exemplified by transformations **6→7** [2] and **8→9** [3].

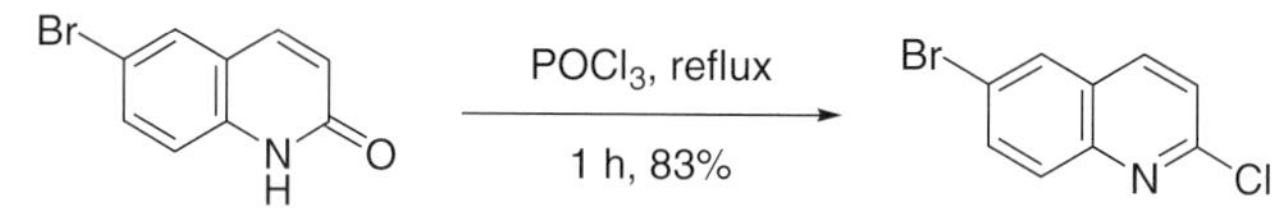

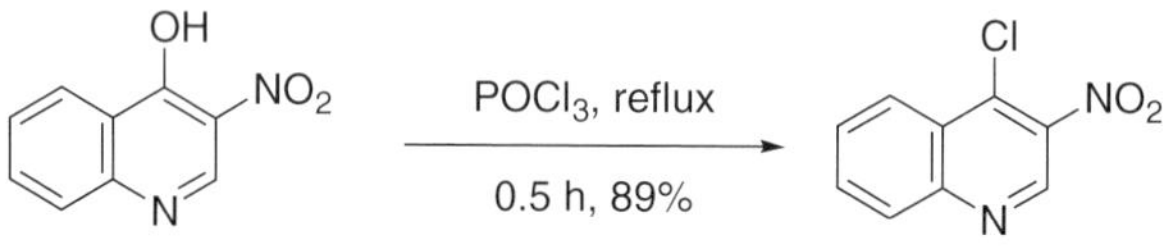

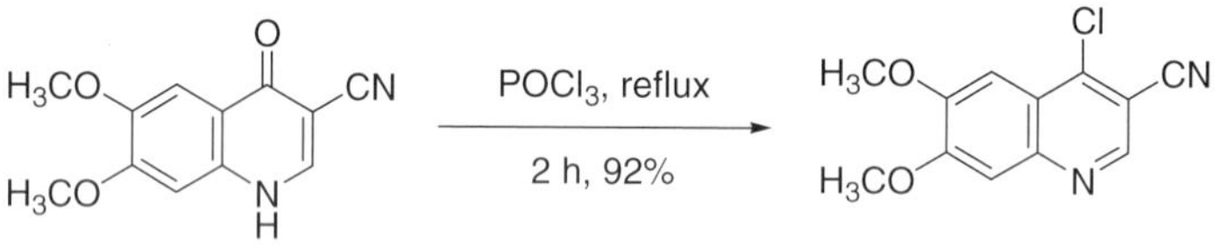

The halogenation can be carried out under more mild conditions. Sugimoto *et al.* [4] reported halogenation of hydroxyheterocycles using *N*-halosuccinimide and triphenylphosphine. Application of the method to 2(1*H*)-quinolone **10** gave 2-bromoquinoline **11** in 90% yield. This method has several advantages over standard halogenating reagents such as phosphorus oxyhalides. It obviates the use of caustic and expensive phosphorus oxybromide and the resulting haloheterocycles are easily isolated via SiO₂ chromatography. It was also found that in order to carry out the halogenation in good yield, 2–5 equivalents of the halogenating reagent were required. Using just one equivalent of triphenylphosphine and N-bromosuccinimide resulted in only 39% yield of 2-bromoquinoline **11**.

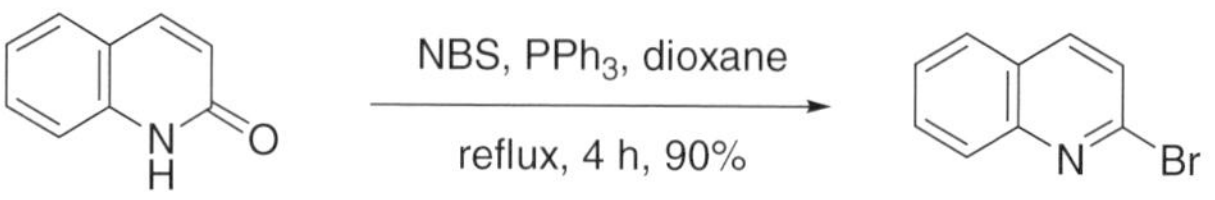

2-Methoxy- and 4-methoxy-quinolines may behave similar to quinolones during halogenation, in some cases they can be converted into the corresponding haloquinolines. For instance, 2- and 4-methoxyquinolines were converted to bromo compounds by Yajima *et al.* [5] using a novel bromination reagent PBr₃–DMF. The reaction proceeded efficiently at 60–80°C with no evolution of HBr which was a disadvantage with using phosphorus oxybromide by making the reaction solution strongly acidic. 2-Methoxyquinoline (**12**) furnished 2-bromoquinoline (**11**) in 78% yield, whereas 4-methoxyquinoline produced 4-bromoquinoline in 68% yield.

12.1.2. Direct halogenation of quinolines

Since the pyridine ring is electron-deficient, direct halogenation of quinolines rarely takes place on the pyridine motif of quinoline unless it is substituted with strong electron-donating groups. Therefore, direct halogenation of quinolines generally occurs at the benzene moiety. Commonly used halogenation reagents include NBS, NCS, and Br₂ etc. For example, treatment of quinolinol **13** with NBS yielded dibromo-quinolinol **14**, along with two minor quinone by-products as a consequence of oxidation by NBS [6]. Bromine, a powerful bromination agent, has found many utilities in preparing bromo-quinolines. It has been known in the literature that substitution on 6-halogenated quinolines gives only 5- or 5,8-substituted compounds [7]. This was applied to the bromination of 6-chloroquinoline **15** which gave only 5-bromo-6-chloroquinoline **16** [8]. Similarly, 6-fluoro-2-methylquinoline **17** was brominated using Br₂ with aluminum chloride as the catalyst and dichloroethane (DCE) as the solvent [9]. The major product was the 5-bromo-6-fluoro-2-methylquinoline **18**, (80% yield) with concurrent formation of a small amount (10%) of the 5,8-dibromo-6-fluoro-2-methylquinoline **19**.

Bromination of 5-hydroxyquinoline (20) normally gives both mono- and dibromides [10]. Under basic conditions, the reaction was not regioselective, producing a mixture of mono- and dibromide, along with recovered starting material. However, under acidic conditions the reaction proceeded regioselectively, affording exclusive formation of 6, 8-dibromo derivative 21 with three equivalents of bromine in acetic acid. The product is unstable but can be isolated by treatment of the reaction mixture with aqueous sodium acetate. Finally, photolysis of 5-azidoquinoline (22) in hydrobromic acid resulted in the formation of both 6-bromo-5-aminoquinoline (23) and 8-bromo-5-aminoquinoline (24) in a 1:1 ratio [11]. The conversion may involve interesting intermediates such as an azirine and/or azacycloheptatetraene.

12.1.3. S$_N$Ar reaction

Despite being activated by the nitrogen atom, 2-chloroquinoline 25 is still a poor substrate for the Stille cross-coupling reactions, though yields are usually improved under Negishi conditions. For instance, the coupling of 2-chloroquinoline 25 and 2-trimethyl-stannyl-6-methylpyridine under standard Pd^0 or Ni^0 catalytic conditions [12] was found to give poor overall yields for the desired coupled product. Thus, via the S$_N$Ar displacement mechanism, 2-chloroquinoline 25 was converted to the more reactive iodo-derivative 26 in excellent yield using sodium iodide in acetyl chloride [13]. Thus the produced 2-iodoquinoline 26 was coupled efficiently with an array of stannanes to give the biaryl products in good yields (see Section 12.3.3 Stille coupling). Likewise, 2-chloro-6-methoxyquinoline was converted into 2-iodo-6-methoxyquinoline in 98% yield.

Nucleophilic aromatic substitution of 4-chloroquinolines is known to proceed in the absence of transition metal catalysts [14, 15]. Thus, electron-deficient 2-substituted-4-chloroquinolines underwent nucleophilic aromatic substitution by amines and thiols under elevated temperature [16]. It was noted, however, that when the palladium-phosphinous acid catalyst POPd (see Section 12.5 Heck reaction) was utilized in the presence of a base (Cy_2NMe or t-BuOK) under the same reaction conditions the yields were improved in most cases. It is thought the POPd-catalyzed amination and thiation of 4-chloroquinoline **27** proceeds via oxidative addition followed by base-promoted nucleophilic displacement of chloride and subsequent reductive elimination.

12.1.4. Halogen-dance reaction

The halogen-dance reaction is a good tactic for moving the halogen substituent around the pyridine ring of quinolines. QuÈguiner *et al.* [17] reported the first halogen-dance reaction in the quinoline series. Thus, treatment of the iodoquinoline **30** with LDA under the standard conditions followed by quenching with iodine led to the corresponding 3-fluoro-iodoquinoline **31** in 90% yield. The mechanism of halogen-dance is proposed to involve an intermolecular crossover deprotonation directed by the fluorine atom.

12.1.5. Vilsmeier–Haack reaction

Meth-Cohn *et al.* [18] demonstrated that treatment of acetanilides with the Vilsmeier–Haack reagent ($POCl_3$/DMF) provided chloroquinoline derivatives. As such, treatment of 2,5-dimethoxyacetanilide **32** with phosphorus oxychloride in DMF led to 5,8-dimethoxy-2-chloroquinoline-3-carbaldehyde **34** in 50% yield [19]. The reaction was presumed to go through a double Vilsmeier–Haack reaction via the putative intermediate **33**, which subsequently cyclized to give **34**. In the same fashion, Korodi and coworkers [20] applied the Vilsmeier–Haack strategy developed by the Meth-Cohn group [21, 22] in the synthesis of differently substituted 2-chloroquinolines. Thus, anilide **35** was converted to 2-chloroquinoline **37** via enamine intermediate **36**.

An interesting application of the Vilsmeier–Haack reaction to 3-phenylisoxazol-5(4*H*)-one **38** using POCl₃/DMF resulted in 2,4-dichloroquinoline-3-carbaldehyde **39** through a novel rearrangement [23]. As the 3-position of the quinoline ring is particularly unreactive, i.e. the electrophilic substitution of a formyl group at C(3) is challenging, the aforementioned method presents a useful route for the preparation of such quinolines-3-aldehydes under easily accessible substrates. Oxidation of the carbaldehyde **39** with aqueous potassium permanganate and sodium carbonate at room temperature led to the corresponding carboxylic acid, which underwent a decarboxylation process to afford 2,4-dichloroquinoline **40**.

12.1.6. Miscellaneous syntheses of haloquinolines

Yamamoto and Yanagi [24] have prepared iodoazines through iodo-destannation of trimethylstannylazines. Trimethylstannyl sodium was prepared *in situ* from chlorotrimethyl-stannane and metallic sodium. Subsequent treatment of 2-chloroquinoline **25** with trimethylstannyl sodium gave 2-trimethylstannylquinoline **41**. Likewise, 3- and 4-trimethyl-stannyl quinolines **42** and **44** were converted to 3- and 4-iodoquinolines **43** and **45** in 96 and 91% yield, respectively via iodo-destannation. In the same fashion, 2,4-bis (trimethylstannyl)quinoline was synthesized from 4-bromo-2-chloroquinoline using two equivalents of trimethylstannyl sodium in 65% yield.

In a more "exotic" approach towards 5-haloquinolines, Uchiyama *et al.* [25] prepared 5-bromo-2-methylquinolin-8-ol **47** from *O*-2,4-dinitrophenyloxime **46** in one-pot. They found that the stereochemistry of the oximes did not affect the outcome of the reaction, obviating the separation of the stereochemical isomers of the oxime.

12.2. Synthesis of quinoline nucleophiles

The S_NAr reaction between 2-chloroquinoline **25** and trimethylstannyl sodium led to 2-trimethylstannylquinoline **41** as shown previously. Similarly, 3-trimethylstannyl quinoline **49** and 4-trimethylstannyl quinoline **51** were prepared from 3-chloroquinoline **48** and 4-chloroquinoline **50**, respectively [26]. Furthermore, the method was extended to sodium triphenylstannane, which was treated with 2-chloroquinoline **25** to afford 2-triphenylstannyl quinoline in 96% yield under irradiation [27].

Stille *et al.* [28] demonstrated that Pd-catalyzed reaction between hexaalkylditin and aryl triflates led to aryl trialkylstannanes. This method has proven to be applicable to both aryl triflates and aryl halides. For example, the Stille coupling between quinolinyl-8-triflate **52** and hexamethylditin offered 8-trimethylstannyl quinoline **53** in 67% yield along with a small amount of the homocoupling product.

The Miyaura coupling reaction involving the pinacol ester of diboron **55** offers a convenient and mild preparation of arylboronic esters. Applying the methodology to 4-bromoquinoline **54**, the Miyaura group [29] obtained the corresponding quinolinyl-4-arylboronic ester **56** in 84% yield. Electron-withdrawing groups on the nucleophiles for such reactions increase the rate of reaction and the use of KOAc is essential to accelerate the reaction as well as to prevent the formation of biaryl by-products. The transmetalation step is accelerated by the use of the base.

Furthermore, Masuda *et al.* have shown that dialkoxyhydroborane can replace diboron **55** in converting aryl halides and triflates into arylboronates [30]. In one example, treatment of 2-methoxyquinolinyl-8-triflate with 4,4,5,5-tetramethyl-[1,3,2]dioxaborolane in the presence of PdCl$_2$•(dppf) and Et$_3$N in dioxane at 80°C for 4 h produced 2-methoxy-8,4,5,5-tetramethyl-[1,3,2]dioxaborolyquinoline in 63% yield.

Finally, functionalized quinolines were prepared by generating lithium tri(quinolinyl)magnesates, such as **60**, from 2-, 3-, and 4-bromoquinolines using Bu$_3$MgLi [31]. The resulting organomagnesium derivatives were utilized in metal-catalyzed coupling reactions with heteroaryl halides or quenched with various electrophiles. Good yields can be obtained at moderate temperatures, with scope for greater complexity by using further functionalized aryl magnesates.

	E	
59a	CH(OH)Ph	28%
59b	CO_2H	39%
59c	SPh	47%
44	I	57%

12.3. Cross-coupling reactions with organometallic reagents

Among the cross-coupling reactions with organometallic reagents that involve a quinoline fragment, the Suzuki and Stille coupling reactions are more prevalent, and only limited precedents exist for the Negishi and Hiyama reactions. Although the Stille coupling reaction is more general and the reaction conditions are neutral, it is not recommended as the first choice simply because of the toxicity of stannanes. If all is equal, the Suzuki coupling reaction should be considered first while keeping in mind that the reaction needs a base (though weak bases such as K_2CO_3, Na_2CO_3, and K_3PO_4 are acceptable) to proceed.

12.3.1. Negishi coupling

The quinolinyl motif has been used in the Negishi coupling reactions as both electrophilic and nucleophilic coupling partners. Imidazol-4-ylzinc reagent **62** was easily generated *in situ* by treating 1-trityl-4-iodoimidazole **61** with ethyl Grignard reagent followed by the addition of $ZnCl_2$. The Negishi reaction of **62** with quinolinyl-4-triflate **63** in the presence of LiCl led to quinolinylimidazole **64** in 72% yield after acidic removal of the trityl protection [32]. The Negishi reaction of **62** and quinolinyl-4-bromide also gave quinolinylimidazole **64** in 74% yield. Furthermore, the authors found that the protection of the iodoimidazole could be avoided if two equivalents of $ZnCl_2$ were employed. In another note, the coupling of quinolinyl-4-triflate **63** and quinolinyl-4-bromide with imidazolylstannane also provided adduct **64**, albeit in a slightly lower yield.

In their formal total synthesis of camptothecin **69**, Murata *et al.* employed a Negishi reaction to establish the A, B, and D ring linkage [33]. The halogen–metal exchange of 2-chloropyridine **65** was achieved using lithium naphthalenide complex, which was superior to *n*-BuLi because nucleophilic addition to the substrate was avoided. Transmetalation of the resulting lithiopyridine with $ZnCl_2$ generated pyridylzinc reagent **66**, which was then coupled with methyl 2-chloro-3-quinolinecarboxylate **67** to provide hetero biaryl **68**, an important intermediate for camptothecin **69** synthesis.

Baston *et al.* prepared quinolinylzinc chloride **71** *in situ* from 6-bromo-2-methoxyquino-line **70** via halogen–metal exchange using two equivalents of *t*-butyllithium followed by quenching with ZnCl$_2$ [1]. The subsequent Negishi coupling reaction of **71** with various 4-bromobenzamides **72** resulted in 6-aryl-substituted 2-methoxyquinolines **73** in 52–83% yield. The R group functionalities were isopropyl, isobutyl and cyclohexyl, and 6-arylquinolines **73** were utilized to synthesize potent inhibitors of steroid 5α reductases types 1 and 2.

12.3.2. Suzuki coupling

Quinoline motifs can be installed from both quinolinyl halides and quinolinyl boranes in the Suzuki coupling reaction. However, quinolinyl halides as the electrophiles are more prevalent because they are readily available. Homogeneous catalysts present a prob-lem of separating the catalysts from the reaction mixture. This shortcoming can be over-come by using polymer-bound palladium catalysts. Further advantages of such a catalyst are the prevention of the contamination of the products by the phosphine ligand and clean and simple work-ups. Fenger and Drian investigated different polymer-supported palla-dium catalysts as an alternative to tetrakis(triphenylphosphine)palladium catalyst in the

Suzuki coupling reaction of phenylboronic acids such as **74** with 4-bromopyridine [34]. Inada and Miyuara [35] have extended the method to 2-chloroquinoline **25**. Therefore, the coupling between **25** and phenylboronic acid **74** led to 2-tolylquinoline **75** in 91% yield. The catalyst was recovered with ease and used in further coupling reactions. Not surprisingly, the couplings of phenylboronic acids with electron-rich chloroarenes were ineffective due to their slow oxidative addition to the palladium(0) complex. This reaction is an example where even quinolinyl chloride is a good substrate for the oxidative addition of Pd(0) if the chlorine atom is at the activated position (α or δ).

Under the phase-transfer catalysis conditions, 2-bromo-8-methylquinoline **76** was coupled with 2-pyridylboronic ester **77** to furnish 2-(2-pyridyl)-8-methylquinoline **78** in 56% yield [36]. At this point, it is opportune to mention that the simple 2-pyridylborane, in contrast to 3- and (4-pyridyl)boranes, is considered an unsuitable Suzuki coupling partner because it forms an unusually stable cyclic dimer resembling a dihydroanthracene. In this case, the obstacle was circumvented by using 2-pyridylboronic ester in place of 2-pyridylborane.

3-Bromoquinolines behave in the Suzuki reaction similarly to simple carbocyclic aryl bromides and the reaction is straightforward. Examples include 3-(3-pyridyl)-quinoline **80** from 3-bromoquinoline **58** and 3-pyridylboronic acid **79** [36]; and 3-phenyl-quinoline **83** from substituted 3,7-dibromoquinoline **81** and (2-pivaloylaminophenyl)boronic acid **82** [37]. Notice that the combination of potassium carbonate and ethanol resulted in debromination at the C(7) position (but the combination of sodium carbonate and methanol left the C(7) bromide intact). It was speculated that the debromination arose from a second palladium insertion, followed by hydrolysis of the intermediate. Further manipulations of biaryl **83** delivered an interesting 11H-indolo[3,2-c]quinoline **84**. The result is particularly intriguing since the cyclization of the azide is regioselective which is not usually the case in the pyridine series [38].

In one case, an unpurified sample of pyrrolyl-bisboronic ester **86**, synthesized using conditions described by Masuda *et al.* [30], was readily engaged in a Suzuki cross-coupling reaction [29] with 5-bromoquinoline **85**, giving the desilylated product **87** in 29% overall yield [39].

Now, we move our attention to quinolinylboron reagents as the nucleophilic coupling partners in the Suzuki coupling. In this case, quinolinylboranes are generally prepared *in situ* from the corresponding quinolinyl halides by taking advantage of the Miyuara reaction. As shown previously, 3-bromoquinoline **58** was transformed to quinolinyl-3-aryl-boronic ester **88** using the pinacol ester of diboron **55**. Subsequent cross-coupling of **88** with 3-bromothiophene **89** provided the 3-(3-thienyl)quinoline **90** in 29% yield [40].

3-Dialkylquinolinyl boranes **91** and **95** were prepared from halogen/metal exchange of 3-bromoquinoline **58** with *n*-BuLi followed by quenching with Et$_2$BOMe and Br-9-BBN

94, respectively. They were then coupled with bromides **92** and **96** to give 3-substituted quinoline derivatives **93** and **97**, respectively [41].

Hiyama's group carried out a Suzuki reaction as part of their synthesis of synthetic analogues of inhibitors of 3-hydroxy-3-methylglutaryl coenzyme A (HMG-CoA) reductase [42]. Hydroboration of terminal alkyne **98** with disiamylborane gave vinylborane **99**, which was used *in situ* to couple with 3-iodoquinoline **100** to produce adduct **101** in 99% yield. It should be noted that although aryl iodide **100** is sterically congested a high yield nevertheless resulted from the coupling reaction.

12.3.3. Stille coupling

Quinolinyl halides as the electrophiles are also more prevalent in the Stille coupling reactions. As stated previously, the Stille coupling of 2- and 4-chloroquinoiline is more sluggish than the corresponding Negishi reaction. However, employing a combination of

Pd(dba)$_2$/PPh$_4$ in a 1:2 ratio, 2-chloroquinoline **25** was successfully coupled with 1-ethoxy-2-tributylstannylethene to produce adduct **102**, which was converted to 2-acetylquinoline **103** by treatment with acid [43]. Using such a tactic, all the acetyl quinoline isomers were synthesized. In the same fashion, 2-chloro-3-cyano-4-methylquinoline **104** was coupled with 4-trimethylstannylpyridine **105** under the standard Stille coupling conditions to assemble 2-(4-pyridyl)quinoline **106** in 39% yield [44].

With regard to 3-bromoquinoline **58**, its coupling with 2-tributylstannyl-6-chloropyridine **107** was straightforward, giving rise to 3-(2-pyridyl)quinoline **108** in 65% yield [45].

The Stille coupling of an aryl triflate normally calls for the addition of at least one equivalent of LiCl. Presumably, the transmetalation is facilitated by replacing triflate with Cl$^-$ at the palladium intermediate generated from oxidative addition. As Stille demonstrated in 1987, 4-quinolinyl triflate **109** was coupled with phenylstannane **110** in the presence of Pd(Ph$_3$P)$_4$ and LiCl in refluxing 1,4-dioxane to furnish biaryl **111**, which was used as an intermediate for the first total synthesis of antibiotic amphimedine [46]. Interestingly, 4-quinolinyl triflate **112** underwent the Stille coupling smoothly with 3-tributylstannylindole **113** to deliver indolylquinoline **114** in 92% yield in the presence of Pd$_2$(dba)$_3$–AsPh$_3$ *in the absence of* LiCl [47]. It is possible that this transmetalation is facilitated by the softer ligand AsPh$_3$.

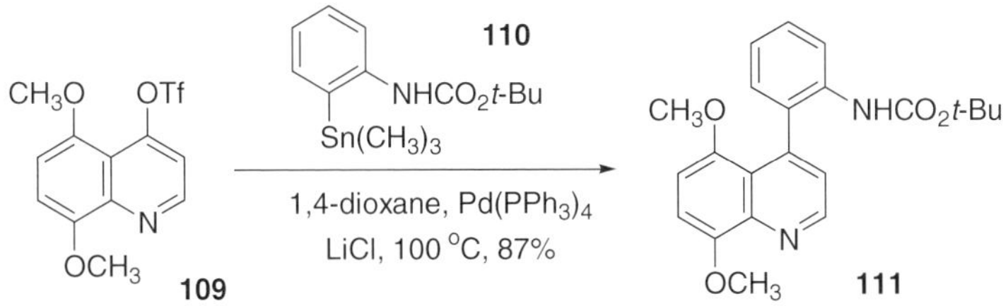

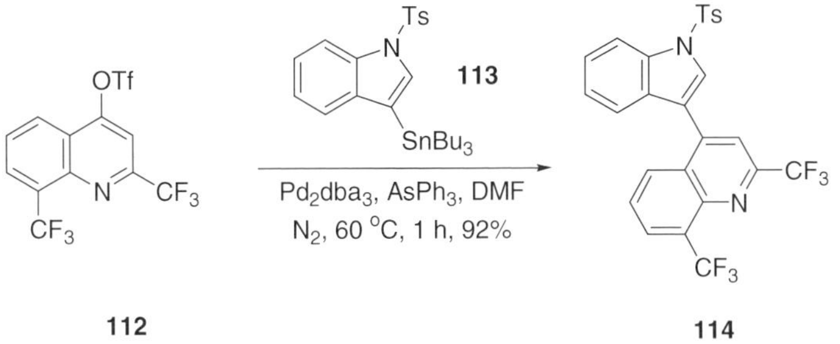

Quinolinylstannanes serve as the nucleophilic coupling partners in the Stille coupling. As illustrated below, the Stille coupling of 3-trimethylstannylquinoline **42** with 2-bromopyridine afforded 3-(2-pyridyl)quinoline **80** in 79% yield [48]. Analogously, the Stille coupling of 2-trimethylstannylquinoline **41** with 2-chloropyridine **116** resulted in 3-substituted quinoline **117** in 78% yield [49].

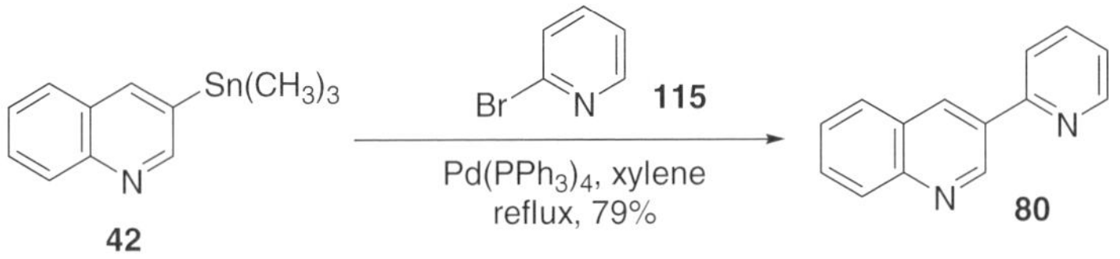

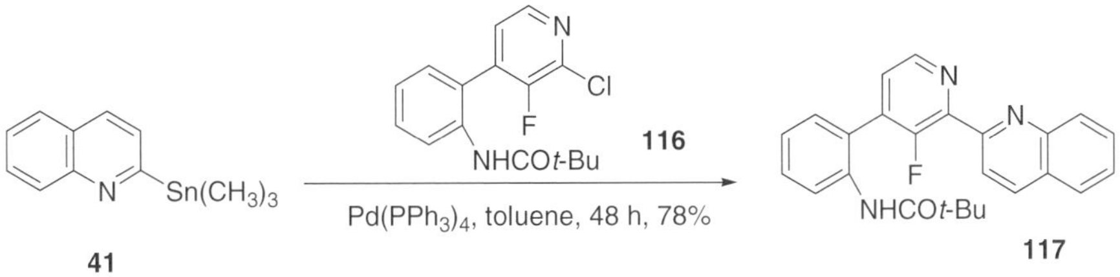

In a model study [50], 2-[2-(4-phenyl-3-amino)-pyridyl]quinoline **119** was derived from the Stille coupling of 2-trimethylstannylquinoline **41** and pyridyl triflate **118**. As an extension of the aforementioned method, decorated 2-trimethylstannylquinoline **120** was coupled with a more intricate pyridyl triflate **121** *in the presence of* LiCl, to assemble adduct **122** in 68% yield [49]. Addition of Cu_2Br_2 is supposed to promote the rate of transmetalation. 2-Aryl-quinoline **122** was an advanced intermediate for the total syntheses of Streptonigrin [51] and Lavendamycin [52], which have been shown to possess antitumor and antiviral activities [51].

The powerful utility of the Stille coupling reaction is showcased by the synthesis of 4-substituted 6-nitroquipazine analogs (e.g. **126**) which are reported to have high affinities to serotonin transporter [53]. As delineated below, 2,4-dibromo-6-nitroquino-line (**123**) underwent an S_NAr displacement by 1-piperazinecarbaldehyde to give *N*-formyl-4-bromo-6-nitroquipazine **124**. With bromide **124** in hand, it was transformed to *N*-formyl-4-tributylstannyl-6-nitroquipazine **125**, which was further converted by iodo-destannation and the resulting iodide was then available for additional Pd-catalyzed reactions. Moreover, bromide **124** was utilized as the nucleophile to couple with an array of stannanes to functionalize the core structure as exemplified by **126**. An operational note is worth-mentioning here for transformation **124→125** via the Pd-catalyzed ditin chemistry. Tributyltin halide, the toxic by-product of the Stille coupling reaction, was removed by first quenching of the reaction mixture with the addition of 10% aqueous KF. After being stirred for three hours, the insoluble products, palladium metal and white tributyltin polymer, were removed via filtration through celite. Such treatment of the reaction mixture alleviated the trouble of dealing with the toxic tin by-product.

125

126

12.3.4. Hiyama coupling

In comparison to the transmetalation of organometallic reagents including Grignard reagents, organozinc reagents, and organostannanes, transmetalation of the organosilicon reagents does not occur under normal Pd-catalyzed cross-coupling conditions. The C—Si bond is much less polarized, possessing more covalent bond properties than aforementioned organometallic reagents. However, a C—Si bond can be activated by a nucleophile such as F⁻ or HO⁻ through formation of a pentacoordinated silicate, which weakens the C—Si bond by enhancing the polarization. As a result, the transmetalation becomes more facile and the cross-coupling proceeds readily. One of the advantages of the Hiyama coupling is that organosilicon reagents are innocuous. Another advantage is better tolerance of functional groups in comparison to other strong nucleophilic organometallic reagents. The combination of these two characteristics makes the Hiyama coupling an attractive alternative to other Pd-catalyzed cross-couplings. As a result, the Hiyama coupling reaction is sometimes called Silicon–Stille coupling. One case that involves the quinolinyl fragment is depicted. Quinolinyl-3-bromide **58** was coupled with functionalized alkyltrifluorosilane **127** to produce adduct **128** [54] in the presence of tetrabutylammonium fluoride and catalytic Pd(PPh$_3$)$_4$. The Hiyama coupling reaction took place on an sp^3 carbon-center of the nucleophile.

Previously, HMG-CoA reductase inhibitor **101** was synthesized via a Suzuki coupling approach (see Section 12.3.3). Hiyama's group also carried out a Hiyama coupling to make the same compound [55]. Vinylsilane **129** was prepared by Pd-catalyzed reaction from terminal alkyne **98**. Palladium-catalyzed cross-coupling between vinylsilane **129** and 3-iodoquinoline **100** then assembled **101** in 83% yield.

98   →   1.2 eq. (CH$_3$)$_2$SiHCl, 0.5 mol%
t-Bu$_3$P•Pt(CH$_2$=CHSi(CH$_3$)$_2$)$_2$O
rt, 1 h   →   **129**

100

TBAF, [(allyl)PdCl$_2$], THF
60 °C, 0.5 h, 83%   →   **101**

12.4. Sonogashira reaction

A simple example of the Sonogashira reaction involving quinolines is illustrated. Installation of 2-trimethylsilylethynylquinoline **130** [56] was achieved from 2-bromo-quinoline **11** under the standard Sonogashira reaction conditions [PdCl$_2$(PPh$_3$)$_2$, CuI, Et$_3$N]. Under these conditions, 2-chloroquinoline **25** gave 87% yield at 80°C for 4 h; however, 4-chloroquinoline gave 85% recovered starting material even after 80 °C for 22 h, echoing the phenomenon observed in transformation **123**→**124**, where we saw less reactivity for halides on the C(4) position compared to the C(2) position in both S$_N$Ar displacement and palladium-mediated reactions.

11   →   Pd(PPh$_3$)$_2$Cl$_2$, CuI, Et$_3$N
70 °C, 3 h, 75%   →   **130**

3-Bromoquinoline **58**, behaving similarly to a simple carbocyclic aryl bromide, was coupled with phenylethyne **131** to provide disubstituted ethyne **132** in 50% yield [57].

58   →   **131**
Pd(OAc)$_2$, CuI, CH$_3$CN, Et$_3$N
PPh$_3$, N$_2$, 80 °C, 24 h, 50%   →   **132**

Although Sonogashira reaction requires terminal alkynes, 2,7-dimethylocta-3,5-diyne-2,7-diol can be readily unmasked in the presence of a base. Diacetylene **133** was obtained from the reaction of 3-bromoquinoline **58** and 2,7-dimethylocta-3,5-diyne-2,7-diol in the presence of NaOH, benzyltrimethylammonium chloride as a phase-transfer catalyst, and palladium(II) and copper(I) catalysts [58]. In addition, by taking advantage of the terminal alkyne on **134** as the anchor, the Sonogashira reaction attached the quinoline fragment to give 2-fluoro-6-O-propargyl-11,12-carbamate ketolide derivative of erythromycin **135** [59].

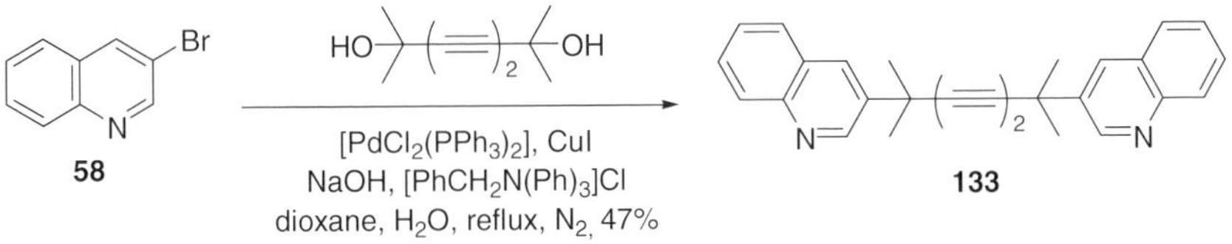

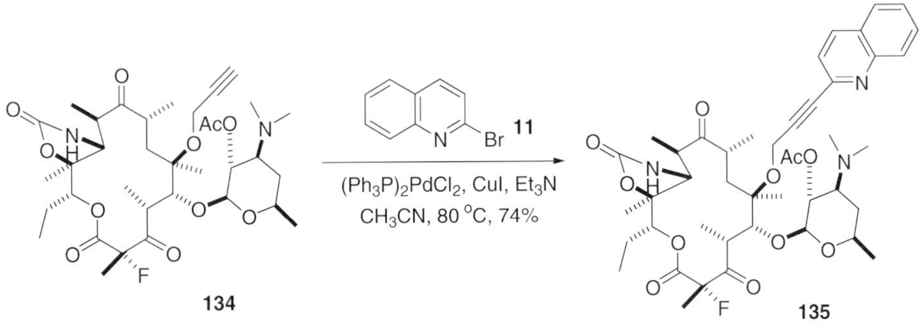

The Pd-catalyzed reaction of o-iodoanilides with terminal acetylenic carbinols provides a facile route to the synthesis of quinolines using readily available starting materials [60]. When o-iodoanilide **136** was stirred with acetylenic carbinol **137** in the presence of bis-triphenyl phosphine palladium(II) chloride in triethylamine at room temperature for 24h, the substituted alkynol **138** was obtained in 65% yield. On cyclization of **138** with sodium ethoxide in ethanol, 2-substituted quinoline **139** was obtained in excellent yield.

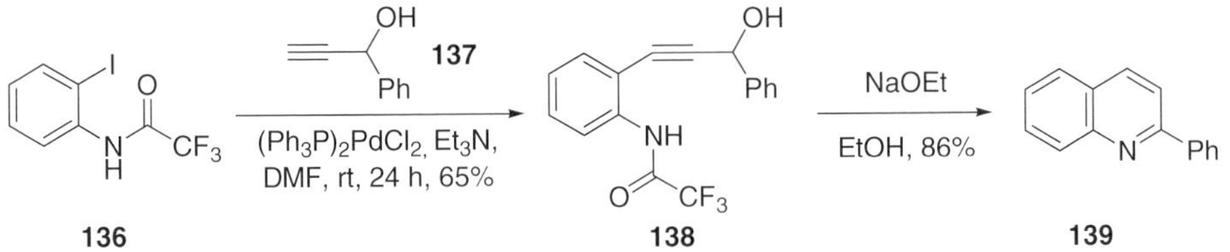

Santelli's group has carried out cross-coupling reactions with heteroaryl halides in the presence of a palladium-tetraphosphine catalyst; these include the Heck reaction [61], Suzuki coupling [62], and more recently, the Sonogashira reaction [63]. Low catalyst loadings are employed to give moderate to good yields of the alkynylated quinolines, as shown below.

12.5. Heck reaction

Heck reaction, Pd-catalyzed cross-coupling reactions between organohalides or triflates with alkenes [64], can take place inter- or intra-molecularly. It is a powerful carbon–carbon bond forming reaction for the preparation of alkenyl- and aryl-substituted alkenes in which only a catalytic amount of a palladium(0) complex is required.

12.5.1. Intermolecular Heck reaction

Legros *et al.* [43] carried out the synthesis of acetylquinolines (e.g. **144** via Heck reaction of 3-bromoquinoline **58** and *n*-butyl vinyl ether employing either Pd(dba)$_2$ or Pd(OAc)$_2$ as the catalyst. In each case it was found that the Heck reaction for this synthesis gave better overall yields than using the Stille reaction (see Section 12.3.3). Additionally, the use of toxic stannane is avoided.

Due to their successful synthesis of 2-(4′-chlorophenyl)-4-iodoquinoline from the corresponding precursor acetylene, Arcadi and group [65] developed a one-step synthesis of 2,4-disubstituted quinolines via Pd-catalyzed coupling reactions. An example is the Heck reaction of 4-iodoquinoline **145** with α-acetamidoacrylate **146**. This one-pot synthesis yielded adduct **147** in 50% overall yield after purification via flash chromatography.

Cacchi and Palmieri [66] investigated a new entry into the quinoline skeleton by Pd-catalyzed Michael-type reactions. They found that phenyl mercurial **148** was a useful intermediate for the synthesis of quinoline derivatives, and that by selecting the reaction conditions the oxidation level of the heterocyclic ring in the quinoline skeleton can be varied. $PdCl_2$-catalyzed coupling between organomercurial reagent **148** and enone **149** delivered adduct **150** which was subsequently cyclized to quinoline **151** under acidic conditions. This reaction is not a *bona fide* Heck reaction *per se* for two reasons; (a) the starting material underwent a Hg–Pd transmetalation first rather than the oxidative addition of an aryl halide or triflate to Pd(0); (b) instead of undergoing an elimination step to give an enone, transformation **148**→**150** terminated the catalytic cycle with a reduction to afford a saturated alkyl ketone. Another case of Heck-like reaction is depicted in reaction **153**→**154** [67]. Thus, *ortho*-cyclopalladation of acetanilide **152** gave organopalladium reagent **153**. The *ortho*-vinylation of **153** afforded enone **154**, which was then cyclized to quinoline **155** under acidic conditions. Notice this reaction requires stoichiometric amounts of $Pd(OAc)_2$.

Larock and Kuo [68] investigated the Pd-catalyzed coupling of allylic alcohols and *o*-iodoaniline which provided a convenient, one-step synthesis of quinolines as represented by the reaction between *o*-iodoaniline and alkenol **156** to give quinoline **157**. The optimal conditions for this reaction were found to be 5 mol% $PdCl_2$, 5 mol% PPh_3, 3 equivalents $NaHCO_3$, 1.5 equivalents alkenol and 10 mL of HMPA per mmol of *o*-iodoaniline at 140°C for 1 day under N_2. The reaction was found to be fairly versatile and the use of a range of allylic alcohols can be utilized although the quinoline could not be isolated easily from using allyl alcohol itself.

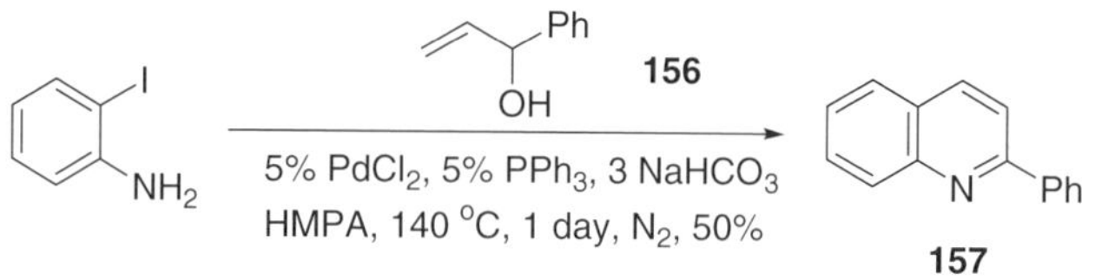

157

The Heck reaction of various aryl chlorides with *tert*-butyl acrylate using highly active palladium-phosphinous acid catalysts (POPd, POPd1, and POPd2) has been employed in the synthesis of 2,4-disubstituted quinolines [69]. The synthesis begins with an unprecedented Ziegler reaction using organolithium reagents resulting in regioselective arylation or alkylation of 4-chloroquinoline followed by treatment with CAN. The Heck additions proceed with diastereoselectivity and in high yield. Similarly, the 4-chloroquinolines **158a-c** were also found to undergo facile Stille coupling with arylstannanes using the aforementioned catalysts. A variety of quinoline derivatives can thus be achieved in just two steps.

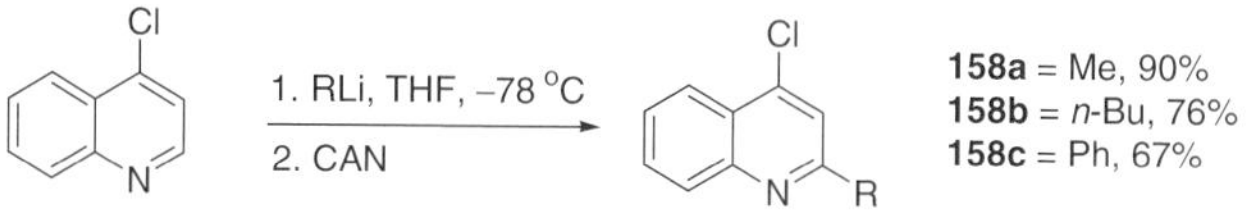

50

158a-c

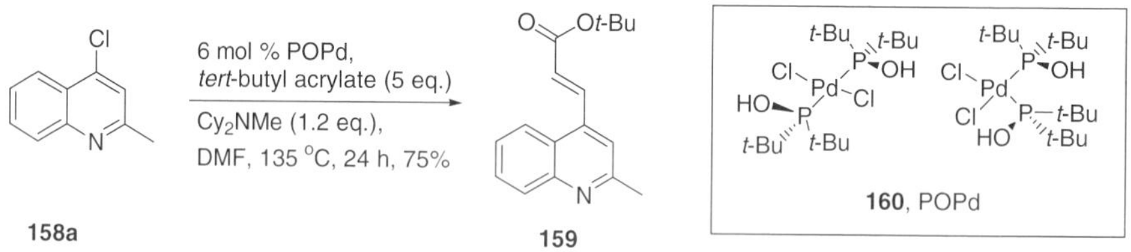

158a

159

160, POPd

12.5.2. Intramolecular Heck reaction

Employing Jefferey's "ligandless" conditions, Larock and Babu [70] synthesized quinolines and other nitrogen-containing heterocycles via the intramolecular Heck reaction strategy as exemplified by reaction **161→162**. This reaction is similar to the Mori–Ban indole synthesis with one additional CH_2 at the olefin moiety. In a six-step synthesis of (*S*)-camptothecin [71], the final step involved a C-ring closure using the Heck reaction. As shown in the following scheme, 2-chloroquinoline **163** was treated with 15% $(PPh_3)_2Pd(OAc)_2$ and 2 equivalents KOAc in CH_3CN at 100°C, affording (*S*)-campotothecin **164** in 64% yield.

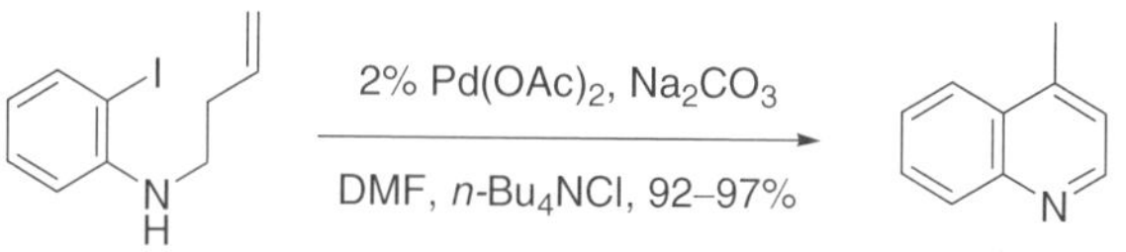

161

162

163 → 164

In their search for new antiplasmodial drugs, Maes and colleagues have synthesized 7*H*-indolo[2,3-*c*]quinoline, a core structure in the alkaloid series cryptolepine, in two steps from commercially available starting materials [72]. The synthesis consists of two consecutive Pd-catalyzed reactions: a selective Buchwald/Hartwig amination followed by an intramolecular Heck-type reaction. Selective methylation of the 7*H*-indolo[2,3-*c*] quinoline **169** gave 5-methyl-5*H*-indolo[2,3-*c*]quinoline **170** (Isoneocryptolepine), an interesting lead compound for novel antiplasmodial drugs.

12.6. Miscellaneous reactions mediated by palladium

12.6.1. Oxidative cyclization

Oxidative cyclizations are generally facilitated by Pd(OAc)$_2$ in acetic acid under reflux. The role of acetic acid in such oxidative cyclization processes is to protonate the acetate ligand, making Pd(II) more electrophilic. The initial step in these oxidative cyclization reactions is electrophilic palladation of the aromatic ring. In organic synthesis, oxidative cyclization offers an expeditious route to those target molecules that may not be easily accessible otherwise. In one case, quindoline **172**, an antimalarial agent isolated from a West African plant *Cryptolepis sanguinolenta*, was synthesized in two steps, a remarkably concise synthesis [73]. The precursor, 3-anilinoquinoline **171**, was prepared by phenylation of 3-aminoquinoline with $Ph_3Bi(OAc)_2$ in the presence of metallic copper. The crucial oxidative cyclization of **171** was then effected by *two equivalents* of $Pd(OAc)_2$ in trifluoroacetic acid under reflux to furnish quindoline **172**.

| 171 | | 172 |

12.6.2. Reductive cyclization

In their quest to synthesize quinolines without the need to involve metalated species as cross-coupling partners, Banwell and coworkers have devised a two-step procedure wherein the first step is the palladium[0]-mediated Ullmann cross-coupling of 1-bromo-2-nitroarenes (e.g. **173**) with β-halo-enals [74]. The resulting β-nitroaryl enal undergoes reductive cyclization, in the style of the Friedländer quinoline synthesis, to give the corresponding quinoline. A wide range of quinolines can be accessed by this method since many 1-halo-2-nitroarenes are commercially available and β-halo-enals such as **174** can be easily prepared by Vilsmeier haloformylation of the appropriate ketones.

| 173 | 174 | 175 | 176 |

In an analogous manner, novel 2-(2-hydroxyaryl)quinolines can be prepared via a one-pot synthesis from intramolecular reductive coupling reactions of 2′-hydroxy-2-nitro-chalcones [75]. Accordingly, 2′-hydroxy-2-nitrochalcones **177a–b** were treated with ammonium formate and Pd/C at room temperature to give a mixture of 2-(2-hydroxy-aryl)quinolines **178a–b** and 2-(2-hydroxyaryl)quinoline-N-oxides **179a–b**.

| 177a-b | 178a-b | 179a-b |

a $R_1 = R_2 = H$
b $R_1 = H, R_2 = Br$

12.6.3. Heteroannulation

As part of their study on gastric (H⁺/K⁺)-ATPase inhibitors, Kang *et al.* developed a simple and convenient synthetic approach to 1,2,3-trisubstituted pyrrolo[3,2-*c*]quinolines by means of Pd-catalyzed heteroannulation of 4-amino-3-iodoquinoline derivatives with internal alkynes [76]. The following scheme shows an example of a reaction using 4-arylamino-3-iodoquinoline derivative **180** with alkyne **181** to provide 1-arylpyrrolo [3,2-*c*]quinoline **182**, illustrating the possibility of introducing diverse substituents to 1-arylpyrrolo[3,2-c]quinolines. In addition, a Pd-catalyzed domino hydroarylation/ cyclization process was reported to form substituted quinolines [77]. Thus, 3-arylquino-lines were prepared in 56–74% yield when 3,3-diethyl-1-phenyl-1-propyne and aryl iodide were refluxed in ionic liquid, 1-butyl-3-methylimidazolium tetrafluoroborate [(bmim)BF₄⁻],in the presence of HCO₂H, Et₃N, and palladium catalyst. Meanwhile, 4-arylquinolines were obtained in 9–21% yield as minor by-products.

12.6.4. Cyanation

The nitrile group is a stable and versatile functional group which can be converted into several other functionalities by relatively simple synthetic routes. Thus the addition of this group into a molecule provides a route by which it can be manipulated into a more complex compound. Sakamoto and Ohsawa [78] investigated the cyanation of various quinolines using copper(I) cyanide as the cyano group source. In the presence of catalytic amount of Pd₂(dba)₃, and dppf as ligand, 3-iodoquinoline **43** was cyanated to give 4-cyanoquinoline **183** in 91% yield. To confirm that their method was indeed catalyzed by palladium, they also ran their reactions without the Pd catalysts and found in most cases that little or none of the desired product was obtained, and mostly the starting material was recovered.

The mechanism of action of the cyanation reaction is considered to progress as fol-lows: An oxidative addition reaction occurs between the aryl halide and a palladium(0) species to form an arylpalladium halide complex which then undergoes a ligand exchange reaction with CuCN thus transforming to an arylpalladium cyanide. Reductive elimination of the arylpalladium cyanide then gives the aryl cyanide.

12.6.5. Homocoupling

Traditionally, the synthesis of symmetrical biaryls was routinely accomplished using the Ullmann reaction. Recently, Pd-catalyzed homocoupling of aryl halides has also been demonstrated to rival the utility of the Ullmann coupling. As illustrated in Scheme 21, using Pd(OAc)$_2$ as the catalyst, K$_2$CO$_3$ as the base in isopropanol [79], 3-bromoquinoline **58** was homocoupled to produce bis-quinoline **184** in 79% yield. Under the same reaction conditions, 2-chloroquinoline was homocoupled to give its corresponding dimer in 79% yield.

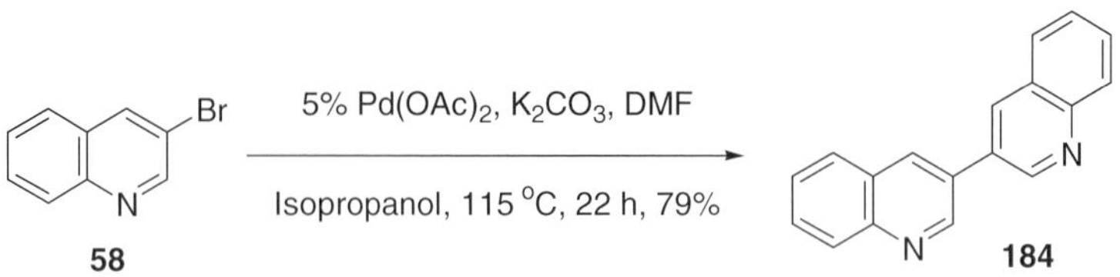

Scheme: 5% Pd(OAc)$_2$, K$_2$CO$_3$, DMF / Isopropanol, 115 °C, 22 h, 79%; **58** → **184**

12.6.6. Phosphination

Tertiary phosphines are an important family of ligands in transition metal-catalyzed reactions. Palladium-catalyzed phosphination with triarylphosphines to produce functionalized tertiary phosphines was carried out by Kwong *et al.* [80]. Using Pd(OAc)$_2$ as the catalyst, PPh$_3$ as both the ligand and phosphinating agent, quinolinyl-8-triflate **185** was converted to quinolinyl-8-diphenylphosphine **186** in 45% yield. The method was compatible with many functional groups such as aldehyde, ketone, nitrile, ester, methoxy, and pyridyl groups. Typical reactions involving simple carbocyclic substrates only took 2–6 h; however, quinolinyl and pyridyl triflates took longer to transform to their corresponding phosphines. This was presumably due to the coordination of the chelating heteroatom to the palladium center, thus causing it to be coordinatively saturated, hence reducing its catalytic activity.

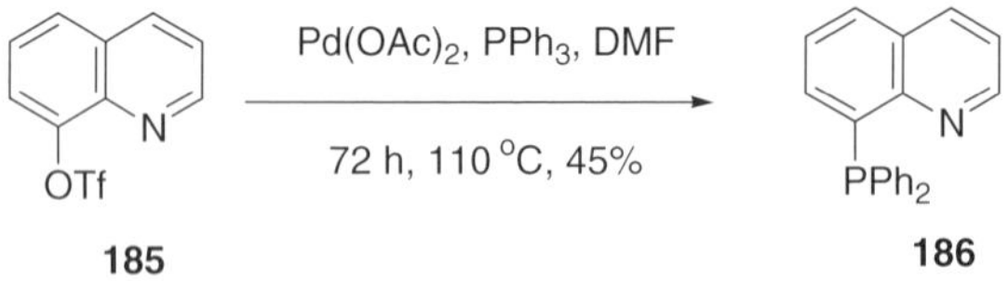

Scheme: Pd(OAc)$_2$, PPh$_3$, DMF / 72 h, 110 °C, 45%; **185** → **186**

12.6.7. Oxidative C–C$_{Ar}$ bond formation

Recently there has been much interest in the development of a strategy which enables C–C$_{Ar}$ bond formation without requiring functionalized reagents such as an aryl halide and an organometallic fragment as is usually the case for Pd-catalyzed reactions [81]. In particular, Pd mediated C–H activation with ensuing functionalization of the Pd-alkyl/aryl species by an arylating agent is deemed attractive [82]. Sanford *et al.* have developed and applied such methodology to quinoline substrates, resulting in facile coupling reactions of heterocycles with iodine-(III) arylating reagents which can be conducted in the presence of ambient air/moisture [83]. Reaction times varied between 8–24 h and a variety of solvents can be employed; for example, AcOH, toluene, and C_6H_6.

In summary, palladium-mediated reactions, especially cross-coupling reactions have found many applications in quinoline synthesis. It is noteworthy that due to the α and δ activation for the C(2) and C(4) positions, even 2-chloro- and 4-chloro-quinolines are viable substrates for Pd-catalyzed reactions under standard conditions. With the advent of palladium chemistry and more commercially available organometallic substrates, more palladium-mediated quinoline syntheses are to be added to the repertoire of quinoline chemistry.

12.7. References

1. Baston, E.; Palusczak, A.; Hartmann, R. W. *Eur. J. Med. Chem.* **2000**, *35*, 931.
2. Van Galen, P. J. M.; Nissen, P.; van Wijngaarden, P. M.; Ijzerman, A. P.; Soudijn, W. *J. Med. Chem.* **1991**, *34*, 1202.
3. Wissner, A.; Berger, D. M.; Boschelli, D. H.; Floyd, M. B., Jr.; Greenberger, L. M.; Gruber, B. C.; Johnson, B. D.; Mamuya, N.; Nilakantan, R.; Reich, M. F.; Shen, R.; Tsou, H.-R.; Upeslacis, E.; Wang, Y. F.; Wu, B.; Ye, F. Zhang, N.; *J. Med. Chem.* **2000**; *43*, 3244.
4. Sugimoto, O.; Mori, M.; Tanji, K-I. *Tetrahedron Lett.* **1999**, *40*, 7477.
5. Yajima, T.; Munakata, K. *Chem. Lett.* **1997**, *8*, 891.
6. Boger, D. L.; Duff, S. R.; Panek, J. S.; Yasuda, M. *J. Org. Chem.* **1985**, *50*, 5782.
7. Gordon, M.; Pearson, D. E. *J. Org. Chem.* **1964**, *29*, 329.
8. Gordon, M.; Hamilton, H.; Adkins, C.; Hay, J.; Pearson, D. E. *J. Heterocycl. Chem.* **1967**, *4*, 410.
9. Bouyssou, P.; Le Goff, C.; Chenault, J. *J. Heterocycl. Chem.* **1992**, *29*, 895.
10. Skraup, Z. H. *Monatsh. Chem.* **1982**, *3*, 531.
11. Sawanishi, H.; Hirai, T.; Tsuchiya, T. *Heterocycles* **1982**, *19*, 1043.
12. Negishi, E. *Acc. Chem. Res.* **1982**, *15*, 340; Stille, J. K. *Angew. Chem. Int. Ed.* **1986**, *25*, 508.
13. Kimber, M.; Anderberg, P. I.; Harding, M. M. *Tetrahedron* **2000**, *56*, 3575.
14. Sicker, D.; Reifegerste, D.; Hauptmann, S.; Wilde, H.; Mann, G. *Synthesis*, **1985**, 331. Galanakis, D.; Davis, C. A.; Ganellin, C. R.; Dunn, P. M. *J. Med. Chem.* **1996**, *39*, 359.
15. Takashiro, E.; Nakamura, Y.; Fujimoto, K. *Tetrahedron Lett.* **1999**, *40*, 5565.
16. Wolf, C.; Lerebours, R. *J. Org. Chem.* **2003**, *68*, 7077.
17. Arzel, E.; Rocca, P.; Marsais, F.; Godard, A.; Queguiner, G. *Tetrahedron Lett.* **1998**, *39*, 6465.
18. Meth-Cohn, O.; Narine, B.; Tarnowski, B. *J. Chem. Soc., Perkin Trans.1* **1981**, *5*, 1520.

19. Alonso, M. M.; del Mar Blanco, M.; Avendano, C.; Menendez, J. C. *Heterocycles* **1993**, *36*, 2315.

20. Korodi, F.; Cziaky, Z. *Org. Prep. Proced. Int.* **1990**, *22*, 579.

21. Meth-Cohn, O.; Rhouati, S.; Tarnowski, B. *Tetrahedron Lett.* **1979**, *20*, 4885.

22. Meth-Cohn, O.; Rhouati, S.; Tarnowski, B.; Robinson, A. *J. Chem. Soc., Perkin Trans. 1* **1981**, 1537.

23. Ashok, K.; Sridevi, G.; Umadevi, Y. *Synthesis*, **1993**, 623.

24. Yamamoto, Y.; Yanagi, A. *Chem. Pharm. Bull.* **1982**, *30*, 1731.

25. Uchiyama, K.; Ono, A.; Hayashi, Y.; Narasaka, K. *Bull. Chem. Soc. Jpn* **1998**, *71*, 2945.

26. Yamamoto, Y.; Yanagi, A. *Heterocycles* **1981**, *16*, 1161.

27. Yammal, C. C.; Podesta, J. C.; Rossi, R. A. *J. Org. Chem.* **1992**, *57*, 5720.

28. Echavarren, A. M.; Stille, J. M. *J. Am. Chem. Soc.* **1988**, *110*, 4051.

29. Ishiyama, T.; Murata, M.; Miyaura, N. *J. Org. Chem.* **1995**, *60*, 7508.

30. Murata, M.; Oyama, T.; Watanabe, S.; Masuda, Y. *J. Org. Chem.* **2000**, *65*, 164.

31. Dumouchel, S.; Mongin, F.; Trecourt, F.; Quequiner, G. *Tetrahedron*, **2003**, *59*, 8629.

32. Jetter, M.; Reitz, A. B. *Synthesis*, **1998**, 829.

33. Murata, N.; Sugihara, T.; Kondo, Y.; Sakamoto, T. *Synlett* **1997**, 298.

34. Fenger, I.; Le Drian, C. *Tetrahedron. Lett.* **1998**, *39*, 4287.

35. Inada, K.; Miyaura, N. *Tetrahedron* **2000**, *56*, 8661.

36. Deshayes, K.; Broene, R. D.; Chao, I.; Knobler, C. B.; Diederich, F. *J. Org. Chem.* **1991**, *56*, 6787.

37. Trécourt, F. Mongin, F.; Mallet, M.; Queguiner, G. *Synth. Commun.* **1995**, *25*, 4011.

38. Smith, P. A. S.; Boyer, J. H. *J. Am. Chem. Soc.* **1951**, *73*, 2626.

39. Banwell, M. G.; Bray, A. M.; Edwards, A. J.; Wong, D. J. *J. Chem. Soc., Perkin Trans. 1* **2002**, 1320.

40. Giroux, A.; Han, Y.; Prasit, P. *Tetrahedron. Lett.* **1997**, *38*, 3841.

41. Ishikura, M.; Oda, I.; Terashima, M. *Heterocycles* **1985**, *23*, 2375.

42. Miyachi, N.; Yanagawa, Y.; Iwasaki, H.; Ohara, Y.; Hiyama, T. *Tetrahedron Lett.* **1993**, *34*, 8267.

43. Legros, J.-Y.; Primault, G.; Fiaud, J.-C. *Tetrahedron* **2001**, *57*, 2507.

44. Bracher, F.; Papke, T.; Legros, J.-Y. *Liebigs Ann. Chem.* **1996**, 115.

45. Choppin, S.; Gros, P.; Fort,Y. *Org. Lett.* **2000**, *2*, 803–05.

46. Echavarren, A. M.; Stille, J. K. *J. Am. Chem. Soc.* **1987**, *109*, 5478.

47. Ciattini, P. G.; Morera, E.; Ortar, G. *Tetrahedron Lett.* **1994**, *35*, 2405.

48. Yamamoto, Y.; Azuma, Y.; Mitoh, H. *Synthesis* **1986**, 564.

49. Godard, A.; Rocca, P.; Fourquez, J.-M.; Rovera, J.-C.; Marsais, F.; Queguiner, G. *Tetrahedron Lett.* **1993**, *34*, 7919.

50. Godard, A.; Rovera, J.-C.; Marsais, F.; Ple, N.; Quequiner, G. *Tetrahedron* **1992**, *48*, 4123.

51. Boger, D. L.; Yasuda, M.; Mitscher, L. A.; Drake, D. D.; Kitos, P. A.; Thompson, S. C. *J. Med. Chem.* **1987**, *30*, 1918.

52. (a) Doyle, T. W.; Balitz, D. M.; Grulich, R. E.; Nettleton, D. E. *Tetrahedron Lett.* **1981**, *22*, 4595; (b) Balitz, D. M.; Bush, J. A.; Bradner, W. T.; O'Herron, F. A.; Nettleton, D. E. *J. Antibiot.* **1982**, *25*, 261.

53. Lee, B. S.; Chu, S.; Lee, B.-S.; Chi, D. Y.; Song, Y. S.; Jin, C. *Bioorg. Med. Chem. Lett.* **2002**, *12*, 811.

54. Mahsuhashi, H.; Kuroboshi, M.; Hatanaka, Y.; Hiyama, T. *Tetrahedron*, **1994**, *35*, 6507.

55. Takahashi, K.; Minami, T.; Ohara, Y.; Hiyama, T. *Tetrahedron Lett.* **1993**, *34*, 8263.

56. Sakamoto, T.; Shiraiwa, M.; Kondo, Y.; Yamanaka, H. *Synthesis* **1983**, 312.

57. Armengol, M.; Joule, J. A. *J. Chem. Soc., Perkin Trans. 1* **2001**, 978.

58. Sarkar, A.; Okada, S.; Nakanishi, H.; Matsuda, H. *Helv. Chim. Acta* **1999**, *82*, 138.

59. Phan, L. T.; Clark, R. F.; Rupp, M.; Or, Y. S.; Chu, D. T. W.; Ma, D. *Org. Lett.* **2000**, *2*, 2951.

60. Kundu, N. G.; Mahanty, J. S.; Das, P.; Das, B. *Tetrahedron. Lett.* **1993**, *34*, 1625.

61. Berthiol, F.; Feuerstein, M.; Doucet, H.; Santelli, M. *Tetrahedron Lett.* **2002**, *43*, 5625.

62. Feuerstein, M.; Doucet, H.; Santelli, M. *J. Organomet. Chem.* **2003**, *687*, 327.

63. Feuerstein, M.; Doucet, H.; Santelli, M. *Tetrahedron Lett.* **2005**, *46*, 1717.

64. Heck, R. F.; Nolley, Jr. J. P. *J. Org. Chem.* **1972**, *37*, 2320.

65. Arcadi, A.; Marinelli, F.; Rossi, E. *Tetrahedron* **1999**, *55*, 13233.

66. Cacchi, S.; Palmieri, G. *Tetrahedron* **1983**, *39*, 3373.

67. Horino, H.; Inoue, N. *Tetrahedron. Lett.* **1979**, *20*, 2403.

68. Larock, R. C; Kuo, M.-Y., *Tetrahedron. Lett.* **1991**, *32*, 569.

69. Wolf, C.; Lerebours, R. *J. Org. Chem.* **2003**, *68*, 7077.

70. Larock, R. C.; Babu, S. *Tetrahedron. Lett.* **1987**, *28*, 5291.

71. Comins, D. L.; Nolan, J. M. *Org. Lett.* **2001**, *3*, 4255.

72. Hostyn, S.; Maes, B. U. W.; Pieters, L.; Lemière, G. L. F.; Mátyus, P.; Hajós, G.; Dommisse, R. A. *Tetrahedron*, **2005**, *61*, 1571.

73. Fan, P.; Ablordeppey, S. Y. *J. Heterocycl. Chem.* **1997**, *34*, 1789.

74. Banwell, M. G.; Lupton, D. W.; Xinghua, M.; Renner, J.; Sydnes, M. O. *Org. Lett.* **2004**, *6*, 2741.

75. Barros, A. I. R. N. A.; Silva, A. M. S. *Tetrahedron Lett.* **2003**, *44*, 5893.

76. Kang, S. K.; Park, S. S.; Kim, S. S.; Choi, J.-K.; Yum, E. K. *Tetrahedron. Lett.* **1999**, *40*, 4379.

77. Cacchi, S.; Fabrizi, G.; Goggiamani, A.; Moreno-Mañas, M.; Vallribera, A. *Tetrahedron Lett.* **2000**, *43*, 5537.

78. Sakamoto, K.; Ohsawa, K. *J. Chem. Soc., Perkin Trans. 1* **1999**, 2323.

79. Hassan, J.; Penalva, V.; Lavenot, L.; Gozzi, C.; Lemaire, M. *Tetrahedron* **1998**, *54*, 13793.

80. Kwong, F. Y.; Lai, C. W.; Tian, Y.; Chan, K. S. *Tetrahedron Lett.* **2000**, *41*, 10285.

81. (a) Lane, B.S.; Sames, D. *Org. Lett.* **2004**, *6*, 2897; (b) Park, C.-H.; Ryabova, V.; Seregin, I. V.; Sromek, A. W.; Gevorgyan, V. *Org. Lett.* **2004**, *6*, 1159; (c) Glover, B.; Harvey, K. A.; Liu, B.; Sharp, M. J.; Tymoschenko, M. F. *Org. Lett.* **2003**, *5*, 301; (d) Campeau, L.-C.; Parisien, M.; Leblanc, M.; Fagnou, K. *J. Am. Chem. Soc.* **2004**, *126*, 9186; (e) Wakui, H.; Kawasaki, S.; Satoh, T.; Miura, M.; Nomura, M. *J. Am. Chem. Soc.* **2004**, *126*, 8658 and references therin; (f) Huang, Q.; Fazio, A.; Dai, G.; Campr, M. A.; Larock, R. C. *J. Am. Chem. Soc.* **2004**, *126*, 7460; (g) Sezen, B.; Franz, R.; Sames, D.; *J. Am. Chem. Soc.* **2002**, *124*, 13372.

82. (a) Zaitsev, V. G.; Daugulis, O. *J. Am. Chem. Soc.* **2005**, *127*, 4165; (b) Boele, M. D. K.; van Strijdonck, G. P. W. N. M. *J. Am. Chem. Soc.* **2002**, *124*, 1586.

83. Kalyani, D.; Deprez, N. R.; Desai, L. V.; Sanford, M. S. *J. Am. Chem. Soc.* **2005**, *127*, 7330.

Chapter 13

Pyridazines

Bert U.W. Maes

Although Taüber synthesized pyridazine (1,2-diazine) already in 1895, it is remarkable that this heterocyclic skeleton attracted almost no interest from medicinal chemists in the following 76 years [1–3]. The contrast with the other diazines is really striking and the question rises "*Why ?*" A reasonable explanation is that medicinal chemistry is often inspired by natural products. The first naturally occurring hexahydropyridazines were only reported in 1971 by Hassall and co-workers [4] and the first natural product containing an aromatic pyridazine ring (Pyridazomycin) [5] was not described before 1988! A medicinal chemistry study performed by Heinisch *et al.* clearly revealed that the pyridazine ring is essential for the antimicrobial activity of this natural product since the synthesis of a deaza analog of Pyridazomycin resulted in a loss of activity [6, 7].

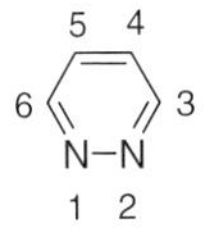

pyridazine

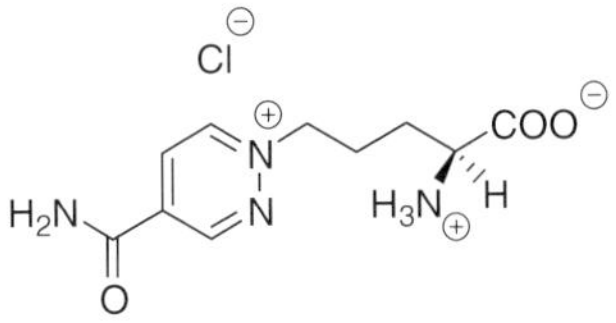

Pyridazomycin

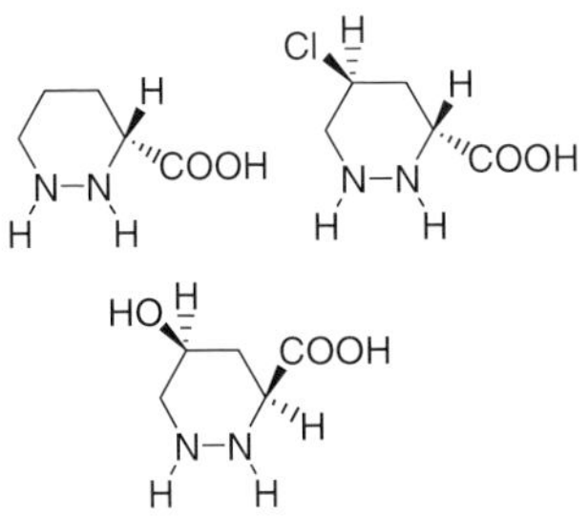

hexahydropyridazine amino acids
produced by *Streptomyces jamaicensis*

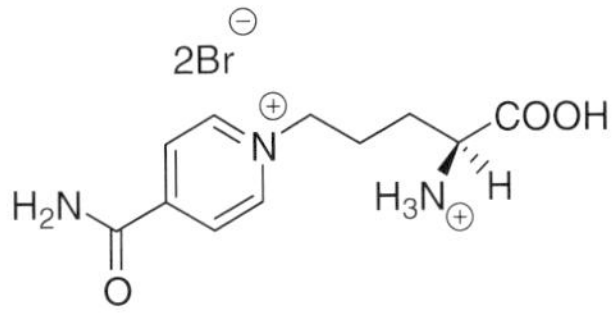

desazapyridazomycin

Since the discovery of hexahydropyridazine containing natural products, a large number of patents and papers has been published dealing with the synthesis of pharmaceutically useful 1,2-diazine derivatives [3, 8–10]. Compounds which have been launched as pharmaceutical products are Amezinium metilsulfate (selective noradrenergic antihypertensive), Azelastine (antiasthmatic, antiallergic, antihistaminic), Cadralazine (antihypertensive, vasodilator), Cefozopran (antibacterial), Cinoxacin (antibacterial), Emorfazone (anti-inflammatory, analgesic), Hydralazine (antihypertensive), Minaprine (antidepressant), and Sulfamethoxypyridazine (antibacterial).

Amezinium metilsulfate   Azelastine   Cadralazine

Cefozopran   Cinoxacin   Emorfazone

Hydralazine   Minaprine   Sulfamethoxypyridazine

Besides the pharmaceutical applications of pyridazines there is also an agrochemical interest for these heterocycles [3, 9]. In contrast to the medicinal relevance, this interest has existed for an even longer time. Thus, in 1964 BASF launched the herbicide Chloridazon. Even today Chloridazon is still used for the cultivation of sugar beet and red table beet. Norflurazon, a structurally closely related compound, was commercialized in 1971. This crop protection agent is used for the cultivation of cotton, tree fruit, and vines. Several other pyridazine herbicides were developed and some of them found the way to the market, such as Pyridate and maleic hydrazide. Besides herbicidal, there are also fungicidal pyridazines; for instance, Diclomezine which was launched in 1987 by Sankyo. There are also reports of insecticidal 1,2-diazines. Important commercial examples are the cholinesterase inhibitor Pyridaphenthion (1974, Mitsui Toatsu Chemicals) and the energy metabolism disruption agent Pyridaben (1991, Nissan Chemicals).

Chloridazon   Norflurazon   Pyridate   Maleic hydrazide

Diclomezine   Pyridaphenthion   Pyridaben

The pyridazine nucleus is a π-deficient aromatic heterocycle. Each nitrogen atom of the 1,2-diazine activates its α- and γ-position for nucleophilic attack. Consequently, all ring carbon atoms are deactivated for electrophilic substitution. All ring carbon atoms of *N*-substituted pyridazin-3(2*H*)-ones are also activated for nucleophilic attack: C(4) and C(6) are in the α- and γ-position of N(1), respectively, and C(5) is part of an enone moiety. The pyridazine and pyridazin-3(2*H*)-one skeleta are usually only sufficiently reactive for electrophilic substitution when one or more electron-donating substituents (e.g. amino, methoxy, hydroxy) are present on these nuclei. Also pyridazine *N*-oxides are sufficiently activated (in the α- and γ-position) for reactions with electrophilic reagents (e.g. nitration and halogenation). Due to the electron deficient nature, halopyridazines and halopyridazin-3(2*H*)-ones, even chlorinated derivatives, undergo carbon–carbon bond forming reactions under standard palladium-catalyzed reaction conditions.

13.1. Synthesis of (pseudo)halopyridazines and (pseudo)halopyridazin-3(2*H*)-ones

13.1.1. Ring synthesis of halopyridazin-3(2*H*)-ones

4,5-Dichloropyridazin-3(2*H*)-one and its 2-substituted analogs can be easily synthesized via the condensation of mucochloric acid (**1**) with hydrazines [11–13]. Similarly, when mucobromic acid (**2**) is used as substrate 4,5-dibromopyridazin-3(2*H*)-ones are obtained. 2-Alkyl substituted 4,5-dihalopyridazin-3(2*H*)-ones are usually not prepared by a direct condensation but via a direct *N*-alkylation on commercially available 4,5-dihalopyridazin-3(2*H*)-ones. In this way the use of toxic substituted hydrazines is avoided [11, 12].

1, X=Cl
2, X=Br

The synthesis of 5-chloro-4-iodo-2-methylpyridazin-3(2*H*)-one (**4**) from the corresponding mixed mucohalic acid (**3**) and methylhydrazine was also reported [14, 15]. The required chloro/iodo mixed mucohalic acid could be smoothly synthesized from commercially available mucochloric acid (**1**) by a transhalogenation with MeMgI [14]. The mechanism consists of a neutralization reaction of the mucochloric acid with the Grignard reagent leading to the halogenomagnesium salt followed by nucleophilic substitution of the α-halogen atom by iodide.

5-Chloro- (**5**) and 5-bromo-6-phenylpyridazin-3(2*H*)-one (**6**) are obtained by Friedel-Crafts alkylation of mucochloric acid (**1**) or mucobromic acid (**2**) with benzene followed by a reaction with hydrazine [11, 16, 17]. In the latter process one halogen atom is selectively eliminated.

5, X = Cl: 63%
6, X = Br: 60%

13.1.2. Direct halogenation of pyridazines and pyridazin-3(2*H*)-ones

2-(4-Chlorophenyl)-5-hydroxypyridazin-3(2*H*)-one (**7a**) and 4-amino-1-(4-chlorophenyl)-6-oxo-1,6-dihydro-3-pyridazinecarboxylic acid (**9**) can be smoothly iodinated in the C-4 and C-5 position, respectively using I_2 in aqueous Na_2CO_3/dioxane at reflux [18]. Chlorination and bromination of methyl 1-(4-chlorophenyl)-4-hydroxy-6-oxo-1,6-dihydro-3-pyridazinecarboxylate (**7c**) with SO_2Cl_2 and Br_2, respectively is also reported [19].

7a, R = H
7b, R = CO$_2$H   **7**
7c, R = CO$_2$Me

I_2, aq. Na$_2$CO$_3$
1,4-dioxane, reflux

SO$_2$Cl$_2$
dioxane, 50 to 65 °C

Br$_2$
EtOH, RT

X=I **8a**, R = H: 77%
     8b, R = CO$_2$H: 75%

X=Cl **8c**, R = CO$_2$Me: 91%

X=Br **8d**, R = CO$_2$Me: 85%

Pyridazines with electron-releasing substituents can also undergo direct halogenation as exemplified by the bromination of 3-(dimethoxymethyl)-6-methylpyridazin-4-amine (**11**) with Br_2 in CCl_4 [20].

13.1.3. From pyridazine *N*-oxides via reaction with $POCl_3$

The reaction of pyridazine *N*-oxides with $POCl_3$ allows the direct introduction of a chlorine atom in the α-position with simultaneous elimination of the *N*-oxide function [21]. If the α-position is blocked, substitution γ to the *N*-oxide occurs [22].

16a, R^1 = MeO, R^2 = Me: 57%
16b, R^1 = Me, R^2 = Me: 28%
16c, R^1 = MeO, R^2 = MeO: 72%

13.1.4. Deoxy-halogenation of pyridazinones

Deoxy-chlorination of pyridazin-3(2*H*)-ones [23, 24], pyridazin-4(1*H*)-ones [25], or pyridazine-3,6(1*H*,2*H*)-diones [26] with $POCl_3$ is the most popular and straightforward route to prepare chloropyridazines. When PCl_5 in $POCl_3$ is used additional C-chlorination is often observed. This additional halogenation occurs especially when the reaction temperature is high and with prolonged reaction times.

Although similar deoxy-brominations with $POBr_3$, or PBr_5 are often described in the literature they are less frequently used [27].

13.1.5. Pseudohalopyridazines and -pyridazin-3(2H)-ones by esterification

The good availability of pyridazin-3(2H)-ones make triflate esters preferred substrates to access bond formation via palladium catalysis. It avoids the use of $POCl_3$, PCl_5, $POBr_3$, or PBr_5 which are often used as reagent and solvent (see Section 13.1.4). Moreover, triflate esters are more reactive towards oxidative addition than the corresponding bromo and chloro derivatives. Pyridazin-(3(2H)-on)e triflates are air stable compounds that can be stored in the cold without decomposition.

The team of Wermuth from Strasbourg was the first to report the synthesis of pyridazine triflates and its use in Pd-catalyzed C—C bond formation [28]. The pyridazine triflates were easily prepared from pyridazin-3(2H)-ones using either triflic anhydride in pyridine or by deprotonation with BuLi followed by the slow addition of triflic anhydride. In 2001,

Aldous and co-workers reported a more convenient procedure based on triethylamine in dichloromethane [29].

$R^1 = Me, R^2 = Ph$: 72%
$R^1 = H, R^2 = Ph$: 82%
$R^1 = H, R^2 = Me$: 90%

Maes and co-workers synthesized 5-chloro-2-methyl-4-trifluoromethanesulfony-loxypyridazin-3(2H)-one (**28**) and its 4-chloro-5-trifluoromethanesulfonyloxy-(**30**) regioisomer from the corresponding chloro-hydroxypyridazin-3(2H)-ones (**27, 29**) by esterification with triflic anhydride using triethylamine as base [30]. The trifluoromethanesulfonyloxypyridazin-3(2H)-ones (**28, 30**) are usually purified by one or more simple extractions (wash steps) followed by drying over MgSO₄.

13.1.6. Direct metalation of pyridazines followed by quenching with halogens

The synthesis of substituted pyridazines via direct metalation followed by reaction with an electrophile is pioneered by Quéguiner and co-workers [31]. The synthesis of halopyri-dazines following this procedure is hitherto only rarely used. Lithiation of 3-chloro-6-methoxypyridazine (**31**) with lithium t-butyl-1-isopropylbutylamide base at −75°C in THF followed by a quench with iodine yielded a 5/95 regioisomeric mixture of 3-chloro-4-iodo-6-methoxy- (**32**) and 6-chloro-4-iodo-3-methoxypyridazine (**33**). Interestingly, the regioselectivity changed depending on the lithium amide used for the metalation step [32].

Applying a lithiation–iodine quench protocol on 4-chlorocinnoline (**34**) as substrate using lithium diisopropylamide (LDA) as base gave 4-chloro-3-iodocinnoline (**35**) in good yield [33].

Metalation of 6-chloro-4-(1-hydroxyethyl)-3-methoxypyridazine (**36**) at 0°C with LTMP followed by reaction of the corresponding 5-lithio derivative with iodine yields 6-chloro-4-(1-hydroxyethyl)-5-iodo-3-methoxypyridazine (**37**) [34].

13.1.7. From metalopyridazines by reaction with an electrophilic halogen source

Although this route has not been used extensively there are several examples. The synthesis of 3-aryl-7-bromo-6-chloro[1,2,4]triazolo[4,3-*b*]pyridazines (**39**) from 3-aryl-6-chloro-7-(trimethylsilyl)[1,2,4]triazolo[4,3-*b*]pyridazines (**38**) was achieved by capturing the anion generated on desilylation with a soft electrophilic source of bromine [35].

Reaction of 3,6-diphenyl-4-tributylstannylpyridazine (**40**) with I$_2$ at room temperature gave 3,6-diphenyl-4-iodopyridazine (**41**) in 77% yield [36]. Likewise, 3-phenyl-5-tributylstannylpyridazine (**42**) gave 5-iodo-3-phenylpyridazine (**43**) [37]. When Br$_2$ was used as the electrophile source, 5-bromo-3-phenylpyridazine (**44**) was obtained [37].

In comparison to the synthesis of other iodopyridazines from trialkylstannylpyridazines a higher temperature and longer reaction time was required for the preparation of 4,5-diiodo-bistrifluoromethylpyridazine (**45**) from 3,6-bistrifluoromethyl-4,5-bistrimethylstannylpyridazine (**44**) [38]. 4,5-Dichloro-3,6-bistrifluoromethylpyridazine (**46**) could be obtained from the same substrate using Cl$_2$ in tetrachloromethane [38].

13.1.8. Diazotization—S$_N$1 reaction of aminopyridazines

The synthesis of halogenopyridazines from the corresponding aminopyridazines is only rarely used. 3-Amino-4-hydroxy-6-methyl- (**47a**) and 3-amino-4-methoxy-6-methylpyridazine (**47b**) have successfully been transformed into the corresponding 3-chloropyridazines (**48**) via diazotization followed by an S$_N$1 reaction with chloride anion [39].

47

47a, R^1 = OH, R^2 = Me
47b, R^1 = OMe, R^2 = Me

48

48a, R^1 = OH, R^2 = Me: 75%
48b, R^1 = OMe, R^2 = Me: 46%

13.1.9. Transhalogenation on chloropyridazines and chloropyridazin-3(2*H*)-ones

Transhalogenation is a frequently used technique to prepare iodopyridazin(-3(2*H*)-on)es from the corresponding chloropyridazin(-3(2*H*)-on)es. 3,6-Diiodopyridazine (**49**) for instance is synthesized from 3,6-dichloropyridazine (**23**) by heating in 50% HI at 150°C [40]. Interestingly, the same substrate can be selectively transformed into 3-chloro-6-iodopyridazine (**50**) using NaI in 57% HI at 40°C [41].

50

23

49

When commercially available 4,5-dichloropyridazin-3(2*H*)-one and its 2-substituted analogs are heated in 57% HI initially both chlorine atoms are replaced by iodine atoms. Subsequently, a selective C-4 hydrodeiodination occurs, via iodide attack on the *O*-protonated pyridazin-3(2*H*)-one, giving easy access to 5-iodopyridazin-3(2*H*)-ones (**52**) in good yields [42, 43]. Heating 4,5-dichloro-2-methyl-6-nitropyridazin-3(2*H*)-one (**53**) with NaI in DMF yields 4-iodo-2-methyl-6-nitropyridazin-3(2*H*)-one (**54**) via a similar reaction pathway [43]. In this case, chemoselective reduction of the C-5-I bond by iodide takes place.

51

R = H, Me, Bn, Ph

52

53

54

13.2. Coupling reactions with organometallic reagents

13.2.1. Negishi coupling

In 1998, Quéguiner and co-workers prepared the first organozinc derivative of a pyridazine. 3,6-Dimethoxypyridazine (**55**) was lithiated in the 4-position and subsequently transmetalated with $ZnCl_2$ yielding the corresponding zinc derivative (**56**) [44]. The organozinc derivative **56** can be stored at room temperature and reacted smoothly with iodobenzene, 2-bromopyridine, and 5-bromopyrimidine under heating. Interestingly, ultrasonic activation significantly decreased the required reaction time.

Ar = Ph, 2-pyridinyl, 5-pyrimidinyl

In 2002, Wonnacott published the synthesis and biological evaluation of a pyridazine analog (**62**) of nicotinic acetylcholine receptor agonist UB-165 [45]. This aza-UB-165 analog (**62**) was synthesized via Negishi cross-coupling reaction on triflate **60** with pyridazin-3-ylzinc halide (**59**). Compound **59** could be obtained from 3-bromopyridazine (**58**) via lithiation and subsequent transmetalation with zinc chloride.

In the frame of a medicinal chemistry program at Merck Sharp & Dohme Ltd, Collins and co-workers reported successful Negishi cross-coupling reaction of 3-aryl-7-bromo-6-[(1-methyl-1H-1,2,4-triazol-5-yl)methoxy][1,2,4]triazolo[4,3-b]pyridazines (**63**) with aliphatic organozinc reagents [35].

63
Ar = Ph, 2-FPh

64
Ar = Ph, R = cyclopentylmethyl
Ar = 2-FPh, R = t-butylmethyl

Negishi cross-coupling reaction of 3,6-dichloropyridazine (**23**) with 1.05 equivalents of aryl and alkyl organozinc derivatives yielded monosubstitution products with more than 98% selectivity [46]. Yields were often low due to an incomplete conversion of **23**. Attempts to increase the loading of catalyst or prolonging the reaction time did not solve the conversion problem. Interestingly, the use of a larger excess of organozinc compound proved to be beneficial and did not really influence the mono/di selectivity in most cases (>92% selectivity). A second Negishi cross-coupling reaction on the remaining C—Cl of the monosubstituted pyridazines **65** at a higher reaction temperature gave 3,6 unsymmetrically C-substituted pyridazines **66**.

In their search for new biologically active pyridazin-3(2*H*)-one derivatives useful as crop protection agents, Stevenson and co-workers of Dupont Crop Protection Products in Newark studied Negishi cross-coupling reaction of 4-bromo-5-methoxy-2-(4-trifluoromethyl-phenyl)pyridazin-3(2*H*)-one (**67**) with 3-trifluoromethylbenzylzinc bromide. When Pd$_2$(dba)$_3$/tri(2-furyl)phosphine was taken as precatalyst a good yield could be obtained, while the use of Pd(PPh$_3$)$_2$Cl$_2$ only gave 15–20% of **68**.

13.2.2. Suzuki coupling

13.2.2.1. (Pseudo)halopyridazines

The first Suzuki cross-coupling reaction on a halopyridazine was not reported before 1993 [47]. In order to obtain 6-aryl-3-chloropyridazines (**71**), commercially available 3,6-dichloropyridazine (**23**) was selected. 6-Aryl-3-chloropyridazines (**71**) can be used for the synthesis of new series of pyridazines by subsequent exchange of the remaining chloro substituent via another cross-coupling or nucleophilic substitution reaction. Unfortunately, the reaction of 3,6-dichloropyridazine (**23**) with phenylboronic acid under standard Suzuki reaction conditions led to a mixture of 3-chloro-6-phenylpyridazine (**69**) and 3,6-diphenylpyridazine (**70**).

Interestingly, Stanforth showed that 6-aryl-3-chloropyridazines (**71**) can be selectively pre-pared if 3-chloro-6-iodopyridazine (**50**) is used as substrate under Gronowitz-type reaction conditions [41].

If specific 6-substituted 3-arylpyridazines are required, selectivity problems can be simply avoided by starting immediately from the corresponding 6-substituted 3-chloropyridazines. This is exemplified by the use of 3-chloro- or 3-iodo-6-methoxypyridazine (**72**) [47–49] and 6-chloro- or 6-iodopyridazin-3-amine (**75**) [50, 51] as substrates. Interestingly, the primary amino group of the 6-halopyridazin-3-amines does not seem to hamper the Suzuki arylation, avoiding protection/deprotection steps. 3-Iodo-6-methoxypyridazine **72** and **75** can be easily obtained from **23** or **49** by a selective nucleophilic substitution of one chlorine or iodine atom, respectively. Also, N-alkyl-or N,N-dialkyl-6-arylpyridazin-3-amines have been prepared in a similar way starting from N-alkyl- or N,N-dialkyl-6-chloropyridazin-3-amines, respectively [49].

3-Methoxy-6-phenylpyridazine (**74**), prepared from 3-chloro-6-methoxypyridazine via Suzuki phenylation, has been utilized as a starting material in an alternative route for the synthesis of the antidepressant drug Minaprine [4-methyl-N-(2-morpholin-4-ylethyl)-6-phenylpyridazin-3-amine] (**78**) [47].

In 1995, Quéguiner reported a new route to antihypertensive 5,6-diarylpyridazin-3(2*H*)-ones starting from 4-acetyl-6-chloro-5-iodo-3-methoxypyridazine (**79**) [34]. **79** was synthesized of **79** from 6-chloro-4-(1-hydroxyethyl)-5-iodo-3-methoxypyridazine (**37**) by oxidation with MnO_2 or PCC. The new route involves a chemoselective Suzuki arylation of a C—I over a C—Cl bond on a pyridazine nucleus. In a first Suzuki reaction, C-5 selective arylation of **79** could be obtained using only a slight excess of arylboronic acid. Subsequently, the obtained 4-acetyl-5-aryl-6-chloro-3-methoxypyridazines (**80**) were transformed into unsymmetrically arylated 4-acetyl-5,6-diaryl-3-methoxypyridazines (**81**) via a second Suzuki reaction. For this second arylation, an excess of arylboronic acid (2 equivalents) was used. By using 4 equivalents of 4-chlorophenylboronic acid on **79** symmetrically arylated 4-acetyl-5,6-bis(4-chlorophenyl)-3-methoxypyridazine was obtained in 85% isolated yield.

Interestingly, when searching for new endogenous bioamine-interfering CNS agents, Bourguignon and co-workers found that the two *ortho*-brominated aminopyridazines 4-bromo-6-phenylpyridazin-3-amine and *N*-benzyl-4-bromo-6-phenylpyridazin-3-amine (**82**) can be used as substrates in Suzuki arylation reactions without the need of any amino group protection [52].

The Suzuki cross-coupling reactions could also be extended to fused pyridazines bearing a halogen atom on the 1,2-diazine nucleus. For example, 7-bromo-3-(2-fluorophenyl)-6-[(1-methyl-1*H*-1,2,4-triazol-5-yl)methoxy][1,2,4]triazolo[4,3-*b*]pyridazine (**84**) was successfully coupled with phenylboronic acid to yield 3-(2-fluorophenyl)-6-[(1-methyl-1*H*-1,2,4-triazol-5-yl)methoxy]-7-phenyl[1,2,4]triazolo[4,3-*b*]pyridazine (**85**) in 51% yield [35]. Compounds like **85** might give access to new anxiolytic drugs since trisubstituted 1,2,4-triazolo[4,3-*b*]pyridazines have been identified as subtype selective ligands for the benzodiazepine binding site of GABA-A receptors.

6-Aryl and 6-methyl-2-(4-fluorophenyl)imidazo[1,2-*b*]pyridazines (**87**) were obtained from 6-chloro-2-(4-fluorophenyl)imidazo[1,2-*b*]pyridazine (**86**) via Suzuki coupling with methylboronic acid and arylboronic acids, respectively [53]. Best yields were obtained with sodium hydroxide as the base in coupling reactions of **86** with arylboronic acids, whereas yields decreased as the strength of the base increased (Na_2CO_3 (27%), NaOH (19%), $Ba(OH)_2$ (11%)) in the reaction with methylboronic acid.

Halogenated benzo fused derivatives have also been successfully used as substrates exemplified by the synthesis of 1-aryl-4-(4-methylpiperazinyl)phthalazines (**89**), possessing 5-HT_3 receptor affinity, via Suzuki reaction of 1-chloro-4-(4-methylpiperazinyl)phthalazine (**88**) with arylboronic acids [54]. The substituents on the arylboronic acid and the reaction conditions (catalyst, base, solvent) significantly influenced the yields of the studied transformations. In the coupling reaction with 4-methylphenylboronic acid, for instance, $Pd(PPh_3)_4$ proved not to be an effective catalyst in the presence of Na_2CO_3 in EtOH/toluene (30% yield), Cs_2CO_3 in toluene (20% yield), K_3PO_4 in DMF (35% yield), or $Ba(OH)_2 \cdot 8H_2O$ in dimethoxyethane (0% yield). However, with $Pd_2(dba)_3/P(t\text{-}Bu)_3$ as the precatalyst and Cs_2CO_3 as the base in 1,4-dioxane, the yield (70%) of 1-(4-methylphenyl)-4-(4-methylpiperazinyl)phthalazine was substantially higher.

In 2002, Bodwell and Li reported an elegant and efficient total synthesis of strychnine based on an intramolecular inverse-electron-demand Diels–Alder reaction that involves a pyridazine as diene. The required cyclophane **91** was built up via hydroboration of *N*-[2-(1-allyl-1*H*-indol-3-yl)ethyl]-6-iodopyridazin-3-amine (**90**) followed by an intramolecular sp^3–sp^2 coupling reaction [55].

In the same year, Wermuth and co-workers published intermolecular examples following the same approach [56]. Hydroboration of alkenes followed by Suzuki cross-coupling reaction with 4- (**92**) or 5-substituted 6-aryl-3-iodopyridazines (**94**) gave the corresponding 3-alkylpyridazines (**93, 95**) in good yield.

The first Suzuki cross-coupling reaction on pyridazinyl trifluoromethanesulfonates was reported in 2001 by Aldous and co-workers [29]. 6-Methylpyridazin-3-yl trifluoromethanesulfonate **96** was readily coupled with arylboronic acids. Also an arylboronic ester and a dialkylarylborane have been used as transmetalating agent. Although the triflate **96** is hydrolytically sensitive in aqueous base, aqueous Na_2CO_3/dimethoxyethane worked fine in most cases. 3-Diethylboranylpyridine was an exception which illustrates the competition between the rate of coupling and the hydrolysis rate. It is worth noting that for the coupling of **96** with 2-thienylboronic acid, a range of anhydrous methods using different bases were investigated. Interestingly, all these led to a lower reactivity in comparison with aqueous conditions. The most effective anhydrous conditions proved to be potassium carbonate in toluene yielding 3-methyl-6-(2-thienyl)pyridazine in only 33% yield together with some unreacted starting material. Generally, it can be concluded that although hydrolysis of triflate is inhibited by using anhydrous conditions, aqueous conditions are better due to higher cross-coupling rates.

Not until 2005 were the first organoboron derivatives of pyridazines reported in the literature [57]. Harrity reported the preparation of pyridazin-4-ylboronic esters (**99**) via cycloaddition of tetrazines (**98**) with alkynyl boronic esters. Some of these boronic sters have been used in Suzuki reactions with iodobenzene. Although the organoboron compounds **99** are electron deficient (and some are also sterically hindered) protodeboronation usually proved no major problem.

R¹ = CO₂Me R₂ = Me, *n*-Bu, Ph, TMS

R¹ = [pyrazolyl] R₂ = Me, Ph, TMS

R¹ = H R² = Ph, H

R¹ = H R₂ = H

R¹ = [pyrazolyl] R² = Ph

R¹ = CO₂Me R² = Me

13.2.2.2. (Pseudo)halopyridazin-3(2H)-ones

The wide range of known biological activities of substituted pyridazin-3(2H)-ones as well as the numerous reported successful Suzuki reactions on chloropyridazines stimulated Maes and co-workers to perform similar studies on easily accessible 2-substituted 4,5-dichloropyridazin-3(2H)-one substrates [58]. When cross-coupling reactions were performed with three equivalents of arylboronic acid, 2-substituted 4,5-diarylated pyridazin-3(2H)-ones (101) could be isolated in excellent yields. Unfortunately, by using a smaller excess of arylboronic acid under standard Suzuki conditions, no regioselective displacement of one chloro substituent could be achieved. A mixture of mono- and diarylated pyridazin-3(2H)-ones was always obtained. Similar observations have been made by Stevenson on the related 2-substituted 4,5-dibromopyridazin-3(2H)-ones using Gronowitz-type reaction conditions [15].

R = Me, Ph
Ar = Ph, 4-FPh, 4-MeOPh

In 2004, Gong and He reported C-5 selective Suzuki arylation of 2-methyl-4,5-dichloropyridazin-3(2H)-one (102) with phenylboronic acid [59]. Best results were obtained with a ratio of 2:1 of 102 and phenylboronic acid. Carefully optimized reaction parameters gave 4-chloro-2-methyl-5-phenylpyridazin-3(2H)-one (103) with a selectivity greater than 300-fold over 5-chloro-2-methyl-4-phenylpyridazin-3(2H)-one (104) and 2-methyl-4,5-diphenylpyridazin-3(2H)-one (105). $Pd(PEt_3)_2Cl_2$ was found to be the most suitable precatalyst for this regioselective Suzuki phenylation reaction, and the reaction temperature and solvent were optimized. Unfortunately, the general applicability of these optimized reaction conditions when other arylboronic acids or 2-substituted 4,5-dichloropyridazin-3(2H)-ones are used remains unknown.

PhB(OH)₂
Pd(PEt₃)₂Cl₂
1M Na₂CO₃
DMF, RT
77%

102   **103**   **104**   **105**

330   <1   1

Other types of catalyst have been tested on 4,5-dichloro-2-methylpyridazin-3(2*H*)-one (**102**). Nájera reported the cross-coupling of **102** with phenylboronic acid (2 equivalents) using a low loading (10⁻¹ mol%) of an oxime-derived palladacycle (**106**) as the precatalyst [60]. The reaction was carried out using K₂CO₃ as base in the presence of tetrabutylammonium bromide in water and afforded the desired 4,5-diphenyl-2-methylpyridazin-3(2*H*)-one (**105**) in 92% isolated yield. Interestingly, 2,4,5-trimethylpyridazin-3(2*H*)-one (**107**) could also be obtained from **102** via Suzuki coupling with an equimolar amount of trimethylboroxine using a high loading of the palladacycle (10 mol%). However, the yield was very low since only 20% of **107** could be isolated. Interestingly, besides **107**, 35% of 4-chloro-2,5-dimethylpyridazin-3(2*H*)-one (**108**) was obtained.

PhB(OH)₂
106
K₂CO₃, TBAB
H₂O, reflux
92%

102   **105**   **106**

102 +   **106**
K₂CO₃, TBAB
H₂O, reflux

48   52
107   **108**
20%   35%

N-unsubstituted 4,5-dichloropyridazin-3(2*H*)-one (**109**) fails to undergo Suzuki cross-coupling reactions. In order to synthesize *N*-2, unsubstituted 4,5-diarylpyridazin-3(2*H*)-ones (**112**) by Suzuki reaction, the temporary protection of the lactam moiety of **109** is essential. To achieve this a simple and efficient retro-ene-assisted Suzuki methodology has been developed by Raviña and co-workers which is based on 4,5-dichloro-2-(hydroxymethyl)pyridazin-3(2*H*)-one (**110**) [61]. This pyridazin-3(2*H*)-one is easily available from **109** via reaction with formaldehyde. Interestingly, **110** reacts smoothly with arylboronic acids to afford directly the deprotected 4,5-diarylated pyridazin-3(2*H*)-ones (**112**) in high yields. The mechanism probably involves the formation of 4,5-diarylated-2-(hydroxymethyl)pyridazin-3(2*H*)-ones (**111**), which subsequently lose formaldehyde by a retro-ene reaction induced by base or heat [62].

35% CH₂O
reflux
62%

ArB(OH)₂
Pd(PPh₃)₄
Na₂CO₃
DME, H₂O
reflux

retro-ene
- CH₂O

109   **110**   **111**   **112**
78-92%

Ar = Ph, 4-MePh, 4-ClPh, 4-MeOPh,
4-N(Me)₂Ph

Based on this methodology, a traceless solid-phase synthesis of 4,5-diarylated pyridazin-3(2H)-ones has also been elaborated employing a dihydropyran-functionalized resin (113) [63].

Ar = Ph, 4-MePh, 4-ClPh, 4-MeOPh

An unsual approach for the protection of the lactam moiety of pyridazin-3(2H)-ones is its transformation into an imidoformyl chloride [24]. 4-Bromo-6-chloro-3-phenylpyridazine (19), synthesized from 4-bromo-6-phenylpyridazin-3(2H)-one (6) via reaction with POCl$_3$, could be chemoselectively arylated in the 4-position to afford 4-aryl-6-chloro-3-phenylpyridazines (116) in 73–96% yields using a variety of arylboronic acids (one equivalent). To prove the concept, 4-phenyl- and 4-furyl-6-chloro-3-phenylpyridazine were subsequently smoothly "deprotected" into the corresponding lactam derivatives using hot acetic acid. The observed selectivity clearly illustrates the higher reactivity of the activated C—Br bond in the oxidative addition reaction of the catalytic cycle in comparison to the activated C—Cl bond of the substrate 19.

Ar = Ph, 4-MePh, 4-ClPh, 4-MeOPh,
2-furyl, 2-thienyl, 4-CHOPh

To deal with the regioselectivity problems observed in Suzuki arylation reactions on 2-substituted 4,5-dichloropyridazin-3(2H)-ones under classical conditions, Mátyus, Maes, and Riedl introduced the concept of "provisionally masked functionalities" (PMF's) [64]. The PMF concept is based on the well-documented regioselective introduction of a methoxy group in the 4 or 5 position of 2-substituted 4,5-dichloropyridazin-3(2H)-ones by carefully controlling the reaction conditions and its inertness in an oxidative addition reaction [58]. The latter guaranties a complete regioselectivity in the subsequent Pd-catalyzed cross-coupling reaction that involves the remaining C—Cl bond. The methoxy group introduced by a nucleophilic substitution reaction of the addition-elimination type could then be utilized for another inter- or intramolecular C—C or C—X bond formation as it is, or by conversion into a better leaving group.

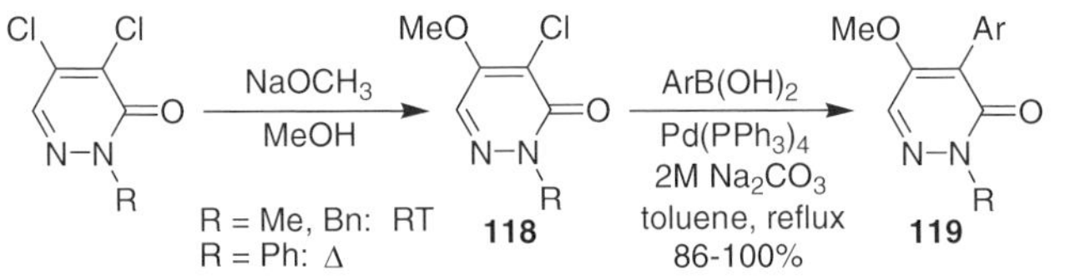

R = Me, Bn, Ph

Ar = Ph, 2-MePh, 3-MeOPh, 4-MeOPh,
4-MeSPh, 4-FPh, 2,4-ClPh, 3-CF₃Ph

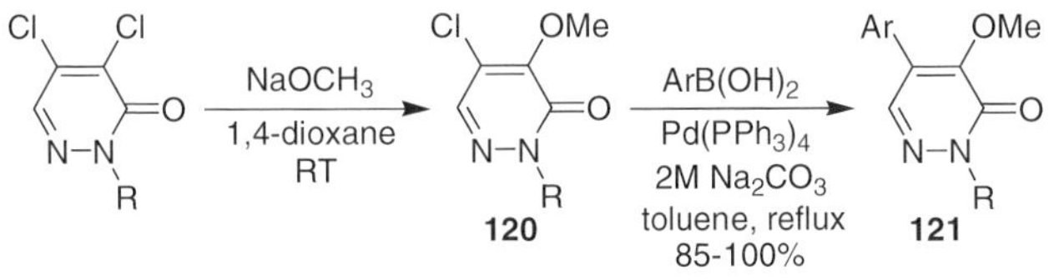

R = Me, Bn, Ph

Ar = Ph, 4-MeOPh, 4-MeSPh, 4-FPh,
3-CF₃Ph, 3-thienyl

The PMF strategy has been applied to prepare 2-substituted 4,5-diarylpyridazin-3(2*H*)-ones with two different aryl groups. For this purpose, 2-substituted 4-aryl-5-methoxy- (**119**) and 5-aryl-4-methoxypyridazin-3(2*H*)-ones (**121**) were converted into the corresponding hydroxypyridazin-3(2*H*)-ones (**122, 126**) by aqueous basic hydrolysis [65]. Subsequently, the aryl-hydroxypyridazin-3(2*H*)-ones were transformed into the corresponding triflate esters (**123, 127**) giving access to a new leaving group for a second Suzuki arylation reaction. Efficient synthetic approaches towards 2-substituted 4,5-diarylpyridazin-3(2*H*)-ones are quite important since some representatives of this class such as 4-(4-fluorophenyl)-2-{[(1*S*,2*S*)-2-methylcyclopropyl]methyl}-5-[4-(methylsulfonyl)phenyl]pyridazin-3(2*H*)-one have been identified as potent and selective COX-2 inhibitors [66].

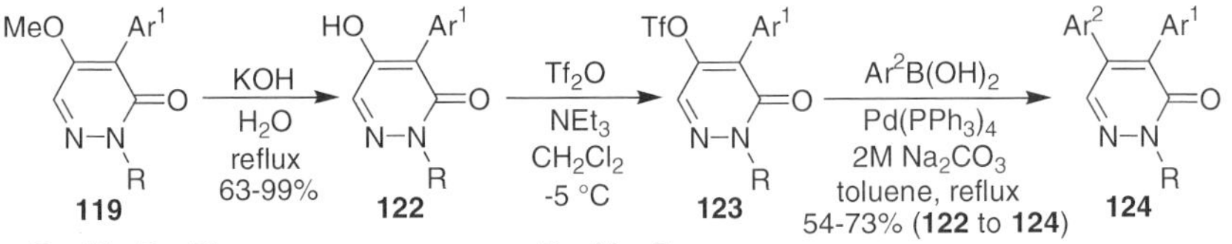

R = Me, Bn, Ph
Ar¹ = Ph, 2-MePh, 4-MeOPh,
4-MeSPh, 2,4-ClPh, 3-CF₃Ph

R = Me, Bn
Ar¹ = 2-MePh, 4-MeSPh
Ar² = Ph, 4-FPh, 3-CF₃Ph

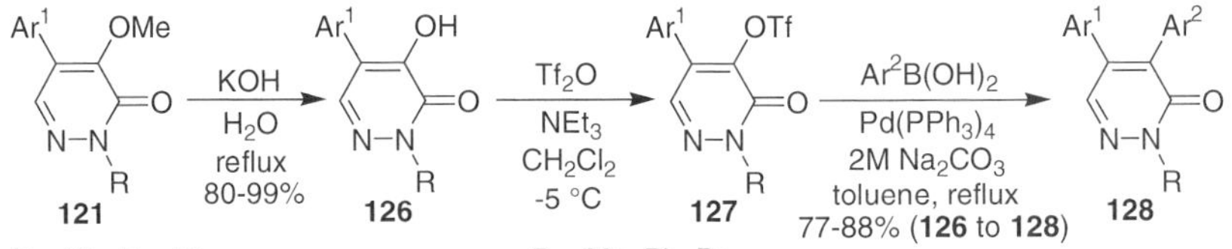

R = Me, Bn, Ph
Ar¹ = 4-MeOPh, 3-thienyl, 4-MeSPh

R = Me, Ph, Bn
Ar¹ = 4-MeOPh, 3-thienyl, 4-MeSPh
Ar² = 4-FPh

As an alternative for the PMF strategy, Stevenson reported the synthesis and use of a pyridazin-3(2*H*)-one with two different halogen atoms; namely, 5-chloro-4-iodo-2-methylpyridazin-3(2*H*)-one (**4**) [15]. While disubstitution could not be completely suppressed, in all cases tested using this substrate only one monoarylated product (**129**) is obtained. This illustrates the greater reactivity of the C—I bond in the oxidative addition reaction of the catalytic cycle.

The usefulness of the PMF strategy has also been demonstrated in the construction of several new polycyclic skeletons that contain a pyridazin-3(2*H*)-one moiety. The dibenzo[*f,h*]phthalazin-1(2*H*)-one skeleton was synthesized via Suzuki cross-coupling of 2-benzyl-5-trifluoromethanesulfonyloxy-4-phenylpyridazin-3(2*H*)-one (**132**) with 2-bromophenylboronic acid, followed by an intramolecular Heck-type reaction on the resulting diarylpyridazin-3(2*H*)-one **133**. Alternatively, the same skeleton could also be prepared from the same substrate (**132**) via coupling with 2-pivaloylaminophenylboronic acid followed by amine deprotection using acid hydrolysis, diazotization, and spontaneous Pschorr-type ring closure [67].

1*H*-isochromeno[3,4-*d*]pyridazine-1,6(2*H*)-dione (**139**) and its regioisomer 3*H*-isochromeno[3,4-*d*]pyridazine-4,6-dione (**143**) have also been constructed using the PMF concept [65]. Alkaline hydrolysis of the methoxy group of **138** gives 2-(2-benzyl-5-hydroxy-3-oxo-2,3-dihydropyridazin-4-yl)benzaldehyde which was not isolated. After cooling the reaction mixture, the aldehyde moiety of the hydroxypyridazin-3(2*H*)-one was oxidized with KMnO₄, and acidification of the filtrate gave directly 2-benzyl-1*H*-isochromeno[3,4-*d*]pyridazine-1,6(2*H*)-dione (**139**) in a one-pot procedure. Surprisingly, attempts to use a similar protocol on the isomeric 2-(1-benzyl-5-methoxy-6-oxo-1,6-dihydropyridazin-4-yl)benzaldehyde (**141**) led to the uncyclized

2-(1-benzyl-5-hydroxy-6-oxo-1,6-dihydropyridazin-4-yl)benzoic acid (142). Lactonization of 142 with a catalytic amount of concentrated H_2SO_4 in DME at 60°C finally gave the desired 3-benzyl-3H-isochromeno[3,4-d]pyridazine-4,6-dione (143). The synthesis of the substrates 138 and 141 required the use of Gronowitz coupling conditions as well as gradual addition of excess of 2-formylphenylboronic acid. Both factors probably suppress the unwanted protodeboronation of the 2-formylphenylboronic acid.

Interestingly, the methoxy group of 138 and 141 can immediately be used as a leaving group as exemplified by the synthesis of regioisomeric pyridazino[4,5-c]isoquinolinones 144 and 145. The mechanism of the ring closure reaction probably involves imine formation followed by an electrocyclic reaction and subsequent aromatization via elimination of methanol [68].

5-Iodo-2-methylpyridazin-3(2H)-one (146) has also been used as a starting material for the preparation of polycyclic pyridazin-3(2H)-ones [69]. Reaction of 146 with 2-pivaloylaminophenylboronic acid under Gronowitz conditions smoothly gave 5-arylpyridazin-3(2H)-one 147. Hydrolysis of the pivaloylanilide 147 and subsequent diazotization of the resulting amine followed by azidation of the corresponding diazonium salt yielded azide 149. Finally, heating 149 in 1,2-dichlorobenzene gave 3-methyl-3,5-dihydro-4H-pyridazino[4,5-b]indol-4-one (150) in 73%. Interestingly, the ring closure occurs in a completely regioselective way since no isomeric 2-methyl-2,9-dihydro-3H-pyridazino[3,4-b]indol-3-one was formed. The mechanism of the cyclization reaction involves the formation of a nitrene. Formally, the 4-H atom of 146 can be regarded as a PMF group since 146 is synthesized from 2-methyl-4,5-dichloropyridazin-3(2H)-one via reaction with 57% HI and the H-4 hydrogen atom acts as a leaving group in the nitrene cyclization reaction.

13.2.3. Stille coupling

The first Stille reactions on halopyridazines were described by Draper and Bailey in 1995 [48]. They successfully coupled 6-substituted 3-iodopyridazines (**151**) with arylstannanes in refluxing THF or DMF at 80°C using $Pd(PPh_3)_2Cl_2$ as the precatalyst.

R^1 = MeO, Me_2N, MeS   Ar = 2-thienyl (R^2 = n-Bu)
5-(1,3-dioxolan-2-yl)-2-furyl (R^2 = Me)

3-Chloropyridazines seem to be preferred over 3-iodopyridazines in this type of cross-coupling reaction [50]. Maes and co-workers compared Stille reactions of 6-chloropyridazin-3-amine and 6-iodopyridazin-3-amine under identical reaction conditions. Remarkably, using the same aryl(tributyl)stannane cross-coupling on the chloropyridazine was substantially faster than the same reaction on the corresponding iodo derivative although oxidative addition on the latter is certainly easier.

X = I: 87-99%
X = Cl: 89-96%   Ar = 2-thienyl, 2-furyl

Besides halopyridazines, pseudohalopyridazines have been used as substrates for Stille reactions [29]. 6-Methylpyridazin-3-yl trifluoromethanesulfonate (**96**) reacted with arylstannanes using a procedure based on Stille's original conditions for aryl triflates. Although 3-methyl-6-(2-thienyl)pyridazine was obtained in a good yield (77%) under these conditions, trialkyl(phenyl)stannanes reacted only very slowly in comparison with tributyl(2-thienyl)stannane. Trimethyl(phenyl)stannane and tributyl(phenyl)stannane gave 3-methyl-6-phenylpyridazine in only 22 and 6% yield, respectively.

Me—[pyridazine]—OSO$_2$CF$_3$ $\xrightarrow[\substack{\text{Pd(PPh}_3)_4 \\ \text{LiCl} \\ \text{1,4-dioxane} \\ 80\text{-}85\,°C \\ 22\text{-}77\%}]{\text{ArSnR}_3}$ Me—[pyridazine]—Ar

96 **97** Ar = Ph (R = Me), 2-thienyl (R = *n*-Bu)

At the end of the nineties, Sauer prepared the first trialkylstannylpyridazines via a regioselective [4+2] cycloaddition of tributylethynylstannane with various 3-aryl-1,2,4,5-tetrazines (**153**) [37, 70]. The studied cycloaddition reactions gave 3-aryl-5-tributylstannylpyridazines (**154**) highly regioselectively since the *o*/*m* ratio varied only from 1:99 to 5:95 depending on the aryl group. Also 4-tributylstannylpyridazine (**156**) was synthesized in a similar manner via cycloaddition with 1,2,4,5-tetrazine (**155**). Stille cross-coupling reactions of 5-tributylstannyl-3-phenylpyridazine (**42**) and 4-tributylstannylpyridazine (**156**) with aryl halides in the presence of Pd(PPh$_3$)$_4$ as the catalyst gave 5-aryl-3-phenylpyridazines (**157**) and 4-arylpyridazines (**158**), respectively, in moderate to good yields.

153 + [SnBu$_3$ alkyne] $\xrightarrow[\substack{\text{or} \\ \text{CH}_2\text{Cl}_2,\ RT \\ 71\text{-}95\%}]{\text{MeCN, 80 °C}}$ **154** + (minor)

Ar = Ph, 4-ClPh, 4-MeOPh, 3-CF$_3$Ph
   2-furyl, 3-furyl, 2-thienyl, 3-thienyl
   2-1*H*-pyrrolyl, 2-(1-methyl-1*H*-pyrrolyl), 2-pyridinyl
   3-pyridinyl, 4-pyridinyl, 2-thiazolyl, 4-thiazolyl

42 $\xrightarrow[\substack{\text{Pd(PPh}_3)_4 \\ \text{toluene} \\ \text{or} \\ \text{1,4-dioxane} \\ \text{reflux} \\ 46\text{-}67\%}]{\text{ArX}}$ **157**

Ar = Ph (X = I), 2-pyridinyl (X = Br)
   2-thienyl (X = Br), 5-pyrimidinyl (X = Br)

155 + [SnBu$_3$ alkyne] $\xrightarrow[76\%]{\text{toluene, RT}}$ **156** $\xrightarrow[\substack{\text{Pd(PPh}_3)_4 \\ \text{toluene} \\ \text{reflux} \\ 51\text{-}93\%}]{\text{ArX}}$ **158**

Ar = Ph (X = I), 2-pyridinyl (X = Br)
   2-thienyl (X = Br), 5-pyrimidinyl (X = Br)

A pyridazine-containing bioisostere of (−)-ferruginine was obtained via the coupling of triflate **159** with 4-(tributylstannyl)pyridazine (**156**) [71]. Several reaction conditions were tried to optimize the yield of this reaction. The best results were obtained using 10 mol% Pd(PhCN)$_2$Cl$_2$, 20 mol% Ph$_3$As, and 20 mol% CuI in combination with three equivalents of LiCl in *N*-methylpyrrolidone. Removal of the carbamate group of **160** finally afforded the bioisostere of (−)-ferruginine (**161**).

(-)-ferruginine

A tributylstannyl group can also be introduced via transmetalation. Collins transmetalated the trimethylsilyl group of the 1,2,4-triazolo[4,3-b]pyridazine **162** to the corresponding tributylstannane **163** using $(Bu_3Sn)_2O$ [35]. Subsequent Stille reaction of **163** with iodobenzene in the presence of $Pd(PPh_3)_4$ in DMF at 100°C resulted in the formation of **164**, albeit in a low overall yield. In the same paper, a reverse approach was presented in which a brominated triazolopyridazine **165** reacted with 2-tributylstannyl-1,3-oxazole to yield 6-[(1-methyl-1H-1,2,4-triazol-5-yl)methoxy]-7-(1,3-oxazol-2-yl)-3-phenyl[1,2,4]triazolo[4,3-b]pyridazine **166** in 37% yield (23% of starting compound was also recovered).

Ditopic bidendate ligands **168**, **169**, and **170**, as well as the dimer of **170** (**171**) were synthesized based on a Stille cross-coupling reaction [72]. Reaction of 3-chloropyridazine **167** with 5-(1,3-dioxolan-2-yl)-2-(tributylstannyl)pyridine in the presence of $Pd(PPh_3)_4$, using CuI as co-catalyst, in refluxing toluene led to **168** in moderate yield. Hydrolysis of the acetal function of **168** gave aldehyde **169** which was transformed into the acetylene **170** using dimethyl (1-diazo-2-oxopropyl)phosphonate. Dimethyl (1-diazo-2-oxopropyl)phosphonate generated the required dimethyl (diazomethyl)phosphonate by methanolysis. The linear dimer ligand **171** was synthesized via homocoupling of **170** under reaction conditions similar to the Eglinton procedure.

Stille reactions on halopyridazin-3(2*H*)-ones were also studied. In the frame of a medicinal program, dealing with the search for new potent antiplatelet agents, Raviña and coworkers investigated the derivatization of the 5-bromo-6-phenylpyridazin-3(2*H*)-one skeleton (**6**) by Stille reactions [73–76]. Cross-coupling of methoxymethylether MOM protected 5-bromo-6-phenylpyridazin-3(2*H*)-one (**172**) with tributyl(vinyl)stannane, tributyl(2-thienyl)stannane, tributyl(ethynyl)stannane, or tributyl(phenylethynyl)stannane gave the desired coupling products [73, 75]. Deprotection of these pyridazin-3(2*H*)-ones was first attempted by acidic hydrolysis (6N HCl, reflux), used typically for removing a MOM group [73, 74]. However, these conditions were unsuitable for compounds with multiple bonds presumably due to electrophilic addition. Therefore, aluminum trichloride or boron tribromide were introduced as the reagents to cleave the MOM group. Interestingly, deprotection of 5-(1-ethoxyvinyl)-2-methoxymethyl-6-phenylpyridazin-3(2*H*)-one, prepared via coupling of **172** with (1-ethoxyvinyl)tributylstannane, with 6N HCl at reflux gave 5-acetyl-6-phenylpyridazin-3(2*H*)-one via additional ether cleavage and tautomerization [75]. When a hydroxymethyl group was used to protect **6**, Stille reaction with tributyl(vinyl)stannane and deprotection (retro-ene reaction) occurred consecutively in a one-pot procedure yielding 6-phenyl-5-vinylpyridazin-3(2*H*)-one in 81% [76]. Similar behavior was observed using 2-hydroxymethyl-4,5-dibromopyridazin-3(2*H*)-one as substrate [61].

R = vinyl, 2-thienyl
ethynyl, phenylethynyl

2-Substituted 5-iodo-pyridazin-3(2*H*)-ones such as 5-iodo-2-methylpyridazin-3(2*H*)-one (**146**) have also been used in Stille reactions as exemplified by the smooth coupling of **146** with tributyl(vinyl)stannane [77].

A trialkylstannylpyridazin-3(2*H*)-one has recently also been reported in the literature by researchers of Dupont Crop Protection Products [15]. Stille reaction of 4-iodo-5-methoxy-2-methylpyridazin-3(2*H*)-one (**176**) with hexamethyldistannane gave 5-methoxy-2-methyl-4-(trimethylstannyl)pyridazin-3(2*H*)-one (**177**) in good yield. **177** was successfully used as transmetalating agent in an agrochemical program dealing with the synthesis of Strobilurin A analogs such as **178**.

13.3. Sonogashira reaction

The first example of a Sonogashira reaction on a pyridazine core was reported in 1978 [78]. Igeta and co-workers initially investigated the coupling of 3-iodo-6-methyl- and 3-chloro-6-methylpyridazine (**179**) with phenylacetylene at room temperature using diethylamine as base and solvent [78, 79]. Only poor results were obtained since starting material could be recovered after a reaction time of 15 h. The iodo- derivative is more reactive than the corresponding chloropyridazine which is in accordance with the expected oxidative addition rate.

Because of their better availability, the less reactive 3-chloropyridazine derivatives were focused on by the authors for further optimization. Better results could be achieved for Sonogashira coupling of 3-chloro-6-methylpyridazine with phenylacetylene (56%) by performing the reaction at a higher temperature with less diethylamine using a higher catalyst loading. Subsequently, several 6-substituted 3-chloropyridazines (**181**) were heated at 50–70°C with several acetylenes under these optimized conditions allowing good access to the corresponding 3-alkynylpyridazines (**182**) [78–80]. However, when a strong electron-releasing substituent was present in the 6-position (e.g. piperidino) of the 3-chloropyridazine, poor results were obtained with significant amounts of recovered starting material [79]. Performing these coupling reactions for a longer reaction time gave acceptable yields.

Furthermore, 3-chloropyridazine 1-oxides (**183**) could be used as substrates in this reaction type [78, 79]. However, 3-chloropyridazine 2-oxides were surprisingly unreactive [79].

In 1983, Yamanaka published smooth ethynylation of several haloazines and -diazines via Sonogashira coupling with trimethylsilylacetylene, using triethylamine as base and solvent, followed by subsequent desilylation with aqueous methanolic potassium hydroxide [81]. Two chloropyridazines were included as substrates in this report; namely, 3-chloro-6-methylpyridazine and 4-chloro-3,6-dimethylpyridazine (**185**).

In 1983 and 1992, the same group successfully coupled methyl 6-chloropyridazine-3-carboxylate (**188**) with several aliphatic acetylenes and phenylacetylene [82]. Reduction of the triple bond of **189** using catalytic hydrogenation (Pd/C, H_2) gave a simple and smooth access to methyl 6-alkylpyridazine-3-carboxylates (**190**).

R = Pr, Bu, *n*-Pen, *n*-Hex, *n*-Hep, Ph

Clearly, chloropyridazines with a strong electron-withdrawing substituent can smoothly react at room temperature since Heinisch and co-workers published the facile coupling of 3-chloro-4-cyanopyridazine (191) with phenylacetylene at room temperature [83].

In 1995, Bailey showed that the harsh reaction conditions often required to prepare 6-substituted 3-alkynylpyridazines from the corresponding 3-chloropyridazines, especially when a strong electron-releasing substituent is present in the 6-position, could be avoided when 3-iodopyridazines (151) are used. These substrates allow efficient coupling at room temperature [48].

Haider and Käferböck developed a route for the preparation of substituted annulated carbazoles (196) based on the Sonogashira reaction of 3-iodopyridazines 193 with 1-(prop-2-yn-1-yl)indoline and 1-(but-3-yn-1-yl)indoline at ambient temperature [84]. Hydrogenation of the triple bond of compounds 194, using Pd/C under a hydrogen gas pressure (60 psi), gave overreduction to dihydropyridazines (195). Subsequent dehydrogenation/ aromatization of the indoline moiety to an indole, using Pd/C as the catalyst in the presence of oxygen in refluxing xylene, simultaneously reoxidized the dihydropyridazines to the required pyridazines. The dehydrogenation reaction yielded the desired diene (pyridazine) and dienophile (indole) for carbazole construction via intramolecular [4+2] cycloaddition.

Heinisch reported efficient dialkynylation when 3,6-diiodopyridazine (49) is used in Sonogashira reactions at room temperature [85]. Interestingly, Stanforth described that selectivity can be realized when 3-chloro-6-iodopyridazine (**50**) is used as the substrate, albeit only low yields of **198** were obtained [41].

R = Ph, C(Me)$_2$OH

R^1 = H, Me   R^2 = Me, Ph
R^3 = CH$_2$OH, SiEt$_3$, C(Me)$_2$OH, CH$_2$Ot-BDPS
C(OH)(CH$_2$)$_5$, CH$_2$NHBOC, CH(OEt)$_2$

Pyridazin-3-yl trifluoromethanesulfonates (**199**) have also been smoothly alkynylated at room temperature [28].

Hitherto, only two papers appeared in the literature in which an alkynyl group was introduced into the 4 position of a monocyclic pyridazine using a Sonogashira reaction [52, 81]. Besides the previously mentioned 4-chloro-3,6-dimethylpyridazine (**185**), Bourguignon showed that electronically rich and sterically hindered *N*-benzyl-4-bromo- and 4-bromo-6-phenyl-pyridazin-3-amine (**82**) are valuable coupling partners for different alkynes.

R^1 = H, Bn
R^2 = TMS, (CH$_2$)$_3$OBn, Ph, CH$_2$NHBOC

The first example of an alkynylation of a bicyclic pyridazine derivative was published in 2000 [35]. Collins studied Sonogashira coupling of 7-bromo-6-[(1-methyl-1*H*-1,2,4-triazol-5-yl)methoxy]-3-phenyl[1,2,4]triazolo[4,3-*b*]pyridazine (**165**) with propargyl alcohol at 90°C in a sealed tube using Pd(PPh$_3$)$_4$ and CuI in triethylamine. Unfortunately, only a low yield of the reaction product (**202**) was obtained.

Sonogashira reactions have also been used to construct pyrrolo fused pyridazines. Gulevskaya described the synthesis of substituted 6,8-dimethyl-1*H*-pyrimido[4,5-*c*]pyrrolo[2,3-*e*]pyridazine-7,9(6*H*,8*H*)-diones (**205**) using a combination of a Sonogashira reaction and an oxidative amination [86, 87]. Besides 3-chloro-6,8-dimethylpyrimido[4,5-*c*]pyridazine-5,7(6*H*,8*H*)-dione, the Russian research team also prepared 6,8-dimethyl-5,7-dioxo-5,6,7,8-tetrahydropyrimido[4,5-*c*]pyridazin-3-yl trifluoromethanesulfonate as substrate for alkynylation. Similar results were obtained starting from both substrates. The resulting 3-alkynyl-6,8-dimethylpyrimido[4,5-*c*]pyridazine-5,7(6*H*,8*H*)-diones (**204**) could be transformed to substituted 6,8-dimethyl-1*H*-pyrimido[4,5-*c*]pyrrolo[2,3-*e*]pyridazine-7,9(6*H*,8*H*)-diones (**205**) via oxidative amination. Alternatively, the same skeleton could be built up via a reverse approach starting from 4-amino-3-chloro-6,8-dimethylpyrimido[4,5-*c*]pyridazine-5,7(6*H*,8*H*)-diones (**206**).

Successful alkynylation of a pyridazin-3(2*H*)-one core has not been reported before 1999 [73]. Raviña and co-workers used *N*-2 MOM protected 5-bromo-6-phenylpyridazin-3(2*H*)-one (**172**) as the initial substrate. The protection was critical for the success of the coupling reactions. Deprotection of *N*-2 of the alkynylpyridazin-3(2*H*)-one reaction products (**207**) with 6N HCl under reflux was unsuitable since anti-Markovnikov addition of the acid to the triple bond was observed [74, 88].

R = TMS, CH$_2$OH, (CH$_2$)$_4$OH, COOH

The same researchers reported that a 2-hydroxymethyl protecting group (**208**) is compatible with Sonogashira cross-coupling reaction conditions [76, 89]. Remarkably, using this protective group, alkynylation occurs first followed immediately by the loss of formaldehyde (deprotection) via a base or heat promoted retro-ene reaction.

R = TMS, CH$_2$OH, CH(OEt)$_2$

Interestingly, the use of 1-phenylprop-2-yn-1-ol as alkyne on *N*-2 protected 5-bromo-6-phenylpyridazin-3(2*H*)-ones gave rearrangement of the initially formed 5-(3-hydroxy-3-phenylprop-1-yn-1-yl)-6-phenylpyridazin-3(2*H*)-ones to the tautomeric 5-[(1*E*)-3-oxo-3-phenylprop-1-en-1-yl]-6-phenylpyridazin-3(2*H*)-ones (**210**, **211**). This *in situ* base-catalyzed isomerization to the *E*-chalcone was observed with the MOM (**172**) as well as with the hydroxymethyl (**208**) as protecting group for the lactam moiety of the pyridazin-3(2*H*)-one core [89, 90]. More recently, the same team found similar behavior on the 4,5-dibromo-2-hydroxymethylpyridazin-3(2*H*)-one core [61].

At the end of 2001, Maes and co-workers described the first examples of Sonogashira reactions on 2-substituted chloropyridazin-3(2*H*)-ones [30]. When 2-methyl- or 2-phenyl-4,5-dichloropyridazin-3(2*H*)-one was coupled with an excess of alkyne using triethylamine as base in THF at 80°C, in combination with Pd(PPh$_3$)$_2$Cl$_2$ as precatalyst and CuI as co-catalyst, symmetrically substituted dialkynylpyridazin-3(2*H*)-ones (**212**) were easily obtained.

Unsymmetrical derivatives could be prepared via sequential alkynylation starting from 5-chloro-2-methyl-4-trifluoromethanesulfonyloxypyridazin-3(2H)-one (**28**) or its regio-isomer 4-chloro-2-methyl-5-trifluoromethanesulfonyloxypyridazin-3(2H)-one (**30**) [30]. To achieve chemoselective alkynylation on the triflates (**28**, **30**), a nearly equimolar amount of alkyne, a room temperature reaction, as well as careful control of the reaction time were required.

4-Alkynyl-5-chloro-2-methylpyridazin-3(2H)-ones (**214**) can also be obtained via chemoselective Sonogashira reaction on 5-chloro-4-iodo-2-methylpyridazin-3(2H)-one (**4**) as exemplified by the coupling of 4-fluorophenylacetylene with **4** [15].

The alkynyl-chloropyridazin-3(2H)-ones (**214**, **216**), formed in the first Sonogashira coupling on **28** and **30**, proved to be very useful building blocks since they could be used to prepare thieno[2,3-*d*]pyridazin-4(5*H*)-one (218), furo[2,3-*d*]pyridazin-4(5*H*)-one (219), thieno[2,3-*d*]pyridazin-7(6*H*)-one (220), furo[2,3-*d*]pyridazin-7(6*H*)-one (221), 1,5-dihydro-4*H*-pyrrolo[2,3-*d*]pyridazin-4-one (222), and 1,6-dihydro-7*H*-pyrrolo[2,3-*d*]pyridazin-7-one (223) skeleta [91, 92].

Interestingly, Sonogashira coupling of 4-bromo-5-methoxy-2-methylpyridazin-3(2*H*)-one (224) with phenylacetylene gave 5-methoxy-2-methyl-4-phenylethynylpyridazin-3(2*H*)-one (226) in good yield, while a similar reaction on the chloro analog (4-chloro-5-methoxy-2-methylpyridazin-3(2*H*)-one) yielded only traces of 226 [15, 91]. Alternatively, 226 could also be synthesized via nucleophilic substitution of the chlorine atom of 5-chloro-2-methyl-4-phenylethynylpyridazin-3(2*H*)-one (225) obtained via chemoselective Sonogashira reaction on 28 [91].

Readily accessible 2-substituted 5-iodopyridazin-3(2*H*)-ones were also used as substrates. For instance, coupling of 5-iodo-2-methylpyridazin-3(2*H*)-one (146) with 1-phenylprop-2-yn-1-ol easily gave 2-methyl-5-[(1*E*)-3-oxo-3-phenylprop-1-en-1-yl]pyridazin-3(2*H*)-one (227) via *in situ* base-catalyzed isomerization of the initially formed tautomeric 5-(3-hydroxy-3-phenylprop-1-yn-1-yl)-2-methylpyridazin-3(2*H*)-one [77].

13.4. Heck and intramolecular Heck reactions

In 1999, Raviña and co-workers described the first example of an intermolecular Heck reaction on a pyridazine derivative [73]. Heating 5-bromo-2-methoxymethyl-6-phenylpyridazin-3(2H)-one (172) with styrene at 120°C for two days in a pressure tube, using Pd(OAc)$_2$/PPh$_3$ as the precatalyst and triethylamine as the base in acetonitrile, yielded 2-methoxymethyl-6-phenyl-5-(2-phenylvinyl)pyridazin-3(2H)-one in moderate yield (40%). Later, the same research group published more examples of Heck alkenylations on the same substrate with alkyl acrylates and acrylonitrile [93]. The catalyst and base were the same, but acetonitrile was replaced by DMF as the solvent. N-2 protection of the pyridazin-3(2H)-one was essential for the alkenylation to occur since no formation of the desired reaction product could be observed on 5-bromo-6-phenylpyridazin-3(2H)-one (6) itself. Instead, Michael addition of the N-2 nitrogen on the alkyl acrylates and acrylonitrile occurred. Besides this 1,4-addition (228), a Heck reaction on the C-5 position of these N-2 blocked pyridazin-3(2H)-ones was also observed. Only activated alkenes gave 1,4-addition and subsequent alkenylation since the use of styrene yielded only recovery of unreacted starting material.

Disappointingly, a Heck reaction on 5-bromo-2-methoxymethyl-6-phenylpyridazin-3(2H)-one (172) afforded only a low yield of 5-alkenyl-2-methoxymethyl-6-phenylpyridazin-3(2H)-ones (232). The major compounds were dehalogenated substrate (230) and 6-substituted 2-methoxymethyl-4-phenylphthalazin-1(2H)-ones (231). The formation of the latter compound depends on the alkene since when styrene was used no phthalazin-1(2H)-one formation occurred. In this case, 65% of 2-methoxymethyl-6-phenylpyridazin-3(2H)-one and 5% 2-methoxymethyl-6-phenyl-5-(2-phenylvinyl)pyridazin-3(2H)-one could be isolated. The formation of alkyl (2-methoxymethyl)-1-oxo-4-phenyl-1,2-dihydrophthalazine-6-carboxylates and (2-methoxymethyl)-1-oxo-4-phenyl-1,2-dihydrophthalazine-6-carbonitrile, using alkyl acrylates and acrylonitrile as alkenes, respectively, can be rationalized by a tandem Pd-catalyzed process.

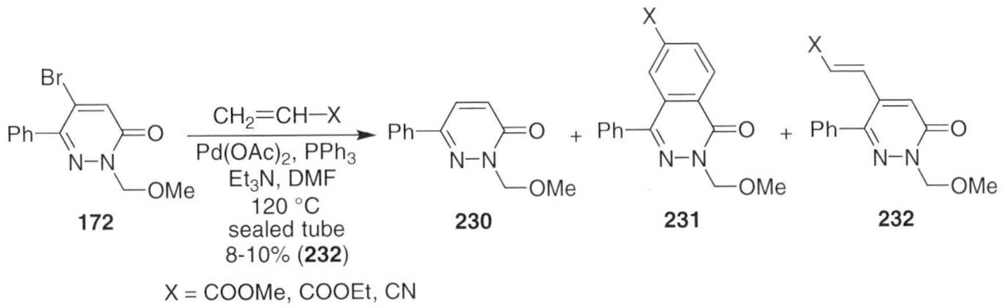

X = COOMe, COOEt, CN

Interestingly, using Pd[P-*o*-tolyl$_3$]$_2$Cl$_2$ as a catalyst dramatically increased the formation of the desired 5-alkenyl-2-methoxymethyl-6-phenylpyridazin-3(2*H*)-ones (232). These compounds were subsequently deprotected using AlCl$_3$ or hydrochloric acid.

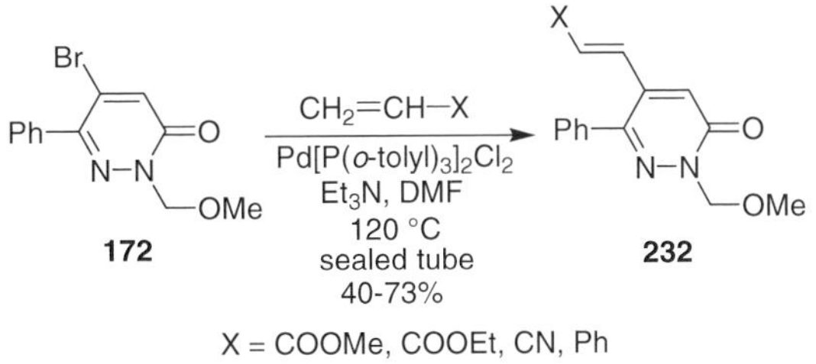

X = COOMe, COOEt, CN, Ph

The optimized Heck conditions could also be used on 5-bromo-2-hydroxymethyl-6-phenylpyridazin-3(2*H*)-one (208) [89]. In this case, a Heck reaction and retro-ene deprotection consecutively occurred in a one-pot procedure.

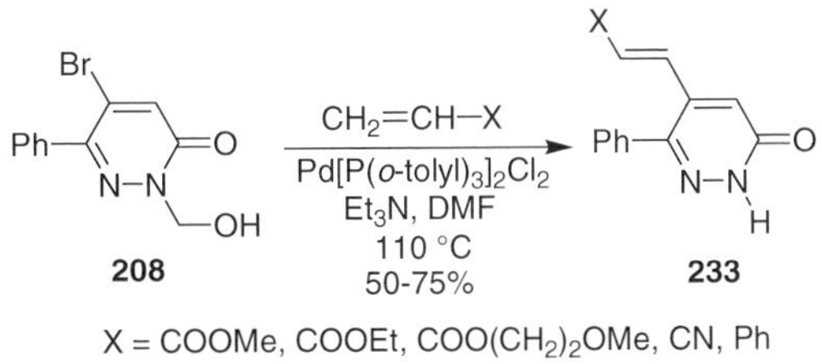

X = COOMe, COOEt, COO(CH$_2$)$_2$OMe, CN, Ph

Heck reaction involving C—H activation of a pyridazin-3(2*H*)-one core has been far less studied. Maes and Mátyus published the first examples in 2004 [42]. Intramolecular Heck-type cyclization of 5-(2-bromophenoxy)-2-methylpyridazin-3(2*H*)-one (234) and 5-[(2-bromophenyl)amino]-2-methylpyridazin-3(2*H*)-one (237), involving C-4 C—H activation of the pyridazin-3(2*H*)-one, yielded 2-methylbenzo[*b*]furo[2,3-*d*]pyridazin-1(2*H*)-one (235) and 2-methyl-2,5-dihydro-1*H*-pyridazino[4,5-*b*]indol-1-one (238), respectively, in good yields. For the construction of 235, a reverse approach in which a bromo atom was present at the C-4 position of the pyridazin-3(2*H*)-one core (236) was also tested but proved to be less efficient.

The required 5-phenoxypyridazin-3(2H)-ones **234** and **236** are easily available through a nucleophilic substitution reaction on halopyridazin-3(2H)-ones **146** and **239**, respectively; whereas the required 5-phenylaminopyridazin-3(2H)-one **237** could be synthesized via a chemoselective Buchwald–Hartwig amination reaction on **146** with 2-bromoaniline.

The 4-phenyl substituted derivatives (**243**, **245**) have also been prepared via this methodology starting from 5-chloro-2-methyl-6-phenylpyridazin-3(2H)-one (**241**). Remarkably, Buchwald–Hartwig reaction of **241** with 2-bromoaniline worked smoothly and no homocoupling of the bromoaniline was observed. The C—Cl bond of **241** seems to be more reactive in the oxidative addition step than the C—Br bond of 2-bromoaniline. The preferential reaction of the C—Cl bond can be explained by taking into account that this C—Cl bond is part of a vinylogous carbamoyl chloride which dramatically increases its reactivity for oxidative addition to Pd(BINAP) in comparison with unactivated C—Cl bonds. In addition, the amino substituent of 2-bromoaniline sterically and electronically deactivates the C—Br bond for the oxidative addition to Pd(BINAP) catalyst. Interestingly, the reaction did not end up with the formation of **244**. In a one-pot process, the intermediate **244** immediately cyclized via an intramolecular Heck-type reaction yielding 2-methyl-4-phenyl-2,5-dihydro-1*H*-pyridazino[4,5-*b*]indol-1-one (**245**) in 55% overall yield. The construction of **245** represents one of the rare examples that involve C—N and C—C bond formation via auto-tandem Pd-catalysis.

There is also a report in the literature in which a pyridazin-3(2*H*)-one core itself is built up via an intramolecular Heck-type reaction [94]. Cyclodehydrobromination of 2-bromo-*N*-1*H*-indol-1-yl-*N*-methylnicotinamide (**248**) yielded tetracyclic 6-methylpyrido[3′,2′:4,5]-pyridazino[1,6-*a*]indol-5(6*H*)-one (**249**) in 92% yield.

13.5. Carbonylation reactions

13.5.1. Alkoxycarbonylation

In 1996, Wermuth reported that various substituted pyridazin-3-yl trifluoromethane-sulfonates could be smoothly transformed into the corresponding methyl pyridazin-3-carboxylates (**251**) using Pd(OAc)$_2$ as palladium source and dppf (1,1′-bis(diphenylphosphino)ferrocene) as the ligand for the *in situ* formed catalyst [95]. A large excess of methanol was used in DMF as the solvent with triethylamine as the base at 50°C. A CO atmosphere was created by simply using a balloon filled with this gas.

Clearly, no high CO pressure is required to get α,α-insertion on the oxidative addition complexes of the pyridazin-3-yl trifluoromethanesulfonates (250).

R^1 = H R^2 = H R^3 = Me, Ph
R^1 = H R^2 = Me R^3 = Ph, 2,4-ClPh
R^1 = Me R^2 = H R^3 = Ph, 4-FPh

In 2003, Gonzáles-Gómez described a similar transformation for the decoration of the pyridazino[4,3-h]psoralen skeleton [96]. Furocoumarins are of pharmacological interest due to their capacity to link covalently to DNA and other macromolecules upon irradiation with long-wavelength UV light.

Interestingly, 3-chloropyridazines (254) have also been used for alkoxycarbonylation [97, 98]. Reactions on these substrates were always executed in an autoclave using a high CO pressure. When 3,6-dichloropyridazine (23) was used as the substrate, dialkyl pyridazin-3,6-dicarboxylates (256) could easily be obtained. 6-Chloropyridazin-3-carboxylates can be synthesized from 3,6-dichloropyridazine (23) by carefully controlling the amount of alcohol. Unfortunately, only a low yield could be obtained in this case since besides recovery of starting material, pyridazin-3,6-dicarboxylates were also isolated.

R = H, morpholino, 3-CF$_3$PhO,

R = Et, i-Pr, n-Bu, c-Hex

Ethoxycarbonylation of 4,5-dibromo-2-methylpyridazin-3(2H)-one (239) smoothly yielded diethyl 1-methyl-6-oxo-1,6-dihydropyridazine-4,5-dicarboxylate (257) when a high CO pressure was used [15].

13.5.2. Aminocarbonylation

In addition to esters, one report also describes the synthesis of amides via a similar methodology [97]. The synthesis of *N,N,N',N'*-tetraethylpyridazin-3,6-dicarboxamide (**259**) proved to be very difficult since the main isolated reaction product was *N,N*-diethyl-6-chloropyridazin-3-carboxamide (**258**). Moreover, starting material was recovered. However, aminocarbonylation of 6-substituted 3-chloropyridazines (**254**) was efficient.

13.6. C–N bond formation

The formation of carbon–nitrogen bonds on a pyridazine nucleus is usually done via direct nucleophilic substitution (addition-elimination) with an amine on the corresponding halopyridazine derivative [2a, 3]. Less frequently, other leaving groups such as alkoxy, alkylsulfanyl, and alkylsulfonyl groups have been used successfully. However, for less nucleophilic amines, such as arylamines, such a strategy usually does not lead to the desired pyridazinamines and the success depends on the electron density of the pyridazine substrate as well as on the nucleophilicity of the arylamine used [99].

In 2000, Košmrlj and Maes described that Buchwald–Hartwig amination with arylamines works well in cases where a classical approach does not work or requires very harsh reaction conditions [64, 100]. Easily accessible and relatively electron-rich chloropyridazin-3(2*H*)-ones (**120, 261**) were used as model substrates yielding a variety of hitherto unknown 4- and 5-arylamino-substituted pyridazin-3(2*H*)-ones (**262, 263**). The standard second generation ligand (±)-BINAP (2,2'-bis(diphenylphosphino)-1,1'-binaphthyl) was found to be suitable for the Pd-catalyzed amination of chloropyridazin-3(2*H*)-ones. Interestingly, a large excess of almost completely insoluble weak carbonate base (K_2CO_3 or Cs_2CO_3) was essential to obtain high coupling rates. Recently, this "base effect" has been investigated in more detail by Maes [101]. A direct interphase deprotonation of the intermediately formed Pd(II)-amine complex explains the observed behavior.

R = Me X = MeO Ar = 4-MePh, 4-ClPh, 4-FPh, 4-CNPh
                4-NO$_2$Ph, 4-MeOPh, 2-pyridinyl

R = Bn X = PhO Ar = 4-FPh, 2-CNPh, 4-EtO$_2$CPh

R = Ph X = EtS Ar = 4-FPh

An alternative way to prepare aminopyridazin-3(2*H*)-ones such as **266** was also described using benzophenone imine as an ammonia equivalent [100].

In 2002, Wermuth showed that the Buchwald–Hartwig reaction can also be used for the amination of 3-iodo-6-phenylpyridazine and 4- or 5-substituted 6-aryl-3-iodopyridazines (**267**) with aliphatic amines [56]. Interestingly, higher and more reproducible yields could be obtained using Pd-catalyzed amination on the 3-iodopyridazines rather than under the more drastic classical aminolysis starting from the corresponding 3-chloropyridazines. Surprisingly, for the synthesis of 6-(2-methoxyphenyl)-5-methylpyridazin-3-amines no reaction product could be obtained following the classical approach. When 3-iodo-6-(2-methoxyphenyl)-5-methylpyridazine (**94**) was used as a substrate in a Pd-catalyzed amination using an excess of primary amine the expected pyridazin-3-amines were obtained. Interestingly, *N*-alkyl-bis[6-(2-methoxyphenyl)-5-methylpyridazin-3-yl]amines (**269**) could easily be synthesized when two equivalents of **94** in combination with one equivalent of primary amine were used. The latter are the first representatives of a hitherto unknown class of bisarylated amino derivatives. When a primary and secondary amino functionality were present in an aliphatic diamine, a preference for the primary over the secondary amine was observed which is in accordance with literature data on other ring systems. Unfortunately, the functional group compatibility of the described protocol is rather limited since a strong alkoxide base was required.

As described in Section 13.4, Mátyus and Maes developed a new synthetic approach for 2-methyl-2,5-dihydro-1H-pyridazino[4,5-b]indol-1-ones (**238**, **245**) in which a chemoselective arylamination of 5-iodo-2-methyl- (**146**) and 5-chloro-2-methyl-6-phenylpyridazin-3(2H)-one (**241**) with 2-bromoaniline was the key step [42]. Interestingly, with substrate **146** a small "base effect" [101] was found. When five equivalents of K_2CO_3 were used 40% of 5-[(2-bromophenyl)amino]-2-methylpyridazin-3(2H)-one (**237**) and a recovery of 52% of **146** were obtained after refluxing for 3 h, whereas the use of one equivalent of the same base under the same reaction conditions gave a yield of 30% of **237** and a recovery of 63% of **146** in the same reaction time.

In summary, the Pd chemistry of pyridazines and pyridazin-3(2H)-ones has its own characteristics when compared to carbocyclic arenes due to the fact that all positions are activated for nucleophilic attack. As a consequence, chlorinated derivatives usually react with standard palladium catalysts. For Negishi, Suzuki, Stille, and Sonogashira reactions on (pseudo)halopyridazines and -pyridazin-3(2H)-ones PPh_3 has been used as the ligand in most of the reported cases. Dppf and P(o-tolyl)$_3$ are suitable ligands for carbonylations. Interestingly, intramolecular Heck-type reactions on pyridazin-3(2H)-ones can be done using PPh_3 while intermolecular Heck reactions seem to require P(o-tolyl)$_3$ as ligand. In Buchwald–Hartwig reactions, dppf and racemic BINAP have proven to give good results. When comparing the availability of published literature examples, Suzuki, Sonogashira, and Stille reactions are reasonably well explored, while data on Negishi, Heck, carbonylation, and Buchwald–Hartwig reactions are far more limited. For C—N bond formation, this is not really surprising since a metal-mediated process makes only sense when classical nucleophilic substitution fails or gives poor yields (less nucleophilic amines and/or electron-rich 1,2-diazine-substrates). Metalated pyridazines are hitherto only rarely reported. This is remarkable since these are preferred for accessing large 1,2-diazine-based libraries for medicinal and agrochemical projects. The few available examples prove that the stability of the metalated pyridazines seems not to be an issue of major concern. Moreover, all the prepared representatives (Zn, Sn, and B) could be efficiently used as organometallic partners in palladium mediated cross-coupling reactions.

13.7. References

1. Taüber, E. *Chem. Ber.* **1895**, *28*, 451–5.
2. For recent reviews on the chemistry of pyridazines see: (a) Haider, N.; Holzer, W. *Product Class 8: Pyridazines* in Science of Synthesis (Houben-Weyl Methods of Molecular Transformations); Yamamoto, Y., Ed.; Thieme: Stuttgart, **2004**; Vol. 16, pp. 125–249. (b) Kolar, P.; Tišler, M. *Adv. in Heterocycl. Chem.* **2000**, *75*, 167–241.
3. For a recent book on the chemistry of pyridazines see: Brown, D. J. *The Pyridazines Supplement 1* in The Chemistry of Heterocyclic Compounds; Taylor, E. C. and Wipf, P., Eds.; J. Wiley and Sons: New York, **2000**; Vol. 57, pp. 1–687.
4. Bevan, K.; Davies, J. S.; Hassall, C. H.; Morton, R. B.; Phillips, D. A. *J. Chem. Soc. (C)* **1971**, 514–22.
5. Grote, R.; Chen, Y.; Zeeck, A.; Chen, Z.; Zähner, H.; Mischnick-Lübbecke, P.; König, W. A. *J. Antibiot.* **1988**, *41*, 595–601.
6. Easmon, J.; Heinisch, G.; Holzer, W.; Matuszczak, B. *Arch. Pharm.* **1995**, *328*, 307–12.
7. Easmon, J.; Heinisch, G.; Holzer, W.; Matuszczak, B. *Pharmazie* **1996**, *2*, 76–83.
8. (a) Heinisch, G.; Frank, H. *In Progress in Medicinal Chemistry*, Ellis G. P.; West G. B., Eds.; Elsevier Science Publishers: Amsterdam, **1990**, Vol. 27, pp. 1–49. (b) Heinisch, G.; Frank, H. *In Progress in Medicinal Chemistry*; Ellis, G. P.; Luscombe, D. K., Eds.; Elsevier Science Publishers: Amsterdam, **1992**, Vol. 29, pp. 141–83.
9. Coates, W. J. *Pyridazines and their Benzo Derivatives* in Comprehensive Heterocyclic Chemistry; Katritzky, A. R., Rees, C. W.; Scriven, E. F. W., Eds.; Pergamon Press: Oxford, **1996**, Vol. 6, pp. 86–9.
10. Kleemann, A.; Engel, J.; Kutscher, B.; Reichert, D. *Pharmaceutical Substances 4th Edition* ; Georg Thieme Verlag: Stuttgart, **2001**, pp. 1–2286.
11. Dury, K. *Angew. Chem. Int. Edit.* **1965**, *4*, 292–300.
12. Tapolcsányi, P.; Mátyus, P. *Recent Developments in the Chemistry of 4,5-Dihalopyridazin-3(2H)-ones* in Targets in Heterocyclic Systems; Attanasi, O. A.; Spinelli, D., Eds.; Societa Chimica Italiana: Rome, **2002**, Vol. 6, pp. 369–98.
13. Mowry, D. T. *J. Am. Chem. Soc.* **1953**, *75*, 1909–10.
14. Beška, E.; Rapoš, P. *J. Chem. Soc., Perkin Trans. 1* **1976**, 2470–1.
15. Stevenson, T. M.; Crouse, B. A.; Thieu, T. V.; Gebreysus, C.; Finkelstein, B. L.; Sethuraman, M. R.; Dubas-Cordery, C. M.; Piotrowski, D. L. *J. Heterocyclic Chem.* **2005**, *42*, 427–35.
16. Sotelo, E.; Fraiz, N.; Yáñez, M.; Terrades, V.; Laguna, R.; Cano, E.; Raviña, E. *Bioorg. Med. Chem.* **2002**, *10*, 2873–82.
17. Raviña, E.; Teran, C.; Santana, L.; Garcia, N.; Estevez, I. *Heterocycles* **1990**, *31*, 1967–74.
18. Schober, B. D.; Megyeri, G.; Kappe, T. *J. Heterocycl. Chem.* **1990**, *27*, 471–7.
19. Schober, B. D.; Kappe, T. *Monatsh. Chem.* **1990**, *121*, 565–9.
20. Turck, A.; Brument, J.-F.; Quéguiner, G. *J. Heterocycl. Chem.* **1981**, *18*, 1465–8.
21. Yanai, M.; Kinoshita, T. *Yakugaku Zasshi* **1965**, *85*, 344–52.
22. (a) Igeta, H. *Chem. Pharm. Bull.* **1960**, *8*, 368–9. (b) Ogato, M.; Kano, H. *Chem. Pharm. Bull.* **1963**, *11*, 29–35. (c) Sako, S. *Chem. Pharm. Bull.* **1963**, *11*, 337–41.
23. Sitamzé, J. M.; Mann, A.; Wermuth, C.-G. *Heterocycles* **1994**, *39*, 271–6.

24. Sotelo, E.; Raviña, E. *Synlett* **2002**, 223–6.

25. Nagashima, H.; Ukai, K.; Oda, H.; Masaki, Y.; Kaji, K. *Chem. Pharm. Bull.* **1987**, *35*, 350–6.

26. Assandri, A.; Bellasio, E.; Bernareggi, A.; Cristina, T.; Perazzi, A.; Odasso, G. *Arzneim.-Forsch.* **1985**, *35*, 508–13.

27. (a) Klinge, D. E.; van der Plas, H. C.; Geurtsen, G.; Koudijs, A. *Recl. Trav. Chim. Pays-Bas* **1974**, *93*, 236–9. (b) Steck, E. A.; Brundage, P.; Fletcher, L. T. *J. Am. Chem. Soc.* **1954**, *76*, 3225–8.

28. Toussaint, D.; Suffert, J.; Wermuth, C. G. *Heterocycles* **1994**, *38*, 1273–86.

29. Aldous, D. J.; Bower, S.; Moorcroft, N.; Todd, M. *Synlett* **2001**, 150–2.

30. R'Kyek, O.; Maes, B. U. W.; Jonckers, T. H. M.; Lemière, G. L. F.; Dommisse, R. A. *Tetrahedron* **2001**, *57*, 10009–16.

31. For a review on the direct metalation of diazines see: Turck, A.; Plé, N.; Mongin, F.; Quéguiner, G. *Tetrahedron* **2001**, *57*, 4489–505.

32. Mojovic, L.; Turck, A.; Plé, N.; Dorsy, M.; Ndzi, B.; Quéguiner, G. *Tetrahedron* **1996**, *52*, 10417–26.

33. Turck, A.; Plé, N.; Tallon, V.; Quéguiner, G. *Tetrahedron* **1995**, *51*, 13045–60.

34. Trécourt, F.; Turck, A.; Plé, N.; Paris, A.; Quéguiner, G. *J. Heterocycl. Chem.* **1995**, *32*, 1057–62.

35. Collins, I.; Castro, J. L.; Street, L. J. *Tetrahedron Lett.* **2000**, *41*, 781–4.

36. Sakamoto, T.; Funami, N.; Kondo, Y.; Yamanaka, H. *Heterocycles* **1991**, *32*, 1387–90.

37. Sauer, J.; Heldmann, D. K. *Tetrahedron* **1998**, *54*, 4297–312.

38. Barlow, M. G.; Haszeldine, R. N.; Pickett, J. A. *J. Chem. Soc., Perkin Trans. 1* **1978**, 378–80.

39. Becker, H. G. O.; Böttcher, H.; Koch, J. *J. Prakt. Chem.* **1969**, *311*, 286–95.

40. Coad, P.; Coad, R. A.; Clough, S.; Hyepock, J.; Salisbury, R.; Wilkins, C. *J. Org. Chem.* **1963**, *28*, 218–21.

41. Goodman, A. J.; Stanforth, S. P.; Tarbit, B. *Tetrahedron* **1999**, *55*, 15067–70.

42. Dajka-Halász, B.; Monsieurs, K.; Éliás, O.; Károlyházy, L.; Tapolcsányi, P.; Maes, B. U. W.; Riedl, Z.; Hajós, G; Dommisse, R. A.; Lemière, G. L. F.; Košmrlj, J.; Mátyus, P. *Tetrahedron* **2004**, *60*, 2283–91.

43. Krajsovszky, G.; Károlyházy, L.; Riedl, Z.; Csámpai, A.; Dunkel, P.; Lernyei, Á.; Dajka-Halász, B.; Hajós, G.; Mátyus, P. *J. Mol. Struct.* **2005**, *713*, 235–43.

44. Turck, A.; Plé, N.; Leprêtre-Gaquère, A.; Quéguiner, G. *Heterocycles* **1998**, *49*, 205–14.

45. Sharples, C. G. V.; Karig, G.; Simpson, G. L.; Spencer, J. A.; Wright, E.; Millar, N. S.; Wonnacott, S.; Gallager, T. *J. Med. Chem.* **2002**, *45*, 3235–45.

46. Chekmarev, D. S.; Stepanov, A. E.; Kasatkin, A. N. *Tetrahedron Lett.* **2005**, *46*, 1303–5.

47. Turck, A.; Plé, N.; Mojovic, L.; Quéguiner, G. *Bull. Soc. Chim. Fr.* **1993**, *130*, 488–92.

48. Draper, T. L.; Bailey, T. R. *J. Org. Chem.* **1995**, *60*, 748–50.

49. Parrot, I.; Rival, Y.; Wermuth, C. G. *Synthesis,* **1999**, 1163–8.

50. Maes, B. U. W.; Lemière, G. L. F.; Dommisse, R. A.; Augustyns, K.; Haemers, A. *Tetrahedron* **2000**, *56*, 1777–81.

51. Guery, S.; Parrot, I.; Rival, Y.; Wermuth, C. G. *Tetrahedron Lett.* **2001**, *42*, 2115–7.

52. Bourotte, M.; Pellegrini, N.; Schmitt, M.; Bourguignon, J. J. *Synlett* **2003**, 1482–4.

53. Enguehard, C.; Hervet, M.; Allouchi, H.; Debouzy, J. C.; Leger, J. M.; Gueiffier, A. *Synthesis* **2001**, 595–600.
54. Guery, S.; Parrot, I.; Rival, Y.; Wermuth, C. G. *Synthesis* **2001**, 699–701.
55. Bodwell, G. J.; Li, J. *Angew. Chem. Int. Ed.* **2002**, *41*, 3261–2.
      For a related article see: Bodwell, G. J; Li, J. *Org. Lett.* **2002**, *4*, 127–30.
56. Parrot, I.; Ritter, G.; Wermuth, C. G.; Hibert, M. *Synlett* **2002**, 1123–7.
57. Helm, M. D.; Moore, J. E.; Plant, A.; Harrity, J. P. A. *Angew. Chem. Int. Ed.* **2005**, *44*, 3889–92.
58. Maes, B. U. W.; R'kyek, O.; Košmrlj, J.; Lemière, G. L. F.; Esmans, E.; Rozenski, J.; Dommisse, R. A.; Haemers, A. *Tetrahedron* **2001**, *57*, 1323–30.
59. Gong, Y.; He, W. *Heterocycles* **2004**, *62*, 851–6.
60. (a) Botella, L.; Nájera, C. *Angew. Chem. Int. Ed.* **2002**, *41*, 179–81. (b) Botella, L.; Nájera, C. *J. Organomet. Chem.* **2002**, *663*, 46–57.
61. Sotelo, E.; Coelho, A.; Raviña, E. *Tetrahedron Lett.* **2003**, *44*, 4459–62.
62. The loss of formaldehyde from a structurally related 2-(hydroxymethyl)pyridazin-3(2*H*)-one derivative was already observed more than a decade ago: Zára-Kaczián, E.; Mátyus P. *Heterocycles* **1993**, *36*, 519–28.
63. Sotelo, E.; Raviña, E. *Synlett* **2003**, 1113–16.
64. Mátyus, P.; Maes, B. U. W.; Riedl, Z.; Hajós, G.; Lemière, G. L. F.; Tapolcsányi, P.; Monsieurs, K.; Éliás, O.; Dommisse, R. A.; Krajsovszky, G. *Synlett* **2004**, 1123–39.
65. Maes, B. U. W; Monsieurs, K.; Loones, K. T. J.; Lemière, G. L. F.; Dommisse, R.; Mátyus, P.; Riedl, Z.; Hajós, G. *Tetrahedron* **2002**, *58*, 9713–21.
66. Li, C. S.; Brideau, C.; Chan, C. C.; Savoie, C.; Claveau, D.; Charleson, S.; Gordon, R.; Greig, G.; Gauthier, J. Y.; Lau, C. K.; Riendeau, D.; Thérien, M.; Wong, E.; Prasit, P. *Bioorg. Med. Chem. Lett.* **2003**, *13*, 597–600.
67. Tapolcsányi, P.; Maes, B. U. W.; Monsieurs, K.; Lemière, G. L. F.; Riedl, Z.; Hajós, G.; Van den Driessche B.; Dommisse, R. A.; Mátyus, P. *Tetrahedron* **2003**, *59*, 5919–26.
68. Riedl, Z.; Maes, B. U. W.; Monsieurs, K.; Lemière, G. L. F.; Mátyus, P.; Hajós, G. *Tetrahedron* **2002**, *58*, 5645–50.
69. Krajsovszky, G.; Mátyus, P.; Riedl, Z.; Csányi, D.; Hajós, G. *Heterocycles* **2001**, *55*, 1105–11.
70. (a) Heldmann, D. K.; Sauer, J. *Tetrahedron Lett.* **1997**, *38*, 5791–4. (b) Sauer, J.; Heldmann, D. K.; Hetzenegger, J.; Krauthan, J.; Sichert, H.; Schuster, J. *Eur. J. Org. Chem.* **1998**, 2885–96.
71. Gündisch, D.; Harms, K.; Schwarz, S.; Seitz, G.; Stubbs, M. T.; Wegge, T. *Bioorg. Med. Chem.* **2001**, *9*, 2683–91.
72. Romero-Salguero, F.; Lehn, J.-M. *Tetrahedron Lett.* **1999**, *40*, 859–62.
73. Estevez, I.; Coelho, A.; Raviña, E. *Synthesis* **1999**, 1666–70.
74. Sotelo, E.; Coelho, A.; Raviña, E. *Tetrahedron Lett.* **2001**, *42*, 8633–6.
75. Sotelo, E.; Coelho, A.; Raviña, E. *Chem. Pharm. Bull.* **2003**, *51*, 427–30.
76. Coelho, A.; Raviña, E.; Sotelo, E. *Synlett* **2002**, 2062–4.
77. Coelho, A.; Sotelo, E.; Novoa, H.; Peeters, O. M.; Blaton, N.; Raviña, E. *Tetrahedron* **2004**, *60*, 12177–89.
78. Abe, Y.; Ohsawa, A.; Arai, H.; Igeta, H. *Heterocycles* **1978**, *9*, 1397–401.
79. Ohsawa, A.; Abey, Y.; Igeta, H. *Chem. Pharm. Bull.* **1980**, *28*, 3488–93.
80. Ohsawa, A.; Abe, Y.; Igeta, H. *Chem. Lett.* **1979**, 241–4.
81. Sakamoto, T.; Shiraiwa, M.; Kondo, Y.; Yamanaka, H. *Synthesis* **1983**, 312–14.

82. (a) Yamanaka, H.; Mizugaki, M.; Sakamoto, T.; Sagi, M.; Nakagawa, Y.; Takayama, H.; Ishibashi, M.; Miyazaki, H. *Chem. Pharm. Bull.* **1983**, *31*, 4549–53. (b) Konno, S.; Sagi, M.; Siga, F.; Yamanaka, H. *Heterocycles* **1992**, *34*, 225–8.

83. Dostal, W.; Heinisch, G.; Lötsch, G. *Monatsh. Chem.* **1988**, *119*, 751–9.

84. Haider, N.; Käferböck, J. *Tetrahedron* **2004**, *60*, 6495–507.

85. Heinisch, G.; Holzer, W. *Can. J. Chem.* **1991**, *69*, 972–7.

86. Gorunenko, V. V.; Gulevskaya, A. V.; Pozaharskii, A. F. *Russ. Chem. Bull. Int. Ed.* **2003**, *52*, 441–6.

87. Gorunenko, V. V.; Gulevskaya, A. V.; Pozaharskii, A. F. *Russ. Chem. Bull. Int. Ed.* **2004**, *53*, 846–52.

88. Coelho, A.; Novoa, H.; Peeters, O. M.; Blaton, N.; Alvarado, M.; Sotelo, E. *Tetrahedron* **2005**, *61*, 4785–91.

89. Coelho, A.; Sotelo, E.; Fraiz, N.; Yáñez, M.; Laguna, R.; Cano, E.; Raviña, E. *Bioorg. Med. Chem. Lett.* **2004**, *14*, 321–4.

90. Coelho, A.; Sotelo, E.; Raviña, E. *Tetrahedron* **2003**, *59*, 2477–84.

91. R'Kyek, O.; Maes, B. U. W.; Lemière, G. L. F.; Dommisse, R. A. *Heterocycles* **2002**, *57*, 2115–28.

92. R'Kyek, O.; Maes, B. U. W.; Lemière, G. L. F.; Dommisse, R. A. *Heterocycles* **2003**, *60*, 2471–83.

93. Coelho, A.; Sotelo, E.; Novoa, H.; Peeters, O. M.; Blaton, N.; Raviña, E. *Tetrahedron Lett.* **2004**, *45*, 3459–63.

94. (a) Melnyk, P.; Gasche, J.; Thai, C. *Tetrahedron Lett.* **1993**, *34*, 5449–50. (b) Melnyk, P.; Legrand, B.; Gasche, J.; Ducrot, P.; Thai, C. *Tetrahedron* **1995**, *51*, 1941–52.

95. Rohr, M.; Toussaint, D.; Chayer, S.; Mann, A.; Suffert, J.; Wermuth, C. G. *Heterocycles* **1996**, *43*, 1459–64.

96. Gonz·les-Gómez, J. C.; Uriarte, E. *Synlett* **2003**, 2225–7.

97. Bessard, Y.; Crettaz, R.; Brieden, W. WO 0107415, (**2001**); *Chem. Abstr.* **2001**, *134*, 147608.

98. Bessard, Y.; Crettaz, R.; Eggel, M. WO 0107416, (**2001**); *Chem. Abstr.* **2001**, *134*, 131545.

99. (a) Dal Piaz, V.; Ciciani, G.; Turco, G.; Giovannoni, M. P.; Miceli, M.; Pirisino, R.; Perretti, M. *J. Pharm. Sci.* **1991**, *80*, 341–8. (b) Schober, B. D.; Megyeri, G.; Kappe, T. *J. Heterocycl. Chem.* **1990**, *27*, 471–7. (c) Maes, B. U. W.; Košmrlj, J.; Lemière, G. L. F. *J. Heterocycl. Chem.* **2002**, *39*, 535–43. (d) Heinisch, G.; Lassnigg, D. *Arch. Pharm.* (*Weinheim, Ger.*), **1987**, *320*, 1222–6.

100. Košmrlj, J.; Maes, B. U. W.; Lemière, G. L. F.; Haemers, A. *Synlett* **2000**, 1581–4.

101. Meyers, C.; Maes, B. U. W.; Loones, K. T. J.; Bal, G.; Lemière, G. L. F.; Dommisse, R. A. *J. Org. Chem.* **2004**, *69*, 6010–17.

102. Hong, Y. P.; Tanoury, G. J.; Wilkinson, H. S.; Bakale, R. P.; Wald, S. A.; Senanayaka, C. H. *Tetrahedron Lett.* **1997**, *38*, 5607–10.

Chapter 14

Industrial scale
palladium chemistry

Marudai Balasubramanian

14.1. Introduction

During the last 30 years, numerous books, monographs, and reviews have been published on organo-palladium chemistry and palladium-catalyzed reactions such as the Heck, Suzuki, Stille, and Sonogashira reactions [1–52]. These represent well-established methods for carbon–carbon bond formation in organic synthesis.

Palladium-catalyzed reactions often offer shorter and more selective synthetic routes to substituted arenes and alkenes when compared to classical stoichiometric organic transformations. Many of these classical reactions need typically 1–5 mol% of the palladium catalyst and suffer from low catalyst efficiency. Hence, in these reactions the catalyst costs dominate over the raw material costs. Relatively, large amounts of palladium catalysts are used in the coupling reactions and it is challenging to keep the level of palladium content low in the pharmaceutical and agrochemical end products. Academics have paid little attention to the industrial availability and price of starting materials. For example, most research groups at universities developed new catalysts and coupling reactions pertaining to aryl iodides and aryl triflates instead of commercially available aryl chlorides [9, 53–61]. In general, organic chemists ignore the problem of catalyst activity (turnover number, TON), which is important for cost-effective manufacturing. Fine chemical production requires catalyst productivities of *ca.* 1000–10,000 TON and catalyst activities of 200–500 h in order to be competitive with noncatalytic routes. For bulk chemicals, the requirements are significantly higher [17].

The interest in Pd-catalyzed coupling reactions has increased since the decline of the precious metal price over the last five years [62]. The price per ounce between 2000 and 2003 was reduced by one-sixth.

This review covers a number of important coupling reactions that are interesting to industries. They are carried out on a kilogram scale in order to be commercialized. It is difficult to get the actual industrial process involving in some cases and an appreciation of the real scale of these reactions. Although a number of industrial fine chemicals are prepared by Pd-catalyzed reactions, relatively few applications have been explored in large scale.

14.2. Pharmaceutical products

14.2.1. Heck coupling

α-Arylpropionic acids are an important class of nonsteroidal anti-inflammatory agents (NSAIDs) with a multi-billion dollar market. The key step in the naproxen process is the alkenylation of 2-bromo-6-methoxynaphthalene (**1**). 2-Bromo-6-methoxynaphthalene is reacted with ethylene in the presence of a homogeneous palladium catalyst [63]. A sterically hindered basic phosphine ligand is used in the synthesis of naproxen [63]. Naproxen was prepared in industrial scale by Albermarle using the Heck reaction strategy [64]. Palladium-catalyzed hydrocarboxylation of **2** was followed by subsequent resolution to provide access to naproxen. Palladacycle is a catalyst used on a kilogram scale by Hoechst AG and Hoechst–Celanese. 2-Methoxy-6-vinylnaphthalene (**2**) was obtained in 89% yield at 20 bar ethylene pressure [65]. High catalyst turnover numbers of *ca.* 10,000 were realized; however, due to a shift in the company's business strategy no further development was made.

Rhône–Poulenc developed ketoprofen [66]. A few synthetic routes have been reported that involving multi-step syntheses [67, 68]. Ketoprofen is produced by similar reaction sequences that were described for naproxen [69]. Transition metal-catalyzed reactions including carbonylations, hydroformylations, and hydrogenations have been applied to the synthesis of ketoprofen. 3-Vinylbenzophenone (**4**) was obtained from 3-bromo-benzophenone (**3**) by a Heck reaction. Palladium-catalyzed carbonylation of **4** provided the isopropyl-α-(3-benzoylphenyl)propionate in 95% yield. Ketoprofen was then obtained in 90% yield by hydrolysis of the isopropyl ester [69].

A practical route to a new LTD4 receptor antagonist **8** was developed at Merck [70]. An aryl bromide or triflate **5** was coupled with a vinylquinoline **6** in DMF at 100°C utilizing 3.0 mol% of palladium(II) acetate as precatalyst. In the case of the bromide, tris-(*o*-tolyl)-phosphine was the optimal ligand yielding 91% of the desired product. In contrast, using the corresponding triflate as the aryl source, triphenylphosphine performed better than tris-(*o*-tolyl)phosphine.

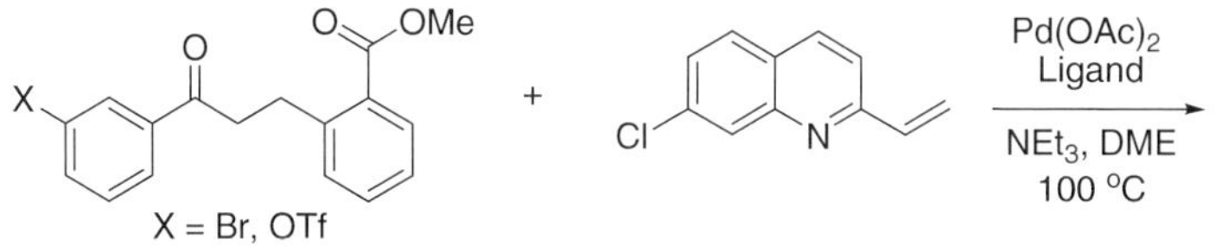
X = Br, OTf
Pd(OAc)2
Ligand
NEt3, DME
100 °C

5 **6**

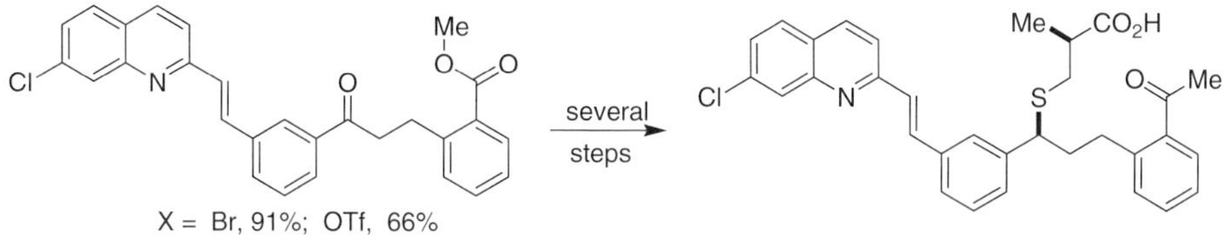
X = Br, 91%; OTf, 66%
several
steps

7 **8**

The synthesis of the nonsteroidal anti-inflammatory drug nabumetone (**9**) was developed by Hoechst–Celanese [71]. It was prepared *via* a Heck coupling of 2-bromo-6-methoxy-naphthalene (**1**) with methyl vinyl ketone in the presence of palladium catalyst [71]. Further reduction of unsaturated ketone provided **9**. Nabumetone was also obtained in a one-step coupling reaction of 2-bromo-6-methoxy-naphthalene with 3-buten-2-ol followed by isomerization of enol [72].

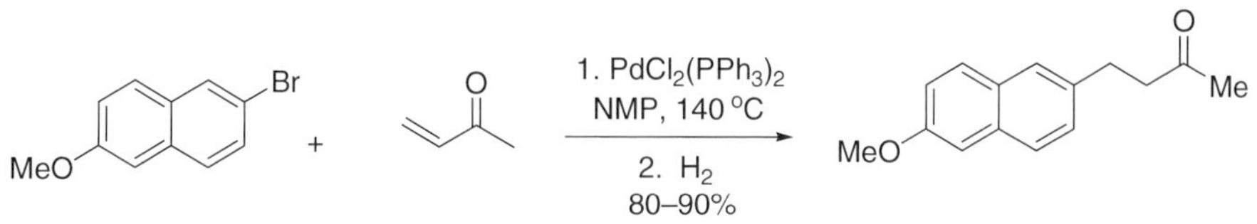
1. PdCl2(PPh3)2
NMP, 140 °C
2. H2
80–90%

1 **9**

The Heck reactions have been explored in the synthesis of the drugs naratriptan, rizatriptan, almotriptan, and eletriptan on laboratory scale; however, none of them have been explored on an industrial scale.

The key step in the synthesis of naratriptan (**12**) involved the Heck reaction of bromoindolyl piperidine with *N*-methylvinylsulfonamide (**11**) at 100–110°C in a sealed tube vessel to afford indolylsulfenamide **12** [73, 74].

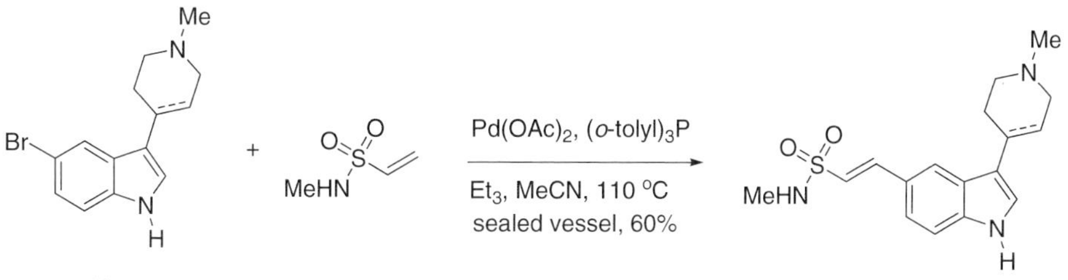
Pd(OAc)2, (o-tolyl)3P
Et3, MeCN, 110 °C
sealed vessel, 60%

10 **11** **12**

Rizatriptan (MK–0462, **16**), an effective antimigraine drug, was synthesized *via* a Pd-catalyzed indolization reaction of iodoaniline **13** with bistrimethylsilyl protected butynol **14** to yield **15** [75–77]. As a catalyst, palladium(II) acetate, although in relatively high concentrations (2 mol%), was used without any ligand. Under these conditions, 80% of the substituted indole was formed in DMF at 100°C [77]. This process was useful to form indoles containing acid labile substitutents such as triazole, acetyl, ketal, cyano, and carbamate or indoles having a leaving group at the benzylic position.

The synthesis of almotriptan (**18**), another triptan in the marketplace as an antimigraine drug, involved the Pd-catalyzed Heck cyclization of **17** with concomitant elimination of a trifluoroacetyl group [78].

An improved synthesis of eletriptan (**20**) was achieved starting from *N*-acylated indole **19** (*N*-protected to avoid addition across the double bond) with phenylvinyl sulfone *via* the Heck reaction [79a,b]. Substituted indole **20** is a potent serotonin (5-HT$_1$) agonist and may be used in the treatment of depression, anxiety, etc.

An alternative synthetic strategy for eletriptan (**20**, R = PhSO₂CH=CH) is reported involving an intramolecular Heck cyclization of appropriately substituted aniline derivative **21** [79b].

21   **20**

14.2.2. Carbonylations

Among the carbonylation reactions of aryl halides, those of heteroaryl halides [80] were of special interest to industrial research groups. The attachment of carbonyl functionalities onto heterocyclic frameworks by replacing a halide provides easy access to valuable intermediates for the manufacture of antifibrotics, herbicides, and other pharmaceuticals.

The Pd-catalyzed carbonylation of aryl halides has proven to be a versatile tool for the synthesis of various benzoic acid and heteroaromatic acid derivatives [81]. Apart from alkenylation and other cross-coupling procedures, carbonylation reactions of aryl and benzyl halides, which make use of the inexpensive reagent carbon monoxide, have attracted industrial interest. By a simple change of the nucleophile used in this reaction, acids, esters, amides, aldehydes, ketones, and other compounds can be synthesized. The synthesis of ibuprofen, one of the most important nonsteroidal anti-inflammatory agents, developed by Hoechst–Celanese in the late 1980s, demonstrates the synthetic utility of this method. *i*-Butylbenzene is acetylated in the *para* position and the resulting acetophenone is reduced to the corresponding benzylic alcohol **22**. Subsequent carbonylation of **22** proceeds in concentrated HCl in the presence of a PdCl₂/PPh₃ to provide **23** [14, 82]. Due to the lower amount of by-products compared to the original Boots process, this catalytic route is now the main industrial production process (3000 tons/year) for ibuprofen.

22   **23**

A carbonylation reaction has been used for the production of lazabemide (**24**), a monoamine oxidase B inhibitor, from commercially available 2,5-dichloropyridine. The original eight-step laboratory synthesis of lazabemide was replaced by a one-step protocol [83, 84]. The product lazabemide (**24**) was isolated in 65% yield. Due to the fact that only small amounts of catalyst need to be used, traces of palladium in the product could be removed by appropriate workup [TON = 3000].

24

14.2.3. Suzuki coupling

The Suzuki reaction of arylboronic acid derivatives with aryl halides is one of the most powerful methods for construction of an unsymmetrically substituted biaryl derivative [11, 23, 85]. Due to the importance of substituted biaryls as building blocks for pharmaceuticals, there is currently a great deal of interest in the coupling of economically attractive aryl halides with arylboronic acids [86–98].

The Suzuki cross-coupling reactions have been explored in the synthesis of drugs and related compounds on the laboratory scale but none have been explored on an industrial scale. A few examples given below are: AT II antagonist losartan, a COX-2 selective inhibitor, rofecoxib, a highly potent HMG–CoA reductase inhibitor, NK–104, and anti-cancer agents epothilones A–B.

The Suzuki reaction was used in the production of intermediates for AT II antagonists developed by Hoechst AG and now is used by Clariant AG in Frankfurt/Main on a multi-ton scale and of some biaryls at E. Merck KG in Darmstadt (Germany) [99a,b]. 2-Cyano-4-methylbiphenyl (**27**) is produced by coupling of 4-tolylboronic acid (**25**) and 2-chlorobenzo-nitrile (**26**) in the presence of a palladium/sulfonated triphenylphosphine (TPPTS) catalyst with yields higher than 90%. The reaction is conducted at 120°C in a polyhydric alcohol solvent (e.g. ethylene glycol), which contains small amounts of sulfoxide or sulfone for catalyst stability reasons. At the end of the reaction two phases form. The catalyst and salts remain in the polar phase, and the product in the organic phase. This biphasic procedure allows for an efficient recycling of the homogeneous catalyst.

25   **26**   **27**

Researchers from Merck elaborated a convergent synthesis for the AT II antagonist losartan **30** with the key biaryl coupling step in the final stage of the protocol [100]. The two reaction partners **28** and **29** were coupled very efficiently (99% yield) utilizing 1 mol% of palladium(II) acetate and 4 mol% triphenylphosphine. As the solvent, a 1:4 mixture of THF and diethoxymethane containing a defined amount of water was crucial for the high reaction rates and yields.

28   **29**   **30**

Rofecoxib (**32**, Vioxx), a COX-2 selective inhibitor was withdrawn in September 2004. A Suzuki coupling strategy was employed for installation of the second aryl group to the furanone. Furanone bromide **31** was coupled with 4-methylthiophenylboronic acid to give diphenylfuranone **32** in 90% yield [101, 102].

31   **SMe**   **32**

NK–104, **35**, a highly potent HMG–CoA reductase inhibitor was prepared by the cross-coupling reaction of iodoquinoline (**33**) with alkenylborane (**34**) [103]. Allylpalladium chloride was an effective catalyst and gave almost quantitative yield. Aryltriflate was found to be another coupling partner for the synthesis of NK–104. The great advantage of this synthetic route was that aryl diversity could be introduced late in the synthesis.

                                                                          NK-104

33   **34**   **35**

Meng *et al.* have recently reported a total synthesis of the promising anticancer agents epothilones A and B using an alkylboranes coupling reaction [104]. In contrast to the stereochemical outcome of the related cross-couplings involving silanes and stannanes, the stereochemistry of the transmetalation of alkylboranes to palladium received little attention, although it was suggested to proceed with the retention of configuration [13, 16, 105, 106]. Meng *et al.* confirmed that primary alkylboranes **36** undergo transmetalation with vinylio-dide **37** with the retention of configuration [104, 107].

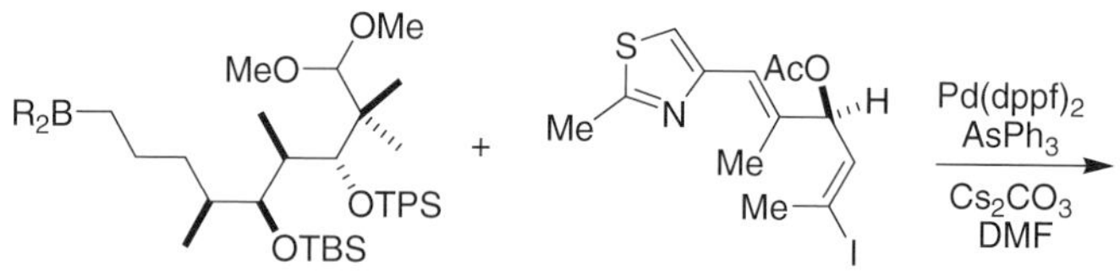

36   **37**

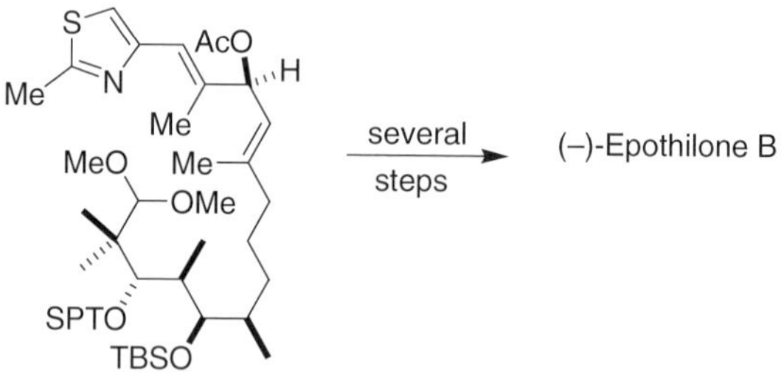

38   several steps → (−)-Epothilone B

14.2.4. Sonogashira coupling

The Sonogashira reaction has been explored on laboratory scale towards the synthesis of fexofenadine and terbinafin. One of the key intermediates, arylalkyn-ol **40** for the synthesis of Fexofenadine (major metabolite of terfenadine) was achieved by the Sonogashira coupling reaction of 4-bromophenylacetic acid (**39**) and 3-butyne-1-ol catalyzed by $Pd(PPh_3)_4$ [108].

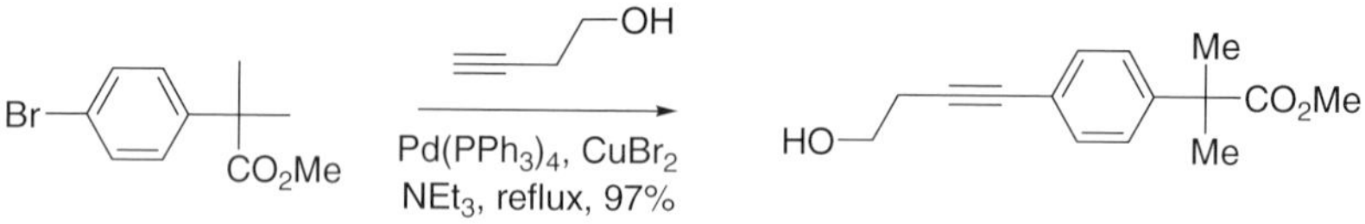

39   **40**

The ecological advantage of Pd-catalyzed reactions compared to a process that relies on stoichiometric reactions is well demonstrated by the Sandoz process for the antifungal terbinafin (**42**) [109]. Terbinafin, the active agent of the broad-spectrum antimycotic Lamisil®, was the first pharmaceutical drug on the market with a 1,3-enyne unit as an integral structural element. An important step is Pd-catalyzed coupling of a substituted alkenyl chloride **41** with *t*-butylacetylene. This coupling reaction proceeded stereoselectively in the presence of less than 0.05 mol% of the precatalyst $PdCl_2(PPh_3)_2$ with CuI as a co-catalyst [109].

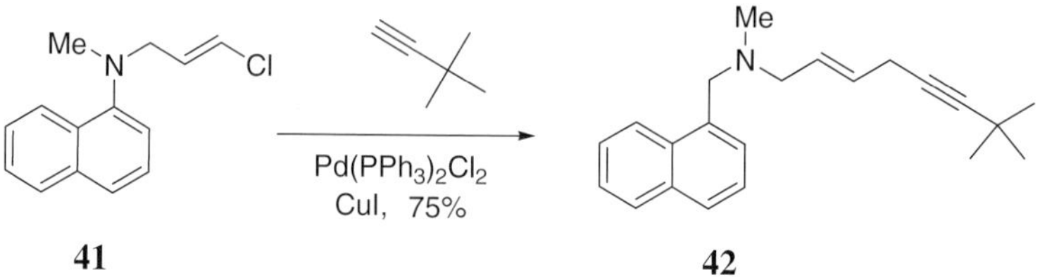

41   **42**

Another example of a cross-coupling process with an alkyne developed in industry is the reaction of an iodo-substituted dideoxynucleoside **45** with trifluoroacetyl *N*-protected propargylamine **46**. By using a homogeneous Pd(0)/Cu(I) catalyst in dimethylformamide, smooth coupling was achieved [110]. DuPont used this Sonogashira methodology in the synthesis of the DNA sequencing agent **47** [111].

14.2.5. Negishi coupling

Allergan explored the synthesis of tazarotene *via* the Negishi coupling on laboratory scale. Tazarotene is a topical prodrug that modulates receptor-selective retinoid and normalizes differentiation and proliferation of keratinocytes. The key step of the Allergan synthesis of tazarotene **51** involved the Negishi coupling of alkynylzinc chloride **48** with ethyl α-chloropicolinate (**49**) using Pd(PPh$_3$)$_4$ [112–114].

Facile synthesis of an intermediate for biotin **53** was achieved through the nonpyrophoric catalyst Pd(OH)$_2$/C promoted coupling reaction of thiolactone **50** with 4-ethoxycarbonyl-butylzinc iodide (**52**) [115, 116]. This synthetic route was advantageous over the previously reported method using Pd/C (yield 30%) [117].

14.2.6. Amination

In the last decade, several research groups have shown considerable interest in Pd-catalyzed C–N coupling reactions [21, 118–135]. The Pd-catalyzed amination reactions have been utilized in the following drug syntheses: aripiprazole, Dup–721, CP–529414, combretastatin A-4, and morphine analogs.

Aripiprazole is used for the treatment of schizophrenia. The key intermediate, phenylpiperazine **56**, was prepared by conventional methodology *via* a ring closure reaction from 2,3-dichlorophenol in six steps with an overall yield of 9%. Alternatively, phenyl-piperazine **56** was prepared *via* a Pd-catalyzed amination of benzyl protected 4-bromo-2,3-dichlorophenol **54** with piperazine. The coupling reactions proceeded regioselectively to provide excellent yield of phenyl piperazine [136–140].

An effective synthesis of structurally unique M_3 antagonist **59** was achieved *via* a Pd-catalyzed amination. Drug candidate **59** is a highly potent, orally active, long acting selective M_3 antagonist and is being investigated for the treatment of chronic obstructive pulmonary diseases (COPD) and urinary incontinence [141]. The following Pd-catalyzed amination has been used to prepare multi-kilogram quantities of the bulk drug.

Benzazepine and benzoxepine ring systems showed high CNS activity and their syntheses have been reported [142]. The key step in this synthesis exploited an alternative route to form oxepine or thiepine ring systems **61** *via* a Pd-catalyzed cyclo-amination of **60**. Overall, the best yields were achieved with $Pd_2(dba)_3$ as the palladium source, $P(t\text{-Bu})_3$ as the ligand, t-BuONa alone or with K_2CO_3, in toluene [142].

Synthesis and structure activity relationship (SAR) of bioisosteric 5-HT$_{1F}$ receptor ligands containing furanopyridine **65** and indole were investigated [143]. The key step involved the conversion of chlorofuranopyridine **62** to the corresponding furanopyridylamine **64** through a Pd-catalyzed amination reaction. Benzophenoneimine **63** was employed as an ammonia equivalent followed by acid hydrolysis to generate primary amine **64**. Acylation of amine **64** subsequently provided target compound **65** [143].

An antibacterial agent in the oxazolidinone class, Dup–721 (**67**) was synthesized *via* a convergent route using Pd-catalyzed amination of oxazolidinones **66** with arylbromides [144]. The amide of arylated oxazolidinone **67** was deprotected with TFA to provide Dup–721 in 65% yield.

Drug candidate CP–529414 (**71**), a novel compound for cholesteryl ester transfer protein (CTEP) inhibitors, was prepared using Pd-catalyzed cyclo-amination of **68** with benzylamine **69** [145, 146]. Amination was very efficient in the presence of dialkyl-phosphinobiphenyl ligand (**70**).

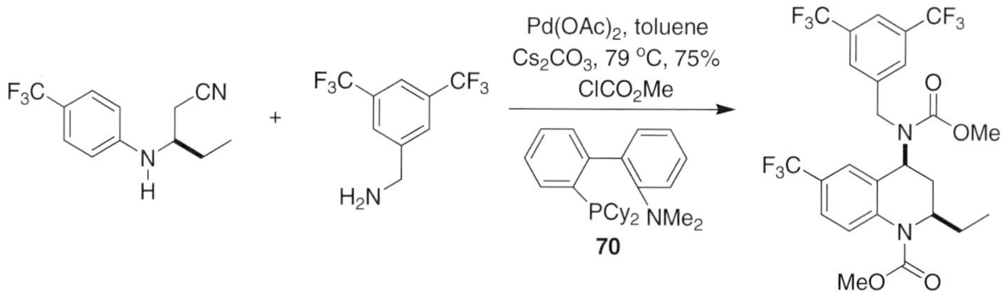

The synthesis and SAR study of pyrazole-based analogs of combretastatin A-4 (**72**) was investigated by Wang *et al.* [147]. The key intermediate hydrazine **74** was prepared from bromotrimethoxybenzene (**73**) and benzophenonehydrazone *via* Pd-catalyzed amination.

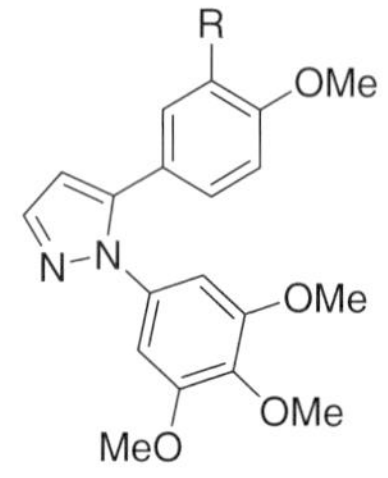

72

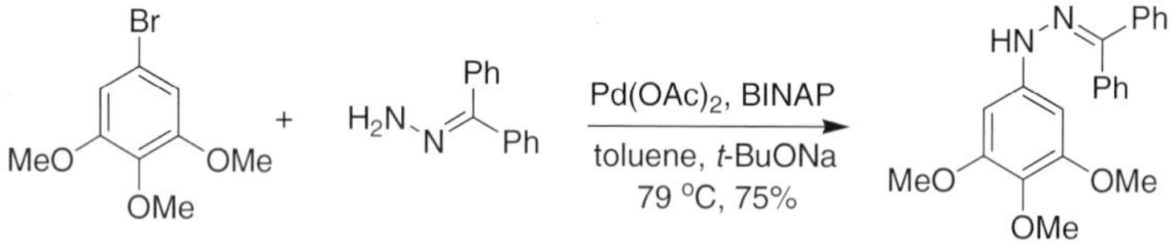

73   **74**

A series of 8-(substituted)-amino analogs of morphine (**76**) were prepared and their opioid receptor binding properties were investigated. The 8-hydroxy group at C(8) was replaced with amino derivatives *via* the Pd-catalyzed amination reaction [148]. Triflate **75** was aminated using Pd(OAc)$_2$ and BINAP as the catalyst source and Pd ligand, respectively. Higher yields were obtained in the presence of *t*-BuONa, which also induced the cleavage of triflate to phenol **76** [148].

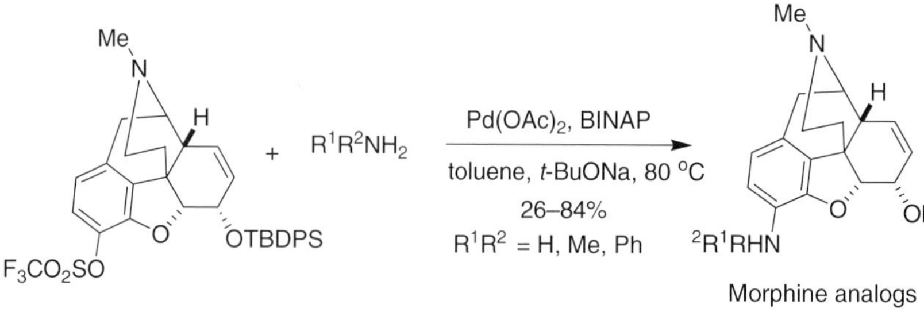

Morphine analogs

75   **76**

14.3. Cosmaceuticals

14.3.1. Heck reaction

A large-scale pilot run Heck reaction has been explored for the synthesis of the fragrance Lilial and of sunscreen agent EHMC but not commercialized. Lilial (**78**), a delightful lily-of-the-valley fragrance, is produced commercially starting from *para-t*-butylbenzene (**77**) through a sequence of reactions: air oxidation, aldol condensation, and hydrogenation. The relatively large volume of Lilial that was needed spurred research on alternative syntheses. Givaudan developed homogeneously catalyzed Heck reactions that have been exploited towards an industrial production of the fragrance Lilial directly from aryl bromide **77** [111, 149]. Palladium catalysis gave 86% conversion and 82% yield. However, the reaction is yet to be commercialized.

The synthesis of the sunscreen agent, 2-ethylhexyl-*p*-methoxycinnamate (EHMC) **80**, was developed by Hoechst AG and other companies. As the most common UV–B sunscreen on the market, EHMC was produced for some time by a heterogeneously catalyzed Heck reaction by an Israeli chemical company [72], which involves bromination of anisole and the Heck coupling with 2-ethylhexylacrylate (**79**) refluxing in *N*-methylpyrrolidone (NMP) in the presence of Pd/C as a catalyst. Pilot plant runs were carried out in a 250 L reactor [20]. By using an optimized concentration of starting materials product yields of 80–90% were realized after high vacuum distillation. The same technology was used for the synthesis of 2-ethylhexyl-*p*-dimethylaminocinnamate (**81**), a potential UV (A–B) filter agent.

14.4. Agrochemical products

14.4.1. Matsuda–Heck reaction

The Matsuda–Heck reaction of an aryl diazonium salt **82** with 1,1,1-trifluoropropene developed at Ciba–Geigy was one of the first examples of a Pd-catalyzed coupling reaction on an industrial scale. Today this process is performed on large scale by Novartis for the synthesis of a sulfonylurea herbicide, prosulfuron **84** [150, 151]. By combining three synthetic steps (diazotization, alkenylation, and hydrogenation) in a one-pot sequence with an overall yield of 93% (i.e. an average yield of 98% per step), this elegant process was made economically feasible. Due to the high stability of the *in situ* generated diazonium salt, safety issues were not a problem in this reaction. Despite diazotization, only 2 kg of waste per kg of product are produced.

14.4.2. Carbonylations

Chlorinated picolinic acids served in the agrochemical chemistry as precursors for herbicides. 5-Substituted pyridine-2,3-dicarboxylic acids **85**, useful as intermediates in the manufacture of a new class of imidazolinone herbicides **86**, are usually prepared by cyclocondensation reactions or by oxidation of 5-substituted quinoline with various oxidizing agents. However, these classical multi-step synthetic methods are unsatisfactory for large-scale production. Bessard and Roduit at Lonza AG explored the synthesis of alkyl 3-chloropicolinates and dialkyl pyridine-2,3-dicarboxylates *via* carbonylation of chloropyridines [152]. Starting from 2,3-dichloro-5-(methoxymethyl)pyridine **87**, both the mono- (**88**) and di- (**89**) carbonylated methoxymethyl pyridines were obtained at low CO pressure (15 atm) with high selectivity and yields. Carbonylation of 2,3-dichloropyridine **87** using PdCl$_2$(PPh$_3$)$_2$ and dppb at 145°C gave 94% of methyl 3-chloropicolinate **88** whereas palladium(II) acetate and dppf at 160°C led to a double alkoxycarbonylation to give **89**.

MeO (structure 88) ← MeOH, CO / Pd₂Cl₂(PPh₃)₂ / dppb / 94% ← (structure 87) → Pd(OAc)₂, dppf / CH₃CO₂Na / EtOH, CO / 90% → (structure 89)

88   **87**   **89**

14.4.3. Amination

Aryl and morpholine substituted analogs of rocagiamide were synthesized and tested for insecticidal and antifungal activity [153]. The Pd-catalyzed amination of bromophenyl derivative **90** with morpholine gave moderate yield of **91**.

(structure 90) → P₂(dba)₃, dioxane / Cs₂CO₃, 80 °C, 41% / morpholine / PCy₂ NMe₂ → (structure 91)

90   **91**

14.5. Material sciences

Amines are useful building blocks for biological or chemical applications, but were also a core element of polymers and materials for the electronics and xerographic industries. Watanabe *et al.* synthesized novel bis-(diarylamino)thiophene oligomers (**92**) and these amines showed intrinsic electronic properties [154]. The employment of the bulky and electron-rich ligand, P(*t*-Bu)₃, aided the Pd-catalyzed amination of 2,5-dibromothiophene to bisdiarylaminothiophene (**92**).

(structure: Br-thiophene-Br) + (Ph₂NH) → Pd(OAc)₂, *t*-Bu₃P / *t*-BuONa, *o*-xylene / 120 °C, 57% → (structure 92)

92

N-Aryl azoles have attracted attention as hole transport molecules for organic light-emitting diodes (LEDs). Watanabe *et al.* used the Pd-catalyzed amination technology to build up tris(*N*-azolyl)triphenylamines (TCTA) **94** [155]. The palladium/P(*t*-Bu)₃ system catalyzed the coupling of aryl halide **93** and carbazole, providing *N*-arylazole **94**.

Queiroz synthesized diarylamines in the benzothiophene series, where the ligand BINAP was used to achieve the Pd-catalyzed amination in medium to high yields [156]. The diarylamines **96** were used in materials with electronic or luminescent properties. These compounds were further cyclized to provide substituted thienocarbazoles, which are bioisosteres of natural antitumoral DNA intercalating compounds. The presence of fluorine atoms increased the solubility of these molecules [156].

Lin and Tao used P(t-Bu)$_3$ for the synthesis of benzo[a]aceanthranylene core compounds (**acen**) **98**. Pure red-emitting devices were fabricated using **acen** as both hole-transporting and emitting materials [157]. Palladium-catalyzed aromatic C−N coupling reactions and cyclization were carried out in one pot in the presence of Pd(OAc)$_2$ and t-Bu$_3$P [157].

97   **98**

Benthocyanin A is a powerful radical scavenger from the mycelium of *Streptomyces prunicolor*. A new route to phenazine was described *via* the Pd-catalyzed intramolecular amination of aryl bromide [158].

Benthocyanin A

14.6. Polymer chemistry

The Pd-catalyzed coupling reaction of an aryl halide and olefin is a very efficient and practical method for making C–C bonds. The Heck alkenylation of aryl bromides with ethylene was used by Dow Chemical to make high-purity 2- and 4-vinyltoluenes, which are of interest as co-monomers in styrene polymers [159]. The monomer, *o*-vinyltoluene (**99**), has a low toxicity and an attractive co-monomer for styrene polymers. *o*-Vinyltoluene improved heat distortion properties of styrene and polymerization rate. It also minimized color formation or cross-linking and it was difficult to make by other routes [159]. Catalyst turnover, rate, and lifetime were significantly improved.

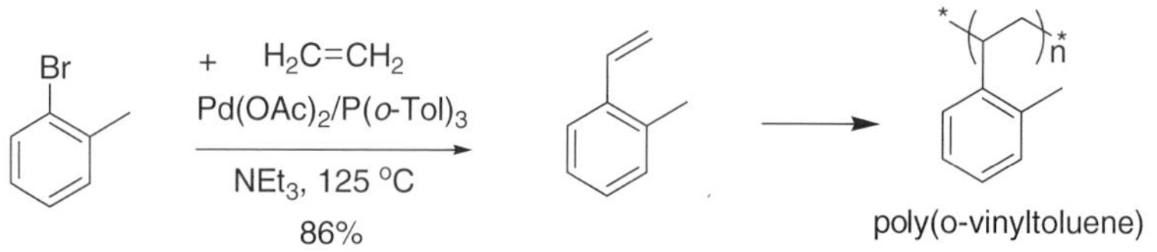

99

Poly *p*-phenylene (**104**) was expected to have good thermal and oxidative stability as well as electrical conductivity in the oxidized or reduced states. Rehahn *et al.* reported the first cross-coupling reaction of dihaloarenes **100** and aryldiboronic acids (**101**) to provide poly (*p*-phenylenes) (**104**) [160]. Homologation was achieved *via* repeating the following sequence of reactions: lithiation, boration, and Suzuki coupling [160].

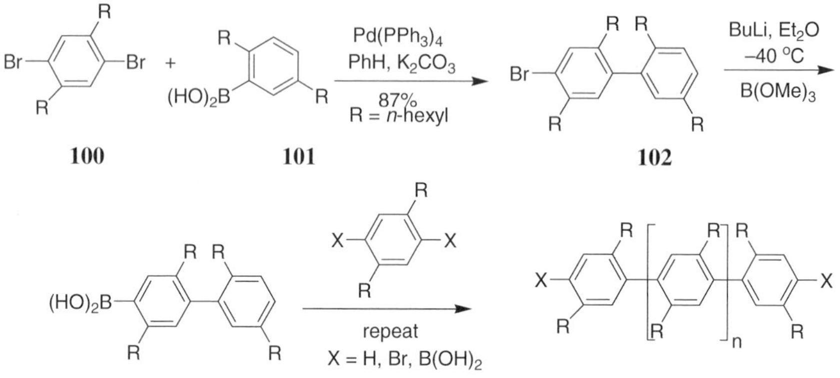

100   **101**   **102**

103   **104**

An all carbon conjugated ladder polymer (graphite ribbon) was synthesized by a novel electrophile-induced cyclization reaction to provide fused benzenoid aromatic hydrocarbon in quantitative yield [161]. Suzuki cross-coupling of dieneyne **105** with 1,4-didodecylbenzene-2,5-diboronic acid (**106**) gave rigid-rod polymers **107**, which was further treated with TFA to produce **108** graphite ribbon as a yellow/orange solid.

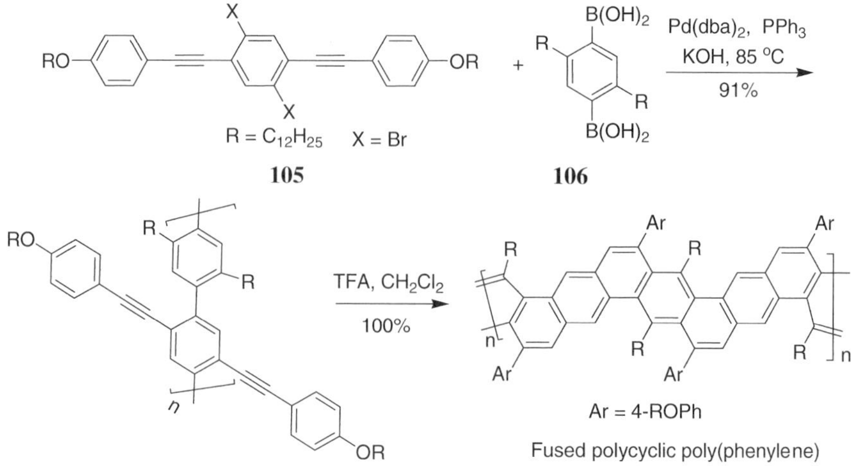

$R = C_{12}H_{25}$ $X = Br$

105   **106**

Ar = 4-ROPh

Fused polycyclic poly(phenylene)

107   **108**

Water-soluble poly(*p*-phenylene) **111** was prepared from dibromobiphenyl **109** and **110** *via* the Suzuki coupling method in water as the free acid formed without the use of highly toxic and corrosive agents [162]. Such polymers exhibited outstanding thermal stability typically found in rigid-chain polymers.

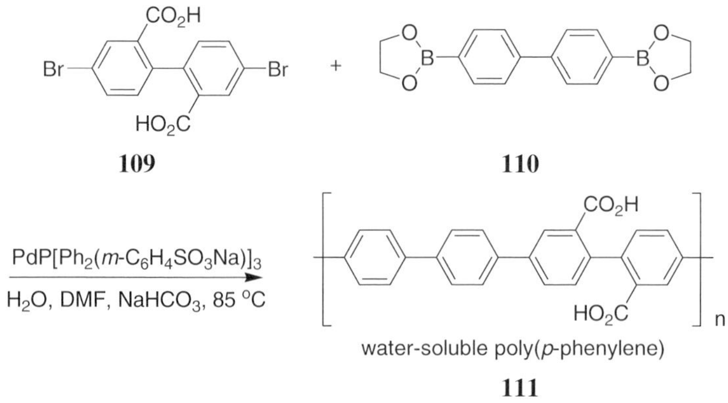

109 **110**

water-soluble poly(*p*-phenylene)

111

Planar poly(*p*-phenylenes) (**115**, PPP), a highly insoluble polymer, has been studied for its possible electronic and photonic applications. The Suzuki coupling of **112** with aryl bromide **113** yielded intermediate polymer **114**, which underwent loss of the Boc protecting group and cyclization to afford flexible free-standing film **115** in 97% yield [163]. Such film is devoid of ketones, carbamates, and amines. The dodecyl groups apparently exert a plasticizing effect so that even planar rigid-rod polymers can possess good film-forming properties.

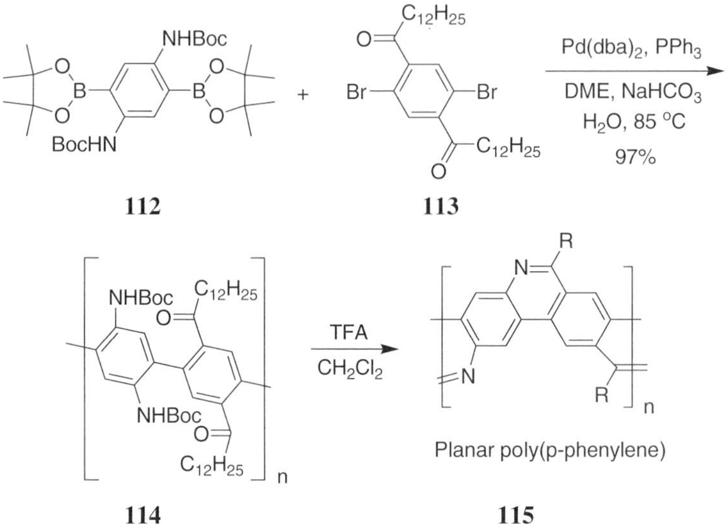

112 **113**

Planar poly(p-phenylene)

114 **115**

Poly *o*-phenylene oligomers **116** having three to nine rings were prepared using a Suzuki coupling strategy [164]. Commercially available 4-methoxyboronic acid was coupled with

1-bromo-2-iodobenzene in the presence of Pd(PPh$_3$)$_4$ to provide the biphenyl. The homologation of biphenyl was carried out *via* a sequence of metal exchange, boration, and Suzuki coupling reaction as shown below [164].

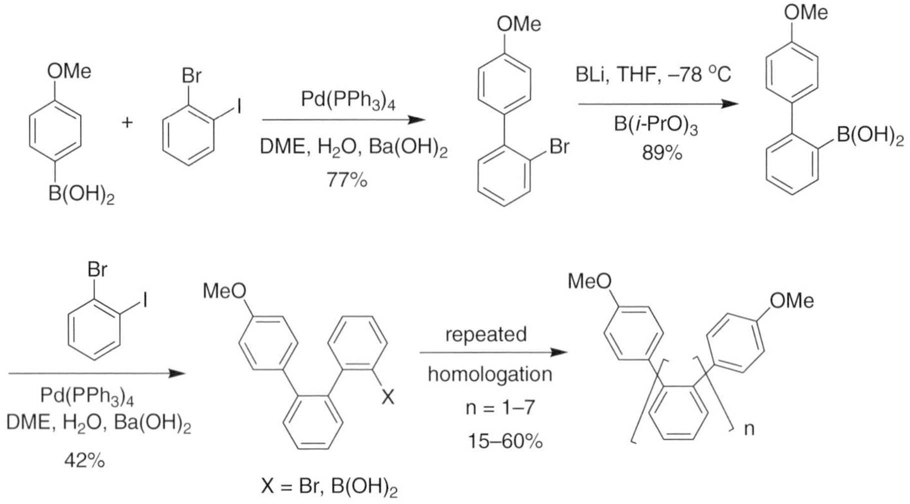

116

Dendrimers have been used as large compartmented hosts for drug delivery, as carriers for catalytically active sites in flow reactors, and for charge or energy transfer purposes. New types of such dendrimers were prepared starting from functionalized poly-*p*-phenylenes as polymeric cores whose functional groups were further used to anchor dendrons [165]. The polymers were synthesized from dibromide **117** and diboronic acid **118** *via* a Suzuki cross-coupling; the dendrons were attached *via* a Williamson ether synthesis [165].

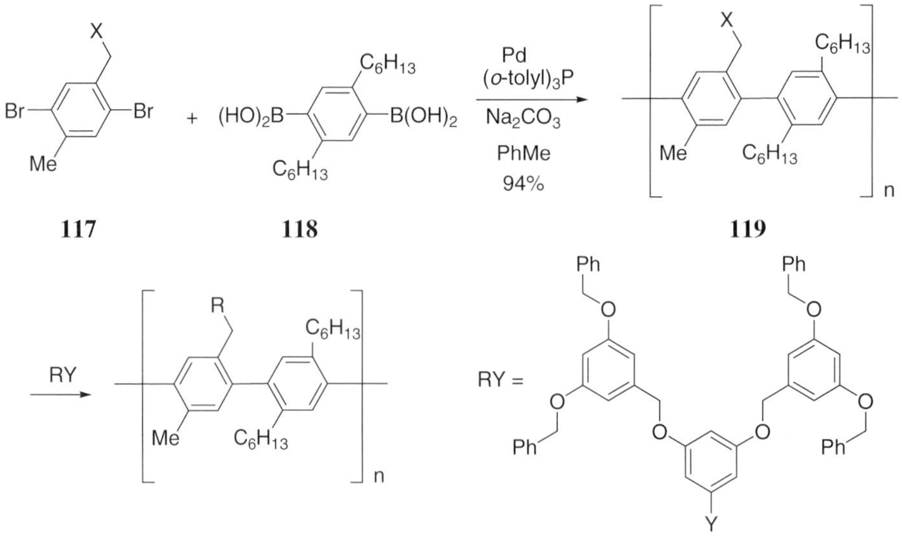

117   **118**   **119**

120

A variety of benzene–furan–alkene/alkyne conjugated oligomers **122** of precise length and constitution were synthesized iteratively by combining furan annulation, Heck reaction, and Sonogashira coupling [166]. Oligoaryls and their vinylene or acetylene homologs of desired conjugation length have been widely investigated due to their potential optoelectronic applications. These oligoaryls were thermally stable and exhibited bright fluorescence in the blue-light region. Treatment of divinyl **121** with 4-bromobenzaldehyde in the presence of Pd(OAc)$_2$ and PPh$_3$ resulted in the formation of dialdehyde **122** in 75% yield.

14.7. New catalyst developments in fine chemical synthesis

In the last decade significant progress has been made towards the development of new catalysts for palladium chemistry [167, 168]. Since the properties of the central metal palladium can be tuned by ligand variation, the introduction of new ligands was the key to success. The refinement of economically attractive aryl–X compounds is of general interest in fine chemical synthesis. As an example, the alkenylation of aryl–X derivatives (Heck reaction) [15, 16, 24, 105, 106, 169, 170, 171–182] has been called "one of the true powerful tools of contemporary organic synthesis" [18].

The following Tables 14.1–14.5 present a list of recent catalyst developments for the Heck, Suzuki, Suzuki–Miyaura, Sonogashira, Negishi, and Kumada reactions.

14.7.1. Heck reactions

Milstein and co-workers were the first to introduce catalysts capable of activating various aryl chlorides in 1992 [175]. By using palladium complexes of highly basic and sterically demanding chelating bisphosphines, for example, dippb [1,4-bis(di-isopropylphosphinyl) butane], even chlorobenzene was coupled with alkenes in high yields (Table 1, 70–95% yield; TON = 70–95) [175]. However, these catalysts are extremely sensitive to air. Herrmann, Beller, and co-workers introduced more robust palladacycles [cyclopalladated complexes of the general formula $Pd_2(-L)_2(P-C)_2$; L = bridging ligand, e.g. OAc, Cl, Br; P-cyclometalated P-donor, e.g. o-{$CH_2C_6H_4P(o$-Tol)$_2$}] as highly active catalyst (TON = 40,000) for Heck reactions of 4-chloroacetophenone (Table 14.1) [88, 177]. Effective Heck couplings between aryl bromides and styrene using bis-triarylphosphines in the presence of Pd(OAc)$_2$ and Cs$_2$CO$_3$ have been reported with low TON [182].

Table 14.1.
Catalyst development for Pd-catalyzed alkenylation of aryl chlorides.

Entry	R^1	R^2	Catalyst	Ligand	Yield (%)	TON	Reference
1	4-CF$_3$	H	Pd/C	–	62	40	[172]
2	H	Ph	Pd(OAc)$_2$/ PPh$_3$	–	49	25	[173]
3	4-CO$_2$Me	CN	Pd(OAc)$_2$/ PPh$_3$	PPh$_3$	51	51	[174]
4	H	4-C$_6$H$_4$OMe	Pd(OAc)$_2$	dippb[a]	77	70–95	[175]
5	H	Ph	Pd(OAc)$_2$	dippp[b]	88	88	[176]
6	4-CHO	Ph	Palladacycle	[c]	99	200,000	[177]
7	H	Ph	Pd(OAc)$_2$/ Ph$_4$PCl	DMG[d]	77	1300	[178]
8	4-CF$_3$	Ph	Pd(OAc)$_2$/ P(OR)$_3$	[e]	84	840	[179]
9	H	Ph	Pd (dba)$_3$	P(t-Bu)$_3$	80	400	[180]
10	Me	CO$_2n$-Bu	Pd(dba)$_2$	t-Bu$_2$PF[c,e]	67	27	[181]
11	H	Ph	Pd(OAc)$_2$	[f]	100	50	[182]

[a]dippb: 1,4-Bis(di-iso-propylphosphinyl)butane; [b]dippp: 1,3-Bis(di-iso-propylphosphinyl)propane; [c]trans-Di(μ-acetato)-bis[o-(di-o-tolylphosphinyl)benzyl]dipalladium(II); [d]N,N-Dimethylglycine; [e]Di-tert-butylphosphinylferrocene; [f][5,17-Dibromo-11,23-bis(diphenylphosphino)-25,26,27,28-tetrapropoxycalix[4]arene.

Apart from palladacycles, a number of catalyst systems are currently known to show TON up to 100,000 for the Heck and Suzuki reactions of all kinds of aryl bromides. It is important to note that coupling reactions of electron-deficient aryl bromides (e.g. 4-bromoacetophenone), which are often used in academic laboratories, are not suitable as test reactions to judge the productivity of a new catalyst, because simple palladium salts without any ligand give turnover numbers up to 100,000 with these substrates.

14.7.2. Suzuki coupling

Palladium complexes in combination with sterically congested basic phosphines (e.g. tri-t-butylphosphine), carbenes, and also phosphates are productive palladium catalysts for the activation of various aryl chlorides (Table 14.2) [183]. Diphosphaferrocene is an efficient catalyst for the coupling reaction between phenylboronic acid and 4-bromoacetophenone in refluxing acetophenone, and a conversion of 98% was obtained with a maximum catalyst turnover number of 9.8×10^5 [184]. Poly(amidoamine)-dendrimer-stabilized Pd nanoparticles as a catalyst for the Suzuki coupling of aromatic halides and p-tolylboronic acid in DMF provided high TON (1200–1800) [185a]. Palladium containing Perovskite $(LaFe_{0.57}Co_{0.38}Pd_{0.05}O_3)$ was an air stable reusable catalyst for the Suzuki cross-coupling reactions under mild conditions with low levels of Pd leaching (Table 14.2) [186]. In the presence of bis-triarylphosphines, the Suzuki cross-coupling of aryl halides with phenyl boronic acid showed somewhat lower TONs (~50) for aryl bromides (Table 14.2) [182]. 1-Phosphabarrelene–phosphine–sulfide (PPS) substituted palladium(II) complexes proved to be very active catalysts in the Suzuki–Miyaura reaction, which allowed

Table 14.2.
Catalyst development for Pd-catalyzed suzuki reactions of aryl chlorides.

Entry	R	X	Catalyst	Ligand	Yield (%)	TON	Reference
1	4-COMe	Cl	–	Palladacycle[a]	82	820	[88]
2	4-Me	Cl	$Pd_2(dba)_3$	$P(t\text{-}Bu)_3$	87	29	[95]
3	4-Me	Cl	$Pd_2(dba)_3$	[b]	96	32	[97]
4	2-Me	Cl	$Pd(dba)_2$	[c]	95	48	[98]
5	4-COMe	Cl	$Pd(OAc)_2$	[d]	96	1×10^8	[96]
6	4-CF$_3$	Cl	$Pd(OAc)_2$	[e]	88	~820,000	[183]
7	4-COMe	Br	–	[f]	98	770,000	[184]
9	H	I	–	[g]	98	1771	[185a]
10	4-Ome	Br	–	[h]	95	–	[186]
11	4-CHO	Br	–	[i]	91	1×10^7	[187]
12	4-NO$_2$	N_2BF_4	$Pd(OAc)_2$	[j]	92	–	[188]
13	H	Br	–	[k]	100	50	[182]
14	H	Br	–	[l]	90	90,000	[189a]

[a]trans-Di(μ-acetato)-bis[o-(di-o-tolylphosphinyl)benzyl]dipalladium(II); [b]N,N-1-Dimesitylimidazolium chloride; [c]2-(2′-Dicyclohexylphosphinylphenyl)-2-methyl-1,3-dioxolane; [d]2-(Dicyclohexylphosphinyl)biphenyl; [e]$P(O-2,4-t\text{-}Bu_2C_6H_3)_3$; [f]bis(octaethyldiphosphaferrocene)palladium(0)complex; [g]Pd_{60}[PAMAMG-4-OH] nanoparticles; [h]$LaFe_{0.57}Co_{0.38}Pd_{0.05}O_3$; [i]furancarbothioamide-based palladacycles; [j]Thiourea-based C_2-symmetric ligand; [k][5,17-bis(diphenylphosphino)-25,26,27,28-tetrapropoxycalix[4]arene; [l][Pd(η^3-C$_3$H$_5$)] [O-trifluromethylsulfonyl].

the synthesis of functionalized biphenyl derivatives from the coupling of arylbromide with phenylboronic acid (TON up to 7×10^6) [189a]. The cationic complex also catalyzes the coupling between allyl alcohol and secondary amines to afford the corresponding N-allylamines in toluene at 70°C.

Palladacycles prepared by the addition of furancarbothioamide to a methanol solution of Li_2PdCl_4 at room temperature are soluble in hexane, chloroform, and moderately soluble in polar solvents DMF and DMSO [187]. These palladacycles are thermally stable, not sensitive to air or moisture, and can be applied effectively in the Heck reaction of aryl halides with terminal olefins and in the Suzuki reaction of aryl halides with arylboronic acids. These reactions were performed under aerobic conditions, leading to turnover numbers

(TONS) up to 1×10^5 [187]. The coupling reactions between arenediazonium salts with styrene, methylacrylate, and aryl boronic acids under aerobic conditions were achieved in moderate yields [188].

Palladium acetate is better than $PdCl_2(dppf)$ in terms of cost and catalyst removal and effectively catalyzes the cross-coupling of a wide variety of aryl halides with bis(pinacolato)diboron to form the corresponding boronates. These boronates can be conveniently isolated or used *in situ* for Suzuki cross-coupling reactions with aryl halides to provide biaryls [189b].

14.7.3. Suzuki–Miyaura coupling

1,4-Diazobicyclo[2.2.2]octane (DABCO) is used as a supporting ligand for the Suzuki–Miyaura cross-coupling reaction. An inexpensive and highly efficient methodology has been developed with $Pd(OAc)_2$/DABCO and turnover numbers up to 950,000 were obtained for the coupling of iodobenzene (PhI) and *p*-chlorophenylboronic acid (Table 14.3) [190]. Excellent yields and high TON up to 960,000 were obtained for the reaction of 1-iodo-4-nitrobenzene with phenylboronic acid using PEG-400 as the solvent [191].

Palladacycles derived from phenone–oximes are efficient precatalysts for the Suzuki–Miyaura reaction of arylboronic acids with aromatic and heteroaromatic bromides and chlorides under refluxing water or water/MeOH [192]. Aryl bromides gave biaryls with TON up to 10^5. Activated and deactivated aryl chlorides require the presence of TBAB for the couplings, showing slightly lower efficiency (TON = 9000). P–S–Bidentate ligands are used as catalysts in the Suzuki–Miyaura cross-coupling reaction of aryl bromides with pinacolborane to yield the corresponding arylboronic esters (TON up to 799,000) (Table 14.3) [193].

The $PdCl_2$–EDTA complex is an efficient catalyst for the Suzuki–Miyaura reactions of aryl and heteroarylhalides with aryl-(heteroaryl)boronic acids in water. Aryl iodides and bromides provide coupled products with TON up to 97,000 (Table 14.3) [194].

Table 14.3.
Catalyst development for Pd-catalyzed Suzuki–Miyaura coupling reactions.

Entry	R	X	Catalyst	Ligand	Yield (%)	TON	Reference
1	4-Me	I	–	a	96	10,000	[185b]
2	4-COMe	Br	–	b	92	92,000	[192]
3	H	I	$Pd(OAc)_2$	DABCO	95	950,000	[190]
4	H	I	–	c	100	100,000	[195]
5	H	Cl, Br	$Pd(OAc)_2$	DABCO	95–98	960,000	[191]
6	4-COMe	Br	–	d	80	799,000	[193]
7	4-CO_2H	Br	–	$PdCl_2$–EDTA	97	97,000	[194]

[a]2-*n*-Butylcyclophosphahexadienylanion; [b]Oximes derived from palladacycles; [c](2,5-Diphenylphospholyl)-2-methylpyridine-$PdCl_2$ complex; [d]P–S–Bidentate ligand.

The palladium(II) complexes of a S–P–S pincer ligand efficiently catalyzed the coupling between pinacolborane and various iodoaryls to yield the corresponding aryl-boronic esters (TON between 5100 and 76,500) [185b].

Reaction of (2,5-diphenylphospholyl)-2-methylpyridine with [(PdCl$_2$)[COD)] afforded a complex which catalyzed the cross-coupling between pinacolborane and iodoaromatics to afford corresponding arylboronic esters with TONs up to 100×10^3 and with TONs up to 90×10^2 in the case of aryl bromide [195].

14.7.4. Sonogashira coupling

Multidentate ferrocenyl phosphine is thermally stable and insensitive to air or moisture (Table 14.4) [196]. The catalytic activity of ferrocenyl aryl/alkyl triphosphine in Sonogashira reactions of alkynes was demonstrated with a variety of halides and allows aryl alkyny-lation with TONs up to 250,000. Efficient Pd-catalyzed homocoupling and Sonogashira reactions of terminal alkynes were achieved under aerobic conditions (Table 14.4) [197]. In the presence of Pd(OAc)$_2$, CuI, and DABCO, homocoupling of various terminal alkynes was carried out efficiently to provide moderate to excellent yields and high TONs (maximum up to 940,000) (Table 14.4) [197]. A mild and efficient Pd(OAc)$_2$/DABCO catalytic system for the Sonogashira cross-coupling reactions of aryl halides with terminal alkynes offered excellent yields. High TON (720,000) for the Sonogashira cross-coupling of 1-iodo-4-nitrobenzene with phenylacetylene was observed (Table 14.4) [198].

Table 14.4.
Catalyst development for Pd-catalyzed Sonogashira coupling reactions.

Entry	R	X	Catalyst	Ligand	Yield (%)	TON	Reference
1	H	I	–	a	95	100,000	[196]
2	4-Me	I, Br	Pd(OAc)$_2$	DABCO	94	940,000	[197]
3	4-NO$_2$	Br	Pd(OAc)$_2$	DABCO	95	720,000	[198]

[a][Pd(C$_3$H$_5$)Cl$_2$]/2Fc(P)$_2$^{t}Bu(P^{i}Pr).

14.7.5. Negishi coupling

Palladium-catalyzed reactions of unsaturated organozincs with arylalkyl bromide gave high yields of alkylbiphenyl with TON over 10^5 [199].

[a]1,1'-Bis(diphenylphosphino)ferrocene

14.7.6. Kumada coupling

The Kumada coupling of aryl halides with phenylmagnesium bromide, in the presence of palladium or nickel, showed high catalyst activity. TON up to 800 was observed for the conversion of chlorobenzene into biphenyl (Table 14.5) [182].

Table 14.5.
Catalyst development for Pd-catalyzed Kumada coupling reactions.

Entry	R	X	Catalyst	Ligand	Yield (%)	TON	Reference
1	H	Br	–	a	100	6900	[200]
2	H	Cl	Pd(OAc)₂	b	98	800	[182]

[a]1,4-Bis(diphenylphosphino)-1,2,3,4-tetraphenyl-1,3-butadiene; [b][5,17-bis(diphenylphosphino)-25,26,27,28-tetrapropoxycalix[4]arene.

14.8. Amination

The first high yielding one-pot tandem Hartwig–Buchwald–Heck cyclization was reported and applied to the synthesis of 2,3-disubstituted indoles [201]. Commercially available enone 123 was coupled with 1,2-dibromobenzene to provide N-aryleneaminone 124 which subsequently cyclized to indole derivative 125 [201]. This reaction was widely applicable to a variety of electron rich, electron poor and neutral aromatic bromides and chlorides as well as heterocyclic halides. Excellent yields were obtained regardless of the substitution pattern on the aromatic halide.

123   124   125

14.9. Carbonylation

Reductive carbonylations of aryl halides to benzaldehydes, amidocarbonylations, and double carbonylations of C–X bonds are of interest in industry.

The formylation of aryl bromides or iodides with HCOONa at an atmospheric pressure of CO readily proceeded in the presence of $PdCl_2(PPh_3)_2$ in DMF to afford the corresponding aldehydes in good yields. The contamination with water promoted a side reaction to form carboxylic acids [202]. The formylation using HCOOK or $HCOONH_4$ as a hydride source was rapid; however, it gave a considerable amount of the reduction product [202a]. Reductive formylation of diaryl halides was carried out under 50-psi pressure of CO in the presence of Pd(PPh$_3$)$_4$ to give mono-aldehydes in excellent yields [202b]. Carbonylation of several aryl chlorides at 200°C has been carried out to provide the corresponding aryl esters in the presence of 5% palladium catalyst. Addition of K$_2$Cr$_2$O$_7$ enhanced the catalytic activity of the palladium catalyst [203]. Electron-rich, chelate-stabilized complex Pd(dppp)$_2$-catalyzed carbonylation of aryl halides to aldehydes in high yield [204]. Aryl, heteroaryl, and vinylic halides were carbonylated under 1200 psi of 1:1 CO−H$_2$ in the presence of a basic tertiary amine and a dihalo-bis-(triphenylphosphine)palladium(II) catalyst at 80–150°C to form aldehydes in good yield. Acid chlorides also worked under similar conditions to form aldehydes [205].

Carbonylation of chloropyrazines in methanol or amines gave the corresponding esters or amides in excellent yields [80a]. The alkoxycarbonylation of various *N*-heteroaryl chlorides was examined with dppb and dppf as ligands [80b]. Several heterocyclic compounds have been prepared with appropriate catalyst and ligand with excellent yields. A catalyst turnover number up to 13,000 was obtained for the carbonylation of heteroaryl chloride [80b].

14.10. Amidocarbonylation

Enantiomerically pure α-amino acids were prepared in excellent yields *via* Pd-catalyzed amidocarbonylations using PdBr$_2$/PPh$_3$, and superior catalyst productivities were achieved (TON = 250,000) [206, 207]. The Pd-catalyzed carbonylation of aldehydes with urea derivatives provided a remarkably simple method for the preparation of 5-,3,5- and 1,3,5-substituted hydantoins with moderate yields [208]. The chemo- and regio-selectivities are an important advantage of this new one-pot multicomponent reaction over classical methods. In this process, low-cost starting materials were used which also eliminated the production of stoichiometric amounts of side products.

14.11. References

1. Maitlis, P. M. *The Organic Chemistry of Palladium,* Academic Press: New York, **1971**; Vols 1 and 2.
2. Tsuji, J. *Organic Synthesis with Palladium Compounds,* Springer: Berlin, **1980**.
3. Trost, B. M.; Verhoeven, T. R. Organopalladium Compounds in Organic Synthesis and in Catalysis. in *Comprehensive Organometallic Chemistry*; Pergamon Press: Oxford, **1982**, Vol. 8, p. 799.
4. Negishi, E.-I. *Acc. Chem. Res.* **1982**, *15*, 340–8.
5. Heck, R. F. *Palladium Reagents in Organic Synthesis*; Academic Press: New York, **1985**.
6. Hegedus, L. S. *Angew. Chem., Int. Ed. Engl.* **1988**, *27*, 1113–26.
7. Hegedus, L. S. *Transition Metals in the Synthesis of Complex Organic Molecules*, 2nd Edn.; University Science Books: Mill Valley, USA, **1999**.
8. Knochel, P.; Singer, R. D. *Chem. Rev.* **1993**, *93*, 2117–88.
9. Grushin, V. V.; Alper, H. *Chem. Rev.* **1994**, *94*, 1047–62.
10. Campbell, I. B. in *Organocopper Reagents,* Taylor, R. J. K. Ed.; IRL Press: Oxford, UK, **1994,** pp. 217–35.

11. Miyaura, N.; Suzuki, A. *Chem. Rev.* **1995**, *95*, 2457–83.

12. Tsuji, J. *Palladium Reagents and Catalysts: Innovations in Organic Synthesis;* Wiley: Chichester, UK, **1995,** pp. 340–5.

13. *Transition Metal Catalyzed Reactions, Chemistry for the 21st Century,* Murahashi, S.; Davies, S. G. Eds.; Blackwell Science: Oxford, **1999**.

14. Beller, M. in *Applied Homogeneous Catalysis with Organometallic Compounds,* Cornils, B. Herrmann, W. A. Eds.; VCH: Weinheim, **1996,** Vol. 1, p. 148.

15. Herrmann, W. A. in *Applied Homogeneous Catalysis with Organometallic Compounds*; Cornils, B.; Herrmann, W. A. Eds.; VCH: Weinheim, **1996**; Vol. 2, p. 712.

16. Jeffery, T. *Adv. Met. Org. Chem.* **1996**, *5*, 153–260.

17. Blaser, H. U.; Pugin, B., Spindler, F. in *Applied Homogeneous Catalysis with Organometallic Compounds*; Cornils, B.; Herrmann, W. A. Eds.; VCH: Weinheim, **1996**, Vol. 2, p. 992.

18. Nicolaou, K. C., Sorensen, E. J. *Classics in Total Synthesis,* VCH: Weinheim, **1996**.

19. Malleron, J.-L.; Fiaud, J.-C.; Legros, J.-Y. *Handbook of Palladium-catalyzed Organic Reactions*; Academic Press, San Diego, USA, **1997**.

20. Eisenstadt, A. in *Catalysis of Organic Reactions*, Herkes, F. E. Ed.; Marcel Dekker: New York, **1998**, p. 415.

21. Beller, M.; Riermeier, T. H. in *Organic Synthesis Highlights III*, Mulzer, J.; Waldmann, H. Eds., Wiley-VCH: Weinheim, **1998**, 126.

22. Farina, V.; Krishnamurthy, V.; Scott, W. J. *The Stille Reaction* Wiley: New York, **1998**.

23. Suzuki, A. in *Metal-Catalyzed Cross-Coupling Reactions*, Diederich, F.; Stang, P. J. Eds., Wiley-VCH: Weinheim, **1998**, 49.

24. Bräse, S.; de Meijere, A. in *Metal-Catalyzed Cross-Coupling Reactions,* Diederich, F.; Stang, P. J., Eds.; Wiley-VCH: Weinheim, **1998**; p. 99.

25. *Cross-Coupling Reactions,* Miyaura, N. Ed.; Spring Verlag: Berlin, **2002**.

26. Tsuji, J. *Perspectives in Organopalladium Chemistry for the XXI Century,* Elsevier: Amsterdam, **1999**.

27. Duncton, M. A. J.; Pattenden, G. *J. Chem. Soc., Perkin Trans. 1* **1999**, 1235–46.

28. Tsuji, J. *J. Org. Chem.* **1999**, 576.

29. Tsuji, J. *Transition Metal Reagents and Catalysts, Innovations in Organic Synthesis*, Wiley: Chichester, **2000**.

30. Li, J. J.; Gribble, G.W. *Palladium in Heterocyclic Chemistry*; Pergamon: Oxford, **2000**.

31. Beller, M.; Zapf, A.; Mägerlein, W. *Chem. Eng. Technol.* **2001**, *24*, 575–82.

32. Chemler, S. R.; Trauner, D.; Danishefsky, S. J. *Angew. Chem. Int. Ed. Engl.* **2001**, *40*, 4544–68.

33. Littke, A. F.; Fu, G. C. *Angew. Chem. Int. Ed. Engl.* **2002**, *41*, 4176–211.

34. Colquhoum, H. M.; Thompson, D. J.; Twigg, M. V. *Carbonylation*, Plenum Press: New York, **1991**.

35. (a) Tsjui, J. in *Handbook of Organopalladium Chemistry for Organic Synthesis,* Negishi, E.-I. Ed.; John Wiley and Sons: Hoboken, **2002**, Vol. 2, pp. 1669–87. (b) Acemoglu, L.; Williams, J. M. J. in *Handbook of Organopalladium Chemistry for Organic Synthesis*, Negishi, E.-I. Ed.; John Wiley and Sons: Hoboken, **2002**, Vol. 1, pp. 1689–705.

36. Negishi, E.-I in *Handbook of Organopalladium Chemistry for Organic Synthesis*, Negishi, E.-I. Ed.; John Wiley and Sons: Hoboken, **2002**, Vol. 1, pp. 229–47.
37. Fujiwara, Y.; Jia, C. in *Handbook of Organopalladium Chemistry for Organic Synthesis*, Negishi, E.-I. Ed.; John Wiley and Sons: Hoboken, **2002**, Vol. 2, pp. 2859–62.
38. Link, J. T. *Org. React.* **2002**, *60*, 157–534.
39. Hassan, J.; Sévignon, M.; Gozzi, C.; Schulz, E.; Lemaire, M. *Chem. Rev.* **2002**, *102*, 1359–469.
40. Suzuki, A. in *Modern Arene Chemistry,* Astruc, D. Ed.; Wiley-VCH: Weinheim, **2002**, pp. 53–106.
41. Denmark, S. E.; Sweis, R. F. in *Metal-Catalyzed Cross-Coupling Reactions,* 2nd Edn.; de Meijere, A.; Diederich, F. Eds.; Wiley-VCH: Weinheim, **2004,** pp. 163–216.
42. Braese, S.; de Meijere, A. in *Metal-Catalyzed Cross-Coupling Reactions*, 2nd Edn.; de Meijere, A.; Diederich, F., Eds; Wiley-VCH: Weinheim, **2004,** pp. 217–315.
43. Tietze, F.; Ila, H.; Bell, H. P. *Chem. Rev.* **2004**, *104*, 3453–516.
44. Echavarren, A. M.; Cardenas, D. J. in *Metal-Catalyzed Cross-Coupling Reactions*, 2nd Edn.; de Meijere, A.; Diederich, F., Eds.; Wiley-VCH: Weinheim, **2004**, pp. 1–40.
45. Miyaura, N. in *Metal-Catalyzed Cross-Coupling Reactions*, 2nd Edn.; de Meijere, A.; Diederich, F., Eds.; Wiley-VCH: Weinheim, **2004**; pp. 41–123.
46. Tsuji, J. *Palladium Reagents and Catalysts New Perspective for the 21st Century*; John Wiley & Sons, Ltd.: Chichester, **2004**.
47. Nakamura, I.; Yamamoto, Y. *Chem Rev.* **2004**, *104*, 2127–98.
48. Zeni, G.; Larock, R. C. *Chem. Rev.* **2004**, *104*, 2285–309.
49. Schlummer, B.; Scholz, U. *Adv. Synth. Cata.* **2004**, *346*, 1599–626.
50. Christmann, U.; Vilar, R. *Angew. Chem. Int. Ed. Engl.* **2005**, *44*, 366–74.
51. Cacchi, S.; Fabrizi, G. *Chem. Rev.* **2005**, *105*, 2873–920.
52. Wolfe, J. P.; Thomas, J. S. *Curr. Org. Chem.* **2005**, *9*, 625–55.
53. Riermeier, T. H.; Zapf, A.; Beller, M. *Top. Catal.* **1997**, *4*, 301.
54. Stürmer, R. *Angew. Chem. Int. Ed. Engl.* **1999**, *38*, 3307–8.
55. Chottard, J. C.; Mulliez, J. S.; Mansuy, D.; Guilhem, J. *Tetrahedron* **1981**, *37*, 31–40.
56. Kikukawa, K.; Nagira, K.; Terao, N.; Wada, F.; Matsuda, T. *Bull. Chem. Soc. Jpn.* **1979**, *52*, 2609–10.
57. Sengupta, S.; Bhattacharya, S. *J. Chem. Soc. Perkin Trans. 1* **1993**, 1943.
58. Kikukawa, K.; Maemura, K.; Kiseki, Y.; Wada, F.; Matsuda, T. Giam, C. *J. Org. Chem.* **1981**, *46*, 4885.
59. Akiyama, F.; Miyazaki, H.; Kaneda, K.; Teranishi, S.; Fujiwara, Y.; Abe, M.; Taniguchi, H. *J. Org. Chem.* **1980**, *45*, 2359.
60. Beller, H.; Fischer, H.; Kühlein, K. *Tetrahedron Lett.* **1994**, *35*, 8773–6.
61. Beller, M.; Kühlein, K. *Synlett* **1995**, 441–2.
62. Quotes on the trading prices of palladium on live – online Kitco Store. http://www.kitco.com/charts/livepalladium.html; November 2005.
63. Wu, T.-C. (Ethyl Corporation). U.S. Patent 5,315,026, **1994**.
64. Wu, T.-C. (Albermarle). U.S. Patent 5,536,870, **1996**.
65. Beller, M.; Tafesh, A.; Herrmann, W. A. (Hoechst AG). DE 19,503,119, **1996**.
66. Brunet, J. P.; Cometti, A. (Rhone-Poulenc). FR 2163875, **1973**.
67. Rieu, J. P.; Boucherle, A.; Cousse, H.; Mouzin, G. *Tetrahedron* **1986**, *42*, 4095–131.
68. (a) Mitra, R.B.; Joshi, V.S. *Synth. Commun.* **1988**, *18*, 2259; (b) Bennetau, B.; Krempp, M.; Dunogués. J. *Synth. Commun.* **1994**, *24*, 77.

69. Ramminger, C.; Zim, D.; Lando, V. R.; Fassina, V.; Monteiro, A.L. *J. Braz. Chem. Soc.* **2000**, *11*, 105–11.

70. Larsen, R. D.; Corley, E. G.; King, A. O.; Carroll, J. D.; Davis, P.; Verhoeven, T. R.; Reider, P. J.; Labelle, M.; Gauthier, J. Y.; Xiang, Y. B.; Zamboni, R. J. *J. Org. Chem.* **1996**, *61*, 3398–405.

71. Aslam, M.; Elango, V. (Hoechst-Celanese). U.S. Patent 5,225,603, **1993**.

72. Eisenstadt, A., in *17th Conference on Catalysis of Organic Reactions,* New Orleans, **1998**; p. 415.

73. Oxford, A. W.; Butina, D.; Owen, M. R. (Glaxo). U.S. Patent 4,997,841, **1991**.

74. Blatcher, P.; Carter, M.; Hornby, R.; Owen, M. R. (Glaxo). WO 09166, **1995**.

75. Chen, C. Y.; Larsen, R. D. (Merck). WO 06725, **1998**.

76. Chen, C. Y.; Larsen, R. D.; Verhoeven, T. R. (Merck). WO 32197, **1995**.

77. Chen, C.-Y.; Lieberman, D. R.; Larsen, R. D.; Reamer, R.A.; Verhoeven, T. R.; Reider, P. J.; Cottrell, I. F.; Houghton, P. G. *Tetrahedron Lett.* **1994**, *35*, 6981–4.

78. Bosch, J.; Roca, T.; Armengal, M.; Fernández-Forner, D. *Tetrahedron* **2001**, *57*, 1041–8.

79. (a) Ogilvie, R. J. (Pfizer). WO 50063, **2002**; (b) Macor, J. E.; Wythes, M. J. (Pfizer). U.S. Patent 5,545,644, **1996**.

80. (a) Takeuchi, R.; Suzuki, K.; Sato, N. *J. Mol. Cat.* **1991**, *66*, 277–88. (b) Beller, M., Magerlein, W.; Indolese, A. F.; Fischer, C. *Synthesis* **2001**, *7*, 1098–109.

81. Beller, M.; Indolese, A.F. *Chimia* **2001**, *55*, 684–7.

82. Jang, E. J.; Lee, K. H.; Lee, J. S.; Kim, Y. G. *J. Mol. Catal. A Chem.* **1999**, *138*, 25–36.

83. Schmid, R. *Chimia* **1996**, *50*, 110–13.

84. Scalone, M; Vogt, P. (Hoffmann-La Roche). EP 385210, **1990**.

85. Geissler, in *Transition Metals for Organic Synthesis*; Beller, M.; Bolm, C. Eds.; Wiley-VCH: Weinheim, **1998**; Vol. 1, 158.

86. Indolese, A. F. *Tetrahedron Lett.* **1997**, *38*, 3513–16.

87. Saito, S.; Oh-Tani, S.; Miyaura, N. *J. Org. Chem.* **1997**, *62*, 8024–30.

88. Beller, M.; Fischer. H.; Herrmann, W.A.; Öfele, K; Broßmer, C. *Angew. Chem. Int. Ed. Engl.* **1995**, *34*, 1848–9.

89. Mitchell, M. B.; Wallbank, P. J. *Tetrahedron Lett.* **1991**, *32*, 2273–6.

90. Reetz, M. T.; Breinbauer, R.; Wanninger, K. *Tetrahedron Lett.* **1996**, *37*, 4499–502.

91. Saito, S.; Sakai, M.; Miyaura, N. *Tetrahedron Lett.* **1996**, *37*, 2993–6.

92. Shen, W. *Tetrahedron Lett.* **1997**, 38, 5575–8.

93. Herrmann, W. A.; Reisinger, C.-P.; Spiegler, M. *J. Org. Chem.* **1998**, *557*, 93–6.

94. Firooznia, F.; Gude, C.; Chan, K.; Satoh, Y. *Tetrahedron Lett.* **1998**, *39*, 3985–8.

95. Littke, A. F.; Fu, G. C. *Angew. Chem. Int. Ed. Engl.* **1998**, *38*, 3387–8.

96. Wolfe, J. P.; Buchwald, S. L. *Angew. Chem. Int. Ed. Engl.* **1999**, 39, 2413–16.

97. Zhang, C.; Huang, J.; Trudell, M. L; Nolan, S. P. *J. Org. Chem.* **1999**, 64, 3804–5.

98. Bei, X.; Turner, H. W.; Weinberg, W. H.; Guram, A. S. *J. Org. Chem.* **1999**, *64*, 6797–803.

99. (a) Poetsch, E., *Kontakte* **1988**, 15. (b) Haber, J. G. S. in *Aqueous Phase Organometallic Catalysis*, Cornils B.; Herrmann, W. A.; Eds.; VCH-Wiley: Weinheim, **1998**, p. 444.

100. Larsen, R. D.; King, A. O.; Chen, C. Y.; Corley, E. G.; Foster, B. S.; Roberts, F. E.; Yang, C.; Lieberman, D. R.; Reamer, R. A.; Tschaen, D. M.; Verhoeven, T. R.;

Reider, P. J.; Lo, Y. S.; Rossano, L. T.; Brookes, A. S.; Meloni, D.; Moore, J. R.; Arnett, J. F. *J. Org. Chem.* **1994**, *59*, 6391–4.

101. Desmond, R.; Dolling, U.; Marcune, B.; Tillyer, R.; Tschaen, D. (Merck). WO 96 08482, **1996**.

102. Hancock, B., Winters, C.; Gertz, B; Ehrich, E. (Merck). WO 97 044028, **1997**.

103. Miyachi, N.; Yanagawa, Y.; Iwasaki, H.; Ohara, Y.; Hiyama, T. *Tetrahedron Lett.* **1993**, *34*, 8267–70.

104. Meng, D.; Bertinato, P.; Balog, A.; Su, D.-S.; Kamenecka, T.; Sorensen, E. J.; Danishefsky, S. J. *J. Am. Chem. Soc.* **1997**, *119*, 10073–92.

105. Mizoroki, T.; Mori, K.; Ozaki, A. *Bull. Chem. Soc. Jpn.* **1971**, *44*, 581.

106. Heck, R. F. *Org. React.* **1982**, *27*, 345.

107. Ridgway, B. H.; Woerpel, K. A. *J. Org. Chem.* **1998**, *63*, 458–60.

108. Kawai, S. H.; Hambalek, R. J.; Just, G. *J. Org. Chem.* **1994**, *59*, 2620–2.

109. Beutler, U.; Mazacek, J.; Penn, G.; Schenkel, B.; Wasmuth, D. *Chimia* **1996**, *50*, 154–6.

110. Robins, M. J.; Barr, P. J. *J. Org. Chem.* **1983**, *48*, 1854–62.

111. Parshall, G. W.; Nugent, W. A. *CHEMTECH* **1988**, 376–83.

112. Chandraratna, R. A. S. (Allergan). EP 284,288, **1988**.

113. Chandraratna, R. A. S. (Allergan). U.S. Patent 5,089,509, **1992**.

114. Chandraratna, R. A. S. (Allergan). WO 011686, **1996**.

115. Mori, Y.; Seki, M. *Heterocycles* **2002**, *58*, 125.

116. Mori, Y.; Seki, M. *J. Org. Chem.* **2003**, *68*, 1571–4.

117. Shimizu, T.; Seki, M. *Tetrahedron Lett.* **2001**, *42*, 429.

118. Kosugi, M., Kameyama, M.; Migita, T. *Chem. Lett.* **1983**, 927–8.

119. Guram, A. S.; Rennels, R. A.; Buchwald, S. L. *Angew. Chem. Int. Ed. Engl.* **1995**, *34*, 1348–50.

120. Louie, J.; Hartwig, J. F. *Tetrahedron Lett.* **1995**, *36*, 3609–12.

121. Beller, M. *Angew. Chem. Int. Ed. Engl.* **1995**, *34*, 1316–17.

122. Wagaw, S.; Buchwald, S. L. *J. Org. Chem.* **1996**, *61*, 7240–1.

123. Wolfe, J. P.; Wagaw, S.; Buchwald, S. L. *J. Am. Chem. Soc.* **1996**, *118*, 7215–16.

124. Driver, M. S.; Hartwig, J. F. *J. Am. Chem. Soc.* **1996**, *118*, 7217–18.

125. Zhao, S.; Miller, A. K.; Berger, J.; Flippin, L. A. *Tetrahedron Lett.* **1996**, *37*, 4463–6.

126. Ward, Y. D.; Farina, V. *Tetrahedron Lett.* **1996**, *37*, 6993–6.

127. Willoughby, C. A.; Chapman, K. T. *Tetrahedron Lett.* **1996**, *37*, 7181–4.

128. Kanbara, T.; Honma, A.; Hasegawa, K. *Chem. Lett.* **1996**, 1135–6.

129. Wolfe, J. P.; Buchwald, S. L. *J. Org. Chem.* **1997**, *62*, 1264–7.

130. Louie, J.; Driver, M. S.; Hamann, B. C.; Hartwig, J. F. *J. Org. Chem.* **1997**, *62*, 1268–73.

131. Marcoux, J.-F., Wagaw, S., Buchwald, S. L. *J. Org. Chem.* **1997**, *62*, 1568–9.

132. Beletskaya, I. P.; Bessmertnykh, A. G.; Guilard, R. *Tetrahedron Lett.* **1997**, *38*, 2287–90.

133. Beller, M.; Riermeier, T. H.; Reisinger, C.-P.; Herrmann, W. A. *Tetrahedron Lett.* **1997**, *38*, 2073–4.

134. Singer, R. A.; Buchwald, S. L. *Tetrahedron Lett.* **1999**, *40*, 1095–8.

135. Hartwig, J. F. *Synlett* **1997**, 329–40.

136. Morita, S.; Kitano, K.; Matsubara, J.; Ohtani, T.; Kawano, Y.; Otsubo, K.; Uchida, M. *Tetrahedron* **1998**, *54*, 4811–18.

137. Oshiro, Y.; Sato S.; Kurahashi, N; Tanaka, T.; Kikuchi, T.; Tottori, K.; Uwahodo, Y.; Nishi, T. *J. Med. Chem.* **1998**, *41*, 658–67.

138. *Drugs Future* **1995**, *20*, 884.

139. Oshiro, Y.; Sato S.; Kurahashi, N. (Otsuka Pharmaceuticals). EP 0367141, **1996**.

140. Oshiro, Y.; Sato, S.; Kurahashi, N. (Otsuka Pharmaceuticals). U.S. Patent 5,006,528, **1991**.

141. Mase, T.; Kato, Y.; Kawasaki, M.; Lang, F.; Lee, J.; Lynch, J.; Maligres, P.; Molina, A.; Nemoto, T.; Okada, S.; Reamer, R.; Song, J. Z.; Tschaen, D.; Wada, T.; Zewge, D.; Volante, R. P.; Reider, P. J.; Tomimoto, K. *J. Org. Chem.* **2001**, *66*, 6775–86.

142. Margolis, B. J.; Swidorski, J. J.; Rogers, B. N. *J. Org. Chem.* **2003**, *68*, 644–7.

143. Mathes, B. J.; Filla, S. A. *Tetrahedron Lett.* **2003**, *44*, 725–8.

144. Madar, D. J.; Kopecka, H.; Pireh, D.; Pease, J.; Pliushchev, M.; Sciotti, R. J.; Wiedeman, P. E.; Djuric, S.W. *Tetrahedron Lett.* **2001**, *42*, 3681–4.

145. Damon, D. B.; Dugger, R. W.; Scott, R. W. (Pfizer). WO 2002088085, **2002**.

146. Damon, D. B.; Dugger, R. W.; Scott, R. W. (Pfizer). WO 2002088069, **2002**.

147. Wang, L.; Woods, K. W.; Li, Q.; Barr, K. J.; McCroskey, R. W.; Hannick, S. M.; Gherke, L.; Credo, R. B.; Hui, Y.-H.; Marsh, K.; Waner, R.; Lee, J. Y.; Zielinski-Mozug, N.; Frost, D.; Rosenberg, S. H.; Sham, H. L. *J. Med. Chem.* **2002**, *45*, 1697–711.

148. Wentland, M. P.; Duan, W.; Cohen, D. J.; Bidlack, J. M. *J. Med. Chem.* **2000**, *43*, 3558–65.

149. Chalk, A. J.; Magennis, S. A. *J. Org. Chem.* **1976**, *41*, 1206–9.

150. Bader, R. R.; Baumeister, P.; Blaser, H.-U. *Chimia* **1996**, *50*, 99–105.

151. Baumeister, P.; Seifert, G.; Steiner, H. (Ciba-Geigy AG). EP-A 584 043, **1994**.

152. Bessard Y.; Roduit, J. P. *Tetrahedron* **1999**, *55*, 393–404.

153. Dobler, M. R.; Bruce, I.; Cederbaum, F.; Cooke, N. G.; Diorazio, L. J.; Hall, R. G.; Irving, E. *Tetrahedron Lett.* **2001**, *42*, 8281–4.

154. Watanabe, M.; Yamamoto, T.; Nishiyama, M. *Chem. Commun.* **2000**, 133–4.

155. Watanabe, M.; Nishiyama, Y.; Yamamoto, T.; Koei, Y. *Tetrahedron Lett.* **2000**, *41*, 481–3.

156. Ferreira, I. C. F. R.; Queiroz, M.-J. R. P., Kirsch, G. *Tetrahedron* **2003**, *59*, 975–81.

157. Huang, T. H.; Lin, J. T.; Tao, Y.-T.; Chuen, C.-H. *Chem. Mater.* **2003**, *15*, 4854–62.

158. Durán, R.; Zubia, E.; Ortega, M. J.; Naranjo, S.; Salvá, J. *Tetrahedron* **1999**, *55*, 13225–32.

159. DeVries, R. A.; Mendoza, A. *Organometallics* **1994**, *13*, 2405–11.

160. Rehahn, M.; Schlüter, A.-D.; Wegner, G.; Feast, W. J. *Polymer* **1989**, *30*, 1054–9.

161. Goldfinger, M. B.; Swager, T. M. *J. Am. Chem. Soc.* **1994**, *116*, 7895–6.

162. Wallow, T. I.; Novak, B. M. *J. Am. Chem. Soc.* **1991**, *113*, 7411–12.

163. Tour, J. M.; Lamba, J. J. S. *J. Am. Chem. Soc.* **1993**, *115*, 4935–6.

164. Blake, A. J.; Cooke, P. A.; Doyle, K. J.; Gair, S.; Simpkins, N. S. *Tetrahedron Lett.* **1998**, *39*, 9093–6.

165. Karakaya, B.; Claussen, W.; Gessler, K.; Saenber, W.; Schlüter, A.-D. *J. Am. Chem. Soc.* **1997**, *119*, 3296–301.

166. Liu, C-Y.; Luh, T.-Y. *Org. Lett.* **2002**, *4*, 4305–7.

167. Farina, V. *Adv. Synth. Catal.* **2004**, *346*, 1553–82.

168. Zapf, A. *Angew. Chem. Int. ed. Engl.* **2003**, *42*, 5394–9.

169. Heck, R. F.; Nolley, J. P. Jr. *J. Org. Chem.* **1972**, *37*, 2320–2.

170. Herrmann, W. A.; Elison, M.; Fischer, J.; Köcher, C.; Artus, G. R. J. *Angew. Chem. Int. Ed. Engl.* **1995**, *34*, 2371–4.

171. de Meijere, A.; Meyer, F. E. *Angew. Chem. Int. Ed. Engl.* **1994**, *33*, 2379–411.

172. Julia, M.; Duteil, M.; Grard, C.; Kuntz, E. *Bull. Soc. Chim. Fr.* **1973**, 2791.

173. Davison, J. B.; Simon, N. M.; Sojka, S. A. *J. Mol. Catal.* **1984**, *22*, 394.

174. Spencer, A. *J. Org. Chem.* **1984**, *270*, 115.

175. Ben-David,Y.; Portnoy, M.; Gozin, M.; Milstein, D. *Organometallics* **1992**, *11*, 1995–6.

176. Portnoy, M.; Ben-David, Y.; Milstein, D. *Organometallics* **1993**, *12*, 4734–5.

177. Herrmann, W. A.; Brossmer, C.; Öfele, K.; Reisinger, C.-P.; Priermeier, T.; Beller, M.; Fischer, H. *Angew. Chem. Int. Ed. Engl.* **1995**, *34*, 1844–8.

178. Reetz, M. T.; Lohmer, G.; Schwickardi, R. *Angew. Chem. Int. Ed. Engl.* **1998**, *37*, 481–3.

179. Beller, M.; Zapf, A. *Synlett* **1998**, 792–3.

180. Littke, A. F.; Fu, G. C. *J. Org. Chem.* **1999**, *64*, 10–11.

181. Shaughnessy, K. H.; Kim, P.; Hartwig, J. F. *J. Am. Chem. Soc.* **1999**, *121*, 2123–32.

182. Sémeril, D.; Lejeune, M.; Jeunesse, C.; Matt, D. *J. Mol. Catal. A: Chemical* **2005**, *239*, 257–62.

183. Zapf, A.; Beller, M. *Chem. Eur. J.* **2000**, *6*, 1830–3.

184. Sava, X.; Richard, L.; Mathey, F.; Le Floch, P. *Organometallics* **2000**, *19*, 4899–903.

185. (a) Pittelkow, M.; Moth-Poulsen, K.; Boas, U.; Christensen, J. B. *Langmuir* **2003**, *19*, 7682–4. (b) Doux, M.; Mezailles, N.; Melaimi, M.; Ricard, L.; Le Floch, P. *Chem. Commun.* **2002**, *15*, 1566–7.

186. Smith, M. D.; Stepan, A. F.; Ramarao, C.; Brennan, P. E.; Ley, S. V. *Chem. Commun.* **2003**, 2652–3.

187. Xiong, Z.; Wang, N.; Dai, M; Li, A.; Chen, J.; Yang, Z. *Org. Lett.* **2004**, *6*, 3337–40.

188. Dai, M.; Liang, B.; Wang, C.; Chen, J.; Yang, Z. *Org. Lett.* **2004**, *6*, 221–4.

189. (a) Piechaczyk, O.; Doux, M.; Ricard, L.; Floch, P. *Organometallics* **2005**, *24*, 1204–13. (b) Zhu, L.; Duquette, J.; Zhang, M. *J. Org. Chem.* **2003**, *68*, 3729–32.

190. Li, J.-H.; Liu, W.-J. *Org. Lett.* **2004**, *6*, 2809–11.

191. Li, J.-H.; Liang, Y.; Wang, D. P.; Liu, W.-J.; Xie, Y.-X. *J. Org. Chem.* **2005**, *70*, 5409–12.

192. Botella, L.; Nájera, C. *J. Org. Chem.* **2002**, *663*, 46–57.

193. Dochnahl, M.; Doux, M.; Faillard, E.; Ricard, L.; Le Floch, P. *Eur. J. Inorg. Chem.* **2005**, *1*, 125–34.

194. Korolev, D. N.; Bumagin, N. A. *Tetrahedron Lett.* **2005**, *46*, 5751–4.

195. Thoumazet, C.; Melaimi, M.; Ricard, L.; Le Floch, P. *Comptes Rendus. Chimie* **2004**, *7*, 823–32.

196. Hierso, J.-C.; Fihri, A.; Amardeil, R.; Meunier, P.; Doucet, H.; Santelli, M.; Ivanov, V. *Org. Lett.* **2004**, *6*, 3473–6.

197. Li, J.-H.; Liang, Y.; Xie, Y.-X. *J. Org. Chem.* **2005**, *70*, 4393–6.

198. Li, J.-H.; Zhang, X.-D.; Xie, Y.-X. *Synthesis* **2005**, *5*, 804–8.

199. Huang, Z.; Qian, M.; Babinski, D. J.; Negishi, E. *Organometallics* **2005**, *24*, 475–8.

200. Doherty, S.; Robins, E. G., Nieuwenhuyzen, M.; Knight, J. G.; Champkin, P. A.; Clegg, W. *Organometallics* **2002**, *21*, 1383–99.

201. Edmondson, S. D.; Mastracchio, A.; Parmee, E. R. *Org. Lett.* **2000**, 2, 1109–12.

202. (a) Okano, T.; Harada, N.; Kiji, J. *Bull. Chem. Soc. Jpn.* **1994**, *67*, 2329–32. (b) Pri-Bar, I.; Buchmann, O. *J. Org. Chem.* **1984**, *49*, 4009–11.

203. Dufaud, V.; Thivolle-Cazat, J.; Basset, J. M. *J. Chem. Soc. Chem. Commun.* **1990**, 426–7.

204. Ben-David, Y.; Portnoy, M.; Milstein, D. *J. Chem. Soc. Chem. Commun.* **1989**, 1816–17.

205. Schoenberg, A.; Heck, R. F. *J. Am. Chem. Soc.* **1974**, *96*, 7761–4.

206. Beller, M.; Eckert, M.; Vollmüller, F.; Bogdanovic, S.; Geissler, H. *Angew. Chem. Int. Ed. Engl.* **1997**, *36*, 1494–6.

207. Beller, M.; Eckert, M.; Geissler, H.; Napierski, B.; Rebenstock, H.-P.; Holla, W. *Chem. Eur. J.* **1998**, *4*, 935–41.

208. Beller, M.; Eckert, M.; Moradi, W. A.; Neumann, H. *Angew. Chem. Int. Ed. Engl.* **1999**, *38*, 1454–7.

Index

Accolate, 82
4-Acetoxyindol, 66
Acetyl chloride, 197
Acetyl fluoride, 193
Acetylenic homocoupling, 365
Acetylpyrazines, 441
3-Acylindoles, 100, 144, 159
Adenosine diphosphate (ADP), 251
Adenosine receptor 2a (A$_{2a}$), 264
Adenosine receptors A$_{2a}$ and A$_1$, 365
Agroclavine, 129
Aklavinone, 307
N-Alkenyl-2-iodoindoles, 134
Alkenylpalladium complex, 330
3-Alkenylpyrroles, 95
Alkoxycarbonylation, 336, 370, 577, 613
3-Alkoxyindoles, 146
Alkylacetylene, 396
Alkylation, 316
N-Alkyl-bis[6-(2-methoxyphenyl)-5-
    methylpyridazin-3-yl]amines, 580
Alkyl (2-methoxymethyl)-1-oxo-
    4-phenyl-1, 2-dihydrophthalazine-
    6-carboxylates, 574
B-Alkyl Suzuki reaction, 315
O-Alkylation, 328, 329
Alkyne hydroamination, 365
Alkynols, 330–333
Alkynones, 333–335
Alkynylcopper, 14
2-Alkynylfuran, 323
Alkynylpyrazine, 440
Allenones, 336
Allenyl ketone intermediate, 333
3-Allenylbenzofuran, 332
Allenylpalladium complex, 328, 329

Alloyohimbone, 166
Allyl-3-haloquinoxalin-2-ylamines, 466
Allylations, 429
Allylic substitution, 353
η^3-Allylpalladium chloride dimmer, 283
π-Allylpalladium complex, 2, 3, 25, 67, 328
Almotriptan, 590
Aluminium chloride, 513
Amaryldaceae alkaloids, 107
Amezinium metilsulfate, 542
Amidic hydrogen transfer, 359
Amination, 71, 155, 237–239, 371, 467, 601
5-Amino-7-azaindole, 158
2-Amino-3-bromoquinoxaline, 463
2-Amino-5-bromopyrazine, 448, 460
2-Amino-5-cyanopyrroles, 71
2-Amino-3-cyano-5-bromopyrazine, 451
2-Amino-3-heteroarylquinoxalines, 463
2-Amino-3-indolyl-5-bromopyrazine, 449
3-Amino-2-phenylpiperidine, 213
6-Aminobenzofuran, 467
Aminocarbonylation, 579
Aminomalononitrile p-toluenesulfonate, 384
o-Aminophenols, 401
2-Aminophenylboronic acid, 389
2-Aminopyrazine, 455
N-Aminopyrroles, 69
Aminothiazoles, 345
Anhydrodehydrolycorine, 137
Anhydrolycorin-7-one, 121, 136
Annonidine A, 129
Annular tautomerism, 410
Annulated isoindoles, 427
Annulation, 364
Anthramycin, 59
Anti-asthmatic, 511

Antifungal, 279, 380, 435
Antihypertensive, 553
Anti-inflammatory agent, 345, 436
Anti-Markovnikov addition, 570
Antimigraine drugs, 82
Antiostatins, 118
Antiplasmodial drugs, 533
Antispychotic sertindole, 412
Antitumor activities, 525
Antiviral activities, 435, 525
Anxiolytic, 553
Anzemet, 82
3-AP, 212, 235
Arbuzov–Michaelis reaction, 429
Arctic acid, 285
Arctium lappa, 285
Arcyriacyanin A, 111, 115, 134
Arenediazonium salts, 610
Argiotoxin 659, 81
Aripiprazole, 596
Aromaticity, 189
Aryl benzylfluoropyrazines, 440
Aryl boronic acids, 52
Aryl halide, 3, 5, 9, 19, 22, 24
Aryl-(heteroaryl)boronic acids, 610
Aryl magnesates, 518
Aryl triflate, 10, 21, 22, 348
Aryl trifluoromethylquinoxalines, 461
5-Aryl-4-methoxypyridazin-3(2*H*)-ones,
    559, 561
2-Aryl-4-oxalone, 391
4-Aryl-5-chloropyridazin-3(2*H*)-ones, 560
4-Aryl-5-methoxypyridazin-3(2*H*)-ones,
    559–561
3-Aryl-6-chloropyridazines, 551, 552
3-Aryl-6-methoxypyridazines, 552, 562
5-Aryl-6-phenylpyridazin-3(2*H*)-ones, 558
Arylaminopyridazin-(3(2*H*)-on)-es, 577, 579
Arylation, 348, 369, 426, 427
O-Arylation, 337
2-Arylbenzoxazole, 388, 389
Arylboranates, 518
Arylboronic esters, 518
Aryldiboronic acids, 604
Aryldifluoromethoxyquinoxalines, 461
N-Arylindoles, 165
2-Aryloxazoloquinolones, 389

6-Aryloxindoles, 112
6-Arylpyridazin-3-amines, 552, 562
4-Arylpyridazines, 563
2-Aryltryptamines, 108
Arynic cyclization, 123
Asperlicin, 162
Aspidophytine, 127
Asterriquinones, 158
Asymmetric induction, 149
Atomic charges, 189
(H$^+$/K$^+$)-ATPase inhibitors, 535
Aurantioclavine, 130
9-Azabicyclononen, 442
Azacycloheptatracene, 514
Azaindoles, 107, 122–124, 126, 129, 145,
    148, 156
5-Azaindolones, 151
7-Azaindolinones, 149
Azaketotetrahydrocarbazoles, 145
Azelastine, 542
5-Azidoquinoline, 514
Azinine, 514
Azobisisobutyronitrile (AIBN), 356

Baeyer–Villiger oxidation, 90
Bartoli indole synthesis, 88
Base effect, 581
Beckmann rearrangement, 70
Benthocyanin A, 437, 603
Benz[*b*]oxazole, 458
Benzaldehyde, 384
Benzannulation, 319
Benzo[4,5]cyclohepta[*b*]indole, 138
Benzo[*a*]aceanthranylene, 602
Benzo[*b*]furan, 306, 457
Benzo[*b*]thiophene, 457
Benzocarbazole-6,11-quinones, 90
3,4-Benzocarbolines, 136
Benzodiazepine receptors, 135
Benzodiazepines, 239, 442
Benzofuran, 305, 328, 336
Benzofuranylthioethers, 441
Benzofuryl-2-boronic acid, 313
Benzopyran, 27
2,1,3-Benzothiadiazole bisindoles, 103
Benzoxazole, 18, 403
2-Benzoxazolyllithium, 395

Benzoxepine ring, 596

Benzoyl chloride, 384

N-Benzoylpyrrole, 39, 42

1-Benzyl-2-chloroindole, 84

1-Benzyl-3-bromopyrrole, 39

3-Benzyl-5-bromo-2-pyrazinamine, 447

9-Benzyl-6-(2-furyl)purines, 319

1-Benzyl-7-bromoindoline, 88

3-Benzyl-5-(4-methoxyphenyl)-2-
    pyrazinamine, 448

Benzylfluoropyrazines, 440, 446

2-Benzylpyrazine, 440

BF$_3$-etherate, 193

2,2′-Bifuran, 306

2,3′-Bifuran, 306

Bi-heteroaryl ketone, 441

3,3′-Biindole, 123

Bi-indole alkaloids, 134, 158, 213

2,2′-Biindoles, 104, 126, 164

BINAP, 292, 293

Bioisosteric analog, 357

3,3′-Bipyrroles, 73

3,4-Bis(alkynyl)pyrroles, 60

Bis-(diarylamino)thiophene oligomers, 601

1,4-Bis(di-isopropylphosphinyl) butane, 607

1,1′-Bis(di-*tert*-butylphosphino)ferrocene
    (dppf), 211

Bis(imidazol-2-ylidine)palladium
    complexes, 239

Bis(pinacolato)diboron, 610

3,4-Bis(tributylstannyl)furan, 317

Bis(triphenylphosphine)palladium(Ii)
    dichloride, 445

3,5-(Bisindolyl)-2-aminopyrazine, 449

Bisindolylmaleimides, 93

Bisindolylpyrazines, 449

4,4-Bisoxazol-2-ylstilbene, 384

Bisoxazolines, 401

Bis-pinacolatoborane, 262

1,2-Bispyrazinylacetylene, 451

Bis-triarylphosphines, 608

N-BOC-2-trimethylstannylpyrrole, 54

Borrerine, 117

Botryllazines, 436

Bright fluorescence, 607

Bromination, 305, 314, 346, 352

Bromine–magnesium exchange, 195, 196

2-(5-Bromo-2-furyl)-5-*t*-butylbenzoxazole, 403

4-Bromo-1-(4-toluenesulfonyl)indole, 129

3-Bromo-1-(phenylsulfonyl)indole, 85, 98

2-Bromo-1-methyl-5-phenylpyrrole, 52

2-Bromo-1-methylpyrrole, 40

3-Bromo-1-methylpyrrole, 40

2-Bromo-1-tosylpyrrole, 40

4-Bromo-2,3-dichlorophenol, 596

5-Bromo-2-furaldehyde, 323

3-Bromo-2-furanylthiophene, 319

1-Bromo-2-iodobenzene, 606

3-Bromo-2-iodoindole, 86

5-Bromo-2-phenylfuran, 304

5-Bromo-2-phenyloxazole, 384

6-Bromo-2-phenyloxazolo[4,5-*b*]pyridine, 389

6-Bromo-3-aminopyrazinoate, 444

2-Bromo-3-iodo-1-(phenylsulfonyl) indole, 86

2-Bromo-3-nitromethylbenzoate, 467

5-Bromo-4-chloropyrrole-2-carboxylic
    acid, 41

5-Bromo-4-methyloxazoles, 396

2-Bromo-5-aminopyrazine, 445

2-Bromo-5-aryloxazole, 388

2-Bromo-5-methylfuran, 336

2-Bromo-6-methoxynaphthalene, 588, 589

4-Bromo-6-phenylpyridazin-3-amines,
    553, 569

5-Bromo-7-methylbenzofuran, 316

4-Bromoalkyl-2,5-diphenyloxazole, 391

8-Bromobenzo[*c*]carbazole, 109

2-Bromobenzofuran, 321

3-Bromobenzofuran, 305, 324

4-Bromobenzofuran, 306

5-Bromobenzofuran, 305, 306, 336

6-Bromobenzofuran, 306

7-Bromobenzofuran, 306, 336

Bromo-bis(triphenylphosphine)-*n*-
    succinimidepalladium (II), 393

2-Bromoethylamine hydrobromide, 384

2-Bromofuran, 304, 314, 323, 336

3-Bromofuran, 98, 303, 314, 320, 322, 325,
    336, 337

Bromofuranylbenzoxazole, 403

3-Bromoindole, 83, 118

4-Bromoindole, 86

5-Bromoindole, 445

6-Bromoindole, 88, 120

7-Bromoindole, 88, 89, 107, 111, 126
4-Bromomethyl-2-chlorooxazole, 395
4-Bromomethyloxazole, 393
3-Bromooxazole, 392
4-Bromooxazole, 385
5-Bromooxazole, 385
2-Bromooxazolines, 279, 319
Bromooxazolopyridines, 400
Bromopyrazine, 444
2-Bromopyridine, 203, 204
3-Bromopyridine, 101, 201
5-Bromopyrimidine, 336
2-Bromopyrrole, 38, 40
Bromopyrroles, 72
3-Bromopyrroles, 57
4-Bromopyrrolo[3,4-*c*]carbazole, 121
3-Bromoquinoline, 521
3-Bromoquinoxalin-2-ylamine, 463
Bromoquinoxaline, 461
7-Bromoquinoxaline, 466
Bromoquinoxaline-*n*-oxide, 461
Bromoquinoxalines, 280, 281
1-Bromo-β-carboline, 101
3-Bromothiophene, 98
Bromotrimethoxybenzene, 598
Bromotrimethylsilane, 197
5-Bromotryptamine, 107
Brotizolam, 284
Buchwald–Hartwig aminations, 21, 162, 165,
        208, 236–239, 291–293, 336, 533,
        579, 612
(*n*-Bu)$_2$(*i*-Pr)MgLi complex, 196
t-BuOK, 193
t-Butylacetylene, 594
t-Butyl-4-ethynylbenzoate, 451
t-Butyldimethylsilyloxy-4-phenyl boronic acid,
        448
para-*t*-Butylbenzene, 599
2-*n*-Butylfuran, 305

Cadralazine, 542
Caerulomycin c, 204
Calycanthine, 144
CAMP phosphodiesterase inhibitors, 120
Camptothecin, 204, 230, 519, 532
CAN, 532
Carazostatin, 109, 118, 126

Carbaldehyde, 516
Carbamoyl group, 369
Carbazole, 12, 91, 145, 163, 165
Carbazole triflates, 109
Carbazolequinones, 91
Carbazolones, 146, 163
Carbazomadurin A, 91
Carbazomycins G, 91
Carbazoquinocins, 91, 109, 118, 126
Carboamination, 24
Carbodepalladation, 334
Carboetherification, 24
Carbohydrate chemistry, 428
Carboline, 112, 145
β-Carbolines, 163
δ-Carbolines, 123, 141
γ-Carbolines, 167
Carbon monoxide, 215, 242
Carbon tetrabromide, 194
Carbonylation, 19, 150, 216, 240, 241, 291,
        330, 336, 370, 400, 458, 577, 590, 612
Carbopalladation of nitriles, 402
Carbosilane dendrimers, 210
1-Carboxy-2-(tributylstannyl)indole, 321
4-Carboxybenzene boronic acid, 447
Cardiovascular atherosclerotic disease, 251
Carquinostatin A, 91
Castro acetylene coupling, 124
Catechol borane, 108
Catharanthine, 166
Catuabine I, 37
CC-1065 analogs, 96, 140, 141, 161
CCK-A antagonist, 151
CDP840, 214
Cefozopran, 542
Cephalosporins, 221, 274, 356
Cesium carbonate, 237–238
Cesium fluoride, 216
CGP-60474, 201
C–H oxidation, 25
Chanoclavine-I, 129, 133, 167
Chemoselective coupling, 411
Chichibabin reaction, 192
Chimonanthine, 144
Chiral phosphines, 349
Chloramine-T, 255
Chloridazon, 543

Chlorination, 305

2-Chloro-3,6-diethylpyrazine, 398, 457

2-Chloro-3,6-diisobutylpyrazine, 400, 443, 451, 453, 457

2-Chloro-3,6-diisopropylpyrazine-4-oxide, 443

2-Chloro-3,6-dimethylpyrazine, 438, 451, 456, 457, 459

2-Chloro-3-alkynylquinoxalines, 464

4-Chloro-3-iodocinnoline, 547

2-Chloro-3-phenylethynylquinoxaline, 464

5-Chloro-4-iodopyridazin-3(2H)-ones, 544, 560, 572

5-Chloro-4-trifluoromethanesulfonyloxypyri-dazin-3(2H)-ones, 547, 572

2-Chloro-5-methoxycarbonyl-3-methylquinox-aline, 464

4-Chloro-5-trifluoromethanesulfonyloxypyri-dazin-3(2H)-ones, 547, 572

3-Chloro-6-iodopyridazine, 549, 552, 569

3-Chloro-6-methoxypyridazine, 552, 567

6-Chloro-7-fluoroindole, 156

3-Chloro-7-methoxy-1-methylquinoxaline-2-one, 464

2-Chlorobenzoxazole, 379, 390, 402

2-Chlorofurans, 337

3-Chlorofuran, 304

3-Chloroindole, 84

2-Chloromethyl-4-vinyloxazole, 390

2-Chlorooxazole, 379

2-Chlorooxazole-4-carboxylate, 386

2-Chlorooxazole-4-ethylcarboxylate, 391, 399

Chloropeptin, 105

2-(4-Chlorophenyl)pyrazine, 441

4-Chlorophenyl-tri-n-butylstannane, 441

2-Chloropyrazine, 98

2-Chloropyrazine 1-oxide, 439

3-Chloropyrazine 1-oxide, 452

Chloropyrazine-n-oxide, 438

6-Chloropyridazin-3-amine, 552, 562

2-Chloropyridines, 303, 337

3-Chloropyridine, 194, 231

4-Chloropyridine, 194

Chloropyridopyrazine, 468

2-Chloroquinoline, 514

1-Chloro-β-carboline, 121

2-Chlorothiophene, 200

2-Chlorozinc benzoxazole, 387

Chronic myeloid leukemia (CML), 201

Chuangxinmycin, 119, 165

Chuchuhuanines, 189

Cinnamyl oxazole, 399

Cinoxacin, 542

Clavicipitic acid, 95, 108, 130, 133

Clopidogrel bisulfate (Plavix®), 251

C−N bond formation, 336, 337, 371

C−O bond formation, 336, 337

Coelenterazine, 437

Coenzyme, 345

Colenterazine, 447

Combinatorial synthesis, 210, 217, 326

Combretastatin A-4, 596, 598

Complete ring bromination, 408

Complex-induced proximity effect (CIPE), 192, 194

Conjugate addition, 331

Copper(I) thiophene-2-carboxylic acid (CuTC), 208, 234, 283, 389, 441

Copper cofactor, 363

Copper iodide, 216, 564

Corticotropin releasing factor (CRF) ligands, 203, 213, 214

Costaclavine, 95, 130, 133

Coupling reactions with organometallic reagents, 347–363, 411–422

COX-2 inhibitor, 559, 593

Cross-coupling, 5–7

Cryptolepine, 533

Cryptolepis sanguinolenta, 534

Crysanthemum macrotum (Dur.) Ball, 276

Cu$_2$Br$_2$, 525

Cy$_2$NMe, 515

Cyanation, 535

4-Cyano-2-chloropyridine, 205

2-Cyano-4-methylbiphenyl, 592

2-Cyano-5,6-diarylpyrazine, 460

N-Cyanoindoles, 160

3-Cyanoindoles, 145

Cyanopyrazine, 460

5-Cyanopyridopyrazine, 468

1-Cyano-β-carboline, 166

Cyclin-dependent kinase 4, 109

Cycloaddition, 563, 568

Cycloamination, 70, 597

Cyclocarbonylation, 66
ortho-Cyclopalladation, 531
Cyclopent[b]indolones, 146
Cyclosexipyridine, 219
Cymbalta®, see duloxetine hydrochloride
Cypridina luciferin, 452
Cytisine, 212, 222
Cytochrome P450, 215

DABCO, 278, 279
Damirones, 162
Debromination, 521
Dechlorination, 460
Deepwater sponges, 435
Dehalogenation, 424
Dehydrotubifoline, 139
Dehydroxy-halogenation, 346
Dendrimers, 606
Deoxy-halogenation, 545
2-Deoxy-β-D-ribofuranosylpyrazine, 456
Desazapyridazomycin, 541
Desoxyeserolin, 141
2,5-Di(chlorozinc)terfuran, 403
Di- and tri-substituted thiazole
        derivates, 373
2,5-Dialkynylfurans, 323
2,6-Diamino-3,5-diaryl-1,4-pyrazine, 447
Diaryl-2,3-dicyanopyrazine, 454
2,3-Diarylindoles, 107
2,5-Diaryloxazoles, 388
Diarylpiperazine, 445
4,5-Diarylpyridazin-3(2H)-ones, 556–560
3,4-Diarylpyrrole, 53
Diazoindoles, 166
Diazonamide A, 109, 379, 392
Diazonium salts, 347
Diazotization, 198, 202, 549
Dibenzo[f,h]phthalazin-1(2H)-one, 560
5,8-Dibromo-2,3-di(pyrrol-2-yl)-
        quinoxaline, 463
4,5-Dibromo-2-furaldehyde, 315, 324
3,5-Dibromo-2-pyrone, 319
2,5-Dibromo-3-octylfuran, 311
Dibromoalkenols, 333
2,3-Dibromobenzofuran, 305
Dibromobiphenyl, 605
2,4-Dibromofuran, 315

2,5-Dibromofuran, 304, 323
2,3-Dibromoindole, 86
4,6-Dibromoindole, 166
2,3-Dibromopyridine, 195
2,5-Dibromopyridine, 195, 205
2,6-Dibromopyridine, 195, 197, 205
2,5-Dibromopyrroles, 39
2,3-Dichloro-5-(methoxymethyl)pyridine, 600
5,6-Dichloroindole, 126
2,6-Dichloropyrazine, 438, 452
3,6-Dichloropyridazine, 549, 551, 578, 579
2,3-Dichloropyridine, 197, 600
2,4-Dichloropyridine, 197
2,5-Dichloropyridine, 197, 216, 591
2,4-Dichloropyrimidine, 201
2,6-Dichloroquinoxaline, 464
Dichloroquinoxalines, 468
2,3-Dichloroquinoxalines, 464
Diclomezine, 543
Dicyclohexyl carbodiimide (DCC), 364
Dicyclohexyl-18-crown-6 (DCH-18-C-6),
        288–290
1,4-Didodecylbenzene-2,5-diboronic acid, 604
Diels–Alder reaction, 143, 317, 420
Dienyl ketene intermediate, 319
2-(Diethylaminocarbonyl)pyrazine, 459
2-Diethylamino-4,5-diphenyloxazole are, 384
Diethyl-(4-isoquinolyl)borane, 314
5,6-Difluoroindole, 158
5,7-Difluoroindole, 126
2,3-Dihaloindoles, 84
4,5-Dihalopyridazin-3(2H)-ones, 543,
        556–559, 561, 565, 578
Dihydrocleavamine, 158
2,3-Dihydropyrroles, 72
2,5-Dihydropyrroles, 72
1,2-Dihydroquinoxaline, 467
3,4-Dihydroquinoxaline, 467
Diimes-HCl, 239
Diindolocarbazoles, 105, 107, 109
2,3-Diiodo-1-(phenylsulfonyl)indole, 85
2,5-Diiodo-1,3,4-trimethylpyrrole, 60
3,5-Diiodo-2,6-dimethoxypyrazine, 452
1,2-Diiodoethane, 384
2,3-Diiodoindole, 85
3,6-Diiodopyridazine, 549, 568
N,N-diisopropylethylamine, 448

1,9-Dilithio-β-carboline, 101

Dimerization, 365

5-(Diethoxymethyl)-2-furylboronic acid, 312

3,4-Dimethoxyphenylboronic acid, 454

2,6-Dimethoxypyrazine, 452

3,6-Dimethoxypyridazine, 550

N-(Dimethylamino)pyrrole, 55

6,8-Dimethyl-1*H*-pyrimido[4,5-*c*]pyrrolo[2,3-*e*]
    pyridazine-7,9(6*H*,8*H*)-diones, 570

4,4-Dimethyl-2-oxazoline, 279, 320

2,5-Dimethyl-3-styrylpyrazine, 451

1,3-Dimethyl-6-propionylpteridine-
    2,4-dione, 469

N,N-dimethylaminoethanol, 194

2,5-Dimethylpyrazine, 438

2,4-Dimethylquinoline, 96

(*E*)-2,5-Dimethyl-3-styrylpyrazine, 456

2,5-Diphenyl-4-tributylstannanyloxazole, 391

2,5-Diphenyl-4-vinyloxazole, 391

2,5-Diphenyloxazole, 400

(2,5-Diphenylphospholyl)-2-methylpyridine, 611

6,7-Diphenylpyrrolopyrazines, 468

4,5-Diphenyl-α-phenylsulfonyl-2-oxazoloace-
    tonitrile, 385

Diphosphaferrocene, 608

2,3-Dipyrrolylquinoxalines, 463

Direct halogenation, 346, 544

Direct metalation, 547

Directed metalation group (DMG), 192, 196

Directed *ortho* metalation (DOM), 192, 201

Disamylborane, 523

Disorazole A1, 379, 394, 397

Disorazole A1 and C1, 397

Disorazole D1, 379, 398

Disorazoles C1, 379

1,3-Dithiole-2-thione stannanes, 462

Ditin chemistry, 356, 357

Ditopic bidentate ligand, 564

DMAP, 212, 441

DNA cross-linking agent, 38

DNA sequencing agent, 595

Domino Heck reaction, 134

Dopamine D$_3$ receptor antagonists, 104

Dopamine receptor antagonist, 38, 49

Dowex® ion exchanger resin, 389

Dragmacidin, 103

Dragmacidin F, 44

Duloxetine hydrochloride (Cymbalta®), 251

Duocarmycin SA, 63, 126, 141, 325

Dupont Crop Protection, 551, 556

EDTA, 208

Elaiolide, 283

Electrochromic polymers, 51

Electrocyclic reaction, 561

Electrocyclic reactions, 420

Electroluminescent polymer, 233

Electrophilic coupling, 418

Electrophilic iodination, 408

Electrophilic palladation, 534

Electrophilic substitution, 190, 252, 279, 294

Electrospray ms, 206

Eletriptan, 590

Ellipticines, 91

Emorfazone, 542

Enamine, 516

Enamine stannane, 275

Endogeneous bioamine-interfering CNS
    agents, 553

Endothelin conversion enzyme-1 (ECE-1)
    inhibitors, 222, 348, 349

Energy transfer, 606

Enones, 335

(±)-Epibatidine B, 221

Epothilones A, 592

Ergot alkaloids, 99, 130, 132, 167, 387

Ergot skeleton, 133

Erythromycin, 529

Ether formation, 22, 23

4-Ethoxycarbonyl-butylzinc iodide, 595

1-Ethoxyprop-1-eneyltin, 469

(*Z*)-1-Ethoxy-2-(tributylstannyl)ethane, 236

Ethyl 2-chlorooxazole-4-carboxylate, 389

Ethyl 5-bromo-2-phenyl-oxazole-4-carboxy-
    late, 389

2-Ethyl-3-methylindole, 96

2-Ethylhexyl-*p*-methoxycinnamate, 599

Ethylpyrazine, 444

Ethyltrichlorosilane, 224

Ethynylquinoxaline, 464

Eudistomin T, 166

(−)-Ferruginine, 563

Fexofenadine, 594

Fibrecat, 207
Film forming, 605
Fluorescence sensor, 142
Fluorescent materials, 103, 384
2-Fluoro-6-tributylstannyl-pyrazine, 450
2-Fluoro-9-oxime ketolides, 454
Fluoroindoles, 86
Fluoropyrazine, 437
Fluorosubstrates, 437
Fluorovinylbenzoxazole, 386
B-Fluoro-β-trifluoromethyl-α-phenylvinyl-
    stannane, 442
2-Formyl-1-(phenylsulfonyl)-1H-indole, 313
2-Formyl-3-furylboronic acid, 313
Formylation, 415
Free-standing film, 605
Friedel–Crafts acylation, 350, 351
Friedlander quinoline synthesis, 534
Fujiwara–Moritani oxidative Heck reaction, 307
Functional group tolerance, 360, 365, 414
Furan, 11, 26, 304, 306, 328, 336
Furan[3,2-b]pyrroles, 314
Furan arylation, 427
Furancarbothioamide, 609
Furanoeudesmanes, 314
Furanopyridine, 597
Furfural, 303
Furo[2,3-c]quinoline, 313
Furo[2,3-d]pyridazinones, 573
Furo[3,2-a]carbazole, 316
Furostifoline, 158, 316
2-Furoyl chloride, 310
Furylboronic acids, 312–314
2-Furylboronic acid, 312, 444
2-(3′-Furyl)-4,4-dimethyl-2-oxazoline, 320
2-Furyllithium, 308
2-Furylzinc chloride, 308, 310, 311
(S)-(+)-Fusarinolic acid, 229

GABA receptor, 162, 374, 389
GABA-A receptor, 553
Geissoschizal, 139
Geissoschizine, 139
Gelsemine, 144, 150
Glivec™, 201
Glucose biosensor, 37
Glycozolidine, 123

GPIIB/GPIIIA antagonists, 400
Gramines, 152
Grignard reagent, 12, 13, 200, 203, 257, 258
Grossularine, 118, 152, 162

4-Halo-3,6-dimethylpyridazine, 567
5-Halo-4-methoxypyridazin-3(2H)-ones, 559,
    561, 580
4-Halo-5-methoxypyridazin-3(2H)-ones, 551,
    559, 561, 566, 573, 580
3-Halo-6-methylpyridazine, 566, 567
5-Halo-6-phenylpyridazin-3(2H)-ones, 544,
    565, 570, 571, 574–576, 581
Halobenzofurans, 305, 306
5-Halobenzoxazoles, 385
Halofurans, 303–305
Halogen shuffling, 196
Halogenated bipyrroles, 37, 41
Halogenated indoles, 83
Halogenated pyrroles, 57
Halogenation, 193, 198, 252–256
Halogen-dance reaction, 515
Halogen–halogen exchange, 191, 197, 198
Halogen–metal exchange, 349, 351, 354–356,
    415, 416, 419, 520
Haloimidazole synthesis, 407–410
5-Haloquinolines, 517
N-Halosuccinimide, 512
Halothiazole synthesis, 345–347
Hantzch–Panek condensation, 399
Hapalindole, 118
Heck reaction, 10, 15–17, 27, 61–63, 128,
    231–236, 240, 287–290, 324–328,
    353, 367–370, 398, 424–427, 530,
    574, 588, 607
Heck reaction, heteroaryl, 327, 328, 368
Heck reaction, intramolecular, 325–327
Hegedus indole synthesis, 27, 155
Helicopodands, 203
Hemolytic substitution, 356
Hennoxazole A, 379
Hermann's catalyst, 326, 335
Hetero cross-coupling, 352
Heteroannulation, 328–337, 535
Heteroaromatics, 361
Heteroaryl Heck reaction, 327, 328, 368
2-Heteroaryl-benzoxazoles, 401

Heteroarylquinoxalines, 281
Heteroatom-Heck, 23
Heterobiaryl phosphates, 417
Heterobiaryls, 352
Hetero-Cope rearrangement, 161
Heterocyclic halides, 358
Heterocyclization, 467
N-Heterocyclic carbene ligands, 237
Heterogeneous catalysts, 207
Hexafluoroacetone, 193
Hexahydropyridazine amino acids, 541
Hexa-methyldistannane, 392, 566
Hexamethylditin, 272, 282
3-(1-Hexynyl)pyrazine-1-oxide, 452
Hippadine, 106, 120, 121, 136
Histidine, 407
HIV protease, 235, 242
HIV-1 reverse transcriptase inhibitor, 365
Hiyama coupling, 6, 13, 224, 283, 284, 322, 527
HMG-CoA reductase inhibitor, 523, 527, 593
Homocoupling reaction, 256, 262, 410, 536
Homotryptamines, 141
5-HT1$_A$ agonist, 387
5-HT$_3$ receptor, 554
5-HT$_{1D}$ receptor agonist, 148
Hunsdiecker reaction, 373
Hydralazine, 542
β-Hydride elimination, 2, 7, 15, 16, 18, 22, 23, 26, 28, 231
Hydroamination, 365
Hydroboration, 213, 315, 554, 555
Hydrocarboxylation, 588
Hydroformylation, 325, 588
Hydrogenation, 467
Hydrogenolysis, 235
Hydrostannation, 274
3-Hydroxy-3-methylglutaryl coenzyme A (HMG-CoA), 523, 527, 593
18-Hydroxygardnutine, 168
4-Hydroxyindoles, 66
3-Hydroxypyridine, 198, 199
18-Hydroxytaberpsychine, 168
5-Hydroxytryptamine, 444
Hyellazole, 109, 118
Hypervalent iodonium salts, 278

Iboga alkaloids, 91, 134
Ibogamine, 91
Ibuprofen, 591
Imatinib, 201
Imidazo[1,2-a]pyridines, 222, 280
Imidazole derivatives, 9, 25, 26, 425
Imidazole halides, 415
Imidazole metalation, 411
Imidazolium ionic liquids, 231
Imidazolylstannanes, 118, 417, 519
Imidazolylzinc chloride synthesis, 411
Imidazopyrazinone, 452
Imine hydrolysis, 367
Indol-2-yl triflate, 313
Indole, 12, 13, 15, 27–29
Indole libraries, 158
Indole-2-carboxylate libraries, 162
Indoleboronic acids, 102, 104, 105
Indolequinones, 66
Indoline, 12, 24, 87
Indolo[1,2-c]quinazolines, 151, 159
Indolo[2,1-a]isoquinolines, 136
Indolo[2,3-a]carbazole, 115, 160, 164
Indolo[2,3-a]pyrrolo[3,4-c]carbazoles, 93
Indolo[2,3-b]quinoxalines, 164
Indolo[3,2-b]benzo[b]thiophenes, 146
Indolo[7,6-g]indole, 126
Indolocarbazoles, 104, 116, 134
Indoloquinones, 140
Indole triflates, 117
3-Indolyl triflate, 90, 102, 125
Indolylborates, 110
5-Indolylboronic acid, 444
Indolylquinolinones, 164, 463
Indolyltributylstannanes, 113
Indolyltriflates, 109
Indolylzinc halides, 97
1-Indolylzinc chloride, 100
Indoxyl acetate, 83
5-Indoylboronic acid, 314
α$_v$β$_3$ Integrin antagonists, 230
Interleukin-8 receptor, 412
Intramolecular Heck reaction, 532, 530
Intramolecular heteroaryl Heck reaction, 427
Intramolecular reductive coupling reactions, 534

Inverto-yuehchukene, 98
Iodide-(III) arylating reagents, 536
Iodination, 198
Iodine, 198, 199
Iodine/magnesium exchange reaction, 351
3-Iodo-1-(phenylsulfonyl)indole, 85
2-Iodo-1-methylindole, 86, 99
3-Iodo-2,6-dimethoxypyrazine, 452
2-Iodo-5-*n*-butylfuran, 305
3-Iodo-6-methoxypyridazine, 552,
        562, 568
4-Iodo-6-nitropyridazin-3(2*H*)-ones, 549
o-Iodoanilide, 529
2-Iodoazaindoles, 86
2-Iodobenzofuran, 305
3-Iodobenzothiophene, 324
2-Iodobenzoxazole, 386
Iodo-destannation, 197, 516
Iodofluorobenzylpyrazines, 440, 446
2-Iodofuran, 323
3-Iodofuran, 303
Iodoimidazole, 519
3-Iodoindole, 83
4-Iodoindole, 87, 89
5-Iodoindole, 87
7-Iodoindole, 90, 129
4-Iodoindole-3-carboxaldehyde, 89
Iodonation, 305
3-Iodo-*N*-TIPS-pyrrole, 41
5-Iodopyridazin-3(2*H*)-ones, 549, 561, 566,
        573, 581
6-Iodopyridazin-3-amine, 552, 562
2-Iodopyridine, 194
3-Iodopyridine, 207, 240
3-Iodopyrrole, 50, 57, 58, 65
Iodopyrroles, 72
2-Iodoquinoxaline, 462
Iodotrimethylsilane, 197
2-Iodotryptamine, 86
2-(5′-Indoyl)furan, 310
Iodozine, 516
IPR, 237
Iron tribromide, 254
Isochanoclavine-I, 129
Isochromeno[3,4-*d*]pyridazinediones, 561
Isocryptolepine, 115
Isogeissoschizal, 139

Isoindolo[2,1-*a*]indoles, 150
Isomerization, 573
Isoneocrytolepine, 533
Isopropylmagnesium chloride, 196
Isopropyl-(3-benzoylphenyl)
        propionate, 588
Isostrychnine, 140

Jeffery's ligandless conditions, 16, 288, 289,
        325, 326, 532
Jellyfish, 437
Jones oxidation, 356

Kalbretorine, 121
KDR kinase inhibitors, 145, 164
Kelly-variation, 220
Ketoprofen, 588
Kistamycin, 105
Komaroine, 109
Konbu'acidin, 37
Kornfeld's ketone, 131
Koumine, 166, 168
Kumada coupling, 10, 12, 46, 96, 200, 201,
        206, 211, 256–258, 285, 308, 311,
        437, 612

L-754, 394, 235
β-Lactamase, 221
Lamellarin alkaloids, 44, 47
Larock indole synthesis, 27, 147, 330
Laughine, 37
Lavendamycin, 161, 525
Lazabemide, 591
LDA, 547
Lego system, 220
Leimgruber–Batcho indole synthesis, 88
Lescol, 82
Leucascandrolide A, 379, 393, 397
Lewis acid, 193
Liebeskind, 208, 214
Lilial, 599
Lintitript, 151
Lipitor, 38
Lipophilic pocket, 361
Liquid phase organic synthesis (LPOS), 210
Lithiation, 193–195, 210, 220, 236, 255,
        259, 260

ortho-Lithiation, 259, 272, 309, 352
2-Lithio-1-methylpyrrole, 65
1-Lithio-3-methylindole, 150
3-Lithiofuran, 304
2-Lithioindoles, 97, 384
Lithium di-*tert*-butyltetramethylpiperidinoz-
    incate (TMP-zincate), 194
Lithium hexamethyldisilazide, 384
Lithium napthalenide, 519, 520
Lophotoxin, 318
Losartan, 592
LTMP, 193, 548
Lumazines, 469
Luminescent, 602
Lysergic acid, 105, 114, 130

Macrocyclization, 389
Magallanesine analogs, 139
Makaluvamine c, 162
Malaria, 511
Maleic hydrazide, 543
MAP kinase inhibitors, 413
Materials synthesis, 350
Maxalt, 82
Maxonine, 137
m-Bromotoluene, 455
Melatonin analogs, 107
Merck Sharp & Dohme Ltd, 550, 553,
    564, 570
Mercuration, 65, 66
Mescengricin, 162
Metabatropic glutamate receptor subtype 5
    (MgluR5) antagonists, 223, 231
Metal–halogen exchange, 252, 260, 261,
    272, 273, 280, 294
Metallated thiazole, 349
Metalation, 193, 203, 352, 354, 411, 412, 423
ortho-Metalation, 350
Metalopyridazines, 548
7-Methoxy-1-methyl-3-(4-methoxy-
    carbonyl)-, 465
2-Methoxy-6-vinylnaphthalene, 588
4-Methoxybenzoylchloride, 443
2-Methoxycarbonyl-3,6-dimethylpyrazine, 459
5-Methoxycarbonyl-3-methyl-2-phenylethynyl-
    quinoxaline but-3-yn-2-ol, 466
(±)-9-Methoxycytisine, 222

11-Methoxykoumine, 168
(2-Methoxymethyl)-1-oxo-4-phenyl-1,2-dihy-
    drophthalazine-6-carbonitrile, 574
7-Methoxymitosene, 144
4-Methoxyphenylboronic acid, 447
2-(4-Methoxyphenyl)pyrazine , 449
2-Methoxypyridine, 194
2-Methoxyquinoline, 513
2-Methyl-1,2,3,4-tetrahydroquinoxaline, 467
1-Methyl-2-(tri-*n*-butylstannyl)pyrrole, 54
2-Methyl-2,5-dihydro-1*H*-pyridazino[4,5-*b*]
    indol-1-ones, 575, 581
1-Methyl-2-indolylzinc chloride, 99
1-Methyl-2-iodolylmagnesium bromide, 97
1-Methyl-2-trimethylstannylpyrrole, 59
2-Methyl-4-phenyl-2,5-dihydro-1*H*-pyri-
    dazino[4,5-*b*]indol-1-one, 577, 581
2-Methyl-6-vinylpyrazine, 435
2-Methylbenzo[*b*]furo[2,3-*d*]pyridazin-1(2*H*)-
    one, 575
Methyleneaziridines, 44
2-Methylfuran, 307
4-Methyloxazole, 390
2-(*n*-Methyl-*n*-phenyl)pyrazine, 455
Oxazolopyridines, Methylpropenoate, 400
3-Methylpropyl-2-methoxypyrazine, 435
6-Methylpyrido[3′,2′:4,5]pyridazino[1,6-*a*]
    indol-5(6*H*)-one, 577
N-Methylpyrrole, 41, 55, 61
3-Methylpyrrolo[2,3-*b*]quinoxaline, 466
3-Methylsalicylate, 441
Methylthiazole ethynyl pyridine (MTEP), 231
Methylthiobenzothiazole, 318
Methylthiooxazoline, 279
N-Methylvinylsulfonamide, 589
Mg–H exchange, 241
Michael acceptor, 331
Michael addition, 367, 531, 574
Michaelis–Arbuzov reaction, 294
Micrococcinic acid, 360
Micro-pH sensors, 83
Microwave irradiation, 201, 207, 225, 286
Microwave-enhanced conditions, 158
Migratory insertion, 2, 4, 15, 23, 24, 29
Minaprine, 542, 552
Mitosene analogs, 66, 144
Mitsunobu conditions, 332

Miyaura boronic ester synthesis, 352
Miyaura reaction, 522
Molybdenum hexacarbonyl, 242
Monoamine transporters, 361
Monodendrons, 125
Mono-*ipso*-iodination, 284
Montelukast, 511, see also Singulair
Moracin M, 319
Mori–Ban indole synthesis, 16, 27, 140, 532
Morphine, 598
2-Morpholinylbenzoxazole, 402
Mouse fibroblast cell line, 364
Mucobromic acid, 543
Münchnones, 73
Murrayaquinone a, 163
Muscarinic agonists, 444
Mycalazol 11, 56

N-(Phenylsulfonyl)pyrrole, 43, 61
Nabumetone, 589
Naltrexone, 122
Naltrindoles, 109, 122
Nanofiltration, 210
Nanoparticles, 233
Naphth[3,2,1-*cd*]indole, 138
1-Naphthalene triflate, 387
Naphthalenylbenzoxazole, 387
Naphthoquinone(imidazolin-2-ylidene)
         palladium complexes, 238
1,4-Naphthoquinone, 43
Naproxen, 588
Naramig, 82
Naratriptan, 589
Nasea, 82
Nauclefine, 166
Naucletine, 152
NBS, 304
Negishi coupling, 6, 7, 9, 46, 48, 97,
         201–206, 210, 211, 226, 236,
         258–261, 280, 295, 308–312,
         347–352, 386, 411–414, 439, 514,
         519, 550, 595, 611
Nematocidal, 279
Neuropathic pain, 251
Nicotine, 189
Nicotinic acetylcholine receptor
         agonist, 550

Nicotinic acetylcholine receptor
         antagonist, 221
Niphatesine C, 229
Nitramarine, 101, 121
Nitrene, 316, 561
Nitrobenzofurans, 306
Nitro-containing biphenyls, 351
Nitroindoles, 158
NK₁ receptor antagonist, 352
Nonaflate functional group, 348
Non-linear optical materials, 233
Non-natural amino acids, 205, 213
Non-peptide inhibitor of ECE, 349
Non-steroidal anti-inflammatory drug
         (NSAID), 251, 589
Norbinaltorphimine, 38
Norchanoclavine, 129
Nordehydrocacalohastine, 331
Norepinephrine transporter, 361
Norflurazon, 543
Norniphatesine C, 229
Nortopsentins, 103, 416
Nosiheptide, 145
Nucleophilic attack, 190
Nucleosidases, 428
Nucleoside chemistry, 428

Octylation, 439
3-Octylfuran, 311
Octylstannane, 439
Olefin metathesis, 143
Oligopyridylimines, 220
Optical brighteners, 384
δ Opioid receptor antagonists, 122
Organoborane, 6–10
Organosilane, 6, 13, 14, 527
Organostannanes, 6, 7, 10, 11, 362
Organozinc reagent, 6, 7, 413, 414
Oroidin alkaloids, 423
Osteoclast inhibitor, 122
Osteoporosis, 251
Otar conditions, 116
Oxazolepiperidine, 400
Oxazolyl triflate, 397
Ω-Oxazolylalkanoic acid, 389
Oxazolylboronic acids, 389
Oxazolylpyrazine, 399

2-Oxazolylzinc chloride, 387

(±)-Oxerime, 234

Oxidative addition, 1, 3, 5, 6, 15, 22–24, 29, 190, 191, 345, 368

Oxidative amination, 570

Oxidative cyclization, 3, 4, 90, 256, 294, 306–308, 533, 536

Oxime, 517

Oxindole, 17, 151, 152, 162

Oxoassoanine, 121

21-Oxogelsemine, 144

21-Oxoyohimbine, 150

Oxypalladation, 334

P450 2A6, 215

Palladacycles, 607, 609

Palladate complex, 412

Palladation, 367

Palladium acetate, 225

Palladium carbene complex, 328

Palladium coupling reaction, 353, 372

Palladium tetrakis(triphenylphosphine), 206, 208

Palladium-catalyzed amination, 402

Paraherquamide B, 91

Parallel synthesis, 326

Parkinson's disease, 82

PBr$_3$-DMF, 513

Pd/C, 207, 226

Pd-catalyzed coupling reactions, 587

PdCl$_2$, 208

Pennigritrem, 81

Pentacoordinated silicate, 291, 527

Pentylation, 439

Peptide mimetics, 235

Perillene, 303

Perlmann's catalyst, 226

Perlolyrine, 109

Phase transfer catalysis, 521, 529

6-Phenyl-2-phenyloxazolo[4,5-b]pyridine, 389

2-Phenyl-4,5-dihydrooxazole, 384

2-Phenyl-5-oxazolecarboxanilide, 399

Phenylacetylene, 227, 451

Phenylalanine boronic acid, 445

2-Phenylbenzofuran, 332, 333

2-Phenylbenzooxazole, 395

Phenylboronic acid, 389, 444

Phenylethynylquinoxaline-2-one, 465

2-Phenylfurano[2,3-b]quinoxaline, 464

Phenylmagnesium bromide, 437

2-Phenyloxazole, 384

4-Phenyloxazole, 317

2-(5-Phenyloxazol-2-yl)-benzoate, 401

2-Phenylpyrazine, 437

2-Phenylpyrrole, 47

3-Phenylpyrrole, 47

N-Phenylsulfonylpyrrole, 458

2-Phenylthieno[2,3-b]quinoxaline, 464

Phenylthiobenzothiazole, 318

Phenylvinyl sulfone, 590

Phorboxazole, 393

Phorboxazole A, 379

Phorboxazole B, 379

Phosphination, 536

Phosphine ligands, 349, 369

Phosphine-free Pd(0), 208

Phosphodiesterase (PDE) IV inhibitor, 214, 241

Phosphonation, 20

Phosphonylation, 429, 430

Phosphorous pentabromide, 199

Phosphorous tribromide, 199

Phosphorous trichloride, 199

Phosphorus esters, 429

Phosphorylases, 428

Phosphorylation, 21

Phosphotyrosine, 222

Photolysis, 514

Photorefractive pyrrole polymers, 37

Physostigmine, 144

Physovenine, 144

Pimprinin, 383

Pinacolborane, 610

Pinocol ester, 518, 522

5-Piperidinoethylquinoxaline-2,3-dione, 461

π-System, 189

o-Pivaloylaminophenyl boronic acid, 445

Plasticizing effect, 605

PMF, 558

POCl$_3$, 512

Poly(3-octylfuran), 311

Poly(amidoamine)-dendrimer, 608

Poly(bipyridines), 219

Poly(p-phenylene), 604, 605

Polyethylene glycol (PEG), 269

Polyketides, 99
Polymer-bound palladium catalysts, 520
Polymeric pyrrolocarbazole, 83
Polymer-incarcerated palladium
　　　(PI-Pd), 208
Polymerization, 603
Polymers, 54
Polymethylhydrosilaoxane, 460
Poly-*p*-phenylenes, 606
Polypyrrole, 37, 38
Popd, 282, 515, 532
Porphyrins, 48, 58, 61, 65, 72, 228
Positron emission tomography (PET), 223
Potassium hydroxide, 238
Potassium phosphate, 206
Potassium trifluoroborates, 111
Pratosine, 121, 136
Proazaphosphatrane ligands, 238, 278
Prodigiosin, 43
N-Propargylamides, 400
Propargyl carbonates, 328–330
Propionylpyrazines, 441
Prosulfuron, 600
Protein kinase C (PKC) inhibitor, 109, 201
Protodeboronation, 209
Protodepalladation, 334
Protoporphyrin IX, 59
Pschorr reaction, 560
Pukalide, 318
Purine biosynthesis, 429
Pyrazine, 15, 16, 18
Pyrazine alkaloids, 436
Pyrazinecarboxamide, 459
Pyrazino[1,2-*a*]indoles, 93, 153
Pyrazino[2,3-*b*]indoles, 164
Pyrazinoindazole, 445
2-Pyrazinyl-(3-chloro-2-pyridyl)amine, 456
2-Pyrazinyl bromide, 442
Pyrazinylacetylenes, 451
Pyrazinylbenzoic acid, 446
Pyrazinylbenzoxazole, 400
2-(Pyrazin-2-yl)indole, 457
5-(2-Pyrazinyl)indole, 445
4-Pyrazinylphenylalanine, 445
Pyrazinyltetralone, 443
2-Pyrazinylzincbromide, 439
Pyridaben, 543

Pyridaphenthion, 543
Pyridate, 543
Pyridazin-(3(2*H*)-on)e triflates, 546, 555, 559,
　　　560, 562, 569, 577, 578
Pyridazine *N*-oxides, 545, 567
Pyridazino[4,3-*h*]psoralen, 578
Pyridazino[4,5-*c*]isoquinolinones, 561
Pyridazinylboronic esters, 555
Pyridazinylzinc halides, 550
Pyridazomycin, 541
Pyridine, 11, 13, 14, 20–22, 24, 384
Pyridine-2,3-dicarboxylic acids, 600
Pyridinium salts, 190
Pyridino[1,2-*a*]indoles, 142
Pyrido[1,2-*a*]benzimidazoles, 145
Pyrido[2′,3′-*d*′]diazepino[1,6,7-*h*,*i*]
　　　indole, 137
Pyrido[2′,3′-*d*′]pyridazino[2,3-*a*]indole, 137
Pyridobenzodiazepinones, 239
Pyridones, 199
Pyridyl triflate, 525
2-Pyridylboronic ester, 521
3-Pyridylboroxin, 209
2-(2-Pyridyl)indoles, 98
2-Pyridylzinc bromide, 386
Pyrimidine, 8
Pyrrole boronate, 51
Pyrrole polymers, 58
Pyrrole-3-boronic acid, 50, 51
Pyrrolidine, 24
Pyrrolnitrin, 37
Pyrrolo[1,2-*a*]indoles, 134, 136, 142
Pyrrolo[1,2-*a*]pyrazin-1-ones, 43
Pyrrolo[2,3-*b*]pyrazines, 149, 158
Pyrrolo[2,3-*b*]quinoxalines, 159, 464, 466
Pyrrolo[2,3-*c*]pyridin-7-ones, 43
Pyrrolo[2,3-*d*]pyridazinones, 573
Pyrrolo[2,3-*e*]indole, 126
Pyrrolo[3,2,1-*ij*]quinolines, 126
Pyrrolo[3,2-*c*]pyridin-4-ones, 43
Pyrrolo[3,2-*c*]quinolines, 148
Pyrrolo-fused steroids, 67
Pyrrolophenanthridine alkaloids, 137
Pyrrolophenanthridines, 136
Pyrrolophenanthridone alkaloids, 92
Pyrrolopyrazine, 468
Pyrroloquinolines, 133, 156

Pyrrolyl-bisboronic ester, 522
N-Pyrrolylzinc chloride, 47
Pyrrolylzinc reagents, 46

Quaterfuran , 403
Quaterpyridines, 211
Quinoline, 25
Quinoline halogenation, 513
Quinolines, 511
Quinolinol, 513
Quinolinyl, 519
Quinolinyl halides, 522
4-Quinolinyl triflate, 524
Quinolinylboron reagents, 522
Quinolinylstannane, 525
Quinolone halogenation, 512
Quinolones, 512
Quinoxaline, 16
Quinoxaline-5,8-dimalononitriles, 462
Quinquifuran, 403
Quinuclidinylpyrazine, 444
Quioline-3-aldehyde, 516

Radical scavenger, 437
Raloxifene hydrochloride (Evista®), 251
Ranitidine, 303
Reductive coupling reactions, 256
Reductive cyclisation, 534, 350
Reductive elimination, 1, 4, 5, 7, 22, 24,
        25, 29
Regioselective bromination, 346, 409, 412,
        415, 418
Regioselective diarylation, 416
Regioselective diiodination, 408
Regioselective functionalization, 424
Regoselective addition, 368
Rescriptor, 82
Resin-bound, 315, 326, 332
Resin-to-resin transfer reactions (RRTR), 226
Retro-ene reaction, 557, 565, 571, 575
Rhizoxin D, 382, 394
Ribonucleotide reductase (RR), 212
Rieke zinc, 201, 260, 311
Rigid-rod polymers, 604
Ring-closing metathesis, 364
Rink amide resin, 142
Rizatriptan, 590

Rofecoxib , 593
Roseophilin, 67
Rutaecarpine, 137
Rutecarpine, 150

Salen ligands, 213
Sandmeyer reaction, 345, 347, 384
Schizophrenia, 596
Sciodole, 81
6,7-Secoagroclavine, 129
Seed germination inhibition, 279
Selective estrogen receptor modulator
        (SERM), 214, 251
Selective heteroarylation, 353
Selective reductive bromination, 408
Selective serotonin and norepinephrine
        reuptake inhibitor (SSNRI), 251
Selectivity, 551–553, 556–558, 560, 572,
        576, 580
Septipyridine, 218
Serdolect, 82
Serotonin, 590
Serotonin reuptake inhibitors, 141
Sertaconazole (Ertaczo®), 251
Shapiro reaction, 275
[3,3]-Sigmatropic rearrangement, 161
Silicon-Stille coupling, 527
Site selective coupling reactions, 371–374
S_N2 displacement, 326
SN-38, 230
S_NAr displacement reactions, 514,
        526, 528
Sodium amide, 194
Sodium carbonate, 198
Sodium hypobromite, 198
Sodium tert-butoxide, 237
Sodium triacetoxyborohydride, 456
Solid phase, 116, 127, 142, 145, 148, 149,
        332, 558
Solid support, 321, 326
Sonogashira reaction, 14, 15, 59–61, 124, 210,
        225–230, 242, 272, 276, 284–286,
        311, 322–324, 334, 337, 363–367,
        396, 422–424, 440, 454, 528, 566,
        594, 611
sp^2 hybridized, 189
Spent palladium, 148

S-Phos, 268
3-Spiro-2-oxindoles, 154
Spiroindoxyls, 164
Spirooxindoles, 143, 144
Squarate, 275
Stannylindoles, 113
2-Stannylindoles, 115, 116
3-Stannylindoles, 117
Stephacidin A, 145
Stephen–Castro reaction, 14, 225
Steroid 5α-reductases type 1 and 2, 520
Stille coupling, 6, 10, 50, 54–59, 113, 210,
        216–223, 225, 272–283, 295,
        314, 316–321, 337, 345,
        354–363, 390, 417–422, 441,
        523, 562
Stille–Kelly reaction, 11, 282, 322, 360
Streptonigrin, 221, 525
Stress urinary incontinence, 251
Strobilurin A, 566
Structure activity relationships (SAR), 191
Strychnine, 139, 554
Styrene, 235
Styrene polymers, 603
4-Substituted 6-nitroquipazine analogues, 526
Sulfamethoxypyridazine, 542
Sulfomycin I, 357
Sulfonylurea herbicide, 600
Sunscreen agent, 599
Suzuki coupling, 6–9, 44, 47–54, 61,
        206–215, 222, 225, 226, 228,
        234–236, 240, 242, 261–272, 280,
        285, 295, 312–316, 333, 337,
        352–354, 388, 414–416, 444, 520,
        551, 608
Suzuki–Miyaura cross-coupling, 389, 610
Synthesis of stannylthiazoles, 354, 356

Tabersonine, 117
Tagetes sp. (marigold), 251
Tamoxifen, 214
Tandem carbonylation–arylation, 335
Tandem cyclization-anion capture, 318
Tautomeric equilibrium, 410
Tazarotene, 595
TBAF, 225
Tedicyp, 226, 270, 285–288, 530

Terminal olefins, 367
Tetra(*p*-methoxyphenyl)stannane, 443
Tetrabutylammonium bromide (TBAB), 199,
        269, 288, 290
1,2,3,4-Tetrahydro-2-vinylquinoxaline, 468
Tetrahydrofuran, 24
Tetralones, 442
Tetramesitylporphyrins, 54
2,2,6,6-Tetramethylpiperidine, 452
Tetraphenyltin, 443
Tetraphosphine ligand, 353
Tetrathiafulvalene (TTF), 279, 280
Thallation, 89, 108, 129
Thallation–iodination, 106
Thermal rearrangement, 359
Thiazole as electrophile, 349–352
[1,2,4]-Triazolo[4,3-*b*]pyridazine, 550, 553,
        564, 570
Thiazole as electrophile, 361–363
Thiazole as nucleophile, 347–349, 356–361
Thieno[2,3-*d*]pyridazinones, 573
Thieno-1,6-naphthyridines, 217
Thienocarbazoles, 602
Thienopyrroles, 64, 71
3-Thienylboronic acid, 53
Thienylpyrazine, 444
2-Thiomethylpyrazine, 440
Thiophene, 12–15, 20
2-Thiopheneboronic acids, 444
3-Thiophenetrifluoroborate, 448
Thiophilicity, 252, 290, 291, 294
Three-component coupling, 330, 331, 334
Thrombin inhibitor, 230
Thyroid receptor ligand, 352
Tiaprofenic acid (Surgam®), 251
N-TIPS-3,4-diiodopyrrole, 60
N-TIPS-3-iodopyrrole, 60
N-TIPS-4-iodogramine, 119
Titanium-isocyanate complex, 164
Tjipanazoles, 164
5-(*p*-Tolyl)oxazole, 384
1-Tosyl-4-indolyltriflate, 119
N-Tosylindolyl-3-boronic acid, 452
Transesterification, 269
Transhalogenation, 549
Transmetalation, 2, 5, 6, 13, 190, 350, 413
2-Trialkylstannylfurans, 316

2,3,5-Tribromobenzofuran, 311
2,3,4-Tribromoindole, 40
3,4,6-Tribromoindoles, 161
Tributyl-(1-ethoxyalkenyl)tin, 441
2-Tributylstannylbenzofuran, 316
2-Tributylstannylfuran, 306, 314, 317–319
3-Tributylstannylfuran, 317
2-Tributylstannylpyrazine, 441
Tributylstannylpyridazines, 548, 563
2-(Tributylstannyl)benzoxazole, 395
2-(Tributylstannyl)-4,4-dimethyl-2-oxazoline, 320
3-(Tributylstannyl)pyridine, 443
3-(Tributylstannyl)-pyridine *n*-oxide, 443
Triethylborane, 438
Trifluoroborate salts, 354
5,6,7-Trifluoroindole, 126
Trifluoromethanesulfonic acid, 448
6-Trifluoromethylindole, 158
meta-Trifluoromethylphenylstannane, 461
α-(Trifluoromethyl)ethenyl boronic acid, 446
Trifuranylphosphine, 387
Trikentrin, 130
Trimethylaluminum, 438
Trimethylorthoformate, 385
Trimethylsilyl protecting group, 355
N-Trimethylsilylpyrrole, 39
Trimethylstannyl sodium, 196
2-Trimethylstannyl-4-methyloxazole, 390
2-Trimethylstannylbenzoxazole, 396
5-Trimethylstannylindole, 120
Trimethylstannylpyridazin-3(2*H*)-ones, 566
2-(Trimethylstannyl)pyrrole, 54
2-Trimethylstannylpyridine, 391
Tri-*n*-butyllithiostannane, 462
2,3,5-Triphenyloxazole, 399
Triphenylphosphine, 199
Tris-(*N*-azolyl)triphenylamines, 601
Tris-(2-furyl)phosphine, 389
Tris-polyethyleneglycol vinyl iodide-biotin ester, 381
Tris-(*o*-tolyl)-phosphine, 589
Tris-(4-trimethylsilylfuran-3-yl) boroxine, 314
2,3,5-Trisubstituted oxazoles, 386
N-Tritylpyrrole, 39

Troger's base, 227
Tryptamine stannane, 114
Tryptamines, 149, 152, 166
Tryptophans, 149
Tryptophols, 149, 152
Tsuji–Trost reaction, 25, 428, 429
Tumor cells, 381
Turn over number (TON), 270, 587

Ullmann reaction, 240, 534, 536
Ultrasonic activation, 310
Ungerimine, 106
USB-165, 439

van Leusen pyrrole ring synthesis, 56
Vascular endothelial growth factor (VEGF) receptor inhibitors, 223
Verrucarin E, 40
Vilsmeier haloformation, 534
Vilsmeier reaction, 346
Vilsmeier–Haack reagent, 515
Vincadifformine, 117
Vindoline, 116
Vinyl stannane, 393
2-Vinylazaindoles, 117
4-Vinylindole, 119
3-Vinylindoles, 160
7-Vinylindoles, 120
2-Vinyloxazole, 399
2-Vinyloxazole-4-ethylcarboxylate, 391, 399
Vinylphosphates, 266, 275, 276
2-Vinylpyrroles, 58
3-Vinylpyrroles, 57
Vinylquinoline, 589
O-Vinyltoluene, 603
Vinyltributyltin, 235, 390
Vinyltrimethylstannane, 391
Virginiamycin m2, 379
Vitamin B1, 345

Wacker oxidation, 2, 4, 26, 28, 307
Wadsworth–Horner–Emmons reaction, 218, 357, 417
Wang resin, 49
Wilkinson's catalyst, 123
Williamson ether synthesis, 606

Wittig, 218

(+)-(*S*)-WS-75624b, 222

XH-14, 332

X-ray crystallography, 56

Yuehchukene analogs, 154

Ziegler reaction, 532

Zinc chloride, 201

Zirconium, 108

Zomig, 82

Zoxazoleamine, 379